PROCEEDINGS OF THE 7th EUROPEAN CONFERENCE ON CONSTITUTIVE MODELS FOR RUBBER, ECCMR, DUBLIN, IRELAND, 20–23 SEPTEMBER 2011

Constitutive Models for Rubber VII

Editors

Stephen Jerrams & Niall Murphy
The Centre of Elastomer Research,
Dublin Institute of Technology, Dublin, Ireland

CRC Press
Taylor & Francis Group
Boca Raton London New York Leiden

CRC Press is an imprint of the
Taylor & Francis Group, an **informa** business

A BALKEMA BOOK

First issued in paperback 2017

CRC Press/Balkema is an imprint of the Taylor & Francis Group, an informa business

© 2012 Taylor & Francis Group, London, UK

Typeset by MPS Limited, a Macmillan Company, Chennai, India

Published by: CRC Press/Balkema
                 P.O. Box 447, 2300 AK Leiden, The Netherlands
                 e-mail: Pub.NL@taylorandfrancis.com
                 www.crcpress.com – www.taylorandfrancis.co.uk – www.balkema.nl

ISBN 13: 978-1-138-11543-9 (pbk)
ISBN 13: 978-0-415-68389-0 (hbk)

Constitutive Models for Rubber VII – Jerrams & Murphy (eds)
© *2012 Taylor & Francis Group, London, ISBN 978-0-415-68389-0*

Table of Contents

Modelling and simulation

Damage mechanisms in elastomers

Stress softening and related phenomena

Design and applications

Fatigue and time dependent behaviour

Test methods and analytical techniques

Constitutive Models for Rubber VII – Jerrams & Murphy (eds)
© 2012 Taylor & Francis Group, London, ISBN 978-0-415-68389-0

Foreword

This book is a collation of all the papers resulting from invitations or selected for Oral and Poster presentations at the Seventh European Conference on Constitutive Models for Rubber (ECCMR VII). The Centre for Elastomer Research, of the Dublin Institute of Technology, hosted this biennial conference in The Radisson-Blu Royal Hotel, Dublin, Ireland and attracted a large number of delegates from a wide variety of disciplines in academia and industry worldwide. Collectively they represented a full range of expertise in the challenging and ever changing field of rubber research.

Since its inauguration in 1999, this conference has become an international forum where world renowned researchers present and discuss the latest trends in topics related to the testing, modelling, characterisation and application of rubber-like materials. In 2003, the emphasis was on dynamic properties; in 2005, fatigue and fracture represented the largest topic and inevitably the emphasis shifted through 2007 and 2009 to dynamic analyses, microstructural observation and biomedical applications. In 2011, in the largest ECCMR programme to date, all the topics of the previous 13 years remained strongly represented. However, there was a need for the organising and scientific committees to timetable two full conference sessions to Modelling and Simulation and similarly there was a discernable increase in presentations devoted to understanding fatigue mechanisms and associated theories, analyses and simulations. Not surprisingly, properties of new adaptive elastomers featured strongly for the first time.

The editors would like to thank their colleagues on the organising committee – Frank Abraham, Marek Rebow, Erwan Verron, Robert Schuster and Jean-Benoît Le Cam and the scientific committee for their hard work in the review process – A. Lion, H. Baaser, P. Buckley, J. Busfield, L. Dorfman, G. Holzapfel, M. Itskov, W. Mars, A. Muhr, P. Wriggers, S. Reese, M. Kluppel & J. Ihlemann. In particular, the additional advice provided by Erwan Verron must be acknowledged and we would also like to extend our gratitude to Michael Kaliske who supported the organisers throughout the two years prior to the event. The editors must single out the contribution of one of their group in bringing the conference to a successful conclusion – Brendan O'Rourke shouldered responsibility for all elements of the conference organisation and worked tirelessly to see it through to completion.

Written by Steve Jerrams on behalf of the Editors – Steve Jerrams & Niall Murphy.

Constitutive Models for Rubber VII – Jerrams & Murphy (eds)
© 2012 Taylor & Francis Group, London, ISBN 978-0-415-68389-0

Sponsors

Invited papers

Constitutive Models for Rubber VII – Jerrams & Murphy (eds)
© 2012 Taylor & Francis Group, London, ISBN 978-0-415-68389-0

Unravelling the mysteries of cyclic deformation in thermoplastic elastomers

C.P. Buckley
Department of Engineering Science, University of Oxford, Oxford, UK

D.S.A. De Focatiis
Department of Chemical and Environmental Engineering, University of Nottingham, Nottingham, UK

C. Prisacariu
Romanian Academy Institute of Macromolecular Chemistry, 'Petru Poni', Iasi, Romania

ABSTRACT: A study has been made of cyclic deformation in a series of model polyurethane thermoplastic elastomers, in which chemical compositions of hard and soft segments were varied systematically. Mechanical tests were supplemented by structural studies. The results reveal significant evidence for series coupling of hard and soft phases, and for deformation-induced conversion from hard phase to soft phase (presumed to occur by chain pull-out), as proposed previously to explain the Mullins effect. The nm-scale structural evidence provides quantitative information on the molecular environment in the two phases, and explains the pull-out and other features of the response. The results suggest a physically-based constitutive model needs to combine series and parallel coupling of the phases, together with phase conversion. Such a model, even when populated with the simplest possible elastic and elastic-plastic representations of soft and hard phases, naturally demonstrates the Mullins effect and other features of the observed cyclic response.

1 INTRODUCTION

1.1 *Preamble*

Thermoplastic elastomers (TPEs) form a class of soft polymeric material of great practical interest. They typically show the high elongations to break and low stiffnesses associated with natural and synthetic rubbery materials, and hence are of value in a wide range of products where flexibility and resilience are required. But whereas in traditional rubbery materials (those available more than 50 years ago) connectivity between the flexible polymer molecules is achieved by chemical crosslinking, this is not the case in TPEs. Instead, in these materials, the chemical architecture of the polymer itself is tailored so as to provide a physical crosslinking effect.

The trick is to employ block copolymers where, in addition to flexible 'soft' segments (SS) providing rubbery characteristics, each molecule also contains 'hard' segments (HS) that, at the service temperature, combine to form relatively rigid junction points between the molecules. Provided there are at least two such hard segments per molecule, there can be percolation of connectivity throughout the material volume, and a macromolecular network achieved, analogous to that in chemically crosslinked polymers.

The great attraction of TPEs comes from their convenience of use in manufacturing. The absence of any permanent molecular connectivity means they can be readily processed as thermoplastics, with all the associated economic advantages for high volume manufacturing. If a TPE is heated to a temperature where the 'hard' segments soften and become fluid, it can be melt processed by usual routes such as injection moulding or extrusion. Moreover, such materials can be readily recycled in similar fashion to other thermoplastics, by remelting and reprocessing.

1.2 *Mysteries of cyclic deformation in TPEs*

Since TPEs are usually employed in products where rubberlike resilience is required, their performance under small and large cyclic deformations is of special relevance. Therefore there is much interest in seeking to understand and to develop methods for predicting their performance, to assist in the design of materials and products. Such work faces serious challenges, as the mechanical response of TPEs remains poorly understood in physical terms, and is undoubtedly highly sensitive to details of their structure on a nm length-scale. Consequently experimental results seem to contain mysterious features, that have impeded the development of physically-based constitutive models. Examples of such mysteries are: unusual relaxation spectra in the SS glass transition region; values of elastic modulus that apparently defy quantitative explanation; pronounced strain-softening (Mullins effect); all of the above being remarkably sensitive to details of chemical composition.

An important sub-class of TPEs is the family of thermoplastic polyurethane elastomers (TPUs). The aim of this paper is to help explain the characteristic features of cyclic response of TPEs generally, by drawing on results from an ongoing study of an extensive series of model TPUs, in which the chemical structure is known, and in many cases the physical structure has been well characterized. Some results from this study were presented earlier: see Buckley et al. (2005, 2010). Here we combine those and other more recent developments to explain the anomalies, unravel the mysteries, and suggest the implications for constitutive modeling.

2 MATERIALS

2.1 Chemical structure

In view of the sensitivity of cyclic deformation of TPEs to their chemical structure, the present work focuses on an extensive series of specially synthesized TPUs, where the chemical structures are known, and vary systematically between the materials. They are all linear block copolymers, consisting of a di-isocyanate (DI), a monodisperse macrodiol (MD) of molar mass 2000, and a chain extender (CE). They were all synthesized to give molar ratios DI:MD:CE = 4:1:3, which leads to an HS volume fraction (assuming complete phase segregation) of ca 35%.

Di-isocyanates employed in this study were 4,4′-methylene bis(phenyl di-isocyanate) (MDI) and 4,4′-dibenzyl di-isocyanate (DBDI). Macrodiols included were poly(ethylene adipate) (PEA), poly(butylene adipate) (PBA) and polytetrahydrofuran (PTHF). Chain extenders included were ethylene glycol (EG), diethylene glycol (DEG) and 1,4-butylene glycol (BG). Most of these TPUs were synthesized by the two-stage pre-polymer route: details were given previously – for example, see Prisacariu et al. (2011). But a subset was also produced via the one-shot route (where DI+MD+CE are reacted in a single stage).

2.2 Physical structure

The physical structures of the materials described above have been studied, primarily by wide-angle X-ray scattering (WAXS) and small-angle X-ray scattering, using the UK Daresbury synchrotron source: for details see Buckley et al. (2010).

SAXS data showed the TPUs are phase segregated. In almost all cases there is a peak in intensity at a scattering vector $|q|$ of ca 0.3 nm^{-1}, indicating density fluctuations with long periods d of ca 20 nm. The chemical compositions of these TPUs ensure that the matrix phase contains predominantly the MD soft segments and is therefore the relatively soft phase. The dispersed phase contains predominantly the DI+CE hard segments and is the relatively hard phase. The degree of phase separation was quantified via the usual scattering invariant Q. It was found to be highest by

far for those PUs where PTHF forms the soft phase. Q and the Porod law coefficient were combined to obtain the surface:volume ratio A/V of the hard domains. See Buckley et al. (2010) for details of the calculations of Q and A/V.

WAXS scans of intensity versus 2θ revealed that in most of the TPUs both hard and soft phases are entirely or nearly amorphous. But in the cases where the DI is DBDI, and the CE is *not* DEG, discrete diffraction peaks are obtained, indicating some crystallization of the hard phase, as has been reported previously. The highest degree of crystallinity χ seen so far is for the composition DBDI+PBA+BG, where $\chi = 0.18$. When DEG is used as CE for DBDI, its central oxygen atom introduces a kink into the DBDI hard segments and inhibits crystallinity.

Thus we see that TPUs, typical of other TPEs, are nanocomposite materials. They each consist of a matrix of rubberlike polymer (T_gs of the soft segments all lie well below room temperature), reinforced by hard domains of width typically 10–16 nm. In addition there is the special feature of molecular connectivity linking the matrix to the hard domains.

Consider what this means for the environment of soft segments in the matrix. We can deduce from either d or A/V that the matrix channels between adjacent hard domains are of average width ca 5 nm. But the unperturbed rms end-to-end distance of a PTHF chain of molar mass 2000 for example is 4.5 nm. Thus, if there is complete phase segregation, such a chain is tethered at its ends at approximately the same separation (on a time rms average basis) as it would have if it were unperturbed. However, the constraint on motion provided by the tethering, the similar tethering of its neighbours, and the excluded volume imposed by the walls of the channel to which it is confined, will restrict its mobility, possibly severely.

Now consider the environment of hard segments within the hard domains. In TPUs the hardness of these domains comes from two sources. Firstly the DIs MDI and DBDI of the present TPUs are bulky, relatively rigid, units at room temperature. Secondly, adjacent units are bound to their neighbours by hydrogen bonding primarily between their urethane linkages. In the present PUs, DBDI is more flexible than MDI at higher temperature when the phase segregation takes place, and hence a higher degree of hydrogen bonding is achieved. As noted above, this may extend to crystallization of the hard segments. However, measured values of A/V indicate that, on average, a large proportion of each hard domain (19%–36% is the range seen so far) is occupied by molecular segments that lie at the surface of the domain, and are therefore bonded to neighbouring hard segments on one side only. This must provide each hard domain with a surface layer that is more weakly bound, and therefore more vulnerable than its core.

This picture of the physical structure is essential information in interpreting the mechanical properties of these materials.

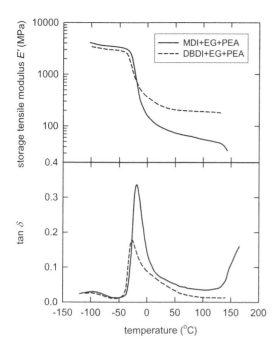

Figure 1. Results of tensile DMA tests at a frequency of 1 Hz, for two TPUs from the present study, with di-isocyanate, macrodiol and chain extender as indicated.

3 LINEAR VISCOELASTIC RESPONSE

3.1 *Dynamic mechanical analysis*

The linear viscoelastic response is readily obtained, for example, from small strain cyclic experiments in the linear region (Dynamic Mechanical Analysis – DMA). Sample data obtained at a frequency of 1 Hz are shown in Figure 1 – see Prisacariu et al. (2011) for experimental details and further examples. In the case of Figure 1, the two materials differ only in the choice of DI employed in the hard segments: MDI or DBDI. Both have a hard segment volume fraction of $\phi_H = 0.36$. Both materials are phase segregated: the DBDI polymer having a higher phase density difference than the MDI polymer. The DBDI polymer is significantly more crystalline than the MDI polymer: $\chi = 0.165$ and $\chi = 0.036$ respectively.

Figure 1 shows traces typical of TPUs. There is a low temperature local relaxation, followed by the prominent α-relaxation associated with the glass transition of the soft phase, and finally at higher temperature the softening of the hard phase (not reached for DBDI in Figure 1). The precise temperature of the α-relaxation damping peak is sensitive to the overall mobility of the soft phase, and hence to the degree of phase mixing, and if there is any phase mixing, to the flexibility of the hard segment.

Here we focus on the performance of the polymers as elastomers at room temperature. First consider the magnitude of the tensile storage modulus at 20°C. As noted above, the TPUs have the structure of a nanocomposite. Therefore, to interpret their macroscopic mechanical properties, our starting point is to consider them as composite materials consisting of two phases each of which is a homogeneous isotropic continuum. For such a system, the simplest combining rule for the storage shear modulus G', consistent with the general bounding rules of Hashin & Shtrikman (1963), is the empirical log law used by Gray & McCrum (1969) in interpreting the modulus of polyethylene:

$$\log G' = \phi_h \log G'_h + (1 - \phi_h) \log G'_s \quad (1)$$

where h and s refer to hard and soft phases. A set of simple assumptions allows us to predict a benchmark value of storage tensile modulus: (i) full phase segregation (hence $\phi_h = \phi_H$); (ii) the soft phase acts as a Gaussian rubber with molar mass 2000 between crosslinks, and hence a shear modulus at room temperature of 1.2 MPa; (iii) the hard phase has the typical stiffness of a glassy polymer, i.e. a shear modulus of 1 GPa; (iv) the macroscopic Poisson's ratio is $v = 0.5$. Equation 1 then predicts a benchmark value for E' of 40 MPa, for all the TPUs in the present study. Thus Figure 1 reveals the first mystery: the room temperature storage tensile modulus is significantly greater than the expected value (by factors of ca 2.4 and 6.6 for MDI and DBDI polymers respectively).

The physical structure identified above provides explanations for this apparent mystery. The soft phase is trapped within narrow nm-scale channels between hard domains, of width similar to the unconstrained rms end-to-end distance of the SS. Two consequences for molecular mobility are to be expected.

(1) The constraint restricts the range of molecular configurations possible, and hence increases the entropy barrier to relaxation of chain segments. Thus the longest relaxation times (corresponding to relaxation of the longest chain portions) are lengthened and the relaxation is broadened on the low frequency (high temperature) side. Therefore both polymers at room temperature are not yet fully relaxed at 1 Hz, i.e. they have not reached their rubbery plateau.

(2) Even when they do reach their respective rubbery plateaus, the benchmark modulus will be exceeded. The constraint will be most severe for the longest modes of chain relaxation (longest Rouse modes) that may be inhibited completely, and the rubbery response of the soft phase will be then *as if* the molar mass between crosslinks is less than 2000 (raising the plateau shear modulus, and hence the tensile modulus). Figure 1 illustrates clearly both these effects. Evidently, the hard domains exert the most extreme constraint in the DBDI polymer. Here the semi-crystallised hard segments in the DBDI-rich domains constrain the soft segments to such a degree that the matrix acts as a highly crosslinked rubber with a molar mass of only ca 100 between crosslinks.

The graphs of tan δ confirm this interpretation. Although the peaks indicating the dominant glass transition of the soft phase occur at ca −20°C or below, both peaks are broader on the high temperature side.

This is an expected consequence of the reduced long range mobility of soft segments. In both cases, room temperature lies under a high temperature shoulder, which is especially pronounced in the case of the DBDI polymer. The other notable features of tan δ in Figure 1 are the small heights of both peaks. This is an expected consequence of the reduced relaxation magnitude caused by increase in the relaxed (plateau) modulus, when the unrelaxed modulus is unchanged.

3.2 Stress relaxation modulus

More precise measurements of small strain elastic moduli are obtained in experiments where extensometers are used to avoid specimen end-effects, not usually possible in commercial DMA instruments. In this way Buckley et al. (2010) determined the short term (0.3s) tensile stress relaxation modulus $E(0.3s)$ for 14 TPUs polymers from the present study, from the stress-strain gradient during constant strain-rate tensile tests at room temperature (20°C) – see the original paper for details. The results corroborated the conclusions above from DMA. Values of $E(0.3s)$ ranged from 39.2 MPa to 324 MPa. It is interesting to note how close the lowest measured modulus is to the predicted benchmark value of 40 MPa. All the polymers with $E(0.3s) < 100$ MPa were based on MDI, while all those with $E(0.3s) > 100$ MPa were based on DBDI, with the highest values shown by the semicrystalline DBDI-based polymers. Again we see evidence for the tighter hydrogen bonding achieved with DBDI, especially when semicrystalline, causing the most severe soft segment constraint.

4 LARGE CYCLIC DEFORMATIONS

4.1 Mullins effect

When cycled to strains outside the linear viscoelastic region, the TPUs exhibit the well-known Mullins effect, widely observed in filled elastomers. Some representative data from tensile cycling of three of the TPUs at a nominal strain rate of $0.03\,s^{-1}$ are shown in Figure 2. Specimens were subjected to a 'pseudo-cyclic' saw-tooth extension history, where a constant nominal extension rate was reversed periodically, allowing the specimen to unload fully before continuation with a positive extension rate. Thus single cycles of deformation were obtained after deformation to various levels of nominal pre-strain e_{max}.

The features characteristic of the Mullins effect can all be observed in Figure 2. Cyclic deformation *within the envelope* of the maximum prior deformation exhibits: (i) lower stiffness than on first loading; (ii) lower hysteresis than on first loading; (iii) on reaching the previous maximum deformation, the material re-joins the first loading stress-strain path. Many previous authors have attempted to explain, and model, the Mullins effect, largely because of its practical importance in the case of particle-filled elastomers. But the literature still shows no firm consensus. Thus

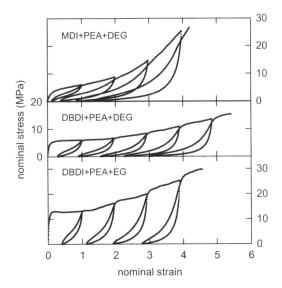

Figure 2. Representative results of pseudo-cyclic tests on three TPUs, where the DI and CE are varied. Top and middle graphs are for amorphous TPUs. The bottom graph is for a TPU with significant crystallinity in the hard domains.

the occurrence of these features in TPUs is a major additional mystery to be explained, and a further challenge for the constitutive modeller.

The extensive series of TPUs included in the present work, where the chemical structures are known, provides insight into the dependencies of (i), (ii) and (iii) on polymer structure.

Figure 3 shows the tensile modulus as measured from data such as shown in Figure 2, at the start of re-loading after each unloading, for a family of 6 TPUs based on PEA as macrodiol, including those in Figure 2. The modulus is expressed here in terms of true stress and true strain at the point at which it is determined. The most prominent feature in Figure 3 is the immediate drop in modulus from the value obtained initially, without any pre-strain. The second point of note is that the drop is followed by an increase in modulus after further straining. Thirdly, the greatest drop and the greatest increase are for the DBDI polymers with crystallinity.

Figure 2 shows that the plastic, unrecovered, strain on unloading is much higher for the semicrystalline DBDI-based TPU, than for the amorphous MDI-based TPU, while the amorphous DBDI-based TPU is intermediate. This is generally observed. A corollary is that the first cycle hysteresis is greater for semicrystalline DBDI-based TPU than for amorphous MDI-based TPU. Nevertheless, an interesting feature visible in Figure 2 is that the *second cycle relative* hysteresis appears similar for all three polymers.

All these features demand explanation.

4.2 Cyclic deformation after pre-strain

Pseudo-cyclic tests such as those illustrated in Figure 2 are a quick, but only approximate, means to determine

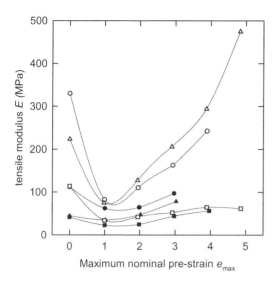

Figure 3. Tensile stress relaxation modulus $E(0.3\,\mathrm{s})$ calculated from data such as shown in Figure 2, based on the current true stress and strain, after previous stretching to the nominal strain shown. Data for six TPUs are shown, all having PEA as macrodiol. Key: closed symbols indicate DI was MDI, open symbols indicate DI was DBDI; circles, squares or triangles indicate the chain extender was EG, DEG or BG respectively.

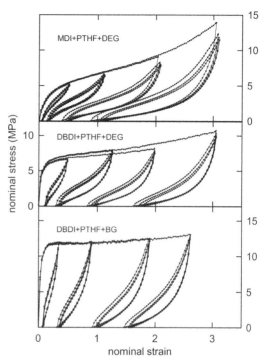

Figure 4. Representative results of experiments in which specimens were cycled 4 times between different maximum strains and zero load. Different specimens were used for different maximum strains. The top and middle graphs are for amorphous TPUs: the bottom is for a TPU with significant crystallinity in the hard domains.

the steady-state cyclic response after a pre-strain, as there is only one unloading/re-loading for each level of pre-strain.

Therefore another set of tests was carried out where each specimen was cycled between zero load and a given strain level four times, again at a nominal strain-rate of $0.03\,\mathrm{s}^{-1}$. Some typical results are shown in Figure 4, for a series of TPUs based on PTHF as macrodiol. In each graph, each set of loops was obtained with a different specimen. It is clear that the cyclic stress-strain curves, after the first to any given strain, converge rapidly onto a common curve. Hence, in each case, the 4th cycle stress-strain loop is a good representation of the steady-state cyclic response to a sawtooth profile of nominal strain, within the envelope of the previous maximum strain.

Considering the role of chemical structure, again the same qualitative pattern can be seen as noted above: the semicrystalline DBDI-based TPU shows the greatest first cycle hysteresis and plastic strain, and the amorphous MDI-based TPU shows the least of both. But on 2nd, 3rd and 4th cycles the relative hysteresis is similar for all three polymers.

This latter observation may be examined quantitatively, as discussed previously – see Buckley et al. (2010). Thus in Figure 5, from a series of experiments such as those of Figure 4, the 4th cycle hysteresis energy ΔW_4 is plotted versus the 4th cycle energy input W_4, for a series of 7 TPUs based on PTHF as macrodiol, for various values of pre-strain (between 4 and 11 for each material: there are 42 data points in total). Thus there were seven permutations of DI, CE and synthesis route (pre-polymer or one-shot).

Four were based on MDI as DI, and three were based on DBDI: four were based on BG as chain extender, three were based on DEG; four were prepared by the pre-polymer route, while three were prepared by the one-shot route.

The remarkable result shown in Figure 5 is that the data for all seven polymers, for all values of pre-strain, lie close to the same straight line with a high degree of consistency ($R^2 = 0.9869$). A similar result was reported by us previously from results of the earlier study (Buckley et al. (2010)) of 10 TPUs, only two of which were common to the set of 7 included in Figure 5. Thus it is now reasonable to consider this simplifying feature of the response of TPUs to be well established. Its existence is another mystery to be explained.

5 UNRAVELLING THE MYSTERIES

5.1 *Mechanical coupling between the phases*

Considering the TPUs as two-phase nanocomposites as suggested above, it follows that their mechanical response is fully determined by three factors: (i) the dispersed phase volume fraction ϕ_h; (ii) the mechanical coupling between the phases; (iii) the mechanical properties of the individual phases (noting that, in

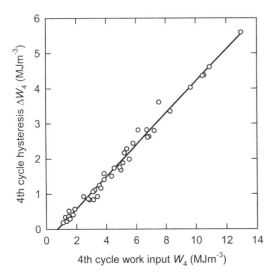

Figure 5. Results from experiments such as those shown in Figure 4, for 7 TPUs from the present study, for various values of maximum pre-strain. The hysteretic energy dissipated on the 4th cycle is plotted versus the energy input on the positive strain-rate portion of the 4th cycle.

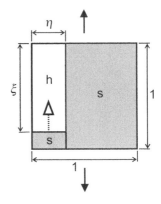

Figure 6. A 1-D Takayanagi representation of series and parallel coupling between the hard (h) and soft (s) phases of a TPE, for modeling tensile deformation in the direction of the solid arrows. The dotted arrow indicates the direction of movement of the s/h boundary, resulting from chain pull-out from the hard phase.

view of the nm scale inhomogeneity of their physical structures, the phase properties may deviate from their macroscopic counterparts).

Mechanical coupling between the phases occurs at the nm length scale particle boundaries: from interphase force equilibrium and displacement compatability. It leads to a complex 3D distribution of stress and strain at the nm scale. To capture the effects of that in a model to describe the macroscopic response, gross simplification of the coupling is necessary. We know that neither simple series nor parallel coupling will be adequate. It is well-known that they under- and over-predict the macroscopic stiffness respectively. We therefore look to experimental data for guidance on whether the coupling can be expressed in any other simple closed form at the macroscopic level, for these materials.

Consider two prominent features of the cyclic response, within the envelope of the previous maximum strain, highlighted above. (1) When the strain continues beyond the previous maximum strain, the stress spontaneously rejoins the original stress-strain curve (Figure 2). (2) The steady state hysteresis varies linearly with the input energy, with only a small intercept: this indicates that the cyclic damping of the material is constant – independent of the maximum pre-strain (and independent of chemistry of DI and CE) – see Figure 4. The occurrence of these two features is the signature of series coupling existing between a relatively soft viscoelastic material, with reproducible cyclic response, and a relatively hard elastic-plastic material. Deformation *within the maximum strain envelope* would then occur only within the soft material (we neglect the small elastic strain in

the hard phase). This explains why the damping is so reproducible as seen in Figure 5. Similarly, the stress cannot rise beyond the flow stress of the hard material. This explains why, on re-loading to the flow stress, the stress spontaneously rejoins the original stress-strain path, as seen in Figure 2.

Notwithstanding these indications of series coupling, we know that the phases are not purely coupled in series – it would require the macroscopic tensile moduli to be much smaller than those observed, by factors of more than 10.

Thus we are led to consider the effective mechanical coupling between the phases as being a combination of series and parallel coupling – as suggested in 1-D form by Takayanagi et al. (1964) for semicrystalline polymers.

For the present purpose we shall test the promise of this approach in simple one-dimensional form – see Figure 6. For consistency with the physical microstructure, we set $\xi\eta = \phi_h$.

The ratio $r = \xi/\eta$ characterizes the relative proportions of soft phase coupled in series and parallel with the hard phase. For example, for a TPE with given ϕ_h and phase tensile moduli E_h and E_s, the tensile modulus would be fully determined by r. The corollary is: from the macroscopic tensile modulus obtained experimentally or by other means, the required value of r may always be determined empirically.

5.2 *Implications for large cyclic deformations*

Note that the 1-D Takayanagi model shown in Figure 6 couples a series s-h combination in parallel to pure s phase. We assume the latter to exhibit reproducible viscoelastic stress-strain response in cyclic straining (see above). It follows that experimental features (1) and (2) above, are exhibited also by the Takayanagi model. Can the model also capture the other mysterious features of the cyclic response of TPEs?

The special challenge is the strain-induced softening visible in, for example, Figures 2 and 4, and associated with the Mullins effect. The Takayanagi model *per se* does not show this. But several previous authors have speculated that the origin of the Mullins effect is pull-out of molecular segments from the hard phase, to create additional soft phase. In particular, it would be energetically favourable for the new soft phase to be coupled preferentially in series with the hard phase, as proposed by Mullins and Tobin (1957) for filler-reinforced vulcanized rubber, and by Enderle et al. (1986) for semicrystalline TPUs.

In the TPEs this is highly plausible. We showed above that a significant fraction of each hard domain lies in a surface layer that must be relatively weakly bound *and* that has molecular connectivity to the soft phase. Therefore a sufficiently high stress in the soft phase is expected to pull off surface segments, that will then add to the series-coupled soft phase. Modelled at the macroscopic level in terms of the Takayanagi-type model in Figure 6, this would be manifest as a shift in the boundary in the series–coupled portion of the model. The boundary would move so as to reduce the proportion of hard phase, as indicated by the dotted arrow in Figure 6.

To test this description, consider the following benchmark application of the model in Figure 6, based on idealised data relevant to a typical material from the present study.

The soft phase is taken to be a Gaussian rubber with tensile modulus $E_s = 3.6$ MPa. The hard phase is taken to be elastic-plastic with $E_h = 3$ GPa and true flow stress in tension $\sigma_{fh} = 9.7$ MPa (here we assume no yield in compression). We include no time-dependence for either phase. The volume fraction ϕ_h is taken equal to $\phi_H = 0.38$. The initial value of the ξ/η ratio is $r = 2.47$, chosen to give quantitative agreement between predicted Youngs' modulus and equation (1). Finally the effect of pull-out is represented empirically via a linear reduction of ξ with hard phase plastic true strain ε_p via an empirical material parameter k:

$$\xi = \xi_0 - k\varepsilon_p. \tag{2}$$

With these simplifying assumptions and values of the parameters, tensile tests similar to those of Figure 4 may be simulated for the 1-D model of Figure 6. Sample results are shown in Figure 7, for two values of the pull-out parameter k. It is immediately clear that for $k > 0$, i.e. with some degree of pull-out, there is progressive strain-induced softening of the cyclic response – i.e. Mullins effect – whereas the case $k = 0$ does not show this. It is also clear that $k > 0$ reduces the accumulation of residual strain with increase in applied strain. From comparison with stress-strain responses of different materials (e.g. Figures 2 and 4), it seems k is a parameter that varies with chemical structure. This is expected, since the ease of pull-out will vary depending on the tightness of binding of segments at hard domain surfaces. Thus it is clear the pull-out-modified Takayanagi model of

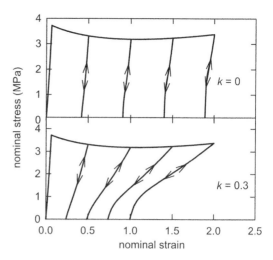

Figure 7. Finite tensile deformation response of the 1-D Takayanagi model of Figure 6, with unloading-reloading cycles, for simplified representations of hard phase and soft phase properties, and incorporating differing degrees of pull-out from the hard phase.

Figure 6 is promising as a physically-based macroscopic constitutive representation of the TPUs.

In detail, of course, the responses shown in Figure 7 deviate substantially from those typically observed. But the remaining discrepancies can be attributed to neglect of other features that are well-known in polymer mechanics – soft phase hysteresis and finite extensibility, time-dependence of both phases, strain-hardening of the hard-phase, multiaxiality of response. These can be added if necessary, to achieve predictive capability.

6 CONCLUSIONS

Results from a study of numerous model TPUs are providing explanations for many of the characteristics seen in cyclic deformation of TPEs generally. They also point to how to structure constitutive models of them, in order to capture naturally the character of their observed behavior.

Their macroscopic responses suggest the soft/hard phase coupling has some series and some parallel character, and the Takayanagi macroscopic model proposed many years ago captures this conveniently. Such a model, when incorporating a representation of chain pull-out from the hard-phase, naturally displays the Mullins effect. The physical structure established by X-ray scattering shows that chain pull-out (from weakly bound hard domain surfaces) is to be expected. Therefore the Mullins effect is an expected consequence.

Similarly, the linear viscoelastic response reveals the effects of constraint on soft segments expected from the nm-scale structure. The measured tensile moduli, via the log-law combining rule, show that in some of the PUs based on MDI the soft segments have approximately the rubbery moduli expected for their chain length. Others, based on DBDI, show significant increase in modulus caused by molecular constraint from being tethered to tightly hydrogen bonded hard domains. The differing chemical structures and nm-scale microstructures are essential information in understanding all these differences.

This work has contributed to unravelling several mysterious features of TPE cyclic response, and has pointed the way to more physically-based constitutive modelling, that naturally captures this response.

ACKNOWLEDGEMENTS

Several collaborators have made important contributions to aspects of the work reviewed here: in particular A.A. Caraculacu, C.M. Martin, and A. De Simone. The UK-Romania collaboration was made possible by financial support from the Royal Society (via a Joint Project Grant) and NATO (via a Collaborative Linkage Grant).

REFERENCES

Buckley, C.P., Prisacariu, C., Caraculacu, A.A. & Martin, C.M. 2005. Inelasticity of hard-phase reinforced elastomers: a study of copolyurethanes with varying hard and soft segments. In P.-E. Austrell & L.Kari (eds), Constitutive Models for Rubber IV; Proc. 4th European conf. on constitutive models for rubber, Stockholm, 27–29 June 2005. Rotterdam: Balkema.

Buckley, C.P., Prisacariu, C. & Martin, C.M. 2010. Elasticity and inelasticity of thermoplastic polyurethane elastomers: sensitivity to chemical and physical structure. Polymer 51: 3213–3224.

Enderle, H.F., Kilian, H.-G., Heise, B., Mayer, J., & Hespe, H. 1986. Irreversible deformation of semicrystalline PUR-elastomers. Colloid & Polymer Sci 264: 305–322.

Gray, R.W. & McCrum, N.G. 1969. Origin of the γ relaxations in polyethylene and polytetraflouroethylene. J. Polymer Sci., Part A-2 7: 1329–1355.

Hashin, Z. & Shtrikman, S. 1963. A variational approach to the theory of elastic behavior of multiphase materials. J. Mech. Phys. Solids 11: 127–140.

Mullins, L. & Tobin, N.R. 1957. Theoretical model for the elastic behavior of filler-reinforced vulcanized rubbers. Rubber Chem. & Technol. 30: 555–571.

Prisacariu, C., Scortanu, E., Stoica, I., Agapie, B. & Barboiu V. 2011. Morphological features and thermal and mechanical response in segmented polyurethane elastomers based on mixtures of isocyanates. Polymer Journal in press.

Takayanagi, M., Uemura, S. & Minami, S. 1964. Application of equivalent model method to dynamic rheo-optical properties of crystalline polymer. J. Polymer Sci., Part C 5: 113–122.

Constitutive Models for Rubber VII – Jerrams & Murphy (eds)
© 2012 Taylor & Francis Group, London, ISBN 978-0-415-68389-0

The role of glassy-like polymer bridges in rubber reinforcement

M. Klüppel, H. Lorenz, M. Möwes, D. Steinhauser & J. Fritzsche
Deutsches Institut für Kautschuktechnologie e.V., Hannover, Germany

ABSTRACT: For a better understanding of the structure and dynamics of elastomer composites, it is convenient to study the dielectric properties of conducting carbon black filled elastomers on a broad frequency scale. By applying combined rheological and dielectric analysis it is demonstrated that during heat treatment (flocculation) a high frequency relaxation transition appears which is traced back to the tunneling of charge carriers over small gaps between adjacent carbon black particles. From the dielectric spectra the gap size can be evaluated which is found to decrease with flocculation time and carbon black loading. The gaps are found in the range of a few polymer layers (4–6 nm) indicating that the polymer in the gaps is strongly immobilized due to strong attractive interactions with the filler surface. Accordingly, from a mechanical point of view the gaps correspond to glassy-like polymer bridges representing quite stiff filler-filler bonds which transmit the stress between adjacent particles of the filler network. The thermo-mechanical response of glassy-like polymer bridges (filler-filler bonds) is found to be closely related to the gap size which plays a major role in understanding the viscoelastic and quasistatic stress-strain properties of filled elastomers. By referring to the structural analysis of elastomer composites, a micro-mechanical model of stress softening and filler-induced hysteresis has been developed. The model is briefly described and shown to be in fair agreement with experimental data. In particular it is demonstrated that the effect of temperature on the stress-strain cycles can be well described based on an Arrhenius like activation behavior of the filler-filler bonds.

1 INTRODUCTION

The field of application of elastomers is determined by their frequency dependent dynamic-mechanical properties, which are connected with characteristic relaxation and energy dissipation mechanisms depending on frequency, temperature and amplitude. They are strongly affected by the bonding of the polymer chains to the filler interface, the polymer-filler interphase dynamics and the filler networking effects. The latter are determined by the morphological arrangement of filler particles to form clusters and the mechanical joints of the particles by filler-filler bonds. For a better understanding and for getting a deeper insight into structure and dynamics of filler reinforced elastomers, it is important to investigate the relations between filler network morphology, interphase dynamics and bulk mechanical properties in dependence of frequency, strain amplitude and temperature.

The role of filler networking in rubber reinforcement has been investigated since more than 50 years by several authors (e.g. Payne 1963, Kraus 1965, Medalia 1973). Later on, more sophisticated approaches have been developed which also considered the disordered nature of the filler network. A review is given in (Heinrich & Klüppel 2002, Klüppel 2003). In particular, it has been demonstrated that, driven by osmotic depletion forces, filler networking takes place during heat treatment or vulcanization. Due to strong polymer-filler couplings the polymer between adjacent filler particles is not squeezed out totally and characteristic gaps remain. The confined polymer in these gaps is strongly immobilized and forms glassy-like bridges between adjacent filler particles transmitting the stress through the filler network. In so far, the viscoelastic response of the filler network results mainly from the stiffness and viscose losses of the filler-filler bonds implying that several dynamic mechanical properties of filled elastomers can be traced back to the specific properties of the filler-filler bonds. In particular, it has been argued that the observed temperature dependence of the small strain storage modulus results from an Arrhenius-like temperature dependence of the stiffness of filler-filler bonds. Accordingly, the changes in the dynamic-mechanical properties of elastomers by the incorporation of fillers can be traced back to the dynamics of the filler network which is governed by the viscoelastic response of glassy-like polymer bridges between adjacent filler particles (Klüppel 2003, Klüppel & Heinrich 2004, Meier & Klüppel 2008, Fritzsche & Klüppel 2011).

In the present paper the effect of filler networking on the mechanical properties of elastomers is analyzed whereby a special focus lies in the role of glassy-like polymer bridges (filler-filler bonds) resulting from the reduced polymer mobility in the gaps between adjacent filler particles. Temperature dependent dynamic-mechanical and dielectric spectra are studied in a wide frequency range. For the first time also the combination of both techniques, i.e. online-dielectric spectroscopy in a plate and plate rheometer, is applied for monitoring the flocculation dynamics of fillers in elastomer melts during heat treatment. In particular, the gap distance corresponding to the length of glassy-like polymer bridges will be estimated in dependence of flocculation time, filler type and filler loading. It will be demonstrated that on the one side these data allow for a physical understanding of the observed Arrhenius dependence of the temperature dependent shear modulus at small strain. On the other side, it will be shown that the temperature dependence of stress-strain cycles at large strain can be described fairly well based on the thermal activation energy of filler-filler bonds found from the Arrhenius plot of the modulus at small strain.

2 EXPERIMENTAL

2.1 *Sample preparation*

The polymers used in this study were a solution styrene-butadien rubber (S-SBR) with 25 wt.% vinyl and 25 wt.% styrene (Buna VSL2525-0) and an ethylene-propylene-diene rubber (EPDM, Keltan 512). Unfilled samples and composites with different filler loadings, i.e. 10, 20, 30, 40, 50 and 60 per hundred rubber (phr) were prepared. As fillers, two types of carbon blacks (CB) with similar structure but different specific surface area (SA) were used: N339 (SA $= 88\,m^2/g$) and HS55 (SA $= 48\,m^2/g$). In addition, 3 phr zinc oxide, 1 phr stearic acid, 2 phr N-isopropyl-N'-phenyl-p-phenylen-diamine (IPPD), 2,5 phr n-cyclohexyl-2-benzothiazole-sulfenamide (CBS) and 1,7 phr soluble sulfur were used as vulcanization system. The curing study was carried out with the help of a rheometer (Monsanto ME 2000) at 160°C. The samples were cured under pressure at 160°C to 2 mm plates in dependence of the determined t_{90} vulcanization time. For uniaxial stress-strain testing, axial-symmetric dumbbells were prepared in a specially designed mould.

For the combined dielectric and rheological investigations we used a standard receipt for EPDM-carbon black composites without curing system. The composite was mixed in a laboratory mixer (Haake Rheomix 300) with Banbury rotors and a volume of 380 cm^3. The starting temperature was 40°C and the rotation speed was 50 rpm/min. To get rid of air bubbles, the samples were pressed with approximately 200 bar at 80°C. The sample shape was disc like with 2 mm height and a diameter of 20 mm.

2.2 *Dielectric and rheological measurements*

The investigation of the dielectric properties of the vulcanized specimens was performed with a broadband dielectric spectrometer, BDS 40 system manufactured by Novocontrol GmbH Germany, providing a bandwidth from 0.1 Hz to 10 MHz. The measurement geometry was a disc shaped plate-capacitor of 40 mm diameter where the specimen was placed between two gold-plated brace electrodes. The thickness of the sample was about 2 mm. Thin gold layers were sputtered onto the flat surfaces of the specimens to ensure electrical contact to the electrode plates. Temperature was controlled better than 0.5°C.

The experimental setup for combined dielectric and rheological measurements consists of the combination of two different machines: a high resolution impedance spectrometer (Novocontrol Alpha-Analyser B 40) and a rotational rheometer (Anton Paar Physica MCR501s). The rheometer has tuned up engines to deliver enough torque (0.3 Nm) and normal force (from −70 N to 70 N) to measure high viscose and elastic polymer composites. Further, it is equipped with a ceramic isolated dielectric chamber and both plates are used as capacitor while rheological measurement are carried out. The measurements were performed at 140°C. The amplitude was kept small ($\gamma = 0.02\%$) while flocculation and recovering. For the strain sweep measurements a ramp from $10^{-3}\%$ to 21% with 5 steps/decade was chosen.

2.3 *Dynamic-mechanical measurements*

The dynamic-mechanical measurements were performed in the torsion-rectangular mode with strip specimen of 2 mm thickness and 30 mm length with an ARES rheometer (Rheometrix). The dynamic moduli were measured over a wide temperature range (−80°C to +80°C) at a frequency of 1 Hz and 0.5% strain amplitude. The temperature dependent measurements.

2.4 *Multi-hysteresis measurements*

Uniaxial multi-hysteresis tests at $\dot{\varepsilon} \approx 0.01/s$ (quasistatic) were carried out at various temperatures by using dumbbell samples, 15 mm in thickness, with a Zwick 1445 universal testing machine. Multi-hysteresis means: at constant velocity up and down – cycles between certain minimum and maximum strains, ε_{min} and ε_{max}, are carried out. This was done 5 times each step, and after every of such steps the boundaries of deformation are successively raised (ε_{max}) or lowered (ε_{min}), respectively. Only every 5th closed cycle was considered for the fitting procedure with the Dynamic Flocculation Model, which can be regarded as being in an equilibrium state in a good approximation. The nominal stress σ_1 was determined by dividing the measured force by the initial cross-section area. To determine the nominal strain ε_1 the displacement between two reflection marks glued onto the surface was measured and divided by the initial distance.

3 RESULTS AND DISCUSSION

3.1 *Dynamics of filler flocculation in polymer melts*

Combined dielectric-rheological investigations of filler flocculation as well as strain induced fracture of filler networks under oscillatory deformations have

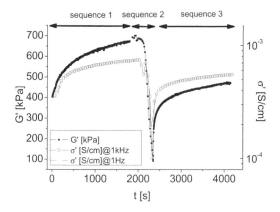

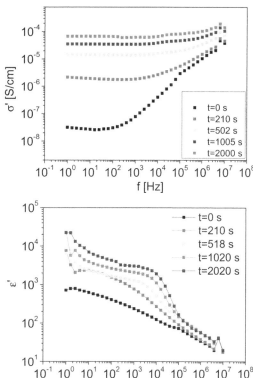

Figure 1. Time dependence of the modulus G' and real part of the conductivity σ' at two frequencies, 1 Hz and 1 kHz, during heat treatment of the EPDM melt with 50 phr N339 at 140°C. Three sequences are shown: first flocculation at 0.02% strain (sequence 1), then a strain sweep up to 21% (sequence 2) and finally recovery at 0.02 % strain (sequence 3).

been applied for uncross-linked rubber melts with various amounts and types of conductive fillers. In particular, a simultaneous analysis of flocculation with rheological methods and dielectric spectroscopy give a deeper insight into the nature and dynamics of filler flocculation.

Fig. 1 shows an example for a measurement of an EPDM melt with 50 phr N339 at 140°C. The shear modulus G', and the real part of the AC-conductivity are presented as sequence of three experiments in a queue: flocculation, strain sweep and recovering. During flocculation (sequence 1), the filler-filler network is build up due to attractive depletion forces acting on the filler particles. The polymer composite becomes stiffer, G' and the conductivity rise. The amplitude is kept small in order that the filler-filler network is not affected. During the strain sweep (sequence 2), the filler-filler network breaks down due to increasing mechanical straining. In the strain sweep, the network is successively destroyed and both, conductivity and G' decrease. This gives direct evidence that the Payne effect results from the breakdown of the carbon black network. During recovery (sequence 3), the amplitude is set back to a smaller value of $\gamma = 0.02\%$ implying that the filler-filler network can re-aggregate again. This leads to a newly rise in modulus and conductivity though the initial levels after flocculation are not reached.

Fig. 2 shows the real part of the conductivity and permittivity, respectively, of the composite with 40 phr HS55 during sequence 1 at different flocculation times. Obviously, the conductivity increases significantly and the permittivity shows in the beginning of the flocculation at $t = 0$ a broad relaxation process, which gets narrower with ongoing time and moves to higher frequencies. We point out that the carbon black loading of the sample is close to the percolation threshold where the strongest changes of

Figure 2. Real part of the conductivity σ' (top) and permittivity ε' (bottom) vs. frequency f of the uncross-linked EPDM melt with 40 phr HS55 at 140°C for different flocculation times. The shift of the observed relaxation transition for the permittivity to higher frequencies indicates a decrease of the gap size between adjacent CB particles with increasing flocculation time.

electrical properties are observed. In particular, the ad-conductivity plateau increases by about four decades from around 10^{-8} S/cm to almost 10^{-4} S/cm and the composite changes from an isolator to a conductor. The shift of the observed relaxation transition of ε' to higher frequencies indicates a decrease of the relaxation time which can be related to a decreasing gap size between adjacent CB particles with increasing flocculation time (Fritzsche & Klüppel 2011). Since the particles become closer, the tunneling distance of the charge carriers is reduced leading to an exponential increasing of the charge transport over the gaps. This explains why the conductivity rises by many decades.

This interpretation of the dielectric spectra relates the characteristic time $\tau_G = 1/\omega_G$ of the observed relaxation transition of ε' to a tunneling process of charge carriers over gaps between adjacent carbon black particles (primary aggregates). Following the model of Kawamoto (Kawamoto 1982), two carbon black aggregates connected by bound rubber can be treated as a micro capacitor with a capacity C_G and a micro resistor R_G in parallel. The resistance can be calculated from the quantum mechanical tunneling

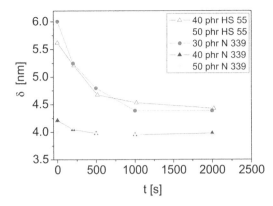

Figure 3. Gap distance δ from online-dielectric spectroscopy vs. flocculation time t for uncross-linked EDM/CB composites at different filler loadings, as indicated.

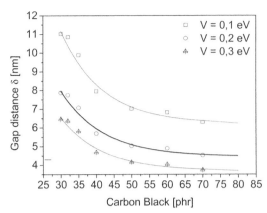

Figure 4. Gap distance δ vs. carbon black loading of cross-linked S-SBR/N339 composites for three different values of the tunneling barrier V, as indicated.

current The characteristic frequency of the tunneling process over a carbon black – carbon black connection can then be related to the gap distance δ as follows (Meier et al. 2007, Meier & Klüppel 2008, Fritzsche & Klüppel 2011):

$$\omega_G = \frac{1}{R_G C_G} = \frac{3e^2}{16\pi^2\hbar\varepsilon_0}\frac{k_0}{\varepsilon}exp(-k_0\delta) \qquad (1)$$

This is obtained from the capacitance of the gaps given by:

$$C_G = \varepsilon_0\varepsilon\frac{A}{\delta} \qquad (2)$$

and the tunneling resistance:

$$R_G = \frac{16\pi^2\hbar}{3e^2}\frac{\delta}{k_0A}exp(k_0\delta) \qquad (3)$$

with

$$k_0 = \frac{2\sqrt{2m_e}}{\hbar}\sqrt{V} \qquad (4)$$

Here, k_0 is a constant that depends on the mean height V of the potential barrier, A is the cross section of the gaps, $\hbar$ the Planck constant, e the charge and m_e the mass of the charge carriers.

With this model and the assumption of a typical potential barrier $V = 0.3$ eV and a dielectric constant of the polymer in the gap of $\varepsilon = 3$, typical nanoscopic gap distances between $\delta = 4$ and 6 nm are estimated with Equ. (1). Thereby the characteristic frequencies ω_G have been obtained from Cole-Cole fits of the permittivity. Fig. 3 shows the behavior of the evaluated gap distances for the HS55 and the N339 carbon black at different filler loadings as a function of flocculation time. It is seen that on the one side, the gaps become smaller with ongoing flocculation time which is most

pronounced for the low filler loadings, i.e. 30 phr N339 and 40 phr HS55. On the other side, the gap distance decreases systematic with filler loading which agrees with former observations made for vulcanized samples (Fritzsche & Klüppel 2011).

3.2 The role of gap size in viscoelastic properties of vulcanized rubbers

The gap distance of carbon black filled elastomers has also been evaluated from the dielectric spectra of the cross-linked samples after vulcanization, by using Equ. (1) has been used and Typical results for S-SBR vulcanizates are depicted in Fig. 4 where the gap distance is plotted in dependence of N339 concentration for three different values of the tunneling barrier V. It is demonstrated that the choice of V has a strong impact on the gap distance. This is due to the fact that the square root of V enters into the exponent of Equ. (1). On the other side, the chosen dielectric constant of the polymer in the gap $\varepsilon = 3$ has no significant effect on the gap distance because it enters as pre-factor in Equ. (1).

We point out that the cross-linked sample have experienced a heat treatment at 160°C during vulcanization for several minutes and therefore the carbon black network should be in a well flocculated state comparing to large flocculation times in Fig. 3. Fig. 4 demonstrates that also for the cross-linked samples the gap distance decreases systematic with filler loading indicating that the driving depletion forced for flocculation increase with filler loading. A possible reason can be the more pronounced multiple-particle interaction for highly filled systems which increases the effective depletion forced between neighboring particles. The evaluated gap distances are listed in Tab. 1. Here, also the gap distances obtained for the vulcanized EPDM/CB composites are included.

The question arises whether the gap distance, corresponding to the length of filler-filler bonds, can be

Table 1. Comparison of the activation energy of cross-linked EPDM and S-SBR composites, determined by temperature dependent measurements of G' (Fig. 5), and gap distance evaluated from dielectric spectra with tunneling barrier $V = 0.3$ eV.

	Activation energy [kJ/mol]		Gap distance [nm]	
	HS55	N339	HS55	N339
EPDM 20 phr	–	3.2	–	–
EPDM 30 phr	3.1	3.3	–	–
EPDM 40 phr	3.8	5.0	4.9	4.4
EPDM 50 phr	4.9	6.2	4.1	4.0
S-SBR 20 phr	–	1.2	–	6.5
S-SBR 40 phr	–	5.3	–	4.7
S-SBR 60 phr	–	9.4	–	4.0

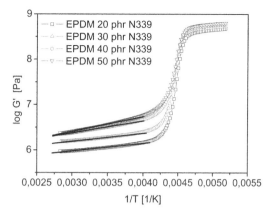

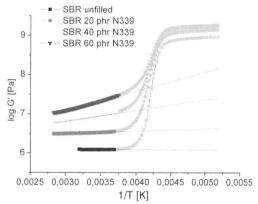

Figure 5. Arrhenius plot of the temperature dependent small strain storage modulus of cross-linked EPDM/N339 composites (top) and S-SBR/N339 composites (bottom) at different filler loadings, as indicated. The slope of the inserted regression lines determines the apparent activation energy of filler-filler bonds listed in Tab. 1.

correlated to the mechanical properties of filled elastomers. An Arrhenius-plot of the temperature dependent small strain storage modulus of cross-linked EPDM/N339 and S-SBR/N339 composites, respectively, at different loadings of carbon black N339 is shown in Fig. 5. As indicated by the inserted regression lines one observes a thermal activation of G' in the high temperature range, which becomes more pronounced with increasing filler content. This effect is typical for elastomers filled with highly reinforcing fillers (Heinrich & Klüppel 2002, Heinrich & Klüppel 2004).

The apparent activation energies obtained from Fig. 5 are summarized in Table 1. They increase systematically with increasing filler loading which correlates with the decreasing gap distance also listed in Tab. 1. This indicates that there is a change of the dynamics of the glassy-like polymer bridges resulting from the decreasing gap size of the glassy-like polymer bridges. A comparison of the two types of CB shows that the gaps are smaller for the more fine black N339 which is accompanied by larger activation energies. In addition, a comparison of the two polymers shows that the gap distance is larger for the S-SBR samples due to a stronger polymer-filler coupling which again results in larger activation energies.

From a mechanical point of view the gaps correspond to glassy-like polymer bridges representing quite stiff filler-filler bonds which transmit the stress between adjacent particles of the filler network. The glassy-like behavior of the polymer in the gaps can be related to strong confinement effects close to the attractive filler surfaces which act from two sides. This correlates with the experimental observation that the glass transition temperature of ultra-thin films between attractive walls increases strongly with film thickness if the thickness falls below 20 nm (Hartmann et al. 2003, Soles et al. 2004, Grohens et al. 2002). These results have been confirmed by simulations and analytical studies concerning the change of dynamics next to filler surfaces (Starr et al. 2002, Douglas &

Freed 1997). The gap size between adjacent filler particles of the filler network typically lies in the range of a few nm (Figs 3 and 4), implying strong effects of the gap spacing on mechanical properties of the filler-filler bonds, e.g. stiffness, strength or activation energy.

3.3 Temperature dependence of filler-filler bonds in quasi-static stress-strain cycles

The quasi-static stress-strain cycles of filler reinforced elastomers have been described successfully by the Dynamic Flocculation Model (DFM). This model combines well established concepts of rubber elasticity with a micro-mechanical approach of dynamic filler cluster breakdown and re-aggregation in strained rubbers at different elongations. The mechanical response of the samples is based on two micro-mechanical mechanisms: (i) hydrodynamic reinforcement of the rubber matrix by a fraction of rigid filler aggregates with strong virgin filler-filler bonds, which have not been broken during previous deformations and (ii) cyclic breakdown and re-aggregation

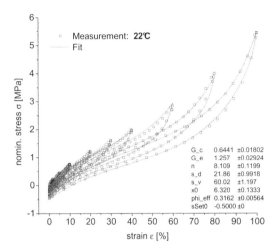

Figure 6. Quasi-static stretching cycles (5th cycle) of a S-SBR sample with 60 phr N339 at 22°C with successive increase of the maximum strain (symbols) and fit with the DFM (lines). Fitting parameters are listed in the legend.

In the figure legend:

G_c	0.6441	±0.01802
G_e	1.257	±0.02924
n	8.109	±0.1199
s_d	21.86	±0.9918
s_v	60.02	±1.197
x0	6.320	±0.1333
phi_eff	0.3162	±0.00564
sSet0	-0.5000	±0

of the remaining fraction of soft filler clusters with damaged and hence weaker filler-filler bonds. The fraction of rigid (unbroken) filler clusters decreases with increasing strain, while the fraction of soft filler-clusters increases. The first effect is responsible for the well known stress softening of filled rubbers and is described by a decreasing strain amplification factor with increasing strain. The second effect results in a filler-induced hysteresis, since the soft clusters that are stretched in the stress field of the rubber store energy that is dissipated when the cluster breaks. We point out that this kind of hysteresis response is present also for quasi-static deformations, where no explicit time dependency of the stress-strain cycles is taken into account.

A detailed mathematical description of the DFM is found in previous publications (Klüppel 2003, Klüppel et al. 2005, Lorenz et al. 2011, Freund et al. 2011). Here we are mainly interested in the basic properties and the physical meaning of the fitting parameters. An example of a fit with the DFM is given in Fig. 6 where the typical quasi-static, multi-hysteresis stretching cycles of the S-SBR/60 phr N339 sample at 22°C are shown. The fit is performed with the 5th cycles that are considered to be in a stationary equilibrium. Fair agreement between fit and experimental data is found with fitting parameters listed in the legend of the plot. The values of all fitting parameters appear physically reasonable. The three upper parameters stand for the polymer matrix the rest are related to the filler clusters. The two moduli $G_c = 0.64$ MPa and $G_e = 1.26$ MPa scale with the cross-link and entanglement density of the rubber and $n = n_e/T_e = 8.1$ is the number of statistical segments between trapped entanglements with T_e being the trapping factor. The parameters $s_v = 60$ MPa and $s_d = 21.9$ MPa are the strength of virgin and damage filler-filler bonds, respectively. As expected it holds

$s_v > s_d$. Since both quantities represent the yield stress of glassy-like polymer bridges it makes sense that they are significantly larger than the polymer moduli G_c and G_e. The parameter $x_0 = 6.3$ is the related mean cluster size (in units of the particle size) indicating that the clusters are quite small. The fitted effective filler volume fraction $\Phi_{eff} = 0.32$ is somewhat larger than the filler ration $\Phi = 0.24$ (60 phr) since it also takes into account the structure of the filler particles. The parameter $s_{set,0} = -0.5$ MPa considers the set behavior of the sample and stands for the setting stress at $\varepsilon_{max} = 1$.

In the following we will focus on the consideration of temperature dependency in the DFM. In the temperature range well above the glass transition temperature the rubber matrix can considered to be independent of temperature (rubber elastic plateau). Then, in agreement with the treatment of small strain viscoelasticity in the previous section, the temperature dependence of mechanical properties results from the thermal activation of filler bonds, i.e. an Arrhenius-like decrease of the stiffness of glassy-like polymer bridges with increasing temperature. This effect is considered in the DFM by an Arrhenius-like temperature dependency of the strength s_v and s_d of virgin and damage filler-filler bonds, respectively:

$$s_v \equiv \frac{Q_v \varepsilon_{b,v}}{d^3} = s_{v,ref} \, e^{\frac{E_v}{R}\left(\frac{1}{T} - \frac{1}{T_{ref}}\right)} \tag{5}$$

$$s_d \equiv \frac{Q_d \varepsilon_{b,d}}{d^3} = s_{d,ref} \, e^{\frac{E_d}{R}\left(\frac{1}{T} - \frac{1}{T_{ref}}\right)} \tag{6}$$

Here, Q_v and Q_d are the respective temperature dependent elastic constants corresponding to the stiffness of virgin and damaged filler-filler bonds, $\varepsilon_{b,v}$ and $\varepsilon_{b,d}$ are the respective yield strains, which are considered to be almost independent of temperature. d is the particle diameter, R is the gas constant, T is temperature and E_v and E_d are the respective activation energies of virgin and damaged filler-filler bonds. $s_{v,ref}$ and $s_{d,ref}$ are reference values of the bond strength determined at $T = T_{ref}$. We note that the strength of the virgin bonds s_v affects the strain amplification factor X which influences the stress softening and also the storage modulus G'. In contrast, s_d affects the hysteresis and loss modulus G''.

Beside the bond strength, also the set behavior of the sample is temperature dependent. This is considered by the empirical relation:

$$\sigma_{set} = s_{set,0} \cdot a(T - T_g) \cdot f(\varepsilon_{min}, \varepsilon_{max}) \tag{7}$$

The temperature dependent factor has the simple form $a = (T_{ref} - T_g)/(T - T_g)$ and the strain dependence of the set stress is best described by a square root function $f = \varepsilon_{max}^{1/2} - |\varepsilon_{min}|^{1/2}$. The fitting parameter $s_{set,0}$ equals the set stress at $\varepsilon_{max} = 1$ (if $\varepsilon_{min} = 0$) at reference temperature $T_{ref} = 22°C$.

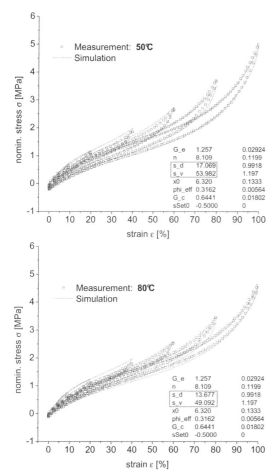

Figure contains two plots. Top plot parameters:

G_e	1.257	0.02924
n	8.109	0.1199
s_d	17.069	0.9918
s_v	53.982	1.197
x0	6.320	0.1333
phi_eff	0.3162	0.00564
G_c	0.6441	0.01802
sSet0	-0.5000	0

Bottom plot parameters:

G_e	1.257	0.02924
n	8.109	0.1199
s_d	13.677	0.9918
s_v	49.092	1.197
x0	6.320	0.1333
phi_eff	0.3162	0.00564
G_c	0.6441	0.01802
sSet0	-0.5000	0

Figure 7. Quasi-static stretching cycles (5th cycle) of the S-SBR sample with 60 phr N339 at 50°C (top) and 80°C (bottom) with successive increase of the maximum strain (symbols) and simulations with the DFM (lines). Thereby, the two marked material parameters, i.e. the strength s_v and s_d of virgin and damage filler-filler bonds, respectively, are modified according to Equs. (5) and (6) with activation energies of $E_v = 3$ kJ/mol and $E_d = 7$ kJ/mol.

Fig. 7 shows the quasi-static, multi-hysteresis stretching cycles of the S-SBR sample with 60 phr N339 at 50°C and 80°C, respectively. Based on the fitting parameters obtained at 22°C, the stress-strain curves were simulated by applying Equ. (7) together with the thermal activation of the bond strengths, Equs. (5) and (6). To simulate the stress-strain curves at 50°C and 80°C, suitable activation energies of $E_v = 3$ kJ/mol and $E_d = 7$ kJ/mol have been chosen to achieve a fair agreement between experiment and simulation. Using these activation energies, the strengths of filler-filler bonds, s_v and s_d, are calculated for the desired temperature T. The resulting parameter values for s_v and s_d are marked in the insets of Fig. 7.

We point out that the activation energies E_v and E_d can in principle be predicted from viscoelastic measurements (compare also Tab. 1). For the same type of S-SBR with 60 phr N339 master curves were created from dynamic mechanical analysis at 3.5% strain amplitude (Le Gal et al. 2005). The thermal activation energies were obtained from vertical shifting factors of G' and G'', respectively, as $E_v = 5.4$ kJ/mol and $E_d = 7.72$ kJ/mol. These values are somewhat higher than the ones necessary to simulate the large-strain behavior but sufficient for strains below $\varepsilon \approx 40\%$. The activation energies at 0.5% strain amplitude are, consequently, even higher: $E_v = 8.81$ kJ/mol and $E_d = 15.26$ kJ/mol.

4 CONCLUSIONS

By applying combined rheological and dielectric analysis it has been shown in the first part of the paper that due to the heat treatment of uncross-linked carbon black filled rubber melts, a pronounced flocculation of filler particles takes place in the rubber matrix leading to an increase of the modulus and a drastic modification of the dielectric spectra. It has been found that the conductivity increases significantly during flocculation and a second high frequency relaxation transition appears which is traced back to the tunneling of charge carriers over small gaps between adjacent carbon black particles (Fig. 2). From the dielectric spectra the gap size has been evaluated which is found to decrease with filler loading and flocculation time (Fig. 3). Obviously, the gap size decreases during heat treatment leading to a stiffening of filler-filler bonds and hence a rise of the modulus. Dielectric investigations with the corresponding cross-linked samples confirm that the gap size decreases with increasing carbon black concentration in the range of 4–6 nm (Fig. 4).

In the second part of the paper, correlations between the gap size and the viscoelastic response of filled, cross-linked rubbers at small strain has been explored. From a mechanical point of view the gaps correspond to glassy-like polymer bridges representing quite stiff filler-filler bonds which transmit the stress between adjacent particles of the filler network. The glassy-like behavior of the polymer in the gaps can be related to strong confinement effects close to the attractive filler surfaces which act from two sides. Based on this picture, the temperature dependence of the small strain modulus above the glass transition temperature can be explained by the thermal response of filler-filler bonds which become softer with increasing temperature. Due to this thermal activation of filler-filler bonds, the modulus above the glass transition temperature shows an Arrhenius-like activation behavior (Fig. 5). The observed increase of the apparent activation energy with increasing filler loading correlates with the observed decrease of the gap size, i.e. the length of filler-filler bonds (Tab. 1) indicating that there is a change of the dynamics of the glassy-like polymer bridges resulting from the decreasing gap size of filler-filler bonds. This observation supports our view about the thermo-mechanical properties of filler-filler bonds which are shown to play a major

17

role in understanding the viscoelastic properties of elastomer composites.

Finally, the thermo-mechanical response of glassy-like polymer bridges (filler-filler bonds) has shown to be closely related to the quasistatic stress-strain properties of filled elastomers up to large strain. By referring to the structural analysis of elastomer composites, a micro-structure based dynamic flocculation model of stress softening and filler-induced hysteresis has been developed (DFM). It is based on a tube model of rubber elasticity together with a micro-mechanical model of stress induced filler cluster breakdown. The evaluation of stress softening is obtained via a pre-strain dependent hydrodynamic amplification of the rubber matrix by a fraction of rigid filler clusters with virgin filler-filler bonds. The filler-induced hysteresis is described by a cyclic breakdown and re-aggregation of the residual fraction of softer filler clusters with already broken, damaged filler-filler bonds. The model is in fair agreement with experimental data obtained with carbon black filled elastomers (Fig. 6). In particular it is shown that the effect of temperature on the stress-strain cycles can be well described by considering an Arrhenius like activation behavior of the filler-filler bonds with activation energies obtained from dynamic-mechanical measurements (Fig. 7).

ACKNOWLEDGEMENTS

This work has been supported by the Deutsche Forschungsgemeinschaft (SPP 1369 and FOR 597) and the Bundesministerium für Bildung und Forschung (BMBF, 03X3533D).

REFERENCES

Douglas, J. F. & Freed, K. F. 1997. *Macromolecules* 30:1813
Freund, M., Lorenz, H., Juhre, D., Ihlemann, J. & Klüppel, M. 2011. *Int. J. Plasticity* 27:902
Fritzsche, J. & Klüppel, M. 2011. *J. Phys.: Condens. Matter* 23:035104
Grohens, Y., Hamon, L., Reiter, G., Soldera, A. & Holl, Y. 2002. *Eur. Phys. J. E*, 8:217
Hartmann, L., Fukao, K. & Kremer, F. 2003. Molecular dynamics in thin polymer films. In Kremer, F., Schönhals, A. (eds.) *Broadband dielectric spectroscopy*, Springer: Berlin, Heidelberg, New York
Heinrich, G. & Klüppel, M. 2002. Recent Advances in the Theory of Filler Networking in Elastomers. *Adv. Polym. Sci.* 160:1
Heinrich, G. & Klüppel, M. 2004. *Kautschuk Gummi Kunststoffe* 57:452
Klüppel M. 2003. The Role of Disorder in Filler Reinforcement of Elastomers on Various Length Scales. *Adv. Polym. Sci.* 164:1
Klüppel, M., Meier, J. & Dämgen M. 2005. Modelling of stress softening and filler induced hysteresis of elastomer materials. In Austrell & Kari (eds), *Constitutive Models for Rubber IV*. London: Taylor & Francis
Kraus G. (ed.) 1965. *Reinforcement of elastomers*. N.Y. London Sydney: Wiley, Interscience Publ.
Le Gal, A., Yang, Y. & Klüppel, M. 2005. *J. Chem. Phys.* 123: 014704
Lorenz, H., Freund, M., Juhre, D., Ihlemann, J. & Klüppel, M. 2011. Macromol. Theory Simul. 20:110
Medalia, A. I. 1973. *Rubber Chem. Technol.* 46:877
Meier, J. G., Mani, J. W. & Klüppel, M. 2007. *Phys. Rev. B* 75: 054202
Meier, J. G. & Klüppel, M. 2008. *Macromol. Mater. Eng.* 293:12
Payne, A. R. 1963. *J. Appl. Polym. Sci.* 7:873
Starr, F. W., Schröder, T. B. & Glotzer, S. C. 2002. *Macromolecules* 35:4481
Soles, C. L., Douglas, J. F. & Wu, W.-L. 2004. *J. Polym. Sci, Part B* 42:3218–3234

Constitutive Models for Rubber VII – Jerrams & Murphy (eds)
© 2012 Taylor & Francis Group, London, ISBN 978-0-415-68389-0

Elastic instabilities in rubber

A.N. Gent

The University of Akron, Akron Ohio, US
Chonbuk National University, Jeonju, Korea

ABSTRACT: Materials that undergo large elastic deformations can exhibit novel instabilities. Moreover, the onset of such highly non-uniform deformations has serious implications for the fatigue life and fracture resistance of rubber components. Several examples are described: development of an aneurysm on inflating a rubber tube; non-uniform stretching on inflating a spherical balloon; formation of internal cracks in rubber blocks when they are subjected to a critical amount of triaxial tension or supersaturated with a dissolved gas; wrinkling of the surface at a critical degree of compression; and the sudden formation of "knots" on twisting stretched cylindrical rods. These various deformations are analyzed in terms of the theory of large elastic deformations [1, 2] and the theoretical results are then compared with experimental measurements of the onset of unstable states [3]. Such comparisons provide new tests of the theory and, at least in principle, critical tests of the validity of proposed strain energy functions for rubber. (They also provide interesting challenges for finite element programmers.)

REFERENCES

[1] R. S. Rivlin, Philos. Trans. Roy. Soc. Lond. Ser. **A241** (1948) 379–397.
[2] O. Lopez-Pamies, M. I. Idiart, T. Nakamura, J. Mech. Phys. Solids **59** (2011) 1464–1505.
[3] 3. A. N. Gent, Internatl. J. Non-Linear Mech. **40** (2005) 165–175.

BIOGRAPHICAL NOTES

Alan N. Gent is Professor Emeritus of Polymer Physics and Polymer Engineering at The University of Akron, where he has been since 1961. He is currently a WCU Professor at Chonbuk National University, Korea. He has received many scientific awards including the Bingham Medal of the Society of Rheology, the Colwyn Medal of the Plastics and Rubber Institute, the International Research Award of the Society of Plastics Engineers, the 3M Award of the Adhesion Society, the Charles Goodyear Medal of the American Chemical Society and the Polymer Physics Prize of the American Physical Society. He received N.A.S.A.'s Public Service Medal for services rendered after the Challenger space-shuttle disaster (1986) and was elected to the U.S. National Academy of Engineering in 1991.

Mechanical characterization of rubber – Novel approaches

Constitutive Models for Rubber VII – Jerrams & Murphy (eds)
© 2012 Taylor & Francis Group, London, ISBN 978-0-415-68389-0

Strain-induced crystallization of natural rubber subjected to biaxial loading conditions as revealed by X-ray diffraction

S. Beurrot, B. Huneau & E. Verron

LUNAM Université, Ecole Centrale de Nantes, GeM, UMR CNRS, Nantes cedex, France

ABSTRACT: Strain induced crystallization (SIC) in carbon black-filled natural rubber (NR) is investigated by wide-angle X-ray diffraction (WAXD) using synchrotron radiation for three deformation states: uniaxial, biaxial (but non-equibiaxial) and equibiaxial tension. The crystallites size is of the same order of magnitude and the lattice parameters are similar for the three states. But the orientation of the crystallites varies: as the crystallites are highly oriented accordingly to the tensile direction for uniaxial loading condition, they are oriented (but with a higher degree of disorientation) along the tensile direction (highest stretch ratio) for biaxial loading condition, and they are not oriented in the plane of tension for equibiaxial loading condition.

1 INTRODUCTION

Natural rubber (NR), cis-1,4-polyisoprene, has remarkable mechanical properties which are generally explained by strain-induced crystallization (SIC). SIC is commonly investigated by wide-angle X-ray diffraction (WAXD) (see Murakami et al. 2002, Toki et al. 2003, Trabelsi et al. 2003a, among others, for a more complete review of these works, the reader can refer to Huneau 2011). The great majority of the studies on SIC focuses on uniaxial tension; but as engineering applications involve multiaxial loading conditions, mechanical models must take into account the multiaxiality of the material response; for example, it is well-known that the efficiency of a given constitutive equation for rubberlike materials requires shear or biaxial experimental data (Marckmann and Verron 2006). In this way, the present paper is devoted to the experimental study of biaxial strain-induced crystallization of NR by X-ray diffraction using synchrotron radiation. To our knowledge, this is the second study on the influence of biaxiality on strain-induced crystallization of natural rubber: Oono et al. (1973) studied the orientation of crystallites in a stretched thin film of unfilled vulcanized natural rubber at $-27°$C.

Here, we compare (i) the orientation and (ii) the size of the crystallites, and (iii) the lattice parameters of the crystal unit cell in filled NR submitted to uniaxial, biaxial and equibiaxial loading conditions, at room temperature and for thick samples.

2 EXPERIMENTAL METHOD

2.1 *Material, sample and local strain measurement*

The material used in this study is a classic carbon black-filled natural rubber. The composition is given in Table 1.

Table 1. Chemical composition of the filled NR used in this study (g per 100 g of rubber).

Rubber	100
Carbon black (N330)	50
Zinc oxide	5
Stearic acid	2
Sulfur	1.2
Accelerator (CBS)	1.2
Antioxydant	1

Figure 1 shows the two samples used in this study. The uniaxial tensile tests were performed on classical flat dumbbell specimens; dimensions are given in Fig. 1a. Obtaining biaxial deformation for soft materials is not an easy task (Demmerle and Boehler 1993). For this study, we developed a cruciform sample, based on a symmetric cross thinned at its center as shown in Fig. 1b and 1c. Under loading, the arms of the sample are uniaxially stretched and the central zone is in a complex deformation state (see Fig. 1d). When the four arms are equally stretched, the only point at which the deformation is equibiaxial is the center point of the specimen. The central part of the sample is made thinner in order to reach higher strain levels, without breaking the arms of the cross sample. Both homogeneity and equibiaxiality were verified by finite element analysis and the zone in which the material is subjected to nearly equibiaxial tension is quiet larger than the beam spot, i.e. about 3 mm^2. To achieve biaxial but non-equibiaxial deformation at the center of the sample, two opposite arms of the cross sample are simply more stretched than the two other ones.

As the cross sample is in a complex state of deformation when the arms are stretched, it is difficult to predict the stretch ratio in the center of the sample from the values of displacement of the actuators only.

Therefore, the biaxial tests have been filmed and a motion analysis system (Tema motion©) has been used to track the displacement of points of a paint pattern applied at the center of the sample. It allowed to determine the deformation gradient **F** from which the two stretch ratios in the directions parallel to the plane of tension are extracted. The general form of the deformation gradient, assuming incompressibility, is:

$$\mathbf{F} = \begin{bmatrix} \lambda & 0 & 0 \\ 0 & \lambda^B & 0 \\ 0 & 0 & \lambda^{-B-1} \end{bmatrix} \quad (1)$$

where B is the biaxiality factor, $B \in [-0.5; 1]$. One can note that $B = -0.5$ for a material subjected to uniaxial tension and $B = 1$ in the case of equibiaxial tension. When the material is subjected to biaxial (non-equibiaxial) tension, $B \in]0; 1[$.

2.2 Synchrotron

Synchrotron measurements have been carried out at the DiffAbs beamline in the French national synchrotron facility SOLEIL. The wavelength used is 1.319 Å and the beam size is 0.3 mm in diameter at half-maximum. The 2D WAXD patterns are recorded by a MAR 345 CCD X-ray detector. In order to make an accurate correction of air scattering, a PIN-diode beam stop was used.

2.3 Tensile testing machine

The experiments have been conducted with a home-made stretching machine shown in Figure 2. It is composed of four electrical actuators, which displacements can be synchronized or not. Their loading capacity is ±500 N and their stroke is 75 mm each. All the experiments are conducted by prescribing the displacement of these actuators. Opposite actuators always have opposite equal displacements in order to keep the central zone of the sample fixed. For uniaxial tests, only two actuators are used; for biaxial tests, the four actuators are used.

2.4 Procedure

The stretching unit is placed on the stand of the diffractometer (see Fig. 2) in order to keep the beam focused on the centre of the samples. The uniaxial experiment is quasi-static (the actuators speed is set to 0.012 mm/s) and scattering patterns are recorded every 100 seconds (due to the reading time of the CCD detector). For the biaxial experiments, the actuators speed is 30 mm/s, and one pattern is recorded at maximum deformation. The exposure time is short (1 second for uniaxial test and 3 seconds for biaxial tests): it permits to reduce the influence of kinetics of crystallization on the results.

An air scattering pattern (without sample) was first collected and has been used to correct the patterns. Moreover, the change in thickness of the sample under extension and the change of intensity of the incident photons have also been considered. All these corrections are performed by following the well-established method of Ran et al. (2001). Both the determination of the pattern center and the calibration of the diffraction angles were achieved by considering the first diffraction ring of ZnO ((100)-plane, $a = 3.25$ Å (Reeber 1970)). Here, small angles scattering is not investigated; the range of diffraction angles is $2\theta \in [8°, 26.7°]$.

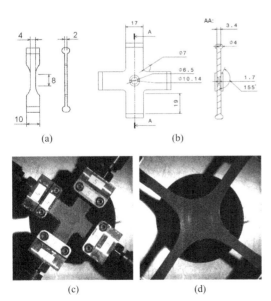

(a)   (b)

(c)   (d)

Figure 1. Dimensions of the (a) uniaxial samples and (b) biaxial samples. Cruciform sample (c) in undeformed state and (d) in equibiaxial deformation state.

Figure 2. Uniaxial and biaxial stretching machine in DiffAbs.

24

2.5 Scattering pattern analysis

The spectra extracted from the diffraction patterns are classically fitted by series of Pearson functions (Trabelsi et al. 2003a, Chenal et al. 2007, Rault et al. 2006, Toki et al. 2000); before deconvolution, the linear baseline of each spectrum is suppressed (see example in Figure 3). The lattice parameters of the crystal cell of the polyisoprene are calculated considering a monoclinic crystal system, as determined by Bunn (1942). The crystallites size is deduced from the Scherrer formula (Guinier 1963):

$$l_{hkl} = \frac{K\lambda}{\text{FWHM}_{2\theta}\cos\theta} \qquad (2)$$

where l_{hkl} is the crystallites size in the direction normal to the hkl plane, K is a scalar that depends on the shape of crystallites (here we adopt 0.78 as Trabelsi et al. (2003a)), λ is the radiation wavelength, θ is the Bragg angle and $\text{FWHM}_{2\theta}$ is the full width at half maximum of the peak hkl in 2θ. Finally, the disorientation ψ_{hkl} (compared to the mean orientation) of the hkl diffraction plane in the cristallites is simply given by half the full width at half maximum (FWHM_{β}) of the peaks, measured on the azimuthal profiles of the reflection.

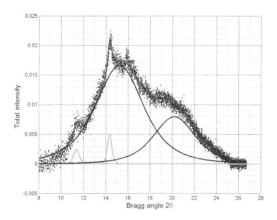

Figure 3. Example of fitting and deconvolution of a spectrum with a series of Pearson functions.

3 RESULTS AND DISCUSSION

3.1 Strain-induced crystallization as revealed by WAXD patterns

Figure 4 shows the diffraction patterns for filled NR in undeformed state (Fig. 4a), uniaxial tension

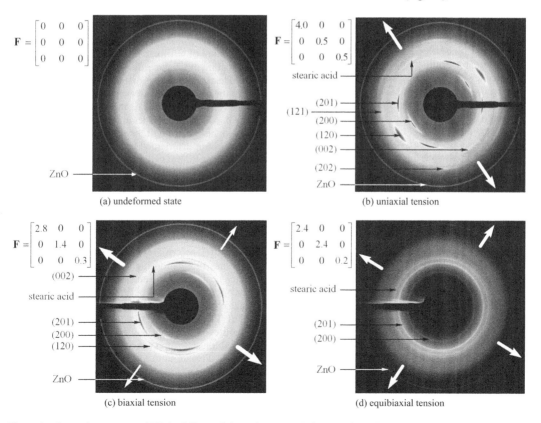

Figure 4. Scattering patterns of NR in different deformation states (**F** is the deformation gradient; the white arrows show the tensile directions).

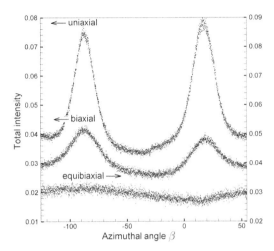

Figure 5. Azimuthal spectra at $2\theta = 12.3°$ (Bragg angle of the (201) arcs and ring) for different deformation states: uniaxial tension, biaxial tension and equibiaxial tension.

(Fig. 4b), biaxial tension (Fig. 4c) and equibiaxial tension (Fig. 4d). For each state, the deformation gradient **F** is given. In the undeformed state, the diffraction pattern consists of a diffuse amorphous halo and a ring due to the (100) Bragg reflection of ZnO. Fig. 4b exhibits the diffraction reflections of crystallized NR in uniaxial tension. The diffraction pattern is of course oriented according to the stretching direction (see the white arrows in the figure). The pattern is composed of eight intense crystalline reflection arcs corresponding to the three different crystallographic planes (200), (201) and (120), ten less intense reflection arcs corresponding to the crystallographic planes (121), (202) and (002), a co-existing amorphous halo, the ZnO reflection ring and stearic acid reflection arcs. The diffraction pattern of crystallized NR in biaxial tension (Fig. 4c) is quite similar to the pattern of NR in uniaxial tension. The pattern is also oriented, according to the main direction of traction (direction of the highest stretch ratio). Some diffraction arcs are not observed in the pattern, and all the arcs (corresponding to the planes (002), (201), (200) and (120)) are much less intense and wider. When filled NR is subjected to equibiaxial tension (Fig. 4d), the diffraction pattern is quite different. As previously, an amorphous halo, a ZnO ring and stearic acid arcs are observed. But the diffraction reflections of the NR crystalline phase are rings and not arcs. Only the reflections corresponding to the planes (201) and (200) are visible.

3.2 *The orientation of the crystallites depends on B ...*

In order to measure the width of the arcs and ring corresponding to the (201) plane, azimuthal spectra are extracted from the three previous diffraction patterns of NR in deformed state and shown in Figure 5; the Bragg angle $2\theta = 12.3°$ corresponds to the position of the (201) arcs and ring in the three patterns. The origin

Table 2. Biaxiality factor and disorientation of the crystallites.

Loading conditions	B	$\psi_{200}(°)$	$\psi_{201}(°)$
Uniaxial	−0.5	11.70	11.09
Biaxial	0.3	15.25	17.03
Equibiaxial	1	90	90

of β is arbitrarily defined as the horizontal direction in the biaxial tension pattern ($\beta = -36°$ is the tension direction and $\beta = 54°$ is the direction perpendicular to the tension direction for all the tests). The width of the reflection arcs in the patterns (i.e. the width at half maximum of the peaks in the spectra) directly stems for the orientation of the crystallites: the wider the arcs, the more disoriented the crystallites. In the case of equibiaxial loading, the reflections of the planes (200) and (201) are rings which means the crystallites are not oriented in the plane of tension. This result extends those of Oono et al. (1973) who showed that the crystallites are oriented in the direction perpendicular to the traction directions in a thin film of NR. The low anisotropy which can be observed in the spectrum of the equibiaxial test is due to experimental settings (it can also be observed in patterns of amorphous undeformed samples). In the case of uniaxial and biaxial tension, all the reflections of the crystalline phase are arcs, which means that the crystallites are oriented in the plane of tension. Furthermore, the (002) arcs are aligned with the direction of tension for the uniaxial test and with the direction of tension of highest stretch ratio for the biaxial test, which means that the c-axis of the crystall cell of the crystallites is in average oriented along the main direction of tension. As the peaks in the spectrum of the biaxial test are wider at half maximum than the peaks in the spectrum of the uniaxial test, the crystallites are more disoriented in NR subjected to biaxial tension than to uniaxial tension. The exact disorientation of the crystallites for each deformation state is measured and given in Table 2, in comparison with the biaxiality factor B. In uniaxial tension, the crystallites are disoriented of about 11°; it is slightly smaller than in Poompradub et al. (2005) and twice smaller than in Trabelsi et al. (2003b). This discrepancy is explained by the difference in cross-link densities: indeed, the crystallites orientation highly depends on cross-link density, i.e. formulation and processing, especially for filled NR (Chenal et al. 2007, Poompradub et al. 2005, Tosaka et al. 2004, Trabelsi et al. 2003b, Trabelsi et al. 2003a). Tab. 2 confirms that the disorientation of the crystallites is higher when the material is in biaxial deformation state than when it is in uniaxial deformation state. Furthermore, it suggests that the larger the biaxiality factor, the higher the disorientation.

3.3 *... but the crystallites are identical*

The Bragg angle spectra corresponding to the three deformed states are shown in Figure 6: the azimuthal

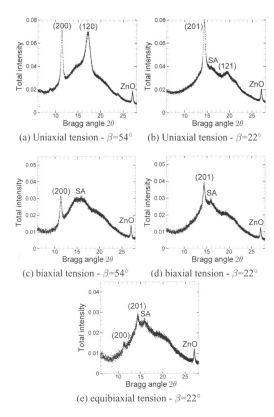

(a) Uniaxial tension - $\beta=54°$ (b) Uniaxial tension - $\beta=22°$

(c) biaxial tension - $\beta=54°$ (d) biaxial tension - $\beta=22°$

(e) equibiaxial tension - $\beta=22°$

Figure 6. Bragg angle spectra for different loading conditions and different azimutal angles β (SA: stearic acid).

Table 3. Lattice parameters.

Loading conditions	a (Å)	c (Å)
Uniaxial	13.38	8.90
Biaxial	13.35	8.93
Equibiaxial	13.33	9.04

angles are chosen in order to observe the (200) peaks ($\beta = 54°$, Fig. 6a and 6c) and the (201) peaks ($\beta = 22°$, Fig. 6b, 6d and 6e). For the equibiaxial tension test, only one spectrum at any azymuthal angle β is necessary to observe both reflections as the pattern is isotropic.

From Fig. 6, we deduce the lattice parameters of the crystal unit cell and the size of the crystallites for the three tests; results are given in Tables 3 and 4 respectively. The crystall cell unit is identical for the three deformation states. Indeed, the diffraction angles of the (200) and (201) planes are very close for the three tests: the corresponding lattice parameters only differ from 1.5% maximum. The lattice parameters of the crystall cell in uniaxial tension are larger than those calculated by other authors in carbon-black filled NR (for example, Poompradub et al. (2004) have found $a = 12.65$ Å, $b = 9.15$ Å and $c = 8.35$ Å). This difference may arise

Table 4. Size of the crystallites.

Loading conditions	l_{200} (Å)	l_{201} (Å)
Uniaxial	127.2	121.2
Biaxial	135.9	122.1
Equibiaxial	122.5	120.7

Table 5. Crystallites size in uniaxial tension for different stretch ratios λ.

λ	l_{200} (Å)	l_{201} (Å)
2.8	161	145
3.4	147	137
4.0	127	121

from our calibration method of Bragg angles with the (100) plane ZnO ring. Indeed, the lattice parameter a of the ZnO used for calibration corresponds to pure ZnO, and may slightly differ from the parameter of industrial ZnO inside rubber. This discrepancy with the bibliography does not influence the previous comparative result as the same method has been used for the three tests. This extends the recent result of Poompradub et al. (2004), who demonstrated that the lattice parameters of filled NR only slightly evolve with strain in uniaxial tension: we demonstrate here that they do not change with the deformation state either.

The crystallites sizes obtained for uniaxial tension are quite different from the results obtained in the rare studies published on crystallization of carbon black-filled NR: $l_{200} = 127.2$ Å and $l_{201} = 121.2$ Å, compared to $l_{200} = 170$ Å or $l_{200} = 220$ Å (depending on the quantity of fillers) found by Poompradub et al. (2005) and $l_{002} = 100$ Å found by Trabelsi et al. (2003b). But this discrepancy is explained by the difference in cross-link densities, similarly as the difference in crystallites disorientation. From the results given in Tab. 4, one may conclude that crystallites in NR subjected to uniaxial, biaxial and equibiaxial deformation states have the same size. But we observe in uniaxial tension that crystallites size depends on strain as shown in Table 5; similar results are given in Poompradub et al. (2005) and Tosaka et al. (2004). In order to precisely compare results in uniaxial, biaxial and equibiaxial tension, it would then be necessary to measure the crystallites size at different stretch ratios in biaxial and equibiaxial deformation states.

4 CONCLUSION

Firstly, this work shows that in equibiaxial tension crystallization of filled NR is isotropic in the plane of tension; it results in rings in the diffraction pattern. It is very different from crystallization in uniaxial tension which is strongly anisotropic, as revealed in the

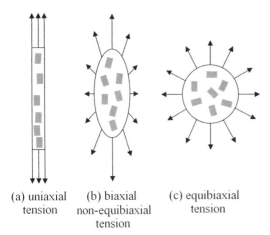

(a) uniaxial tension   (b) biaxial non-equibiaxial tension   (c) equibiaxial tension

Figure 7. Schematic representation of crystallites in stretched filled NR: the more biaxial the loading conditions are, the more disoriented the crystallites are; but crystallites are identical (in size and lattice parameters) for all experiments.

diffraction pattern by narrow reflection arcs. In biaxial (non-equibiaxial) tension, crystallites of filled NR have an intermediate orientation: they are oriented in average along the direction of highest stretch ratio but the level of disorientation is higher than in uniaxial tension; indeed the reflection arcs in the diffraction pattern are wider for biaxial tension than for uniaxial tension. It seems that the larger the biaxial factor B calculated from the deformation gradient is, the more disoriented crystallites are. Secondly, crystallites are similar in uniaxial, biaxial and equibiaxial deformation states: the lattice parameters of the cristal unit cell are identical and the size of the crystallites is of the same order of magnitude. This result is illustrated in Figure 7.

ACKNOWLEDGEMENT

The authors thank Dr D. Thiaudière, Dr C. Mocuta and Dr A. Zozulya from the DiffAbs line in the synchrotron facility Soleil for their great help during the experiments.

REFERENCES

Bunn, C. W. (1942). Molecular structure and rubber-like elasticity i. the crystal structures of beta gutta-percha, rubber and polychloroprene. *Proceedings of the Royal Society of London Series A-Mathematical and Physical Sciences 180(A980)*, 40–66.

Chenal, J.-M., L. Chazeau, L. Guy, Y. Bomal, & C. Gauthier (2007). Molecular weight between physical entanglements in natural rubber: A critical parameter during strain-induced crystallization. *Polymer 48*, 1042–1046.

Chenal, J.-M., C. Gauthier, L. Chazeau, L. Guy, & Y. Bomal (2007). Parameters governing strain induced crystallization in filled natural rubber. *Polymer 48*, 6893–6901.

Demmerle, S. & J. P. Boehler (1993). Optimal design of biaxial tensile cruciform specimens. *Journal of the Mechanics and Physics of Solids 41*, 143–181.

Guinier, A. (1963). *X-ray Diffraction*. W. H. Freeman & Co.

Huneau, B. (2011). Strain-induced crystallization of natural rubber: a review of X-ray diffraction investigations. *Rubber Chemistry And Technology 84*, to appear.

Marckmann, G. & E. Verron (2006). Comparison of hyperelastic models for rubberlike materials. *Rubber Chemistry And Technology 79*, 835–858.

Murakami, S., K. Senoo, S. Toki, & S. Kohjiya (2002). Structural development of natural rubber during uniaxial stretching by in situ wide angle X-ray diffraction using a synchrotron radiation. *Polymer 43*, 2117–2120.

Oono, R., K. Miyasaka, & K. Ishikawa (1973). Crystallization kinetics of biaxially stretched natural rubber. *Journal of Polymer Science 11*, 1477–1488.

Lattice defvormation of strain-induced crystallites in carbon-filled natural rubber. *Chemistry Letters 33*, 220–221.

Poompradub, S., M. Tosaka, S. Kohjiya, Y. Ikeda, S. Toki, I. Sics, & B. S. Hsiao (2005). Mechanism of strain-induced crystallization in filled and unfilled natural rubber vulcanizates. *Journal of Applied Physics 97*, 103529/1–103529/9.

Ran, S., D. Fang, X. Zong, B. S. Hsiao, B. Chu, & P. F. Cunniff (2001). Structural changes during deformation of kevlar fibers via on-line synchrotron SAXS/WAXD techniques. *Polymer 42*, 1601–1612.

Rault, J., J. Marchal, P. Judeinstein, & P. A. Albouy (2006). Chain orientation in natural rubber, part II: 2H-NRM study. *The European Physical Journal E 21*, 243–261.

Reeber, R. R. (1970). Lattice parameters of ZnO from 4.2 degrees to 296 degrees K. *Journal of Applied Physics 41*, 5063–5066.

Toki, S., T. Fujimaki, & M. Okuyama (2000). Strain-induced crystallization of natural rubber as detected real-time by wide-angle X-ray diffraction technique. *Polymer 41*, 5423–5429.

Toki, S., I. Sics, S. Ran, L. Liu, & B. S. Hsiao (2003). Molecular orientation and structural development in vulcanized polyisoprene rubbers during uniaxial deformation by in-situ synchrotron X-ray diffraction. *Polymer 44*, 6003–6011.

Toki, S., I. Sics, S. F. Ran, L. Z. Liu, B. S. Hsiao, S. Murakami, K. Senoo, & S. Kohjiya (2002). New insights into structural development in natural rubber during uniaxial deformation by in situ synchrotron X-ray diffraction. *Macromolecules 35*), 6578–6584.

Tosaka, M., S. Murakami, S. Poompradub, S. Kohjiya, Y. Ikeda, S. Toki, I. Sics, & B. S. Hsiao (2004). Orientation and crystallization of natural rubber network as revealed by WAXD using synchrotron radiation. *Macromolecules 37*, 3299–3309.

Trabelsi, S., P. A. Albouy, & J. Rault (2003a). Crystallization and melting processes in vulcanized stretched natural rubber. *Macromolecules 36*, 7624–7639.

Trabelsi, S., P. A. Albouy, & J. Rault (2003b). Effective local deformation in stretched filled rubber. *Macromolecules 36*, 9093–9099.

Effects of strain field and strain history on the natural rubber matrix

K. Bruening & K. Schneider

Department of Mechanics and Structure, Leibniz-Institut für Polymerforschung Dresden, Germany

ABSTRACT: The effects of the strain field and the strain history on the matrix morphology of carbon black filled natural rubber has been investigated using synchrotron wide-angle X-ray scattering. It is shown that strain-induced crystallization occurs analogously in pure shear and simple tensile tests despite the different strain fields. Comparison between quasistatic cyclic tensile tests and mechanically aged samples showed that strain-induced crystallization and chain orientation are completely reversible and do not contribute to aging. To evaluate the filler and crystallite contributions to the mechanical reinforcement, we compared graphitized and untreated carbon black filled samples and found that for the case under investigation the filler contribution dominates over the crystallite contribution.

1 INTRODUCTION

In order to understand the long-term behavior of elastomers subjected to cyclic mechanical loads, and in order to model the material performance in their respective application, tear fatigue analysis (TFA) experiments are frequently carried out in addition to classical quasistatic tensile tests.

Strain-induced crystallization (SIC), which is the key for many of the oustanding mechanical properties of natural rubber (Grellmann & Seidler 2001), has been investigated by X-ray scattering for more than seven decades (Gehman & Field 1939). Thanks to the high brilliance of modern third generation synchrotron sources it now has become possible to bring dynamics and X-ray scattering together, e.g. acquire scattering images of propagating cracks or during dynamic mechanical experiments (DMA) in real-time and thus correlate results from mechanical tests with structural information gained by X-ray scattering.

In the last years, numerous studies on in-situ quasi-static cyclic tensile experiments were reported (Poompradub et al. 2005, Goppel 1949, Gehman & Field 1939, Tosaka et al. 2004, Toki et al. 2002, Trabelsi et al. 2003a). It is well-known that SIC is reversible but shows some hysteresis, comparable to the mechanical hysteresis. However, the role of fatigue loading, i.e. several thousand strain cycles, on the matrix morphology has not yet been investigated.

Also, the effect of the strain field on the degree of crystallinity (DOC) has thus far not been systematically investigated with modern-day methods (Mering & Tchoubar 1968). On the other hand, several works dealing with a spatial structural resolution around the crack tip have recently been published (Trabelsi et al. 2002, Zhang et al. 2009, Lee & Donovan 1987). However, the correlation of these findings with the mechanical fields in the crack tip zone remains to be done.

To understand the fundamental influences of the strain field, we compared the morphology evolution under strain in simple tensile and pure shear modes.

To gain better insight into the role of strain history, we performed in-situ cyclic tests and tests on ex-situ mechanically aged samples.

2 EXPERIMENTAL SECTION

2.1 Materials and specimens

The materials consist of carbon black-filled natural rubbers (NR-F*) and unfilled rubbers (NR-S and NR-P). NR-F1 and NR-F1g contain 40 phr untreated and graphitized N339, respectively. NR-F2 contains 60 phr N234. NR-S is sulfur-vulcanized, whereas NR-P is peroxide-vulcanized (Table 1).

The shouldered tensile bars are of 1 mm thickness, 6 mm parallel length and 2 mm width. Pure shear specimens are 30 mm wide, 4 mm high and 1 mm thick. Both geometries have beads at the ends for better clamping.

The objective of the large width-to-height ratio in pure shear mode is to reduce lateral contraction and thus to generate a planar stress state (Treloar 1975).

Table 1. Rubber formulations. The sulfur-cured formulations additionally contain 3 phr zinc oxide, 1 phr stearic acid, 2.5 phr accelerator (CBS) and 1.5 phr antioxidant (IPPD).

	Filler	Crosslinker
NR-S	0	1.7 phr sulfur
NR-F1	40 phr N339	1.7 phr sulfur
NR-F1g	40 phr N339 graphitized	1.7 phr sulfur
NR-F2	60 phr N234	1.7 phr sulfur
NR-P	0	2.0 phr DCP

Mechanical aging of the tensile samples was performed on a dynamic mechanical testing machine at a frequency of 1 Hz and a strain amplitude of 70% for 10^4 cycles and consecutively at a strain amplitude of 120% for 2000 cycles, which is very close to fatigue failure.

2.2 Experimental setup

The stretching experiments were performed on a selfmade miniature tensile machine at the BW4 and P03 beamlines at DESY, Germany, using Pilatus 300K CCD detectors (Schneider 2010). The optical strain was evaluated following a pattern sprayed onto the sample surface and is hereafter simply referred to as strain.

2.3 Data processing

After background correction and normalization with respect to sample thickness, the degree of crystallinity was evaluated in 2θ-φ space by subtracting the amorphous contribution I_{am} from the scattering image I_{total} and setting the obtained crystalline contribution in relation to the total scattering intensity:

$$\text{DOC [\%]} = 100 \cdot \frac{I_{total} - I_{am}}{I_{total}} \tag{1}$$

The strain amplification factor A_ψ is taken as the ratio of the strain ($\varepsilon = \Delta l/l_0$) of the unfilled reference NR-S versus the strain of the filled sample at which the property ψ (stress σ or DOC) has the same value (Trabelsi et al. 2003b):

$$A_\psi = \frac{\epsilon_{unfilled}\left(\psi = \psi^*\right)}{\epsilon_{filled}\left(\psi = \psi^*\right)} \tag{2}$$

3 RESULTS AND DISCUSSION

3.1 Influence of strain field on strain-induced crystallization

It was found that the different strain fields in pure shear and simple tensile mode do not induce differences in SIC, as is exemplified for NR-P in Figure 1. Also, the stress-strain curves for simple tensile and pure shear generally coincide, except that the strain at break is lower for pure shear.

The results show that the SIC mechanism is insensitive to small variations in the strain field and that it is valid to deduce the strain field around a crack tip from a WAXS scan. However, this finding does not permit to draw any general conclusions regarding the effects of larger deviations from the uniaxial strain state (e.g. biaxial strain).

3.2 Cyclic stretching of natural rubber

As has been reported in the literature (e.g. Trabelsi, et al. 2003b), SIC is fully reversible upon complete

a)
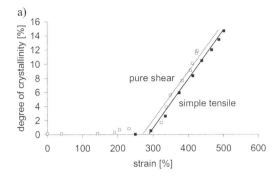

b)
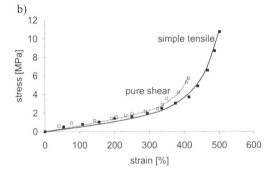

Figure 1. a) stress-strain curve and b) degree of crystallinity vs. strain for NR-P in pure shear and simple tensile geometry.

retraction. As shown in Figure 2, this is also valid for multiple cyclic stretching. The hysteresis in the DOC curves resembles the stress-strain curves, representing the Mullins effect. Making the step from several in-situ cycles to ex-situ mechanical aging, no significant permanent changes were observed in terms of SIC or amorphous orientation, as is exemplified in Figure 3.

3.3 Correlation between crystallinity and mechanics

The definition of strain amplification factors (eq. 2) for quantities obtained by WAXS can serve as connection between the macroscopic mechanical behavior and the microscopic morphology. Graphitization of carbon black is a commonly applied method to change its surface structure and to reduce the interaction with the rubber matrix (Fröhlich et al. 2005, Klüppel 2003). Therefore, the stress-strain curve of NR-F1g is significantly lower than that of the ungraphitized counterpart NR-F1, which is reflected in a smaller mechanical strain amplification factor A_σ (Fig. 4). Nevertheless, the degree of crystallinity and thus A_{DOC} is the same in both samples. This shows that SIC is mainly strain-driven (not stress-driven) and the larger DOC in a more highly filled samples is due to the hydrodynamic reinforcement and not due to the higher stress level (Poompradub, et al. 2005). It also shows that the mechanical properties in NR-F1 and NR-F1g are dominated by the filler rather than by SIC.

a)

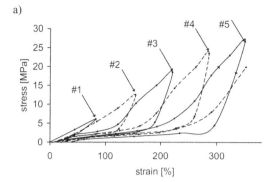

b)

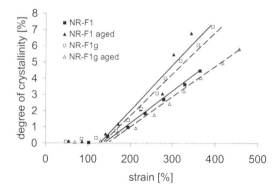

Figure 2. a) stress-strain curves and b) degree of crystallinity in cyclic tensile test on NR-F2 with increasing strain amplitude (#1: 80%; #2: 150%; #3: 220%; #4: 290%; #5: 350%).

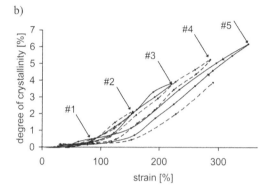

Figure 3. Degree of crystallinity before and after mechanical aging.

4 CONCLUSION

Wide-angle X-ray scattering studies on filled and unfilled natural rubbers revealed that the difference in strain fields between simple tension and pure shear modes has no effect on the extent of strain-induced crystallization.

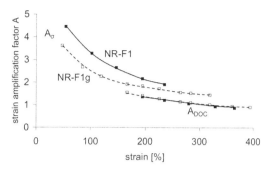

Figure 4. Strain amplification factors A_σ and A_{DOC} for NR-F1 and NR-F1g, based on NR-S.

Performing fatigue tests (application of several thousand strain cycles) does not induce any permanent changes in terms of amorphous orientation or crystallization.

Comparison between samples filled with graphitized and non-graphitized carbon black underlines the fact that the crystallization is mainly strain-driven and that the contributions of filler and crystallites to the mechanical reinforcement can be distinguished.

With the gained understanding of the influence of the strain field and the effects of cyclic stretching, the foundations for the in-situ study of dynamic mechanical experiments and crack propagation have been established.

REFERENCES

Fröhlich, J., W. Niedermeier, & H.-D. Luginsland (2005). *The effect of filler-filler and filler-elastomer interaction on ruber reinforcement*. Composites Part A: Applied Science and Manufacturing 36(4), 449–460. Filled and Nano-Composite Polymer Materials.

Gehman, S. & J. Field (1939). *An X-Ray Investigation of Crystallinity in Rubber*. Journal of Applied Physics 10, 564.

Goppel, J. (1949). On the degree of crystallinity in natural rubber. Applied Scientific Research 1(1), 3–17.

Grellmann, W. & S. Seidler (2001). *Deformation and fracture behaviour of polymers*. Springer Verlag.

Klüppel, M. (2003). *The role of disorder in filler reinforcement of elastomers on various length scales*. Advances in Polymer Science 164, 1–86.

Lee, D. & J. Donovan (1987). *Microstructural changes in the crack tip region of carbon-black-filled natural rubber*. Rubber chemistry and technology 60(5), 910–923.

Mering, J. & D. Tchoubar (1968). *Interpretation de la diffusion centrale des rayons X par les systemes poreux*. I. Journal of Applied Crystallography 1(3), 153–165.

Poompradub, S., M. Tosaka, S. Kohjiya, Y. Ikeda, S. Toki, I. Sics, & B. Hsiao (2005). *Mechanism of strain-induced crystallization in filled and unfilled natural rubber vulcanizates*. Journal of Applied Physics 97, 103529.

Schneider, K. (2010). *Investigation of structural changes in semi-crystalline polymers during deformation by synchrotron X-ray scattering*. Journal of Polymer Science Part B: Physics 48(14), 1574–1586.

Toki, S., I. Sics, S. Ran, L. Liu, B. Hsiao, S. Murakami, K. Senoo, & S. Kohjiya (2002). *New insights into structural development in natural rubber during uniaxial deformation by in situ synchrotron X-ray diffraction.* Macromolecules 35(17), 6578–6584.

Tosaka, M., S. Murakami, S. Poompradub, S. Kohjiya, Y. Ikeda, S. Toki, I. Sics, & B. Hsiao (2004). *Orientation and crystallization of natural rubber network as revealed by WAXD using synchrotron radiation.* Macromolecules 37(9), 3299–3309.

Trabelsi, S., P. Albouy, & J. Rault (2002). *Stress-induced crystallization around a crack tip in natural rubber.* Macromolecules 35(27), 10054–10061.

Trabelsi, S., P. Albouy, & J. Rault (2003a). *Crystallization and melting processes in vulcanized stretched natural rubber.* Macromolecules 36(20), 7624–7639.

Trabelsi, S., P. Albouy, & J. Rault (2003b). *Effective local deformation in stretched filled rubber.* Macromolecules 36(24), 9093–9099.

Treloar, L. (1975). *The physics of rubber elasticity (third ed.).* Oxford Classic Series, 2005.

Zhang, H., J. Niemczura, G. Dennis, K. Ravi-Chandar, & M. Marder (2009). *Toughening effect of strain-induced crystallites in natural rubber.* Physical review letters 102(24), 245503.

Constitutive Models for Rubber VII – Jerrams & Murphy (eds)
© 2012 Taylor & Francis Group, London, ISBN 978-0-415-68389-0

Parameter identification based on multiple inhomogeneous experiments of practical relevance

D. Schellenberg & D. Juhre
German Institute for Rubber Technology, Hannover, Germany

J. Ihlemann
Professorship of Solid Mechanics, Chemnitz University of Technology, Chemnitz, Germany

ABSTRACT: Existing procedures to identify material parameters are based on experiments with simple specimens. Additionally, the load distribution is approximately homogeneous. But there are only few feasible experiments which produce homogeneous or almost homogeneous load distributions. Furthermore, deviations from homogeneity and their consequences cannot be avoided, but are often ignored. We present a solution algorithm which takes several different experimental results into account. Thereby, the experiments are performed on specimens which respond with inhomogeneous distributions of strains and stresses. The restriction to homogeneous loads is not necessary. Thus, it is possible to use different measured data of multiple load cases on one and the same test specimen. The component-oriented design of the specimen permits to consider the specific properties of product groups, the load types and the effect of the manufacturing process on the final material properties already during the identification process.

1 INTRODUCTION

Many material models used in industry are chosen because of their suitability for daily use. In the majority of cases it is not the most accurate reproduction of the real material behavior that has highest priority. Moreover, the adequacy is determined by finding the right compromise between specific demands and economic restrictions due to the available calculating capacity and time. Consider, for example, different components using the same material and the same constitutive model. As a matter of fact, the material properties depend on both the manufacturing process and the geometry of the components. Then, it is to be expected that the optimal set of material parameters is different in each case. Consequently, the identification process for the specific problem should to be performed separately. This is only possible in parts using common standard specimens.

However, identifying material parameters by means of those specimens is already state of the art (Hohl 2007). They usually show a simple geometry. Additionally, the experiments are performed in such a way that the load is approximately homogeneous. Hence, the parameter identification requires rather little calculation capacity and the material behavior can be characterized straight forward. Therefore, that method was chosen as the basis for this paper and is further explained in section 2.

The geometry of standard specimens, on the one side and real components, on the other side is in most cases very different. Nevertheless, the former are used

to obtain the material parameters characterizing the latter. In addition, they are not even manufactured under the same process conditions. These circumstances can lead to significant differences concerning the material behavior. As a consequence, component simulations using material parameters fitted to standard specimens are a priori afflicted with errors. To overcome these drawbacks, an alternative method is presented in section 3. Therefore, experiments have been performed on specimens with a geometry close to the real component shape. Because the geometry is still simplified, this sort of specimens will be called component-oriented specimens or simplified component in the following. In contrast to the standard method, there is no restriction to experiments with homogeneous stress and strain distributions. On the basis of such measured data the material parameters were then identified. At the end of this paper in section 4 the proposed method and the standard approach are compared with the help of reference measurements on real components.

2 IDENTIFICATION USING STANDARD TEST SPECIMENS

Parameter identification poses a second order inverse problem (Moritz 1993). In general, an analytical solution does not exist. This is because material parameters cannot be directly obtained from measurements. Mostly, it is only possible to analyze them indirectly by their effect on the experiment. A common approach to handle this task is to restate it as an optimization problem. In particular, the method of least squares is suitable to determine parameters $\boldsymbol{p} = (p_1, \ldots, p_n)^T$ of constitutive equations. Therefore, the residua are calculated at certain instants of time t_j. The value of each residuum is then given by the difference between measured data $\bar{y}_j = \bar{y}_j(t_j)$ of the component and results of the respective model $\bar{m}(\boldsymbol{p}, t_j) = y_j(\boldsymbol{p})$ (Nocedal 1999):

$$r_j(\boldsymbol{p}) \; = \; \bar{y}_j - \bar{m}(\boldsymbol{p}, t_j) \; = \; \bar{y}_j - y_j(\boldsymbol{p}, t_j)$$
$$\text{with} \quad j \; = \; 1, \ldots, m \tag{1}$$

Now the set of parameters $\boldsymbol{p}^*$ has to be found which minimizes the target function $\Phi(\boldsymbol{p})$:

$$\Phi(\boldsymbol{p}) \; = \; \frac{1}{2} \sum_{j=1}^{m} [r_j(p_1, \ldots, p_n)]^2 \; \rightarrow \; \min_{p} \tag{2}$$

In the present example the residua are always calculated as differences between the measured data and the respective numerically obtained values from the material model.

Assuming that the target function is sufficiently smooth a gradient-based optimization method is employed. Starting with the initial set of parameters the next one is calculated as follows:

$$\boldsymbol{p}_{k+1} \; = \; \boldsymbol{p}_k + \omega \, \boldsymbol{s}_k, \tag{3}$$

at which ω represents the step length along the step direction $\boldsymbol{s}_k$. To solve the optimization problem the Levenberg-Marquardt method (Levenberg 1944) is adopted. It combines the advantages of the steepest decent method with the benefits of the Newton procedure.

To calculate the necessary gradients displacements and forces have to be differentiated with respect to material parameters. This is usually done numerically using finite differences (Schnur 1992). Since this has to be accomplished for each material parameter the calculation effort rises dramatically with increasing number of parameters. Hence, finite forward differences are applied in this work. For this purpose, the target function is expanded into a first order Taylor series:

$$\Phi(p_j + \triangle p_j) \; = \; \Phi(p_j) + \frac{\partial \Phi}{\partial p_j} \triangle p_j +$$
$$O\left((\triangle p_j)^2\right), \quad j \in [1, n], \tag{4}$$

at which n is the number of parameters. The remaining term O can be neglected and accordingly the derivative with respect to material parameters reads as follows:

$$\frac{\partial \Phi}{\partial p_j} \; = \; \frac{\Phi(p_j + \triangle p_j) - \Phi(p_j)}{\triangle p_j} \quad j \in [1, n] \tag{5}$$

From this it follows that for each varied parameter a single FEM simulation has to be carried out.

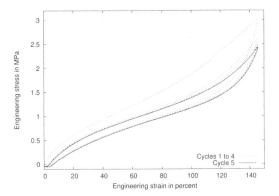

Figure 1.   Stress-strain diagram due to uniaxial tensile test.

Table 1.   Results of the identification of material parameters based on homogeneous experiments.

	C_{10} [MPa]	C_{20} [MPa]	C_{30} [MPa]	Φ
Initial set	1.75	1.4	1.25	2.4e+6
Target set	0.359	−0.0115	0.00454	3.847

In the present work four different, nearly homogeneous experiments were performed on a NR 50 (N330):

- uniaxial tensile test,
- biaxal tensile test,
- plain strain tensile test,
- simple shear test.

Each test was carried out by displacement controlled amplitudes with five repeating cycles. Since the material parameters were not to be identified for the virgin state the transient cycles of the tests were not taken into account. Solely the fifth cycle which is nearly stationary is considered. As an example the stress-strain diagram of the uniaxial tensile test is shown in Figure 1.

The FEM simulations in the present work were carried out using the commercial program *MSC.MARC*. Furthermore, the hyperelastic Yeoh model is used. Like in the most non-linear material models for rubber the material parameters of the Yeoh model have no physical meaning and thus, their values are not predictable in advance. However, due to the dependence of the Yeoh model only on the first invariant of the Cauchy-Green tensor, its characteristics does not change that much on a variation of the material parameters. This is reflected in the optimization process, too. It detects the same minimum point of the target function for varying initial sets of parameters. Table 1 shows the determined material parameters starting from a randomly chosen initial set.

The iteration steps shown in Figure 2 demonstrate the good convergence of the optimization process.

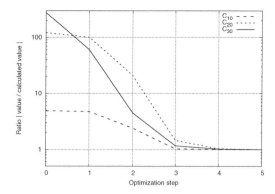

Figure 2. Convergence ratio of homogeneous identification.

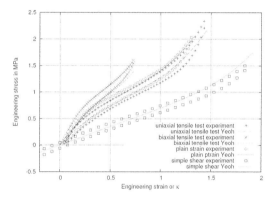

Figure 3. Stress-strain diagram with experiments and simulations using Yeoh parameters identified with homogeneous standard tests.

The stress curves obtained from the experiments for the parameter identification are depicted in Figure 3. Likewise the results from the respective FEM simulations using the identified set of parameters are shown.

3 IDENTIFICATION USING SIMPLIFIED COMPONENTS

In the present work the procedure to determine material parameters is exemplified for a typical component of practical relevance. To be concrete a bushing was chosen which consists of an outer and an inner sleeve made of metal and a rubber layer in between. The corresponding FEM model of the rubber part is illustrated in Figure 4.

Deploying standard test specimens and homogeneous comparative calculations leads to several disadvantages:

- The material properties diverge.
- Only very few load types can be investigated in the experiments.
- Discrepancies from homogeneity in the experiments are neglected in comparative calculations.

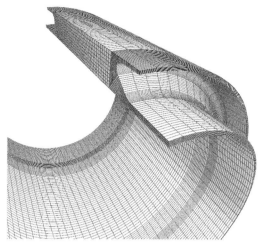

Figure 4. Rubber layer of the original bushing in the undeformed state.

- Different load types require different specimens. Hence, a combined load is not feasible.

To overcome these drawbacks the experiments could be performed on the original bushing. But then respective FEM simulations would also have to use the complex geometry of the component. This again would drastically increase the calculation time which is, in turn, not acceptable regarding the multitude of required simulations. Furthermore, it is necessary to ensure the convergence of FEM simulations even though the set of parameters is somewhat "nonsensical". Thus, using component-oriented specimen is an appropriate way to reproduce the properties of the real component and, at the same time, to significantly reduce computational costs. The advantages can be summarized as follows:

- The material properties are nearly identical.
- A multitude of typical load cases can be realized.
- Discrepancies from homogeneity are not relevant.
- Different load types can be performed on one and the same specimen.
- Boundary conditions can be chosen such that a good reproducibility is achieved.

The FEM model of the rubber part of the component-oriented specimen is shown in Figure 5.

The experiments needed for the identification were adjusted to typical loading conditions of the original bushing. For the chosen example they were determined from the component's specification sheet and read as follows:

- axial load,
- radial load,
- torsion load,
- cardanic load (Bending around the rotational axis).

The actual identification only takes the first three load cases into account. The measured data from the cardanic load serve as possibility for the validation

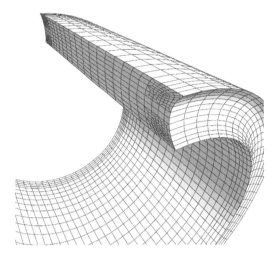

Figure 5. Rubber layer of the component-oriented specimen in the undeformed state.

Table 2. Results of the identification of material parameters based on experiments with simplified components.

	C_{10} [MPa]	C_{20} [MPa]	C_{30} [MPa]	Φ /10^4
Initial set	0.359	−0.0115	0.00454	1.77
Target set	0.372	0.00988	0.00001	1.06

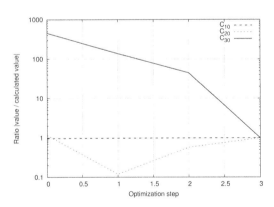

Figure 6. Convergence ratio of inhomogeneous identification.

of the obtained parameters. Likewise to the previous identification method the Yeoh material model is adopted. The initial set of parameters is given by the set identified from the tests on the standard specimens. This ensures a good comparability of the two identification methods. As can be seen in table 2 the initial value of the target function can be significantly reduced in the course of the new identification.

Due to the good initial set of parameters only few optimization steps are necessary to find the minimum of the target function (Figure 6)

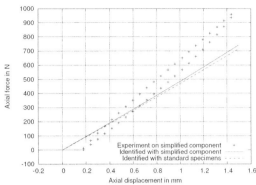

Figure 7. Comparison of the resulting forces due to axial loading (experiment on component-oriented specimen and simulation).

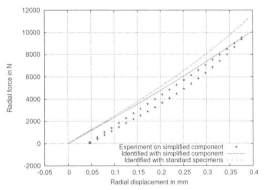

Figure 8. Comparison of the resulting forces due to radial loading (experiment on component-oriented specimen and simulation).

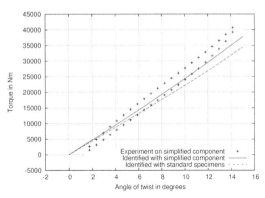

Figure 9. Comparison of the resulting torque due to torsional loading (experiment on component-oriented specimen and simulation).

The decline of the target function value is also noticeable in Figure 7–Figure 9. For all load cases the result curves using the new identified set of parameters fit the measured data much better than those curves using the old set (fitted to the homogeneous tests).

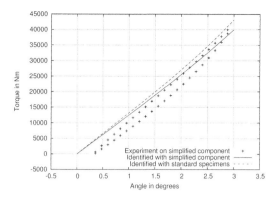

Figure 10. Comparison of the resulting torque due to cardanic loading (experiment on component-oriented specimen and simulation).

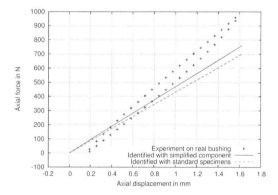

Figure 11. Comparison of the resulting forces due to axial loading (experiment on original bushing and simulation).

To evaluate the reliability of the identified parameters the test with cardanic load is simulated using the initial set of parameters and the new identified set, respectively. The comparison of both results is shown in Figure 10. It is obvious that the material response to cardanic load can be well reproduced. Of course, the quality is limited by the abilities of the chosen material model. Still, this agreement is remarkable since the cardanic test causes a very complex loading of the component.

In the last step the experiments performed on the component-oriented specimen were repeated using the original bushing. Again, parallel FEM simulations were carried out by using the two sets of material parameters, respectively. The comparison of the results by applying axial loading is depicted in Figure 11 and for torsion in 12. Again, the discrepancy between simulation and experiment is significantly reduced by using the newly identified parameter set. Additionally, it is obvious that the curves for the component-oriented specimen and the original bushing are quite similar,

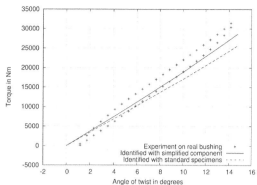

Figure 12. Comparison of the resulting torque due to torsional loading (experiment on original bushing and simulation).

even though there are considerable geometrical differences. Therefore, it is likely that FEM simulations of other bushing geometries belonging to the same product group will also produce reliable results.

4 CONCLUSIONS AND OUTLOOK

The present paper shows the advantages of identification methods using component-oriented specimen and inhomogeneous experiments compared to the conventional method. Using the identified material parameters in FEM simulations leads to a significantly better reproduction of the component's behavior. Furthermore, the global response of the component-oriented specimen is quite similar to the real component in spite of considerable geometrical differences. Thus, the identified set of parameters can also be applied for the simulation of different components of the same product group.

Independent of the chosen identification method all the figures comparing measured data with numerical results show that the accuracy of the Yeoh material model is limited. Neither it is possible to reproduce the hysteresis of the material nor other phenomena such as remaining strains in the unloaded state. In the future the proposed identification method is to be applied to more complex and thus more accurate material models.

ACKNOWLEDGEMENTS

We would like to thank the following companies for the support of our project.

- Anvis Deutschland GmbH
- Freudenberg Forschungsdienste KG, Vibracoustic GmbH & Co. KG
- Henniges Automotive GmbH & Co. KG
- KKT Frölich Kautschuk und Kunststofftechnik GmbH

REFERENCES

Hohl, C. (2007). *Anwendung der Finite-Elemente-Methode zur Parameteridentifikation und Bauteilsimulation bei Elastomeren mit Mullins-Effekt.* Fortschritt-Berichte VDI, Reihe 18, Nr. 310.

Levenberg, K. (1944). A method in the solution of certain non-linear problems in least squares. Q. Appl. Math. 2, 164–168.

Moritz, H. (1993). General considerations regarding inverse and related problems. *Inverse Problems: Principles and Applications in Geophysics, Technology and Medicine 74,* 11.

Nocedal, J., W. S. J. (1999). *Numerical Optimization.* Springer-Verlag, New York.

Schnur, D. S., Z. N. J. (1992). An inverse method for determining elastic material properties and a material interface. *Int. J. Num. Eng. 33,* 2039–2057.

Constitutive Models for Rubber VII – Jerrams & Murphy (eds)
© 2012 Taylor & Francis Group, London, ISBN 978-0-415-68389-0

New insights about strain-induced crystallization of natural rubber thanks to in situ X-rays measurements during uniaxial cyclic deformation at high velocity

Nicolas Candau, Catherine Gauthier, Laurent Chazeau & Jean-Marc Chenal
INSA-Lyon, MATEIS CNRS UMR5510, Villeurbanne, France

ABSTRACT: In this study, real time X-ray measurements are performed during cyclic tensile tests at high velocity (much higher than $1\,s^{-1}$). Both unfilled and filled crosslinked NR have been prepared to analyse physical parameters governing the strain induced crystallisation (SIC) in these conditions. Prior to crystallisation study, mechanical behaviour of NR and filled NR has been investigated with a special attention paid to temperature and strain rate effects. The X-rays coupled to tensile test measurements are carried out thanks to a cyclic tensile test machine equipped with a stroboscope developed in the laboratory. The crystallinity, the crystallisation rate, the evolution of the size and orientation of the crystallite, during loading and active unloading, are analysed to characterize the development of the microstructure according to strain ratio.

1 INTRODUCTION

The natural rubber (NR) is an atypical polymer, because of its natural origin, and its isoprene mer. NR is essentially constituted of cis-1,4 polyisoprene whose stereoregularity allows its crystallization at low temperature ($-25°C$) or under strain. When NR is stretched more than 300% of its original length, a rapid crystallisation occurs. This peculiarity gives to NR a self-reinforcement character with regard to synthetic products (Trabelsi et al. 2003). In many applications, fillers (carbon black, silica…) strengthen natural rubber. It is well established that the incorporation of some kinds of fillers with natural rubber can strongly increase its rigidity, its abrasion resistance, tear strength and stress at break (Kraus 1965, Gent 1992). These improvements depend on the nature of the fillers-fillers and fillers-NR interactions, but also on the effect of fillers on the strain induced crystallization of polymer matrix.

Rubbers under deformation have been the subject of intensive research since the 1940s. Several major topics in polymer science, including rubber elasticity theory and structure-property relationships in polymeric solids, were initiated from this subject. It has been well demonstrated that the roles of both orientation and strain-induced crystallization (SIC) are crucial for determining the final properties of rubbers. However all the studies in the literature about SIC of NR (filled and unfilled) have been carried out in quasi-static state. The aim of the presented work is to present new insights about strain-induced crystallization of natural rubber thanks to in situ X-Rays measurements during uniaxial cyclic deformation at high velocity.

2 BACKGROUND

Strain induced crystallisation studies of natural rubber have been carried out thanks to the use of several techniques of investigation.

In the first studies (before 2000) the crystallization of natural rubber have been carried out after a fast strain step (Mitchell & Meier 1968) and using various techniques of investigation such as volume-change measurements (Gent 1954, Gent 1966, Gent et al. 1998), stress relaxation (Luch D 1973, Gent et al. 1998, Gent & Zhang 2001), electron microscopy (Andrews 1966, Shimizu et al. 1998, Toshiki et al. 2000), infrared absorption (Siesler 1986), birefringence (Shimomura et al. 1982), X-ray diffraction (Mitchell 1984). The morphology of strain-induced crystallites has been reported to be various: fibrils (Andrews 1966), lamellae and fibrils (Shimizu et al. 1998), and shish-kebabs (Andrews 1963).

More recently (after 2000), real time X-rays studies using synchrotron radiation have been carried out during in situ cyclic tensile tests (strain rates close to $10^{-3}\,s^{-1}$) (Toki et al. 2004, Chenal et al. 2007, Toki et al. 2009) and in situ tests under constant strain (Trabelsi et al. 2003, Tosaka 2009) at room temperature where structural development (SIC) and stress-strain relationship were measured simultaneously. The synchrotron studies revealed new insights about behaviour of molecular orientation and strain-induced crystallization in natural rubber.

The influence of deformation rate is still an issue. For instance, if NR is stretched more than 300% of its original length, a rapid crystallization occurs. Mitchell (Mitchell & Meier 1968) has estimated that 50 ms are necessary to promote SIC at 400% elongation.

Table 1. Recipe of vulcanized NR sample and network chain density

Sample code	NR	IR	F-NR
Rubber	100	100	100
N234, phr			50
Stearic acid, phr	2	2	2
ZnO, phr	1.5	1.5	1.5
6PPD, phr	3	3	3
CBS[a], phr	1.9	1.9	1.9
Sulfur, phr	1.2	1.2	1.2
ν[b]	1.56	1.56	2.20

[a] N-Cyclohexyl-2-benzothiazole sulfonamide.
[b] Network chain density ($\times 10^4$ mol/cm^3) determined from swelling ratio of the polymer itself.

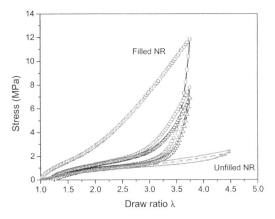

Figure 1. Stress-strain curves of filled and unfilled NR measured during mechanical cycles at 25°C (strain rate 0.25 min^{-1}). Reproduced from Polymer 48 (2007) 6893–6901.

It means that if the rate of deformation is high enough, even at deformation ratio of 400%, the SIC could not take place. Therefore it is interesting to study the critical rate (v_c) at which SIC disappears during tensile tests since it will impact mechanical properties of the material. In particular it could influence the material wear resistance.

The two following sections present the series of materials we have prepared and some of the experimental techniques used for this study.

3 MATERIALS

Vulcanized unfilled natural (NR), synthetic (IR) and filled (50 phr) rubber samples were prepared. The materials have been obtained by sulfur vulcanization of natural rubber according to the recipes given in Table 1. Prior to curing (170°C), each cure time was determined from torque measurement as a function of temperature performed with a Monsanto analyser. Network chains densities (ν) presented in Table 1 have been estimated from swelling ratio in toluene and the Flory Rehner equation (Flory & Rehner 1943):

$$-[\ln(1-v_2) + v_2 + \chi_1 v_2^2] = V_1 \nu [v_2^{1/3} - \frac{v_2}{2}] \quad (1)$$

where v_2 is the volume fraction of polymer in the swollen mass. $v_2 = 1/Q_{pol} = V/V_0$, where Q_{pol} is the swelling ratio, V and V_0 are the volumes of the polymer network, respectively, at swelling equilibrium and after swelling and drying. V_1 (106.3 cm^3/mol) is the molar volume of the solvent (toluene), and χ_1 is the Flory-Huggins polymer-solvent dimensionless interaction term (χ_1 is equal to 0.39 for the system NR-toluene).

The calculus of the filled rubber network chain density takes into account the introduction of additional crosslinks due to the rubber-filler interactions. With the assumption of a perfect rubber-filler interface, the relation between the swelling ratio of the polymer itself (Q_{pol}) and the equilibrium swelling ratio of the composite (Q) should be given by the formula:

$$Q_{pol} = \frac{Q - \varphi_f}{1 - \varphi_f} \quad (2)$$

The molecular weight between crosslinks is deduced from ν as following:

$$M_c = \rho/\nu \quad (3)$$

where ρ is the density and M_c the molecular weight between crosslinks.

Tensile test specimens are dumbbell shaped with the following dimensions: length 6 mm, section 10×0.8 mm^2.

4 EXPERIMENTS

4.1 Mechanical characterization

Prior to X-rays experiments, all the samples have been characterized mechanically in order to estimate the most interesting levels of pre-deformation and amplitudes of the cyclic tensile test. Typical stress-strain curves for NR and filled NR are presented in Figure 1. Stress is nominal i.e. it is the force divided by the initial section of the sample.

4.2 Volume change measurement

Measurements of volume change can be a way to evaluate microstructural and crystalline evolution in a stretched rubber. Such measurements have been performed using a specific apparatus developed in the laboratory based on the Farris apparatus (Fig. 2). The equipment is based on the following principle: a specimen chamber and a reference chamber filled with gas are connected by a differential pressure sensor. When the sample is stretched, a volume change

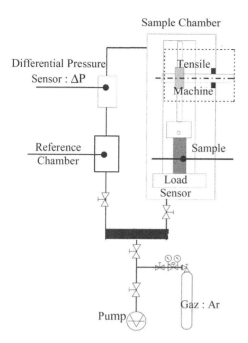

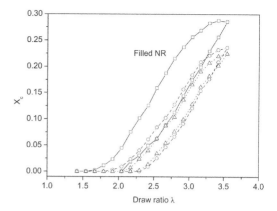

Figure 4. Crystallinity (X_c) change during successive stress-strain curves of filled and unfilled NR measured during mechanical cycles at 25°C (strain rate 0.25 min^{-1}) as a function of the draw ratio. Reproduced from Polymer 48 (2007) 6893–6901.

to a volume increase) and crystallisation (leading to a volume decrease). When active unloading is applied, a fast recovery of decohesion occurs and a negative volume variation observed is associated to the amount of strain induced crystallisation.

4.3 X-Rays measurement

The SIC of NR and filled NR samples have been first characterized at low strain rate (10^{-3} s^{-1}) (Chenal, et al. 2007, Toki et al. 2009). To do so, a homemade stretching machine allowing the symmetric deformation of the sample was used to probe by X-ray the same sample position during stretching at 0.25 min^{-1} strain rate. High resolution is necessary for real-time measurements; thus our in situ wide-angle X-ray scattering (WAXS) studies are carried out on the D2AM beamline of the European Synchrotron Radiation Facility (ESRF). The wavelength of the X-ray is 0.54 Å. The two-dimensional (2D) WAXS patterns can be recorded every 10 s by a CCD Camera (Princeton Instrument). During stretching, the thickness and then the absorption of the sample decrease. The in situ measurements of the absorption by photomultiplicators located ahead and behind the sample, are used to normalize the scattered intensities.

Typical results obtained during mechanical cycles at 25°C (strain rate 0.25 min^{-1}) for filled NR are presented in Figure 4. The crystallinity shows a marked difference between the first cycle and the other cycles, i.e. the Mullins effect plays a major role during stress-induced crystallization.

We can also observe that the onset strain of NR crystallization is shifted to a smaller value when the material is filled by carbon black (un-presented curve). This fact confirms that fillers act as amplifiers of local draw ratio.

In order to avoid sample microstructure evolution due to Mullins effect, the study of SIC is performed

Figure 2. Diagram of volume-change measuring apparatus and strain curve depends on decohesion and crystallisation.

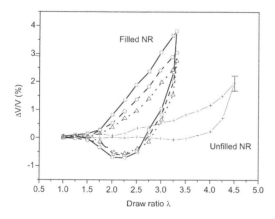

Figure 3. Volume change/strain curve of filled and unfilled NR measured during mechanical cycles at 25°C (strain rate 0.25 min^{-1}). Reproduced from Polymer 48 (2007) 6893–6901.

of the sample causes a pressure change relative to the reference chamber. Thus the pressure difference can be converted into a volume change of the sample. A pressure difference will also be recorded if the sample adsorbs gas from the surroundings during deformation. In order to avoid this phenomenon, argon, which is not absorbed by samples, is used to fill specimen and reference chambers.

Typical in situ measurements of volume variation during mechanical cycling solicitations of the NR are presented on Figure 3. During the stretching process, positive volume variation is recorded. It is assumed to be due to the superposition of decohesion (which leads

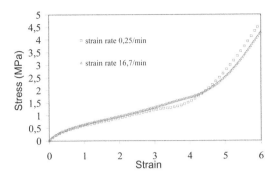

Figure 5a. Stress-strain curves of unfilled NR measured during mechanical cycles at 0°C (strain rate 0.25/min and 16.7/min).

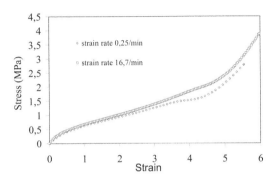

Figure 5b. Stress-strain curves of unfilled NR measured during mechanical cycles at 12°C (strain rate 0.25/min and 16.7/min).

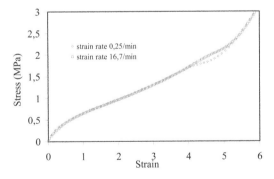

Figure 5c. Stress-strain curves of unfilled NR measured during mechanical cycles at 23°C (strain rate 0.25/min and16.7/min).

during the fourth cycle. In that case, samples are said to be de-mullinized.

5 RESULTS AND PROSPECTS

5.1 *Monotonic tests*

Tensile tests performed at different temperatures and different strain rates on NR are presented on Figure 5a, 5b and 5c.

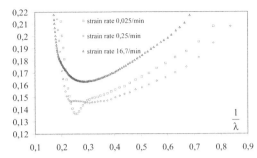

Figure 6a. Mooney-Rivlin plots of stress-strain data for natural rubber (strain rate 0.025/min, 0.25/min and 16.7/min).

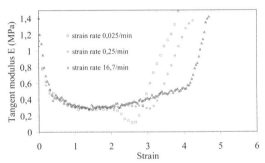

Figure 6b. Tangent modulus plots for natural rubber (strain rate 0.025/min, 0.25/min and 16.7/min).

In these figures, one can notice a stress relaxation phenomenon that occurs at a strain that depends on both temperature and strain range. For instance, at 12°C and 23°C, it appears at a strain of 3.45 and 4.05 for the lowest strain rate. This phenomenon is associated to SIC: when a crystallite grows parallel to the axis of elongation, the distance between the remaining amorphous segments is decreased. Consequently these segments can accept a greater number of configurations (Flory 1947). This theoretical prediction of Flory has been confirmed by recent experimental studies combining stress-strain curves and X-rays measurements (Albouy et al. 2005).

The strain curve of NR measured during the fourth cycle at different strain rates (0.15 mm/min, 1.5 mm/min and 100 mm/min à −25°C) are plotted in a Mooney-Rivlin representation, i.e.

$$\frac{\sigma}{2\left(\lambda - \frac{1}{\lambda^2}\right)} = f\left(\frac{1}{\lambda}\right) \qquad (4)$$

$$\frac{\sigma}{2(\lambda - \frac{1}{\lambda^2})}$$

The stress relaxation associated to the strain induced crystallisation is also clearly evidenced on such representations. We note that at −25°C, witch is the optimal temperature of crystallization (Chenal et al. 2007), the

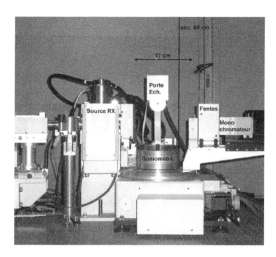

Figure 7. X-Ray in situ facility permitting WAXS and SAXS analysis of samples stretched monotonically or dynamically.

strain rate has an important impact on the value of the incipient strain of crystallization, much more than at 23°C. This result suggests the coexistence of thermal and strain induced crystallites when the temperature is sufficiently closed to −25°C.

5.2 Dynamic tests

From a physical point of view, we want to establish the correlations between the change of v_c according to the level of pre-deformation (100%, 200%, etc.) and the amplitude of deformation (50%, 150%, etc.) during tensile tests. The aim is to characterize and to understand the kinetic as well as the crystalline microstructure that appears during such tensile tests.

To address strain rate influence on SIC phenomenon, a second generation of homemade stretching machine has been developed. It is based on stroboscopic acquisition of X-rays enabling to "observe" the sample always at the same strain ratio, with a reduced exposure time. The conception of the apparatus has been made thanks to SolidEdge, taking into account two opposite requirements: the reduction of mass in motion to reduce the inertial forces at high frequency (bending dynamics, vibration) and the optimization of the mechanical strength to minimize undesirable transverse forces (friction, static bending, etc.). Some constraints also derive from the topology of the X-rays source (See for instance Figure 7), as the machine should be used with both laboratory and synchrotron sources.

Laboratory source is obviously more available, but long image acquisition times are required to accommodate the weak intensity. Rubbers tested in drastic conditions during a long time could be damaged. The synchrotron radiation enable us to maximize the intensity recorded during each cycle, to reach an excellent time resolution (which is essential to study the phenomenon of strain-induced crystallization from

a dynamic point of view) and lower the time of each experiment. Furthermore, the usefulness of a very powerful X-ray source is strengthened by the weak scattering character of our crystallized samples, related to the relatively weak ratio of crystallinity.

More results will be presented and discussed during the conference in particular X-Rays data acquired at high velocity.

REFERENCES

Albouy, P. A. et al. 2005. "Chain orientation in natural rubber, Part 1: The inverse yielding effect." European physical journal 17(3): 247–259.

Andrews, E. H. 1963. "Crystalline morphology in thin films of natural rubber." Proceedings of the royal society of London. Series a, Mathematical and Physical Sciences 277(1371): 562–570.

Andrews, E. 1966. "Microfibrillar Textures in Polymer Fibers." Journal of Polymer Science 4: 668–72.

Chenal, J. M. et al. 2007a. "New insights into the cold crystallization of filled natural rubber." Journal of Polymer Science Part b-Polymer Physics 45(8): 955–962.

Chenal, J. M. et al. 2007b. "Molecular weight between physical entanglements in natural rubber: a critical parameter during strain-induced crystallization." Polymer 48(3): 1042–1046.

Chenal, J. M. et al. 2007c. "Parameters governing strain induced crystallization in filled natural rubber." Polymer 48(23): 6893–6901.

Flory, P.J. & Rehner, J. 1943. "Statistical Mechanics of Cross-Linked Polymer Networks." Journal of Chemical physics 11: 512–520.

Flory, P. J. 1947. "Thermodynamics of crystallization in high polymers. I. Crystallization induced by stretching." J. Chem. Phys 15: 397–408.

Gent, A. N. 1954. "Crystallization and the relaxation of stress in stretched natural rubber vulcanizates." Transactions of the Faraday Society 50: 521–533.

Gent, A. N. 1966. "Crystallization in Stretched Polymer Networks. II. Trans-Polyisoprene." Journal of Polymer Science Part A-1 4: 447–64.

Gent, A. N. 1992. Engineering with rubber. Oxford, England: Hanser Publishers, Oxford University Press.

Gent, A. N. et al. 1998. "Crystallization and strength of natural rubber and synthetic cis-1,4-polyisoprene." Rubber Chemistry and Technology 71(3): 668–678.

Gent, A. N. &. Zhang, L. Q. 2001. "Strain-induced crystallization and strength of elastomers. I. cis-1,4-polybutadiene." Journal of Polymer Science Part B-polymer physics 39(7): 811–817.

Kraus, G. 1965. Reinforcement of elastomers. New York, Interscience Publishers.

Luch, D. 1973. "Strain-induced crystallization of natural rubber. III. Reexamination of axial-stress changes during oriented crystallization of natural rubber vulcanizates." Journal of Polymer Science 11(3): 467–86.

Mitchell, J. C. & Meier, D. J. 1968. "Rapid stress induced crystallization in natural rubber." Journal of Polymer Science 6: 1689–1703.

Mitchell 1984. "A wide-angle x-ray study of the developpement of molecular orientation in crosslinked natural rubber." Polymer 25.

Shimizu, T. M. et al. 1998. "A tem study on natural rubber thin films crystallized under molecular orientation." materials science research international 4(2): 117–120.

Shimomura, Y. et al. 1982. "A comparative-study of stress-induced crystallization of guayule, hevea, and synthetic polyisoprenes." Journal of Applied Polymer Science **27**(9): 3553–3567.

Siesler, H. W. 1986. "Rheooptical Fourier-transform infrared-spectroscopy of polymers 12. Variable temperature studies of strain-induced crystallization in sulfur-cross-linked natural-rubber." Makromolekulare Chemie-macromolecular symposia **5**: 151–155.

Toki, S. et al. 2004. "Strain-induced molecular orientation and crystallization in natural and synthetic rubbers under uniaxial deformation by in-situ synchrotron x-ray study." Rubber Chemistry and Technology **77**(2): 317–335.

Toki, S. et al. 2009. "New insights into the relationship between network structure and strain-induced crystallization in un-vulcanized and vulcanized natural rubber by synchrotron x-ray diffraction." polymer **50**(9): 2142–2148.

Tosaka, M. 2009. "A route for the thermodynamic description of strain-induced crystallization in sulfur-cured natural rubber." Macromolecules **42**(16): 6166–6174.

Toshiki, S. et al. 2000. Rubber Chemistry and Technology **73**: 926–37.

Trabelsi, S. et al. 2003. "Effective local deformation in stretched filled rubber." Macromolecules **36**(24): 9093–9099.

Constitutive Models for Rubber VII – Jerrams & Murphy (eds)
© 2012 Taylor & Francis Group, London, ISBN 978-0-415-68389-0

Application of full-field measurements and numerical simulations to analyze the thermo-mechanical response of a three-branch rubber specimen

X. Balandraud, E. Toussaint, J.-B. Le Cam & M. Grédiac
Laboratoire de Mécanique et Ingénieries (LaMI), Université Blaise Pascal (UBP),
Institut Francais de Mécanique Avancée (IFMA), Clermont-Ferrand, France

R. Behnke & M. Kaliske
Institute for Structural Analysis (ISD), Technische Universität Dresden, Germany

ABSTRACT: The study deals with the characterization of the thermo-mechanical behavior of rubber. First of all, experiments are performed with a three-branch-shaped rubber specimen previously proposed by Guélon et al. (2009) used for this purpose. This heterogeneous test induces simultaneously the three types of strain states classically employed to identify mechanical properties of rubber (uniaxial and equibiaxial tension, and pure shear), as well as some intermediary states. Recent experimental studies that investigated such heterogeneous tests only considered the deformation field, but not the corresponding thermal field. This aim of the present study is to push forward the idea by measuring both the displacement and thermal fields during this test, and to compare the experimental results with their numerical counterparts predicted by a thermo-mechanical numerical simulation performed at ISD using a novel model for thermo-viscoelasticity – an improved version of the extended-tube model presented by Kaliske & Heinrich (1999) in the case of isothermal hyperelasticity. The comparison between predicted and experimental fields, as well as the correlation between thermal and kinematic fields, will be discussed in the presentation.

1 INTRODUCTION

Classically, constitutive relations for rubber are identified and validated from several tests, assumed to be homogeneous, i.e. uniaxial tension, pure shear and equibiaxial tension. In a previous work, (Guélon et al. 2009) have proposed a new method to characterize and to identify the constitutive parameters in the framework of hyperelasticity. For this purpose, the authors used a three-branch sample under tension that induces simultaneously the three homogeneous and intermediate deformation states. In the present study, both thermal and kinematic full field measurements are performed to characterize the thermo-mechanical response of the three-branch-specimen and to compare the experimental results to the predictions of a thermo-viscoelastic model developed at ISD.

2 PRESENTATION OF THE THREE-BRANCH TEST

The test was carried out with a MTS 858 Elastomer Test System testing machine. Its loading capacity is ± 15 kN and the loading cell is ± 1 kN. In order to apply biaxial loading, i.e. to generate equibiaxial tension and pure shear at the sample center, a new tensile apparatus has been designed to be adapted to the uniaxial tensile machine (Guélon et al. 2009). In practice, the two vertical branches of the sample were fastened to

the grips of the conventional testing machine, whereas the branch perpendicular to the previous ones was fastened in the grip.

The sample geometry, which corresponds to a three-branch sample, was 2 mm thick and 52 mm high, and the branches were 20 mm in width. It is presented in Figure 4. The bottom branch was clamped, the loading was carried out in two steps by applying prescribed displacements along the axis of the two other perpendicular branches. The first step consisted in applying a 12 mm displacement along the y-direction and a 30 mm displacement along the x-direction. During this step, images were stored only with a CCD camera. This first step was necessary to generate fields which were sufficiently heterogeneous on the sample surface. The configuration obtained at the end of the first step is called intermediate configuration in the following. The second step consisted in applying a 47 mm displacement along the y-direction. The displacement rate was equal to 15.66 mm/s. At the end of the test, the corresponding global stretch ratios were 2.13 and 1.85, respectively. During this second step, images were stored by both the CCD and IR cameras. Each camera was positioned on each side of the specimen.

2.1 Measurement of kinematic fields

During the test, images for kinematic fields were stored using a cooled 12-bit dynamic Sensicam

45

camera. The images correspond to different stretch ratio levels. The charge-coupled device (CCD) of the camera had 1.4×10^6 connected pixels (1376×1040). Uniform lighting at the sample surface was ensured by a set of fixed lamps. Digital Image Correlation (DIC) was used to measure the displacement field on the sample surface. DIC is a full-field measurement technique developed at the beginning of the 1980s (Sutton et al. 1983). This method consists in matching, before and after displacement, the brindled pattern in a physical part of the observed surface of the specimen, called "Region Of Interest" (ROI). The SeptD software was used (Vacher et al. 1999) for this purpose. This optical technique enables us to reach a resolution of 0.03 pixels, corresponding to 2.9 μm and a spatial resolution (defined as the smallest distance between two independent points) of 10 pixels, corresponding to 975 μm. A set of sub-images is considered to determine the displacement field of a given image with respect to a reference image. This set is referred to as "Zone of Interest" (ZOI). A suitable correlation function was used to calculate the displacement of the center of a given ZOI between two images captured at different stages of an experiment. To improve image contrast, white paint was sprayed on the sample surface before testing. The DIC technique is well-suited for measuring large strains (Sasso et al. 2008, Chevalier et al. 2001).

2.2 Measurement of thermal fields

Temperature measurements were carried out with a Cedip Jade III-MWIR infrared (IR) camera, which features a local plane array of 320×240 pixels and detectors with a wavelength range of 3.5–5 μm. The integration time was 1500 μs and the acquisition frequency was set to 150 Hz. The noise equivalent temperature difference (NETD) of the measurement, i.e. the thermal resolution, is equal to about 20 mK for the used temperature range 5–40 degrees C. In order to ensure that the internal temperature of the camera was optimal for performing the measurements, it was set up and switched on for one hour before the experiment. The stabilization of the temperature in the camera was necessary to avoid any drift of the measurements during the test.

3 EXTENDED TUBE MODEL

3.1 Isothermal hyperelasticity

The extended tube model, introduced by Kaliske & Heinrich (1999), is a material model which is well suited for the characterization of the mechanical material behavior of particle-reinforced rubber. The isochoric free energy function $\bar{\Psi}_0$ can be derived from statistical mechanics and represents the isochoric material response for the case of isothermal hyperelasticity. A comparison of several material models for hyperelasticity in Marckmann & Verron (2006)

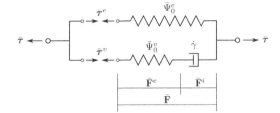

Figure 1. Rheology of the viscoelastic material model, $\mathbf{F} = \bar{\mathbf{F}}^e \, \bar{\mathbf{F}}^i$.

highlights the capacity of the extended tube model to fit the experimental data in uniaxial, equibiaxial as well as pure shear strain states. For this reason, the extended tube model is chosen as the starting point for the derivation of a finite visco-elastic material model which is briefly described in the next section.

3.2 Isothermal viscoelasticity

For many practical applications, the material response of rubber can be assumed to be purely hyperelastic and often volume changes under deformation are neglected, i.e. the rubber is regarded as incompressible. However, experiments with rubber test samples reveal viscous behavior, time dependent (aging) and load dependent (Mullins effect, damage effects) material characteristics. In addition, dissipative phenomena cannot be taken into account by a purely hyperelastic material model. As a result, the aforementioned hyperelastic extended tube model is further developed to a finite non-linear visco-elastic material model. Its rheology is shown in Figure 1. The rheological Zener element contains an elastic (equilibrium – EQ) and viscous (non-equilibrium – NEQ) branch which is modelled by a Maxwell element. In order to describe finite viscoelasticity, a volumetric-isochoric split of the free energy Ψ_0 is introduced where the deformation gradient $\mathbf{F}$ is multiplicatively decomposed into a volumetric $\mathbf{F}_{\text{vol}}$ and isochoric $\bar{\mathbf{F}}$ contribution, $\mathbf{F} = \bar{\mathbf{F}} \, \mathbf{F}_{\text{vol}}$ and $\bar{\mathbf{F}} := J^{-1/3} \, \mathbf{F}$ with $J := \det \mathbf{F}$. The isochoric part of the deformation gradient is further split into an elastic and inelastic part, $\bar{\mathbf{F}} = \bar{\mathbf{F}}^e \, \bar{\mathbf{F}}^i$. The viscosity of the material is assumed to be purely isochoric. The dashpot is governed by the effective creep rate

$$\dot{\gamma} := \dot{\gamma}_0 \left[\frac{\lambda}{\lambda_e} - 1 \right]^c \left(\frac{\tau_v}{\hat{\tau}} \right)^m, \quad \text{with} \tag{1}$$

$$\lambda := \sqrt{\frac{\bar{I}_1}{3}}, \quad \lambda_e := \sqrt{\frac{\bar{I}_1^e}{3}}, \tag{2}$$

where $\dot{\gamma}_0$ denotes the reference effective creep rate, λ/λ_e is the inelastic network stretch and $c \leq 0$ as well as $m > 0$ are model parameters which control the inelastic network kinematics and the energy activation, respectively. The first invariants can be calculated by $\bar{I}_1 = \text{tr} \, \bar{\mathbf{C}}$ and $\bar{I}_1^e = \text{tr} \, \bar{\mathbf{C}}_e$ from the elastic Cauchy-Green tensor $\bar{\mathbf{C}}_e$. The stress term τ_v can be obtained from the isochoric viscous stress $\boldsymbol{\tau}_{\text{iso}}^v$,

$$\tau_v := \frac{\sqrt{\boldsymbol{\tau}_{\text{iso}}^v : \boldsymbol{\tau}_{\text{iso}}^v}}{\sqrt{2}}, \qquad \boldsymbol{\tau}_{\text{iso}}^v := \mathbb{P} : \bar{\boldsymbol{\tau}}^v. \qquad (3)$$

The algorithmic setting is provided in detail in Dal & Kaliske (2009) and Behnke et al. (2011).

3.3 Thermo-mechanical coupling

During cyclic loading, elastomeric material shows also a significant heat build up which is due to dissipative phenomena like irreversible network rearrangements, micro cracking or viscous characteristics. All the phenomena of mechanical energy dissipation lead to a critical heating of cyclic loaded elastomeric components which can cause alteration and deterioration. Due to their entropy and high extensibility, elastomers are also characterized by an observable temperature change upon deformation (heating during stretching, cooling during unloading). In the following, mainly these latter effects are studied.

The concept of the thermo-mechanical modelling follows the approach presented in Reese & Govindjee (1998). The kernel of this approach consists of the evaluation of the free energy functions of the elastic and viscous branch as well as the inner energy e_0 and the heat capacity $\bar{c}$ at a reference temperature Θ_0. At the Gauss point level, the stresses are always evaluated using the reference potentials $\bar{\Psi}_0^e$ and $\bar{\Psi}_0^v$ which are marked therefore by a subscript 0. The temperature dependency of the material and its properties are then constructed by multiplying the reference energy and heat capacity terms by dimensionless temperature functions $f(\Theta)$, $t(\Theta)$,

$$\Psi = f_{\text{EQ}}(\Theta) \left[\bar{\Psi}_0^e(\bar{b}_i) + U_0(J) \right] + t_1(\Theta) \, e_{0\text{EQ}}$$

$$+ f_{\text{NEQ}}(\Theta) \left[\bar{\Psi}_0^v(\bar{b}_{e_i}) \right] + t_2(\Theta) \, \bar{c}. \qquad (4)$$

Θ, $\bar{b}_i$ and U_0 are the absolute temperature, the eigenvalues of the left Cauchy-Green tensor $\bar{\mathbf{b}}$ and the volumetric free energy function, respectively. For the FE implementation, the governing equations of coupled thermo-mechanics (balance of linear and angular momentum, the first and second law of thermodynamics in form of the balance of energy and the dissipation inequality as well as a constitutive law for the heat flux) are written in their weak form. For more details, we refer to Reese & Govindjee (1998).

4 MODEL PARAMETER IDENTIFICATION

For parameter identification, a cyclic uniaxial extension test is carried out at LaMI. A rubber specimen of cuboidal shape ($l = 22$ mm, $b = 4$ mm, $t = 2$ mm) is placed in a testing machine. The mechanical material response is characterized by the reaction force which is

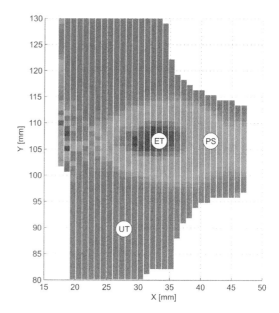

Figure 2.   Stretch states in the gauge section.

recorded during the test. The test consists of a sawtooth wave motion of the upper grip of the testing machine. At the beginning, 3 cycles with a positive strain amplitude of $\varepsilon = 1.0$ are carried out. They are followed by 3 cycles with a positive strain amplitude of $\varepsilon = 1.75$. The absolute value of the displacement rate during the test is kept constant (500 mm/min). The thermal material response – which is dominated by the thermo-elastic coupling for this loading case – is measured by an IR camera during the experiment. The thermal material properties are physically measurable quantities (heat conductivity, heat capacity at reference temperature) which are taken from Pottier et al. (2009).

The recorded mechanical and thermal data are used in the following for the parameter identification. For this automated task, a parameter identification algorithm has been developed at ISD. It is based on least square minimization of the measured and computed test data. The result is one locally optimized parameter set, specified in Behnke et al. (2011), which is used for the simulation of the three-branch rubber specimen.

5 EXPERIMENTAL RESULTS

5.1 Displacement distribution measured by DIC

Figure 2 shows the stretch states as functions of the (X, Y) coordinates in the reference configuration using a suitable color code at the end of the first step.

The Equibiaxial Tension (ET), Pure Shear (PS) and Uniaxial Tension (UT) states correspond to the blue, green and red colors, respectively. Intermediate states are defined by a color that is a weighted average of two colors, namely, blue and green for the states between ET and PS, green and red for the states between PS and

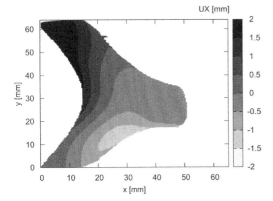

a)

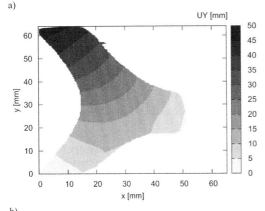

b)

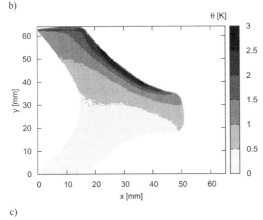

c)

Figure 3. Experimented displacements and temperature variation; a) horizontal displacement; b) vertical displacement; c) θ-temperature variation.

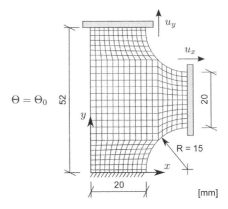

Figure 4. FE mesh and boundary conditions, $\Theta_0 = 292$ K.

the geometry of the coupon at any stage of the loading and the reference geometry. It will then be possible to correctly analyze and process the temperature maps at different loading amplitudes. This is explained in the following section.

5.2 Thermal distribution measured by IR thermography

Since the specimen undergoes large deformations, any given material point at its surface clearly moves during the loading cycle, while the zone captured by the detector array of the IR camera remains unchanged. The smaller the distance to a fixed grip, the greater the amplitude of this displacement. Consequently, a given pixel of the IR detector matrix does not correspond to the same material point plotted during the deformation of the specimen surface. The objective is now to explain how to track the material points.

Constructing a suitable motion compensation technique in the context of mechanical testing and temperature measurement using an IR camera has only seldom been addressed in the literature (Pottier et al. 2009). In the present study, the kinematic field measurement is used to track the material point in the present study. A suitable algorithm was developed.

Two configurations must be distinguished: a reference geometry and the current one. For the present application, the reference geometry corresponds to the configuration obtained at the end of the first step of the loading, the so-called intermediate configuration. Considering kinematic and thermal fields at a same time, the suitable compensation technique is built in practice with the following procedure:

- kinematic fields are calculated at the center of each ZOI. In the present case, the number of ZOIs is lower than the number of points where temperatures are measured. The first step consists in interpolating the values of the displacements on the thermal grid;
- using these values of displacement, temperatures are then interpolated on a regular grid defined on the reference configuration.

UT. Figure 2 shows that the central zone of the specimen is mainly in an ET state. The zones corresponding to the three branches are mainly in a UT state. The PS zone is located on a ring-shaped zone located between the ET and UT zones.

A typical view of the displacement fields obtained at the end of the second step is shown in Figure 3 a) and b). The displacement fields during the second step are necessary to establish the correspondence between

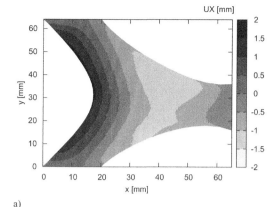

a)

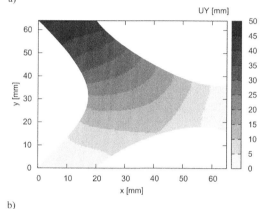

b)

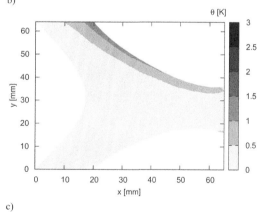

c)

Figure 5. Simulated displacements and temperature variation; a) horizontal displacement; b) vertical displacement; c) θ-temperature variation.

Then temperature variations θ are obtained by subtracting the temperature field corresponding to the reference configuration. Figure 3 c) gives the temperature variation θ at the end of the loading procedure. The temperature variation field obtained highlights a high temperature gradient in the two branches between the moving grips. The temperature variation reaches 3 degree C in the zone mainly submitted to uniaxial tension (see Figure 2).

6 PREDICTION OF THE THERMO-MECHANICAL RESPONSE

The aim of the simulation is twofold. First, the identified model parameters can be validated by a non-homogeneous test. Second, the simulation capacity of the thermo-mechanical coupled material model and its finite element implementation are demonstrated and used to predict the experimental results. The FE mesh, the prescribed displacements and the boundary conditions are illustrated in Figure 4. The FE mesh consists of 384 isoparametric finite solid and 436 finite surface elements to model the heat conduction through the surface. Only one half of the specimen is modelled with one layer of solid elements in z-direction. The simulation results at the end of the loading are illustrated in Figure 5. They are projected to the intermediate configuration by an in-house post-processing algorithm. The temperature variation field prediction is in good agreement with the temperature variation field measurement.

7 CONCLUSIONS

This study deals with the characterization of the thermo-mechanical behavior of rubber. From an experimental point of view, the heterogeneity induced by the three-branch test has been characterized related to the three classical homogeneous deformation states. Coupled kinematic and thermal full field measurements allowed us to associate at each point of the sample surface a deformation state and a temperature variation. This experimental test has been simulated using the thermo-viscoelastic model developed at ISD. Numerical simulations satisfactorily predict the thermo-mechanical response measured.

ACKNOWLEDGEMENTS

Part of the research reported herein is carried out within the DFG research unit 597. The coauthors R. Behnke and M. Kaliske would like to acknowledge gratefully the financial support of the DFG under grant KA 1163/3-2.

REFERENCES

Behnke, R., H. Dal, & M. Kaliske (2011). An extended tube model for thermo-viscoelasticity of rubberlike materials: Theory and numerical implementation. In *Constitutive Models for Rubber VII.*

Chevalier, L., S. Calloch, F. Hild, & Y. Marco (2001). Digital image correlation used to analyze the multiaxial behavior of rubber-like materials. *Eur. J of Mech. A-Solid 2,* 169–187.

Dal, H. & M. Kaliske (2009). Bergström-Boyce model for nonlinear finite rubber viscoelasticity: Theoretical aspects and algorithmic treatment for the FE method. *Computational Mechanics 44,* 809–823.

Guélon, T., E. Toussaint, J.-B. Le Cam, N. Promma, & M. Grédiac (2009). A new characterization method for rubber. *Polymer Testing 28*, 715–723.

Kaliske, M. & G. Heinrich (1999). An extended tube-model for rubber elasticity: Statistical-mechanical theory and finite element implementation. *Rubber Chemistry and Technology 72*, 602–632.

Marckmann, G. & E. Verron (2006). Comparison of hyperelastic models for rubberlike materials. *Rubber Chemistry and Technology 79*, 835–858.

Pottier, T., M.-P. Moutrille, J.-B. Le Cam, X. Balandraud, & M. Grédiac (2009). Study on the use of motion compensation techniques to determine heat sources. Application to large deformations on cracked rubber specimens. *Experimental Mechanics 49*, 561–574.

Reese, S. & S. Govindjee (1998). Theoretical and numerical aspects in the thermo-viscoelastic material behaviour of rubber-like polymers. *Mechanics of Time-Dependent Materials 1*, 357–396.

Sasso, M., G. Palmieri, G. Chiappini, & D. Amodio (2008). Characterization of hyperelastic rubber-like materials by biaxial and uniaxial stretching tests based on optical methods. *Polymer Testing 27*, 995–1004.

Sutton, M. A., W. J. Wolters, W. H. Peters, W. F. Ranson, & S. R. McNeil (1983). Determination of displacements using an improved digital correlation method. *Image and Vision Computating 1*, 133–139.

Vacher, P., S. Dumoulin, F. Morestin, & S. Mguil-Touchal (1999). Bidimensional strain measurement using digital images. In *Proceedings of the Institution of Mechanical Engineers. Part C: Journal of Mechanical Engineering Science*, p. 213.

Constitutive Models for Rubber VII – Jerrams & Murphy (eds)
© 2012 Taylor & Francis Group, London, ISBN 978-0-415-68389-0

Effect of thermal cycles on the kinematic field measurement at the crack tip of crystallizable natural rubber

Evelyne Toussaint*

Clermont Université, Université Blaise Pascal, Laboratoire de mécanique et Ingénieries, Clermont-Ferrand, France

Jean-Benoît Le Cam*

Clermont Université, Institut Français de mécanique Avancée, Laboratoire de mécanique et Ingénieries, Clermont-Ferrand, France

Fédération de Recherche TIMS CNRS FR 2856, Complexe scientifique des Cézeaux, Aubière Cedex, France

ABSTRACT: This paper deals with crack growth in rubbers, especially in crystallizable rubbers. In crystallizable rubbers, the high deformation level encountered at the crack tip engenders the formation of crystallites. Thus, the crack tip is reinforced and resists the crack growth. Moreover, it is well-known that temperature affects the crack propagation. This is observed at the macroscopic scale in terms of crack propagation rate and path. In this study, the effect of temperature is studied at the local scale, by measuring the change in the kinematic field at the crack tip during thermal cycles. Results show that, in such crystallizable rubbers, the effect of temperature depends of the stretch ratio reached in the considered zone. In fact, the temperature variation accounts for the competition between variation of internal energy and of entropy. Such a study is an interesting way to improve and validate thermomechanical models physicaly motivated, taking into account the effect of crystallization and crystallite melting in the thermomechanical response of rubbers.

1 INTRODUCTION

For sixty years, crack propagation in rubbers has remained an area of intensive research. Before the 1950s, no mechanical quantity available in the literature was considered as intrinsic to rubber failure. Indeed, the classical approach that consists in determining the elastic energy release rate is not appropriate; it is formulated within the framework of small deformations and requires the knowledge of either the stress field at the crack tip (available only from the 1970s) or the surface production (which remains difficult to measure without the full-field measurement method available only from the 1990s, see for instance (Sutton, Wolters, Peters, Ranson, and McNeil 1983). This motivated the theoretical work of (Rivlin and Thomas 1953), who proposed an energy-based approach to formulate tearing energy as the large strain counterpart to the elastic energy release rate. Then, from several kinds of notched sample geometry (Rivlin and Thomas 1953; Thomas 1960) tearing energy was measured using various methodologies (see the works of (Rivlin and Thomas 1953; Greensmith and Thomas 1956) and (Thomas 1958) under quasi-static loading and those of (Thomas 1958) and (Gent, Lindley, and Thomas 1964) for cyclic loading). Therefore, using tearing energy, it became possible to study the influence of material formulation and loading conditions on crack propagation (Gent and Pulford 1984; Hamed,

Kim, and Gent 1996; Gent, Razzaghi-Kashani, and Hamed 2003). Even if such an approach is now widely used, it remains a global approach and does not account for the thermophysical phenomena involved in crack growth. We consider that the improvement of crack growth prediction requires the identification of the physical mechanisms that occur and modify the kinematic field at the crack tip and in its vicinity during crack growth. For this purpose, the measurement of the mechanical fields in the crack tip vicinity seems to be an interesting line of investigation.

In this study, we focus on the effect of temperature on the kinematic field at the crack tip in crystallizable rubbers. In fact, the effect of the temperature is classically studied through crack propagation curves and the post-mortem analysis of fracture surfaces. It is well known that temperature affects the relative contribution of phenomena such as knotty tearing, stick-slip and steady tearing (Greensmith and Thomas 1956; Greensmith 1956). The effect of temperature on crack propagation is therefore classically studied through variations in tearing energy and crack propagation rate. We believe that the understanding of temperature effects can be improved by analyzing them at the local scale through the variations in the kinematic field at the crack tip, and by linking them to phenomena due to thermoelastic couplings such as thermoelastic inversion, i.e. the competition between isentropic and entropic effects, and crystallite melting. To the best of our knowledge, no study has yet investigated quantitatively the effect of temperature on the kinematic field at the crack tip, and this is particularly the case for crystallizable material.

The aim of the present study is therefore to carry out such an analysis by measuring the displacement field at the crack tip during stretching using Digital Image Correlation (DIC) techniques, and then to apply thermal cycles. The first part of this paper presents the experimental set-up, the material formulation and mechanical properties, the testing machine and the measurement of the kinematic and thermal fields. The second part deals with the results obtained. Conclusions and perspectives close the paper.

2 EXPERIMENTAL SETUP

2.1 Material and sample geometry

The material considered here is a 34 phr carbon black-filled natural rubber. It is denoted NR in the following. The stretch ratios at the beginning of crystallization and at the end of melting at 23°C were previously determined using volume change measurements (Le Cam and Toussaint 2008; Le Cam and Toussaint 2009) and equal 1.64 and 1.44, respectively. A schematic view of the sample, whose geometry is 32 mm wide, 2 mm thick and 38 mm high, is presented in Figure 2. It was notched at its centre using a razor blade. The crack length was about 10 mm. Once the crack was stretched, a heat source was applied to its interior. Further details are given in the next section.

2.2 Loading conditions

The tests were carried out at ambient temperature under prescribed displacement using a MTS 858 Elastomer Test System testing machine. Its loading capacity is 15 kN and the loading cell is 1 kN.

Samples were first stretched to a nominal stretch ratio equal to 1.88 at a strain rate set to 1.3 min^{-1}. This level of stretch ratio ensured that we would obtain crack shape close to that of a circular hole of 10 mm diameter. A metallic cylinder of the same diameter, surrounded by a heater, was then inserted into the crack. A PT100 temperature sensor linked to a Proportional-Integral-Derivative (PID) temperature controller enabled us to regulate the temperature precisely. Two thermal cycles were then performed between ambient temperature and 95°C. This maximum temperature value was chosen in such a way that it led to the melting of all the crystallites formed in the bulk material (Marchal 2006).

2.3 Measurement of thermal and kinematic fields

The kinematic field at the sample surface was obtained using the DIC technique. This consists in correlating the grey levels between two different images of a Region Of Interest (ROI) recorded using a cooled 12-bit dynamic Sensicam camera. Each image corresponds to different stretch ratio levels. This optical

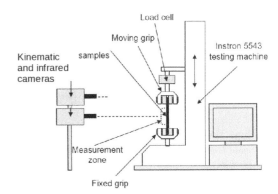

Figure 1. Schematic view of the experimental setup with the two cameras.

technique enabled us to reach a resolution of 0.03 pixels, corresponding to 1.02 μm and a spatial resolution (defined as the smallest distance between two independent points) of 10 pixels, corresponding to 341 μm. Uniform lighting at the sample surface was ensured by a set of fixed lamps. The charge-coupled device (CCD) of the camera had 1.4×10^6 connected pixels (1376 × 1040). The software used for the correlation process was SeptD (Vacher, Dumoulin, Morestin, and Mguil-Touchal 1999).

Temperature measurements were carried out with a Cedip Jade III-MWIR infrared camera, which features a local plane array of 320 × 240 pixels and detectors with a wavelength range of 3.5–5 μm. The integration time was 1500 μs and the acquisition frequency was set to 150 Hz. The noise equivalent temperature difference (NETD) of the camera, i.e. the thermal resolution, equals 20 mK for a temperature range of 5–40°C. Here, as the temperature varied between about 23°C and 95°C, two temperature ranges were used (5–60°C and 60–125°C) and the NETD of the camera does not exceed 30 mK. The temperature measurements were recorded in two files. Therefore, the variation of the temperature over the course of the test had to be reconstructed from these two files. In order to ensure that the internal temperature of the camera was stabilized to perform the measurements, it was set up and switched on for one hour before the experiment started. This camera temperature stabilization was necessary to avoid any drift in the measurements during the test.

The infrared camera is located above the kinematic one. So the two kinematic and thermal fields are obtained at the same time. The fact that the surface of measurement is considered suffciently flat and plane, and the fact that carbon black fillers make surfaces naturally black lead to consider the material emissivity close to 1. This assumption was also checked under deformation by comparing simultaneously the temperature value measured at the sample surface and at the surface of an aluminium sheet recovered by a black painting, which gave the reference temperature. Figure 1 presents a shematic view of the experimental setup with the two cameras.

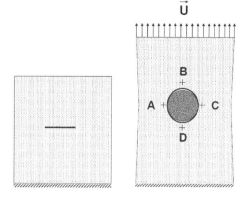

$\vec{U}$

B
+
A + ○ + C
+
D

Figure 2. Schematic view of sample geometry in the unde-formed (left) and deformed (right) states.

3 RESULTS

As mentioned above, the test was composed of two steps. The first was the stretching of the notched sample at room temperature in order to obtain crystallized crack tips. The second consisted in applying a heat source inside the crack. During the test, images were recorded by kinematic and thermal cameras in order to link the changes in the kinematic field to the temperature at each point of the sample surface. In the following, the effect of temperature on the kinematic field is discussed with respect to the variations in stretch ratio and in temperature at points A, B, C and D (see Figure 2).

3.1 Kinematic field at the stretched crack tip

In order to generate crystallites at the crack tips, the sample is first stretched at a nominal stretch ratio of 1.88. Figure 5 shows the kinematic fields obtained using the DIC technique in terms of displacement and stretch ratio in the loading direction. The reference image corresponds to the undeformed state and the current image corresponds to the sample geometry at the end of stretching. The results presented relate to the deformed sample geometry. This figure highlights the fact that the local stretch ratio at the crack tip exceeds 1.64, i.e. the stretch ratio at which crystallization begins. *A contrario*, the stretch ratio obtained at points B and D is close to 1. This is explained by the fact that these zones are located on either side of the stretched crack tips and are therefore relaxed.

3.2 Effect of temperature

Once the sample is stretched, the displacement of the moving grip is halted and remains fixed. A heat source is then applied to the circular crack. Figure 4 gives the temperature cartography at the sample surface when the temperature reaches a maximum value of 95°C at the crack tips. This figure shows that the temperature

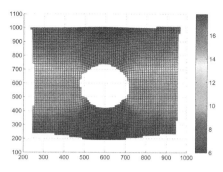

(a) displacement field in the loading direction (mm)

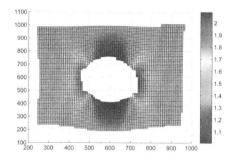

(b) stretch ratio in the loading direction

Figure 3. Kinematic fields after stretching the sample.

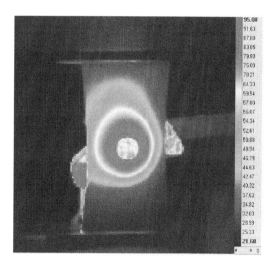

Figure 4. Cartography of sample surface temperature.

iso-values correspond to quasi-circular lines. The temperature gradient between the crack tip and the edge of the specimen is about 50°C (from 95 to 45°C). This is a good illustration of the poor thermal conductivity of rubbers. It should be noted that the maximum temperature value in the vicinity of the crack during each thermal cycle exceeds or is close to 80°C, i.e. 93.9°C at point A and 78°C at point B. This means that all the crystallites have melted in this zone, whatever the stretch ratio level (Marchal 2006).

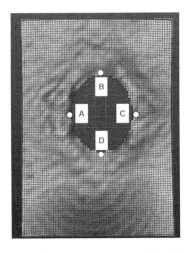

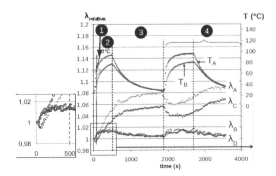

Figure 7. Relative stretch ratio in zone A, B, C, and D, and temperature in zone A and B versus time.

Figure 5. Example of the stretch ratio in the loading direction.

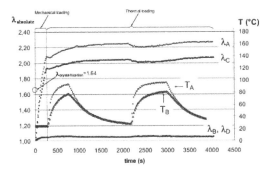

Figure 6. Absolute stretch ratio in zone A, B, C, and D, and temperature in zone A and B versus time.

Figure 5 gives an example of the stretch ratio in the loading direction when the temperature increases inside the hole. Gradients of stretch ratio are well distinguised.

In order to evaluate the effect of temperature variation on the change in the kinematic field, the temperature variation at points A and B and the stretch ratio at points A, B, C and D are considered. In the following, the stretch ratio considered is that calculated in the loading direction. Figure 6 gives the results obtained.

First, the temperature at point B is lower than that at point A because the hole is not exactly circular and also because in this zone the metallic cylinder is not in contact with the crack lip. Thus, a part of the heat brought by the heated cylinder is lost by convection. Second, analyzing precisely the variation in the kinematic field related to the variation in temperature by using the absolute value of the stretch ratio seems difficult. This is the reason why the relative stretch ratio is used in the following. This relative stretch ratio is calculated by considering the end of the stretching step as the reference deformation state. Thus, the relative stretch ratio at the beginning of the thermal cycles equals 1. Figure 7 presents the diagram obtained.

First of all, in the relaxed zones B and D the relative stretch ratio increases with the increase in temperature and decreases with the decrease in temperature, whatever the thermal cycle considered. This phenomenon highlights the effect of the variation in internal energy, which is of the first order, compared to that of the variation in entropy. Let us now consider the zones whose stretch ratio is higher and more particularly higher than 1.64 (points A and C), which is the stretch ratio at the beginning of crystallization. At this level of stretch ratio, the previous phenomenon is reversed and the increase in temperature is expected to decrease the stretch ratio. As shown by the magnification of the outlined zone, this is observed at the beginning of the first thermal cycle for temperatures lower than approximately 63°C. However, for higher temperatures, the stretch ratio increases. In fact, this phenomenon is due to the crystallites' melting. Indeed, crystallites act as fillers and tend to reinforced material stiffness by the contraction and orientation of the polymer chains (Chenal, Gauthier, Chazeau, Guy, and Bomal 2007). The melting of crystallites thus increases the stretch ratio, and at the end of the temperature increase, the stretch ratio at points A and C is higher than that at the end of stretching, because all the crystallites have melted. When heat generation is stopped, the temperature at points A and C decreases and the stretch ratio increases due to entropic effects. As a consequence, the stress level also decreases in this zone, the test being performed under controlled displacement. Without additional stretching or mechanical cycles, the same quantity of crystallites can not be regenerated when the thermal cycle is reconducted. Thus, the increase in temperature leads to a decrease in stretch ratio and vice-versa. It should be noted that the stretch ratio level is not exactly the same at points A and C. This is explained by the fact that the rubber cut, and more particularly in the zone corresponding to the crack tip, is not perfectly symmetrical. Moreover, in terms of pixels, points A and C are not at exactly the same distance from the crack tips.

4 CONCLUSION AND PERSPECTIVES

This study has investigated the effect of temperature variation at the crystallized crack tip in carbon black-filled natural rubber, and more particularly the changes in the kinematic field with respect to the temperature. To this end, kinematic and thermal fields were simultaneously measured. In such a material, with a heterogeneous kinematic field, temperature variation accounts for the competition between energy and entropy effects. This was clearly observed in terms of evolution in the kinematic field. Indeed, in relaxed zones, the stretch ratio increases with the increase in temperature, whereas it decreases in highly stretched crack tips. Moreover at crack tips and more generally in crystallized zones, crystallite melting increases the stretch ratio. This is explained by the fact that crystallites act as fillers, concentrating the stress and therefore increasing the apparent stiffness of the material. *In fine*, such a study could be an interesting way to improve and validate physical thermomechanical models, taking into account the effect of crystallization and crystallite melting in the thermomechanical response of rubbers.

ACKNOWLEDGEMENTS

The authors thank Dr. X. Balandraud, N. Blanchard and H. Perrin for their collaboration.

REFERENCES

Chenal, J.-M., C. Gauthier, L. Chazeau, L. Guy, & Y. Bomal (2007). Parameters governing strain induced crystallization in filled natural rubber. *Polymer 48*, 6893–6901.

Gent, A., P. Lindley, & A. Thomas (1964). Cut growth and fatigue of rubbers. i. the relationship between cut growth and fatigue. *Journal of Applied Polymer Science 8*, 455–466.

Gent, A., M. Razzaghi-Kashani, & G. Hamed (2003). Why do cracks turn sideways? *Rubber Chemistry and Technology 76*, 122–131.

Gent, A. N. & C. Pulford (1984). Micromechanics of fracture in elastomers. *Journal of Materials Science 19*, 3612–3619.

Greensmith, H. W. (1956). Rupture of rubber – iv – tear properties of vulcanized containig carbon black. *Journal of Polymer Science 21*, 175–187.

Greensmith, H. W. & A. G. Thomas (1956). Rupture of rubber – iii – determination of tear properties. *Rubber Chemistry and Technology 29*, 372–381.

Hamed, G. R., H. J. Kim, & A. N. Gent (1996). Cut growth in vulcanizates of natural rubber cis-polybutadiene and a 50/50 blend during single and repeated extension. *Rubber Chemistry and Technology 69*, 807–818.

Le Cam, J.-B. & E. Toussaint (2008). Volume variation in stretched natural rubber: competition between cavitation and stress-induced crystallization. *Macromolecules 41*, 7579–7583.

Le Cam, J.-B. & E. Toussaint (2009). Cyclic volume changes in rubbers. *Mechanics of Materials 41*, 898–901.

Marchal, J. (2006). *Cristallisation des caoutchoucs chargés et non chargés sous contrainte: Effet sur les chaînes amorphes*. Phd thesis, Université Paris XI Orsay, France.

Rivlin, R. S. & A. G. Thomas (1953). Rupture of rubber. I. Characteristic energy for tearing. *Journal of Polymer Science 10*.

Sutton, M. A., W. J. Wolters, W. H. Peters, W. F. Ranson, & S. R. McNeil (1983). Determination of displacements using an improved digital correlation method. *Image and Vision Computing 1*, 133–139.

Thomas, A. G. (1958). Rupture of rubber – v – cut growth in natural rubber vulcanizates. *Journal of Polymer Science 31*, 467–480.

Thomas, A. G. (1960). Rupture of rubber – vi – further experiments on the tear criterion. *Journal of Polymer Science 3*, 168–174.

Vacher, P., S. Dumoulin, F. Morestin, & S. Mguil-Touchal (1999). Bidimensional strain measurement using digital images. In *Proceedings of the Institution of Mechanical Engineers. Part C: Journal of Mechanical Engineering Science*, p. 213.

Processing heterogeneous strain fields with the virtual fields method: A new route for the mechanical characterization of elastomeric materials

N. Promma
Laboratoire de Mécanique et Ingénieries, Université Blaise Pascal & IFMA, France

B. Raka
Laboratoire de Mécanique et Technologie, ENS Cachan/CNRS UMR 8535/Université Paris 6/PRES UniverSud Paris, France

M. Grédiac, E. Toussaint, J.-B. Le Cam & X. Balandraud
Laboratoire de Mécanique et Ingénieries, Université Blaise Pascal & IFMA, France

F. Hild
Laboratoire de Mécanique et Technologie, ENS Cachan/CNRS UMR 8535/Université Paris 6/PRES UniverSud Paris, France

ABSTRACT: This paper deals with the identification of the constitutive parameters governing a hyperelastic law. These parameters are identified using a multiaxial mechanical test that gives rise to heterogeneous stress/strain fields. Since no analytical relationship is available between the measurements and the unknown parameters, a suitable tool is used to identify these unknowns: the virtual fields method. Results obtained with experiments performed on rubber specimens illustrate the approach.

1 INTRODUCTION

Modeling the mechanical response of elastomeric materials is commonly carried out within the framework of hyperelasticity (Ward & Hadley 1993, Holzapfel 2000). Identifying constitutive parameters that govern a type of law is classically carried out with homogeneous tests, namely uniaxial tensile extension, pure shear and equibiaxial extension. These three types of loading conditions completely describe the domain of possible loading paths (Ward & Hadley 1993, G'Sell & Coupard 1994). It is well known that the values of the constitutive parameters that are identified with those three types of test are generally different in practice. A trade-off between these three sets of values must therefore be found to obtain parameters that can reasonably be considered as intrinsic.

The aim of the present work is to identify the parameters of a given model from only one single heterogeneous test in which the three different types of strain states exist. In that case, the parameters obtained are directly a weighted average of those that would be obtained from the three different tests described above. The challenge here is in retrieving the parameters in a situation for which no closed-form solution exists for the stress, strain and displacement distributions as functions of the applied load and the material properties. Consequently, inverse techniques have to be used. Among these, the finite element model updating

technique (Kavanagh & Clough 1971, Molimard, Riche, Vautrin, & Lee 2005), the constitutive equation gap method (G. Geymonat & Pagano 2007, Feissel & Allix 2007, Latourte et al. 2008), the equilibrium gap method (Claire et al. 2004), the virtual fields method (Grédiac 1989) and the reciprocity gap method (Ikehata 1990) are currently developed. The main features of these different methods are presented in (Avril et al. 2008). The equilibrium gap method has been recently used for identification purposes in case of large deformations on steel (Medda et al. 2007) but the finite element model updating technique is the most commonly used in this case (Genovese et al. 2006, Giton et al. 2006, Drapier & Gaied 2006). Updating finite element models however entails large numbers of calculations. Another technique is used in the present study, namely, the virtual fields method. This method leads to a direct identification of the constitutive parameters in case of linear elasticity (Grédiac 1989, Grédiac et al. 2006).

The main features of the virtual fields method and its extension to large deformations are presented in the first part of the paper. Some numerical examples then illustrate the relevance of the approach with a special emphasis on the influence of measurement noise on identified parameters. A biaxial test that gives rise to heterogenous strain fields is analyzed. Kinematic fields are provided by an image correlation code suitable for large deformations. Typical displacement/strain fields are shown and results obtained in terms of identified parameters are discussed.

2 THE VIRTUAL FIELDS METHOD

Assuming a plane stress state in a solid, the principle of virtual work (PVW) is written as follows (Holzapfel 2000)

$$t \times \int_{S_0} \Pi : \frac{\partial \mathbf{U}^*}{\partial X} dS_0 = t \times \int_{\partial S_0} (\Pi.\mathbf{n}).\mathbf{U}^* dl \quad (1)$$

where Π is the first Piola-Kirchhoff (PK1) stress tensor, $\mathbf{U}^*$ a kinematically admissible virtual field, $\mathbf{n}$ the normal to the external boundary where the load is applied, and t the thickness of the solid. $\mathbf{X}$ are the Lagrangian co-ordinates. It must be emphasized that the above equation is valid for *any* admissible virtual field $\mathbf{U}^*$. The virtual fields method consists of two main operations:

– The first operation consists in expressing the stress components as functions of the measured strain components by introducing the constitutive equations. A Mooney law is considered in this work (Mooney 1940). The strain energy W reads in this case

$$W = C_1(I_1 - 3) + C_2(I_2 - 3) \quad (2)$$

where I_1 and I_2 are the first two invariants of the left Cauchy-Green tensor $\mathbf{B}$. The Cauchy stress components are deduced using the following equation

$$\sigma_i = -p + \lambda_i \frac{\partial W}{\partial \lambda_i}, i = 1 \cdots 3 \quad (3)$$

The problem here is to identify the parameters that govern the Mooney law, namely C_1 and C_2 in Equation 2, assuming that the displacement field is measured and that the gradient tensor is deduced by space-differentiation over a grid of measurements regularly distributed. In practice (see Section 3 below), these quantities are obtained in the reference configuration. Assuming that the problem is plane and that the material is incompressible, the two in-plane Cauchy stress components can be written as follows

$$\sigma_i = 2\left[C_1(\lambda_i^2 - \lambda_3^2) - C_2\left(\frac{1}{\lambda_i^2} - \frac{1}{\lambda_3^2}\right)\right], i = 1, 2 \quad (4)$$

where the $\lambda_i, i = 1, 2, 3$ are the principal in-plane stretch ratios. The PK1 stress tensor Π is derived from the Cauchy stress tensor σ by using the following expression

$$\Pi = J\sigma\mathbf{F}^{-t} \quad (5)$$

where J is the Jacobian of the transformation. This leads to the PK1 stresses written as functions of the Mooney parameters C_1, C_2, the principal stretch

ratios $\lambda_i, i = 1, 2, 3$, the gradient tensor components and the transition matrix components

$$\Pi = C_1\Theta(\lambda_1, \lambda_2, \lambda_3) + C_2\Lambda(\lambda_1, \lambda_2, \lambda_3) \quad (6)$$

where Θ and Λ are expressions, which can be found in (Promma et al. 2009). The fact that Cauchy stresses are used here is due to the fact that numerical simulations performed to check the relevancy of the method were carried with a finite element package that provided this type of stress components only, and not the PK1 components, thus leading to the approach described above (see (Promma et al. 2009) for further details). Introducing the above expression in Eq. (1) leads to the following equation

$$C_1 \int_{S_0} \Theta : \frac{\partial \mathbf{U}^*}{\partial \mathbf{X}} dS_0 + C_2 \int_{S_0} \Lambda : \frac{\partial \mathbf{U}^*}{\partial \mathbf{X}} dS_0 = \cdots$$
$$\cdots \int_{\partial S_0} (\Pi.\mathbf{n}).\mathbf{U}^* dl \quad (7)$$

It is worth noting that Eq. (7) is a *linear* function of the sought parameters C_2 and C_1.

– The second operation is to choose a suitable set of kinematically admissible virtual fields ($\mathbf{U}^*$ in Equation 1) and to write the principle of virtual work with these particular virtual fields. The underlying idea is to obtain as many different and independent equations of the same type of Equation 7. Since two parameters have to be determined here, two different and independent virtual fields are sufficient. These equations lead to a linear system, which provides the unknown parameters after inversion. It can be shown that the linear system is invertible if the stress field in the specimen is heterogeneous and if the virtual fields are independent. Choosing at best the virtual field is a major issue, since this choice directly influences the robustness of the method. In this study, the idea was to choose virtual fields that minimized the effect of measurement noise on the identified parameters. The strategy proposed in (Grédiac et al. 2002), which is based on a heuristic method, was used here for the sake of simplicity. It was here easier to define the virtual fields piecewise (with "virtual elements") rather than with continuous functions defined over the whole gauge section (Toussaint et al. 2006). This virtual mesh was defined by 18 nodes and 12 rectangular elements within which the virtual displacement was expressed with the classic shape functions of four-noded in-plane element. The virtual fields were therefore defined by the virtual displacement of the 18 nodes that defined the mesh.

3 EXPERIMENTS

3.1 *Test and specimen*

Biaxial tests were performed on a cross-shaped specimen made of carbon black filled natural rubber (34 parts of carbon black per hundred parts of rubber in

58

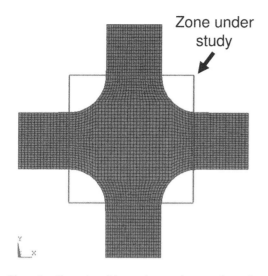

Figure 1. Geometry of the specimen and zone under study.

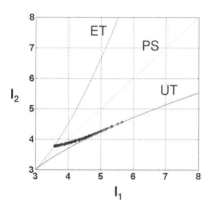

Figure 2. State state distribution in the I_1-I_2 plane, simulated data.

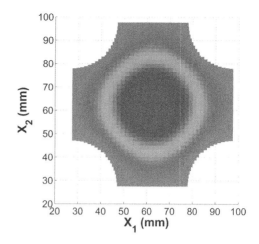

Figure 3. Strain state distribution in the gauge section of the specimen, simulated data.

Figure 4. Mechanical setup.

weight). The geometry of the specimen is shown in Figure 1. The total length is 125 mm and its thickness is 2 mm. The gauge section is a $42 \times 42\,\mathrm{mm}^2$ square.

The central zone of the specimen was subjected to a heterogeneous state of stress, going from a uniaxial state of stress (UT) in the branches to a equibiaxial state of stress (ET) at the center, with a pure shear state of stress (PS) in between. Since combinations of these states of stresses also took place in the specimen, it was necessary to define a suitable representation showing the corresponding "degree of heterogeneity". Assessing this heterogeneity can in fact be obtained by plotting the maximum value of the stretch ratio ($\lambda_{\max}$) with the following color code: red (ET), blue (ET) and green (PS). In the same way, it is useful to plot the distribution of the points in the gauge section in the $I_1 - I_2$ plane. This gives a point cloud, which can be considered as the "signature" of the heterogeneous state of stress, which takes place in the gauge section of the specimen. Both types of representation are reported in Figure 2 and 3. These figures are obtained

with numerical simulations performed with the Ansys package.

3.2 Testing machine

The tests were carried out with the multiaxial testing machine ASTRÉE. Of the six servohydraulic actuators, four of them were displacement-controlled. Consequently, the center of the sample was motionless, which made the displacement measurements easier. External load cells (of capacity equal to 2500 N) were installed. The specimen stretched between the grips of the testing machine is shown in Figure 4.

3.3 Testing conditions

Specimens were tested under equibiaxial loading conditions (Fig. 4). The maximum global stretch ratio λ_g was equal to 1.70 in both directions. In order to avoid the Mullins effects (Mullins 1948), five loading cycles were performed prior to testing up to the same value of λ_g in order to stabilize the mechanical response of the specimen. This maximum stretch ratio was obtained in practice by prescribing a displacement

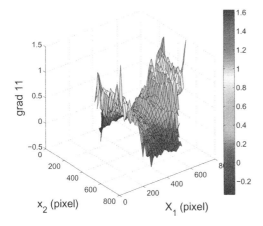

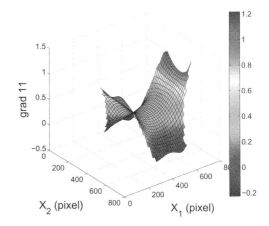

Figure 5. Displacement gradient $\partial U_1/\partial X_1$ for $\lambda_g = 1.70$, unfiltered data.

Figure 6. Displacement gradient $\partial U_1/\partial X_1$ for $\lambda_g = 1.70$, filtered data.

to each grip equal to 40 mm, with a displacement rate of 0.19 mm/s. Then, measurements were performed during three stabilized mechanical cycles.

3.4 Displacement field measurements

For each loading cycle, forty images were shot by a digital camera (DALSA, resolution: 1024×1024 pixels, dynamic range: 12 bits) located on one side of the specimen. Digital image correlation (DIC) was performed to measure displacement fields for successive strain states. Such a technique has already been successfully used for measuring large strains while testing elastomeric materials (Chevalier et al. 2001) or glass wool samples (Hild et al. 2002). The CORRELI DIC software was used in practice to measure the displacement field. Full detail concerning the features of this software and the calculations performed here to obtain large displacements can be found in (Hild et al. 2002), (Bergonnier et al. 2005) and (Promma et al. 2009).

3.5 Results

A typical view of the displacement gradient field $\partial U_1/\partial X_1$ obtained for $\lambda_g = 1.70$ is shown in Figures 5 and 6. It is deduced from the displacement maps using a centered finite differences scheme.

The gradient component is noisy in Figure 5 since the displacement field is not filtered prior to differentiation, contrary to the results shown in Figure 6 for which the displacement gradient is derived from displacements smoothed with a 9th degree polynomial fit. The choice of this degree is justified by the fact that the standard deviation between actual and smoothed data is higher for lower values of this degree, but does not significantly evolve when choosing greater values.

Experimental results are reported in the $I_1 - I_2$ plane (Fig. 7). The points are very scattered. The points obtained from smoothed measurements are distributed around the same curve as the one obtained with finite element results (see Figures 2 and 3 for comparison purposes). In the same way, the top view of the gauge section shown in Figure 8 is very similar to

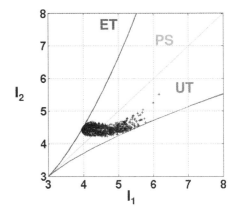

Figure 7. State of strain in the I_1-I_2 plane, filtered data.

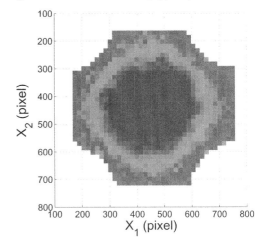

Figure 8. Strain state distribution in the gauge section of the specimen, filtered data.

its numerical counterparts (see Figure 3). The patch corresponding to the ET zone is not rigorously a circle. It seems moreover to be wider in the experimental results. This is probably due to the fact that the border

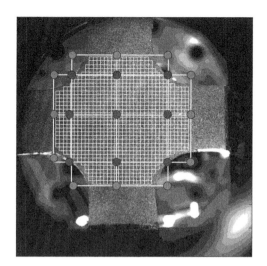

Figure 9.   Virtual mesh and specimen under test.

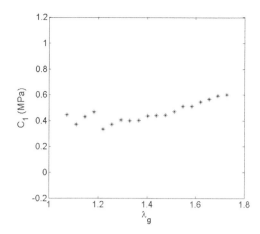

Figure 10.   Identified value of C_1 vs. λ_g.

of the specimen cannot be precisely defined, namely, interrogation windows of the DIC software astride the boundary of the specimen are removed, thus leading to a smaller UT zone.

3.6   Identification

The identification of C_1 and C_2 was performed by using the procedure described above with filtered data. In Figure 9, the virtual mesh used for describing the virtual fields (the 12 "large" rectangles) is superimposed with the interrogation windows used by the CORRELI DIC software (the small squares) and the reference picture of the specimen. This view illustrates the fact that 1033 experimental points were processed in practice to identify the constitutive parameters.

Typical curves showing the evolution of the identified value of C_1 and C_2 vs. λ_g are shown in Figures 10 and 11, respectively. Similar results were obtained with other tests, thus illustrating reproducibility of the results. It can be observed that C_1 slightly increases during the test whereas C_2 slightly decreases. It is also worthy of emphasis that the scatter is higher at the beginning of the test, for the lowest values of λ_g. This is certainly due to the fact the heterogeneity of the stress distribution is lower than for higher values of λ_g, so it is more difficult to retrieve accurate values of the identified parameters in this case. This scatter is taken into account for the calculation of the average values of the parameters denoted $\bar{C}_1$ and $\bar{C}_2$. These quantities are defined by

$$\bar{C}_i = \frac{\sum_{j=1}^{n_t} \left(\dfrac{\lambda_{g_j} - 1}{\sqrt{stdx_{ij}^2 + stdy_{ij}^2}} \right) \times C_{ij}}{\sum_{j=1}^{n} \left(\dfrac{\lambda_{g_j} - 1}{\sqrt{stdx_{ij}^2 + stdy_{ij}^2}} \right)} \qquad (8)$$

where $stdx_{ij}$ ($stdy_{ij}$ respectively) is the standard deviation of the difference between raw and smoothed u_x

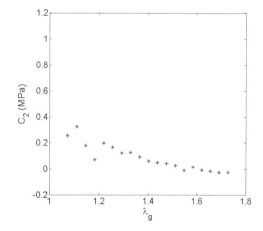

Figure 11.   Identified value of C_2 vs. λ_g.

(u_y respectively) distributions, $i = 1, 2$, $j = 1, \ldots, n_t$, where n_t is the number of images. The higher scatter in the displacement field for a given stretch ratio, the lower confidence in the experimental results. The values of $\bar{C}_1$ and $\bar{C}_2$ are equal to 0.411 MPa and 0.139 MPa, respectively. The identified values are very close to the classical ones that characterize this type of material.

The beneficial effect of smoothing the u_x and u_y fields can finally be illustrated as follows. Applying the complete identification procedure with raw data (as shown in Figure 5 for instance), instead of smoothed data, leads to a $\bar{C}_1$ value which remains reasonable: 0.51 MPa. $\bar{C}_2$ becomes however much lower than with smoothed data: 0.041 MPa. This clearly illustrates that C_2 is much more sensitive to noise than C_1. This is certainly due to the fact that its influence on the strain distribution is lower than C_2.

4   CONCLUSION

The extension of the virtual fields method from small to large deformations is addressed in this paper in case of hyperelasticity. A linear equation is established between parameters governing the Mooney model, actual stretch ratio distribution within the specimen and virtual fields. This equation is written with different virtual fields that are optimized with respect to their independence. This finally leads to a linear system that provides the unknown parameters after inversion. Numerical simulations illustrate the feasibility and the robustness of the method.

Experiments were then carried out on a cross-shaped rubber specimen. Strain fields were deduced from displacements provided by a digital image correlation system whose measurement quality is discussed in this particular case of large deformations. Results obtained are in agreement with theoretical and numerical expectations, thus confirming the feasibility of the present approach. The identification of parameters governing other types of hyperelastic models will be addressed in the near future.

ACKNOWLEDGEMENTS

The support of this research by the "Agence Nationale pour la Recherche" is gratefully acknowledged (PHOTOFIT project).

REFERENCES

Avril, S., M. Bonnet, A. Bretelle, M. Grédiac, F. Hild, P. Ienny, F. Latourte, D. Lemosse, S. Pagano, S. Pagnacco, & F. Pierron (2008). Overview of identification methods of mechanical parameters based on full-field measurements. *Experimental Mechanics 48*(4), 381–402.

Bergonnier, S., F. Hild, & S. Roux (2005). Digital image correlation used for mechanical tests on crimped glass wool samples. *Journal of Strain Analysis 40*, 185–197.

Chevalier, L., S. Calloch, F. Hild, & Y. Marco (2001). Digital image correlation used to analyze the multiaxial behavior of rubber-like materials. *European Journal of Mechanics. A-Solids 20*(2), 169–187. Elsevier.

Claire, D., F. Hild, & S. Roux (2004). A finite element formulation to identify damage fields. *International Journal for Numerical Methods in Engineering 61*, 189–208. Elsevier.

Drapier, S. & I. Gaied (2006). Identification strategy for orthotropic knitted elastomeric fabrics under large biaxial deformations. *Inverse problems in Science and Eniigneering 15*(8), 871–894. Taylor and Francis.

Feissel, P. & O. Allix (2007). Modified constitutive relation error identification strategy for transient dynamics with corrupted data: The elastic case. *Computer Methods in Applied Mechanics and Engineering 196*, 1968–1983.

G. Geymonat, F. H. & S. Pagano (2007). Identification of elastic parameters by displacement field measurement. *Comptes Rendus de l'Académie des Sciences 330*, 403–408.

Genovese, K., L. Lamberti, & C. Pappalettere (2006). Mechanical characterization of hyperelastic materials with fringe projection and optimization techniques. *Optic and Lasers in Engineering 44*, 423–442. Elsevier.

Giton, M., A. Caro-Bretelle, & P. Ienny (2006). Hyperelastic behaviour identification by a forward problem resolution: application to a tear test of a silicon-rubber. *Strain 2006*, 291–297. Blackwell.

Grédiac, M. (1989). Principe des travaux virtuels et identification. principle of virtual work and identification. *Comptes Rendus de l'Académie des Sciences*, 1–5. Gauthier-Villars. In French with abridged English version.

Grédiac, M., F. Pierron, S. Avril, & E. Toussaint (2006). The virtual fields method for extracting constitutive parameters from full-field measurements: a review. *Strain 42*, 233–253.

Grédiac, M., E. Toussaint, & F. Pierron (2002). Special virtual fields for the direct determination of material parameters with the virtual fields method. 1-Principle and definition. *International Journal of Solids and Structures 39*, 2691–2705. Else-vier.

G'Sell, C. & A. Coupard (1994). *Génie Mécanique des caoutchoucs et des élastomères thermoplastiques*. Apollor.

Hild, F., B. Raka, M. Baudequin, S. Roux, & F. Cantelaube (2002). Multi-scale displacement field measurements of compressed mineral wool samples by digital image correlation. *Applied Optics 41*(32), 6815–6828.

Holzapfel, G. A. (2000). *Nonlinear Solid Mechanics: A Continuum Approach for Engineering*. JohnWileyand Sons.

Ikehata, M. (1990). Inversion formulas for the linearized problem for an inverse boundary value problem in elastic prospection. *SIAM Journal for Applied Mathematics 50*, 1635–1644. Elsevier.

Kavanagh, K. T. & R. W. Clough (1971). Finite element applications in the characterization of elastic solids. *International Journal of Solids and Structures 7*, 11–23.

Latourte, F., A. Chrysochoos, S. Pagano, & B. Wattrisse (2008). Elastoplastic behavior identification for heterogeneous loadings and materials. *Experimental Mechanics*. Sage. In press.

Medda, A., G. Demofonti, S. Roux, F. Hild, F. Bertolino, & A. Baldi (2007). Sull'identificazione del comportamento plastico di un'accaio a partire da misure a campo intero ottenute tramite correlazione digitale di immagini. In *Proceedings of the XXXVI Convenio Nazionale, AIAS*.

Molimard, J., R. L. Riche, A. Vautrin, & J. Lee (2005). Identification of the four orthotropic plate stiffnesses a single open-hole tensile test. *Experimental Mechanics 45*, 404–411. Sage.

Mooney, M. (1940). A theory of large elastic deformation. *Journal of Applied Physics 11*, 582–592.

Mullins, L. (1948). Effect of stretching on the properties of rubber. *Rubber Chemistry and Technology 21*, 281–300.

Promma, N., B. Raka, M. Grédiac, E.Toussaint, J. B. L. Cam, X. Balandraud, & F. Hild (2009). Application of the virtual fields method to mechanical characterization of elastomeric materials. *International Journal of Solids and Structures 46*, 698–715.

Toussaint, E., M. Grédiac, & F. Pierron (2006). The virtual fields method with piecewise virtual fields. *International Journal of Mechanical Sciences 48*, 256–264. Elsevier.

Ward, I. M. & D. W. Hadley (1993). *An introduction to the mechanical properties of solid polymers*. John Wiley and Sons.

Constitutive Models for Rubber VII – Jerrams & Murphy (eds)
© 2012 Taylor & Francis Group, London, ISBN 978-0-415-68389-0

Energy losses at small strains in filled rubbers

L.B. Tunnicliffe, A.G. Thomas & J.J.C. Busfield

Department of Materials, Queen Mary University of London, London, UK

ABSTRACT: The dynamic mechanical properties of particulate-filled rubbers are examined within the small strain linear viscoelastic region using a free oscillation technique. Values of tanδ derived from experimental damping oscillations were found to be independent of strain as would be expected for a linear viscoelastic response. However, a significant non-linear dependence of tanδ upon filler loading was observed. This filler dependence demonstrated that a loss mechanism other than that of the inherent rubber viscoelasticity occurs during sample deformation even at very small strains. Potential mechanisms behind this energy loss such as interfacial slippage between the rubber matrix and filler particle and the absorbed rubber layer theory are discussed in the context of these findings.

1 INRODUCTION

Particulate reinforced rubbers are important industrial materials. The physics of particle reinforcement in rubbers has been an area of intense scientific investigation for many years. Despite these efforts, the subject is still not yet fully understood. When a particulate-filled rubber is deformed the energy required to do so is either stored (described by the complex storage modulus G′ or E′) or dissipated (described by the complex loss modulus G″ or E″) depending on the degree of filler breakdown and the nature of the strain response in the matrix rubber. Considerable effort has been made to measure and model the deformation of filled rubbers and in particular the breakdown of the filler structure upon application of increasing strain known as the Payne or Fletcher-Gent effect (Fletcher & Gent, 1953). At large strains not only might the filler structure breakdown but also crosslinks might rupture and significant alignment of the rubber macromolecules arises all of which makes the physical response characteristically non-linear and difficult to interpret.

When the deforming strains are small enough (typically below 1%) the mechanical response is linear and subsequently easier to interpret with the ratio of the storage and loss moduli, tan δ, being independent of strain within this region.

Attempts to model the effects of reinforcing fillers in rubbers within the linear viscoelastic region have been based on the hydrodynamic approach developed by Einstein (1926) for particle-filled Newtonian fluids (equation 1).

$$\eta = \eta_0(1 + 2.5\phi) \tag{1}$$

where η = viscosity of the filled liquid; η_0 = viscosity of the unfilled liquid and ϕ = volume fraction of filler.

Guth & Gold (1938) adapted equation 1 to describe the increase in modulus in reinforced rubbers within the linear, small strain region; the effective Newtonian response of the filled rubber. They incorporated a higher expansion term in an attempt to account for filler networking effects at high loadings (equation 2).

$$E = E_0(1 + 2.5\phi + 14.1\phi^2) \tag{2}$$

where E = small strain modulus of the filled rubber and E_0 = unfilled modulus.

Guth (1945) later introduced an anisotropy term to account for non-spherical fillers such as highly branched carbon blacks and high aspect ratio clays (equation 3).

$$E = E_0(1 + 0.67a\phi + 1.62a^2\phi^2) \tag{3}$$

where a = aspect ratio of the filler particle.

These models all share the initial assumption made by Einstein of no interfacial slippage between the matrix and the filler particles. If this were the case in reality then at small strains the energy of deformation can only be stored in the elastic response of the rubber and filler and any dissipation of energy can only occur within the rubber matrix. Therefore measurements of dynamic losses in filled rubbers within the small strain region should be equal to the viscoelastic characteristics of the rubber matrix (i.e. tan δ should be independent of filler loading). Recent studies of carbon black-filled rubbers (Suphadon et al., 2010) over a wide range of strains seem to indicate that this is in fact not the case with the tan δ values at the smallest strains reported instead being highly dependent on the carbon black filler loading.

Possible mechanisms to account for this relative increase in loss modulus are the absorbed rubber layer theory and the possibility of frictional losses occurring at the filler-matrix interface as a result of rubber slippage.

In the case of the absorbed rubber theory, rubber chains are physisorbed onto the surface of the filler particles resulting in a reduction of their molecular energetics. The result of this is to effectively raise the glass transition temperature of the bound layer (Wang, 1998). The subsequent increase in the tan δ values of this layer may possibly account for some of the increase in loss.

The slippage mechanism is simple; friction associated with the sliding of the rubber matrix over the filler particles upon applied strain results in an energetic loss at the filler-rubber interface (Jha et al., 2008). It is worthwhile noting that interfacial slippage is contrary to the fundamental assumptions of the hydrodynamic models described earlier.

In this work we have developed highly precise dynamic measurements that work within the small strain linear viscoelastic region on rubber-silica composites. Silica was chosen as the filler for this experiment because the use of a chemical surface treatment allows the compatibility of the silica surface with the rubber to be easily tailored (Ou & Yu, 1994). Measurements were made using a free torsion technique as opposed to forced oscillation (the Dynamic Mechanical Analysis method) as this method has previously been shown to give considerably more reliable results within the small strain region (Akutagawa et al., 1996).

2 MATERIALS AND METHODS

2.1 Rubber compounding and curing

The base rubber for this work was natural rubber (NR) grade CV60 obtained from the Tun Abdul Razak Research Centre (TARRC), Hertfordshire, UK. The filler was industrial tyre grade precipitated silica Z115G from Rhodia. The silica was incorporated into the rubber by internal mixing in a 1 litre Banbury. Curing chemicals were then added on a two roll open mill. Samples were prepared with filler loadings between 0 and 40 phr at 10 phr intervals. Each sample was cured with 1 phr elemental sulphur and 5 phr accelerator CBS (N-(1,3-Dimthylbutyl)-N'-Phenyl-P-Phenylenediamine). Curing times were determined by a Monsanto 2000E rheometer at plate temperature of 140°C. Cylindrical samples for dynamic torsion analysis were cured for the T_{90} times at 140°C on a hot press with plate pressure of roughly 11 MPa.

In order to effectively determine if there is any non-matrix loss effect the resilience of the base matrix (determined by rubber type and crosslinking) must remain consistent between samples. In order to achieve this an efficient vulcanizing (EV) system was used

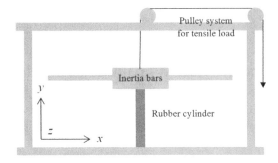

Figure 1. Side view of experimental setup with torsional movement in the x-z plane.

with an accelerator/sulphur ratio of 5. It has previously been demonstrated (Porter, 1963) that this type of cure has two benefits when attempting to control rubber resilience between samples;

1. EV reduces the amount and sulphur rank variety of polysulphidic crosslinks in favour of monosulphidic crosslinks. Typically >80% monosulphides are obtained from EV systems.
2. EV chemistry is relatively insensitive to the incorporation of filler particles (carbon black in the case of Porter's work).

By using an EV system, the variation of matrix resilience between samples was kept to a minimum.

Additionally, the cure system used was zinc-free. This was because solubilized zinc (from stearate or oxide sources) has been shown to preferentially absorb onto the surface of untreated silica filler particles (Hewitt, 2007). This effect would add a complex surface non-linearity into the experiment so zinc sources were omitted from the cure.

2.2 Free torsion analysis

The dynamic experimentation was based upon the free oscillation principal. Much work has been published using this technique for the analysis of complex loadings, damping characteristics and small strain behaviours in rubbers (see for examples; Deeprasertkul et al., 2000 and Suphadon et al., 2009). The data obtained for this study was taken from the amplitude decay of the small strain free torsion of a rubber cylinder. The experimental setup was simple; a cylindrical sample of rubber was bonded using Loctite 480 adhesive to the base of a rigid frame and an inertia bar which in turn was attached via pulley system to a load allowing variable tensile forces to be applied along the y axis of the cylinder. Free oscillations in the x-z plane were initiated by slightly tapping one side of the inertia bar (see figure 1). Resulting strains were found to reach a maximum of 1% torsional strain at the cylindrical sample surface and decayed to much smaller values. In the case of all the experiments performed, the tensile load on the samples was minimal being only enough to keep the sample perpendicular to test frame.

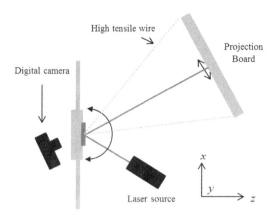

Figure 2. Top view of experimental setup including data collection apparatus. Torsional movement is in the x-z plane.

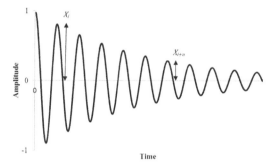

Figure 3. Damped oscillation.

The damped oscillation data was collected by video camera recording of sample motion projected by laser beam reflection from a mirror mounted at the center of the inertia bar onto a white board (see figure 2). A high tensile wire attached to the center of the z axis of the cylinder allowed the white board to be kept exactly perpendicular to the projected laser beam. The video camera sampled at 33 Hz while the frequency of oscillation was controlled by the configuration of the inertia bar to ensure a maximum of 1.5 Hz.

Considering the damped oscillation of a cylinder in free torsion (figure 3) a damping constant, the logarithmic decrement (Δ) of sequential amplitudes is defined by equation 4 (Ferry, 1961).

$$\Delta = \frac{1}{n} \ln \left(\frac{x_i}{x_{i+n}} \right) \tag{4}$$

where $n =$ number of oscillations and $x_i =$ amplitude of specific oscillations.

tan δ can be approximated easily from this decrement value (equation 5).

$$\tan\delta \cong \left(\frac{\Delta}{\pi \left(1 + \frac{\Delta^2}{4\pi^2} \right)} \right) \tag{5}$$

The torsional stiffness of the cylinder can be defined by Δ in terms of in and out-of phase stiffness (equations 6 and 7).

$$K_\theta' = I\omega^2 \left(1 + \frac{\Delta^2}{4\pi^2} \right) \tag{6}$$

$$K_\theta'' = I\omega^2 \frac{\Delta}{\pi} \tag{7}$$

where K_θ' and $K_\theta'' =$ in-phase and out of phase torsional stiffness respectively; $\omega =$ frequency of oscillation and $I =$ inertia of the system.

The relationship between torsional stiffness and static shear modulus (equation 8) then allows definition of the out of phase stiffness with the out of phase shear modulus (equation 9) and thus by equating equations 7 and 9, calculation of G'' from the experimentally determined Δ (equation 10) (Brown, 2006).

$$K_\theta = \frac{\pi r^4 G}{2l} \tag{8}$$

$$K_\theta'' = \frac{\pi r^4 G''}{2l} \tag{9}$$

$$G'' = \frac{2l\omega^2 I\Delta}{\pi^2 r^4} \tag{10}$$

where G = bulk shear modulus; G' and G'' = in-phase and out-of-phase shear moduli respectively, l = length of the rubber cylinder and r = radius.

From this derivation it is possible to determine the dissipation of energy from small strain damping. In practice, the maxima of the successive oscillation peaks were found by measuring the laser beam displacement from a baseline mark made when the cylinder was at rest within the corresponding video frames. Five repeats were performed per sample.

3 RESULTS AND DISCUSSION

3.1 Free torsion oscillation

An example of the collected video frame data is shown in figure 4. The frames selected from the videos at the times of a peak maximum for successive oscillations. As Δ is defined by a ratio of amplitudes, scale determination for these images was not required. Frequency of oscillation was found from peak-to-peak timescale from the camera internal timer.

Values of Δ were calculated from the experimental data. The Δ and therefore tan δ values were found to be independent of the maximum in oscillating strain cycles indicating that the measurements had been taken within the material's linear region (figure 5).

Values of Δ averaged over the linear region and similarly tan δ values were found to vary with the filler loading – demonstrating that an additional loss process other than that of the base rubber viscoelasticity was occurring (figure 6). The calculated G'' values reflect this trend (figure 7).

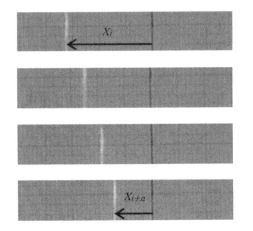

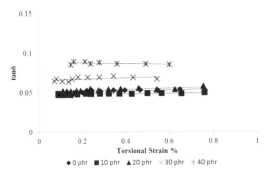

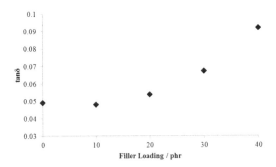

Figure 4. Examples of video frame determination of oscillation amplitude.

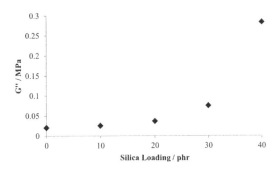

Figure 7. G'' values as a function of silica filler loading.

3. Using more resilient rubbers, for example BR, to emphasise the loss mechanism over rubber viscoelastic effects.

4 CONCLUSIONS

Measuring the very small strain dynamic mechanical properties of filled rubbers in the linear viscoelastic region using a free oscillation technique reveals significant non-rubber energy losses in silica-filled NR. The logarithmic decrement of the damped oscillation data was found to be independent of strain over the range of strains examined while the average tan δ values varied with filler loading. This filler dependence was indicative of a significant increase in the dissipation of energy during the damping process. This loss and associated loss mechanism was discussed in terms the rubber layer theory and interfacial slippage between rubber and filler – the implications of the latter for established hydrodynamic models was also highlighted. It is clear that using the linear region free oscillation technique will in the future help determine the physics of the loss mechanisms.

Figure 5. tan δ values as a function of maximum torsional strains.

Figure 6. tan δ values as a function of silica filler loading.

From this limited data set it is not possible fully understand the viscoelastic loss mechanism (absorbed layer, interfacial slippage or a combination of both). However, the new small strain technique provides a way to unpick the details of the loss mechanism by:

1. Evaluating how chemical coupling agents on the filler surface impact on the loss modulus and reduce potential interface slippage effects.
2. Observing the effect of temperature on the loss mechanism allowing calculation of the activation energy for potential interfacial slippage.

ACKNOWLEDGEMENTS

The authors would like to thank Rhodia for the supply of the precipitated silica filler and TARRC for use of their internal mixer facilities. One of the authors, LB Tunnicliffe would like to thank Sibelco for the funding to support his studies.

REFERENCES

Akutagawa K., Davies C.K.L., Thomas A.G. 1996. The Effect of Low Molar Mass Liquids on the Dynamic Mechanical Properties of Elastomers Under Strain, *Progress in Rubber and Plastics Technology*, 12, 3, 174–190.

Brown R. 2006. *Physical Testing of Rubber*, New York, Springer.

Deeprasertkul C., Thomas A.G., Busfield J.J.C. 2000. The effect of liquids on the dynamic properties of carbon black filled natural rubber as a function of pre-strain, *Polymer*, 41, 9219–9225.

Einstein A. 1926. *Investigations on the Theory of the Brownian Movement*, New York, Dover.

Ferry J.D. 1961. *Viscoelastic properties of polymers*, New York, John Wiley and Sons Inc.

Fletcher W. P. and Gent A. N. 1953. Non-Linearity in the Dynamic Properties of Vulcanised Rubber Compounds, *Trans. Inst. Rubber Ind.* **29**, pp 266–280.

Guth E. and Gold O. 1938. On the hydrodynamic theory of the viscosity of suspensions, *Phy. Rev.*, 53, 322.

Guth E. 1945. Theory of filler reinforcement. *Journal of Applied Physics*, 16, 20–25.

Hewitt N. 2007. *Compounding Precipitated Silica in Elastomers*, New York, William Andrew Publishing.

Jha V., Hon A.A., Thomas A.G., Busfield J.J.C. 2008. Modeling of the effect of rigid fillers on the stiffness of rubbers. *Journal of Applied Polymer Science*, 107, 4, 2572–2577.

Porter M. 1963. Structural Characterization of filled vulcanizates part 1. Determination of the concentration of chemical crosslinks in natural rubber vulcanizates containing high abrasion furnace black, *Rubber Chemistry and Technology*, 36, 547–558.

Suphadon N., Thomas A.G., Busfield J.J.C. 2009. The viscoelastic behaviour of rubber under a complex loading, *Journal of Applied Polymer Science*, 113, 693–699.

Suphadon N., Thomas A.G., Busfield J.J.C. 2010. The Viscoelastic Behavior of Rubber Under a Complex Loading. II. The Effect Large Strains and the Incorporation of Carbon Black Volume, *Journal of Applied Polymer Science*, 117, 3, 1290–1297.

Ou Y.C., Yu Z.Z. 1994. Effects of alkylation of silica filler on rubber reinforcement, *Rubber Chemistry and Technology*, 67, 834–844.

Wang M.J. 1998. Effect of polymer-filler and filler-filler interactions on dynamic properties of filled vulcanizates, *Rubber Chemistry and Technology*, 71, 520–590.

Constitutive Models for Rubber VII – Jerrams & Murphy (eds)
© 2012 Taylor & Francis Group, London, ISBN 978-0-415-68389-0

The mechanism of fatigue crack growth in rubbers under severe loading: The effect of stress-induced crystallization

Jean-Benoît Le Cam*

Clermont Université, Institut Français de Mécanique Avancée, Laboratoire de mécanique et Ingénieries, Clermont-Ferrand, France

Evelyne Toussaint*

Clermont Université, Université Blaise Pascal, Laboratoire de mécanique et Ingénieries, Clermont-Ferrand, France

*Fédération de Recherche TIMS CNRS FR 2856, Complexe scientifique des Cézeaux, Aubière Cedex, France

ABSTRACT: This study deals with the mechanism of fatigue crack growth in natural rubber submitted to severe relaxing loading conditions. In one mechanical cycle under such loading conditions, the high level of stress at the crack tip engenders high crystallinity, which halts crack growth in the plane perpendicular to the loading direction. Consequently, the crack bifurcates. Then the fracture surfaces tear, slide and relax simultaneously along a highly crystallized crack tip to form striations. The higher the stress level, the lower the crack growth in the plane perpendicular to the loading direction and the greater the bifurcation phenomenon. This explains why the striation shape evolves from triangular to lamellar during crack propagation. This study was carried out using a novel experimental approach based on full-field measurement techniques.

1 INTRODUCTION

Crack propagation in elastomers is mostly studied through theoretical approaches which do not take into account the physical phenomena involved during the crack propagation process. Thus, these approaches do not satisfactorily predict the effect of microstructure changes due to aging or stress-induced crystallization, and consequently do not adequately predict the crack path. For this purpose, numerous studies investigate the physical phenomena of crack growth. Under quasi-static and repeated loadings (approximately ten cycles for instance), the works of Gent et al. give some important answers concerning the physical mechanisms involved during the crack growth (Gent and Pulford 1984; Hamed et al. 1996; Gent et al. 2003). Under fatigue loadings, crack growth has been studied at the microscopic scale (Fukahori 1991; Bhowmick 1995; Le Cam et al. 2004; Hainsworth 2007; Wang and G. 2002) using Scanning Electron Microscopy (SEM). These studies are carried out under relaxing loading conditions, i.e. the loading returns to zero at the end of each mechanical cycle. Moreover, the maximum stress level is moderate in the sense that it does not involve oligocyclic fatigue. Here, the term "oligocyclic fatigue" is used for severe fatigue loadings leading to a duration life inferior to 10^4 cycles. In such a material and loading conditions, micro-mechanisms of fatigue crack growth have already been proposed in carbon black filled natural rubber (Le Cam et al. 2004; Hainsworth 2007; Beurrot et al. 2010). Even if

these mechanisms proposed by the authors is suitable for moderate loading, it is not sufficient to explain the morphology of the fracture surface when the stress level increases significantly at the crack tip, i.e. for olygocyclic fatigue or for the end of crack propagation under moderate cyclic loading. In these cases, no wrenching is observed at the fracture surface; only striations are observed. The last remark suggests that the mechanism of fatigue crack growth changes when the stress increases significantly at the crack tip. The aim of this paper is to establish the mechanism of fatigue crack growth under severe cyclic loading conditions and to evaluate the contribution of stress-induced crystallization to this mechanism. Even if the phenomenon of stress-induced crystallization has been widely studied under quasi-static loading (Trabelsi et al. 2002; Trabelsi and Rault 2003; Rault et al. 2006; Toki et al. 2002), its influence on the mechanism of fatigue crack growth under severe fatigue loading has never been investigated at the microscopic scale.

2 EXPERIMENTAL SECTION

2.1 Materials and samples

The material considered here is a 34 phr carbon black-filled natural rubber. It is referred to as F-NR in the following. Two sample geometries were used. The first one was a classic sample geometry for fatigue tests. It is axisymmetrical and is usually called a diabolo sample. It is adhered on both sides to metallic inserts in order to be clamped in the grips of the testing machine. The second geometry corresponds to a plate of 20 mm height, 5 mm width and 2 mm thick. It is referred to as the "flat" sample in the following and was used to observe the real-time change in the mechanism of fatigue crack growth when the stress level increases at the crack tip.

2.2 *Fatigue loading conditions*

The uniaxial fatigue tests were first performed with the diabolo samples under uniaxial cyclic prescribed force with a MTS 858 Elastomer Test System testing machine. under relaxing loading conditions, i.e. the minimum value of the cyclic force is equal to zero. Four levels of force were applied using a sinusoidal signal: 300, 500, 600 and 750 N. For each test, three samples were tested and the number of cycles at crack initiation (the occurrence of a self-initiated crack of 2 mm at the sample surface) and at failure were stored. Three tests were also halted during crack propagation under 500 N maximum force in order to investigate the crack tip morphology when a change in the mechanism of fatigue crack growth was observed. The frequency was set to limit the rise in surface temperature to 20°C and to avoid creating thermal damage in addition to the mechanical damage. These tests were carried out at 23°C regulated temperature. The flat samples were pre-cut with a razor blade and were then tested under the same conditions as the diabolo sample but with a lower prescribed force (between 0 and 15 N) to account for the difference in cross-section area between the two sample geometries. When the crack tip becomes similar to that of a typical fatigue crack tip obtained under moderate loading, the sample is clamped in the grips of a micro-tensile machine. It is then stretched to an elongation higher than the maximum measured during the fatigue test in order to observe in real-time the effect of the increase in stress on the crack propagation mechanism.

2.3 *Scanning electron microscopy*

Photomicrographs were performed with a HITACHI S-3200N model SEM using secondary electrons. Samples were stretched with a rudimentary static tensile apparatus (Le Cam et al. 2004) to a stretch ratio equal to that measured at the last mechanical cycle and the crack tip was coated with a gold layer to ensure electrical conduction.

2.4 *Measurement of the relative displacement field at the crack tip*

The measurement was carried out at the crack tip of the notched flat sample while it was stretched using a DEBEN micro-tensile machine. This machine allows the sample to be stretched symmetrically. Figure 1 presents the experimental set-up.

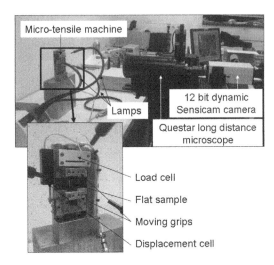

Figure 1. Real-time measurement of the relative displacement field at the crack tip: experimental set-up.

It consisted of a cooled 12-bit dynamic Sensicam camera with a Questar long distance microscope which enables the observation of zones of 2.7×2 mm^2 area. Uniform lighting at the sample surface was ensured by lamps. The charge-coupled device (CCD) of the camera has 1.4×10^6 joined pixels (1376×1040). The camera was fixed on a multidirectional adjustable support. The relative displacement field at the crack tip was obtained using the Digital Image Correlation (DIC) technique (Sutton et al. 1983). This consists in correlating the grey levels between two different images of a given zone. Each image corresponds to different stretch ratio levels. This optical technique offers a resolution of 0.03 pixel corresponding to 0.06 µm and a spatial resolution of 16 pixels corresponding to 31 µm. The software (Hild 2002) used for the correlation process was CorreliLMT.

3 RESULTS

3.1 *Typical fracture surface*

At the macroscopic scale, cracks always propagate in a plane perpendicular to the applied force direction in a zone close to the middle plane of the sample. Figure 2 summarizes the fracture surface morphology observed. Figure 2(a) is the top view of a typical fracture surface. Five zones are considered to describe it: (i) zone A is the initiation zone. It is always located close to the sample surface. This zone is identified by the orientation of wrenchings on the fracture surface, which describe ellipses around this zone. Further investigations using SEM coupled with an Energy Dispersive X-ray Spectrometer (EDXS) have shown that crack initiation is due either to the failure of carbon black agglomerates or to the cavitation in the rubber matrix in the vicinity of the agglomerate poles (Saintier 2000; Le Cam 2005); (ii) zone B contains;

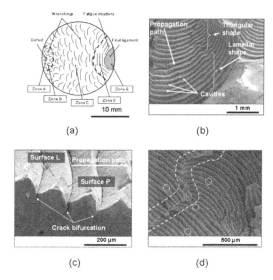

(a)   (b)

(c)   (d)

Figure 2. Description of the fracture surface obtained under relaxing loading conditions: (a) schematic view; (b) triangular and lamellar shapes; (c) crack bifurcation between triangular striations; (d) cavities located on surface L.

zone A and corresponds to the area where the fatigue crack begins to grow from the critical defect. It forms an elliptical crack tip (due to the surface vicinity) which propagates to the outer surface and simultaneously through the bulk of the sample. The crack grows and generates wrenchings formed by the shrinking of ligaments located at the crack tip; (iii) zone C is the zone corresponding to crack propagation through the bulk of the sample. It is covered by wrenchings which increase in size with the increase in stress at the crack tip; (iv) zone D is composed of striations. No wrenching is observed in this zone. It should be noted that the size of this zone increases with the maximum stress at the crack tip. Thus, zone B is not observed on the fracture surface and striations begin to form around the crack initiation zone. These observations show that fatigue striations are the signature of the increase in stress at the crack tip. Moreover, their shape evolves as the stress increases. As shown in Figure 2(b), two types of striation shape are observed: triangular and lamellar striation shapes. They occur successively with the increase in stress. Each of them is smooth. As shown in Figure 2(c), the triangular shape is composed of two perpendicular surfaces, denoted L and P in the following. As the stress level increases, the area of surface L increases and that of the perpendicular surface P decreases until it becomes a line, consequently forming lamellar striations. Moreover, cracks are observed between the striations. This phenomenon resembles crack bifurcation or micro-branching. The fact that this crack is the continuation of surface P indicates that this surface corresponds to crack propagation. At this stage of the present observations, a question of importance arises: do these surfaces form in one or several cycles? This question is discussed in the following. Figure 2(d) presents striations that do not form

in the same plane. The frontier between them is highlighted with dotted lines in this figure. This indicates that the crack can propagate independently in several zones by forming striations along the tip. When the stress increases at the crack tip, these zones can coalesce, i.e. crack propagates simultaneously through them. Finally, cavities are observed on surface P (Figure 2(b) and Figure 2(d)), but contrary to the case of moderate loading, they do not seem to contribute to the mechanism of crack growth: the crack propagates through them; (v) zone E corresponds to the final fracture surface. This zone is smooth and is comparable to the fracture surface obtained for static crack propagation. That is explained by the fact that the stress level is as great as that necessary for static fracture. As a summary, the observation of fracture surfaces provides relevant information about crack growth, especially the fact that fatigue striations take place under severe loading conditions, but it is not sufficient either to establish the mechanism of fatigue striation formation or to explain the evolution from a triangular to a lamellar striation shape. This last remark motivates the fatigue tests which are halted during crack propagation, especially when fatigue striations begin to form. Thus, using an apparatus to stretch the sample, the crack tip can be observed in-situ. The aim of this experiment is to link the morphology of the crack tip to that of the fracture surface, and more particularly that of the fatigue striations. Thus, it could be possible to explain the change in the fatigue crack growth mechanism between moderate and severe loading.

3.2 Crack tip observation

The observations were carried out with samples from halted fatigue tests for which striations begin to form. In order to observe the crack tip morphology, the sample is stretched using a basic apparatus and a gold layer is vapor-deposited at its surface. Figure 4 shows successive magnifications of the crack font. Because of its size, the sample stretched using a basic apparatus undergoes rotation in the SEM chamber.

This is the reason why the sample is not exactly aligned with the photomicrograph borders. The four photomicrographs of Figure 3 correspond to successive magnifications of the zone where fatigue striations initiate. Figure 3(a) shows the frontier (dotted line) between the crack tip and the relaxed failed surfaces that form the crack lip. This frontier corresponds to that between Zones C and D in Figure 2(a). Figure 3(b) is the magnification of the outlined zone in Figure 3(a) and shows that the crack tip is composed of large ligaments and elliptical zones. Moreover, fatigue striations are observed at the relaxed failed surface. This seems to indicate that ligaments and elliptical zones take part in the beginning of fatigue striation formation and that the change in the fatigue crack growth mechanism is a continuum phenomenon. The two last photomicrographs highlight the feet of the ligaments and of the elliptical zones whose morphology clearly differs from that observed during the

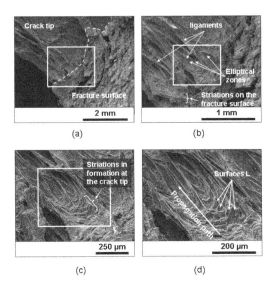

(a)　　　　　　　　　　(b)

(c)　　　　　　　　　　(d)

Figure 3.   Description of the crack tip: (a) frontier between the crack tip and the fracture surface; (b) magnification of the outlined zone of Figure 3(a); (c) magnification of the outlined zone of Figure 3(b); (d) magnification of the outlined zone of Figure 3(c).

formation of wrenchings (?): no micro-cracks are observed but fatigue striations begin to form. This shows that the mechanism of fatigue crack growth under moderate loading is not suitable to describe the formation of fatigue striations under high loading: under moderate loading, cavities grow and weaken the crack tip but under severe loading, the analysis of Figure 2(b) and Figure 2(d) shows that cavities do not grow significantly to allow the crack to propagate. Indeed, the crack propagates through cavities whose size (about 10 m) does not exceed significantly that of zinc oxide agglomerates (5 m maximum). Thus, due to the high stress level, the crack seems to propagate rapidly through the elliptical zones and consequently cavities have no time to grow. Figure 3(d) shows the magnification of the striation formation zone. Striations are identified by the smooth surfaces perpendicular to the plane of crack propagation. In fact, they correspond to the L surfaces. Here, questions of importance arise: Why are smooth surfaces observed perpendicular to the plane of crack propagation? What is the effect of stress-induced crystallization on the mechanism of crack growth? Is the change of mechanism between moderate and severe loading continuous? The next experiment aims to answer these questions.

3.3   *Real-time observation of crack growth under high stress levels*

As mentioned previously, striations are generated under high stress levels at the crack tip. The aim of the present experiment is therefore to investigate the effect of the increase in stress on the mechanism of fatigue crack growth. The experiment consists of cutting the flat sample with a razor-blade and then applying moderate relaxing cyclic loading to propagate the crack. In order to generate a similar stress level at the crack tip as that applied to the diabolo sample, the force varies between 0 and 15 N. As a consequence, only wrenchings are generated on the fracture surface. It should be noted that the number of cycles before stopping the test is set in such a way that the effect of the razor-blade cutting is eliminated. In fact, compared to a self-initiated crack, the crack tip obtained with a razor-blade generates a higher stress concentration. Consequently, the crack rapidly propagates and generates smooth fracture surfaces until a crack tip shape and morphology is obtained which is similar to that obtained without the pre-cut with a razor blade (Thomas 1958; Lake and Lindley 1965), i.e. with a self-initiated crack. Figure 4(a) presents the crack tip obtained for the maximum force applied during the fatigue test. As expected by applying such moderate loading conditions, it is composed of ligaments and elliptical zones. It is flat but rough. It should be noted that the stretch ratio level measured at the crack tip equals 2.25. This was established by observing the side view of the crack during stretching and by using the DIC technique to determine the stretch ratio. As previously explained, in such a material, the stretch ratio at which crystallization begins equals 1.64 at ambient temperature. This means that under moderate loading, stress-induced crystallization occurs at the crack tip but does not significantly affect the crack path. Hence, progressively increasing the force applied, and consequently the stress concentration at the crack tip, by increasing the enforced displacement at a strain rate corresponding to 2 mm/min leads to an increase in the crystallinity. During stretching, images of the crack tip are stored in order to measure the relative kinematic field and to calculate the deformation level at the crack tip using the Digital Image Correlation (DIC) technique.

Figure 4(b) shows the evolution of the crack tip morphology during stretching. Ligaments and the elliptical zones are highly stretched and they join the relaxed zone. None of them are regenerated and the crack tip, previously rough, becomes smooth. Figure 4(c) shows the same experiment performed with the diabolo sample. The crack tip obtained is similar. This indicates that the phenomena observed with the flat sample do not depend on the geometry of the sample and are intrinsic to the microstructure of the natural rubber. It should be noted that this image is obtained by SEM analysis. As SEM induces image distortion, optical microscopy is preferred here to estimate the kinematic field by image correlation. Figure 4(d) shows the crack tip morphology obtained at the end of the stretching. To explain the change in the morphology of the crack tip between moderate and severe loadings, it is necessary to establish the relative deformation of the zones that compose the crack tip. For this purpose, the DIC technique is used. Two images are considered: the reference image in Figure 4(b) and the more stretched image in Figure 4(d), obtained by increasing the displacement of the

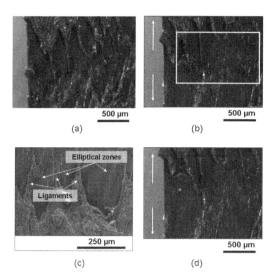

(a) 500 μm

(b) 500 μm

Elliptical zones

Ligaments

250 μm

(c)

(d) 500 μm

Figure 4. Evolution in crack tip morphology when the stress increases: (a) crack tip morphology obtained under the moderate cyclic loading; (b) Morphology of the crack tip obtained with the diabolo sample under moderate cyclic loading (SEM photomicrograph); (c) the morphology of the crack tip evolves with the increase in stress. This image is the reference for the image correlation to obtain the kinematic field in the outlined zone; (d) A higher level of stress. This is the last image for the image correlation.

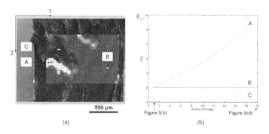

(a) 500 μm

(b)

Figure 5. Image correlation in the outlined zone in Figure 4(b): (a) Cartography of the relative deformation in terms of the Green-Lagrange strain indirection 2; (b) variation of the relative deformation at points A, B, and C during stretching.

grips. By observing the evolution of the smooth, flat surfaces between the two images, it clearly appears that their area is higher when the sample is stretched and that no new ligament is formed. The question is to know if this is due to the crack that propagates through it or to another phenomenon. For this purpose, the DIC technique is used to measure the relative displacement (and therefore the relative deformation) of the different zones that form at the crack tip.

Figure 5(a) presents the result of the image correlation. The correlation area corresponds to the boxed area in Figure 4(b). The results, in terms of the components of the Green-Lagrange tensor (Holzapfel 2000) in the stretching axis (left), are given relative to this boxed area. First of all, three zones can be distinguished in terms of deformation. They correspond to

the smooth surfaces, to the ligaments and to the surface between the relaxed zone and the ligaments. To compare the relative deformation (deformation obtained by considering that Figure 4(b) is the reference image) of these zones during stretching, three points are considered, one in each zone; point A for the ligament, point B for the smooth surface and point C for smooth surface between the ligament and relaxed zone. Figure 5(b) gives the relative deformation at these points. By considering point B, the deformation of the flat, smooth zone remains close to zero between the two deformation states. In fact, the maximum deformation level is obtained at point A in the zone that contains the ligament. This zone joins the relaxed zone by tearing and sliding along the smooth, flat surface. This is the reason why a compression is detected at point C between the ligament and the relaxed zones. The fact that the crack no longer propagates from the smooth surfaces and that no deformation is measured for the smooth, flat zone, indicates that the microstructure has changed, i.e. crystallinity has increased and has reinforced the crack tip. This is the reason why the crack does not propagate through it. In the present natural rubber, crystallization begins at a 1.64 stretch ratio (Le Cam and Toussaint 2008; Le Cam and Toussaint 2009) i.e. at a stretch ratio lower than that at the crack tip. This zone is therefore similar to a wall that stops crack propagation and induces crack bifurcation. Moreover, ligaments and elliptical zones are not regenerated. This explains why wrenchings, which are due to the successive shrinking of ligaments, are not observed in the striation zone at the fracture surface (see zone D in Figure 2(a)). This also explains why the surface perpendicular to the direction of crack propagation is smooth: it corresponds to the smooth and flat surface observed.

3.4 Mechanism of fatigue crack growth under severe loading

In this section, the previous observations are considered to establish the scenario of fatigue crack growth under severe relaxing loading conditions. It is described in Figure 6 and Figure 7 through two chronological sketches corresponding to the font and side views of the crack tip, respectively. These two views are necessary to describe precisely the three-dimensional nature of the phenomenon.

It should be noted that the chronology of the two figures is different: the step shown in Figure 6(a) does not correspond to that of Figure 7(a). Contrary to Figure 6(a), Figure 8(a) shows the crack in the non-deformed state. In most cases, the formation of striations follows the formation of wrenchings. As the transition between the two fatigue crack growth mechanisms is a continuum phenomenon (see Figure 3(b)), it seems relevant to begin the description of the crack growth mechanism under severe loading by the crack tip composed of elliptical zones separated by ligaments, i.e. obtained under moderate loading. It is presented in Figure 6(a).

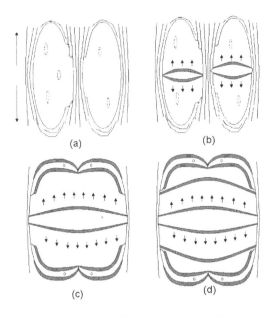

(a) (b)

(c) (d)

Figure 6. Mechanism of fatigue crack growth: front view: (a) crack tip morphology under moderate loading; (b) transition between the mechanism of fatigue crack growth under moderate loading and that under severe loading (c and d) under severe fatigue loading, each striation is formed in one cycle.

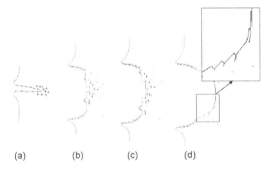

(a) (b) (c) (d)

Figure 7. Mechanism of fatigue crack growth: side view.

In this case, cavities located behind the crack tip (dotted lines) weaken the material in the elliptical zones by forming thin membranes between them and the crack tip which allow the crack to propagate. Ligaments, which do not resist crack propagation, fail and shrink to form wrenchings at the fracture surface. When the stress increases at the crack tip, the crack propagates rapidly in the centre of the elliptical zones (Figure 6(b)) and cavities do not have enough time to grow and to weaken the material. The surface generated by the crack growth is perpendicular to the loading direction and corresponds to surfaces P (the grey surfaces in Figure 6(c)). When the stretch ratio is higher at the end of stretching, the crack encounters zones which are much more crystallized than at the beginning of the stretching, and the direction perpendicular to the loading direction is no longer that of lesser energy.

Consequently, the crack bifurcates (see Figure 2(c)). Once this new mechanism is established, one cycle is sufficient to form one striation (see Figure 6(c) and Figure 6(d)). Figure 7(b), Figure 7(c) and Figure 7(d) illustrate the fact that with a decrease in the sample cross-section, which leads to an increase in stress at the crack tip and in crystallinity in this zone, the part of the cycle dedicated to propagation in the direction perpendicular to that of loading becomes less and less significant. In other words, the area of surfaces P is lower. Moreover, as highlighted by the second experiment described in section 3.1, the formation of surface L is due to both crack bifurcation and the tearing and sliding of the striations along the smooth, flat surface which is highly crystallized. Then, successively generated P surfaces join the relaxed zones and reorient themselves in the loading direction. This mechanism explains the change in the striation shape, the transition between triangular and lamellar striations. Finally, Figure 6(c) and Figure 6(d) illustrate the fact that elliptical zones are not regenerated. This explains why the flat, rough crack tip under moderate loading becomes flat and smooth under severe loading.

It should be noted that in the case of oligocyclic fatigue, striations form around the crack initiation zone and no wrenching is observed. This indicates that only one mechanism is activated: the one of crack growth under severe loading previously described in Figure 6(c) and Figure 6(d).

4 CONCLUSION AND PERSPECTIVES

The mechanism of fatigue crack growth under severe relaxing loading conditions was established using two complementary experiments. The first one consists of observing the stretched crack tip morphology using SEM. The second one enables the relative deformation of each zone at the crack tip to be established using the DIC technique. Results show that under severe loading, the mechanism strongly differs from that of fatigue crack growth under moderate loading. Under severe fatigue loading, the crack propagates by generating only striations; no ligament is regenerated. This phenomenon is due to the fact that the high cristallinity at the crack tip reinforces the material in such a way that the crack can not propagate through it and bifurcates at the microscopic scale. Then the striations formed tear, slide and relax along the smooth surface of the crack tip. This smooth surface joins the relaxed zone and reorients in the direction perpendicular to that of the loading. Thus, striations and consequently micro-bifurcation are the characteristic phenomena of fatigue cracks. The results obtained in the present study at the microscopic scale present some interesting ways to analyze crack bifurcation occurring under non-relaxing conditions (the minimum stress level remains positive) at the macroscopic scale (Le Cam et al. 2008). The fact that the force does not remain equal to zero at the end of each cycle prevents crystallites from melting at the crack tip and therefore prevents cracks from propagating. Finally, it is well-known that cristallinity depends on both the stretch ratio and the temperature. This last remark shows the relevancy of investigating the influence of temperature on the mechanism of fatigue crack growth. Further work in this field is currently being envisaged by the authors of this paper.

REFERENCES

Beurrot, S., B. Huneau, and E. Verron (2010). In situ sem study of fatigue crack growth mechanism in carbon black-filled natural rubber. *Journal of Applied Polymer Science 117*, 1260–1269.

Bhowmick, A. K. (1995). *Rubber Chemistry and Technology 68*, 132–135.

Fukahori, Y. (1991). *Fractography of Rubber Materials, pp. 71*. Elsevier Applied Science: London and New York.

Gent, A. N. and C. Pulford (1984). Micromechanics of fracture in elastomers. *Journal of Materials Science 19*, 3612–3619.

Gent, A. N., M. Razzaghi-Kashani, and G. R. Hamed (2003). Why do cracks turn sideways? *Rubber Chemistry and Technology 76*, 122–131.

Hainsworth, S. V. (2007). An environmental scanning electron microscopy investigation of fatigue crack initiation and propagation in elastomers. *Polymer Testing 26*, 60–70.

Hamed, G. R., H. J. Kim, and A. N. Gent (1996). Cut growth in vulcanizates of natural rubber cis-polybutadiene and a 50/50 blend during single and repeated extension. *Rubber Chemistry and Technology 69*, 807–818.

Hild, F. (2002). Correli[LMT]: a software for displacement field measurements by digital image correlation. Internal report n 254, LMT.

Holzapfel, G. (2000). *Nonlinear Solid Mechanics: A Continuum Approach for Engineering*. Wiley, New York.

Lake, G. J. and P. B. Lindley (1965). *Journal of Applied Polymer Science 9*, 2031–2045.

Le Cam, J.-B. (2005). *Endommagement en fatigue des elastomères*. Phd thesis, Université de Nantes, École Centrale de Nantes.

Le Cam, J.-B., B. Huneau, E. Verron, and L. Gornet (2004). *Macromolecules 37*, 5011–5017.

Le Cam, J.-B. and E. Toussaint (2008). Volume variation in stretched natural rubber: competition between cavitation and stressinduced crystallization. *Macromolecules 41*, 7579–7583.

Le Cam, J.-B. and E. Toussaint (2009). Cyclic volume changes in rubbers. *Mechanics of Materials 41*, 898–901.

Le Cam, J.-B., E. Verron, and B. Huneau (2008). *Fatigue and Fracture of Engineering Material and Structures 31*, 1031–1038.

Rault, J., Marchal, J., P. Judeinstein, and P.-A. Albouy (2006). *Macromolecules 39*, 83568368.

Saintier, N. (2000). *Prévisions de la durée de vie en fatigue du NR, sous chargement multiaxial*. Phd thesis, École Nationale Supérieure des Mines de Paris.

Sutton, M. A., W. J. Wolters, W. H. Peters, W. F. Ranson, and S. R. McNeil (1983). Determination of displacements using an improved digital correlation method. *Image and Vision Computating 1*, 133–139.

Thomas, A. G. (1958). *Journal of Polymer. Science 31*, 467–480.

Toki, S., S. Sics, I., Ran, L. Liu, B. Hsiao, S. Murakami, K. Senoo, and S. Kohjiya (2002). *Macromolecules 35*, 65786584.

Trabelsi, S., Albouy, P.-A. and J. Rault (2003). *Macromolecules 36*, 90939099.

Trabelsi, S., P.-A. Albouy, and J. Rault (2002). *Macromolecules 35*, 10054–10061.

Wang, B. L. H. and K. G. (2002). *Mechanics of Materials 34*.

Influence of thermal ageing on mechanical properties of styrene-butadiene rubber

I. Petrikova, B. Marvalova & P.T. Nhan
Technical University of Liberec, Czech Republic

ABSTRACT: The mechanical properties, the temperature dependence of dynamic behaviour of SBR and post-thermal ageing properties of styrene-butadiene rubber (SBR) were investigated. The rubber specimens were heated at a temperature of 100°C in a ventilated air oven with natural convection for 1, 3, 7, 14 and 21 days, respectively. The specimens were then tested in the tensile strength, elongation at break and their dynamic mechanical properties were determined by means of DMA at different frequencies and amplitudes. The dependence of the hardness ShA on the ageing was also investigated.

1 INTRODUCTION

In order to determine rate-dependent properties of examined rubber we previously performed experimental measurements of the time dependent response and of damping properties of rubber materials consisting of uniaxial creep and stress relaxation tests which were convenient for studying material response at long times. The behaviour at different strain levels was examined in detail through quasistatic cyclic tests and in simple and multistep relaxation tests. The viscosity-induced rate-dependent effects were described and parameters of the material model were determined. The model was implemented into FE code (Marvalova 2008). Next we focused on the dynamic mechanical analysis of filler-reinforced rubber. The dependence of the storage and dissipation moduli on the static pre-strain, on the deformation amplitude and on the frequency was investigated (Petrikova & Marvalova 2010).

The dynamic mechanical analysis (DMA) is well suited for the identification of the short-time range of rubber response. DMA consists of dynamic tests, in which the force resulting from a sinusoidal strain controlled loading is measured.

Payne (1965) first pointed out that the moduli of carbon black filled rubber decrease with increasing deformation amplitudes. By means of further tests he reached the conclusion that this behaviour has to be attributed to a thixotropic change. Lion (1998) observed that both the storage and the dissipation modulus depend on the frequency of the deformation process. This variation is weakly pronounced and it is of power series type approximately. In terms of the theory of linear viscoelasticity this behaviour corresponds to a continuous relaxation time distribution. With increasing temperatures Lion (1998) observed both a decrease in moduli and a lessening of the frequency dependence. The dependence of the dynamic moduli on the filler content and the static pre-strain is investigated in detail by Namboodiri and Tripathy (1994).

When a viscoelastic material is subjected to a sinusoidally varying strain after some initial transients the stationary stress-response will be reached in which the resulting stress is also sinusoidal, having the same angular frequency but advanced in phase by an angle δ. Then the strain lags the stress by the phase angle δ. The axial displacement $u(t)$ consists of a static pre-strain u_0 under tension which is superimposed by small sinusoidal oscillations:

$$u(t) = u_0 + \Delta u \sin(2\pi f t). \tag{1}$$

Stresses and strains are calculated with respect to the reference geometry (Lion & Kardelky 2004) of the pre-deformed specimen

$$\varepsilon_0 = u_0 / (L_0 + u_0), \quad \Delta\varepsilon = \Delta u / (L_0 + u_0), \tag{2}$$

where L_0 is the undeformed length of the specimen. The force response $F(t)$ of the specimen is a harmonic function and can be written as:

$$F(t) = F_0 + \Delta F \sin(2\pi f t + \delta). \tag{3}$$

F_0 is the static force depending only on the pre-deformation u_0. The force amplitude ΔF and the phase angle δ depend, in general, on the pre-deformation, the frequency and the strain amplitude (Lion & Kardelky 2004, Hofer & Lion 2009).

If the incompressibility of the rubber is assumed $A_0 L_0 = A(L_0 + u_0)$, where A_0 is the cross-sectional area of the undeformed specimen, we can relate the force to the cross-sectional area A of the pre-deformed specimen:

$$\sigma(t) = \frac{F(t)}{A} = \sigma_0 +$$

$$+\Delta\sigma \left[\cos\delta \sin(2\pi f t) + \sin\delta \cos(2\pi f t) \right]. \tag{4}$$

The dynamic stress-response $\sigma(t)$ normalized by the deformation amplitude $\Delta\varepsilon$ can be written:

$$\sigma(t) = \sigma_0 + \Delta\varepsilon\Big[E'(\varepsilon_0, f, \Delta\varepsilon)\sin(2\pi f t) +$$

$$+ E''(\varepsilon_0, f, \Delta\varepsilon)\cos(2\pi f t)\Big], \tag{5}$$

where

$$E'(\varepsilon_0, f, \Delta\varepsilon) = \frac{\Delta\sigma}{\Delta\varepsilon}\cos(\delta), \tag{6}$$

and

$$E''(\varepsilon_0, f, \Delta\varepsilon) = \frac{\Delta\sigma}{\Delta\varepsilon}\sin(\delta) \tag{7}$$

are the storage and dissipation moduli respectively and δ is the phase angle. In general, carbon black-reinforced rubber has fairly weak frequency dependence in conjunction with a pronounced amplitude dependence (Hofer & Lion 2009). If the strain amplitude $\Delta\varepsilon$ increases, the storage modulus E' lessens and the dissipation modulus E'' shows a more or less pronounced sigmoidal behaviour – Payne effect. If the material is linear viscoelastic, then these two moduli depend neither on the deformation amplitude nor on the static pre-strain. The damping factor or loss tangent (tan δ) which is the ratio E''/E' is the measure of mechanical energy dissipated as heat during the dynamic cycle. If the dynamic strain amplitude is constant in time, we can observe time-independent moduli (Lion 1998). These phenomena are frequently interpreted as a dynamic state of equilibrium between breakage and recovery of physical bonds linking adjacent filler clusters. The most common model of this state is Kraus model (Kraus 1984, Ulmer 1996) which describes the amplitude dependence of dynamic moduli. The influence of static pre-deformation ε_0 is included in different models (Kim et al. 2004, Cho & Youn 2006) and the uniaxial form of the frequency, amplitude and pre-strain dependent dynamical moduli is proposed by Lion (2004).

The purpose of present paper is to summarize the results of experimental research of the behaviour of rubber under dynamic loading conditions in harmonic strain-controlled tests under tension and to show the dependence of the storage and dissipation moduli on the frequency, on the deformation amplitude, on the static pre-strain and different temperatures.

2 EXPERIMENTAL MEASUREMENT

The carbon-black-filled rubbers investigated were obtained from commercial sources. They are based on common formulations containing 30–40% carbon-black for the manufacturing of O-rings, seals and other products. The materials examined were styrene-butadiene (SBR) rubbers which have been characterized previously (Marvalova et al. 2010).

Table 1. Parameters of testing at 22°C.

$\Delta\varepsilon$	0.014	0.028	0.042	0.056	0.070
f [Hz]	1.0	2.5	5.0	7.5	10.0
ε_0	0.17	0.17	0.21	0.21	0.25

DMA tests under sinusoidal tension mode were carried out on an electro-dynamic testing machine Instron ElectroPuls E3000 equipped with an environmental chamber and with WaveMatrix software. The specimens were thin rectangular strips of length 160 mm, width 25 mm and thickness 2.75 mm. The basic sampling frequency was 100 Hz and was increased up to 500 Hz when needed according to the testing frequency used.

The dynamic properties of SBR were investigated at different temperature levels. The tensile loading was strain-controlled. Every test was performed on a virgin specimen. At the chosen static pre-deformation ε_0, the frequency and the strain amplitude $\Delta\varepsilon$ were changed in order to determine their influence to the storage (SM) and loss moduli (LM) and to hysteretic losses. Before each test the virgin specimens were preconditioned in order to exclude the Mullins effect. The preloading process started on the static pre-strain ε_0 and consisted of 10 cycles with the maximum strain amplitude to be reached in the subsequent experiment. After that specimens relaxed 15 min at static pre-strain ε_0. After this preconditioning, the mean stress σ_0 changed only little in the subsequent cyclic loading.

Raw test data were recorded by a PC and evaluated in the Matlab Signal Processing Toolbox. The discrete Fourier transform was used to determine the frequency content of force and displacement signals and to calculate the phase delay δ between them. Furthermore, we determined the complex dynamic modulus as the ratio between the amplitudes of stress and strain and dynamic moduli were calculated according to the Eqn. (5–7).

2.1 Testing at ambient temperature

In order to determine the dependence of the dynamic moduli we carried out the tests at the temperature 22°C with five frequencies and amplitudes $\Delta\varepsilon$ and with three static pre-strains ε_0 as shown in Tab. 1.

After the preconditioning at the given pre-strain ε_0 the test started at the smallest amplitude and the frequency sweep in the chosen range was performed then the amplitude was raised to the next value. The number of cycles executed at each frequency step was between 200 and 300 and was adapted to achieve a steady state. The results of tests are represented by synoptic graphs in Figures 1–3 where the storage and dissipation moduli and the loss angle (LA) are plotted as a function of frequency and amplitude for the static pre-deformation ε_0 as a parameter. We can make the three following essential conclusions:

• The storage and dissipation moduli increase and the loss angle decreases with increasing static pre-strain ε_0.

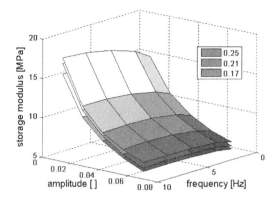

Figure 1. SM amplitude and frequency dependence.

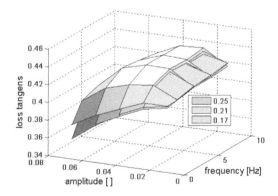

Figure 2. LM amplitude and frequency dependence.

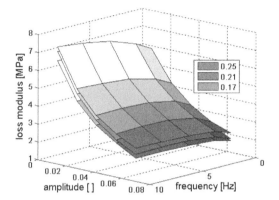

Figure 3. LA amplitude and frequency dependence.

- The storage modulus and the loss modulus increase slightly with increasing frequency i.e. increasing frequencies lead to an increase in stiffness and an increase in energy loss. The graph of the loss angle has a convex shape and shows a slight maximum in the range of applied frequencies and amplitudes.
- Both moduli show a pronounced decrease with an increasing strain amplitudes – so called Payne effect.

The Payne effect is explained by a concept (Lion 1998, Lion & Kardelky 2004, Drozdov & Dorfmann

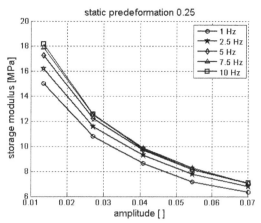

Figure 4. SM amplitude and frequency dependence.

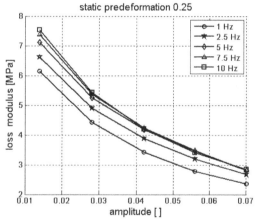

Figure 5. LM amplitude and frequency dependence.

2000) that during cyclic deformations the weak physical bonds between molecules of rubber and clusters of filler are breaking and recovering continually. The rate of breakage is assumed to be an increasing function of the strain amplitude and the rate of recovery is a decreasing function. The storage modulus is assumed to be proportional to the total number of intact bonds and the dissipation modulus to the rate of breakage per unit of time.

The detailed dependence of the storage and dissipation moduli on amplitude for different frequencies is shown in Figures 4–5. The both moduli decrease monotonically with increasing strain amplitudes. In our range of amplitudes the loss modulus does not show any sigmoidal behaviour which was reported by Lion & Kardelky (2004).

2.2 Temperature dependency of dynamic properties

In order to determine the influence of temperature on dynamic properties of examined SBR rubber another

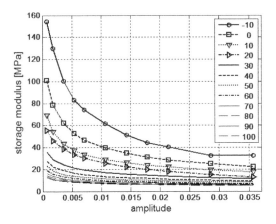

Figure 6. SM amplitude and temperature dependence.

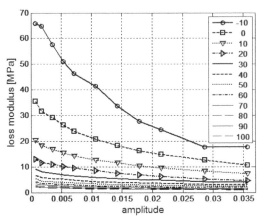

Figure 7. LM amplitude and frequency dependence.

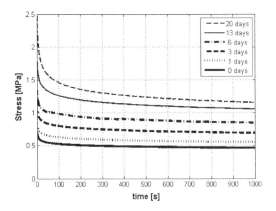

Figure 8. Stress relaxations.

series of tests was carried on with the temperature sweep for different frequencies and amplitudes and with a sole static pre-strain $\varepsilon_0 = 0.29$. The tests were accomplished in the Instron 3119 Environmental Chamber suitable for a temperature range from $-70°$ to $+250°C$.

The first series of tests was lead with the temperature sweep from $-10°C$ to $100°C$ with a step $10°C$. The frequency of the strain controlled loading was fixed at 5 Hz.

After each temperature step the specimens relaxed 15 min at the static pre-strain. Then the cyclic loading started and the amplitude sweep was performed in a range from 0.001 to 0.036. The number of cycles was 200 at each amplitude.

We can see a similar dependency of the storage and loss moduli on the temperature and strain amplitude. Both moduli are rising sharply with decreasing temperature but their dependence on the strain amplitude (Payne effect) is much more pronounced in the low temperature zone as can also be seen on the detailed graphs in Figures 6 and 7 where the amplitude dependency is displayed with the temperature as a parameter.

All dynamic properties deteriorate considerably at higher temperatures. Rubber loses its elasticity and damping properties. Similar results are reported by other authors (Lion et al 2009).

Described tests were conducted also at frequencies 2.5 Hz and 7.5 Hz. The values of investigated quantities did not show substantial differences in this frequency range.

2.3 Thermal ageing of rubber samples

Rubber samples (25×160 mm) were thermally aged in an air-circulated oven at $100°C$. Samples were removed from the oven after given periods of time up to 21 days. Tests were performed at room temperature. Intervals of ageing were 24 hours, 72 hours = 3 days, 7, 14 and 21 days. Mechanical properties as hardness, storage and loss moduli, loss tangent

and stress relaxation in dependency of ageing were investigated.

The relaxation behaviour of new samples and ageing samples is examined in tension relaxation test with displacement 15 mm. The stress relaxation was recorded for 1200 s. The Figure 8 shows the loss of stresses for different ageing time of rubber samples. All curves reveal the existence of a very fast stress relaxation during the first 10 seconds followed by a very slow rate of relaxation that continues in an asymptotic sense.

The stress-strain diagram from tensile tests of ageing rubber is shown in Figure 9. The breaking of the samples occurs at similar values of stress but the elongations are different. Rubbers with longer ageing period show significantly lower values of strain.

The hardness ShA of samples increases with time of thermal ageing as shown in Figure 10. The hardness of specimens increases with increasing of number days of ageing.

Dynamic properties were measured by means of the DMA on artificial ageing samples. The storage modulus, loss modulus and loss tangent were investigated at samples with different time of ageing at the same regime as in Section 2.1, see Tab. 1.

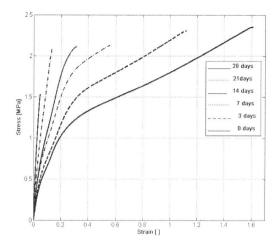

Figure 9. Stress-strain diagram.

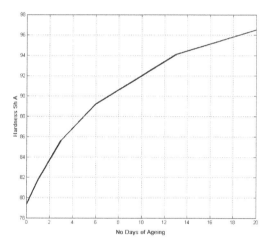

Figure 10. Hardness ShA.

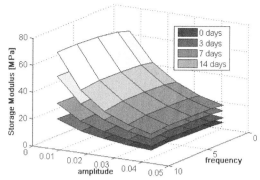

Figure 11. SM amplitude and frequency dependence.

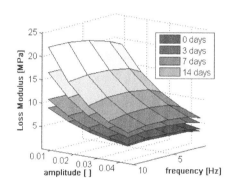

Figure 12. LM amplitude and frequency dependence.

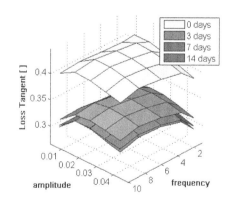

Figure 13. LA amplitude and frequency dependence.

Variation of storage moduli, loss moduli and loss tangent are displayed as a function of amplitudes and frequencies in the Figures 11–13. The storage modulus and the loss modulus increase slightly with increasing frequency i.e. increasing frequencies lead to an increase in stiffness and an increase in energy loss. The graph of the loss angle has a convex shape and shows a slight maximum in the range of applied frequencies and amplitudes.

The detailed dependence of the storage and dissipation moduli on amplitude for different frequencies is shown in Figures 14, 15. The both moduli decrease monotonically with increasing strain amplitudes.

Comparison of storage moduli of specimens with different ageng time is on Figure 14, loss moduli dependency on ageing time is on Figure 15. We see a considerable fall of storage moduli of all samples with increasing values of amplitude. The values of the storage moduli and loss moduli of samples are increased with time of ageing.

The loss tangent (Fig. 16) rapidly falls at the first days of ageing. The difference of tangent loss values after 7 days of ageing is small.

The storage modulus and the loss modulus increase slightly with increasing frequency i.e. increasing frequencies lead to an increase in stiffness and an increase in energy loss. The graph of the loss angle has a convex shape and shows a slight maximum in the range of applied frequencies and amplitudes. All dynamic properties deteriorate considerably at higher temperatures. Rubber loses its elasticity and damping properties.

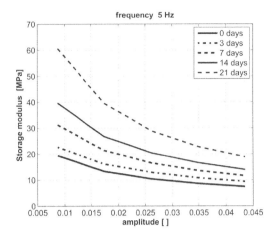

Figure 14. SM amplitude dependency.

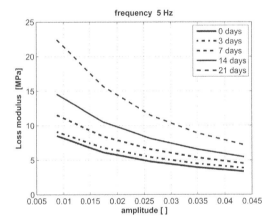

Figure 15. LM amplitude dependency.

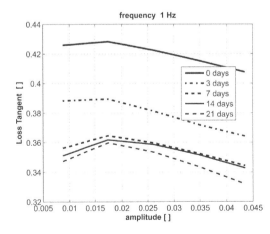

Figure 16. LA amplitude dependency.

3 CONCLUSIONS

In this experimental essay we present the results of dynamic testing of SBR rubber under different conditions. In order to investigate the internal damping of rubber a complex experimental research of dynamic properties was lead by DMA at different frequencies, strain amplitudes and temperatures.

The dependency of storage and loss moduli and of loss angle on these quantities was identified and displayed synoptically.

We should emphasize that material properties of rubber are affected by the temperature to a great extent. Results show that the response of rubber changes at temperatures even slightly different than the ambient temperature which leads to a drastic change in the material properties.

Influence of artificial ageing on mechanical properties of rubber is also considerable.

Another salient property of rubber is its phenomenal memory. Rubber remembers almost all that happened to it from the beginning and the slightest change in the experimental procedure leads to the scattering of results.

ACKNOWLEDGEMENTS

This work was supported by a grant from Ministry of Education of Czech Republic under Contract Code MSM 4674788501.

REFERENCES

Cho, J. H. & Youn, S. K. 2006. A Viscoelastic Constitutive Model of Rubber under Small Oscillatory Load Superimposedon Large Static Deformation Considering the Payne Effect. *Arch. Appl. Mech* 75: 275–288.

Drozdov, A. & Dorfmann, A. 2000. The Payne effect for particle-reinf. elastomers. arXiv:cond-mat/0011223v1.

Hofer, P. & Lion, A. 2009. Modelling of Frequency- and Amplitude-dependent Material Properties of Filler-reinforced Rubber. *J. Mech. Phys. Solids* 57: 500–520.

Kim, B. K., Youn, S. K. & Lee, W. S. 2004. A Constitutive Model and FEA of Rubber under Small Oscillatory Load Superimposed on Large Static Deformation. *Arch. Appl. Mech.* 73: 781–798.

Kraus, G. 1984. Mechanical Losses in Carbon Black Filled Rubbers. *J. Appl. Polym. Sci.* 39: 75–92.

Lion, A. 1998. Thixotropic Behaviour of Rubber under Dynamic Loading Histories: Experiments and Theory. *J. Mech. Phys. Solids* 46(5): 895–930.

Lion, A. 2004. Phenomenological Modelling of the Material Behaviour of Carbon Black-filled Rubber. *Kautschuk Gummi Kunststoffe* 57(4): 184–190.

Lion, A. & Kardelky, C. 2004. The Payne Effect in Finite Viscoelasticity: Constitutive Modelling Based on Fractional Derivatives and Intrinsic Time Scales. *Int. J. Plasticity* 20: 1313–1345.

Lion, A., Retka, J. & Rendek, M. 2009. On the Calculation of Pre-strain-dependent Dynamic Modulus Tensors in Finite Nonlinear Viscoelasticity. *Mechanics Research Communications* 36(6): 653–658.

Marvalova, B. 2008. Viscoelastic properties of filled rubber – Experimental observations and material modelling, In A. Boukamel, L. Laiarinandrasana, S. Meo &

E. Verron (eds), *Proceedings of the 5th European Conference on Constitutive Models for Rubber, Paris 4–7 September 2009,* London: Taylor & Francis.

Marvalova, B., Petrikova, I. & Cirkl, D. 2010. Tribological and Viscoelastic Behaviour of Carbon Black Filled Rubber. In *ASME 10th Biennial Conference on Engineering Systems Design and Analysis. ESDA* 2010. 12–14 July 2010, Istanbul.

Namboodiri, C. & Tripathy, D. 1994. Static and Dynamic Strain-Dependent Viscoelastic Behaviour of Black-Filled EPDM Vulkanisates. *Journal of Applied Polymer Science* 5: 877–889.

Payne, A. R. 1965. Dynamic Properties of Natural Rubber Containing Heat Treated Carbon Black. *Journal of Applied Polymer Science* 9: 3245–54.

Petrikova, I. & Marvalova, B. 2010. Experimental determination of the mechanical properties of naturally aged rubber. In G. Heinrich, M. Kaliske, A. Lion and S. Reese (eds), *Proceedings of the 6th European Conference on Constitutive Models for Rubber, Dresden, 7–10 September 2009.* Taylor & Francis: London.

Ulmer, J. D. 1996. Strain Dependence of Dynamic Mechanical Properties of Carbon Black-filled Rubber Compounds. *Rubber chem. technol.* 69(1): 15–47.

Modelling and simulation

Constitutive Models for Rubber VII – Jerrams & Murphy (eds)
© 2012 Taylor & Francis Group, London, ISBN 978-0-415-68389-0

An extended tube model for thermo-viscoelasticity of rubberlike materials: Theory and numerical implementation

R. Behnke, H. Dal & M. Kaliske
Institute for Structural Analysis (ISD), Technische Universität Dresden, Germany

ABSTRACT: This contribution is devoted to the further development of the original purely hyperelastic extended tube model, presented by Kaliske & Heinrich (1999), to finite nonlinear thermo-viscoelasticity using a nonlinear evolution law and a thermo-mechanical coupled free energy formulation. Particular emphasis is given to an appropriate parameter identification technique for the thermal field. For the latter, a uniaxial extension test is carried out where the recorded data of the temperature field of the rubber specimen under cyclic loading is used for parameter identification.

1 INTRODUCTION

Elastomers are widely used in industry and everyday life's application. Their outstanding mechanical extensibility and damping characteristics are the reason for their use as vibration dampers in large structures, engine mounts and machine components. Their mechanical behavior has been described in the context of hyperelasticity, viscoelasticity as well as in the context of damage and fracture mechanics. However, due to their low heat conductivity in comparison to metal components, internal dissipation phenomena lead to a high temperature increase, especially for geometrically compact and cyclic loaded parts. Therefore, one aim of this work is to include the thermal response of elastomeric materials by a thermo-mechanical material model as well as its application in the context of the Finite Element Method (FEM). Nowadays, the FEM is a standard method for engineers to investigate and predict the behavior of structural components of arbitrary geometry and loading conditions. Hence, a second aim of this contribution is the identification of the model parameters which is a prerequisite for its practical application to design tasks.

2 MATERIAL MODEL

The extended tube model, introduced by Kaliske & Heinrich (1999), is a material model for the description of the mechanical behavior of filled elastomers in the context of hyperelasticity. Its isochoric strain energy function can be derived from statistical mechanics and contains four model parameters. The comparison of several material models for hyperelastic material behavior, reported in Marckmann & Verron (2006), reveals the excellent predictive and fitting capabilities of the extended tube model. As a result, its extension to finite nonlinear viscoelasticity in the context

of a thermo-mechanical formulation seems to be very promising.

2.1 Continuum mechanical basis of isothermal viscoelasticity

The mapping $\varphi : \mathbf{X} \mapsto \mathbf{x}$ of the deformation at time $t \in \mathbb{R}_+$ of a continuous and isotropic body relates points $\mathbf{X} \in \mathfrak{B} \subset \mathbb{R}^3$ of the reference configuration to points $\mathbf{X} \in \mathfrak{b} \subset \mathbb{R}^3$ in the current configuration by the mapping function $\mathbf{X} = \varphi(\mathbf{X}, t) \in \mathfrak{b}$. The deformation gradient $\mathbf{F}$ is defined as the gradient of the mapping φ by $\mathbf{F} := \nabla_X(\varphi(\mathbf{X}, t))$ and its determinant $J := \det \mathbf{F} > 0$ represents the volume changes. The operator ∇_X or Grad $(\cdot)$ indicates the gradient operator with respect to the reference configuration while ∇_x or grad $(\cdot)$ is devoted to the spatial configuration. The deformation gradient $\mathbf{F}$ is multiplicatively split into isochoric and volumetric parts,

$$\mathbf{F} = \bar{\mathbf{F}} \mathbf{F}_{\mathrm{vol}}, \quad \bar{\mathbf{F}} := J^{-\frac{1}{3}} \mathbf{F}, \quad \mathbf{F}_{\mathrm{vol}} := J^{\frac{1}{3}} \mathbf{1}. \tag{1}$$

The unimodular part $\bar{\mathbf{F}}$ is further split into elastic and inelastic contributions, $\bar{\mathbf{F}} = \bar{\mathbf{F}}^e \bar{\mathbf{F}}^i$. A decoupled volumetric-isochoric structure of isothermal finite viscoelasticity is obtained by the additive composition of the Helmholtz free energy function Ψ_0 for a unit reference volume,

$$\Psi_0 = U_0(J) + \bar{\Psi}_0(\bar{\mathbf{F}}^i, \bar{\mathbf{F}}^e), \tag{2}$$

where the isochoric part contains elastic and viscous contributions,

$$\bar{\Psi}_0(\bar{\mathbf{F}}^i, \bar{\mathbf{F}}^e) = \bar{\Psi}_0^e(\bar{\mathbf{C}}) + \bar{\Psi}_0^v(\bar{\mathbf{C}}_e). \tag{3}$$

The subscript 0 characterizes the evaluation of the isothermal quantities at a fixed reference temperature.

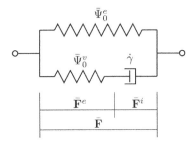

Figure 1. Rheology of the viscoelastic material model.

$\overline{\mathbf{C}} = \overline{\mathbf{F}}^\mathrm{T} \overline{\mathbf{F}}$ and $\overline{\mathbf{C}}_e = \overline{\mathbf{F}}^{e\mathrm{T}} \overline{\mathbf{F}}^e$ are the unimodular part of the right Cauchy-Green tensor and the elastic right Cauchy-Green tensor, respectively. The additive structure of the Helmholtz free energy function yields an additive decomposition of the Kirchhoff stress $\boldsymbol{\tau}$ into spherical and deviatoric contributions,

$$\boldsymbol{\tau} = \boldsymbol{\tau}_{\mathrm{vol}} + \boldsymbol{\tau}_{\mathrm{iso}}, \quad \boldsymbol{\tau}_{\mathrm{vol}} = p\,\mathbf{1}, \quad \boldsymbol{\tau}_{\mathrm{iso}} = \mathbb{P} : \bar{\boldsymbol{\tau}}. \quad (4)$$

$p := J\,U_0'(J)$ denotes the volumetric pressure and $\bar{\boldsymbol{\tau}} := 2\,\partial_{\overline{\mathbf{b}}} \overline{\Psi}_0(\overline{\mathbf{F}}^e, \overline{\mathbf{F}}^i)\,\overline{\mathbf{b}}$ the Kirchhoff stress of the unimodular part. With the help of the fourth order projection tensor $\mathbb{P}_{abcd} = 1/2\,[\delta_{ac}\,\delta_{bd} + \delta_{ad}\,\delta_{bc}] - 1/3\,\delta_{ab}\,\delta_{cd}$, the isochoric Kirchhoff stresses can be obtained. With this definitions at hand, the unimodular Kirchhoff stress $\bar{\boldsymbol{\tau}}$ can be further split into $\bar{\boldsymbol{\tau}} := \bar{\boldsymbol{\tau}}^e + \bar{\boldsymbol{\tau}}^v$, with

$$\bar{\boldsymbol{\tau}}^e := 2\,\partial_{\overline{\mathbf{b}}} \overline{\Psi}_0^e(\overline{\mathbf{b}})\,\overline{\mathbf{b}}, \quad \bar{\boldsymbol{\tau}}^v := 2\,\partial_{\overline{\mathbf{b}}_e} \overline{\Psi}_0^v(\overline{\mathbf{b}}^e)\,\overline{\mathbf{b}}^e. \quad (5)$$

2.2 Free energy function

The free energy function $U_0(J)$ of the volumetric part is described by

$$U_0(J) = \kappa_0\,(J - \ln J - 1). \quad (6)$$

κ_0 denotes the bulk modulus and is chosen large enough to enforce the rubberlike incompressibility. The isochoric free energy functions $\overline{\Psi}_0^e(\overline{\mathbf{C}}) := W^e(\bar{I}) + L^e(\bar{\lambda}_a)$ and $\overline{\Psi}_0^v(\overline{\mathbf{C}}_e) := W^v(\bar{I}_1^e)$ are defined for the equilibrium branch by

$$W^e(\bar{I}_1) = \frac{G_c}{2} \left[\frac{(1-\delta^2)\,(\bar{I}_1 - 3)}{1 - \delta^2\,(\bar{I}_1 - 3)} + \ln\left(1 - \delta^2\,(\bar{I}_1 - 3)\right) \right],$$

$$L^e(\bar{\lambda}_a) = \frac{2\,G_e}{\beta^2} \sum_{a=1}^{3} \left(\bar{\lambda}_a^{-\beta} - 1 \right) \quad (7)$$

and for the non-equilibrium branch by

$$W^v(\bar{I}_1^e) = \frac{G_c^v}{2} \left[\bar{I}_1^e - 3 \right]. \quad (8)$$

$\bar{I} = \mathrm{tr}\,\overline{\mathbf{C}} = \bar{\lambda}_1^2 + \bar{\lambda}_2^2 + \bar{\lambda}_3^2$ and $\bar{I}_1^e = \mathrm{tr}\,\overline{\mathbf{C}}_e = \lambda_1^{e2} + \lambda_2^{e2} + \lambda_3^{e2}$ are the first invariants of the total and elastic right Cauchy-Green tensors, respectively. G_c, G_e, β

and δ are material parameters of the equilibrium (EQ) part $\overline{\Psi}_0^e(\overline{\mathbf{C}})$. The non-equilibrium (NEQ) part contains G_c^v as material parameter. $\bar{\lambda}_a$ is the principal unimodular stretch corresponding to the eigenvector $\mathbf{N}_a$ in the reference configuration of the unimodular right Cauchy-Green tensor $\overline{\mathbf{C}}$. The corresponding eigenvector in the current configuration is noted $\mathbf{n}_a$. Viscous effects are assumed to be purely isochoric.

2.3 Evolution law

Employing a similar approach as proposed by Bergström & Boyce (1998), an evolution for the inelastic rate of deformation tensor in the current configuration is given by

$$\tilde{\mathbf{d}}_i := \dot{\gamma}\,\mathbf{N}, \quad \mathbf{N} = \frac{\boldsymbol{\tau}_{\mathrm{iso}}^v}{\|\boldsymbol{\tau}_{\mathrm{iso}}^v\|}, \quad (9)$$

with

$$\boldsymbol{\tau}_{\mathrm{iso}}^v := \mathbb{P} : \bar{\boldsymbol{\tau}}^v, \quad \|\boldsymbol{\tau}_{\mathrm{iso}}^v\| := \sqrt{\boldsymbol{\tau}_{\mathrm{iso}}^v : \boldsymbol{\tau}_{\mathrm{iso}}^v}, \quad (10)$$

and

$$\dot{\gamma} := \dot{\gamma}_0 \left[\frac{\lambda}{\lambda_e} - 1 \right]^c \left(\frac{\tau_v}{\hat{\tau}} \right)^m, \quad (11)$$

where the creep process is also assumed to be energy activated. Therefore, the term $(\tau_v/\hat{\tau})^m$ is added to the creep rate expression with $m > 0$, where

$$\tau_v := \frac{\|\boldsymbol{\tau}_{\mathrm{iso}}^v\|}{\sqrt{2}}, \quad \lambda := \sqrt{\frac{\bar{I}_1}{3}}, \quad \lambda_e := \sqrt{\frac{\bar{I}_1^e}{3}}. \quad (12)$$

The ratio λ/λ_e represents the inelastic finite network stretch, $\dot{\gamma}_0$ denotes the reference effective creep rate and $\hat{\tau}$ is a parameter for dimensional purposes. $c \leq 0$ controls the influence of the inelastic network deformation on the creep process.

2.4 Stress and moduli terms for the elastic part

The derivation of the Kirchhoff stresses and the Eulerian moduli terms corresponding to the volumetric part of the deformation is standard and not addressed here. In order to obtain the deviatoric stresses and moduli, the projection of the unimodular tensors is carried out,

$$\boldsymbol{\tau}_{\mathrm{iso}} = \mathbb{P} : \bar{\boldsymbol{\tau}}, \quad (13)$$

$$\mathbb{c}_{\mathrm{iso}} = \mathbb{P} : \left[\bar{\mathbb{c}} + \frac{2}{3}\,\mathrm{tr}\,[\bar{\boldsymbol{\tau}}]\,\mathbb{I} - \frac{2}{3}\,(\bar{\boldsymbol{\tau}} \otimes \mathbf{1} + \mathbf{1} \otimes \bar{\boldsymbol{\tau}}) \right] : \mathbb{P}, \quad (14)$$

$$\bar{\mathbb{c}} := \bar{\mathbb{c}}^e + \bar{\mathbb{c}}_{\mathrm{algo}}^v. \quad (15)$$

$\mathbb{I}_{abcd} = 1/2[\delta_{ac}\delta_{bd} + \delta_{ad}\delta_{bc}]$ is the fourth order identity tensor, $\overline{\mathbb{C}}^e$ denotes the elasticity moduli of the equilibrium response and $\overline{\mathbb{C}}^v_{\text{algo}}$ the algorithmic consistent moduli for the non-equilibrium response. In the following, the isochoric contributions of the equilibrium branch are considered. The stress and moduli expressions of the equilibrium response are

$$\bar{\boldsymbol{\tau}}^e := 2 \frac{\partial \bar{\Psi}^e_0(\bar{\mathbf{b}})}{\partial \bar{\mathbf{b}}} \bar{\mathbf{b}} = \bar{\boldsymbol{\tau}}^e_W + \bar{\boldsymbol{\tau}}^e_L, \tag{16}$$

$$\bar{\mathbb{C}}^e := 4 \bar{\mathbf{b}} \frac{\partial^2 \bar{\Psi}^e_0(\bar{\mathbf{b}})}{\partial \bar{\mathbf{b}} \, \partial \bar{\mathbf{b}}} \bar{\mathbf{b}} = \bar{\mathbb{C}}^e_W + \bar{\mathbb{C}}^e_L, \tag{17}$$

where $\bar{\boldsymbol{\tau}}^e$ and $\mathbb{C}^e$ denote the Kirchhoff stress and the elasticity moduli of the equilibrium branch, respectively.

2.5 Stress and moduli terms of the viscous part

The viscous part of the Kirchhoff stress is defined as

$$\bar{\boldsymbol{\tau}}^v := 2 \frac{\partial \bar{\Psi}^v_0(\bar{\mathbf{b}}_e)}{\partial \bar{\mathbf{b}}_e} \bar{\mathbf{b}}_e, \tag{18}$$

where the derivatives of the viscous energy function $W^v(\bar{I}^{e}_1)$ can be obtained by a straight forward exploitation. To be able to compute the Kirchhoff stresses, the current value $\bar{\mathbf{b}}_e$ is needed at time $t = t_{n+1}$. The computation of $\bar{\mathbf{b}}_e$ depends on the treatment of the evolution law. The integration of the evolution law is based on an operator split of the material time derivative of $\bar{\mathbf{b}}_e$ into an elastic predictor (PRE) and an inelastic corrector step (ICOR)

$$\dot{\bar{\mathbf{b}}}_e := \underbrace{\mathbf{l}^{\text{iso}} \bar{\mathbf{b}}_e + \bar{\mathbf{b}}_e \mathbf{l}^{\text{iso T}}}_{\text{PRE}} + \underbrace{\mathcal{L}_v \bar{\mathbf{b}}_e}_{\text{ICOR}}. \tag{19}$$

$\mathcal{L}_v \bar{\mathbf{b}}_e := \overline{\dot{\bar{\mathbf{F}} \mathbf{C}^{-1}_i \bar{\mathbf{F}}^T}}$ is a unimodular operator. During the elastic trial step, the time derivative of the inverse inelastic strains is equal to zero,

$$\overline{\dot{\mathbf{C}^{-1}_i}} = 0. \tag{20}$$

Therefore, the inverse inelastic trail strains of the time step t_{n+1} are equal to the inverse inelastic strains of the previous time step t_n,

$$\left(\bar{\mathbf{C}}^{-1}_i \right)_{tr} = \left(\bar{\mathbf{C}}^{-1}_i \right)_{t_n} \quad \rightarrow \quad \bar{\mathbf{b}}^{tr}_e = \mathbf{F} \left(\bar{\mathbf{C}}^{-1}_i \right)_{t_n} \mathbf{F}^T. \tag{21}$$

In the inelastic corrector step, $\mathbf{l} = \mathbf{l}^{\text{iso}} + \mathbf{l}^{\text{vol}}$ is set to zero, yielding $\mathcal{L}_v \bar{\mathbf{b}}_e = \dot{\bar{\mathbf{b}}}_e$. The consequences for the evolution law can be noted as follows

$$\dot{\bar{\mathbf{b}}}_e = [-2\dot{\gamma} \, \mathbf{N}] \, \bar{\mathbf{b}}^{tr}_e. \tag{22}$$

Eq. (22) is solved by the so-called exponential mapping,

$$\bar{\mathbf{b}}_e = \exp \left[-2 \int_{t_n}^{t_{n+1}} \dot{\gamma} \, \mathbf{N} \, dt \right] \bar{\mathbf{b}}^{tr}_e. \tag{23}$$

Subsequently, the integral equation (23) is expressed as a linear approximation with respect to time. Due to the assumption of an isotropic body, $\boldsymbol{\tau}^v_{\text{iso}}$ and, as a result, $\mathbf{N}$ commute with $\bar{\mathbf{b}}_e$ and also with $\bar{\mathbf{b}}^{tr}_e$. As a consequence, Eq. (23) can be written in principal stretch directions and iteratively solved by a Newton-Raphson scheme.

2.6 Algorithmic moduli for the viscous part

The third part of the consistent tangent moduli, which is called algorithmic moduli, will be briefly addressed in this subsection and concerns the viscous part. The trial elastic deformation can be noted in its spectral decomposition as

$$\bar{\mathbf{F}}^{tr}_e = \sum_{a=1}^{3} \bar{\lambda}^{e \, tr}_a \, \mathbf{n}_a \otimes \mathbf{N}_a. \tag{24}$$

A fictious second Piola-Kirchhoff stress tensor can be defined in principal directions as well,

$$\tilde{\mathbf{S}}^v := \bar{\mathbf{F}}^{tr \, -1}_e \, \bar{\boldsymbol{\tau}}^v \, \bar{\mathbf{F}}^{tr \, -T}_e, \tag{25}$$

and the incremental rate equation yields finally

$$\Delta \tilde{\mathbf{S}}^v = \bar{\mathbb{C}}^v_{\text{algo}} : \Delta \bar{\mathbf{C}}^{tr}_e, \quad \bar{\mathbb{C}}^v_{\text{algo}} = 2 \frac{\partial \tilde{\mathbf{S}}^v}{\partial \bar{\mathbf{C}}^{tr}_e}. \tag{26}$$

The expression for the moduli in principal directions can be found by a straight forward evaluation of Eq. (26). Finally, this expression of the fictitious intermediate configuration has to be formulated in the current configuration by a push-forward operation using $\bar{\mathbf{F}}^{tr}_e$. For more details, we refer to Dal & Kaliske (2009).

3 THERMO-MECHANICAL COUPLING

The temperature acts as an influence parameter on the behavior of rubberlike materials. Their mechanical response depends on the temperature field of the body. Therefore, a coupled thermo-mechanical analysis enables a more realistic representation of the rubber behavior.

3.1 Constitutive equations for thermo-mechanics

The Clausius-Duhem inequality

$$-\dot{\Psi} + \mathbf{S} : \frac{1}{2} \dot{\mathbf{C}} - \eta \dot{\Theta} - \frac{1}{\Theta} \mathbf{Q} \cdot \text{Grad} \, \Theta \geq 0 \tag{27}$$

is the starting point for deriving the coupled thermo-viscoelastic behavior. In the Clausius-Duhem inequality, Ψ denotes the Helmholtz free energy per unit reference volume, $\mathbf{S}$ the second Piola-Kirchhoff stress tensor and $\mathbf{C}$ the right Cauchy-Green tensor, η the entropy per unit reference volume and Θ the absolute temperature. The heat flux $\mathbf{Q} = J\,\mathbf{F}^{-1} \cdot \mathbf{q} = -J\,k\,\mathbf{F}^{-1} \cdot$ Grad$\Theta \cdot \mathbf{F}$ is the heat flux in the reference configuration with the heat conductivity coefficient k. The further evaluation of Eq. (27) leads to

$$\left(\mathbf{S} - 2\frac{\partial\Psi}{\partial\mathbf{C}}\right) : \frac{1}{2}\dot{\mathbf{C}} - \frac{\partial\Psi}{\partial\overline{\mathbf{C}}_i^{-1}} : \dot{\overline{\mathbf{C}}_i^{-1}}$$

$$- \left(\frac{\partial\Psi}{\partial\Theta} + \eta\right)\dot{\Theta} - \frac{1}{\Theta}\mathbf{Q}\cdot\mathrm{Grad}\,\Theta \geq 0, \qquad (28)$$

where the inverse of the inelastic right Cauchy-Green tensor $\overline{\mathbf{C}}_i^{-1}$ is used as internal variable. To satisfy the dissipation inequality and the thermodynamic consistency, the terms in brackets in Eq. (28) have to vanish for arbitrary strain and temperature evolution.

3.2 Helmholtz free energy

The Helmholtz free energy is mainly responsible for the thermo-mechanical behavior of the material model. For the functional formulation, an approach based on a reference energy Ψ_0 evaluated at a given reference temperature Θ_0 is proposed in Reese & Govindjee (1998). In this case, the Helmholtz free energy is

$$\Psi = f_{\mathrm{EQ}}(\Theta)\left[\bar{\Psi}_0^e(\bar{b}_i) + U_0(J)\right] + t_1(\Theta)\,e_{0\mathrm{EQ}}$$

$$+ f_{\mathrm{NEQ}}(\Theta)\left[\bar{\Psi}_0^v(\bar{b}_{e_i})\right] + t_2(\Theta)\,\bar{c}, \qquad (29)$$

where temperature evolution functions $f_{\mathrm{EQ}}(\Theta)$ and $f_{\mathrm{NEQ}}(\Theta)$ are multiplied with the reference free energy functions $\bar{\Psi}_0^e$ and $\bar{\Psi}_0^v$. According to Reese & Govindjee (1998), the function for the equilibrium part can be found to be

$$f_{\mathrm{EQ}}(\Theta) = \frac{\Theta}{\Theta_0} + g_{\mathrm{EQ}}(\Theta) - g_{\mathrm{EQ}}(\Theta_0)$$

$$+ \left.\frac{\partial g_{\mathrm{EQ}}}{\partial\Theta}\right|_{\Theta_0}(\Theta_0 - \Theta), \qquad (30)$$

where for the function of the non-equilibrium part $f_{\mathrm{NEQ}}(\Theta) = f_{\mathrm{EQ}}(\Theta)$ is assumed in the following. The remaining temperature expressions have the form

$$t_1(\Theta) = 1 - \frac{\Theta}{\Theta_0}, \qquad (31)$$

$$t_2(\Theta) = \Theta - \Theta_0 - \Theta\ln\frac{\Theta}{\Theta_0}. \qquad (32)$$

In Eq. (29), the constant $\bar{c}$ can be regarded as the reference heat capacity of the material, evaluated at the temperature Θ_0. As dimensionless functions $g_{\mathrm{EQ}}(\Theta)$ and $g_{\mathrm{NEQ}}(\Theta)$, various functions of different type can be

chosen in order to model the temperature dependency of the investigated material, for example

$$g_{\mathrm{EQ}}(\Theta) = g_{\mathrm{NEQ}}(\Theta) = b\left(\frac{\Theta}{\Theta_0}\right)^a. \qquad (33)$$

For the reference internal energy $e_{0\mathrm{EQ}}$, the form

$$e_{0\mathrm{EQ}} = \kappa_0\,\alpha_0\,\Theta_0\ln J \qquad (34)$$

is chosen, where κ_0 and α_0 are the bulk modulus and the thermal expansion coefficient of the reference temperature Θ_0, respectively. For a more detailed description, we refer to Reese & Govindjee (1998).

3.3 Strong form of the balance equations

Investigating a thermo-mechanical problem, the entropy inequality and the balance equations for energy, mass and momentum have to be satisfied within each point of time. The balance of mass is fulfilled by the relation $J\rho = \rho_0$, where ρ is the density in the current configuration and ρ_0 the density in the reference configuration. The balance of linear momentum is given by

$$J\,\mathrm{div}\left(\frac{\boldsymbol{\tau}}{J}\right) + \mathbf{p}_v - \rho_0\ddot{\mathbf{u}} = 0, \qquad (35)$$

where $\mathbf{p}_v$ and $\ddot{\mathbf{u}}$ are the referential unit volume body force and the acceleration of a material point, respectively. The balance of energy reads

$$J\,\mathrm{div}\,\mathbf{q} - \boldsymbol{\tau}:\mathbf{d} + \dot{e} - \mathcal{R} = 0, \qquad (36)$$

where $\mathcal{R}$ represents a heat source per unit reference volume and $\mathbf{d}$ is the symmetric strain rate tensor. After several mathematical manipulations, Eq. (36) can be written as

$$J\,\mathrm{div}\,\mathbf{q} + \underbrace{\left(\boldsymbol{\tau}_{\mathrm{NEQ}} - \Theta\frac{\partial\boldsymbol{\tau}_{\mathrm{NEQ}}}{\partial\Theta}\right):\frac{1}{2}\mathcal{L}_v\left(\bar{\mathbf{b}}_e\right)\cdot\bar{\mathbf{b}}_e^{-1}}_{-w_{\mathrm{int}}}$$

$$- \underbrace{\Theta\left(\frac{\partial\boldsymbol{\tau}_{\mathrm{EQ}}}{\partial\Theta} + \frac{\partial\boldsymbol{\tau}_{\mathrm{NEQ}}}{\partial\Theta}\right):\mathbf{d}}_{w_{\mathrm{ext}}} + c\dot{\Theta} - \mathcal{R} = 0. \qquad (37)$$

The term w_{int} represents the inner dissipation which stems from the power of the viscous branch. The expression w_{ext} is the thermo-elastic and thermo-viscoelastic coupling term of the system.

3.4 Finite element formulation

The implementation of the thermo-mechanical coupled material model into the FEM necessitates a coupled finite element formulation. For a simultaneous solution scheme, a fully coupled solid finite element has been developed. In addition, the heat boundary

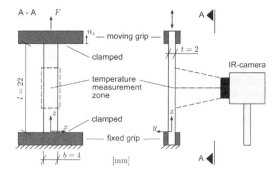

Figure 2. Experimental test set-up.

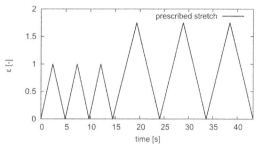

Figure 3. Prescribed longitudinal stretch.

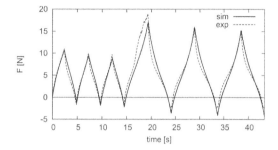

Figure 4. Reaction force.

conditions are modelled by fully coupled surface elements which include the phenomena of static heat conduction as well as dynamic heat convection for a surface displacement relative to its surrounding. The phenomenological heat transfer law for the heat flux trough the boundary is

$$q_{ext} = q_0 + \kappa_{stat}\,\Delta\Theta + \kappa_{conv}\,\left|\dot{u}_{\|surf}\right|\Delta\Theta. \qquad (38)$$

$q_0, \kappa_{stat}, \kappa_{conv}$ denote the prescribed heat flux, the static and the convective heat exchange coefficient, respectively. $\dot{u}_{\|surf}$ is the tangential relative surface velocity of the considered surface point. The temperature difference between the surface Θ_{surf} and the ambient air Θ_{amb} is given by $\Delta\Theta = \Theta_{surf} - \Theta_{amb}$. A more detailed presentation of the finite element formulation and implementation is in preparation.

4 UNIAXIAL EXTENSION TEST

In Figure 2, the test set-up of the uniaxial extension test is illustrated. The test has been carried out by J.-B. Le Cam and his research group at LaMI, France. A rubber specimen with a rectangular cross-section ($a = 2$ mm, $b = 4$ mm, $l = 22$ mm) is clamped between the grips of a testing machine. During the test, the reaction force is measured with the help of a load cell and the surface temperature in the middle of the specimen by an IR-camera (see Figure 2).

4.1 Parameter identification

The recorded data of the reaction force and the surface temperature variation are used to identify the model parameters. For this task, an identification tool has been developed at ISD which is based on the solution of least square minimization of the experimentally obtained and the simulated data curves. Starting from an initial material parameter set, the parameters are iteratively adapted in order to represent best the measured results. The gradient based procedure makes use of a batch processing of FE simulations with slight variations of the initial parameter set to obtain an numerical tangent. The result is a locally optimized parameter set. The obtained parameters are $G_c = G_c^v = 0.35$ MPa,

$\delta = 0.245$, $\dot{\gamma}_0/\hat{\tau}^m = 0.27\,\mathrm{s}^{-1}\mathrm{MPa}^{-m}$, $m = 1.65$ for $\Theta_0 = 292$ K. Certain assumptions regarding the model parameters have been made in order to simplify the identification process. Due to the lack of additional experimental data, the shear modulus G_e of the cross-linking part is set to zero. The remaining shear modulus of the elastic G_c and viscous branch G_c^v are assumed to be equal. The contribution of chain relaxation kinetics to the evolution law is omitted by setting $c = 0$. The bulk modulus is chosen to be $\kappa_0 = 100$ MPa and $\alpha_0 = 16 \cdot 10^{-5}\,\mathrm{K}^{-1}$. The thermal material properties are taken from Pottier et al. (2009) where the material formulation of the rubber is also provided. The parameters of the equilibrium and non-equilibrium temperature function are chosen to be $a = 2.0$ and $b = 0.5$. The heat transfer over the boundary is modelled by setting $\kappa_{stat} = 20\,\mathrm{W\,K}^{-1}\,\mathrm{m}^{-2}$ and $\kappa_{conv} = 20\,\mathrm{kWs\,K}^{-1}\,\mathrm{m}^{-3}$.

4.2 Results of simulation and experiment

The displacement controlled test is carried out with a displacement rate of 500 mm/min and consists of 6 sawtooth waves as shown in Figure 3. The results of the extension test and the simulation are plotted for comparison in Figures 4 and 5. The results agree qualitatively and quantitatively very well. During unloading, buckling of the specimen occurs which is due to the fact that the specimen is still elongated when the moving grip returns in its initial position. This is an indicator for viscous effects during the experiment. The buckling causes a non-plane surface where the IR rays are deflected. This is the reason for the lack of experimental data regarding the surface temperature evolution during unloading as depicted in Figure 5.

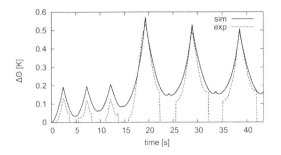

Figure 5. Surface temperature variation.

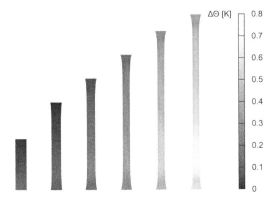

Figure 6. Simulation of surface temperature evolution for a displacement rate of 500 mm/min.

As a result, it was one aim of the simulation to predict the temperature evolution in the periods of time where no measured data is available. In Figure 5, temperature peaks are visible which can be correlated to the points in time where the specimen is completely unloaded and the maximum compressive force appears. The surface temperature variation is illustrated again in Figure 6 for the beginning of the fourth cycle up to the stretch amplitude of 175 %. Due to the modelling of convection phenomena, the surface temperature varies significantly as a function of the displacement of a surface point relativ to the ambient air. For this reason, the upper part of the test specimen near the moving grip is cooler than the lower part near the fixed grip.

5 CONCLUSIONS

Starting from the hyperelastic free energy function of the extended tube model, a further development of this model suitable for filled elastomers to finite thermo-viscoelasticity is derived. The kernel of the viscoelastic formulation is the notation of the inelastic rate of deformation in the eigenspace of the elastic strains. As a result, the inversion of a fourth order tensor during the local Newton iteration can be avoided and an operator split combined with an exponential mapping scheme is carried out. The temperature dependency of the material is taken finally into account by adding additional terms coming from the balance of energy and the balance of entropy. The coupled formulation is based on two steps. First, the terms and expressions containing a contribution from the free energy function are evaluated at a chosen reference temperature. Second, the evaluation of the temperature function allows for the computation of the derivations with respect to temperature. The presented formulation is applied to a uniaxial extension test which reveals mainly thermo-elastic coupling effects between the mechanical and thermal field. The model and material parameters have been identified and the simulation results agree very well with the experimental data. The presented material model, in combination with the identified model parameters, is finally applied to predict the thermo-mechanical response of a three-branch rubber specimen, which is documented in Balandraud et al. (2011).

ACKNOWLEDGEMENTS

The research reported herein is carried out within the DFG research unit 597. The authors would like to acknowledge gratefully the financial support of the DFG under grant KA 1163/3-2. The uniaxial extension test has been carried out by J.-B. Le Cam and his research group at LaMI (IFMA, France). The authors thank J.-B. Le Cam and his research group for providing the test data and the fruitful collaboration.

REFERENCES

Balandraud, X., E. Toussaint, J.-B. Le Cam, M. Grédiac, R. Behnke, & M. Kaliske (2011). Application of full-field measurements and numerical simulations to analyze the thermo-mechanical response of a three-branch rubber specimen. In S. Jerrams, N. Murphy, M. Rebow, F. Abrahams, E. Verron, R. Schuster, and J.-B. Le Cam (Eds.), *Constitutive Models for Rubber VII.*

Bergström, J. & M. Boyce (1998). Constitutive modeling of the large strain time-dependent behavior of elastomers. *Journal of the Mechanics and Physics of Solids 46,* 931–954.

Dal, H. & M. Kaliske (2009). Bergström-Boyce model for nonlinear finite rubber viscoelasticity: Theoretical aspects and algorithmic treatment for the FE method. *Computational Mechanics 44,* 809–823.

Kaliske, M. & G. Heinrich (1999). An extended tube-model for rubber elasticity: Statistical-mechanical theory and finite element implementation. *Rubber Chemistry and Technology 72,* 602–632.

Marckmann, G. & E. Verron (2006). Comparison of hyperelastic models for rubberlike materials. *Rubber Chemistry and Technology 79,* 835–858.

Pottier, T., M.-P. Moutrille, J.-B. Le Cam, X. Balandraud, & M. Grédiac (2009). Study on the use of motion compensation techniques to determine heat sources. Application to large deformations on cracked rubber specimens. *Experimental Mechanics 49,* 561–574.

Reese, S. & S. Govindjee (1998). Theoretical and numerical aspects in the thermo-viscoelastic material behaviour of rubber-like polymers. *Mechanics of Time-Dependent Materials 1,* 357–396.

Constitutive Models for Rubber VII – Jerrams & Murphy (eds)
© *2012 Taylor & Francis Group, London, ISBN 978-0-415-68389-0*

Characterization and identification of the memory decay rates of carbon black-filled rubber

J. Ciambella, A. Paolone & S. Vidoli

Dipartimento di Ingegneria Strutturale e Geotecnica, SAPIENZA Università di Roma, Roma, Italy

ABSTRACT: This paper is concerned with the identification of the dynamic moduli of viscoelastic constitutive models on the basis of a nonlinear optimization method by fitting data on carbon black-filled rubber.

The models considered can be categorized in two different classes according to the definition of the relaxation function: models whose kernel decays exponentially fast, *e.g.*, Debye model, and non-Debye relaxation spectra with a slower decay rate of the viscoelastic kernel, *e.g.*, Cole-Cole model. On the basis of a recent result (Ciambella, Paolone, & Vidoli 2011), we show that all the kernels whose decay rate is too fast lead to identification problems which are, in general, ill-conditioned. This shortcoming, in turn, manifests itself in the presence of multiple sets of optimal parameters for the same data sets. Moreover, this multiplicity can lead to very different constitutive responses when extrapolating material behaviour outside the range of the experimental data.

1 INTRODUCTION

The problem of correctly identifying the sophisticated behaviour of carbon black-filled elastomers is influenced by the demand of testing procedures both highly-repeatable and not time consuming. Oscillatory strain histories at different frequencies are usually preferred over other experimental procedures because they allow the dynamic properties of the material to be assessed more accurately (Knauss, Emri, & Lu 2008). In this context the storage, G', and loss, G'', moduli are introduced to measure the components of the stress in-phase and in-quadrature with the applied deformation, respectively: for instance, tire manufacturers and rubber producers specify the mechanical characteristics of their compounds in terms of frequency by supplying the dynamic moduli of the material.

The experimental evidence collated from the literature and the experiments carried out by the authors on carbon black-filled rubber have shown a high sensitivity of both dynamic moduli with respect to frequency variations. In this regard, Fig. 1 shows the frequency dependence of a carbon black filled-rubber compound (Osanaiye 1996) in the range 1–10 Hz, which is relevant in many engineering applications: rolling tyre at 100 km/h, vibration absorbers, etc.

In (Ciambella, Paolone, & Vidoli 2011) an extensive analysis of the relationship between the frequency behaviour of the dynamic moduli and the time behaviour of the viscoelastic kernel, *i.e.*, the material's memory, has been carried out for both linear and nonlinear viscoelastic models. As a result we have shown that, if the rate of decay of the material memory is too fast, *e.g.*, exponential decay's rate as in the Prony series, the resulting storage modulus has a vanishing

sensitivity for $\omega \to 0$, in open contrast with the data in Fig. 1.

Most of models used in the literature show this behaviour and, consequently, a large number of coefficients is employed to achieve an adequate match of the experimental curves: 5, 10 (Knauss & Zhao 2007) and even 20 (Antonakakis, Bhargava, Chuang, & Zehnder 2006) material parameters have been used.

Here, we show that the use of models with a viscoelastic kernel whose decay rate is too fast leads, in general, to identification problems of the constitutive coefficients which are strongly ill-conditioned; therefore, multiple sets of optimal parameters could appear.

2 FORMULATION OF THE PROBLEM

Since we restrict our attention to small amplitude oscillatory experiments, the linear viscoelastic constitutive model is considered, *viz.*,

$$\sigma(t) = G(0)\,\varepsilon(t) + \int_{-\infty}^{t} \dot{G}(t-s)\varepsilon(s)\,\mathrm{d}s, \tag{1}$$

where σ is the stress, ε is the strain and G is the relaxation function, or *viscoelastic kernel*.

In this linear context, the storage and loss moduli are defined as (Lockett 1972)

$$G'(\omega) = G_0 + \int_0^{+\infty} \dot{G}(s)\cos(\omega s)\,\mathrm{d}s,$$

$$G''(\omega) = -\int_0^{+\infty} \dot{G}(s)\sin(\omega s)\,\mathrm{d}s. \tag{2}$$

with $G_0 = G(0)$.

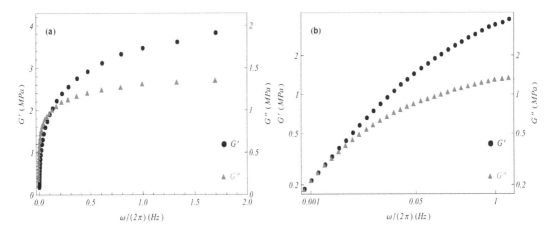

Figure 1. Frequency dependence of the storage and loss moduli for carbon black-filled rubber (experimental results from (Osanaiye 1996)) in linear-linear (a) and log-log (b) plots.

Different forms of $G(t)$ produce a different frequency dependence of dynamic moduli; in particular, it has been proved that (Ciambella, Paolone, & Vidoli 2011), if the viscoelastic kernel satisfies the inequality

$$\int_0^{+\infty} t\,|\dot{G}(t)|\,\mathrm{d}t < +\infty , \tag{3}$$

then the resulting storage modulus G' has an horizontal tangent for $\omega \to 0$, *i.e.*,

$$\lim_{\omega \to 0} \frac{\partial G'}{\partial \omega} = 0 . \tag{4}$$

Incidentally, the previous condition is satisfied by most viscoelastic kernels employed in the literature. Among the different contributions we have compared four models: two satisfying Eq. (3), *i.e.*, Debye and Davidson-Cole models (Hilfer 2002), and two models not satisfying (3), *i.e.*, Cole-Cole and fractional Kelvin models (Lewandowski & Chorazyczewski 2010). All these are listed in Tab. 1 where the sensitivity of the corresponding storage modulus is highlighted.

2.1 Debye model

In the literature, the linear viscoelastic model 1 is often used in conjunction with an exponential viscoelastic kernel expressed as

$$G_D(t) = G_0 \left(1 + g(\mathrm{e}^{-t/\tau} - 1)\right) , \tag{5}$$

with $g < 1, \tau > 0$ (Lockett 1972). The linear viscoelastic model endowed with this kernel was extensively studied by Debye, thus is often referred to as Debye relaxation function. With such a choice, the integrability condition (3) is clearly satisfied, thus the resulting storage modulus has an horizontal tangent for $\omega \to 0$. In particular, from Eqs. (2), the well-known expression

of the storage and loss moduli of the linear viscoelastic model is obtained:

$$G'_D = G_0 \left(1 - \frac{g}{1 + \tau^2\omega^2}\right) ,$$
$$G''_D = G_0 \frac{g\tau\omega}{1 + \tau^2\omega^2} , \tag{6}$$

and, therefore, $\partial G'_D / \partial \omega = 0$ for $\omega \to 0$. Often a larger number of parameters is needed to match accurately the experimental data, thus Eq. (5) is replaced by a series, known as Prony's series, *i.e.*,

$$G_D(t) = G_0 \left(1 + \sum_{i=1}^N g_i(\mathrm{e}^{-t/\tau_i} - 1)\right) . \tag{7}$$

2.2 Davidson-Cole model

Another form of relaxation function used in the literature is the one introduced by Davidson and Cole (Hilfer 2002). In this case, the exponential function in (5) is replaced by the Gamma function, *viz.*,

$$G_D(t) = G_0 \left(1 - g + g\,\frac{\Gamma(\delta, t/\tau)}{\Gamma(\delta)}\right) , \tag{8}$$

with the corresponding expression of the storage and loss moduli

$$G'_D = G_0 \left(1 - g\,(1 + \tau^2\omega^2)^{-\delta/2} \cos(\delta\,\mathrm{atan}(\omega\,\tau))\right) ,$$
$$G''_D = G_0\,g\,(1 + \tau^2\omega^2)^{-\delta/2} \sin(\delta\,\mathrm{atan}(\omega\,\tau)) , \tag{9}$$

where atan is the arctangent function.

If $0 < \delta \leq 1$, the relaxation function G_D satisfies the integrability condition and, hence, $\partial G'_D / \partial \omega = 0$ as $\omega \to 0$.

Table 1. Comparison of the different kernels considered in terms of the sensitivity of the storage modulus $\partial G'/\partial\omega$.

$G(t)/G_0$		$\int_0^{+\infty} t\lvert\dot{G}(t)\rvert\,\mathrm{d}t$	$\left.\left\lvert\dfrac{\partial G'}{\partial\omega}\right\rvert\right\rvert_{\omega\to 0}$
$1 + \sum_{i=1}^{N} g_i\left(e^{-t/\tau_i} - 1\right)$	Debye	$< +\infty$	0
$1 + g\left(\dfrac{\Gamma(\delta, t/\tau_\delta)}{\Gamma(\delta)} - 1\right)$	Davidson-Cole	$< +\infty$	0
$1 + g\left(E_\delta(-(t/\tau)^\delta) - 1\right)$	Cole-Cole	$+\infty$	$+\infty$
$1 + g\dfrac{(t/\tau)^{-\delta}}{\Gamma(1-\delta)}$	fractional Kelvin	$+\infty$	$+\infty$

2.3 Cole-Cole model

Many materials, such as polymers, show a non-Debye relaxation spectrum. In this case, fractional differential models have been increasingly used in recent years. The fractional generalization of the Debye model (5) leads to a hereditary constitutive equation where the kernel $G_C(t)$ is as follows

$$G_C(t) = G_0\left(1 + g\left(E_\delta[-(t/\tau)^\delta] - 1\right)\right), \quad (10)$$

with $g < 1, 0 < \delta \le 1$ and $\tau > 0$. Here E_δ is the δ-order Mittag-Leffler function, i.e.,

$$E_\delta(u) = \sum_{j=0}^{\infty} \frac{u^j}{\Gamma(1 + j\delta)}, \quad (11)$$

which interpolates between a purely exponential-law (for $\delta \to 1$) and a power-law like behavior (for $\delta \to 0$). For each time instant, the evaluation of $G_C(t)$ requires the assessment of the infinite number of terms in the series (11).

In (Metzler & Nonnenmacher 2003), the asymptotic behavior of the kernel (10) has been found for $t >> \tau$:

$$G_C(t) \sim G_0\left(1 - g + g\frac{(t/\tau)^{-\delta}}{\Gamma(1-\delta)}\right). \quad (12)$$

If $0 < \delta < 1$, this expression does not satisfy the integrability condition (3); thus "strictly" fractional kernels, the ones having $\delta < 1$, are good candidates to match the observed high sensitivities at low frequency. For $\delta = 1$, the fractional kernel reduces instead to the standard viscoelastic one.

Equation (10) leads to the following expression of the storage and loss moduli,

$$G_C' = G_0\left(1 - g\frac{1 + (\tau\omega)^\delta\cos(\pi\delta/2)}{1 + (\tau\omega)^{2\delta} + 2(\tau\omega)^\delta\cos(\pi\delta/2)}\right),$$

$$G_C'' = G_0 g\frac{(\tau\omega)^\delta\sin(\pi\delta/2)}{1 + (\tau\omega)^{2\delta} + 2(\tau\omega)^\delta\cos(\pi\delta/2)}, \quad (13)$$

and, being $0 < \delta < 1$, $\partial G_F'/\partial\omega \to \infty$ for $\omega \to 0$.

2.4 Fractional Kelvin model

Another generalization to fractional order of a rheological element is that introduced for the Kelvin element whose fractional counterpart leads to the following viscoelastic kernel (Lewandowski & Chorazyczewski 2010):

$$G_K(t) = G_0\left(1 + g\frac{(t/\tau)^{-\delta}}{\Gamma(1-\delta)}\right), \quad (14)$$

which through Eqs. (2) gives

$$G_K' = G_0\left(1 + \tau^\delta\omega^\delta\cos\left(\frac{\pi\delta}{2}\right)\right), \quad (15)$$

$$G_K'' = G_0\tau^\delta\omega^\delta\sin\left(\frac{\pi\delta}{2}\right). \quad (16)$$

with $\tau > 0, 0 < \delta \le 1$ and, hence, the inequality (3) does not hold.

3 OUTLINE OF THE IDENTIFICATION METHOD

The identification has been carried out by considering data reported in Fig. 1 (Osanaiye 1996); the models are those listed in Tab. 1.

The objective function to be minimized is:

$$F(\mathbf{p}) = \alpha\left\lVert\mathbf{G}'(\Omega, \mathbf{p}) - \widehat{\mathbf{G}'}\right\rVert_2^2 + \beta\left\lVert\mathbf{G}''(\Omega, \mathbf{p}) - \widehat{\mathbf{G}''}\right\rVert_2^2, \quad (17)$$

where $\mathbf{p}$ is the set of material parameters to be identified, e.g., $\mathbf{p} = \{G_0, g, \tau\}$ for the Debye model, $\widehat{\mathbf{G}'}$, $\widehat{\mathbf{G}''}$ are the vectors containing the experimental results, $\Omega = \{\omega_1, \ldots, \omega_N\}$ are the experimental frequencies

95

considered and α, β are two positive weights. Here $\| \cdot \|_2$ is the standard Euclidean norm.

The identification of material parameters is, hence, obtained through the minimization of $F(\mathbf{p})$:

$$\mathbf{p}^* = \operatorname{argmin}_{\mathbf{p}} F(\mathbf{p}) \tag{18}$$

subjected to the constraints

$$0 < g < 1, \qquad \tau > 0, \qquad 0 < \delta \le 1. \tag{19}$$

Since the dependence of G' and G'' on τ and δ is nonlinear for all the models considered, the problem (18) is nonlinear, presumably non-convex; much effort should, hence, be devoted to avoid to be trapped in local minima of the functional F. In the literature, different algorithms have been used to solve nonlinear identification problems (Ciambella, Paolone, & Vidoli 2010) and all of them are based on modification of Newton algorithm (Trust-Region, Interior Point, ...). Here we have used a Trust-region method already implemented in the Matlab function *fmincon*.

4 RESULTS AND DISCUSSION

We report on the identification results obtained for all models listed in Tab. 1. To make a fair comparison, initially, only one term in the Prony's series was chosen, so that all the models had the same number of parameters: 3 coefficients for Debye model, G_0, g and τ, and 4 coefficients for all the others, G_0, g, τ and δ. Thereafter, we will show the effects of increasing the number of coefficients.

Figure 2 shows the fitting curves compared to the experimental data. Having an horizontal tangent for $\omega \to 0$, both Debye and Davidson-Cole models have difficulties in matching the experimental data at the lowest frequencies reflecting in higher relative errors especially for $\omega < 0.1$ Hz. With these models, one faces the problem of fitting smoothly increasing ramp with an "s-shaped" function which badly match the experimental data.

On the contrary, the Cole-Cole and the fractional Kelvin models, for which $\partial G' / \partial \omega \to 0$ as $\omega \to 0$, produce lower relative errors in all the frequency range.

To obtain an adequate match of the experiments also with the Debye model, the number of coefficients has been increased by adding terms in the series (7). The results of the fitting are shown in Fig. 3 against the Cole-Cole model and the experimental data. It is evident that, passing from $N = 1$ to $N = 4$, a more accurate match of the data is achieved and the quality of the fitting is comparable to that obtained with the Cole-Cole model. However, some oscillations in the fitting curves appear which are symptomatic of the difficulties of the model to adapt to the experimental data. The relative errors which were larger than 90%, for the one-term series, drop to less than 10% in the case of $N = 4$ terms; as a consequence, the value of the objective function at the minimum point passes from 1.502 to 0.269. All

Table 2. Comparison between Cole-Cole and Debye models for an incrasing number of parameters in terms of the value of the objective function at minimum $F(\mathbf{p}^*)$ and of the condition number ρ of the hessian.

Kernel Model	Number of Coef.	$F(\mathbf{p}^*)$	ρ
Cole-Cole	4	0.302	454
Debye (N = 2)	5	1.502	1.51×10^8
Debye (N = 3)	7	0.278	1.72×10^{11}
Debye (N = 4)	9	0.269	3.87×10^{16}

these positive effects, however, are counterbalanced by the fact that the identification problem becomes strongly ill-conditioned.

In Tab. 2 the value of the objective function achieved after minimization for the different models considered are displayed, while the last column shows the condition number of the hessian of the functional F around the minimum. A high value of the condition number implies that F is "low" and "flat"; this is an unwanted situation as one can move away from the minimum point without significant changes in the value of the objective function. Therefore, within the numerical tolerance assigned in the minimization algorithm, one could find multiple set of minima.

To exploit further this issue one has to plot the multidimensional functional F along the different directions in the parameters space; to this end, the plot of $F(\mathbf{p}^*(1+\gamma))$ for γ ranging in $[-0.2, 0.2]$ is shown in Fig. 4. It is evident from the figure that objective function of Debye model have multiple directions along which the functional is flat.

Finally, the sensitivity of the minimum point of the Debye model with $N = 3$ terms in the series was assessed for variations of the tolerance of the minimization algorithm. With a tolerance of $TolF = 10^{-5}$ on the objective function value, the parameters of the relaxation function found were

$$g_1 = 0.4773, \qquad \tau_1 = 0.0610,$$
$$g_2 = 0.1755, \qquad \tau_2 = 15.6075,$$
$$g_3 = 0.3006, \qquad \tau_3 = 0.98023;$$

by changing the value of $TolF$ to 10^{-16}, corresponding to the machine precision, a different minimum was found, *i.e.*,

$$g_1 = 0.4605, \qquad \tau_1 = 0.0576,$$
$$g_2 = 0.1947, \qquad \tau_2 = 12.9046,$$
$$g_3 = 0.3003, \qquad \tau_3 = 0.8205.$$

Despite producing a similar value of the objective function, the previous minima result in a different frequency dependence of the storage and loss moduli outside the experimental range observed. This behaviour is symptomatic of the strong ill-conditioning of the identification problem.

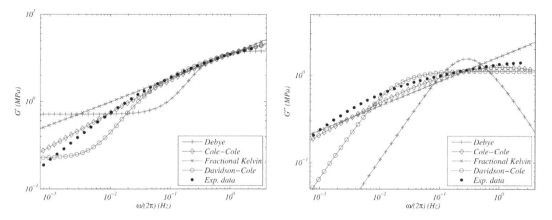

Figure 2. The fitting results for the different models considered and listed in Tab. 1.

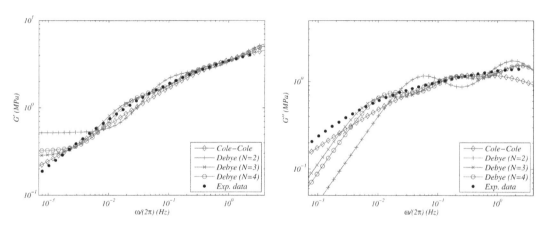

Figure 3. The fitting results for an increasing number of parameters in the Debye model.

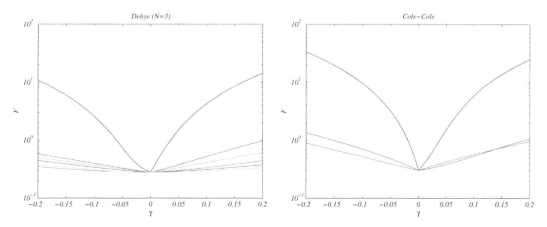

Figure 4. Plot of the objective functions around the minimum found ($\gamma = 0$) for the Debye model with N = 3 and the Cole-Cole model along the different directions in the parameters space.

5 CONCLUSIONS

In this paper, we have compared different linear viscoelastic models in terms of the ability of the dynamic moduli to match the experimental data on carbon black-filled rubber at low frequency (less than 10 Hz).

In general, those models whose viscoelastic kernel decreases in time too fast, as in the Prony's series, lead to identification problem strongly ill-conditioned. With reference to the Debye relaxation function, we have shown that an high condition number implies an high sensitivity of the minimum with respect either to errors in the experimental data or to variations of parameters in the identification algorithm. This, in turn, could reflect on multiple sets of optimal parameters.

On the contrary, the models with a slower rate of decay of the relaxation function produce a more accurate match of the data and result in a higher stability of the minimum point with respect to variations (or errors) in the experiments. Among the different models considered, the Cole-Cole model has produced the better fitting results.

REFERENCES

Antonakakis, J. N., P. Bhargava, K. C. Chuang, & A. T. Zehnder (2006, May). Linear viscoelastic properties of HFPE-II-52 polyimide. *J. Appl. Polym. Sci. 100*(4), 3255–3263.

Ciambella, J., A. Paolone, & S. Vidoli (2010, October). A comparison of nonlinear integral-based viscoelastic models through compression tests on filled rubber. *Mech. Mater. 42*(10), 932–944.

Ciambella, J., A. Paolone, & S. Vidoli (2011). Memory decay rates of viscoelastic solids: not too slow, but not too fast either. *Rheologica Acta.*

Hilfer, R. (2002, July). Analytical representations for relaxation functions of glasses. *Journal of Non-Crystalline Solids 305*(1–3), 122–126.

Knauss, W. G., I. Emri, & H. Lu (2008). *Handbook of Experimental Solid Mechanics*, Chapter 3, Mechani, pp. 49–93. Springer.

Knauss, W. G. & J. Zhao (2007, December). Improved relaxation time coverage in ramp-strain histories. *Mech. Time-Depend. Mat. 11*(3–4), 199–216.

Lewandowski, R. & B. Chorazyczewski (2010, January). Identification of the parameters of the KelvinVoigt and the Maxwell fractional models, used to modeling of viscoelastic dampers. *Computers & Structures 88*(1–2), 1–17.

Lockett, F. (1972). *Nonlinear Viscoelastic Solids.* Boston: Academic Press.

Metzler, R. & T. F. Nonnenmacher (2003, July). Fractional relaxation processes and fractional rheological models for the description of a class of viscoelastic materials. *Int. J. Plast. 19*(7), 941–959.

Osanaiye, G. J. (1996). Effects of temperature and strain amplitude on dynamic mechanical properties of EPDM gum and its carbon black compounds. *J. Appl. Polym. Sci. 59*(4), 567–575.

Constitutive Models for Rubber VII – Jerrams & Murphy (eds)
© 2012 Taylor & Francis Group, London, ISBN 978-0-415-68389-0

Theoretical and numerical modelling of unvulcanized rubber

Hüsnü Dal, Michael Kaliske & Christoph Zopf

Institut für Statik und Dynamik der Tragwerke, Technische Universität Dresden, Germany

ABSTRACT: The forming process of unvulcanized rubber is of great interest. However, classical hyperelastic models developed for cross-linked rubber do not apply to unvulcanized rubber due to the lack of crosslinks giving the material its elasticity. Experiments show that unvulcanized rubber exhibits strong viscoplastic flow without a distinct yield point accompanied with hardening. In this contribution, we propose a new constitutive model suited for unvulcanized rubber. The kinematic structure of the model is based on the micro-sphere model (see Miehe et al., JMPS 52:2617–2660, 2004). The computation of the stretch in the orientation direction follows the Cauchy-Born rule. The micro-sphere enables numerical integration over the sphere via finite summation of the orientation directions corresponding to the integration points over the sphere. This structure replaces the complex three-dimensional formulations, e.g. finite inelasticity models based on multiplicative split of the deformation gradient, by a simpler and more attractive one-dimensional rheological representation at the orientation directions. The rheology of the model consists of two parallel branches. The first branch consists of a spring connected to a Kelvin element where the latter spring models the kinematic hardening. The dashpot describes a time-independent endochronic flow rule based solely on the deformation history. The second branch consists of a spring connected to a Maxwell element in parallel to a dashpot. The two dashpots in the latter branch model the ground-state viscoelasticity and rate-dependent hardening phenomenon. Albeit its complexity, the proposed rheology and the numerical implementation show promising results suitable for large scale FE-based simulations.

1 INTRODUCTION

Rubber elasticity and inelasticity are active fields of research since decades. However, to the authors' knowledge, no substantial effort exists on the modelling of unvulcanized rubber. Expansion of the simulation based design in all engineering processes increased the interest not only on the constitutive behaviour of the end products but also on the material behaviour of raw materials in parallel to the enhancements in the process simulation techniques. Within this context, we propose a one-dimensional rheology for unvulcanized rubber and its three-dimensional continuum extension within the context of the micro-sphere model, see Miehe et al. (2004).

1.1 *Rheology*

The mechanical response of unvulcanized rubber is quite dissimilar to vulcanized rubber. Experiments demonstrate clearly that unvulcanized rubber is a highly deformable material which shows yield surface free equilibrium and non-equilibrium hysteresis with hardening at large strains. Unlike the classical viscoelastic behaviour of crosslinked rubber, it is observed that the amount of hardening is also strongly rate dependent. The rheology of the model consists of two parallel branches with endochronic and viscoelastic dashpots accommodated in parallel to nonlinear springs responsible for the hardening. In the latter branch, the spring responsible for hardening is serially connected to a viscous dashpot in order to account for the rate dependent post-yield hardening phenomenon. In contrast to the crosslinked rubber, unvulcanized rubber does not show ground-state elasticity and material flows at very small strain rates without a distinct yield surface. Three different characteristics can be observed from experiments. Firstly, the material shows an equilibrium hysteresis up to moderate stretches which can be idealized by a serially connected elastic spring and a dashpot element converted from viscoelastic flow rule via the correspondence principle, see Haupt (2000). Further loading leads to post-yield hardening which is modeled via a spring connected in parallel to the friction element. Furthermore, at elevated loading rates, one observes strong rate effects around the thermodynamical equilibrium. The rheology representing the complexities inherent to the model is depicted in Fig. 2. Vulcanized rubber consists of a network connected via cross-links which gives the material the memory for the undeformed state upon removal of the mechanical loading. Although unvulcanized rubber also posssesses a degree of elasticity due to the entanglements, the effect is small. The behaviour is nonlinear, showing significant equilibrium and non-equilibrium hysteresis. The cyclic loading of the material at low strain rates (cf. Fig. 5) shows that flow occurs even at very small deformation levels without a distinct yield surface. Further loading leads to kinematic hardening. Unvulcanized rubber shows significant rate dependency as depicted in the uniaxial tension tests, see Fig. 3a. The multistep relaxation test (cf. Fig. 4a) shows that the stresses do not fully relax reinforcing the hypothesis of existence of an equilibrium elastoplastic branch. Moreover, uniaxial tension experiments (Fig. 3a) clearly demonstrate a post-yield hardening regime which is strongly rate-dependent. Within this context, two discrete mechanisms are proposed for the description of the rheology of the material. For the equilibrium hysteresis observed at very small loading rates, a spring is connected serially to a Kelvin element, where an endochronic dashpot replaces the classical viscous dashpot via *correspondence principle*, see Haupt (2000). The spring in the Kelvin element models the rate-independent post-yield kinematic hardening regime. The latter branch consists of a non-linear spring connected serially to a system consisting of a Maxwell element in parallel to a viscous dashpot. The dashpot parallel to the Maxwell branch is responsible for the rate dependence of the initial loading whereas the dashpot in the Maxwell branch is responsible for the rate dependent kinematic hardening.

2 A MICROMECHANICAL MODEL FOR UNVULCANIZED RUBBER

2.1 *Kinematic description*

The key aspect of the micro-sphere based modelling is to link the deformation of a single point on the sphere to the macroscopic isochoric deformation $\bar{F}$. Let r denote a Lagrangean orientation unit vector with $|r| := \sqrt{r_\flat \cdot r} = 1$, where $r_\flat := r$ is the co-vector of r obtained by mapping with the standard metric $G = \delta_{AB}$. It can be described in terms of spherical coordinates

$$r = \cos \varphi \sin \vartheta e_1 + \sin \varphi \sin \vartheta e_2 + \cos \vartheta e_3 \quad (1)$$

in Cartesian frame $\{e_i\}_{i=1,2,3}$, where $\varphi \in [0, 2\pi]$ and $\vartheta \in [0, \pi]$, see Fig. 1. Mapping of the unit orientation vector r by the isochoric deformation of the continuum $\bar{F}$ leads to the spatial orientation vector

$$t = \bar{F}r . \quad (2)$$

The *affine-stretch* of a material line element in the orientation direction r reads

$$\bar{\lambda} := \sqrt{t_\flat \cdot t} \quad \text{where} \quad t_\flat := gt \quad (3)$$

is the co-vector of t obtained by a mapping with the current metric $g = \delta_{AB}$. Before we introduce the averaging operator on the unit micro-sphere $\mathcal{S}$, the infinitesimal area $dA = \sin \vartheta d\varphi d\vartheta$ and the area $A(\varphi, \vartheta) = \int_0^\vartheta \int_0^\varphi \sin d\varphi d\vartheta$ are defined. The total area reads $\mathcal{S} = 4\pi$. To this end, we define the averaging operator over the unit sphere $\mathcal{S}$

$$\langle \cdot \rangle = \frac{1}{|\mathcal{S}|} \int_{\mathcal{S}} (\cdot) \, dA \quad (4)$$

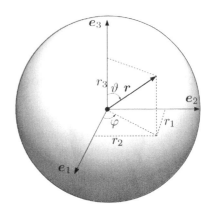

Figure 1. The unit micro-sphere and the orientation vector $r = r_1 e_1 + r_2 e_2 + r_3 e_3$ where $r_1 = \cos \varphi \sin \vartheta, r_2 = \sin \varphi \sin \vartheta$ and $r_3 = \cos \vartheta$ in terms of spherical coordinates $\{\varphi, \vartheta\}$, respectively.

which can be considered as the homogenization of the state variable $(\cdot)$ over the unit micro-sphere $\mathcal{S}$. For the affine micro-sphere model, the macroscopic free energy function is linked to the microscopic free energies

$$\bar{\Psi}(g; F, \mathcal{I}) := n \langle \psi(\lambda, \mathcal{I}) \rangle \quad (5)$$

in terms of continuos integration on the sphere. The set of microscopic internal variables of are denoted by $\mathcal{I}$. We propose, the *affinity assumption* $\lambda = \bar{\lambda}$ holds for the linkage between micro-stretches λ and macro-stretches $\bar{\lambda}$. The affinity assumption corresponds to the Cauchy-Born rule in crystal elasticity, stating that for crystals undergoing small deformations, stretch in micro-orientation direction is equivalent to the macroscopic stretch $\bar{\lambda}$.

2.2 *Free energy and the dissipation function*

A general internal variable formulation of finite inelasticity based on two scalar functions: the energy storage function and the dissipation function, will be constructed. The general set up of this generic type of models in the context of multiplicative split of the deformation gradient dates back to the works of Biot (1965), Maugin (1990) among others. A similar framework is applied to rubber viscoelasticity by Miehe and Göktepe (2005) in the context of micro-sphere model. They used logarithmic stretches and internal variables in the orientation directions of the sphere with their work-conjugatetes for the description of the flow rule and/or the dissipation function. Such an ansatz circumvents the complexities inherent to the multiplicative split of the deformation gradient which entails additional assumptions concerning the inelastic rotation and inelastic spin. In what follows, we propose microscopic free energy functions and dissipation potentials for the constitutive description of the unvulcanized rubber material consistent with the rheological description depicted in Fig. 2.

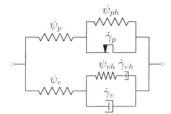

Figure 2. Rheological representation of the proposed model.

The microscopic free energy additively decomposes into *elasto-plastic* and *visco-elastic* parts

$$\psi(\lambda, \mathscr{I}) = \hat{\psi}^p(\lambda, \mathscr{I}^p) + \hat{\psi}^v(\lambda, \mathscr{I}^v) . \tag{6}$$

The storage and dissipation functions describing each part will be considered as follows.

Endochronic plasticity + kinematic hardening:
The first branch of the rheology consists of two storage functions

$$\hat{\psi}^p(\lambda, \mathscr{I}^p) = \psi_p(\lambda_{ep}) + \psi_{ph}(\lambda_p), \tag{7}$$

where the latter expression is responsible for the post-yield kinematic hardening. A generic power-type expression is adopted for the free energy functions

$$\psi_p(\lambda_{ep}) := \frac{\mu_p}{\delta_p}\left((\lambda_{ep})^{\delta_p} - 1\right) \quad \text{and} \tag{8}$$

$$\psi_{ph}(\lambda_p) := \frac{\mu_{ph}}{\delta_{ph}}(\lambda_p - 1)^{\delta_{ph}} . \tag{9}$$

μ_p, δ_p, μ_{ph} and δ_{ph} are the material parameters. The stretch expression is split into elastic and plastic contributions

$$\lambda = \lambda_{ep}\lambda_p \quad \text{and} \quad \varepsilon := \varepsilon_{ep} + \varepsilon_p , \tag{10}$$

where $\varepsilon = \ln(\lambda)$, $\varepsilon_{ep} = \ln(\lambda_{ep})$ and $\varepsilon_p = \ln(\lambda_p)$, respectively. Hence, the internal state of the material due to plastic deformation is described by

$$\mathscr{I}^p := \{\varepsilon_p\} . \tag{11}$$

The micro-dissipation due to the internal variable reads

$$\mathcal{D}_{loc}^{ep} := \hat{\beta}_p \dot{\varepsilon}_p \geq 0 \tag{12}$$

where we introduce the thermodynamic force driving the endochronic dashpot

$$\hat{\beta}_p := -\partial_{\varepsilon_p}\psi_p - \partial_{\varepsilon_p}\psi_{ph} = \beta_p - \beta_{ph} \tag{13}$$

as work-conjugate to the logarithmic internal variable ε_p. $\beta_{ph} = \partial_{\varepsilon_p}\psi_{ph}$ is the back stress. The model of inelasticity must be supplemented by additional constitutive equations which determine the evolution of the internal variables ε_p in time. A broad spectrum of

inelastic solids is covered by the so-called standard dissipative media where the evolution $\dot{\varepsilon}_p$ is governed by a smooth dissipation function ϕ^{ep} which is related to the free energy function through Biot's equation

$$\partial_{\varepsilon_p}\psi_p(\varepsilon_e) + \partial_{\dot{\varepsilon}_p}\phi^{ep}(\dot{\varepsilon}_p, \varepsilon_p) = 0 \tag{14}$$

with $\varepsilon_p(0) = 0$. We propose a power-type generic expression for the dissipation function

$$\phi^{ep}(\dot{\varepsilon}_p, \varepsilon_p) := \frac{\dot{z}\, m_p}{\eta_p(1 + m_p)}\left(\eta_p|\dot{\varepsilon}_p|\right)^{\frac{m_p}{1+m_p}} \tag{15}$$

governed by the material parameters m_p and η_p. Insertion of (15) into (14) leads to the evolution of the inelastic logarithmic strain after some manipulations

$$\dot{\varepsilon}_p := \dot{\gamma}_p(\hat{\beta}_p)\hat{\beta}_p \quad ; \quad \dot{\gamma}_p(\hat{\beta}_p) := \frac{\dot{z}}{\eta_p}|\hat{\beta}_p|^{m_p}. \tag{16}$$

A special choice of the evolution of the arclength

$$\dot{z} := |\dot{\varepsilon}| = \frac{|\dot{\lambda}|}{\lambda} \tag{17}$$

renders the formulation (16) rate-independent. Hence, the deformation of the plastic strain is controlled solely by the magnitude of the deformation history and plastic flow occurs without a distinct yield surface.

Viscoelasticity + kinematic hardening: The second branch of the rheology also consists of two storage functions

$$\hat{\psi}^v(\lambda, \mathscr{I}^v) = \psi_v(\lambda_{ve}) + \psi_{vh}(\lambda_{vh}), \tag{18}$$

where the latter expression is responsible for the post-yield kinematic hardening. A generic power-type expression is adopted for the free energy functions

$$\psi_v(\lambda_{ve}) := \frac{\mu_v}{\delta_v}\left((\lambda_{ve})^{\delta_v} - 1\right) \quad \text{and} \tag{19}$$

$$\psi_{vh}(\lambda_{vh}) := \frac{\mu_{vh}}{\delta_{vh}}(\lambda_{vh} - 1)^{\delta_{vh}} . \tag{20}$$

μ_v, δ_v, μ_{vh} and δ_{vh} are the material parameters. The stretch expression is split into elastic and viscous contributions

$$\lambda = \lambda_{ve}\lambda_v \quad \text{and} \quad \varepsilon := \varepsilon_{ve} + \varepsilon_v , \tag{21}$$

where $\varepsilon = \ln(\lambda)$, $\varepsilon_{ep} = \ln(\lambda_{ep})$ and $\varepsilon_p = \ln(\lambda_p)$, respectively. In order to model the rate-dependency of the hardening stress, a second dashpot is introduced. Accordingly, the viscous stretch and strain is decomposed as

$$\lambda_v = \lambda_{eh}\lambda_{vh} \quad \text{and} \quad \varepsilon_v := \varepsilon_{eh} + \varepsilon_{vh} . \tag{22}$$

The internal state of the material due to viscous deformations is described by

$$\mathscr{I}^v := \{\varepsilon_v, \varepsilon_{vh}\} . \tag{23}$$

Table 1. Material parameters of the specified model

Param.	Description	Eqn.
μ_p, δ_p	Elas. const., elasto-plastic branch	(8)
μ_{ph}, δ_{ph}	Elas. const., back-stress in pl. branch	(9)
μ_v, δ_v	Elas. const., visco-elastic branch	(19)
μ_{vh}, δ_{vh}	Elas. const. back-stress in visc. branch	(20)
η_p, m_p	Evol. par., plastic flow	(15, 16)
η_v, m_v	Evol. par., viscous flow	(27, 29)
η_{vh}, m_v	Evol. par., hardening viscosity	(28, 30)

The micro-dissipation due to the internal variables reads

$$\mathcal{D}^{ve}_{loc} := \hat{\beta}_v \dot{\varepsilon}_v + \beta_{vh} \dot{\varepsilon}_{vh} \geq 0 \,, \qquad (24)$$

where we introduce the thermodynamical forces driving the viscous dashpots

$$\hat{\beta}_v := -\partial_{\varepsilon_v} \psi_v - \partial_{\varepsilon_v} \psi_{vh} = \beta_v - \beta_{vh} \qquad (25)$$

as work-conjugate to the logarithmic internal variable ε_v and

$$\beta_{vh} := -\partial_{\varepsilon_{vh}} \psi_{vh} \qquad (26)$$

as work-conjugate to the logarithmic internal variable ε_{vh}, respectively. The identity $\partial_{\varepsilon_v} \psi_{vh} = -\partial_{\varepsilon_{vh}} \psi_{vh}$ is utilised in the previous equation. $\beta_{vh} = -\partial_{\varepsilon_{vh}} \psi_{vh}$ is the rate-dependent back stress.

The power-type generic expressions for the dissipation functions read

$$\phi^{ve}(\dot{\varepsilon}_v, \varepsilon_v) := \frac{m_v}{\eta_v(1+m_v)} \left(\eta_v |\dot{\varepsilon}_v|\right)^{\frac{m_v}{1+m_v}} \,, \qquad (27)$$

$$\phi^{vh}(\dot{\varepsilon}_{vh}, \varepsilon_{vh}) := \frac{m_{vh}}{\eta_{vh}(1+m_{vh})} \left(\eta_{vh} |\dot{\varepsilon}_{vh}|\right)^{\frac{m_{vh}}{1+m_{vh}}} \qquad (28)$$

and are governed by the material parameters m_v, m_{vh}, η_v and η_{vh}, respectively. The dissipation potentials (27) and (28) lead to the evolution laws

$$\dot{\varepsilon}_v := \dot{\gamma}_v(\hat{\beta}_v)\hat{\beta}_v \quad ; \quad \dot{\gamma}_v(\hat{\beta}_v) := \frac{1}{\eta_v}|\hat{\beta}_v|^{m_v} \quad \text{and} \quad (29)$$

$$\dot{\varepsilon}_{vh} := \dot{\gamma}_{vh}(\beta_{vh})\beta_{vh} \quad ; \quad \dot{\gamma}_{vh}(\beta_{vh}) := \frac{1}{\eta_{vh}}|\beta_{vh}|^{m_v}, \quad (30)$$

respectively. In (16), (29) and (30), $|(\cdot)| := \{[(\cdot)/\text{unit}(\cdot)]^2\}^{1/2}$ is the norm operator with neutralization of the units. In addition, the material parameters of the specified model are listed in Table 1 along with their brief description and the equation numbers where they appear.

2.3 Stresses and moduli

Endochronic plasticity + kinematic hardening:
(i) Stresses:
Having the free energy function and the dissipation function at hand, stress expression and the numerical

tangent necessary for the finite element implementation will be derived. The Kirchhoff stresses $\bar{\tau} := 2\partial \mathbf{g} \bar{\Psi}(\mathbf{g}; \bar{F}, \mathcal{I})$ are computed via

$$\bar{\tau}^{ep} := \langle f^p \lambda^{-1} \mathbf{t} \otimes \mathbf{t} \rangle \quad ; \quad f^p := \frac{\partial \hat{\psi}^p}{\partial \lambda} \,, \qquad (31)$$

with the help of the identity

$$2\partial_g \lambda = \lambda^{-1} \mathbf{t} \otimes \mathbf{t} \,. \qquad (32)$$

Applying the chain rule, the relation between the micro-stresses $\sigma_p = \partial_{\varepsilon_e} \psi_p$ and f^p can be established

$$f^p = \sigma_p / \lambda \,. \qquad (33)$$

$(31)_2$ and (33) into (8) leads to

$$f^p = \mu_p \frac{\lambda_{ep}^{\delta_p}}{\lambda} \quad ; \quad \sigma_p = \mu_p \lambda_{ep}^{\delta_p} \,. \qquad (34)$$

Computation of σ_p at an orientation direction $\mathbf{r}$ entails the description of the current state of the history variable λ_p or ε_p. In order to compute ε_p for a given time-step t_{n+1}, we recall the flow rule expression (16) and recast it into a discrete residual form by backward Euler scheme

$$r^p = \varepsilon_p - \varepsilon_p^n - \dot{\gamma}_p(\hat{\beta}_p)\hat{\beta}_p \Delta t = 0 \,. \qquad (35)$$

The index n denotes the previous time-step t_n, whereas the index $n+1$ is dropped from ε_p and $\hat{\beta}_p$ expression for the sake of convenience. The thermodynamical force driving the endochronic dashpot can be derived by incorporating $(13)_2$ into (8) and (9)

$$\hat{\beta}_p = \sigma_p - \beta_{ph} \quad ; \quad \beta_{ph} = \mu_{ph}(\lambda_p - 1)^{\delta_{ph}-1} \lambda_p \,. \qquad (36)$$

(35) is a nonlinear equation and cannot be solved analytically for ε_p. Linearization of the residual around ε_p^k yields

$$\text{Lin } r^p := r^p|_{\varepsilon_p^k} + \frac{\partial r^p}{\partial \varepsilon_p}|_{\varepsilon_p^k} \Delta \varepsilon_p^{k+1} = 0 \,. \qquad (37)$$

Setting $\varepsilon_p^0 = \varepsilon_p^n$, the incremental plastic strain in (37) can be obtained

$$\Delta \varepsilon_p^{k+1} = -\mathcal{K}_p^{-1} r^p|_{\varepsilon_p^k} \quad ; \quad \mathcal{K}_p = \frac{\partial r^p}{\partial \varepsilon_p}|_{\varepsilon_p^k} \,. \qquad (38)$$

Update of the plastic strain reads

$$\varepsilon_p^{k+1} = \varepsilon_p^k + \Delta \varepsilon_p^{k+1} \,. \qquad (39)$$

(35) is to be solved repeating the steps (38) and (39) until a certain residual tolerance $|r^p| < TOL$ is obtained. The tangent term $\mathcal{K}_p$ in (38) reads

$$\mathcal{K}_p = 1 - (m_p + 1)\frac{\dot{z}}{\eta_p}\Delta t|\hat{\beta}_p|^{m_p}\frac{\partial \hat{\beta}_p}{\partial \varepsilon_p} \qquad (40)$$

Table 2. Local Newton update of the internal variable ε_p

1. Set initial values	$k = 0,\ \varepsilon_p^0 = \varepsilon_v^n$
DO	
2. Residual equation	$r^p = \varepsilon_p - \varepsilon_p^n - \dot{\gamma}_p(\hat{\beta}_p)\hat{\beta}_p \Delta t = 0$
3. Linearization	$\text{Lin } r^p := r^p|_{\varepsilon_p^k} + \partial r^p / \partial \varepsilon_p\ |_{\varepsilon_p^k}$
	$\Delta\varepsilon_p^{k+1} = 0$
4. Compute	$\mathcal{K}_p := \partial r^p / \partial \varepsilon_p\big|_{\varepsilon_p = \varepsilon_p^k}$
5. Solve	$\Delta\varepsilon_p^{k+1} = -\mathcal{K}_p^{-1} r_p$
6. Update	$\varepsilon_p^{k+1} \leftarrow \varepsilon_p^k + \Delta\varepsilon_p^{k+1}$
	$k \leftarrow k + 1$
WHILE	$TOL \leq |r^p|$

where

$$\frac{\partial \hat{\beta}_p}{\partial \varepsilon_p} = -\left(\sigma_p' \lambda_{ep} + \beta_{ph}' \lambda_p\right), \tag{41}$$

$$\sigma_p' = \delta_p \mu_p \lambda_{ep}^{\delta_p - 1}, \tag{42}$$

$$\beta_{ph}' = (\delta_{ph}-1)\mu_{ph}(\lambda_{p-1})^{\delta_{ph}-2}\lambda_p + \mu_{ph}(\lambda_{p-1})^{\delta_{ph}-1}. \tag{43}$$

The Newton iteration scheme for the computation of the current value of ε_p is summarized in Table 2.

(ii) Algorithmic moduli:

Further derivation of the stress expression (31) with respect to the Eulerian metric $\boldsymbol{g}$ yields the spatial algorithmic tangent

$$\bar{\mathbb{C}}_{algo}^p := 2\partial_{\boldsymbol{g}} \bar{\tau}^{ep}(\boldsymbol{g}, \mathcal{I}^p, \bar{\boldsymbol{F}}; \bar{\boldsymbol{F}}_n, \mathcal{I}_n^p). \tag{44}$$

Use of the results $\partial_{\boldsymbol{g}}(\boldsymbol{t} \otimes \boldsymbol{t}) = \boldsymbol{0}$ and (32) and incorporation (44) into (31) leads to the compact representation

$$\bar{\mathbb{C}}_{algo}^p := \langle (c^p - f^p\lambda^{-1})\lambda^{-2}\boldsymbol{t} \otimes \boldsymbol{t} \otimes \boldsymbol{t} \otimes \boldsymbol{t}\rangle \tag{45}$$

where

$$c^p := \frac{df^p}{d\lambda} = \frac{1}{\lambda^2}\frac{d\sigma_p}{d\varepsilon} - \frac{f^p}{\lambda} \quad \text{and} \tag{46}$$

$$\frac{d\sigma_p}{d\varepsilon} = \delta_p \mu_p \lambda_{ep}^{\delta_p}\left(1 - \frac{d\varepsilon_p}{d\varepsilon}\right). \tag{47}$$

In order to derive the expression $d\varepsilon_p/d\varepsilon$, we will make use of the *implicit function theorem*. Recalling the identity $dr^p/d\varepsilon = 0$ and applying the chain rule, one finally ends up with

$$\frac{d\varepsilon_p}{d\varepsilon} = \frac{\Delta t \frac{\dot{z}}{\eta_p}|\hat{\beta}_p|^{m_p}\sigma_p'\lambda_{ep}}{\mathcal{K}_p}. \tag{48}$$

Viscoelasticity + kinematic hardening:
(i) Stresses:

The viscoelastic part of the Kirchhoff stresses can be derived similar to what has been proposed for the endochronic branch

$$\bar{\tau}^{ve} := \langle f^v \lambda^{-1}\boldsymbol{t} \otimes \boldsymbol{t}\rangle \quad ; \quad f^v := \frac{\partial \hat{\psi}^v}{\partial \lambda}, \tag{49}$$

along with the definitions

$$f^v = \sigma_v/\lambda = \mu_v \frac{\lambda_{ve}^{\delta_v}}{\lambda} \quad \text{and} \quad \sigma_v = \mu_v \lambda_{ve}^{\delta_v}. \tag{50}$$

Computation of σ_v at an orientation direction $\boldsymbol{r}$ entails outer and inner iterations for the viscous strain ε_v and the hardening viscous strain ε_{vh}, respectively. The residual equation for the outer iteration step is written as

$$r^v = \varepsilon_v - \varepsilon_v^n - \dot{\gamma}_v(\hat{\beta}_v)\hat{\beta}_v \Delta t = 0. \tag{51}$$

The thermodynamical force driving the endochronic dashpot can be derived by incorporating $(26)_2$ into (19) and (20)

$$\hat{\beta}_v = \sigma_v - \beta_{vh} \quad ; \quad \beta_{vh} = \mu_{vh}(\lambda_{eh} - 1)^{\delta_{vh}-1}\lambda_{eh}. \tag{52}$$

(51) is a nonlinear equation and cannot be solved analytically for ε_v. Linearization of the residual around ε_p^k yields

$$\text{Lin } r^v := r^v|_{\varepsilon_v^k} + \frac{\partial r^v}{\partial \varepsilon_v}\ |_{\varepsilon_v^k} \Delta\varepsilon_v^{k+1} = 0. \tag{53}$$

Setting $\varepsilon_v^0 = \varepsilon_v^n$ and updating (53) iteratively via

$$\Delta\varepsilon_v^{k+1} = -\mathcal{K}_v^{-1} r^v|_{\varepsilon_v^k} \quad ; \quad \mathcal{K}_v = \frac{\partial r^v}{\partial \varepsilon_v}|_{\varepsilon_v^k} \tag{54}$$

and

$$\varepsilon_v^{k+1} = \varepsilon_v^k + \Delta\varepsilon_v^{k+1}, \tag{55}$$

one obtains the current state of the history variable ε_v. The tangent term $\mathcal{K}_v$ in (54) reads

$$\mathcal{K}_v = 1 - (m_v + 1)\frac{1}{\eta_v}\Delta t|\hat{\beta}_v|^{m_v}\frac{\partial \hat{\beta}_v}{\partial \varepsilon_v} \tag{56}$$

along with the following definitions

$$\frac{\partial \hat{\beta}_v}{\partial \varepsilon_v} = -(\sigma_v'\lambda_{ve} + \beta_{vh}'\lambda_{eh}), \tag{57}$$

$$\sigma_v' = \delta_v \mu_v \lambda_{ve}^{\delta_v - 1}, \tag{58}$$

$$\beta_{vh}' = (\delta_{vh}-1)\mu_{vh}(\lambda_{eh}-1)^{\delta_{vh}-2}\lambda_{eh} + \mu_{vh}(\lambda_{eh}-1)^{\delta_{vh}-1}. \tag{59}$$

The tangent $\mathcal{K}_v$ and the residual r^v are dependent on the elastic strain ε_{eh} and the viscous strain ε_{vh} of the hardening part. This requires the solution of the residual expression

$$r^{vh} = \varepsilon_{vh} - \varepsilon_{vh}^n - \dot{\gamma}_{vh}(\beta_{vh})\beta_{vh}\Delta t = 0 \qquad (60)$$

prior to the update of (59). Linearization of the residual expression (60) around ε_{vh}^k yields

$$\text{Lin } r^{vh} := r^{vh}|_{\varepsilon_{vh}^k} + \frac{\partial r^{vh}}{\partial \varepsilon_{vh}}|_{\varepsilon_{vh}^k}\Delta\varepsilon_{vh}^{k+1} = 0. \qquad (61)$$

Setting $\varepsilon_{vh}^0 = \varepsilon_{vh}^n$, the incremental viscous-hardening strain in (61) can be obtained

$$\Delta\varepsilon_{vh}^{k+1} = -\mathcal{K}_{vh}^{-1}r^{vh}|_{\varepsilon_{vh}^k} \quad ; \quad \mathcal{K}_{vh} = \frac{\partial r^{vh}}{\partial \varepsilon_{vh}}|_{\varepsilon_{vh}^k}. \qquad (62)$$

Update of the viscous-hardening strain reads

$$\mathcal{K}_{vh} = 1 - (m_{vh}+1)\frac{1}{\eta_{vh}}\Delta t|\beta_{vh}|^{m_{vh}}\frac{\partial\beta_{vh}}{\partial\varepsilon_{vh}} \qquad (64)$$

(60) is to be solved iteratively, updating (62) and (63) until a certain residual tolerance $|r^{vh}| < TOL$ is reached. The tangent $\mathcal{K}_v$ of the internal Newton iteration reads

$$\mathcal{K}_{vh} = 1 - (m_{vh}+1)\frac{1}{\eta_{vh}}\Delta t|\beta_{vh}|^{m_{vh}}\frac{\partial\beta_{vh}}{\partial\varepsilon_{vh}} \qquad (64)$$

where

$$\frac{\partial\beta_{vh}}{\partial\varepsilon_{vh}} = -\beta'_{vh}\lambda_{eh}. \qquad (65)$$

Once the solution of the internal iteration is achieved, the update of the outer residual (53) follows straightforwardly through (54). The Newton iteration scheme for the update of ε_v and ε_{vh} is summarized in Table 3.

(ii) Algorithmic moduli:
Further derivation of the of the stress expression (49) with respect to the Eulerian metric g yields the spatial algorithmic tangent

$$\bar{\mathbb{C}}^v_{algo} := 2\partial_g\bar{\tau}^{vc}(g,\mathcal{I}^v,\bar{F};F_n,\mathcal{I}^v_n). \qquad (66)$$

Incorporation of (66) into (49) leads to the representation

$$\bar{\mathbb{C}}^v_{algo} := \langle(c^v - f^v\lambda^{-1})\lambda^{-2}t\otimes t\otimes t\otimes t\rangle \qquad (67)$$

where

$$c^v := \frac{df^v}{d\lambda} = \frac{1}{\lambda^2}\frac{d\sigma_v}{d\varepsilon} - \frac{f^v}{\lambda} \quad \text{and} \qquad (68)$$

Table 3. Local Newton update of the internal variables $\{\varepsilon_v, \varepsilon_{vh}\}$

1. Set initial values	$k = 0, \varepsilon_v^0 = \varepsilon_v^n$
DO	
(i) Set initial values	$i = 0, \varepsilon_{vh}^0 = \varepsilon_{vh}^n$
DO	
(ii) Residual equation	$r^{vh} = \varepsilon_{vh} - \varepsilon_{vh}^n - \dot{\gamma}_{vh}(\beta_{vh})\beta_{vh}\Delta t = 0$
(iii) Linearization	$\text{Lin } r^{vh} := r^{vh}|_{\varepsilon_{vh}^i} + \partial r^{vh}/\partial\varepsilon_{vh}|_{\varepsilon_{vh}^i}$
	$\Delta\varepsilon_{vh}^{i+1} = 0$
(iv) Compute	$\mathcal{K}_{vh} := \partial r^{vh}/\partial\varepsilon_{vh}|_{\varepsilon_{vh}=\varepsilon_{vh}^i}$
(v) Solve	$\Delta\varepsilon_{vh}^{i+1} = -\mathcal{K}_{vh}^{-1}r_{vh}|_{\varepsilon_{vh}^i}$
(vi) Update	$\varepsilon_{vh}^{i+1} \leftarrow \varepsilon_{vh}^i + \Delta\varepsilon_{vh}^{i+1}$
	$i \leftarrow i+1$
WHILE	$TOL \leq |r^{vh}|$
2. Residual equation	$r^v = \varepsilon_v - \varepsilon_v^n - \dot{\gamma}_v(\hat{\beta}_v)\hat{\beta}_v\Delta t = 0$
3. Linearization	$\text{Lin } r^v := r^v|_{\varepsilon_v^k} + \partial r^v/\partial\varepsilon_v|_{\varepsilon_v^k}$
	$\Delta\varepsilon_v^{k+1} = 0$
4. Compute	$\mathcal{K}_v := \partial r^v/\partial\varepsilon_v|_{\varepsilon_v=\varepsilon_v^k}$
5. Solve	$\Delta\varepsilon_v^{k+1} = -\mathcal{K}_v^{-1}r_v|_{\varepsilon_v^k}$
6. Update	$\varepsilon_v^{k+1} \leftarrow \varepsilon_v^k + \Delta\varepsilon_v^{k+1}$
	$k \leftarrow k+1$
WHILE	$TOL \leq |r^v|$

$$\frac{d\sigma_v}{d\varepsilon} = \delta_v\mu_v\lambda_{ve}^{\delta_v}(1 - \frac{d\varepsilon_v}{d\varepsilon}). \qquad (69)$$

The derivation of the expression $d\varepsilon_v/d\varepsilon$ is also carried out by applying the *implicit function theorem* and yields

$$\frac{d\varepsilon_v}{d\varepsilon} = \frac{\Delta t\frac{1}{\eta_v}|\hat{\beta}_v|^{m_v}\sigma'_v\lambda_{ve}}{\mathcal{K}_v}. \qquad (70)$$

Finally, the algorithmic Eulerian tangent moduli $\bar{\mathbb{C}}_{algo} := 2\partial_g\bar{\tau}(g,\mathcal{I},\bar{F};\bar{F}_n,\mathcal{I}_n)$ read

$$\bar{\mathbb{C}}_{algo} := \langle(c - f\lambda^{-1})\lambda^{-2}t\otimes t\otimes t\otimes t\rangle \qquad (71)$$

with

$$c := c^p + c^v \quad \text{and} \quad f = f^p + f^v. \qquad (72)$$

The isochoric part of the Eulerian tangent moduli reads

$$\mathbb{C}^{iso}:=\mathbb{P}:[\bar{\mathbb{C}}_{algo}:\frac{2}{3}(\bar{\tau}:g)\mathbb{I} - \frac{2}{3}(\bar{\tau}\otimes g^{-1}+g^{-1}\otimes\bar{\tau})]:\mathbb{P} \quad (73)$$

where $\mathbb{I}^{abcd} = [\delta_c^a\delta_d^b + \delta_d^a\delta_c^b]/2$ is the fourth order symmetric identity tensor. The total Eulerian tangent is obtained by summing up the volumetric and isochoric parts

$$\mathbb{C} := \mathbb{C}^{vol} + \mathbb{C}^{iso}. \qquad (74)$$

2.4 Micro-macro transition

Numerical implementation of the continuous integral over the unit sphere requires discretization in terms of finite summation

$$\frac{1}{|\mathcal{S}|}\int_{\mathcal{S}}(\cdot)\,dA \approx \sum_{i=1}^m(\cdot)^iw^i. \qquad (75)$$

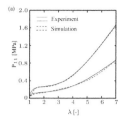 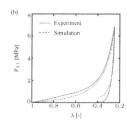

Figure 3. Comparison of the proposed model with (a) uniaxial tensile experiments on unvulcanized natural rubber at stretch rates $\dot{\lambda} = 0.1/35\,[\mathrm{s}^{-1}]$ and $\dot{\lambda} = 10/35\,[\mathrm{s}^{-1}]$ and (b) nonhomogeneous compression tests, respectively.

Here, $\{w^i\}_{i=1\ldots m}$ are the weight factors in the discrete orientation directions $\{r^i\}_{i=1\ldots m}$. The weights w^i and the discrete orientations r^i should satisfy certain normalization conditions in order to preserve isotropy and stress-free reference configuration. Analytically, it can be easily shown that

$$\langle r \rangle = 0 \quad \text{and} \quad \langle r \otimes r \rangle = \frac{1}{3}\mathbf{1} \,. \tag{76}$$

Similarly, the discrete representation should preserve the identities $(76)_1$ and $(76)_2$

$$\sum_{i=1}^{m} r^i w^i = 0 \quad \text{and} \quad \sum_{i=1}^{m} r^i \otimes r^i w^i = \frac{1}{3}\mathbf{1} \,. \tag{77}$$

Set of orientation vectors $\{r^i\}_{i=1\ldots m}$ and associated weight factors $\{w^i\}_{i=1\ldots m}$ which satisfy the constraints $(77)_1$ and $(77)_2$ are given in Bažant and Oh (1986). The set of $m = 21$ integration points for the half sphere are used for the current investigations.

3 MODEL VALIDATION AND NUMERICAL EXAMPLES

In what follows, the proposed constitutive model will be compared to the experimental data. The modelling capacity will be assessed according to the homogeneous uniaxial tension tests at two different loading rates, and the identified parameters from uniaxial tension tests will be used for the demonstration of the nonhomogeneous compression tests. Furthermore, multistep relaxation tests and cyclic compression tests will be used for the verification of the constitutive model.

3.1 Comparison of homogeneous test results

Fig. 3a depicts the uniaxial tension tests and the model predictions with the identified material parameters given in Table 4. During the identification process, more attention is given for the stretches beyond $\lambda = 2$. The proposed model shows slight underestimation upto stretch level $\lambda = 2$ and excellent fitting capability from $\lambda = 2$ to $\lambda = 7$ at different stretch levels. The identified parameters are used for validation of nonhomogeneous compression tests and the results

Table 4. Identified material parameters

Parameter	Figure 3	Figure 4	Figure 5	Dimension
μ_p	0.11	0.055	0.22	[MPa]
δ_p	6.0	6.0	6.0	[–]
μ_{ph}	0.45	0.45	0.23	[MPa]
δ_{ph}	6.0	6.0	6.0	[–]
η_p	0.4	0.15	0.10	[MPa]
m_p	1.0	1.0	1.0	[–]
μ_v	0.10	0.20	0.20	[MPa]
δ_v	7.1	10	7.1	[–]
μ_{vh}	0.36	0.36	0.18	[MPa]
δ_{vh}	7.1	7.1	7.1	[–]
η_v	2.0	$1.0 \cdot 10^2$	1.0	[MPa·s]
m_v	2.0	1.0	2.0	[–]
η_{vh}	$9.2 \cdot 10^2$	$9.2 \cdot 10^2$	$9.2 \cdot 10^2$	[MPa·s]
m_{vh}	1.0	1.0	1.0	[–]

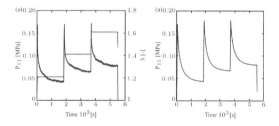

Figure 4. Comparison of (a) the experimental results of uniaxial tension stress vs. time, (b) the model prediction for tensile relaxation test. The loading rate is $\dot{\lambda} = 0.2\,[\mathrm{s}^{-1}]$, and intermittent relaxations are applied at stretch levels $\lambda = 1.2, \lambda = 1.4$ and $\lambda = 1.6$, respectively.

are shown in Fig. 3b. The compression tests show an inital underestimation in comparison to the experiments similar to the uniaxial tensile tests. However, the qualitative description of the material behaviour is successfully captured.

In Fig. 4a, the multistep tensile relaxation curve for unvulcanized natural rubber is depicted. Fig. 4b shows the model prediction with modified parameters given in Table 4. Fig. 4 is especially significant in order to assess the form of the dissipation potential governing the viscous flow. The reason for using modified parameters compared to the identified parameters for uniaxial tension and compression tests is to capture long-term relaxation effects which are lumped to the endochronic branch in the previous identification process. Besides, the parameters identified for the tension-compression tests underestimate the initial behaviour where the relaxation experiments are carried out and prestressing with an amount of $\sigma_{pre} \approx 0.05\,\mathrm{MPa}$ is applied to the uniaxial test specimens before loading. In reality, a spectrum of visco-elastic branches is needed in order to capture the range of relevant loading frequencies and the number of parallel viscoelastic branches can be increased for improving the accuracy. Fig. 5a depicts cyclic tension test results and the model prediction

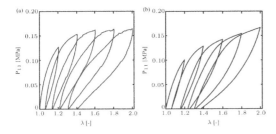

Figure 5. Cyclic tension test on unvulcanized natural rubber: stress vs. stretch plot (a) experiment, (b) model prediction. The loading rate is $\dot{\lambda} = 0.2[\mathrm{s}^{-1}]$.

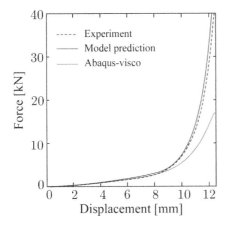

Figure 6. Comparison of the load-displacement curves for the drawing simulation of unvulcanized natural rubber specimen.

is shown in Fig. 5b. Permanent setting due to plastic deformations is captured successfully with slight underestimation of the initial stresses. In general, the proposed theory clearly captures the distinct characteristics of the unvulcanized rubber qualitatively. Considering the difficulties in preparing the test specimens due to the sticky nature of the unvulcanized rubber and the damage caused to the specimens during cutting procedure, the discrepancy observed between theory and the experiments is tolerable. However, it is to be mentioned that more experimental information is needed on vulcanized rubber in order to assess the quality of the material model proposed.

3.2 Numerical example: Forming simulation of an unvulcanized rubber specimen

The algorithm proposed in the previous section is implemented into ABAQUS as a UMAT subroutine. A three-dimensional mould is constructed as a frictionless contact surface and a rubber sample is pressed into the mould. The three-dimensional geometry of the specimen is hidden due to copyright restrictions of our industrial partner. The rubber sample is prepared from identical compound used for homogeneous experiments and under identical conditions. The measured pressing force versus tip displacement curves

of the experiment are depicted in Fig. 6. For control purposes, a generalized Maxwell model with three Maxwell branches in parallel to a Yeoh type hyperelasticity model is used. The formulation is based on the linear finite viscoelasticity framework of Kaliske and Rothert (1997a, 1997b) and documented in ABAQUS Theory Manual. The material parameters are identified in discarding the kinematic hardening region in the identification process according to the homogeneous experiments presented in Figs. 3, 4, 5. The hardening part is captured by including a small amount of equilibrium elasticity. Both simulations are carried out with an explicit FE-method due to extreme distortions rendering implicit FE-simulation impossible. The results with the proposed constitutive model and the ABAQUS simulations based on the finite viscoelasticity combined with Yeoh type hyperelastic formulation are compared to the experimental results in Fig. 6. The proposed model is quite successful in capturing the load displacement characteristics of the unvulcanized rubber material.

4 CONCLUSION

We proposed a new constitutive model for unvulcanized rubber and a novel algorithmic setting based on the kinematics of affine micro-sphere model. The model uses logarithmic strains in the orientation direction of a unit-sphere as state variables. The complex three-dimensional kinematics based on the multiplicative split of the deformation gradient is a priori eliminated. The power type free energy and dissipation potentials are used for the construction of the rheology. Fitting capacity of the proposed constitutive model is demonstrated to be succesful in capturing the observed material behaviour.

REFERENCES

Bažant, Z.P. & B.H Oh (1986). Efficient numerical integration on the surface of a sphere. *Zeitschrift für angewandte Mathematik und Mechanik 66*, 37–49.

Biot, M. A. (1965). *Mechanics of Incremental Deformations*. John Wiley & Sons, Inc., New York.

Haupt, P. (2000). *Continuum Mechanics and Theory of Materials*. Springer-Verlag, Berlin.

Kaliske, M. & H. Rothert (1997a). Formulation and implementation of three-dimensional viscoelasticity at small and finite strains. *Computational Mechanics 19*, 228–239.

Kaliske, M. & H. Rothert (1997b). On the finite element implementation of rubber-like materials at finite strains. *Engineering Computations 14*, 216–232.

Maugin, G. (1990). Internal Variables and Dissipative Structures. *Journal of Non-Equilibrium Thermodynamics 15*, 20.

Miehe, C. & S. Göktepe (2005). A micro-macro approach to rubber-like materials. Part II: The micro-sphere model of finite rubber viscoelasticity. *Journal of the Mechanics and Physics of Solids 53*, 2231–2258.

Miehe, C., S. Göktepe, & F. Lulei (2004). A micro-macro approach to rubber-like materials. Part I: The non-affine micro-sphere model of rubber elasticity. *Journal of the Mechanics and Physics of Solids 52*, 2617–2660.

Constitutive Models for Rubber VII – Jerrams & Murphy (eds)
© 2012 Taylor & Francis Group, London, ISBN 978-0-415-68389-0

Thermomechanical material behaviour within the concept of representative directions

C. Naumann & J. Ihlemann

Professorship of Solid Mechanics, Chemnitz University of Technology, Chemnitz, Germany

ABSTRACT: The material behaviour of rubber-like materials is often significantly influenced by temperature. In order to take thermal effects in numerical simulations into account a three-dimensional thermomechanically coupled material model is required. The development of such a general constitutive relation is very challenging. A much easier task is to formulate a one-dimensional material model. (Freund and Ihlemann 2010) proposed a technique, the so-called concept of representative directions, which can be used to generalise one-dimensional mechanical material models to complete three-dimensional constitutive relations. In the presented work this concept is extended to cover thermomechanical effects. Starting from an one-dimensional material model, this concept is able to compute the second PIOLA-KIRCHHOFF stress tensor, the entropy, the dissipation, and the heat flux vector as a reaction to an arbitrary thermal and mechanical loading. It is shown that for thermodynamically consistent one-dimensional material models this generalisation technique leads to a thermodynamically consistent constitutive relation. Simulations show that typical thermomechanical effects can be modelled.

1 INTRODUCTION

The concept of representative directions is able to generalise one dimensional mechanical material models for uniaxial tension to complete three dimensional material models. A related generalisation technique, the so called micro-sphere-model, was developed by (Miehe et al. 2004). In this model the stresses are assumed to be a derivative of the free energy ψ. Hence a dashpot, which provides a stress but does not contribute to the free energy, cannot be modelled. No such assumption is made in the concept of representative directions.

In some applications the interactions between the temperature field and the deformation have to be taken into account. The temperature influences the deformation via thermal expansion and temperature dependence of material parameters. Due to dissipative and thermoelastic effects the temperature is affected by the deformation. To simulate this coupling using the finite element method a thermomechanically coupled material model is required. In the presented work the concept of representative directions is extended to generalise one dimensional thermomechanically coupled material models.

2 FUNDAMENTALS

2.1 *Decomposition of the deformation gradient*

The dominant influence of the temperature field on the deformation arises mostly from thermal expansion. To take this effect into account (Lu and Pister 1975)

proposed a decomposition of the deformation gradient $\underline{\underline{F}}$ into a stress-free expansion $\underline{\underline{F}}_\theta$ due to a temperature change and a stress producing mechanical part $\underline{\underline{F}}_m$. The thermal expansion is assumed to be volumetric:

$$\underline{\underline{F}} = \underline{\underline{F}}_m \cdot \underline{\underline{F}}_\theta, \quad \text{with} \quad \underline{\underline{F}}_\theta = \varphi \underline{\underline{E}} \quad . \tag{1}$$

$\underline{\underline{E}}$ denotes the identity tensor and φ is a material function that depends on the temperature θ:

$$\varphi = \exp\left[\alpha_t \left(\theta - \theta_0\right)\right] \quad . \tag{2}$$

The material parameter α_t is the coefficient of thermal expansion. It takes a value of approximately $2 \cdot 10^{-4}\,\mathrm{K}^{-1}$ for rubber materials. The temperature θ_0 is set to room temperature (296K).

Another widely used decomposition is the distortional-volumetric decomposition. A detailed derivation of the depicted relations is given in (Ihlemann 2003). The mechanical part of the deformation gradient is decomposed into a volumetric part $\overset{V}{\underline{\underline{F}}}$ and an isochoric part $\overset{G}{\underline{\underline{F}}}$:

$$\underline{\underline{F}}_m = \overset{G}{\underline{\underline{F}}} \cdot \overset{V}{\underline{\underline{F}}}, \quad \text{with} \quad \overset{G}{\underline{\underline{F}}} = J_3^{-\frac{1}{3}} \underline{\underline{F}}, \quad \overset{V}{\underline{\underline{F}}} = J_{3m}^{\frac{1}{3}} \underline{\underline{E}} \quad . \tag{3}$$

J_3 and J_{3m} are the third invariants of the deformation gradient $\underline{\underline{F}}$ and its mechanical part $\underline{\underline{F}}_m$, respectively.

2.2 *The Clausius-Duhem-inequality*

The deformation influences the temperature field mainly due to thermoelastic and dissipative effects.

The amount of dissipated and stored energy in the material due to a deformation process can be determined by interpreting the CLAUSIUS-DUHEM-inequality. This inequality results from the principle of irreversibility. For a detailed derivation see for example (Greve 2003) and (Haupt 1999).

The CLAUSIUS-DUHEM-inequality states that the dissipation $\mathcal{D}$ has to be positive for all conceivable thermomechanical processes:

$$\mathcal{D} = \frac{1}{\tilde{\rho}} P_{sp} - \dot{\psi} - \eta\dot{\theta} - \mathcal{D}_q \geq 0 \qquad (4)$$

The quantity P_{sp} is the stress power, η is the entropy, ψ is the free energy, $\tilde{\rho}$ is the density in the undeformed configuration and $\mathcal{D}_q$ denotes the dissipation due to heat conduction.

In the framework of rational thermodynamics material models have to be formulated in such a way that no thermomechanical process exists which violates the inequality 4. Material models that satisfy the CLAUSIUS-DUHEM-inequality for all thermomechanical processes are called thermodynamically consistent (Haupt 1999).

Usually the dissipation due to heat conduction $\mathcal{D}_q$ is demanded to be positive by itself:

$$\mathcal{D}_q := -\frac{1}{\tilde{\rho}\theta} \underline{Q} \cdot \tilde{\nabla}\theta \qquad (5)$$

$$= -\frac{1}{\rho\theta} \underline{q} \cdot \nabla\theta \geq 0 \qquad (6)$$

The quantities $\underline{Q}$ and $\underline{q}$ denote the KIRCHHOFF and CAUCHY heat flux vector, $\tilde{\nabla}\theta$ and $\nabla\theta$ the material and spatial temperature gradient, respectively.

The relation $\mathcal{D}_q \geq 0$ leads to the mechanical dissipation $\mathcal{D}_m$, which has to be positive:

$$\mathcal{D}_m = \frac{1}{\tilde{\rho}} P_{sp} - \dot{\psi} - \eta\dot{\theta} \geq 0 \quad . \qquad (7)$$

3 THREE DIMENSIONAL MATERIAL MODELS

Three dimensional material models are able to predict the material behavior for arbitrary loading processes. For these models the stress power P_{sp} can be expressed as (Ihlemann 2003)

$$P_{sp} = \frac{1}{2} \tilde{\underline{T}} \cdot\cdot \overset{\triangle}{\underline{C}} \qquad (8)$$

where $\tilde{\underline{T}}$ denotes the second PIOLA-KIRCHHOFF stress tensor and $\overset{\triangle}{\underline{C}}$ the material time derivative of the right CAUCHY-GREEN-Tensor. Based on the decomposition of the deformation gradient into a volumetric and an isochoric part (eq. 3) (Ihlemann 2003) derives the following relations:

$$\underline{C}^{-1} \cdot \overset{\triangle}{\underline{C}} = \left(\underline{C}^{-1} \cdot \overset{\triangle}{\underline{C}}\right)' + \frac{1}{3}\left(\underline{C}^{-1} \cdot\cdot \overset{\triangle}{\underline{C}}\right)\underline{E} \qquad (9)$$

$$= \overset{G}{\underline{C}}{}^{-1} \cdot \overset{\overset{\triangle}{G}}{\underline{C}} + \frac{2}{3}\frac{\dot{J_3}}{J_3}\underline{E} \quad . \qquad (10)$$

Hence the stress power can be expressed as

$$P_{sp} = \frac{1}{2}\tilde{\underline{T}} \cdot \underline{C} \cdot\cdot \left(\underline{C}^{-1} \cdot \overset{\triangle}{\underline{C}}\right)' + \tilde{\underline{T}} \cdot \underline{C} \cdot\cdot \frac{1}{3}\frac{\dot{J_3}}{J_3}\underline{E} \qquad (11)$$

$$= \frac{1}{2}\left(\tilde{\underline{T}} \cdot \underline{C}\right)' \cdot\cdot \overset{G}{\underline{C}}{}^{-1} \cdot \overset{\overset{\triangle}{G}}{\underline{C}} + \tilde{\underline{T}} \cdot \underline{C} \frac{1}{3}\frac{\dot{J_3}}{J_3} \quad . \qquad (12)$$

By defining the distortional second PIOLA-KIRCHHOFF-stress tensor $\overset{G}{\underline{T}}$ and the hydrostatic pressure p as

$$\overset{G}{\underline{T}} := \left(\tilde{\underline{T}} \cdot \underline{C}\right)' \cdot \overset{G}{\underline{C}}{}^{-1}, \qquad p := -\frac{1}{3J_3}\tilde{\underline{T}} \cdot\cdot \underline{C} \qquad (13)$$

the stress power can be rewritten as

$$P_{sp} = \overset{G}{\underline{T}} \cdot\cdot \overset{\overset{\triangle}{G}}{\underline{C}} - p\dot{J_3} = \overset{G}{P}_{sp} + \overset{V}{P}_{sp} \quad . \qquad (14)$$

A three dimensional material model is supposed to provide equations for the distortional part of the second PIOLA-KIRCHHOFF-stress tensor $\overset{G}{\underline{T}}$, the hydrostatic pressure p, the entropy η, the mechanical dissipation $\mathcal{D}_m$ and the heat flux vector $\underline{Q}$. These quantities can depend on the deformation, the deformation velocity, inner variables, the temperature, and the temperature gradient.

A detailed derivation of the following thermodynamically consistent thermoviscoelastic material model can be found in (Lion 2000). As an inner variable an inelastic right CAUCHY-GREEN-tensor $\underline{C}_i$ is introduced. The constitutive equations are as follows:

$$\overset{G}{\underline{T}} = 2C_{10}\frac{\theta}{\theta_0}\left[\overset{G}{\underline{C}}{}' \cdot \overset{G}{\underline{C}}{}^{-1} + c\left(\overset{G}{\underline{C}}_i{}^{-1} \cdot \overset{G}{\underline{C}}\right)' \cdot \overset{G}{\underline{C}}{}^{-1}\right]$$

$$p = -K\frac{1}{\varphi^3}\left(J_{3m} - 1\right)$$

$$\eta = -C_{10}\frac{1}{\theta_0}\left[\left(\overset{G}{\underline{C}} \cdot\cdot \underline{E} - 3\right) + c\left(\overset{G}{\underline{C}}_i{}^{-1} \cdot\cdot \overset{G}{\underline{C}} - 3\right)\right]$$

$$\qquad + 3\alpha_t K J_{3m}\left(J_{3m} - 1\right) + \eta_\theta$$

$$\mathcal{D}_m = 4\zeta c^2 C_{10}^2\left(\overset{G}{\underline{C}}_i{}^{-1} \cdot \overset{G}{\underline{C}}\right)' \cdot\cdot \left(\overset{G}{\underline{C}}_i{}^{-1} \cdot \overset{G}{\underline{C}}\right)'$$

$$\overset{\overset{\triangle}{G}}{\underline{C}}_i = 4\zeta c C_{10}\overset{G}{\underline{C}}_i \cdot \left(\overset{G}{\underline{C}}_i{}^{-1} \cdot \overset{G}{\underline{C}}\right)'$$

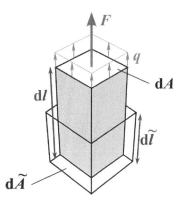

Figure 1. A cube under uniaxial loads.

The material parameters are C_{10}, c, K and ζ.

The heat conduction inequality 6 is satisfied by the FOURIER model of isotropic heat conduction (Haupt 1999):

$$\underline{q} = -\overline{k}\,\underline{\nabla}\theta \quad . \tag{15}$$

The material parameter $\overline{k}$ is called the coefficient of heat conduction.

4 ONE DIMENSIONAL MATERIAL MODELS

The concept of representative directions is able to generalise one dimensional material models. For mechanical material models one dimensional material models provide a stress for uniaxial tension processes. For thermomechanical material models it has to be clarified, which quantities have to be provided by the one dimensional material model.

Therefor a cube with length $d\tilde{l}$, area $d\tilde{A}$ and density $\tilde{\rho}$ in the unloaded configuration is considered.

The stress-free thermal expansion of the cube is accompanied by a change of the density. The expanded cube has the density $\hat{\rho}$ which can be computed by using the material function φ (eq. 2):

$$\hat{\rho} = \frac{1}{\varphi^3}\tilde{\rho} \tag{16}$$

The cube is loaded with a force F which leads to an elongation to length dl and a change of the density to ρ. As independent variables the density ratios J_3, J_{3m} and the distortional stretch λ are defined:

$$J_3 := \frac{\tilde{\rho}}{\rho}, \qquad J_{3m} := \frac{\hat{\rho}}{\rho}, \qquad \lambda := J_3^{-\frac{1}{3}}\frac{dl}{d\tilde{l}} \quad . \tag{17}$$

The power density P of the force is given by

$$P = \frac{1}{d\tilde{V}}F\dot{dl} = \frac{F}{d\tilde{A}}\frac{\dot{dl}}{d\tilde{l}} = \frac{F}{d\tilde{A}}\left(J_3^{\frac{1}{3}}\lambda\right)^{\cdot} \quad . \tag{18}$$

By defining the stress $\tilde{T}$ and the hydrostatic pressure p as

$$\tilde{T} := J_3^{\frac{1}{3}}\frac{1}{2\lambda}\frac{F}{d\tilde{A}}, \qquad p := -\frac{1}{3}J_3^{-\frac{2}{3}}\lambda\frac{F}{d\tilde{A}} \tag{19}$$

the power density can be written as

$$P = \frac{F}{d\tilde{A}}\left(J_3^{\frac{1}{3}}\lambda\right)^{\cdot} = \tilde{T}\left(\lambda^2\right)^{\cdot} - p\dot{J}_3 =: \overset{G}{P} + \overset{V}{P} \quad . \tag{20}$$

Obviously the power of the force can be (similarly to the three dimensional models) decomposed into a distortional part $\overset{G}{P}$ and a volumetric part $\overset{V}{P}$.

The lateral faces of the cube do not exchange any heat with the environment, i.e. they are adiabatic. Only the surface area exchanges heat by a heat flow density q which can be regarded as a component of the heat flux vector $\underline{q}$. Between the top and the bottom there is a temperature difference $d\theta$. By dividing $d\theta$ by the deformed length dl a one dimensional temperature gradient g_θ is obtained:

$$g_\theta := \frac{d\theta}{dl} \tag{21}$$

A one dimensional material model provides a stress $\tilde{T}$, a hydrostatic pressure p, an entropy η and a dissipation $\mathcal{D}_m$. These quantities can depend on the deformation (which is uniquely described by the stretch λ and the density ratio J_3), the deformation velocity, inner variables, the temperature and the temperature gradient. These constitutive equations are said to be thermomechanically consistent if they fulfill the inequalities

$$\mathcal{D}_m = \frac{1}{\rho}\tilde{T}\left(\lambda^2\right)^{\cdot} - \frac{1}{\rho}p\dot{J}_3 - \left(\dot{\psi} + \eta\dot{\theta}\right) \geq 0 \tag{22}$$

$$\mathcal{D}_q = -\frac{1}{\tilde{\rho}\theta}q\,g_\theta \qquad\qquad \geq 0 \quad . \tag{23}$$

The following thermoviscoelastic material model is equivalent to the three dimensional material model in section 3 for uniaxial tension:

$$\tilde{T} = 2C_{10}\frac{\theta}{\theta_0}\left[\left(1 - \frac{1}{\lambda^3}\right) + c\left(\frac{1}{\xi^2} - \frac{\xi}{\lambda^3}\right)\right]$$

$$p = -K\frac{1}{\varphi^3}\left(J_{3m} - 1\right)$$

$$\eta = 2C_{10}\frac{1}{\theta_0}\left[\left(\lambda^2 + \frac{2}{\lambda} - 3\right) + c\left(\frac{\lambda^2}{\xi^2} + 2\frac{\xi}{\lambda} - 3\right)\right]$$

$$+ 3\alpha_t K J_{3m}\left(J_{3m} - 1\right) + \eta_\theta$$

$$\mathcal{D}_m = \frac{8}{3}\zeta C_{10}^2 \frac{1}{c}\left(\frac{\lambda^2}{\xi^2} - \frac{\xi}{\lambda}\right)^2$$

$$\dot{\xi} = \frac{4}{3}\zeta C_{10}\xi\left(\frac{\lambda^2}{\xi^2} - \frac{\xi}{\lambda}\right)^2$$

The material parameters are C_{10}, c, K and ζ.

The inequality 23 is fulfilled by the one dimensional FOURIER model of heat conduction:

$$q = -k g_\theta \quad . \tag{24}$$

5 THE CONCEPT OF REPRESENTATIVE DIRECTIONS

The basis of the concept is the consideration of an infinitesimal neighborhood $d\tilde{\Omega}$ of a material point. A material line $d\tilde{r}$ of this volume element is mapped to a deformed material line dr by the deformation gradient:

$$dr = \underline{\underline{F}} \cdot d\tilde{r} \quad . \tag{25}$$

The distortional elongation of this material line can be defined as (cf. (Freund and Ihlemann 2010))

$$\lambda := J_3^{-\frac{1}{3}}\frac{|dr|}{|d\tilde{r}|} = \sqrt{\frac{d\tilde{r}}{\sqrt{d\tilde{r}\cdot d\tilde{r}}}\cdot\underline{\underline{G}}\cdot\frac{d\tilde{r}}{\sqrt{d\tilde{r}\cdot d\tilde{r}}}} \quad . \tag{26}$$

The normalized material lines $d\tilde{r}$ can be uniquely identified by an ordered pair of solid angles $\alpha := (\vartheta, \phi)$. This motivates the following definitions:

$$\overset{\alpha}{\underline{e}} := \frac{d\tilde{r}}{\sqrt{d\tilde{r}\cdot d\tilde{r}}} \quad\Rightarrow\quad \lambda(\alpha) =: \overset{\alpha}{\lambda} = \sqrt{\overset{\alpha}{\underline{e}}\cdot\overset{G}{\underline{\underline{C}}}\cdot\overset{\alpha}{\underline{e}}}$$

Furthermore a one dimensional temperature gradient of the deformed material line dr is introduced:

$$\overset{\alpha}{g}_\theta := \overset{\alpha}{\underline{e}}\cdot\nabla\theta \quad . \tag{27}$$

The volume element $d\tilde{\Omega}$ is divided into volume segments which are characterized by an ordered pair of solid angles α. Each of the volume segments is stretched by $\overset{\alpha}{\lambda}$ according to eq. 26 and is subjected to a temperature gradient $\overset{\alpha}{g}\theta$ according to eq. 27. All of the volume elements are subjected to the same temperature θ and the same density ratio J_3. Hence, for every volume element the one dimensional material model can provide a stress $\overset{\alpha}{\tilde{T}}$, a hydrostatic pressure $\overset{\alpha}{p}$, an entropy $\overset{\alpha}{\eta}$, a dissipation $\overset{\alpha}{\mathcal{D}}_m$, and a heat flux $\overset{\alpha}{q}$ that satisfy the inequalities 22 and 23.

For mechanical material models the concept of representative directions demands the equivalence of stress powers (cf. (Freund and Ihlemann 2010)). For thermomechanical material models the generalisation of the one dimensional material model arises from

the postulation that the mechanical dissipation of the unknown three dimensional material model is equivalent to the integral over all mechanical dissipations of all volume segments, i.e.

$$\mathcal{D}_m = \frac{1}{4\pi}\int_0^{2\pi}\int_0^\pi \overset{\alpha}{\mathcal{D}}_m \sin\vartheta d\vartheta d\phi \quad . \tag{28}$$

The insertion of the CLAUSIUS-DUHEM-Inequalities 7 and 22, using the relation

$$\left(\overset{\alpha}{\lambda}^2\right)^{\cdot} = \overset{\alpha}{\underline{e}}\circ\overset{\alpha}{\underline{e}}\cdot\cdot\overset{\triangle}{\underline{\underline{C}}}^G \tag{29}$$

and reordering leads to

$$0 = \left(\overset{G}{\underline{\underline{T}}} - \frac{1}{4\pi}\int_0^{2\pi}\int_0^\pi \overset{\alpha}{\tilde{T}}\overset{\alpha}{\underline{e}}\circ\overset{\alpha}{\underline{e}}\sin\vartheta d\vartheta d\phi\right)\cdot\cdot\overset{\triangle}{\underline{\underline{C}}}^G$$

$$+ \left(-p + \frac{1}{4\pi}\int_0^{2\pi}\int_0^\pi \overset{\alpha}{p}\sin\vartheta d\vartheta d\phi\right)\dot{J}_3$$

$$+ \tilde{\rho}\left(-\dot{\psi} + \frac{1}{4\pi}\int_0^{2\pi}\int_0^\pi \overset{\alpha}{\dot{\psi}}\sin\vartheta d\vartheta d\phi\right)$$

$$+ \left(-\eta + \frac{1}{4\pi}\int_0^{2\pi}\int_0^\pi \overset{\alpha}{\eta}\sin\vartheta d\vartheta d\phi\right)\dot{\theta} \tag{30}$$

By demanding that for all conceivable thermomechanical processes equation 30 holds, the following relations can be deduced:

$$\overset{G}{\underline{\underline{T}}} = \left(\underline{\underline{\tilde{T}^*}}\cdot\overset{G}{\underline{\underline{C}}}\right)'\cdot\overset{G}{\underline{\underline{C}}}^{-1} \tag{31}$$

with

$$\underline{\underline{\tilde{T}^*}} = \frac{1}{4\pi}\int_0^{2\pi}\int_0^\pi \overset{\alpha}{\tilde{T}}\overset{\alpha}{\underline{e}}\circ\overset{\alpha}{\underline{e}}\sin\vartheta d\vartheta d\phi$$

$$p = \frac{1}{4\pi}\int_0^{2\pi}\int_0^\pi \overset{\alpha}{p}\sin\vartheta d\vartheta d\phi \tag{32}$$

$$\eta = \frac{1}{4\pi}\int_0^{2\pi}\int_0^\pi \overset{\alpha}{\eta}\sin\vartheta d\vartheta d\phi \quad . \tag{33}$$

The postulation 28 leads to thermomechanical consistency. Due to the thermomechanical consistency of the one dimensional material model the integrand $\overset{\alpha}{\mathcal{D}}_m$ is nonnegative for all α. The dissipation $\mathcal{D}_m$ of the three dimensional material model, defined as an integral over a nonnegative function, is therefore always nonnegative, i.e. the generalised material model is thermomechanically consistent.

The constitutive model for the heat conduction can be deduced similarly: the dissipation due to heat transfer of the three dimensional model is defined as the integral over all dissipations of all volume segments, i.e.

$$\mathcal{D}_q = \frac{1}{4\pi} \int_0^{2\pi} \int_0^\pi \overset{\alpha}{\mathcal{D}}_q \sin\vartheta \mathrm{d}\vartheta \mathrm{d}\phi \quad . \qquad (34)$$

The insertion of the one dimensional temperature gradient 27 and the FOURIER-ansatz 24 leads to the following relation:

$$\underline{q} = -\frac{k}{3}\underline{\nabla}\theta \quad . \qquad (35)$$

Obviously the generalisation of the one dimensional FOURIER-ansatz leads to the three dimensional FOURIER-model of heat conduction.

The integrals 28, 31, 32 and 33 are generally not analytically computable, so that a numerical integration has to be performed. The integral is approximated as a weighted sum of the integrand at discrete solid angles α_i. For an arbitrary function $\overset{\alpha}{g}$ the integration algorithm is

$$\frac{1}{4\pi} \int_0^{2\pi} \int_0^\pi \overset{\alpha}{g} \sin\vartheta \mathrm{d}\vartheta \mathrm{d}\phi \approx \sum_{i=1}^{n_\alpha} \overset{\alpha_i}{g} \overset{\alpha_i}{w} \qquad (36)$$

Nodes α_i and corresponding weighting factors $\overset{\alpha i}{w}$ are given for example in (Bažant and Oh 1986) and (Freund and Ihlemann 2010).

6 EXAMPLES

The concept of representative directions is implemented in the commercial FE-code MSC.MARC via the user subroutine *Hypela2*. The thermomechanically coupled problem is solved by a staggered solution algorithm with an isothermal split (see (Armero and Simo 1992) for details).

(Freund and Ihlemann 2010) point out that the mechanical material response of three dimensional material models and their generalised one dimensional versions show remarkable similarities if the material parameters are chosen appropriately. To assure that the thermomechanical material responses of the three dimensional material model and the generalised one dimensional material model are similar a parameter fit has been performed. Because of the small number of material parameters only a qualitatively similar material response can be achieved.

To compare the material models a simulation of an adiabatic stretching of a thermoelastic material

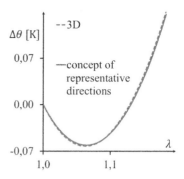

Figure 2. Gough-Joule-Effect.

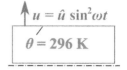

Figure 3. Model of a cyclically loaded disk.

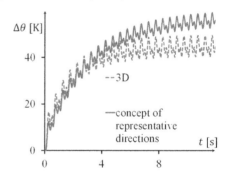

Figure 4. Dissipative heating of a disk.

has been performed. The typical temperature response is known as the GOUGH-JOULE-effect. For small stretches the change in temperature is negative, for larger deformations the material heats due to entropy elastic effects. For a detailed description of this effect see e.g. (Lion 2000).

The graph 2 shows the temperature change for the three dimensional material model and the generalised one dimensional model at different stretches λ.

It can be seen that the GOUGH-JOULE-effect is well described by the concept of representative directions. Due to the small deformation even a good quantitative agreement is achieved.

To consider inelastic effects a disk with thermo-viscoelastic material behavior is loaded by a cyclic displacement. The boundary conditions are illustrated in Fig. 3.

Fig. 4 shows the temperature change at the symmetry plane for the three dimensional material model and the generalised one dimensional model.

Due to the dissipative effects the disk heats up. After a transient start the heat conduction due to the temperature boundary condition at the top of the disk and the dissipative heating are in balance and a stationary state arises. Obviously the generalised one dimensional material model is able to describe qualitatively the same effects as the three dimensional material model. The dissipative heating is as well reproduced as the thermoelastic effects which lead to local temperature extrema. Due to the large deformations a quantitative agreement can not be achieved.

7 CONCLUSION

In the presented work the concept of representative directions, which is able to generalise one dimensional mechanical material models, is adapted to thermomechanically coupled material models. It is shown that thermomechanical consistency of the one dimensional model leads to thermomechanical consistency of the three dimensional model.

Although only a one dimensional material model is required as an input this generalisation technique provides all necessary quantities for arbitrary loading conditions: the second PIOLA-KIRCHHOFF stress tensor, the heat flux vector, the entropy, and the dissipation.

The concept of representative directions is implemented in the commercial FE-program MSC.Marc. Simulations show that typical thermomechanical effects are reproduced.

REFERENCES

Armero, F. and J. C. Simo (1992). A new unconditionally stable fractional step method for non-linear coupled thermomechanical problems. *International Journal for Numerical Methods in Engineering 35*, 737–766.

Bažant, P. and B. H. Oh (1986). Efficient numerical integration on the surface of a sphere. *ZAMM – Journal of Applied Mathematics and Mechanics 66*, 37–49.

Freund, M. and J. Ihlemann (2010). Generalization of one-dimensional material models for the finite element method. *ZAMM – Journal of Applied Mathematics and Mechanics 90*, 399–417.

Greve, R. (2003). *Kontinuumsmechanik*. Berlin/Heidelberg: Springer.

Haupt, P. (1999). *Continuum Mechanics and Theory of Materials*. Springer.

Ihlemann, J. (2003). *Kontinuumsmechanische Nachbildung hochbelasteter technischer Gummiwerkstoffe*. Düsseldorf: VDI-Verlag.

Lion, A. (2000). *Thermomechanik von Elastomeren*. Habilitation, Universität Gesamthochschule Kassel.

Lu, S. C. H. and K. S. Pister (1975). Decomposition of deformation and representation of the free energy function for isotropic thermoelastic solids. *International Journal of Solids and Structures 11*, 927–934.

Miehe, C., S. Göktepe, and F. Lulei (2004). A micro-macro approach to rubber-like materials. part i: the non-affine micro-sphere model of rubber elasticity. *Journal of the Mechanics and Physics of Solids 52*, 2617–2660.

Constitutive Models for Rubber VII – Jerrams & Murphy (eds)
© 2012 Taylor & Francis Group, London, ISBN 978-0-415-68389-0

Chemical ageing of elastomers: Experiments and modelling

Michael Johlitz, Johannes Retka & Alexander Lion

Institute of Mechanics, Faculty of Aerospace Engineering, Universität der Bundeswehr München,
Neubiberg, Germany

ABSTRACT: If typical elastomer components like tires, belts, sealings, suspension or engine mounts are exposed to environmental conditions, complicated ageing phenomena take place which can lead to embrittlement, swelling or shrinkage, stiffening or softening or to changes in other material properties. While physical ageing is a so-called thermoreversible phenomenon when the aged polymer is heated above the glass transition temperature, the current project investigates irreversible chemical ageing behaviour. In this case, the elastomer degenerates and changes its chemical structure in the aged regions. As model material, a commercial elastomer which is applied in the shipping industry is investigated. In order to reproduce marine environmental conditions in the laboratory, a series of long-term experiments under different temperatures with respect to the medium sea water with and without deformation were carried out. To examine the mechanical behaviour of the aged elastomer specimens in terms of their stress-strain characteristics, continuous and intermittent relaxation tests were performed. In order to represent the experimentally observed material behaviour in the field of a phenomenological material theory, a continuum mechanical approach has been applied which is able to describe the ageing phenomena with respect to the different boundary conditions. The constitutive equations and evolution equations are applied in the framework of a finite thermoelastic theory. Based on this approach, a series of finite element simulations is shown and the results are physically interpreted in the context of the experimental data.

1 INTRODUCTION

With the advance of technology, products become more and more complex. If we consider a modern product like a car or an aircraft, one can observe that not only the formerly widely separated areas of engineering (mechanical engineering, electrical engineering, electronics, control technology, etc.) move together and interlock, but also the materials and their development play a crucial role. Cars and aircrafts shall become lighter but saver without impacting the environment too much and without endangering the occupants.

In particular, the manufacturer has to guarantee the desired properties of the product over the whole application period, the so-called life time. In order to develop the best product in this respect, the material behaviour must be known with sufficient accuracy. Especially, ageing studies are increasingly becoming the focus of attention.

Referred to the term ageing, one understands all the chemical and physical changes of the material with time that determine the measurable properties. In the case that the mechanical loads are of primary interest and the number of loading cycles appropriates for the ageing of the material, one speaks of fatigue (Flamm et al. 2011). In contrast to this, there is the chemical and physical ageing of materials. This implies, for example, the influence of temperature profiles, weather and other environmental effects like radiation (Tobolsky 1967; Shaw et al. 2005; Duarte and Achenbach 2007) on the mechanical behaviour of materials during their period of application.

Physical ageing processes are thermally reversible and relevant for thermoplastic polymers usually from below the glass transition temperature T_g. These materials are in a non-equilibrium state resulting from their production process, and relax to an equilibrium state over time (Schöenhals and Donath 1986; Perez et al. 1991; Hodge 1995; Perera 2003). Thus, they change their mechanical properties as well as their thermal expansion behaviour. A full healing occurs when the material is again heated above its glass transition temperature (Struik 1976; Perera 2003). Likewise, also reversible relaxation processes that are rooted in the viscoelastic behaviour of polymers are referred to as physical ageing phenomena.

In comparison to this, chemical ageing is an irreversible degenerative process. This process changes the chemical structure of the molecules and can not be undone by heating (Hutchinson 1995; Ehrenstein and Pongratz 2007). On the one hand, chemical bonds in the form of cross-linked nodes are destroyed and also chain scission can occur. On the other hand new chemical bonds are created. These processes are described, particularly in view of the thermo-oxidative ageing, under the terms of network degradation and network formation that can run simultaneously. That means thermo-oxidative ageing conceives the temperature-dependent ageing in the presence of oxygen which includes the above mentioned chain scission and cross-linking processes, see (Blum et al. 1951; Tobolsky 1967) or (Shaw et al. 2005). Since elastomers are generally used above T_g, the corresponding chemical ageing processes take place on much larger time scales than the physical relaxation processes (Budzien et al. 2008). This makes it clear that ageing of components of finite thickness depends on the diffusion process and the associated oxygen transport. To be able to assume a homogeneous distribution of oxygen and in order to neglect the diffusion in a first step only thin specimens are investigated and modelled in this paper, which are saturated with oxygen, i.e. (Blum et al. 1951). Otherwise, the diffusion of oxygen would have to be taken into account (Shaw et al. 2005).

The present work contributes to the large field of experimental-based material modelling with respect to the estimation of suitability and life time.

2 EXPERIMENTAL INVESTIGATIONS

To detect the chemical ageing behaviour of materials as a function of temperature and exposure in various media, there are well established experiments and detailed descriptions which can be found in literature: the continuous and intermittent relaxation tests. The central idea and the physical motivation of these tests is going back to the contributions of (Andrews et al. 1946; Scalan and Watson 1957; Dunn et al. 1959; Ore 1959; Tobolsky 1967; Smith 1993).

The continuous relaxation test corresponds to the classical relaxation test, as practiced in viscoelastic materials. Here it takes place on much longer timescales, typically several weeks or months. Temperature and mechanical strain are kept constant over the entire testing period. The experiment is used to detect the network degradation described in the introduction. This process can be thermally accelerated and later extrapolated by using the well-known Arrhenius function (Tobolsky 1967; Shaw et al. 2005; Duarte and Achenbach 2007). The stress increases monotonically during the whole experiment, except that the network formation process implies shrinkage (Andrews et al. 1946).

The intermittent relaxation test is more complicated. First of all, the examined samples need to be stored stress-free and isothermally over the testing period in the ageing medium. At defined intervals, the samples are subjected to mechanical relaxation tests on much shorter time scales. Thereby, the distribution of the overstress caused by the transient network and the evolution of the equilibrium stress, which reflects a measure of the basic elasticity of the material, are of interest. At this both the network formation and the network degradation play a role. Since the two phenomena chain scission and formation of new chemical bonds run simultaneously, the material can become stiffer or softer while performing the experiments (Andrews et al. 1946; Tobolsky et al. 1967).

 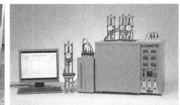

Figure 1. Establishment of the equipment for measuring continuous long-time relaxation tests in sea water under isothermal conditions (right) and detailed view of a rig (left).

2.1 Experimental setup

To carry out the ageing studies in various media and under different thermal conditions, a testing device from the company Elastocon was purchased by the Institute of Mechanics. This device consists of two cell ovens, each with three chambers. Therein different temperatures can be adjusted over a testing period with an accuracy of 0.5°C (Figure 1, right). The chambers are constructed so that they are able to contain a number of S2-dogbone specimens for stress-free ageing. The temperature range is between room temperature and 300°C. The chambers are sealed fairly airtight and can be filled with various media, such as water, sea water, air or something else. Alternatively, the chambers of the cell oven can be equipped with so-called "rigs" (see Figure 1, left), making it possible to investigate the ageing behaviour in the medium under mechanical loads. Normally, the device is able to perform both tensile deformations and compressive deformations. During the testing period, both the force signal and the temperature signal are read and recorded via a control computer which is connected to the device.

2.2 Continuous relaxation tests

Carrying out the continuous relaxation tests makes it possible to investigate the ageing behaviour of the specimens under mechanical loads. Therefor, the S2 samples are clamped on mechanical sample holders incorporated in the rigs. The rigs are filled with the already preheated sea water. Then, they are sealed and placed in the cell oven. An additional sensor in the fluid ensures the temperature control of the sample chamber, such that isothermal conditions can be guaranteed. The experiment and data acquisition for three concurrent continuous ageing tests is started by software. The record of the data is implemented in such way that it fits into the semilogarithmic scheme. Now, a deformation of 25% is applied over a screw mechanism to the specimen and both the force-time signal and the temperature-time signal of the various isothermal ageing experiments are recorded over a period of 12 weeks.

With respect to Figure 2 there are clearly two regions to distinguish. In the first region (I), i.e. up to 10^4 s, an ageing mechanism is predominant, leading to a

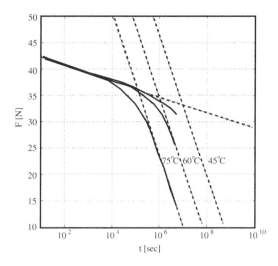

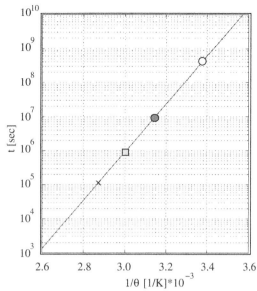

Figure 2. Long-term relaxation tests of NBR 55 in sea water with respect to different isothermal temperatures.

Figure 3. Assessment of the lifetime of NBR 55 in sea water based on the activation points at different isothermal temperatures.

slow change in properties. In the relaxation test, it is obvious that this is a physical ageing process due to the new-arrangement and rearrangement of molecular chains. In the second region (II), i.e. over 10^4 s, an ageing mechanism is superimposed that occurs after an induction time. In our case this may be interpreted as an oxidation reaction. This instance leads to chain scission and results in a reduction of the stress which can be absorbed by the network. It is possible to determine the chemical activation time t_s with respect to Figure 2. The activation point represents the time where the second ageing mechanism is starting to be more pronounced than the first. It is clear to see that the ageing mechanism in the region (II) is a thermally activated reaction, since the activation time t_s decreases by using higher temperatures. The thermal-based activation times are plotted in Figure 3. These can be related by the Arrhenius relationship

$$t_s(\theta) = t_{s0} \exp\left(\frac{k}{\theta}\right). \tag{1}$$

Thus, the measurement results of the different activation times can be extrapolated to a range of temperatures, as visible in Figure 3. In doing so $t_s(\theta)$ describes the relation between the reciprocal of the thermodynamic temperature θ and the activation time t_s with the constants t_{s0} and k. Taking this approach into account, and the previously mentioned restrictions, a time scaling can be achieved. The validity of this extrapolation depends on the experimental and theoretical foundations, or prior knowledge, working in the present case properly.

2.3 Intermittent relaxation tests

The intermittent relaxation tests are taken in order to study the chemical ageing behaviour of the load-free

specimens. The procedure can be described as follows: The S2-samples are measured in cross-section and then stored isothermally in sea water at 45°C, 60°C and 75°C over a period of 10 weeks. Now, for each temperature a sample is taken out of the chambers of the cell oven at defined time intervals. The samples are tested in the Zwick testing machine by performing relaxation experiments under the same isothermal conditions as in the sea water. For this purpose, a constant deformation of 25% is applied to the specimen by using a displacement rate of 10 mm/min. Then, the stress-time behaviour is recorded over a period of one hour. The duration of the experiment is chosen as short that only physical ageing processes in the form of stress relaxation occur. Chemical ageing does not proceed during this short time and may be neglected. As stress measurement, the engineering stress $\sigma = F/A_0$ is taken into account. From the recorded data, the non-equilibrium stress (viscoelastic component) σ_{neq} and the equilibrium stress (basic elasticity) σ_{eq} are separated and plotted in a diagram as a function of the ageing time and the temperature, see Figure 4.

The results show a temperature- and ageing-dependent increase in the equilibrium stress while the overstress remains almost unaffected.

3 MODELLING APPROACH

In the literature there are only a few works which deal with the modelling of the chemical ageing behaviour. Many of them are based on the idea of (Andrews et al. 1946) and take the acceptance of simultaneously running network degradation and creation into account. In addition, incompressible material behaviour is

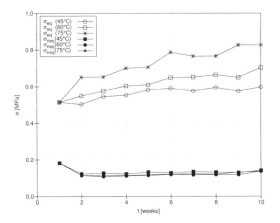

Figure 4. Equilibrium stress σ_{eq} and non-equilibrium (viscoelastic) stress σ_{neq} of the intermittent relaxation tests at different ageing times and temperatures.

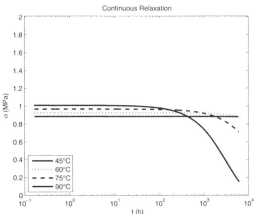

Figure 5. Long-term relaxation experiments at different isothermal temperatures.

assumed, i.e. shrinkage is neglected. These approaches were enhanced, e.g. by (Septanika and Ernst 1998a; Septanika and Ernst 1998b), in order to describe the rate-independent ageing behaviour. Therefor, the stress was formulated as a rate-independent functional of the continuous time variable. To capture the network formation process with respect to a finite strain theory, a rate-independent hypoelasticity approach is used in analogy to (Hossain et al. 2008; Hossain et al. 2009).

3.1 Continuous ageing tests

The derivation of the constitutive equations for a three-dimensional, thermomechanical consistent material model concerning ageing can be found in (Lion and Johlitz 2011). In the present paper, only the equations of the continuous uniaxial relaxation tests are provided:

$$
\begin{aligned}
\sigma(t) &= \rho_R \frac{\theta}{\theta_R}(1 - q(t))\left[c_{10}\left(\lambda_0^2 - \frac{1}{\lambda_0}\right)\right] \\
&+ \rho_R \frac{\theta}{\theta_R}(1 - q(t))\left[c_{20}\left(\lambda_0 - \frac{1}{\lambda_0^2}\right)\right]
\end{aligned}
\tag{2}
$$

Here, a material model of the Mooney-Rivlin type is recognised as a stress-strain relationship. The constant stretch $\lambda_0 = 1.3$ is applied on the sample at the beginning of the experiment, $\rho_R = 1000\,\text{kg/m}^3$ and θ_R stand for the reference density and reference temperature and θ is the value of the current temperature. The parameters c_{10} and c_{20}, respectively, characterise the virgin Mooney-Rivlin material. The network degradation during the continuous tests is described as a function of time using the equation

$$
q(t) = 1 - e^{-\left(\nu_q e^{-\frac{E_q}{R\theta}}\right)t}
\tag{3}
$$

In addition, the model parameters ν_q, E_q and the gas constant $R = 8.314$ J/molK are introduced. The creation of a second network can not be captured with

the continuous experiment, since it remains free of stress. In Figure 5 various simulations of the continuous ageing test at different isothermal temperatures in the range of $\theta = \theta_R = 45°$C up to $\theta = 90°$C are shown. The corresponding model parameters can be found in Table 1.

It can be clearly seen that the network degradation at constant deformation and the resulting decrease of the stress are temperature dependent. Thus, the model is able to reflect the experimentally observed effects in a right manner and allows to predict the operating time of the elastomer through an extrapolation with the Arrhenius function.

3.2 Intermittent relaxation tests

In the case of stress-free ageing and the associated intermittent relaxation tests, the model is used in its full dimension, i.e. the stress-strain relationship is formulated in that way that both the degradation and the assembling of the polymer network can be described.

$$
\begin{aligned}
\sigma(t_a) &= \rho_R \frac{\theta}{\theta_R}\left[c_1(t_a)\left(\lambda_0^2 - \frac{1}{\lambda_0}\right)\right] \\
&+ \rho_R \frac{\theta}{\theta_R}\left[c_2(t_a)\left(\lambda_0 - \frac{1}{\lambda_0^2}\right)\right]
\end{aligned}
\tag{4}
$$

This equation again describes a material of the Mooney-Rivlin type. The corresponding equations of the parameters c_1 and c_2 are formulated as functions

Table 1. Model parameters used in this contribution.

c_{10} [J/kg]	c_{20} [J/kg]	E_q [J/mol]	E_p [J/mol]
0.0005	0.0005	30000	30000
g_0^A [J/kg]	g_0^B [J/kg]	ν_q [1/s]	ν_p [1/s]
0.0007	0.0007	10^{11}	10^{11}

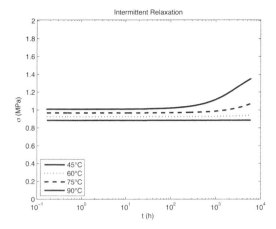

Figure 6. Intermittent relaxation tests at constant temperatures.

of the ageing time t_a with respect to the network creation $p(t_a)$ and the network degradation $q(t_a)$ as follows (Lion and Johlitz 2011):

$$
\begin{aligned}
c_1(t_a) &= (1 - q(t_a))c_{10} + p(t_a)g_0^A \\
c_2(t_a) &= (1 - q(t_a))c_{20} + p(t_a)g_0^B
\end{aligned}
\tag{5}
$$

Therein the parameters c_{10} and c_{20} stand again for the virgin material, the parameters g_0^A and g_0^B describe the creation of the network architecture in conjunction with $p(t_a)$ in terms of the Mooney-Rivlin model. Depending on the choice of parameters, both stiffening and softening effects can be described. If one assumes that the deformation application process is short compared with the ageing process during the measurement period, then the functions of the network formation and degradation are considered to be independent of time and satisfy the following equations:

$$
\begin{aligned}
q(t) \approx q(t_a) &= 1 - e^{-\left(\nu_q e^{-\frac{E_q}{R\theta}}\right)t_a} \\
p(t) \approx p(t_a) &= 1 - e^{-\left(\nu_p e^{-\frac{E_p}{R\theta}}\right)t_a}
\end{aligned}
\tag{6}
$$

Within this goal, the additional parameters ν_p, E_p are introduced. To carry out the simulation, the model parameters from the continuous experiment are now complemented by the additional model parameters. A suitable choice of the additional model parameters, see Table 1, now ensures that the experimentally observed stiffness increase can be described as a function of temperature θ as can be seen in Figure 6.

4 SUMMARY AND CONCLUSIONS

In this paper the reader is familiarised with the fundamentals of the thermo-oxidative ageing of elastomers. After a brief motivation and literature review different testing procedures to capture the chemical

ageing behaviour of elastomers are presented. Here particularly the difference between continuous and intermittent experiments is explained. After describing the experimental testing device, experimental data is provided and discussed, which show the different mechanisms of the thermo-oxidative ageing processes taking place during the investigations. The temperature shift principle according to Arrhenius allows for an extrapolation of the data in order to predict the associated lifetime of the material. On the theoretical side, the equations of a material model to describe the chemical ageing are presented (Lion and Johlitz 2011). The model is able to reproduce the experimentally observed effects. As an outlook, the expansion of the model according to the finite thermoviscoelastic ageing behaviour of the materials has to be applied. Therefore, also the execution of the related experiments has to be done. Finally, another important aspect is the coupling of the thermal mechanisms with the diffusion process.

REFERENCES

Andrews, R., A. Tobolsky, and E. Hanson (1946). The theory of permanent set at elevated temperatures in natural and synthetic rubber vulcanizates. *Journal of Applied Physics 17*, 352–361.
Blum, G., J. Shelton, and H. Winn (1951). Rubber oxidation and ageing studies. *Industrial and Engineering Chemistry 43* (464–471).
Budzien, J., D. Rottach, J. Curro, C. Lo, and A. Thompson (2008). A new constitutive model for the chemical ageing of rubber networks in deformed states. *Macromolecules 41*, 9896–9903.
Duarte, J. and M. Achenbach (2007). On the modelling of rubber ageing and performance changes in rubbery components. *Kautschuk Gummi Kunststoffe 60*, 172–175.
Dunn, J., J. Scalan, and W. Watson (1959). Stress relaxation during the thermal oxidation of vulcanized natural rubber. *Transactions of the Faraday Society 55*, 667–675.
Ehrenstein, G. and S. Pongratz (2007). *Beständigkeit von Kunststoffen.* Carl Hanser Verlag.
Flamm, M., J. Spreckels, T. Steinweger, and U. Weltin (2011). Effects of very high loads on fatigue life of NR elastomer materials. *International Journal of Fatigue*, in press.
Hodge, I. (1995). Physical ageing in polymer glasses. *Science 267*, 1945–1947.
Hossain, M., G. Possart, and P. Steinmann (2008). A small-strain model to simulate the curing of thermosets. *Computational Mechanics 43*, 769–779.
Hossain, M., G. Possart, and P. Steinmann (2009). A finite strain framework for the simulation of polymer curing. Part I: elasticity. *Computational Mechanics 44*, 621–630.
Hutchinson, J. (1995). Physical ageing of polymers. *Prog. Polym. Sci. 20*, 703–760.
Lion, A. and M. Johlitz (2011). On the representation of chemical ageing of rubber in continuum mechanics. *Int. J. Solids Structures*, under review.
Ore, S. (1959). A modification of the method of intermittent stress relaxation measurements on rubber vulcanisates. *Journal of Applied Polymer Science 2*, 318–321.
Perera, D. (2003). Physical ageing of organic coatings. *Progress in Organic Coatings 47*, 61–76.
Perez, J., J. Cavaille, R. Calleja, J. Ribelles, M. Pradas, and A. Greus (1991). Physical ageing of amorphous

polymers – theoretical analysis and experiments on polymethylmethacrylate. *Die Makromolekulare Chemie 192*, 2141–2161.

Scalan, J. and W. Watson (1957). The interpretation of stress relaxation measurements made on rubber during ageing. *Transactions of the Faraday Society 54*, 740–750.

Schönhals, A. and E. Donth (1986). Analyse einiger Aspekte der physikalischen Alterungv on amorphen Polymeren. *Acta Polymerica 37*, 475–480.

Septanika, E. and L. Ernst (1998a). Application of the network alteration theory for modelling the time-dependent constitutive behaviour of rubbers. Part I. General theory. *Mechanics of Materials 30*, 253–263.

Septanika, E. and L. Ernst (1998b). Application of the network alteration theory for modelling the time-dependent constitutive behaviour of rubbers. Part II. Further evaluation of the general theory and experimental verification. *Mechanics of Materials 30*, 265–273.

Shaw, J., S. Jones, and A. Wineman (2005). Chemorheological response of elastomers at elevated temperatures: experiments and simulations. *Journal of the Mechanics and Physics of Solids 53*, 2758–2793.

Smith, L. (1993). *The language of rubber: an introduction to the specification and testing of elastomers*. Butterworth-Heinemann publication house.

Struik, L. (1976). Physical ageing in amorphous glassy polymers. *Annals of the New York Academy of Sciences 279*, 78–85.

Tobolsky, A. (1967). *Mechanische Eigenschaften und Struktur von Polymeren*. Berliner Union Stuttgart.

Constitutive Models for Rubber VII – Jerrams & Murphy (eds)
© 2012 Taylor & Francis Group, London, ISBN 978-0-415-68389-0

A stable hyperelastic model for foamed rubber

M.W. Lewis & P. Rangaswamy
Los Alamos National Laboratory, Los Alamos, New Mexico, USA

ABSTRACT: A hyperelastic strain energy function for foamed rubber that is based on the physical features of a typical uniaxial compression curve for foams and the behavior of a spherical pore in a spherical, incompressible Mooney-Rivlin matrix is presented. The model is unconditionally stable when positive moduli are used, and can represent most foam test data, including variable Poisson behavior during compression. This most general model has six parameters, including four moduli, a dimensionless parameter associated with buckling or plateau strain, and an initial porosity. The model will be described and its ability to fit compression response data for PDMS foams over a range of relative densities will be evaluated. Several potential extensions of this model will be discussed, including representation of the Mullins effect, representation of aging behavior, inclusion of matrix compressibility, and modeling the effect of pore gas response.

1 INRODUCTION

Elastomeric foams, or foamed rubbers, comprise a class of materials that are useful in many engineering applications. Because of their relatively low densities, they offer benefits at low component weight. Because of low initial stiffness and ability to undergo large, fully recoverable elastic deformations, the materials work well as impulse limiters and cushions that lower stresses that result from differential thermal expansion or contraction of nested components.

Current efforts at mechanical modeling of foamed rubbers can be divided into two approaches. The first of these approaches consists of attempts to understand the mechanics of foam deformation by modeling realizations of unit cells or larger volumes of foam with mesh discretizations that represent actual porosity and matrix morphology. The second of these approaches consists of phenomenologically-based constitutive model development, fitting, and evaluation.

The first of these approaches, which may be seen in work by Kraynik *et al.* (1998), Braydon *et al.* (2005), and a few others, will be referred here to as Direct Numerical Simulation (DNS) investigations. These DNS investigations are often limited by mesh resolution, mesh distortion, and contact mechanics issues. Additionally, the process of moving from DNS to component- or system-level modeling is not yet developed well.

The second of these approaches, of which the current work is an example, is represented by the work of Jemiolo and Turtletaub (2000). The most commonly used hyperelastic model for foamed rubbers in finite element analysis is the model proposed by Jemiolo and Turtletaub, a generalization of the Ogden (1984) strain energy function in isochoric principal stretches. That

model has the advantage that it can reproduce test data with good accuracy. It has disadvantages that include possible instability and high sensitivity of model fit parameters to data perturbation and fitting procedure. As a result, it is impractical to represent variability in mechanical behavior with model parameter distributions. The work of Danielsson *et al.* (2004) is also an example of a developed hyperelastic model for these materials, but is limited to lower porosity foams as it does not capture the classic plateau behavior observed in foams with low relative density.

This work describes a proposed new model for foamed rubber. The model was developed to provide both accurate fits to test data and model parameters that can be varied sensibly to represent material variability. The model consists of a strain energy function that was manufactured to reproduce foam compression behavior at moderate porosities, namely an initial Poisson's ratio, a nearly linear region at small strains, an inflection point where the material tangent compressive modulus drops significantly, and a strongly stiffening region as porosity is eliminated. These first three features were introduced phenomenologically, that is without a micromechanical motivation.

The stiffening region response was derived using an approach similar to Danielsson *et al.* (2004) for evaluating the strain energy of a void in an incompressible material subjected to far-field hydrostatic compression and a volume-conserving deformation. Our only extension to this part of the model is the calculation of the integrated second isochoric invariant for this deformation.

This model has been coded as a material model subroutine for use with a non-linear static finite element program and has been used to model a validation problem involving large compressive strains combined with substantial torsional deformation.

The following sections of this paper consist of a model description section detailing the form of the model and the parameter fitting process used here, an example section in which the model is shown to reasonably fit uniaxial strain compression test data, a model extensions section in which extensions to the model are considered and simple aging and pore gas compression extensions are demonstrated, and a conclusions section.

2 MODEL DESCRIPTION

Before we begin the discussion of the phenomenology of foamed rubber, a few words about nomenclature and conventions are appropriate. When foams are compressed, large compressive strains can obtain and significant volume change occurs. We therefore need to be very specific about our choices of stress and strain measures. Note that if an experimentalist refers to "true stress" when working with foams in compression, he or she had better have measured lateral strain, because standard assumptions of incompressibility at large deformations associated with metals and solid elastomers do not apply to foams. For our purposes, we will refer to engineering stress and engineering strain, both used here as positive in compression. It should also be noted that the present work is primarily focused on foam response under combined loads of compression and shear. The theory developed extends to tension, but most foams are not useful in tension as they are very weak because of their porosity.

2.1 Phenomenology

In much of the literature on foams, foam response is characterized in terms of the relative density of the foam, ρ^*. The relative density of a foam is defined as the ratio of the density of the foam, ρ_f, to the density of the parent solid, ρ_s. The initial porosity of the foam, ϕ_o, is related to the relative density as follows:

$$\phi_o = 1 - \rho^* = \frac{\rho_s - \rho_f}{\rho_s}. \tag{1}$$

A plot of uniaxial stress in compression vs. axial strain for monotonic loading of foam with a relative density of approximately 0.37 is shown in Figure 1. The linear, plateau-like, and densification regions are clearly indicated. A hyperelastic material subjected to this loading would unload along the same curve to a state of no strain when unstressed. A real foamed elastomer would exhibit some hysteresis and possibly cyclic softening. For our purposes, we ignore these dissipative mechanisms, with the understanding that they can be added to a stable, energy-conserving model.

Figure 2 is a plot of the nonincremental form of Poisson's ratio, here defined as the ratio of the lateral engineering strain to the compressive axial engineering strain in a uniaxial stress compression

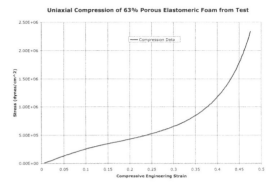

Figure 1. Sample uniaxial compression test data on PDMS foam with an initial porosity of approximately 0.63.

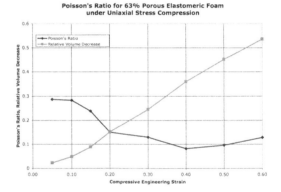

Figure 2. A plot showing the nonincremental version of Poisson's ratio and the relative volume decrease measured in a lubricated uniaxial stress compression test of a PDMS foam with an approximate initial porosity of 0.63.

test, as measured by Mooday (2002). These data suggest that the volumetric stiffness drops significantly in the plateau-like region of the compression stress-strain curve. The relative volume change, $1 - J$, where J is the relative volume or the determinant of the deformation gradient, is also plotted in Fig. 2. During densification, the volumetric response stiffens, until the nearly incompressible behavior of the parent elastomer is approached.

2.2 The model

We have chosen to break the total strain energy function into two main terms to capture the salient features discussed previously.

2.2.1 Initial stiffness and abrupt stiffness decrease

In order to describe the initial abrupt decrease in volumetric stiffness discussed previously, we have chosen to use a simple hyperelastic model based on linear uncoupled stress response in principal stretches added to a volumetric strain response that is quadratic in $(J - 1)$ until the relative volume hits a critical point,

J_b, at which point the function becomes linear. This part of the strain energy function is as follows:

$$U_{lp} = \frac{\hat{E}}{2}\sum_{i=1}^{3}(\lambda_i - 1)^2$$

$$+\hat{K}\left\{(J_b - 1)\left(J - \frac{J_b + 1}{2}\right) + H[J - J_b]\left[\frac{(J-1)^2}{2} - (J_b - 1)\left(J - \frac{J_b + 1}{2}\right)\right]\right\} \tag{2}$$

In Equation 2 above, $\hat{E}$ is an uncoupled version of Young's modulus (i.e. if the uncoupled bulk modulus were zero), λ_I are the principal stretches, and $\hat{K}$ is the uncoupled bulk modulus (i.e. if the uncoupled Young's modulus were zero). The expression H[x] is the unit step function at x. The other terms in Equation 2 have been defined previously in this article.

2.2.2 Micromechanically-based hyperelastic model for compaction

In order to describe the substantial increase in stiffness and stress as the compressive strain approaches the initial porosity of a foam, we borrow from a strain energy approach used by Danielsson et al. (2004) for porous rubber. We have generalized their approach to include both first and second isochoric invariants in the strain energy function for an isolated void enclosed in a spherical shell of incompressible, Mooney-Rivlin material. The strain energy expression is as follows:

$$U_{pc} = C_{10}\left(\hat{\bar{I}}_1 - 3\right) + C_{01}\left(\hat{\bar{I}}_2 - 3\right). \tag{3}$$

In Equation 3, C_{10} and C_{01} are the Mooney-Rivlin coefficients of the parent material, and the other two unfamiliar symbols represent the generalized isochoric invariants as follows:

$$\hat{\bar{I}}_1 = \bar{I}_1 f_1(J) + 3\phi_o \tag{4a}$$

where

$$f_1(J) = \frac{2J - 1}{J^{\frac{1}{3}}} + (2 - 2J - \phi_o)\left[\frac{\phi_o}{J - (1 - \phi_o)}\right]^{\frac{1}{3}}, \tag{4b}$$

and

$$\hat{\bar{I}}_2 = \bar{I}_2 f_2(J) + 3\phi_o. \tag{5a}$$

where

$$f_2(J) = J^{\frac{1}{3}}(2 - J) + \frac{(J - 1 - \phi_o)[J - (1 - \phi_o)]^{\frac{1}{3}}}{\phi_o^{\frac{1}{3}}}$$

In Equations 4a and 5a, the first and second isochoric invariants have been used. These invariants are defined as follows:

$$\bar{I}_1 = \bar{B}_{ii} \tag{6}$$

and

$$\bar{I}_2 = \bar{B}_{ii}^{-1}, \tag{7}$$

where

$$\bar{\mathbf{B}} = J^{-\frac{2}{3}}\mathbf{B}, \tag{8}$$

and

$$\mathbf{B} = \mathbf{F} \cdot \mathbf{F}^T, \tag{9}$$

or

$$\bar{B}_{ij} = J^{-\frac{2}{3}}F_{ik}F_{kj}. \tag{10}$$

In Equations 9 and 10, $\mathbf{F}$ is the deformation gradient, or the derivative of the current configuration with respect to the reference configuration.

2.2.3 Model summary

The full constitutive model proposed, then, can be summarized as a hyperelastic model with a strain energy function that is expressed as follows:

$$U = U_{lp} + U_{pc}. \tag{11}$$

The Cauchy stress, σ, can be derived as follows:

$$\sigma = \frac{1}{J}\frac{\partial U}{\partial \mathbf{F}} \cdot \mathbf{F}^T. \tag{12}$$

More explicitly, the Cauchy stress is as follows:

$$\sigma = \frac{\hat{E}}{J}\sum_{i=1}^{3}\lambda_i(\lambda_i - 1)\mathbf{p}_i \otimes \mathbf{p}_i + \hat{K}\{J_b - 1 + \langle J - J_b\rangle\}\mathbf{i}$$

$$+ C_{10}\left[\frac{2}{J}f_1(J)dev(\bar{\mathbf{B}}) + \bar{I}_1 f_1'(J)\mathbf{i}\right] \tag{13}$$

$$+ C_{01}\left[\frac{2}{J}f_2(J)dev(\bar{I}_1\bar{\mathbf{B}} - \bar{\mathbf{B}}^2) + \bar{I}_2 f_2'(J)\mathbf{i}\right]$$

In Equation 13, the vectors $\mathbf{p}_i$ are unit principal directions vectors associated with the corresponding principal stretches λ_i, $dev(\mathbf{A})$ signifies the deviatoric (traceless) part of the second order tensor $\mathbf{A}$, and $\mathbf{i}$ is the second order identity tensor. The prime symbol, $()'$, denotes differentiation of a function with respect to its argument. The other symbols in Equation 13 should have been adequately explained at this point, except for the Macaulay brackets, $< \ldots >$, which provide the value of the term inside the brackets if it is positive, and zero otherwise.

3 EXAMPLE

In this section, we consider uniaxial strain compression. This consideration is motivated by the fact that for many cushioning applications, foams are made

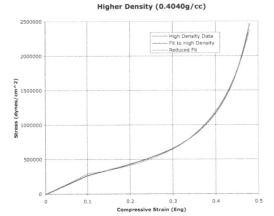

Higher Density (0.4040g/cc)

Figure 3. Uniaxial strain compression data and fits to PDMS foam with a density of 0.4040 g/cc, an approximate initial porosity of 0.63. The reduced fit is one for which C_{01} has been set to zero.

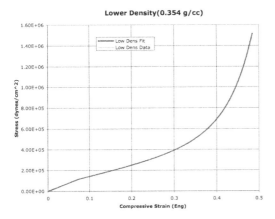

Lower Density(0.354 g/cc)

Figure 4. Uniaxial strain compression data and fits to PDMS foam with a density of 0.354 g/cc, an approximate initial porosity of 0.68.

very thin. As a result, compression tests on these materials tend to be in the thin direction and end effects are not negligible. The specimens tend to be in a condition of near uniaxial strain.

In uniaxial strain, only one stretch is not unity. Let the axial stretch be λ. The relative volume, J, is then also λ. The only direction in which the stress can be easily measured is the axial direction, and it can be shown that the axial stress can be represented by a combination of linearly independent functions of axial stretch with the model moduli as coefficients.

Provided that one has values for the initial porosity, ϕ_o, and for the buckling point, J_b, one may fit the moduli using a linear least squares approach.

In Figures 3 and 4, model fits to uniaxial strain compression data for foams with porosities of 0.63 and 0.68, respectively, are presented. It should be noted that the tests used to generate these two figures were not true uniaxial strain tests, but were rather

unconfined compression tests with very low L/D ratios, which were approximately 0.025.

4 MODEL EXTENSIONS

In this section we consider two main model extensions. The first is associated with material aging in the form of changing crosslinks under a defined deformation. The second is associated with the effects of pore gas compression.

4.1 Material aging

We consider here the effect aging under an imposed deformation. The approach here is based on work of Tobolsky (1960). We consider that the material in the reference state is initially stress free and its network of polymer chains and that the mechanical response of its physical and chemical crosslinks is well described by a particular strain energy function (here chosen as our model for foams).

The material is then deformed to some new configuration. In this new configuration, the storage configuration, new crosslinks are formed during an aging process. Old crosslinks may break, too. The new set of crosslinks are envisioned as forming a second network that is stressless in the configuration in which they formed. The mechanical response of this second network is assumed to be well represented by a strain energy function similar to that of the original material, but the reference state for the second network is the one under which it was formed.

Let us call the deformation gradient that describes the deformation of the material under which this type of aging occurs $\mathbf{F}_s$. The relative deformation gradient, $\mathbf{F}_r$, for the second network is found using a multiplicative decomposition of the total deformation gradient as follows:

$$\mathbf{F} = \mathbf{F}_r \cdot \mathbf{F}_s, \qquad (14)$$

or

$$\mathbf{F}_r = \mathbf{F} \cdot \mathbf{F}_s^{-1}. \qquad (15)$$

Similarly, one can construct all the relevant tensors and invariants pertinent to the strain energy function based on this relative deformation tensor. The relative volume as measured from the storage state is as follows:

$$J_r = \frac{J}{J_s}. \qquad (16)$$

Based on incompressibility of the parent material, we can derive a new initial porosity for the storage state as follows:

$$\phi_o^r = 1 - \frac{1 - \phi_o}{J_s}. \qquad (17)$$

4.1.1 Application of aging model to uniaxial strain

Let us consider using the approach outlined above to describe a material subjected to a uniaxial strain compression storage condition at a strain well beyond the buckling point (stress-strain slope decrease point). Furthermore we will consider that the virgin material is well described with $C_{01} = 0$.

The mechanical response of the second network about the storage configuration will be assumed for demonstration purposes to be described well by the first isochoric invariant part of the strain energy function as follows:

$$U_2 = C_{10}^r \left(\hat{\bar{I}}_1^r - 3 \right), \tag{18}$$

where

$$\hat{\bar{I}}_1^r = I_1^r f_1^r (J_r) + 3\phi_o^r \tag{19a}$$

where

$$f_1^r (J_r) = \frac{2J_r - 1}{J_r^{\frac{1}{3}}} + \left(2 - 2J_r - \phi_o^r \right) \left[\frac{\phi_o^r}{J_r - \left(1 - \phi_o^r \right)} \right]^{\frac{1}{3}}. \tag{19b}$$

We could consider that the moduli for the strain energy network of the initial network change also, but for demonstration purposes we will consider them constant. The resulting strain energy density function is as follows:

$$U = U_{lp} + U_{pc} + J_s C_{10}^r \left(\hat{\bar{I}}_1^r - 3 \right). \tag{20}$$

We now consider the uniaxial strain mechanical response of the new material as follows. The Cauchy stress for this aged material is then as follows:

$$\sigma = \frac{\hat{E}}{J} \sum_{i=1}^{3} \lambda_i (\lambda_i - 1) \mathbf{p}_i \otimes \mathbf{p}_i + \hat{K} \{ J_b - 1 + \langle J - J_b \rangle \} \mathbf{i}$$

$$+ C_{10} \left[\frac{2}{J} f_1 (J) dev(\overline{\mathbf{B}}) + \overline{I}_1 f_1' (J) \mathbf{i} \right] \tag{21}$$

$$+ C_{10}^r J_s \left[\frac{2}{J_r} f_1^r (J_r) dev(\overline{\mathbf{B}}_r) + \overline{I}^r \left(f_1^r \right)' (J_r) \mathbf{i} \right]$$

The axial stress under conditions of uniaxial strain is again a combination of linearly independent functions multiplied by the moduli of the model.

A sample plot showing the unaged and aged uniaxial deformation response for a fictitious 62% porous rubber material aged under an imposed compressive strain of 22% is shown in Figure 5. In this case, the C_{10} moduli for the second network and the initial network are identical.

4.2 Pore gas effects

Let us consider the two different relationships for the compression of gases. The first is an isothermal

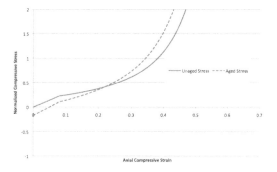

Figure 5. Example normalized compressive stress curves for an unaged material and one stored at 22% axial strain compression.

treatment where it is assumed that the ambient environment can easily exchange heat with the gas while it is compressed so that the gas is always at the ambient temperature. Another way of saying this is that the compression is slow compared to heat transfer processes and that the heat generated in compression is not enough to raise the temperature of the environment.

The pressure under these isothermal conditions is as follows:

$$p_i = p_o \frac{v_o}{v}, \tag{22}$$

where p_i is the current pressure, v is the current volume, p_o is the initial pressure, and v_o is the initial volume. We can express this in terms of the relative volume of the gas, J_g, as follows:

$$p_i = \frac{p_o}{J_g}. \tag{23}$$

J_g is defined as follows:

$$J_g = \frac{v}{v_o}. \tag{24}$$

The other relationship that can be used is that for adiabatic gas compression, in which no heat transfer is allowed to occur. The pressure for this case is as follows:

$$p_a = \frac{p_o}{J_g^\gamma}. \tag{25}$$

It can be shown that if a strain energy density per initial unit volume of gas is $W(J_g)$, then the relationship between W and the current pressure p is as follows:

$$p = -\frac{\partial W}{\partial J_g}. \tag{26}$$

Both the adiabatic and isothermal compression laws may be integrated to provide the following strain energy functions for pore gas:

$$W_i = W_0 - p_o \ln \left(J_g \right) \tag{27}$$

and

$$W_a = W_0 + \frac{p_o}{\gamma - 1}\left(J_g^{1-\gamma} - 1\right). \tag{28}$$

We can use the incompressible matrix assumption to derive an expression for the relative volume of pore gas in a foam as follows:

$$J_g = \frac{J - 1}{\phi_o} + 1. \tag{29}$$

Once can substitute the expression in Equation 31 into the strain energy expressions of Equations 27 and 28 to calculate the strain energy per initial unit volume of gas in terms of the initial porosity and the macroscopic relative volume. To develop an expression for the pore gas strain energy per initial unit volume of the foam, one simply multiplies this expression by the initial porosity as follows:

$$U_{g(a,i)} = \phi_o W_{(a,i)}\left(J_g\left(J,\phi_o\right)\right). \tag{30}$$

Similar but less rigorous developments have been made by others, but this model extension is rigorously tied to the initial foam porosity. This provides a stress contribution to the foam as follows:

$$\sigma_g = -p_o \mathbf{i} \begin{cases} \dfrac{\phi_o}{J - \left(1 - \phi_o\right)} & \textit{isothermal} \\[2em] \dfrac{\phi_o^{\gamma}}{\left[J - \left(1 - \phi_o\right)\right]^{\gamma}} & \textit{adiabatic} \end{cases} \tag{31}$$

5 CONCLUSIONS

A strain energy density function for foamed rubber has been developed and demonstrated. It is a hybrid model, combining rigorously developed strain energy functions for isolated pores in an incompressible Mooney-Rivlin parent material with more phenomenological *ad hoc* strain energy function to capture initial stiffness, Young's modulus, and Poisson's ratio along with buckling phenomena.

The model has a feature that it becomes singular when the foam is compressed to the point where the relative volume is equal to the foam relative density. This is an approximation to observed physical response and ignores effects of matrix material compressibility which would serve to remove this singularity while preserving a very stiff behavior at large compressions.

Given that the developed constitutive model is only a strain energy density function, several extensions are possible. Two have been considered here.

The first model extension considered is associated with material aging and represents the effects of crosslink density change in a deformed storage state. This approach can produce observed permanent set and stiffness changes.

The second model extension considered here is the inclusion of pore gas compression effects. This concept has been considered by others, but is more rigorously developed here.

Several other model extensions are possible, including the use of developed equivalent isochoric invariants in more physically-based strain energy functions like the Arruda-Boyce model proposed in Arruda and Boyce (1993), viscoelasticity, and the Mullins effect.

REFERENCES

Arruda, E.M., and Boyce, M. C., 1993, A three-dimensional constitutive model for the large stretch behavior of rubber elastic materials, *JMPS*, 41(2):389–412.

Braydon, A. D., Bardenhagen, S. G., Miller, E. A., and Seidler, G. T., December 2005, Simulation of the densification of real open-celled foam microstructures, *JMPS*, 53(12): p. 2638–2660.

Danielsson, M., Parks, D.M., and Boyce, M. C., 2004, Constitutive modeling of porous hyperelastic materials, *Mech. of Mat.* 36, 347–358.

Jemiolo, S. and Turteltaub, S., June 2000, Parametric model for a class of foam-like isotropic hyperelastic materials, *J. Appl. Mech.*, v. 67, no. 2, p. 248–254.

Kraynik, A.M., Nielsen, M.K., Reinelt, D.A., and Warren, W.E., 1998, "Foam Micromechanics," NATO Institute for Advanced Studies, Sandia Report SAND98-2454J, Albquerque, New Mexico.

Mooday, R., 2002, private communication, Los Alamos National Laboratory, Los Alamos, New Mexico.

Ogden, R.W., 1984, Non-linear elastic deformations, published by Ellis Harwood, Ltd., New York.

Tobolsky, A.V., 1960, *Properties and Structure of Polymers.* John Wiley and Sons, New York.

Constitutive Models for Rubber VII – Jerrams & Murphy (eds)
© 2012 Taylor & Francis Group, London, ISBN 978-0-415-68389-0

Simulation of self-organization processes in filled rubber considering thermal agitation

H. Wulf & J. Ihlemann

Professorship of Solid Mechanics, Chemnitz University of Technology, Chemnitz, Germany

ABSTRACT: The molecular scale mechanisms responsible for time-dependent phenomena in rubber behavior like creep and relaxation are still subject of discussion. Here, an explanation based on the theory of self-organized linkage patterns (SOLP) is presented. The central claim of this theory is that the rubber behavior is a result of a self-organization process based on connections created by Van-der-Waals interactions. A simulation program has been developed based on an abstract model of the molecular structure of rubber with special attention to the Van-der-Waals forces. Time-dependent behavior was modeled by destruction of such bonds due to thermal agitation. Typical loading schedules were simulated and compared to experimental results. Although the used model is an extreme simplification of the real rubber structure, several properties including Mullins effect, creep and relaxation could be reproduced.

1 INTRODUCTION

In cyclic tension experiments using filled rubber a significant reduction of stress at equal strain in the course of cycles is observed. In addition, the stress is lower if the strain is beneath the maximum strain applied of the loading history. Both phenomena are usually referred to as Mullins effect [11]. Moreover, hysteresis and permanent set always occur.

The strain-stress response of filled rubber is also time-dependent. An increase in strain rate yields increased stresses. When keeping the stress constant over a period of time, the material slowly stretches. If strain is kept constant, the resulting stress drops over time. These two effects, which are called creep and relaxation respectively, were first described by Gent [8]. However, the claim, that relaxation causes a stress reduction, is only true if strain is kept constant after a loading phase. If the relaxation phase follows an unloading, the resulting stress, surprisingly, increases over time.

Concerning the mechanical behavior of rubber, a vast variety of material models has been developed. While some of them are purely phenomenological, others are motivated by molecular structure of filled rubber. For those physically motivated models, there is a wide range of theories, which components and which processes on molecular scale are considered as critical to the mechanical behavior. For instance, *Bergstrøm und Boyce* consider molecule slipping and disentanglement to be most important [3]. On the contrary, *Klüppel* and *Heinrich* emphasize the stiffening role of a filler network. According to them, the stress-induced breakdown of filler agglomerates is critical to Mullins

softening [9]. A third theory, presented by *Dargazany and Itskov*, claims that polymer-filler debonding is responsible for the Mullins effect [4]. For an excellent review on both experimental results and physical explanations, refer to *Diani* [6].

The relevance of weak physical interactions (Van-der-Waals bonds) is usually acknowledged by the authors of physically motivated material models. However, a minor impact on material behavior is attributed to them. Moreover, the average density of interactions is considered only. The possible existence and importance of meso-scale patterns in bond density is therefore neglected. The essential role of emergent patterns has been shown in other areas of material science. For instance, *Devincre et al.* examined plastic deformation of crystalline matter by discrete dislocation dynamics [5]. Here, modelling the dislocations in an explicit way instead of taking an average density was decisive. An overview over applications of such simulation techniques in material science is provided by *Raabe* [12]. Concerning rubber, the previous work by *Besdo* and *Ihlemann* is the only known approach considering self-organized development of structures [2].

First, in section 2 of this paper a brief presentation of the theory of self-organized linkage patterns is given. Then, the extension by thermal agitation is described. Specifically, explanations of effects like speed-dependent stress response and relaxation according to the theory are given. The simulation model used is presented in section 3. Again, special attention is paid to the extensions made to cover time-dependent behavior. The model is evaluated by comparing experimental data to simulated strain-stress response curves in section 4.

2 THEORY OF SELF-ORGANIZED LINKAGE PATTERNS

The common approach for developing physically motivated material models of rubber is to search for molecular scale components and interactions that reproduce the macroscopic material behavior. However, there are various examples of systems exhibiting completely new, complex properties, which are not present for the simple components they consist of. According to *Ebeling* such properties are emergent properties [7]. One prominent example are metals, which form crystallites. This is clearly the emergence of a structure with critical influence on the macroscopic behavior – a structure that could hardly be expected when considering the behavior and interaction of single atoms only. It should be noted that the pattern occurs at a typical scale. Furthermore, metal behavior can be described well with average values over crystallites, but not with average values over atoms. In a similar way, it is claimed, that rubber properties are emergent properties based on the formation of a pattern, rather than average values over its molecular components.

The central idea of the theory of Self-Organized Linkage Patterns (SOLP) is that the structure is based on an non-uniform distribution of the weak physical linkages. It can be assumed that small inhomogenities in linkage density always exist. The development of such a pattern requires a local amplification of initial differences. As regions with decreased bond density will be weaker, they will bear the majority of an external deformation. Due to the network structure of the rubber, this deformation will incur a multitude of movements in various directions on molecular level. Van-der-Waals bonds are easily established as molecules get close enough and they are easily broken up when stretched beyond a critical length. Hence, linkages will be destroyed rather in the soft regions with considerable relative movement of the linkage partners. Simultaneously, new linkages are established in both areas, leading to growth of the stable areas. Overall, the positive feedback separates regions with high from regions with low linkage density. Whereas the positive feedback allows the growth of the structures, there is also negative feedback which limits their size. Here, the essential property of rubber is that polymer molecules can not be further stretched after reaching the linear state. When the soft areas adjacent to a compound of high linkage density are fully stretched, additional external deformation must strain the hard area. As soon as the first links break up, the structure is weakened leading to a chain reaction which might ultimately dissolve the whole compound.

The Mullins effect and hysteresis can be explained by considering the evolution of the linkage pattern. As already explained, regions with high link density are permanently dissolved with others being created. The pattern existing at a certain deformation consists of compounds of different age serving both as memory and characterization of the current state. In general, structures are created which exhibit small resistance to the current deformation. When the strain changes, these structures exert a force tending to their initial deformation. Hence, the stress response differs considerably between loading and unloading path leading to hysteresis. Especially intense reorganization processes occur when a new upper bound in strain is reached. During this loading cycle compounds emerge that are especially adapted to the strain values achieved for the first time. While most of these compounds are destroyed during unloading, some of them persist, reducing the stress during the next loading cycle considerably. During further loading cycles, the new patterns establish and become more pronounced, leading to additional small reductions in maximum stress. This explains the softening by cyclic loading. Finally, in a stationary cycle a closed loop of pattern states exists, with each state implying the next one.

The influence of temperature and time on the self-organization process is taken into account by considering thermal agitation. According to their temperature, all material components carry out small, irregular movements also referred to as Brownian motion. This also applies to molecule and filler parts participating in a physical link. If the distance of the linkage partners grows beyond the critical length of Van-der-Waals bonds, the link will break up. In contrast to free Brownian motion, the link exerts a pullback force directed to the initial position. This means that the expected displacement of a linkage partner is rather constant than growing with time. Obviously, the probability to observe a certain displacement drops with the displacement value. In consequence, links which are prestrained close to the critical length have a higher chance to be expanded above this length by thermal agitation and be destroyed. As a limit case, links exactly strained to the critical length break up immediately.

According to the SOLP-theory, the removal of few physical links can cause the reorganization of wide parts of the linkage pattern. Especially, entire areas of high linkage density might dissolve while other such regions with many new links are created. Such chain-reactions mainly occur when links in strongly stressed structures are destroyed. An important fact to note is, that the influence of thermal agition in the context of the self-organization process should not be considered as mere destruction of links. Instead it should be viewed as initiation of reorganizations in the linkage distribution.

Thermal agitation preferably destroys strongly stressed, instable structures. It can be expected that these are older compounds which hinder the current deformation. They are replaced by new structures, which are adapted to the current deformation and exhibit less resistance. In consequence, the yield stress is reduced. The effect is more pronounced the higher the temperature and the lower the deformation speed. This prediction matches with experimental results. Relaxation after loading can be explained in a similar way: Older compounds are replaced by structures adapted to the current deformation, which induce smaller stresses. However, relaxation after unloading can be explained as well: Here, some structures remaining from previous loading belong to higher strain values. They are actually yielding a negative stress contribution. Replacing those patterns results in a stress increase. In both cases, relaxation is explained as fading of the memory stored in the linkage pattern.

3 SIMULATION MODEL

Verifying the theory of self-organized linkage patterns is very difficult due to the complex processes it predicts and the different scales involved. Therefore, a simulation program called Trial program was developed to test the theory. This program is neither intended to be used as material law for numerical analysis nor attempts to deliver an exact atomistic simulation. Instead, all processes are modeled on a meso-scale which matches the size of the expected patterns. All physical and chemical interactions are abstracted as far as possible without loosing the properties which are considered essential for the self-organization process. The central purpose of the Trial program is to test, whether self-organization based on weak physical links is capable of qualitatively reproducing several typical rubber properties.

The model used for simulation is two-dimensional with the only allowed deformation being uniaxial tension. The first model element are nodes, which are initially placed in a grid layout. They may move in vertical direction. The node position are the degrees of freedom of the model. All other model elements are attached to these nodes.

All nodes on a horizontal line in the initial configuration belong to a chain. The chains possess a bending stiffness and a maximum bending. The aim is to model the stretching stiffness of the polymer chains as well as the maximum length achieved in linear configuration. The application of bending was necessary to keep the stress state uniaxial. Here, extreme abstraction was applied keeping only the most basic characteristics. The important property that was to be preserved is the long-range transfer of forces provided by the polymer network.

Next, an element similar to the sulfur cross-links created by vulcanization is introduced. The purpose is to connect the chains to a mesh structure and ensure a multitude of local movements generated by global deformation. A random selection of vertically aligned pairs of nodes is selected and equipped with a very rigid connection. Essentially, the distance of the nodes in a pair is fixed. These connections enforce the bending of the chains due to external deformation. The randomization process is influenced in such a way, that each chain is attached to at least two crosslinks, ensuring that the system is statically defined.

On the upper and lower edge, cyclic boundary conditions are applied. The model is elongated in vertical direction. By doing so, the node positions can be computed from an equation system. Finally, the resulting

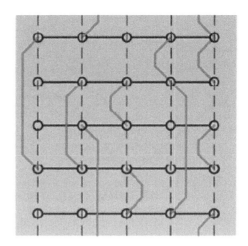

Figure 1. Simulation model with nodes, chains (horizontal) and cross-links (vertical).

force is computed by summing up the internal forces over an arbitrary horizontal cut.

However, in the current state the model behavior is perfectly elastic. The most important component is still missing: Some model element representing the physical links. If two nodes on the vertical line get closer than a critical distance, a spring is inserted between them. These links have an unstretched length of zero, which means they are inserted in stretched state. As soon as the distance of the nodes increases above the critical length, the link is removed. The characteristic of the springs is simply linear with a rather low stiffness. To regulate the influence of the physical links, nodes are divided in active and inactive nodes. Physical links are established between two active nodes only. The activity state of the nodes is set at random before the start of the simulation and remains constant thereafter.

The influence of thermal agitation is introduced by randomized removal of physical links. In each timestep of length Δt, each link is destroyed with the following probability:

$$p_{decay}(\Delta t) = \left(1 - e^{\ln(1/2) * \frac{\Delta t}{t_h}}\right) \tag{1}$$

The formula is similar to those describing radioactive decay probability. The lifetime of a spring is exponentially distributed with half-value period t_h. The specialty of this distribution is, that the expected remaining lifetime is independent of the previous lifetime. As there is no reason to belief in aging of Van-der-Waals bonds, it is a very obvious choice. The value of t_h depends on the length of the physical link:

$$t_h(l) = a * \left(1 - \frac{l}{l_{crit}}\right) \tag{2}$$

Therein, l_{crit} denotes the critical length. Note that the half-value period approaches zero as the length approaches l_{crit}. Hence, this formula implements the

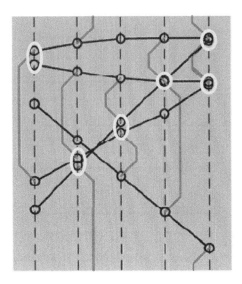

Figure 2. Simulation model in deformed state with inserted physical links (ellipses).

rule, that links strained to the limit are destroyed immediately. The parameter a describes the half-value period of an unstrained link. Its value can be associated with the temperature T, but also material properties.

4   SIMULATION RESULTS

Simulation models as small as depicted in Fig. 2 are dominated by random effects and deliver a very unstable strain-stress response. All results presented here are obtained from models consisting of 1600 Nodes on 40 chains, with 90% of the nodes being active. In addition, 200 models are calculated in parallel and the average is taken.

The first experiment simulated is a cyclic tension test with four subsequent upper bounds in strain. Very good results simulating this experiment have already been presented by *Ihlemann* [2]. Mullins effect, hysteresis and permanent set could be successfully reproduced. Here, the influence of strain rate is examined. The same simulation was executed with thermal agitation deactivated, with high and with low speed. The results are shown in Fig. 3. Therein, transient cycles have been omitted to keep the figure clean. First of all, Mullins effect and hysteresis are not harmed by activation of thermal decay of physical links. Permanent set is still present and almost equal for all experiments. In general, lower loading speed leads to reduced stress. However, the two simulations differ in strain rate by a factor of 10. Moreover, inactive thermal decay could be interpreted as loading speed so high, that decay probabilities are close to zero. So the influence of loading speed is much smaller and different from the characteristics one could expect when assuming viscous material behavior. An interesting detail is, that at lower speed the tips of the curves are rounded off. This is an effect which is frequently observed in measurements.

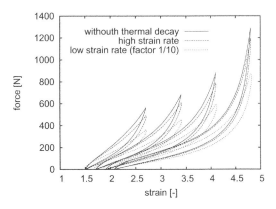

Figure 3. Simulation of cyclic uniaxial tension test at different strain rates.

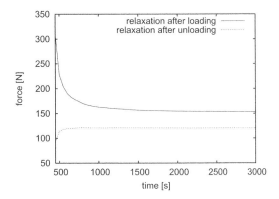

Figure 4. Relaxation after loading and after unloading.

As second experiment, relaxation was simulated. The loading schedule consisted of several stretching cycles up to stretch $\lambda = 2$ followed by a holding phase with $\lambda = 1.5$ kept constant. In two simulation runs the relaxation phase followed a loading branch in one case and an unloading branch in the other. The stress response in the holding phase for both experiments is shown in Fig. 4. In both cases, relaxation can be observed clearly. The most interesting result is of course, that the stress actually increases in the holding phase after unloading. This is especially surprising, as it is achieved by a mechanism removing springs from the model. This result can be considered as a strong indication that the pattern of physical linkages is important and undergoing significant reorganization processes. Both relaxation curves converge, but not to a common equilibrium stress. Although many material models are based on this assumption, existence of a history-free equilibrium state is unclear. Measurements rather seem to support the concept of an equilibrium hysteresis as proposed by *Lion* [10].

Another interesting property of filled rubber recently discovered by *Ahmadi et al.* is, that relaxation speed also depends on the strain rate in the previous loading cycles [1]. They performed shear

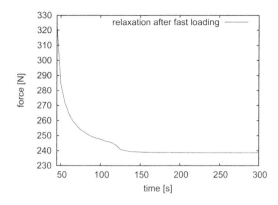

Figure 5. Relaxation after fast loading. Strain rate was increased by factor 10.

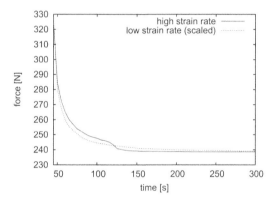

Figure 6. Combined relaxation curves from Fig. 4 and 5. Stress response for the slow test was scaled by 10 and shifted.

experiments with cyclic loading at different frequencies and observed significantly different behavior in a subsequent relaxation phase. To examine this effect, the simulation of relaxation after loading was repeated with the initial loading cycles speed up by factor 10. The resulting stress is depicted in Fig. 5. The similarity of the relaxation curve with Fig. 4 is striking, but note that the time axis is scaled down by 10. Apparently, the relaxation is much faster when previous loading cycles are performed at high strain rate.

Ahmadi et al. also pointed out, that when the time axis is scaled according to the difference in strain rate, the resulting curves can be superposed. Applying an additional vertical shift, they obtained common relaxation curves for a multitude of experiments. Similarly, the result in Fig. 5 was matched to the upper curve of Fig. 4. The common relaxation curve shown in Fig. 6 demonstrates the excellent reproduction of the observations made by *Ahmadi et al.*

This relaxation behavior could be considered as second-order memory effect: The strain-rate history affects how fast some memory inside the material fades. Judging from the simple model components, a similar behavior of the model could never be expected. Obviously, none of the single parts has a characteristic such as relaxation, not to mention history-dependent

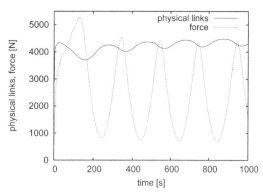

Figure 7. Stress response and evolution of total number of physical links during sinusodial loading.

relaxation. It is therefore a completely new, emergent property. This is a clear sign that the simulation behavior is heavily influenced by complex processes inside the model. As the only dynamic model element are the physical links, whose characteristics are identical to each other, the new properties must emerge from a self-organization process based on these links.

Yet another experiment was conducted to underline that the linkage structure is important rather than the sheer number. The loading schedule applied consisted of a linear stretch to $\lambda = 2.5$ followed by a sinusodial oscillation around $\lambda = 2.5$ with amplitude of 0.5. Again, 200 parallel systems were used for computation. In addition to the stress response, the average number of physical linkages was calculated. The results are shown in in Fig. 7. Apparently, the number of linkages oscillates as well, but relative amplitude is much smaller compared to the stress curve. Moreover, the maxima do not coincide with the maxima of the stress response. The guess that overall stress is tightly coupled with the total number of linkages is clearly refuted. Finally, the softening by repeated loading can be observed as maximum stress decreases with each cycle. The maximum number of physical links, on the contrary, slightly increases. One can conclude, that this type of softening is not induced by the destruction but rather advancing establishment of structures. Again, such an evolution of patterns can be explained with a self-organization process only.

5 CONCLUSIONS

1. The theory of Self-Organized Linkage Patterns (SOLP) delivers an explanation for rubber behavior based on the molecular structure. The central claim is, that the rubber properties are emergent properties arising from a self-organization process based on the pattern of physical linkages.

2. Time-dependent properties are explained by considering that thermal agitation may destroy physical linkages, initiating a reorganization of the linkage pattern.

3. To verify the theory, a simulation program called Trial program was developed. It relies on an extreme abstraction of the molecular structure of rubber. Thermal agitation was added by implementing an exponential decay of the model element corresponding to physical linkages.

4. Uniaxial tension tests and relaxation tests at different strain rates have been simulated. The results qualitatively reproduce several properties of rubber. Moreover, it could be shown that the model behavior is based on self-organization.

5. Considering the abstraction and simplicity of the model and the variety of positive results, one can conclude that essential processes have been covered in the simulation. Hence, self-organization might be crucial for understanding rubber behavior.

REFERENCES

[1] Ahmadi, H.R., Ihlemann, J., Muhr, A.H., 2007: *Time-dependent effects in dynamic tests of filled rubber.* Constitutive Models for Rubber V, Taylor & Francis, London: pp. 305–310

[2] Besdo, D., Ihlemann, J., 2003: *Properties of rubberlike materials under large deformations explained by self-organizing linkage patterns.* Internat. J. of Plasticity 19, London: pp. 305–310

[3] Bergstrøm, J.S., Boyce, M.C., 1998: *Constitutive Modelling of the Large-Strain Time-dependent Behavior of Elastomers.* J. Mech. Phys. Solids 46, S. 931–954

[4] Dargazany, R., Itskov, M., 2009: *A network evolution model for the anisotropic Mullins effect in carbon black filled rubbers.* Internat. J. of Solids and Structures 46, S. 2967–2977

[5] Devincre B., Kubin, L.P., Lemarchand, C., Madec, R., 2001: *Mesoscopic simulations of plastic deformation.* Materials Science and Engineering A309–310, S. 211–219

[6] Diani, J., Fayolle, B., Gilormini, P. 2009: *A review on the Mullins effect.* European Polymer Journal 45, S. 601–612.

[7] Ebeling, W., Freund, J., Schweitzer, F., 1998: *Komplexe Strukturen: Entropie und Information.* Teubner Verlag, Leipzig

[8] Gent, A.N. 1962: *Relaxation processes in vulcanized rubber.* J. Appl. Polym. Sci.

[9] Lorenz, H., Freund, M., Juhre, D., Ihlemann, J., Klüppel, K., 2010: *Constitutive Generalization of a Microstructure-Based Model for Filled Elastomers* Macromol. Theory Simul. 2010, 19, DOI: 10.1002/mats. 201000054

[10] Lion, A., 1996: *A constitutive model for carbon black filled rubber. Experimental investigation and mathematical representation.* Continuum Mech. Thermodyn. 8, S. 153–169

[11] Mullins, L., 1948: *Effect of stretching on the properties of rubber.* Rubber Chem. Tech. 21, S. 281–300

[12] Raabe, D. 1998: *Computational Material Science.* Wiley Verlag Weinheim

Constitutive Models for Rubber VII – Jerrams & Murphy (eds)
© 2012 Taylor & Francis Group, London, ISBN 978-0-415-68389-0

On finite strain models for nano-filled glassy polymers

M. Hossain & P. Steinmann
Chair of Applied Mechanics, University of Erlangen-Nuremberg, Erlangen, Germany

A.R. Sanchez
Corporate Sector Research and Advance Engineering, Robert Bosch GmbH, Waiblingen, Germany

ABSTRACT: The recently proposed constitutive framework based on logarithmic strain space by Miehe et al. (3) is a successful model for the modelling of amorphous glassy polymers. The modular structure of this model makes it attractive especially for the numerical implementation. In this contribution, the Miehe model will be adapted to model the influence of different types of nano-fillers mixed into bulk amorphous glassy polymers. Here, different types of amorphous glassy polymers with various filler content, orientation of nanofillers and loaded by different strain rates have been investigated experimentally. Then, the Miehe model will be extended towards modelling the nano-particle influence in the amorphous glassy polymer which can be conceptualized by adapting the appropriate material parameters. The proposed modified model will finally be validated with experimental data.

1 INTRODUCTION

The reinforcement of polymers with nanofillers results an advanced class of composite materials that demonstrate a significant enhancement in strength, stiffness, and thermal properties compared to homopolymers. This property advantages and relative low cost of the nano-fillers make such materials interesting for a number of applications, e.g. especially in automotive and construction industry, electronics, optical devices and medical technology (3). In the case of silica-reinforced polymer nanocomposites, almost all of the researches have involved investigations into synthesis techniques, structure-property relationships, and mechanical behaviour enhancements (5). Nanoscale particles, e.g. silica, alumina, and hematite used inside the polymer matrix have been studied to manipulate the polymeric material properties such as internal structure, dynamics, stiffness, and end macroscopic properties. Although enormous efforts that are presented in the literature in the recent years to understand and characterize the changes in mechanical behaviour induced by nanoparticles, only one constitutive model in the large deformation region, to the authors knowledge, for this class of polymer nanocomposites has been suggested so far, cf. (5).

A continuum-based micromechanically-motivated model is developed for silica-filled nanocomposites with nanoparticle interfacial treatments which was mainly based on the Mori-Tanaka approach (4; 6) for multiscale modelling. The model incorporates the molecular structures of the nanoparticle, and interfacial regions, which are determined using a molecular modeling method that involves coarse-grained and reverse-mapping techniques. The main limitation of these careful works is that the developed models are in the small deformation regime and did not have a clear-cut strategy on viscoelastic or viscoplastic extensions while amorphous glassy polymers are typically elasto-viscoplastic.

Mulliken & Boyce (5) proposed a finite strain constitutive model for elasto-plastic deformations of amorphous glassy polymers from low to high strain rates. Later they modified the constitutive model for polycarbonate and polycarbonate polyhedral oligomeric silsesquioxanes (POSS) nanocomposites at high rate of deformation (5). It is northworthy to mention here that both models were originally based on the constitutive framework proposed by Boyce in her dissertation and developed further by her co-workers for amorphous glassy homopolymers. The modifications proposed by Mulliken & Boyce are to incorporate the nanofiller influences, i.e. influence of POSS in PMMA homopolymer. Since the Mulliken & Boyce model is based on the multiplicative decomposition of the deformation gradient, the approach makes the numerical implementation somehow complicated. Additionally, there are two viscoplastic spring and dash-pots in the rheology of this model which make the number of the material parameters large, as a result, more efforts will be required to determine these parameters.

In this study, we investigate the rate-dependent mechanical properties of some particular silica-filled nanocomposites and their homopolymer counterparts by a set of experimental investigations. To model the mechanical behaviour, the model development is carried out in accordance with the theoretical framework

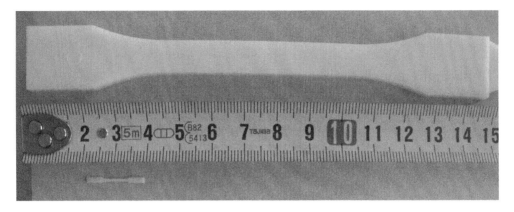

Figure 1. Tensile specimens. The mechanical characterization was performed on micro-samples that are displayed below the measurement scale. For comparison also the standard ISO 527-2 sample is depicted.

(constitutive model) of Miehe et al. (3) for amorphous glassy homopolymers. The Miehe model has a clear modular structure, i.e. the total model structure can be decomposed into three distinct parts, e.g. pre-processing, constitutive core and the post-processing which will be discussed shortly later on. Then, this constitutive model is fitted to the experimental results obtained in our works for the both polymer matrix (homopolymer) and the (nano) silica-reinforced polymer composites.

The paper is organized as follows: In section 2, the experimental setup to determine the mechanical behaviour of some amorphous glassy homopolymers and their nano-filled counterparts is presented along with the experimental detail descriptions and experimental results. Section 3, the main framework of the Miehe constitutive model is revisited briefly. In section 4, the experimental results have been validated with the modified model. Finally, concluding remarks close the paper.

2 EXPERIMENTAL RESULTS

The nanocomposite materials characterized in this work have a monodisperse polystyrene matrix filled with spherical polystyrene (PS) grafted silica nano-particles (PS-silica) and a PAMM matrix with ellipsoidal TMP-silica-hematite nanocomposites in different mass percentages. These materials were prepared by the partners of the European project "NanoModel" (project number: 211778).

The material development was performed on laboratory scale, so that only small quantities of material in the range of some grams could be provided at this stadium. The mechanical characterization was thus performed on micro-samples (tensile specimens), which were prepared by injection moulding. The geometry of the samples corresponds to 1/12th of the dimensions of the standard sample defined in ISO 527-2, see Figure 1.

One of the challenging aspects in the mechanical tests is to perform an accurate strain-measurement on the micro-samples. For this purpose a special extensometer was used for the mechanical strain measurement at room temperature, which allows a strain measurement taking an initial length of 5 mm (as comparison: the initial length for the strain measurement on standard samples is 50 mm). These tests were performed on a universal testing machine (manufacturer Zwick/Roell).

In order to check the quality of the mechanical strain measurement on the micro-samples, further tensile tests with optical strain measurement by means of digital image correlation were performed. These tests were performed on a servohydraulic testing machine (manufacturer Schenck) using the digital image correlation software ARAMIS. The strain is measured by tracking the gray value pattern in small neighbourhood facets. In order to obtain a random speckle pattern on the surface of the measured object, the samples were varnished and sprayed with graphite. Figure 2 (top) shows the results of tensile tests on unfilled polystyrene (PS) at room temperature comparing the data obtained from the mechanical and the optical strain measurement. A good agreement between both strain measurement devices can be observed, thus approving the reliability of the mechanical strain measurement on the micro-samples.

The testing procedure presented was applied to measure the mechanical behaviour of the new nanocomposites developed. Figure 2 (bottom) shows the results of tensile tests on nanocomposite materials with a mass percentage of nanofillers of 5.0% and on the corresponding unfilled polymer. The tensile tests were performed at a strain-rate of 1.0%/min and at room temperature (23°C). The strain was measured mechanically as described before. Before testing the samples were annealed for 5 hours at 80°C in a convection oven in order to remove internal stresses introduced by the injection moulding process. The nanocomposites show a Young's modulus comparable to the one of the unfilled material. The tensile strength and the elongation at break of the nanocomposites are lower than those of the unfilled material. Thus, the nanofillers provoke an embrittlement of the polymer.

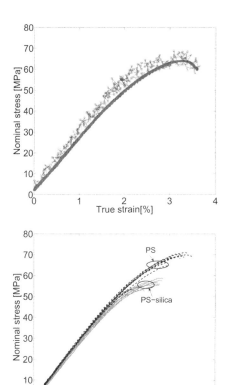

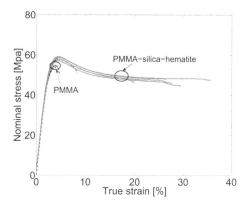

Figure 3. Tensile tests at 23°C temperature and 1%/min strain rate for PMMA + TPM-silica-hematite composites prepared by solvent integration method. Comparison of nanocomposites with unfilled PMMA (**dotted lines**).

It has been observed that in most cases, at least in the room temperature and low strain rate, there is very little influence of nano-fillers on the enhancement of the mechanical behaviours of the polymers under consideration. This assertion is also mentioned by other literature, e.g. Mulliken & Boyce (5). They observed and reported that the influence of nano-fillers will be magnificant at very high rate of strains. Spathis & Kontou (2; 8) demonstrated experimentally that the nano-fillers influence can be visible at higher deformations and at elevated temperature, e.g., near the glass transition temperature. Our experimental results also echoed their observations.

Figure 2. (**Top**) Optical (scattered dots) versus mechanical true strain measurements. Tensile tests (nominal stress versus true strain) on micro-sample made out of polystyrene (PS). Tests performed at 0.14 mm/min (clamping length 14 mm). (**Bottom**) Tensile tests (nominal stress versus true strain) at room (23°C) temperature and 1%/min strain rate on micro-sample of nanocomposite material filled with spherical silica nanoparticles at 5.0% (by mass). Comparison with the results obtained on the corresponding unfilled material (**dotted lines**).

The macroscopic behaviour observed is strongly influenced by the nanostructure of the materials, i.e. by the geometry of the nanofillers, their dispersion in the matrix as well as their interaction with the surrounding polymer.

The stress-strain behaviours of the nanocomposites prepared by solvent integration with PMMA matrix and ellipsoidal TPM-silica-hematite nanoparticles are presented in Fig (3). The results of the nanostructure characterization showed a homogeneous dispersion of the nanoparticles in the matrix. In the mechanical measurements, the nanocomposites with higher filler content (2.7 and 6.3%) showed an increased Young's modulus compared to the unfilled material. Compared to the previous nanocomposites (PS-silica), this one shows a better ductile behaviour with large elongation at break compared to its homopolymer counterpart. The data presented in this work will serve as a validation of the model described in Sections 3 and 4.

3 CONSTITUTIVE MODEL

Following the typical structure of linear elasto (visco) plasticity, the additive decomposition of the total strain tensor in the logarithmic strain space into elastic and viscoplastic parts yields

$$\varepsilon = \varepsilon^e + \varepsilon^p. \tag{1}$$

One important aspects in constructing a framework of finite (visco) plasticity is the definition of an elastic strain measure ε^e. Going back to Seth and Hill (3), it is assumed that this strain measure will be a function of the total and the plastic Cauchy-Green tensors, i.e.

$$\varepsilon^e = \varepsilon^e(C, C^p) \tag{2}$$

where by the definition of the total strain tensor and plastic part of the total strain tensor is as follows

$$\varepsilon = \frac{1}{2}\ln C, \quad \text{and} \quad \varepsilon^p = \frac{1}{2}\ln C^p, \tag{3}$$

respectively. The key to model the material behaviour at large strains is the potential function, i.e. the Helmholz free energy which can be decomposed into elastic and viscoplastic parts as

$$\Psi = \tilde{\Psi}(\varepsilon, \varepsilon^p) = \Psi^e(\varepsilon - \varepsilon^p) + \Psi^p(\varepsilon^p). \tag{4}$$

Then, $\sigma^\star := -\partial_{\varepsilon^p}\Psi = \sigma - \beta$ with $\sigma := \partial_\varepsilon \Psi^e$ and $\beta := \partial_{\varepsilon^p}\Psi^p$, denotes the thermodynamic driving stress tensor conjugate to the viscoplastic strains ε^p. A widely used evolution framework which has been used in viscoelastic and viscoplastic modelling as proposed by Boyce et al. (1), Wu et al. (10), can be rewritten in the formulation below as

$$\dot{\varepsilon}^p := \dot{\gamma}^p \boldsymbol{N} \quad \text{with} \quad \boldsymbol{N} := \frac{\text{dev}[\sigma^\star]}{||\text{dev}\sigma^\star||} \tag{5}$$

where $\dot{\gamma}^p \geq 0, \text{dev}(\sigma^\star) := \sigma^\star - 1/3\text{tr}(\sigma^\star)I, \text{tr}(\bullet) = (\bullet) : I$. The nonlinear evolution equation (5) can be solved by any Newton-like iterative scheme, see Miehe et al. (3). A quadratic form for the elastic free energy has been chosen in terms of elastic strains, e.g. $\Psi^e = \kappa/2\text{tr}^2[\varepsilon - \varepsilon^p] + \mu\text{dev}[\varepsilon - \varepsilon^p]: \text{dev}[\varepsilon - \varepsilon^p]$ where, κ, μ are the bulk modulus and the shear modulus, respectively. Following the Coleman & Gurtin argumentation (1), this particular form of Ψ^e immediately yields the stress expression as

$$\sigma := \frac{\partial\Psi^e}{\partial\varepsilon} = -p\boldsymbol{I} + 2\mu\text{dev}[\varepsilon - \varepsilon^p]$$

$$\text{with} \quad p := -\kappa\text{tr}[\varepsilon - \varepsilon^p] \tag{6}$$

where p is termed as the hydrostatic pressure. Based on non-Gaussian statistics, the plastic network free energy is specified in terms of the network plastic stretch as $\Psi^p = \Psi^p(\varepsilon^p) = \mu_p N_p[\lambda_r^p \mathcal{L}^{-1}(\lambda_r^p) + \ln(\mathcal{L}^{-1}(\lambda_r^p)/\sinh\mathcal{L}^{-1}(\lambda_r^p))]$. In this expression, μ_p, N_p stand for the plastic shear modulus and the number of segments in a polymer chain, respectively. The function $\mathcal{L}^{-1}(\lambda_r^p)$ denotes the inverse of the well-known Langevin function defined by $\mathcal{L}(\bullet) := \coth(\bullet) - (1/(\bullet))$. Using the definition of the back stress, it follows

$$\beta := \frac{\partial\Psi^p(\varepsilon^p)}{\partial\varepsilon^p} = \mu_p\lambda^p\frac{3N_p - \lambda^{p2}}{N_p - \lambda^{p2}}\frac{\partial\lambda^p}{\partial\varepsilon^p}. \tag{7}$$

In the derivation of Ψ^p, the Padé approximation of the inverse Langevin function, i.e. $\mathcal{L}^{-1}(\lambda_r^p)$ has been used. Using the spectral representation for the plastic Cauchy-Green tensor, the relation between plastic strain and plastic deformation tensor is as follows

$$\boldsymbol{C}^p = \exp(2\varepsilon^p) = \sum_{a=1}^{3}(\lambda_a^p)^2\boldsymbol{n}_a^p \otimes \boldsymbol{n}_a^p \tag{8}$$

where C^p is co-axial to the plastic strain tensor, $\varepsilon^p = \sum_{a=1}^3 \epsilon_a^p n_a^p \otimes n_a^p$, and $\epsilon_a^p := \ln(\lambda_a^p)$. Following the overall structure of the eight-chain model (1), the plastic network stretch λ_{ec}^p is obtained in terms of the first invariant of the reference plastic deformation tensor C^p, i.e. $\lambda_{ec}^{p2} = 1/3\text{tr}[C^p] = 1/3[(\lambda_1^p)^2 + (\lambda_2^p)^2 + (\lambda_3^p)^2] = \sum_{a=1}^3 1/3\exp[2\epsilon_a^p]$, where $\lambda_1^p, \lambda_2^p, \lambda_3^p$ are eigenvalues of the plastic Cauchy-Green tensor C^p and λ_{ec}^p

refers to the continuum-stretch or network-stretch of the eight-chain (ec) model. Once the expression of continuum-stretch λ_{ec}^p has been established, the tensorial part of the back stress (7) can be derived using the result $\partial_{\varepsilon^p}\epsilon_a^p = n_a^p \otimes n_a^p$. For the eight-chain model, the closed-form expression for the back stress yields eventually

$$\beta = \mu_p\frac{3N_p - \lambda_{ec}^{p2}}{3[N_p - \lambda_{ec}^{p2}]}\boldsymbol{C}^p. \tag{9}$$

The evolution law for the plastic strain rate $(\dot{\gamma}^p)$ can be obtained following the Arrhenius type ansatz as proposed in Argon et al. (1)

$$\dot{\gamma}^p := \dot{\gamma}_0 \exp\left[-\frac{A\tilde{s}}{\theta}\left[1 - \left(\frac{\tau}{\tilde{s}}\right)^{5/6}\right]\right] \geq 0 \tag{10}$$

where $\tau := \sqrt{\text{dev}(\sigma^\star):\text{dev}(\sigma^\star)/2}$ and $\dot{\gamma}_0, A$ are the material parameters. The modified athermal shear strength is expressed by $\tilde{s} := s + \alpha p$ while θ is the absolute temperature. A phenomenological expression for the evolution of s is proposed as

$$\dot{s} = h[1 - s/s_{ss}]\dot{\gamma}^p \quad \text{with} \quad s(t=0) = s_0 \tag{11}$$

where h, s_0 and s_{ss} denote additional material parameters describing the slope of the softening, the initial and steady state (saturation) values of the athermal shear strength s, respectively. For further details on this model and its numerical implementation, see Miehe et al. (3).

4 SIMULATION AND VALIDATION

In this section, the modelling capacity of the Miehe model for bulk and nano-filled amorphous glassy polymers has been assessed by comparing its fitting capacity to homogeneous experimental data. The homogeneous test data involve the stress-true strain curves (strain expressed in logarithmic scale) obtained from experiments conducted under different conditions, e.g. different strain rates, various amount of nano-fillers (called filler content, normally expressed in percentage) etc on some well-known glassy polymers, e.g. PMMA (Polymethylmethacrylate), PS (Polystyrene). In this study, the main ingredient is silica as nanofiller.

As it was discussed earlier the constitutive modelling of the nano-filled polymer composite will be obtained by adapting the proper set of material parameters to the model proposed for the bulk glassy polymer. In order to determine the material parameters, initially, the parameter sets will be optimized for unfilled polymer. Then, the same algorithm will be used to determine the parameter sets for nanocomposites.

The validation results of silica-filled polystyrene (PS) and its nanocomposite counterpart are demonstrated in Fig (4) where the filler content is 5.0%. The strain rate for these experiments is 1%/min while

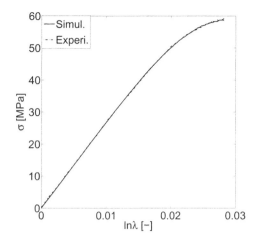

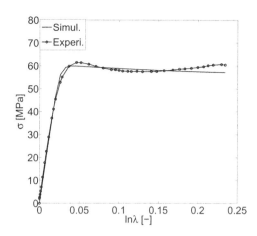

Figure 5. Comparison between simulation and experiment of PMMA-silica-hematite composite at 23°C temperature and 1%/min strain rate. The material parameter identified in the simulations are $\kappa = 133$ MPa, $\mu = 1181$ MPa, $\mu_p = 15.5$ MPa, $N_p = 36.4$, $\dot{\gamma}_0 = 1.13 \times 10^2 s^{-1}$, $A = 104.5$ K/MPa, $h = 10$ MPa, $s_0 = 94.4$ MPa, $s_{ss} = 20$ MPa, $\alpha = 0.1$.

The comparison between experimental and simulation results for PMMA-silica-hematite composite is presented in Fig (5). As it is mentioned earlier, although this nanocomposite is tested at room temperature which is quite below of its glass transition temperature but it shows considerable ductile behaviour compared to its homopolymer counterpart. The Fig (5) shows a good agreement between experiment and simulation.

5 CONCLUSION

In an attempt to adapt a finite strain modular, easy to implement, constitutive model for nano-filled amorphous glassy polymers, the suitability of the logarithmic strain space-based Miehe model has been investigated. The adaptibility of the Miehe model is validated by experimental results of silica-filled and silica-hematite-filled amorphous glassy nanopolymers. It is clear from the previous discussions that the Miehe model can be utilized in case of different types of nano-filler modelling within the same framework as for the bulk glassy polymers by adapting the material parameter sets properly.

ACKNOWLEDGMENTS

Financial support by the European Union (EU) project NanoModel is gratefully acknowledged.

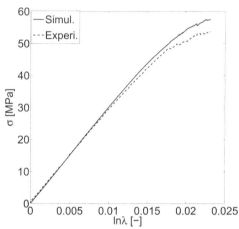

Figure 4. Performance of the model on the experimental data. Bulk and silica-filled polystyrene (PS) at room (23°C) temperature with 1%/min strain rate. (a) The material parameters identified for bulk polystyrene in the simulation are $\kappa = 133$ MPa, $\mu = 1526.2$ MPa, $\mu_p = 4.38$ MPa, $N_p = 6.14$, $\dot{\gamma}_0 = 4.07 \times 10^{-1} s^{-1}$, $A = 28.7$ K/MPa, $h = 269.5$ MPa, $s_0 = 141$ MPa, $s_{ss} = 269.5$ MPa, $\alpha = 0.1$. (b) Uniaxial tensile stress-stretch diagrams of the polystyrene (PS) with 5.0% nano-filler concentration. The material parameter identified in the simulations are $\kappa = 134$ MPa, $\mu = 1308$ MPa, $\mu_p = 21.2$ MPa, $N_p = 1.6$, $\dot{\gamma}_0 = 3.88 \times 10^{-1} s^{-1}$, $A = 30.01$ K/MPa, $h = 5.6$ MPa, $s_0 = 129.9$MPa, $s_{ss} = 256.63$ MPa, $\alpha = 0.1$.

both experiments are performed on annealed microsamples at room temperature. At this temperature, the experimental results, neither for bulk polymer nor for nano-filled polymer did confirm any softening and hardening behaviours. Subsequently, the materials is brittle of around 3% deformation which can be considered in the linear strain domain. Despite the fact that the Miehe model is well-suited for modelling the softening and kinematic hardening behaviours in large deformations, the same constitutive framework can be fitted for small strain results (without well-defined hardening and softening) with quite good agreement between experiments and simulations, cf. Fig (4).

REFERENCES

[1] Boyce MC, Parks DM, Argon AS, *Large Inelastic Deformation of Glassy Polymers. Part I: Rate Dependent Constitutive Model*. Mechanics of Materials 7: 15–33, 1988.

[2] Kontou E, Anthoulis G, *The Effect of Silica Nanoparticles on the Thermomechanical Properties of Polystyrene*. Journal of Applied Polymer Science 105:1723–1731, 2007.

[3] Miehe C, Goektepe S, Diez M, *Finite Viscoplasticity of Amorphous Glassy Polymers in the Logarithmic Strain Space*. International Journal of Solids and Structures 46:181–203,2009.

[4] Mori T, Tanaka K, *Average Stress in Matrix and Average Elastic Energy of Materials with Misfitting Inclusions*. Acta Metallurgica 21: 571–574,1973.

[5] Mulliken AD, Boyce MC, *Polycarbonate and a Polycarbonate-POSS Nanocomposite at High Rates of Deformation*. Journal of Engineering Materials and Technology 128:543–550, 2006.

[6] Odegard GM, Gates TS, Nicholson LM, Wise KE, *Equivalent-Continuum Modeling of Nano-Structred Materials*. Composites Science and Technology 62 : 1869–1880, 2002.

[7] Rubin MR (1994), *Plasticity Theory Formulated in Terms of Physically Based Microstructural Variables-Part I. Theory*. Internatonal Journal of Solids and Structures 31(19):2615–2634, 1994.

[8] Spathis G, Kontou E, *Mechanism of Plastic Deformation for Polycarbonate Under Compression by a Laser Extensometer Technique*. Journal of Applied Polymer Science 79:2534–2542, 2002.

[9] Tomita Y, *Constitutive Modelling of Deformation Behaviour of Glassy Polymers and Applications*. International Journal of Mechanical Sciences 42: 1455–1469, 2000.

[10] Wu PD, Giessen E van der, *Computational Aspects of Localized Deformations in Amorphous Glassy Polymers*. Eur. J. Mech. A/Solids 15(5): 799–823, 1996.

Constitutive Models for Rubber VII – Jerrams & Murphy (eds)
© *2012 Taylor & Francis Group, London, ISBN 978-0-415-68389-0*

Aspects of crack propagation in small and finite strain continua

Michael Kaliske, Kaan Özenç & Hüsnü Dal

Institute for Structural Analysis, Technische Universität Dresden, Germany

ABSTRACT: The contribution presents an r-adaptive crack propagation scheme for the description of the elastic and viscoelastic response of rubber-like materials at large strains and considers details of its three-dimensional numerical implementation within the context of finite element method (FEM). The eight-chain model and the Bergstörm-Boyce model are considered for the description of hyperelastic and finite viscoelastic behaviour of rubber-like bulk material, respectively. However, the methodology can be generalized to any hyperelastic and viscoelastic formulation at finite strains. The proposed FEM based computational framework for the fracture process is thermodynamically consistent in the sense of Coleman's method. We start with the work of Miehe and Gürses (CMAME 198: 1413–1428, 2009) and extend the approach for a generalized finite inelastic continuum. The key feature of this procedure is restructuring the overall system by duplication of crack front degrees of freedom based on the minimization of the overall energy via the Griffith criterion. The crack driving force and the crack direction are predicted by the material force approach. The predictive capability of the proposed method is demonstrated by representative numerical examples.

1 INTRODUCTION

The physical force concept, which is traced back to Newton and Galilei, describes the relation between physical forces and deformations, which engineers deal with in order to investigate quantities during continuum and structural mechanical investigation. However, in fracture mechanics, issues addressing discontinuities, flaws and inhomogeneities can only be described with additional physical and mathematical concepts. Based on the mechanics in material space, a concept of generalized forces acting on imperfections of crystals such as dislocations, foreign atoms and grain boundaries etc. is first introduced by Eshelby (1951). Although, he used the term forces in singularity in his time, nowadays they are commonly called as material or configurational forces. Eshelby stress can be interpreted as the representation of the negative gradient of total energy with respect to the position of an imperfection. In this manner, in a cracked body, this explanation coincides with the J-integral of Rice (1968) in vectorial setting where its tangential component to crack surface represents the total strain energy release rate. Different derivations of the Eshelby stress tensor are given in literature which is also known as energy-momentum tensor (see Braun (1997)). In the work of Maugin and Trimarco (1997), this theory is developed within the context of large strain. A general application for finite element implementation of material forces is presented by Braun (1997), Steinmann (2000), Steinmann et al. (2001). The extension of the material forces approach to nonlinear and inelastic materials is treated by Maugin (1993) and Gurtin (1995). In addition, the finite element

implementation of the material force concept to small strain inelastic and finite inelastic materials is introduced by Nguyen et al. (2005), Näser et al. (2007) and Näser et al. (2009), respectively.

The thermodynamical consistency of brittle crack propagation based on energy minimization algorithm and Griffith's laws with r-adaptive crack propagation by using numerical evaluation of material forces in finite element discretization is outlined by Miehe and Gürses (2007) and Gürses and Miehe (2009). In this contribution, we elaborate with the work of Miehe and Gürses (2007) a general framework applicable to large strain inelastic continua. In this sense, the history of internal variables and nodal displacement fields have to be correctly considered during numerical solution for the consistent formulation of the balance of the material motion at frozen time step for crack propagation. Therefore, an additional step is implemented to expand the developed algorithm in order to project and map history variables.

The current path of crack propagation is modeled by an r-adaptive alignment of the critical nodes at the crack tip by the help of the material force resultant around them. To this end, the data structure is constituted with a dynamic memory allocation (see Ortiz and Pandolfi (1999)). In a typical incremental or frozen loading step, the trial value of the material forces at the crack tip is obtained for the enhanced domain integral value around the crack-tip nodes. This information provides a robust failure criterion. In other words, whenever this value exceeds the critical state, the energy of the system will start to be absorbed by developing a crack surface and the release of an aligned crack faces will take place. This scheme provides a natural setting of fracturing in the finite element method. Therefore, the released face and duplicated node will provide a new stress free zone as it is in the nature of brittle fracture theory. In general circumstances, this fracture process can cause a decrease in stiffness of the system upon mesh adaptivity due to an increase of the crack size.

2 CONSTITUTIVE MODEL

In order to describe more accurate elastic or viscoelastic behavior, many material models have been published. In this sense, Marckmann and Verron (2006) discussed capabilities of various rubber elasticity models and their performance under uniaxial tension, equibiaxial tension and pure shear. According to their fitting capabilities to experimental data, the extended tube model of Kaliske and Heinrich (1999), the micro-sphere model of Miehe et al. (2004) and Ogden's model are rated as the most successful approaches among other 21 models. In general, the accuracy of a viscoelastic model depends not only on the constitutive law for the evolution of the internal variables, but also on the constitutive model describing the ground state elastic response. In the contribution, the Bergstörm-Boyce model with time-dependent viscoelastic behavior is considered by using the parallel assembly of an elastic spring and a Maxwell-element.

Let $\varphi : \mathbf{X} \mapsto \mathbf{x}$ be the deformation map at time $t \in \mathcal{R}_+$ of a body. φ maps points $\mathbf{X} \in \mathcal{B}$ of the reference configuration $\mathcal{B} \subset \mathcal{R}^3$ onto points $\mathbf{x} = \varphi(\mathbf{X}; t) \in \mathcal{S}$ of the current configuration $\mathcal{S} \subset \mathcal{R}^3$. Let $\mathbf{F} := \nabla \varphi(\mathbf{X}; t)$ denote the deformation gradient with the determinant $J := \det \mathbf{F} > 0$. The material time derivative of the deformation gradient is $\dot{\mathbf{F}} = \boldsymbol{l}\mathbf{F}$, where $\boldsymbol{l} := \nabla \boldsymbol{v}$ is the spatial velocity gradient. The boundary value problem for a general inelastic body is governed by the balance of linear momentum

$$\rho \, \ddot{\varphi} = \operatorname{div}[\boldsymbol{\tau}/J] + \boldsymbol{B} \,, \tag{1}$$

along with prescribed displacement boundary conditions $\boldsymbol{\varphi} = \bar{\boldsymbol{\varphi}}(\mathbf{X}; t)$ on $\partial \mathcal{B}_t$ and the traction boundary conditions $[\boldsymbol{\tau}/J]\boldsymbol{n} = \bar{\boldsymbol{t}}$ with the outward normal $\boldsymbol{n}$. ρ is the density and $\boldsymbol{B}$ is the prescribed body force with respect to unit volume of the current configuration. Subsequently, we consider a volumetric and isochoric decomposition of the elastic part of the Helmholtz free energy function with $\bar{\mathbf{F}} = J^{-1/3}\mathbf{F}$. Furthermore, the unimodular part of deformation gradient is split into elastic and inelastic contributions as $\bar{\mathbf{F}} = \mathbf{F}^e \mathbf{F}^i$. A decoupled volumetric-isochoric representation of finite viscoelasticity is obtained by the special form of the Helmholtz free energy function for a unit reference volume

$$\Psi = U(J) + \bar{\Psi}(\boldsymbol{g}, \mathbf{F}^i, \mathbf{F}^e) \,, \tag{2}$$

where the volumetric part of the free energy is described by the simple form

$$U(J) = \kappa(J - \ln J - 1) \,. \tag{3}$$

κ is the bulk modulus and will be taken large enough in order to enforce incompressibility. The isochoric part of the free energy is further decomposed into elastic and viscous contributions

$$\bar{\Psi} = \bar{\Psi}^e(\bar{\mathbf{C}}) + \bar{\Psi}^v(\mathbf{C}_e) \,, \tag{4}$$

where

$$\bar{\Psi}^e(\lambda_r) := \mu N \left(\lambda_r \mathcal{L}^{-1}(\lambda_r) + \ln \frac{\mathcal{L}^{-1}(\lambda_r)}{\sinh \mathcal{L}^{-1}(\lambda_r)} \right) \tag{5}$$

and

$$\bar{\Psi}^v(\lambda_r^e) := \mu_v N_v \left(\lambda_r^e \mathcal{L}^{-1}(\lambda_r^e) + \ln \frac{\mathcal{L}^{-1}(\lambda_r^e)}{\sinh \mathcal{L}^{-1}(\lambda_r^e)} \right) \,. \tag{6}$$

$\lambda_r = \sqrt{I_1/3N}$ and $\lambda_r^e = \sqrt{I_1^e/3N_v}$ denote the relative network stretches for the equilibrium and nonequilibrium deformations. $I_1 := \operatorname{tr}\bar{\mathbf{C}}$ and $I_1^e := \operatorname{tr}\mathbf{C}_e$ are the first invariants of the total and elastic right Cauchy-Green tensors, respectively. μ and μ_v are the shear moduli of the elastic chain network and superimposed free chain network, whereas N and N_v are the segment numbers of the elastic and superimposed networks, respectively. For further information regarding the finite element implementation of the Bergström-Boyce model, we refer to Dal and Kaliske (2009).

3 MATERIAL FORCE APPROACH IN FINITE INELASTICITY

As it is briefly outlined, the concept of material forces can be interpreted as the driving force acting on every kind of inhomogeneities in the material manifold. Elastomeric materials show significant amount of hysteresis. Moreover, energy is dissipated in the fracture process zone due to viscous effects. Accordingly, this effect must be taken into account in fracture mechanical investigations. The subsequent discussion is devoted to the derivation of the Eshelby and Mandel stress tensor for a general finite viscoelastic continuum. To this end, the local form of the momentum balance in the reference configuration is multiplied from the left by $\mathbf{F}^T$ in order to perform a pull back operation

$$\mathbf{F}^T \operatorname{DIV} \mathbf{P} + \mathbf{F}^T \mathbf{B} = 0 \,. \tag{7}$$

For the sake of simplicity, inertia effects are neglected. Applying the chain rule to the first term, Eq. 7 can be rewritten as

$$\operatorname{DIV}(\mathbf{F}^T \mathbf{P}) - \mathbf{P} : \nabla_X \mathbf{F} = \mathbf{F}^T \mathbf{B} \,. \tag{8}$$

Subsequently, the material gradient of the strain energy density yields

$$\nabla_X \Psi = \nabla_X \Psi^e(\mathbf{F}) + \nabla_X (J^i \bar{\Psi}^i(\mathbf{F} \mathbf{F}^{i-1})) \,. \tag{9}$$

By assembling the partial derivatives of the strain energy function together into Eq. 9, one obtains

$$\nabla_X \Psi = \frac{\partial \nabla_X \Psi^e}{\partial \boldsymbol{F}} : \frac{\partial \boldsymbol{F}}{\partial \boldsymbol{X}} + \frac{\partial \Psi^e}{\partial \boldsymbol{X}}\bigg|_{exp}$$

$$= +\frac{\partial J^i}{\partial \boldsymbol{X}} + \frac{\partial \bar{\Psi}^i}{\partial \boldsymbol{F}^e} : \frac{\partial \boldsymbol{F}^e}{\partial \boldsymbol{X}} + \frac{\partial \Psi^i}{\partial \boldsymbol{X}}\bigg|_{exp} . \quad (10)$$

Combination of Eq. 8 and Eq. 10 will yield the material balance

$$\mathrm{DIV}(\boldsymbol{\Sigma}) = \bar{\boldsymbol{M}}^i : \nabla_X \boldsymbol{F}^i + \boldsymbol{F}^T \boldsymbol{B} + \frac{\partial \Psi}{\partial \boldsymbol{X}}\bigg|_{exp} \quad (11)$$

with the Eshelby stress tensor $\boldsymbol{\Sigma} = \Psi \boldsymbol{1} - \boldsymbol{F}^T \boldsymbol{P}$ in the material configuration and Mandel stress tensor $\bar{\boldsymbol{M}}^i = \boldsymbol{F}^{i-T}(\boldsymbol{F}^T \boldsymbol{P}^i)$ in the intermediate configuration, respectively. For further information on the above mentioned formulation and the finite element implementation, we refer to Näser et al. (2007) and Näser et al. (2009).

4 ALGORITHMIC TREATMENT OF THREE DIMENSIONAL ADAPTIVE CRACK PROPAGATION OF VISCOELASTIC RUBBER

In this section, we briefly comment on an algorithm of crack propagation in a medium and prediction of its failure path. Hereof, the data structure required for simulation of failure of element faces by node duplication is outlined by Ortiz and Pandolfi (1999). Recently, it is generalized for energy minimization techniques by using the material force approach by Miehe and Gürses (2007) for elastic continua. In this contribution, a tetrahedral element $\mathcal{T}$ is introduced as an object which is composed of four nodes $\{\mathcal{N}_i\}_{i=1...4}$, six faces $\{\mathcal{F}_i\}_{i=1...6}$, and six lines $\{\mathcal{L}_i\}_{i=1...6}$. Similarly, a face $\mathcal{F}$ is an object and consists of three nodes $\{\mathcal{N}_i\}_{i=1...3}$ and three lines $\{\mathcal{L}_i\}_{i=1...3}$. The third class of objects is a line $\mathcal{L}$ and it is related to two nodes $\{\mathcal{N}_i\}_{i=1...2}$. Additionally, every node is also an object in the data structure. The schematic view of the data structure can be seen in Figure 1. Initially, the data structure remains unchanged unless node duplication, in other words crack propagation, does occur. An advancement of the crack requires three steps. First, the standard connectivity information and numerical solution of the finite elements is needed. Secondly, a new data structure from the last topology has to be formed. As last step, history data belonging to internal variables and displacement fields needs to be stored and mapped for the numerical solution of the next time step.

In Figure 2, a general view of an adaptive algorithm of doubling the critical faces and the critical nodes can be seen. Here, $n_{\mathcal{F}_c}$ is the crack face normal belonging to $\mathcal{F}_c$. $\mathcal{N}_c$ is a node at the crack tip and $\mathcal{G}_c$ is a material force resultant at node $\mathcal{N}_c$. After having found critical node $\mathcal{N}_c$, the critical face $\mathcal{F}_c$ needs to be determined. This procedure can be archived by a loop over the faces

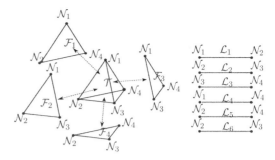

Figure 1. Data structure of a tetrahedral object.

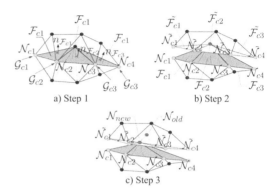

a) Step 1   b) Step 2

c) Step 3

Figure 2. Adaptive element alignement algorithm and node doubling procedure.

which are related to the critical node $\mathcal{N}_c$. When this is accomplished, the crack tip nodes can be duplicated by their mutual ones $\tilde{\mathcal{N}}_c$. In the last step, the new crack surface is aligned to the required direction by moving the node which is not an object of the crack tip node components.

The computation of a boundary value problem by the finite element method demands a global representation of the internal variable data which is only available at the integration points. Since the data structure and the complete numerical problem are changed during mesh adaptivity, the data has to be stored and projected from old Gauss points to the new ones. This transformation can be achieved by projection of these variables onto node level. The L_2-projection introduced by Zienkiewicz and Taylor (2000) is used for the minimization of the functional

$$\Pi = \int_{\mathcal{B}} (\hat{\boldsymbol{\mathcal{I}}}^{sm} - \hat{\boldsymbol{\mathcal{I}}}^{or})^T (\hat{\boldsymbol{\mathcal{I}}}^{sm} - \hat{\boldsymbol{\mathcal{I}}}^{or}) dV , \quad (12)$$

where $\hat{\mathcal{I}}^{sm}$ and $\hat{\mathcal{I}}^{or}$ are smoothed and original internal variables at Gauss points, respectively. Additionally, we can introduce smoothed internal variables as a function of the shape functions $\boldsymbol{N}$ and nodal internal variables $\mathcal{I}$

$$\hat{\boldsymbol{\mathcal{I}}}^{sm} = \sum_{i=1}^{E} \boldsymbol{N}^i \boldsymbol{\mathcal{I}}_i . \quad (13)$$

By minimizing of $\Pi \rightarrow Min$, we obtain the following equation which provides the projection of the variables from Gauss point to nodes in the global system

$$\int_{\mathscr{B}} \boldsymbol{N}^{iT} \boldsymbol{N}^i \boldsymbol{\mathcal{I}} = \int_{\mathscr{B}} \boldsymbol{N}^i \hat{\boldsymbol{\mathcal{I}}} . \qquad (14)$$

As soon as the internal history variables are available at the nodal points, they can be mapped with the moving node as described in Figure 2. We can introduce this mapping as

$$\tilde{\boldsymbol{\mathcal{I}}} = \sum_{i=1}^{E} \boldsymbol{N}^i(\xi) \boldsymbol{\mathcal{I}} . \qquad (15)$$

In the above equation, $\tilde{\boldsymbol{\mathcal{I}}}$ is the tensor of the new nodal values of internal variables while $\boldsymbol{\mathcal{I}}$ is the tensor of the old nodal values of internal variables. As it is described in Eq. (15), this mapping can be done by the shape functions. In this procedure, $\xi \in [-1; 1]$, which is the location of new node in the isoparametric element, is unknown. After the transfer, the history variables are provided at the nodal point of the new mesh. In order to evaluate the next step of the general solution, inverse mapping has to be done from nodal points to new Gauss points.

5 NUMERICAL EXAMPLES

In the following section, we will assess the capability of the proposed algorithmic setting and the material force approach with a couple of examples. First, we validate the proposed model with experimental results taken from the published paper by Pidaparti et al. (1990) which uses Styrene Butadiene Rubber without viscosity and crystallization. Then, fracture parameters are computed in order to give a general idea of the success of the material force approach. Following, a simulation of crack propagation of an inclined side crack specimen from this reference paper is numerically investigated. In the second subsection, the mixed mode crack examples by using the same parameters are simulated in order to show further the capability of the proposed algorithmic setting. And finally, stability of the proposed algorithm will be discussed using the viscoelastic rubber model.

5.1 Example 1

In the first part of the numerical examples, we made a parameter characterization study to evaluate mechanical properties of the rubber which we obtain from the reference paper (1990). For this purpose, fitting of the considered material model with respect to the experimental data, which are obtained from a tension test with specimen of $2.54 \times 10.16 \times 0.21$ cm rubber sheets, is carried out. In Figure 3, the numerical solution of the test by finite element method of the material with $\kappa = 1000$ MPa, $\mu = 1.95$ MPa and $N = 4.65$ and its correspondence to experimental data can be seen.

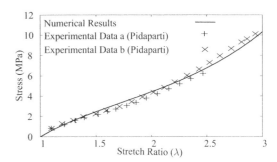

Figure 3. Engineering stress versus stretch curves.

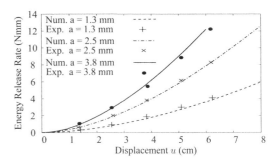

Figure 4. Energy release rates measured by material force approach and their fit to experimental data with various crack lengths.

Subsequently, rubber sheets with dimensions $2.54 \times 10.16 \times 0.21$ cm and single edge cracks are subjected to displacement based loading. Similarly, they are numerically evaluated and compared to the experimental data with various crack sizes $a = 0.13$, 0.25, 0.38 cm, respectively. Energy release rates are evaluated by a domain integral around the crack tip. In Figure 4, the accordance of the material force approach with respect to experimental results is described. Figure 4 successfully justifies the assumption made for the evaluation of the energy release rate by the material force approach in these examples. After validation of the material model and the material force approach for given examples, we investigate the proposed algorithm for the simulation of crack propagation. To this end, we investigate an inclined side crack example under mode I loading. Figure 5 presents the failure path of the inclined side crack specimen and its mesh alignment. In this example, the fracture criterion is employed as $\mathcal{G}_{critical} = 9.5$ Nmm from the reference paper (see Pidaparti et al. (1990)).

5.2 Example 2

In the following subsection, we extend the numerical crack propagation examples by mixed mode loading. Accordingly, two asymmetric test specimens with the same material parameter as in the previous examples are considered. The asymmetry is implemented to single edge crack specimens with cutting circles from the

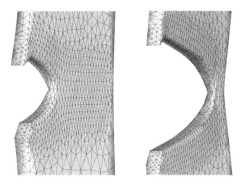

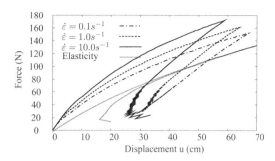

Figure 5. Tensile test of inclined crack specimen with a 45°. angle.

Figure 7. Displacement versus force diagram for visco-elastic crack propagation.

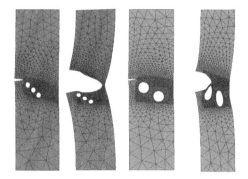

Figure 6. Failure mechanism of mixed-mode test specimen with tetrahedral elements under tension loading.

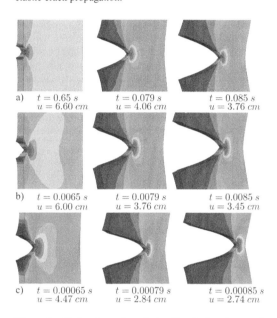

a) $t = 0.65\ s$ $t = 0.079\ s$ $t = 0.085\ s$
    $u = 6.60\ cm$ $u = 4.06\ cm$ $u = 3.76\ cm$

b) $t = 0.0065\ s$ $t = 0.0079\ s$ $t = 0.0085\ s$
    $u = 6.00\ cm$ $u = 3.76\ cm$ $u = 3.45\ cm$

c) $t = 0.00065\ s$ $t = 0.00079\ s$ $t = 0.00085\ s$
    $u = 4.47\ cm$ $u = 2.84\ cm$ $u = 2.74\ cm$

Figure 8. Finite element analysis of a single edge crack specimen under longitudinal displacement based loading with strain rates a) $\dot{\varepsilon} = 0.1\ \mathrm{s}^{-1}$, b) $\dot{\varepsilon} = 1.0\ \mathrm{s}^{-1}$, c) $\dot{\varepsilon} = 10.0\ \mathrm{s}^{-1}$.

half symmetric part of the subject which are below the crack. Figure 6 demonstrates two examples in order to represent the capability of the algorithmic setting under mixed mode loading and finite elasticity. In this Figure, geometries and crack propagation paths can be seen. Since a dynamic data structure update is inevitable during solution step in mixed mode examples, history data belonging to displacement fields need to be projected to its new positions. The details of this procedure is widely explained in previous sections.

5.3 Example 3

In this example, the modeling capabilities of the proposed algorithm with the material force approach and the viscoelastic material model are investigated. In this example, similar to the previous one, due to the dynamic data structure, a projection procedure is required. However, not only history data of displacement fields have to be projected but also history data of internal variables belonging to inelasticity. In this example, a specimen with a single edge crack $a = 0.20$ cm is considered and subjected to the same loading at three different displacement rates as $\dot{\varepsilon} = 0.1 s^{-1}$, $\dot{\varepsilon} = 1.0\ s^{-1}$ and $\dot{\varepsilon} = 10.0\ \mathrm{s}^{-1}$. Material parameters are taken as $\kappa = 1000\ \mathrm{MPa}$, $\mu = 1.95\ \mathrm{MPa}$, $\mu_v = 3.12\ MPa$, $N = 4.65$, $N_v = 4.65$ $\dot{\gamma}_0/\hat{\tau}^m = 7\ \mathrm{s}^{-1} MPa^{-m}$ $c = -1$ and $m = 4$. The same

fracture criteria is considered as in the previous examples, however, the after first step of crack propagation, we start to decrease the loading. The reason why we use this kind of loading is that we experience unstable crack propagation with r-adaptivity which can cause inaccurate numerical solutions with the Newton-Raphson method. However, by using decreasing loading, we avoid this problem. In Figure 7, corresponding curves to these solutions and its comparison with the curve without viscous response can be found.

The deformed configurations and the crack paths of the single edge crack examples under tension loading are depicted in Figure 8 at different loading rates.

6 CONCLUSIONS AND DISCUSSIONS

We extended the crack propagation scheme proposed by Miehe and Gürses (2007), Gürses and Miehe (2009) to a general finite inelastic continuum. Since

141

the consideration of inelastic material depends on history fields, we implement a projection approach and an appropriate data structure to the r-adaptive crack propagation algorithm. This algorithmic treatment provides adequate solution steps even in unstable crack propagations.

Particularly, we used the developed algorithmic setting for the analysis of crack propagation in finite strain continua with a viscous response. The Bergström-Boyce model is used for the numerical examples by the finite element method.

The numerical study has been carried out in order to assess the influence of material behavior on the fracture process. The solution method provides a stable solution for elastic and viscoelastic problems. The predicted fracture path is shown to fit well to published experimental data in the nonlinear elastic case. Unfortunately, due to the lack of experimental data for mixed mode fracture in viscoelastic systems, we only made a numerical study with different strain rates. Consequently, the convergence and stability of the proposed algorithm is elaborated which demonstrates the efficiency of the algorithmic setting. Future work will be devoted to a deeper comparison of numerical and experimental studies.

REFERENCES

Braun, M. (1997). Configurational forces induced by finite-element discretization. *Proceedings of the Estonian Academy of Sciences, Physics, Mathematics 35*, 379–386.

Dal, H. & M. Kaliske (2009). Bergström-Boyce model for nonlinear finite rubber viscoelasticity: Theoretical aspects and algorithmic treatment for the FE method. *Computational Mechanics 44*, 809–823.

Eshelby, J. D. (1951). The force on an elastic singularity. *Philosophical Transactions of the Royal Society London A 244*, 87–112.

Gürses, E. & C. Miehe (2009). A computational framework of three-dimensional configurational-force-driven brittle crack propagation. *Computer Methods in Applied Mechanics and Engineering 198*, 1413–1428.

Gurtin, M. E. (1995). The nature of configurational forces. *Archive for Rational Mechanics and Analysis 131*, 67–100.

Kaliske, M. & G. Heinrich (1999). An extended tube-model for rubber elasticity: Statistical-mechanical theory and finite element implementation. *Rubber Chemistry and Technology 72*, 602–632.

Marckmann, G. & E. Verron (2006). Comparion of hyperelastic models for rubber-like materials. *Rubber Chemistry and Technology 12*, 835–858.

Maugin, G. A. (1993). Material inhomogeneities in elasticity. Chapman & Hall, London.

Maugin, G. A. & C. Trimarco (1997). Pseudomomentum and material forces in nonlinear elasticity: Variational formulations and application to brittle fracture. *Acta Mechanica 94*, 1–28.

Miehe, C., S. Göktepe, & F. Lulei (2004). A micro-macro approach to rubber-like materials. Part I: The non-affine micro-sphere model of rubber elasticity. *Journal of the Mechanics and Physics of Solids 52*, 2617–2660.

Miehe, C. & E. Gürses (2007). A robust algorithm for configurational-force-driven brittle crack propagation with r-adaptive mesh alignment. *International Journal for Numerical Methods in Engineering 72*, 127–155.

Näser, B., M. Kaliske, H. Dal, & C. Netzker (2009). Fracture mechanical behaviour of visco-elastic materials: application to the so-called dwell-effect. *Zeitschrift für Angewandte Mathematik und Mechanik 89*, 677–686.

Näser, B., M. Kaliske, & R. Müller (2007). Material forces for inelastic models at large strains: Application to fracture mechanics. *Computational Mechanics 40*, 1005–1013.

Nguyen, T. D., S. Govindjee, P. A. Klein, & H. Gao (2005). A material force method for inelastic fracture mechanics. *Journal of the Mechanics and Physics of Solids 53*, 91–121.

Ortiz, M. & A. Pandolfi (1999). Finite-deformation irreversible cohesive elements for three-dimensional crack-propagation analysis. *International Journal for Numerical Methods in Engineering 44*, 1267–1282.

Pidaparti, R. M. V., T. Y. Yang, & W. Soedel (1990). Plane stress finite element prediction of mixed-mode rubber fracture and experimental verification. *International Journal of Fracture 45*, 221–241.

Rice, J. R. (1968). A path independent integral and the approximate analysis of strain concentration by notches and cracks. *Journal of Applied Mechanics 35*, 379–386.

Steinmann, P. (2000). Application of material forces to hyperelastostatic fracture mechanics. I. Continuum mechanical setting. *International Journal of Solids and Structures 37*, 7371–7391.

Steinmann, P., D. Ackermann, & F. J. Barth (2001). Application of material forces to hyperelastostatic fracture mechanics. II. Computational setting. *International Journal of Solids and Structures 38*, 5509–5526.

Zienkiewicz, O. & R. Taylor (2000). *The Finite Element Method, Volume I, The Basis*. Butterworth Heinemann, Oxford.

Testing, modelling and validation of numerical model capable of predicting stress fields throughout polyurethane foam

C. Briody, B. Duignan & S. Jerrams

Dublin Institute of Technology, Dublin, Ireland

ABSTRACT: Wheelchair seating systems are specialised for a number of reasons as users can have impaired mobility, which increases the possibility of pressure build up. These areas of high pressure frequently occur in the trunk region under the bony prominences known as the Ischial Tuberosities (IT), pressure ulcers may occur consequently. Polyurethane foam has been in use for some time in wheelchair seating systems as it exhibits good pressure relieving capabilities in most cases. However, little quantitative research has gone into foamed polymers, in comparison with conventional elastomeric materials. This lack of knowledge can ultimately lead to more time being spent in fitting, increased possibility of refitting and potentially an increase in trunk region pressures leading to the development of ulcers. Test results were used to accurately validate a Visco-Hyperfoam material model. Accurately simulating an indentation procedure using FE software verified the validation of the material model.

1 INRODUCTION

A pressure ulcer can be defined as a localised injury to the skin or underlying tissue, usually over a bony prominence, as a result of pressure, or pressure in combination with shear forces or friction (Black 2007). Pressure ulcers have the potential to diminish physical, psychological and social wellbeing and cause serious pain and discomfort which drastically decreases quality of life (Medicine 2001; Voss et al. 2005). There are two types of pressure ulcers, superficial and Deep Tissue Injury (DTI). Upon application of bodyweight, high pressure radiates outwards from bony prominences called the Ischial Tuberosities which are located on the pelvis. This high pressure can cause the damage termed DTI (Bouten, Oomens et al. 2003). Superficial ulcers occur on the outer layers of the skin tissue, although generally the extent of a superficial pressure sore is not as serious as a DTI. Improved understanding of the behaviour of the materials used in wheelchair seating can enable superior designs with improved pressure distribution. This will enhance comfort and support and potentially reduce the onset of pressure ulcers.

Polyurethane foam is an open celled elastomeric polymer and its constituent elastomer, polyurethane rubber, can undergo large and reversible deformations. Foamed polymeric material is known to exhibit three regions of different stress-strain behaviour in simple uniaxial compression: (i) approximately linear behaviour for strains less than about 0.05; this linear elasticity arises from the bending of the cell edges, (ii) a plateau region in which strain increases at constant or nearly constant stress up to a strain of roughly 0.6; this plateau arises from elastic buckling of the cell edges and (iii) a densification of the collapsed cell edges causing the foam to act as would its elastomeric constituent material. In this final region, known as the densification region, the slope of the stress-strain curve increases exponentially with strain as the crushed foam's cell struts and vertices come into contact (Gibson 1997). When the material reaches this level of compression, it is clinically referred to as 'bottomed out'.

Viscoelastic materials can be idealised as an intermediate combination of elastic solids and viscous liquids (Ward 2004). All polyurethane foams exhibit some degree of viscoelasticity (Mills 2007). Viscoelastic polyurethane foam is widely used in clinical seating as it offers excellent comfort and support due to its particular polymeric properties which are dependent on time, temperature and strain rate. The work presented in this paper is part of an ongoing investigation into improvements in the methodologies in the specification of viscoelastic polyurethane foams in wheelchair seating.

2 MATERIAL TESTING

2.1 *Materials tested*

Open-celled polyurethane seating foam with a density of $40 \, \text{kg/m}^3$ was tested in this piece of work. Several empirical tests were conducted on foam samples and the results were used to represent the behaviour of the material using prediction based numerical material models, which would later be used for seating design optimisation.

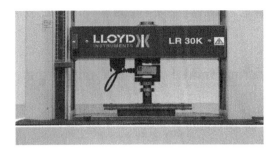

Figure 1. Uni-axial compression testing set-up.

2.2 Uni-axial compression testing

Uni-axial compression testing was conducted on the selected material in accordance with 'ISO 3386: Polymeric materials, cellular flexible – Determination of stress-strain characteristic in compression'. The compression tests were performed on a Lloyd LR 30 K materials testing machine which incorporated a calibrated 3 kN Lloyd instruments load cell as shown in Figure 1. The test-piece was inserted centrally between two horizontal platens in the testing machine. For the first test, the sample was compressed by 70% of its initial height at a strain rate of 5 mm/min. This continuous cycle was repeated immediately three times and on the fourth compression cycle, load-deflection data was recorded. The initial 3 cycles applied to the virgin foam sample removed most of the Mullins effect (Mullins 1969). After recording the load-deflection data, the sample was decompressed. This test procedure was then repeated at strain rates of 50, 100, 250 and 500 mm/min.

2.3 Constant displacement stress relaxation testing

The same test set-up as described in section 2.2 was used to conduct constant displacement stress relaxation testing following the guidelines proposed in 'ISO 3384 – Rubber, vulcanized or thermoplastic – Determination of stress relaxation in compression – Part 1: Testing at constant temperature'. This test procedure measured the decrease in counterforce exerted by a test piece of polymer foam which was compressed to a constant deformation. Samples were compressed at a strain rate of 250 mm/min and held at 80% compressive strain for extended time periods of up to 8 hours. This time was chosen to replicate a typical daily occupancy of a wheelchair user. The dissipating force was monitored over the entire period of the test.

2.4 Simple shear testing

Shear (rigidity) modulus testing was conducted in accordance with 'ISO 1827: Rubber, vulcanized or thermoplastic – Determination of modulus in shear'. Samples were bonded with cyanoacrylate adhesive on both sides to the rigid plates during testing. The shear load was applied at a rate of 4 mm/min until sample failure. A minority of the shear tests failed at

Figure 2. Shear test set-up, sample on left failed due to shear.

relatively low strain values due to adhesive failure – any test that failed at less than 100% shear strain was regarded as unrepresentative of material behaviour and disregarded.

2.5 Indentation Force Deflection testing

Indentation Force Deflection (IFD) tests were conducted on the foam. A circular indenter based on 'ISO:2439, Flexible Cellular Polymeric Materials-Determination of Hardness (Indentation Technique)' but scaled down to 81.2 mm in diameter, to be compatible with the 150 mm square test-pieces. This indenter was axially indented into the foam samples up to 65% of sample height using the Lloyd Instruments testing machine. The result of this test is presented in section 4.1 and compared to the results suggested by a Finite Element (FE) simulation.

3 DEVELOPMENT OF MATERIAL MODEL FROM TEST DATA

3.1 Uni-axial test data fit

Nominal uniaxial compression test data sets, obtained from the procedure described in section 2.2, were used to calculate the constants for the 2nd order form of Ogden's Hyperfoam material model (Ogden 1972; Simulia 2010) described in Equation 1.

$$U = \sum_{i=1}^{N} \frac{2\mu_i}{\alpha_i^2} \cdot [\widehat{\lambda_1}^{\alpha_i} + \widehat{\lambda_2}^{\alpha_i} + \widehat{\lambda_3}^{\alpha_i} - 3 +$$
$$\frac{1}{\beta_i} \cdot ((J^{el})^{-\alpha_i \beta_i} - 1)] \qquad (1)$$

where N is the order of fitting, μ_i, α_i, and β_i are temperature-dependent material parameters to be determined by curve-fitting test data to the model and J^{el} is given by:

$$\hat{\lambda}_i = (J^{th})^{-\frac{1}{3}} \xrightarrow{yields} \widehat{\lambda_1}\widehat{\lambda_2}\widehat{\lambda_3} = J^{el} \qquad (2)$$

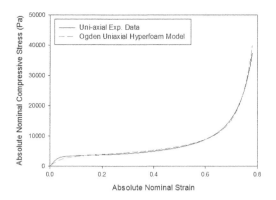

Figure 3. Uniaxial compression data for a sample compressed at a crosshead speed of 50mm/min, to a compressive strain of 80% at 20°C compared with Ogden Hyperfoam Model curve-fit.

Table 1. Coefficients of Ogden Hyperfoam model for uniaxial compression case

N	μ (Pa)	α	β
1	44185.6	21.4556	0
2	3.7050	-6.8900	0

where λ_i is the principal stretch ratio, J_{el} and J_{th} are the elastic and thermal volume ratios respectively and can be defined by the following equations $J_{el} = J/J_{th}$ and $J_{th} = (1 + \varepsilon_{th})^3$. J is the total volume ratio and the thermal strain, ε_{th}, is calculated from the temperature and the isotropic thermal expansion coefficient.

The test data modelled here is taken from a uniaxial compression test conducted on a sample at a strain rate of 5 mm/min. It was assumed that Poisson's ratio (ν) = 0 and that the lateral principal stretches, λ_1 and λ_3 can be considered to be zero. Equation 3 is used to calculate the nominal engineering stress, σ_2, in the λ_2 direction.

$$\sigma_2 = \frac{\partial U}{\partial \lambda_2} = \frac{2}{\lambda_2} \sum_{i=1}^{N} \frac{\mu_i}{\alpha_i} \left[\lambda_2^{\alpha_i} - J_{el}^{-\alpha_i \beta_i} \right] \qquad (3)$$

Abaqus automatically fits the parameters μ, α and β using a non-linear least squares optimisation procedure. The parameters used in the curve fit shown in figure 3, are given in Table 1.

Overall the model fits accurately to the experimental data in Figure 3. Some slight error is noticeable in the initial elastic region as the test data stiffer material than the model predicts. The shape of this initial elastic region is strongly dependent on the μ_1 material constant. This error can be eradicated by weighting the data towards the lower values of strain; however this would introduce error in the higher strain range. Error at lower values of strain was deemed less important than error at the more critical higher strain values.

Table 2. Coefficients of Ogden Hyperfoam model for simple shear case

N	μ (Pa)	α	β
1	7242.15	5.99916	0
2	7242.15	-5.99916	0

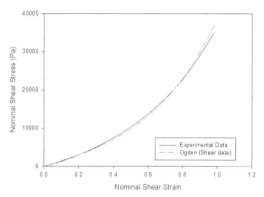

Figure 4. Simple shear data for a sample sheared at a crosshead speed of 4 mm/min at 20°C compared to Ogden Hyperfoam Model curve-fit.

3.2 Simple shear test data fit

Simple shear test data, extracted from the procedure described in Section 2.3, was also curve-fitted to the material model as during service the material will deform in both compression and shear. Simple shear was fitted using Equation 4

$$T_S = \frac{\partial U}{\partial \gamma} = \sum_{j=1}^{2} \left[\frac{2\gamma}{2(\lambda_j^2 - 1) - \gamma^2} \sum_{i=1}^{N} \frac{\mu_i}{\alpha_i} \left(\lambda_j^{\alpha_i} - 1 \right) \right] \qquad (4)$$

where γ is the shear strain and λ_j are the two principal stretches in the plane of shearing and are related to the shear strain by

$$\lambda_{1,2} = \sqrt{1 + \frac{\gamma^2}{2} \pm \gamma \sqrt{1 + \frac{\gamma^2}{4}}} \qquad (5)$$

Abaqus used the same curve-fit procedure to calculate representative parameters of the Ogden hyperfoam model to best fit the curve to the experimental simple shear data. The parameters calculated and shown in Table 2, give a very accurate curve fit for the shear loading mode as can be seen in Figure 4.

3.3 Combination of uniaxial compression data and simple shear data

A compromise was made in the accuracy of the two curve fits, uniaxial compression and simple shear to ensure the model could predict stress fields under complex combinations of both modes of deformation.

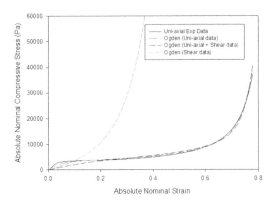

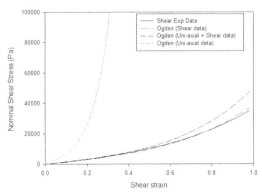

Figure 5. Uni-axial compression data for foam sample compressed at a crosshead speed of 50 mm/min to a compressive strain of 80% at 20°C compared with an Ogden Hyperfoam Model curve-fit for the uniaxial, shear and combination cases.

Figure 6. Simple shear data for a sample sheared at a crosshead speed of 4mm/min, at 20°C compared with an Ogden Hyperfoam Model curve-fit for shear, uni-axial and combination cases.

Table 3. Coefficients of Ogden Hyperfoam model for combination case

N	μ (Pa)	α	β
1	12740.4000	7.2810	0
2	2.7459	−5.7311	0

Data sets from both forms of deformation were used in the calculation of the Ogden hyperfoam constants. This meant that while some accuracy was lost when compared to the fits for each separate mode, the model was more robust. The derivation of model constants followed best practise guidelines (Simulia 2010), constants were derived from test modes which were the most relevant to the materials in-use mode of deformation. The accuracy of the model in simulating a uniaxial test is demonstrated graphically in Figure 5. Three different curve fits, uniaxial, shear and uniaxial plus shear, are plotted against the uniaxial compression test data curve in Figure 5. A compromise can clearly be seen in the accuracy of the curve-fit made in using both modes of deformation to derive material parameters when compared with using the uniaxial mode of deformation by itself. The inaccurate curve fit derived using simple shear test data demonstrates the importance of using more than one mode of deformation when simulating complex modes of deformation. The material parameters derived using both deformation mode data sets are shown in Table 3.

The accuracy of the model when a shear test is undertaken was also studied. The material coefficients were determined with uniaxial data, shear data and a combination of both data sets. The accuracy of each of the curve fits is compared in Figure 6. Again the model parameters derived from the mode of deformation in question generate the most accurate model, as was the case for the uni-axial test data modelling.

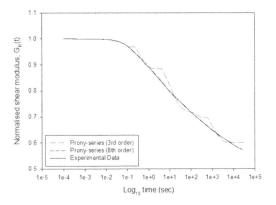

Figure 7. Normalised shear modulus, $g_R(t)$ plotted against $\log_{10}$ time.

3.4 Modelling viscoelastic behaviour

The foam modelled in this research displayed viscoelastic behavior which required modelling to simulate the foam's behaviour accurately. Stress relaxation was a prominent viscoelastic phenomenon noted during the compression and hold tests. Since bulk modulus is quite weak in this type of foamed polymer, the viscoelastic portion of the material model was dominated by the shear modulus, $G_R(t)$. Viscoelasticity was added to the model in the form of time based Prony-series constants based on the shear modulus of the foam, $g_R(t)$:

$$g_R(t) = 1 - \sum_{i=1}^{N} \bar{g}_i^P \left(1 - e^{-\frac{t}{\tau_i^G}}\right) \qquad (6)$$

where g_j^{-P} is the relaxation modulus, τ_j^G is the relaxation time. Both are material dependent properties and N is the order of the Prony-series.

The normalised shear modulus $g_R(t)$ is plotted against the $\log_{10}$ time (Figure 7) and ten data points (each decade of the log time plot) are extracted for use in the curve fitting procedure (3rd order). More data

Table 4. Coefficients of Prony-series model for 3rd order case, used to model viscoelasticity

N	G(i)	Tau(sec)
1	0.0973	0.30639
2	0.1740	11.21
3	0.1290	1011

Table 5. Coefficients of Prony-series model for 8th order case, used to model viscoelasticity

N	G(i)	Tau(sec)
1	6.17E-04	1.01E-03
2	−1.27E-03	1.89E-03
3	8.99E-02	0.2928
4	1.15E-01	4.7441
5	8.03E-02	55.234
6	7.72E-02	629.87
7	6.86E-02	8656
8	−3.01E-02	1.74E+08

Figure 8. IFD physical test set-up, 1/2 size sample compressed 50% of initial height.

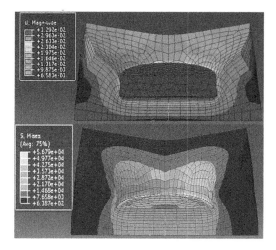

Figure 9. (a): Deformation plot of IFD test 1/2 size simulation in Abaqus compressed by 65% of initial height (m); (b) Von Mises stress plot of IFD test 1/2 size simulation in Abaqus compressed 65% of initial height (Pa).

points were then used to try and improve the curve fit, over 300 data points resulted in the model converging, with minimal error, after 8 iterations. The data is fitted using a non-linear least squares procedure to define the Prony-series parameters, (g_j^{-P}, τ_i), which are shown in Tables 4 and 5 for the respective model orders.

To incorporate viscoelasticity into the material model, the shear modulus of the foam, $g_R(t)$, is multiplied by the material constant, μ_i^0, in Ogden's strain energy function (Equation 1) giving μ_i^R (Equation 7).

$$\mu_i^R = \mu_i^0 (1 - \sum_{k=1}^{N} g_k^{-P} (1 - e^{-\frac{t}{\tau_k}}))$$ (7)

The introduction of viscoelasticity using the Prony series also enables the accurate prediction of loads with variable strain rate as well as the prediction of stress relaxation at constant strain.

4 FINITE ELEMENT SIMULATION OF STANDARD TESTS

Standard testing procedures on polyurethane foam sample were simulated using Abaqus FE software. As previously described, Ogden's material model (Equation 1) for describing the behaviour of compressible rubber-like materials (Ogden 1972) was chosen as a suitable strain energy function. Material test data sets were fitted to the material model and material constants were extracted that gave the most accurate and robust fit available. These coefficients were examined thoroughly as their accuracy was paramount to creating accurate simulations; their stability was ensured as they passed Drucker's criterion (Simulia 2010). These

modes of deformation were chosen as they were representative of the deformation undergone during seating. Only the loading curve was considered when evaluating the material parameters for the material model. Hysteretic effects were not simulated in the model presented here.

4.1 Simulation of IFD testing

An IFD (Indentation Force Deflection) test was conducted to demonstrate the accuracy of the material model. It can be seen from Figure 9(b) that the highest stress values were in tension along the side of the indenter. The accuracy of the model was initially validated by visually comparing material from tests and simulations at the sides of the sample and the grid deformation on the front face of the sample shown in Figure 9. The mesh used in this simulation was optimized by undertaking convergence testing. Refinement was conducted on the foam material mesh around the edge of where the indenter came into contact, as this is where mesh distortion was most likely to occur. The friction coefficient for the contact region was set to 0.75 (Mills 2000).

The simulated force in Figure 10 was the sum of the reaction forces from the top of the indenter; this

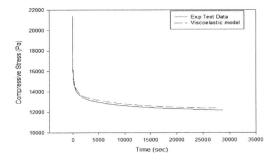

Figure 10. IFD test results compared to simulation results.

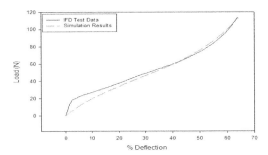

Figure 11. Graph showing accuracy of viscoelastic results over an extended time period.

force was compared to the force from the experimental IFD test procedure. The model demonstrated good accuracy over the majority of the load curve, with some initial elastic region inconsistencies attributed to minor inaccuracies in the material model that is described by the coefficients in Table 3. This curve validated the hyperelastic section of the material model. Figure 11 indicates the accuracy of the viscoelastic model. The foam samples response to the stress relaxation procedure described in section 2.3 is demonstrated. The model predicts an instantaneous stress value which is less than that from testing. The shape of the relaxation curve closely correlates with the predicted relaxation over the extended time period of 8 hours. Hence, the Prony series model is capable of predicting the viscoelastic response of the polyurethane foam over an extended time period.

5 CONCLUSIONS

Polyurethane foam samples were tested in uniaxial compression and simple shear modes. Test results were used to obtain suitable parameters for a second-order hyperelastic material model. This model was implemented in a simulation of an ISO indentation test. Good correlation was found between test results and simulation. The hyperelastic section of the validation process illustrated some inaccuracies in the initial strain region; this was due to similar inaccuracies within the material model. Constitutively modelling the initial elastic region for elastomeric materials is problematic and the initial region was not of in-service importance, therefore this error was not significant. The material also displayed some inherent viscoelastic properties, with stress relaxation being the most noticeable of these. Results of a compression-hold test were used to model the long term reduction of stress, which was modelled with good accuracy using a 7th order Viscoelastic Prony Series. With both the hyperelastic and viscoelastic sections of the model validated, the user can accurately interpret displacements and stresses throughout the material during loading while also being able to monitor stress dissipation over a longer time scale.

Future work will focus on the development of this model to incorporate temperature effects, ultimately this model will provide information to aid in the prescription of wheelchair seating systems.

REFERENCES

Black, J. et al (2007). "National Pressure Ulcer Advisory Panel's updated pressure ulcer staging system." *Urol Nurs.* **2**(27): 144–150.

Bouten, C. V. et al. (2003). "The etiology of pressure ulcers: Skin deep or muscle bound?" *Archives of Physical Medicine and Rehabilitation* **84**(4): 616–619.

Gibson, L. J. & Ashby, M. F. (1997). "Cellular Solids-Structures and Properties." Second Ed. (Cambridge University Press).

Medicine, C. o. S. C. (2001). "Pressure ulcer prevention and treatment following spinal cord injury: a clinical practice guideline for health-care professionals." *J Spinal Cord Med.* **24**: S40–101.

Mills, N. (2007). *Polymer Foams Handbook.* Oxford, Elsevier.

Mills, N. & Gilchrist, A (2000). "Modelling the indentation of low density Polymers." *Cellular Polymers* **19**: 389–412.

Mullins, L. (1969). "Softening of Rubber by Deformation." *Rubber Chemistry and Technology* **42**(1): 339–362.

Ogden, R. W. (1972). "Large Deformation Isotropic Elasticity: On the Correlation of Theory and Experiment for Compressible Rubberlike Solids." *Proceedings of the Royal Society of London. A. Mathematical and Physical Sciences* **328**(1575): 567–583.

Simulia (2010). "Analysis User's Manual." *Abaqus/CAE* **6.1**.

Siriruk, A., Y. et al. (2009). "Polymeric foams and sandwich composites: Material properties, environmental effects, and shear-lag modeling." *Composites Science and Technology* **69**(6): 814–820.

Voss, A. C. et al. (2005). "Long-Term Care Liability for Pressure Ulcers." *Journal of the American Geriatrics Society* **53**(9): 1587–1592.

Wada, A. et al. (2003). "A method to measure shearing modulus of the foamed core for sandwich plates." *Composite Structures* **60**(4): 385–390.

Ward, I. M. & Sweeney, J. (2004). *An introduction to the mechanical properties of solid polymers*, Wiley.

Constitutive Models for Rubber VII – Jerrams & Murphy (eds)
© *2012 Taylor & Francis Group, London, ISBN 978-0-415-68389-0*

Comparison of two approaches to predict rubber response at different strain rates

J.-C. Petiteau, E. Verron & R. Othman
LUNAM Université, École Centrale de Nantes, GeM, UMR CNRS 6183, Nantes cedex, France

H. Le Sourne
Laboratoire Énergétique Mécanique et Matériaux (LE2M), Institut Catholique des Arts et Métiers (ICAM), Carquefou, France

J.-F. Sigrist
DCNS Research, La Montagne, France

B. Auroire
DGA Techniques Navales, Toulon Cedex, France

ABSTRACT: The aim of this work is the choice of relevant constitutive equations for rubber-like materials subjected to intermediate strain rates loading conditions (from 1 to $100\,s^{-1}$). Classically, these materials are considered as large strain elastic solids and the constitutive equations are hyperelastic with more or less complicated free energy densities. In order to phenomenologically take into account the change of response (from rubber-like to glassy) of the polymers and investigate their damping properties at different strain rates, incompressible hyper-viscoelastic models must be considered. Two different approaches have been proposed in the bibliography: (i) the internal variable approach in which the strain is multiplicatively split into an elastic and a viscous part, the latter being driven by an evolution equation (similarly as for elastoplastic materials) and (ii) the integral approach that states that the response at a given time explicitly depends on the strain history of the material. We compare simple models derived from both approaches; their elastic parts are neo-Hookean and their viscous parts are (i) a Maxwell model and (ii) a basic time integral. It finally leads to two three-parameter constitutive equations. Simulating uniaxial tension-compression cycles at different strain rates, we demonstrate that the two approaches are not equivalent for intermediate strain rates, even if both are able to predict material strengthening and hysteresis; we investigate their differences and peculiarities.

1   INTRODUCTION

Elastomers are often used for damping parts in different industrial applications because of their remarkable dissipative properties. Indeed, they can undergo severe mechanical loading conditions: large strain and strain rates. Nevertheless, their mechanical response can vary from purely rubber-like to glassy depending on the strain rate undergone, as shown by Yi et al. (2006), and Sarva et al. (2007).

In order to simulate the behaviour of such non-linear materials, different types of model have been used:

- hyperelastic models predict the static response at finite strain (and also the response at very high strain rates),
- linear viscoelastic models describe the vibration response for small strain over a wide range of strain rates,
- hyper-viscoelastic constitutive equations have been derived to extend viscoelasticity theories to large strain.

The different zones of the strain rate-strain domain in which these different approaches can be used are schematized in Figure 1.

In the bibliography, hyper-viscoelastic constitutive equations are derived considering two different approaches: (i) the internal variable approach and (ii) the time-integral approach. Quantitatively, these two viscoelastic approaches lead to similar results. Nevertheless, the explicit relationships between them and their possible equivalence is not fully established. In the following, we compare them by considering two simple models: one of the internal variable models proposed by Huber and Tsakmakis (2000) and a simple integral model based on the K-BKZ theory (Kaye 1962, Bernstein et al. 1963). Both models only involve three material parameters. The relationship between these parameters will be investigated by considering three limiting loading cases: the strain rate tends to zero, the strain rate tends to infinity and the small strain response.

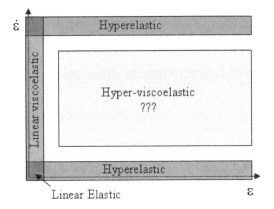

Figure 1. Applicability ranges of constitutive equations.

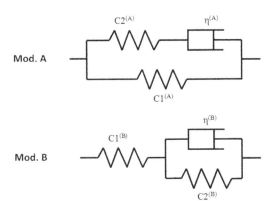

Figure 2. Rheological models A and B as proposed in Huber and Tsakmakis (2000).

2 THE TWO DIFFERENT APPROACHES

2.1 Internal variable model

2.1.1 Constitutive equation
Several different internal variable viscoelastic constitutive equations found in literature are based on the models proposed by Huber and Tsakmakis (2000). Their two models can be described by the rheological sketches presented in Figure 2.

Here, we only consider the model A and we will then refer to it as the H&T(A) model. We do not consider the model B because, as the strain rate tends to zero (i.e. for static loading conditions), the dashpot does not bring any rigidity and then an equilibrium between each spring has to be reached. Although both springs are hyperelastic, the result of this equilibrium state is not representable by a hyperelastic constitutive equation.

The internal variable models are based on the multiplicative split of the deformation gradient $\mathbf{F}$ into an elastic $\mathbf{F_e}$ and an inelastic $\mathbf{F_i}$ parts (Sidoroff 1974), the latter taking into account the viscous response:

$$\mathbf{F} = \mathbf{F_e}\mathbf{F_i}. \tag{1}$$

If rubber is supposed to be incompressible, we assume that both parts are incompressible, i.e. $\det \mathbf{F} = \det \mathbf{F_e} = \det \mathbf{F_i} = 1$.

Then, the total strain energy density of the material is defined as the sum of two strain energy densities corresponding to the two springs in Fig. 2 (Mod. A). In the single spring branch:

$$\Psi_1(\mathbf{B}) = \chi_1(I_B, II_B) \text{ with } \mathbf{B} = \mathbf{F}\mathbf{F}^T \tag{2}$$

in which $\mathbf{B}$ is the left Cauchy-Green strain tensor, and I_B and II_B are its two first invariants; and in the Maxwell part:

$$\Psi_2(\mathbf{B_e}) = \chi_2(I_{B_e}, II_{B_e}) \text{ with } \mathbf{B_e} = \mathbf{F_e}\mathbf{F_e}^T \tag{3}$$

where $\mathbf{B_e}$ is the left Cauchy-Green strain tensor of the elastic part of the deformation, and I_{B_e} and II_{B_e} are its two first invariants.

Finally, introducing the viscosity of the dashpot η and using the classical method that consists in applying the second Principle of Thermodynamics and the Coleman-Noll method (see Huber and Tsakmakis (2000) and Holzapfel (2000) for details), the set of constitutive equations that drives the model reduces to

$$\boldsymbol{\sigma} = -p\mathbf{I} + \boldsymbol{\sigma}_E, \tag{4a}$$

$$\boldsymbol{\sigma}_E = \boldsymbol{\sigma}_E^{(E)} + 2\frac{\partial \chi_2^{(A)}}{\partial I_{B_e}}\mathbf{B_e} + 2\frac{\partial \chi_2^{(A)}}{\partial II_{B_e}}\mathbf{B_e}^{-1}, \tag{4b}$$

$$\boldsymbol{\sigma}_E^{(E)} = 2\frac{\partial \chi_1}{\partial I_B}\mathbf{B} + 2\frac{\partial \chi_1}{\partial II_B}\mathbf{B}^{-1}, \tag{4c}$$

$$\dot{\mathbf{B}}_e = \mathbf{B_e}\mathbf{L}^T + \mathbf{L}\mathbf{B_e} - \frac{2}{\eta}\mathbf{B_e}\left(\boldsymbol{\sigma}_E - \boldsymbol{\sigma}_E^{(E)}\right)^D. \tag{4d}$$

where $\boldsymbol{\sigma}$, $\boldsymbol{\sigma}_E$ and $\boldsymbol{\sigma}_E^{(E)}$ are respectively the total Cauchy stress tensor, its extra part and its elastic-branch contribution; p is the hydrostatic pressure due to incompressibility; $\mathbf{L}$ is the velocity gradient; and the exponent $.^D$ refers to the deviatoric part of a tensor. The three first equations define the stress-strain relationship and the fourth one drives the evolution of the internal variable $\mathbf{F_e}$.

2.1.2 Limiting loading cases
In order to compare the two approaches, three limiting responses of the H&T(A) model are investigated: when the strain rate tends to zero, when the strain rate tends to infinity and for small strain.

1. For quasi-static experiments i.e. when the strain rate tends to zero:

$$\dot{\mathbf{F}} = \dot{\mathbf{F}}_e\mathbf{F_i} + \mathbf{F_e}\dot{\mathbf{F}}_i \to 0. \tag{5}$$

As $\mathbf{F}$ is not null, all strain rate tensors $(\dot{\mathbf{F}}_e, \dot{\mathbf{F}}_i, \mathbf{L})$ tend to zero. So, the deviatoric part of the stress becomes:

$$\left(\boldsymbol{\sigma}_E - \boldsymbol{\sigma}_E^{(E)}\right)^D \to 0 \tag{6}$$

which means that the complete tensor tends to a spherical tensor spherical

$$\sigma_E - \sigma_E{}^{(E)} \to p'\mathbf{I}, \tag{7}$$

p' being a scalar. Finally the quasi-static limiting response of the H&T(A) model reduces to the following hyperelastic equation:

$$\sigma = -(p+p')\mathbf{I} + 2\frac{\partial \chi_1}{\partial I_B}\mathbf{B} + 2\frac{\partial \chi_1}{\partial II_B}\mathbf{B}^{-1}. \tag{8}$$

2. When the strain rate tends to infinity, $\mathbf{F_e}$ tends to $\mathbf{F}$ (and then $\mathbf{B_e}$ tends to $\mathbf{B}$). So, the limiting response is hyperelastic:

$$\sigma = -p\mathbf{I} + 2\left(\frac{\partial \chi_1}{\partial I_B} + \frac{\partial \chi_2}{\partial I_B}\right)\mathbf{B}$$

$$+ 2\left(\frac{\partial \chi_1}{\partial II_B} + \frac{\partial \chi_2}{\partial II_B}\right)\mathbf{B}^{-1} \tag{9}$$

3. Finally, we examine the case of small strain. As it is more complicated than the two previous cases, we limit our investigation to the simplest model for which the springs are neo-Hookean. So, the constitutive equation only involves three material parameters: $C_1 = \partial \chi_1/\partial I_B$, $C_2 = \partial \chi_1/\partial I_{B_e}$ and η. Thus, the constitutive equation reduces to

$$\sigma = -p\mathbf{I} + 2C_1\mathbf{B} + 2C_2\mathbf{B_e}, \tag{10}$$

the evolution equation being unchanged. We consider the uniaxial response of this model in the small strain context. In this case, the deformation gradient is

$$\mathbf{F} = \begin{bmatrix} \lambda & 0 & 0 \\ 0 & \frac{1}{\sqrt{\lambda}} & 0 \\ 0 & 0 & \frac{1}{\sqrt{\lambda}} \end{bmatrix}. \tag{11}$$

Denoting ε the strain in the extension direction, the deformation gradient and the strain rate tensor reduce to

$$\mathbf{F} = \begin{bmatrix} 1+\varepsilon & 0 & 0 \\ 0 & 1-\frac{\varepsilon}{2} & 0 \\ 0 & 0 & 1-\frac{\varepsilon}{2} \end{bmatrix} \tag{12}$$

and

$$\dot{\mathbf{F}} = \begin{bmatrix} \dot{\varepsilon} & 0 & 0 \\ 0 & -\frac{\dot{\varepsilon}}{2} & 0 \\ 0 & 0 & -\frac{\dot{\varepsilon}}{2} \end{bmatrix}. \tag{13}$$

And the constitutive equation simplifies to the following linear viscoelastic equation

$$\sigma = 6C_1\varepsilon + 6C_2\varepsilon_e \text{ and } \dot{\varepsilon}_e = \dot{\varepsilon} - \frac{4C_2}{\eta}\varepsilon_e. \tag{14}$$

Introducing the classical complex notations of linear viscoelasticity, i.e. $\varepsilon = \exp(i\omega t)$, we calculate the storage modulus

$$E' = 6C_1 + 6C_2\frac{\omega^2}{\omega^2 + \left(\frac{4C_2}{\eta}\right)^2}, \tag{15}$$

and the loss modulus

$$E'' = 6C_2\frac{\omega\frac{4C_2}{\eta}}{\omega^2 + \left(\frac{4C_2}{\eta}\right)^2}. \tag{16}$$

2.2 Integral model

2.2.1 Constitutive equation

Number of integral viscoelastic models are based on the theory proposed by Green and Rivlin (1956). The corresponding constitutive equation relates the Cauchy stress tensor to a polynomial expansion of the Green-Lagrange strain tensor. Nevertheless, this approach is difficult to use practically (Verron 2008). Other authors proposed phenomenological models to overcome these difficulties; this is the case of the famous K-BKZ model (Kaye 1962, Bernstein et al. 1963, Tanner 1988), which has been formulated for both fluid and solid polymers.

Here, we consider a solid hyper-viscoelastic model that is composed by a hyperelastic part and a fluid K-BKZ model for the viscous part. It is to note that this model is different from the solid K-BKZ model; in fact, at our knowledge the solid model has not been used in the bibliography whereas the fluid model has proved its efficiency (Tanner 1988). Thus, the key quantity is the deformation gradient at the current time τ with respect to the final deformed configuration at time t

$$\mathbf{F}_t(\tau) = \mathbf{F}(\tau)\mathbf{F}^{-1}(t). \tag{17}$$

Introducing the elastic strain energy function $U_0(\mathbf{B})$ and the "viscous" strain energy $U_1(t-\tau, \mathbf{C}_t(\tau))$ (with $\mathbf{C}_t(\tau) = \mathbf{F}_t(\tau)^T\mathbf{F}_t(\tau)$), the stress-strain relationship (in the case of incompressibility) is given by

$$\sigma(t) = -p\mathbf{I} + 2\frac{\partial U_0}{\partial I_B}\mathbf{B} - 2\frac{\partial U_0}{\partial II_B}\mathbf{B}^{-1}$$

$$+ \int_{-\infty}^t \left[2\frac{\partial U_1(t-\tau, I_{C_t^{-1}}, II_{C_t^{-1}})}{\partial I_{C_t^{-1}}}\mathbf{C}_t^{-1}(\tau) \right.$$

$$\left. - 2\frac{\partial U_1(t-\tau, I_{C_t^{-1}}, II_{C_t^{-1}})}{\partial II_{C_t^{-1}}}\mathbf{C}_t(\tau) \right] d\tau. \tag{18}$$

Similarly to the H&T(A) model, we define the simplest K-BKZ solid model by considering the neo-Hookean elastic strain energy

$$U_0(\mathbf{B}) = g_0(I_B - 3), \tag{19}$$

and a viscoelastic counterpart of the neo-Hookean model

$$U_1 = \frac{g_1}{\tau_R}\exp\left(-\frac{t-\tau}{\tau_R}\right)(I_{C_t^{-1}} - 3). \tag{20}$$

Thus, the constitutive equation involves three parameters, two stiffness parameter g_0 and g_1, and a relaxation time τ_R, and reduces to

$$\boldsymbol{\sigma}(t) = -p\mathbf{I} + 2g_0\mathbf{B}(t)$$

$$+ \int_{-\infty}^{t} 2\frac{g_1}{\tau_R}\exp\left(-\frac{t-\tau}{\tau_R}\right)\mathbf{C}_t^{-1}(\tau)d\tau. \quad (21)$$

or, after integration by parts

$$\boldsymbol{\sigma}(t) = -p\mathbf{I} + 2g_0\mathbf{B}(t) - \mathbf{F}(t)\left[\int_{-\infty}^{t} 2g_1\right.$$

$$\left.\exp\left(-\frac{t-\tau}{\tau_R}\right)\frac{\partial\mathbf{C}^{-1}(\tau)}{\partial\tau}d\tau\right]\mathbf{F}^T(t). \quad (22)$$

2.2.2 Limiting loading cases

For this model and in order to investigate the limiting cases presented above, we use the method initially introduced by Christensen (1980) that consists in considering an accelerated/decelerated strain history through a time scale $\mathbf{C}(\alpha t)$ ($\alpha > 1$: accelerated and $\alpha < 1$: decelerated strain histories). The Cauchy stress tensor is then

$$\boldsymbol{\sigma}\left(\frac{t}{\alpha}\right) = -p\mathbf{I} + 2g_0\mathbf{B}(t) - \mathbf{F}(t)\left[\int_{-\infty}^{\frac{t}{\alpha}} 2g_1\right.$$

$$\left.\exp\left(-\frac{t/\alpha-\tau}{\tau_R}\right)\frac{\partial\mathbf{C}^{-1}(\alpha\tau)}{\partial\tau}d\tau\right]\mathbf{F}^T(t). \quad (23)$$

With a change of variable in the integral, it leads to

$$\boldsymbol{\sigma}\left(\frac{t}{\alpha}\right) = -p\mathbf{I} + 2g_0\mathbf{B}(t) - \mathbf{F}(t)\left[\int_{-\infty}^{t} 2g_1\right.$$

$$\left.\exp\left(-\frac{t-\gamma}{\alpha\tau_R}\right)\frac{\partial\mathbf{C}^{-1}(\gamma)}{\partial\gamma}d\gamma\right]\mathbf{F}^T(t). \quad (24)$$

1. The quasi-static loading case is obtained by considering that α tends to zero, i.e. for an infinitely decelerated strain history. Thus, the strain rate also tends to zero and we obtain the limiting hyperelastic model

$$\boldsymbol{\sigma}\left(\frac{t}{\alpha}\right)\Bigg|_{\alpha\to 0} = -p\mathbf{I} + 2g_0\mathbf{B}(t). \quad (25)$$

2. When the strain rate tends to infinity, i.e. for an infinitely accelerated strain history $\alpha \to \infty$, the constitutive equation reduces to

$$\boldsymbol{\sigma}\left(\frac{t}{\alpha}\right)\Bigg|_{\alpha\to\infty} = -p\mathbf{I} + 2(g_0 + g_1)\mathbf{B}(t). \quad (26)$$

Table 1. Material parameters of the models

Model H&T(A)		Model K-BKZ	
C_1 (MPa)	1	g_0 (MPa)	1
C_2 (MPa)	9	g_1 (MPa)	9
η (MPa.s)	1	τ_R (s)	1/36

3. Finally, as for the H&T(A) model, we examine the limiting model for small strain in uniaxial extension and we obtain:

$$\sigma(t) = 6g_0\varepsilon(t) + \int_{-\infty}^{t} 6g_1\exp\left(\frac{t-\tau}{\tau_R}\right)\frac{d\varepsilon(\tau)}{d\tau}d\tau, \quad (27)$$

and the storage and loss moduli are respectively

$$E' = 6g_0 + 6g_1\frac{\omega^2\tau_R^2}{1+\omega^2\tau_R^2}, \quad (28)$$

and

$$E'' = 6g_1\frac{\omega\tau_R}{1+\omega^2\tau_R^2}. \quad (29)$$

3 COMPARATIVE STUDY OF THE MODELS

3.1 Values of the material parameters

It is now simple to relate the material parameters of the models. Considering the stiffnesses under very low and very large strain rates, and the storage and loss moduli, leads to the following relationships between the material parameters

$$g_0 = C_1 \; ; \; g_1 = C_2 \; ; \; \tau_R = \frac{\eta}{4C_2}. \quad (30)$$

C_1 (g_0) drives the stiffness of the response for very low strain rates, C_2 (g_1) represents the additional stiffness exhibited at very high strain rates, and finally η (τ_R) accounts for the dependence of the response on the strain rate and leads to both strengthening at large strain, and size and shape of the hysteresis loop between loading and unloading.

In the following, we adopt the values given in Table 1.

3.2 Comparison of the mechanical response for different strain rates

The aim of this analysis is the comparison of the behaviour of both models with respect to the strain rate. For large strain, the strain rate can be defined in various ways. The simplest and most common choice is the nominal strain rate that is defined as the time derivative of the deformation gradient $\dot{\mathbf{F}}$. However, in order to study what is really experienced by the material, we

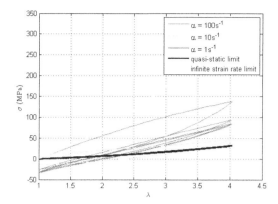

Figure 3. Cauchy stress vs. stretch ratio curve for the H&T(A) model at different strain rates.

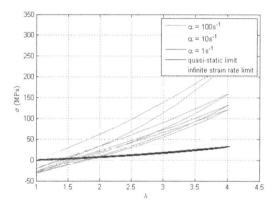

Figure 4. Cauchy stress vs. stretch ratio curve for the K-BKZ model at different strain rates.

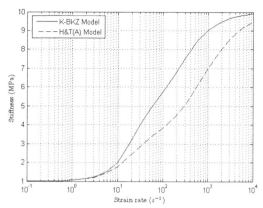

Figure 5. Stiffness of the models vs. strain rate.

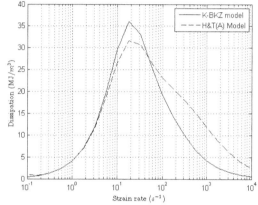

Figure 6. Dissipation depending on strain rate.

must consider the true (Eulerian) strain rate tensor $\mathbf{D}$ defined as follow

$$\mathbf{D} = \frac{1}{2}(\dot{\mathbf{F}}\mathbf{F}^{-1} + \mathbf{F}^{-T}\dot{\mathbf{F}}^{T}). \tag{31}$$

In the case of uniaxial tensile loading conditions, it reduces to

$$\mathbf{D} = \begin{bmatrix} \frac{\dot{\lambda}}{\lambda} & 0 & 0 \\ 0 & -\frac{1}{2}\frac{\dot{\lambda}}{\lambda} & 0 \\ 0 & 0 & -\frac{1}{2}\frac{\dot{\lambda}}{\lambda} \end{bmatrix}. \tag{32}$$

So, in order to simulate tensile-compressive loading cycles at constant strain rate, the stretch ratio λ is chosen as $\lambda = e^{\alpha t}$ where α is the prescribed true strain rate. The corresponding stress-strain curves are shown in Figure 4 for the H&T(A) model and in Figure 5 for the K-BKZ model. Both models exhibit strengthening as the strain rate increases as shown in Figure 5. The stiffness of both model increases. The corresponding limits for static and infinitely fast loading conditions are given by the values of the material parameters. It is to note that the stiffness of the internal model increases faster than the one of the internal variable model.

Moreover, the hysteresis loop evolves with the strain rate. For very low and very high strain rates, we recover our previous analytical results: dissipation is negligible and the response is nearly hyperelastic. For intermediate strain rates, dissipation takes place. The change in dissipation, i.e. the size of the hysteresis loop, with respect to strain rate is presented in Figure 6. The maximum dissipation corresponds to the same strain rate for both models. This maximum value is higher for the integral model and the dissipation decrease is faster for it than for the H&T(A) model. For small strain rates (from 0.1 to $10\,s^{-1}$), dissipation of the two models are similar. They differ for larger strain rates; as an example, the dissipation is equal to 10% of the maximum value for $\alpha = 10^{3}\,s^{-1}$ for the integral model and for $\alpha = 10^{4}\,s^{-1}$ for the internal variable model.

4 FINAL REMARKS

In this paper, we compare two approaches to derive constitutive equations for large strain viscoelasticity of elastomers. Two simple models, based on the neo-Hookean strain energy density, have been considered: an internal variable model developed by Huber and

Tsakmakis (2000) and an integral model based on the K-BKZ constitutive equation (Kaye 1962, Bernstein et al. 1963). For low strain rates, the models are revealed equivalent. This similarity can explain the fact that internal variable models are now widely used for the viscoelasticity of rubberlike materials (Verron 2008), because these models are easier to implement in implicit finite element softwares. Nevertheless, the approaches differs as the strain rate becomes higher. As an example, Hoofatt and Ouyang (2008) carried out dynamic experiments and showed that the stiffness increases as the strain rate increases. Then, for a similar internal variable model as the one considered here, they replace the constant viscosity by a material function of the invariants of $\mathbf{B}$ and $\mathbf{B_e}$ and consequently the viscosity increases as the strain rate increases. In regards with the present work, it appears that allowing the change in viscosity with the strain level makes the model response tend to the response of an integral model. In fact, it seems that the integral viscoelastic models are more appropriate to reproduce, with a small number of material parameters, the response of elastomers for large strain and intermediate strain rates. Moreover, recalling that the implementation of such models in dynamic explicit finite element softwares is not a very difficult task (Feng 1992, Verron et al. 2001), we think that the integral viscoelastic models, which were classically used in the 70's, must be reconsidered.

REFERENCES

Bernstein, B., E. A. Kearsley, and L. J. Zapas (1963). A study of stress relaxation with finite strain. *Trans. Soc. Rheol. 7*, 391–410.

Christensen, R. (1980). A nonlinear theory of viscoelasticity for application to elastomers. *J. Appl. Mech. 47*, 762–768.

Feng, W. W. (1992). Viscoelastic behavior of elastomeric membranes. *J. Appl. Mech. 59*, S29–S34.

Green, A. E. and R. S. Rivlin (1957). The mechanics of non-linear materials with memory, Part 1. *Arch. Ration. Mech. and Anal. 1*, 1–21.

Holzapfel, G. A. (2000). *Nonlinear Solid Mechanics. A continuum approach for engineering*. Chichester: J. Wiley and Sons.

Hoofatt, M. and X. Ouyang (2008). Three-dimensional constitutive equations for styrene butadiene rubber at high strain rates. *Mech. Mater. 40*, 1–16.

Huber, N. and C. Tsakmakis (2000, January). Finite deformation viscoelasticity laws. *Mech. Mater. 32*, 1–18.

Kaye, A. (1962). Non-newtonian flow in incompressible fluids. *College of Aeronautics Press, Cranford, U.K. Note 134*.

Sarva, S. S., S. Deschanel, M. C. Boyce, and W. Chen (2007). Stress-strain behavior of a polyurea and a polyurethane from low to high strain rates. *Polymer 48*, 2208–2213.

Sidoroff, F. (1974). Un modèle viscoélastique non linéaire avec configuration intermédiaire. *J. de Mécanique 13*, 679–713.

Tanner, R. I. (1988). From A to (BK)Z in constitutive relations. *J. Rheol. 32*, 673–702.

Verron, E., G. Marckmann, and B. Peseux (2001). Dynamic inflation of non-linear elastic and viscoelastic rubber-like membranes. *Int. J. Numer. Meth. Engng. 50*, 1233–1251.

Verron, E. (2008). Mechanical behaviour of rubber-like materials. *Journal Club Theme of iMechanica*, http://imechanica.org/node/4167.

Yi, J., M. C. Boyce, G. F. Lee, and E. Balizer (2006). Large deformation rate-dependent stress-strain behavior of polyurea and polyurethanes. *Polymer 47*, 319–329.

Constitutive Models for Rubber VII – Jerrams & Murphy (eds)
© 2012 Taylor & Francis Group, London, ISBN 978-0-415-68389-0

Numerical analysis of the heterogeneous ageing of rubber products

L. Steinke, U. Weltin & M. Flamm
Institute for Reliability Engineering, Technical University Hamburg-Harburg, Germany

A. Lion
Institute for Mechanics, Faculty of Aerospace Engineering, University of the Federal Armed Forces, Munich, Germany

M. Celina
Sandia National Laboratories, Albuquerque, NM, USA

ABSTRACT: Ageing of elastomeric materials is an important subject in product development concerning life-time prediction. Because of environmental influences like temperature, oxygen, radiation and similar conditions rubber and elastomers will change their material properties and hence affect the lifetime. The reason for this material change is the modification of the polymer network. In case of sulphur cured natural rubber it can result in hardening of the material and pronounced surface hardening for thick-walled products. This is known as the Diffusion-Limited Oxidation effect (DLO) when oxidation is the key environmental influence. This paper describes a model which is designed to calculate the change in network density caused by thermo-oxidative ageing and the required measurements for the parameterisation.

1 INTRODUCTION

Degradation of elastomers plays a crucial part in product development. The lifetime of most elastomeric products depends on the relaxation or compression set behaviour and hence on the condition of the elastomeric network. For seals for example it is very important to know the set behaviour to predict the point of time when leakage will occur. The degradation of the elastomeric network is mainly influenced by the environmental factors such as temperature, oxygen, ozone and radiation. In case of fluids the degradation process is depending on the diffusion behaviour of the elastomer which can lead to a heterogenious degradation of the material, known as Diffusion-Limited-Oxidation Effect (DLO). The degradation process thus is a chemical reaction controlled by diffusion with influence on the mechanical properties (see Figure 1). The waisted specimen was aged for 100 h at 130°C and then stressed at room temperature once. Due to the DLO effect the elastomeric network changed mainly in the surface area and the damage will first occur in this area while the inner part of the specimen will still be intact.

Over the past years different simulation models were developed which partially give answers to the previously mentioned issues. Chemical modeling of the degradation process including the diffusion behaviour was recently accomplished by Colin et al. (Colin et al. 2007, Colin et al. 2007). From the mechanical viewpoint lots of investigations were done like

Figure 1. Stressed waisted specimen aged for 100 h at 130°C.

Achenbach et al. who developed a degradation model based on a hyperelastic material model (Achenbach 2000, Achenbach 2007). Wise et al. developed a model which links the chemical reaction with a mechanical parameter, which is the tensile compliance (Wise et al. 1997). Although much work has been accomplished in this area both on the experimental and the simulation aspect the direct link between the chemical reaction and the impact from chain scission and cross linking on the mechanical behaviour is missing. This paper provides a contribution to the missing link between the degradation chemistry part which describes the DLO effect and the mechanical part modelling the

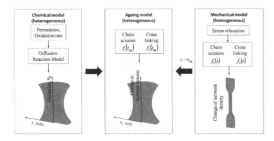

Figure 2. Concept of the degradation model.

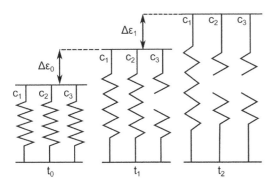

Figure 3. Linear spring model for chain scission.

cross linking and chain scission based on a Neo-Hooke model.

2 FORMULATION OF THE DEGRADATION MODEL

In this section, the concept of the degradation model is introduced. It consists of a chemical model which takes the diffusion and the oxidation due to oxygen into account and a mechanical model which describes the chain scission and cross linking of elastomeric chains based on a hyperelastic formulation (see Figure 2). The material law is formulated in a total Lagrange formulation in which the stress state is defined by the second Piola-Kirchhoff stress tensor. The material model was implemented to the FE-software MSC.MARC via the subroutine HYPELA2.

2.1 Mechanical model

The mechanical model describes the chain scission and cross linking of the elastomer chains based on a hyperelastic formulation. We briefly describe the essential derivations for each process which are motivated by a linear spring model.

2.1.1 Chain scission
The chain scission is based on the model pictured in figure 3.

The elastomeric body consists at time t_0 of the linear springs with the stiffness c_1, c_2 and c_3. The stress

which arises from the strain increment $\Delta\varepsilon_0$ at time t_0 is defined by

$$\sigma(t_0) = (c_1 + c_2 + c_3)\Delta\varepsilon_0. \tag{1}$$

Due to chain scission the stiffness of the system is reduced at time t_1 by the stiffness c_3. A new strain increment $\Delta\varepsilon_1$ at the same time defines the state of stress with

$$\sigma(t_1) = (c_1 + c_2)(\Delta\varepsilon_0 + \Delta\varepsilon_1). \tag{2}$$

If at this state the stiffness is again reduced the new state of stress follows:

$$\sigma(t_2) = (c_1)(\Delta\varepsilon_0 + \Delta\varepsilon_1). \tag{3}$$

The stiffness of the body is therefore only a function of time which leads to the following equation:

$$\sigma(t) = c(t)\varepsilon(t) \tag{4}$$

This can be transfered to a hyperelastic formulation. In case of an incompressible neo-Hookean model the potential reads

$$W_1 = C_{10}\left(III_C^{-\frac{1}{3}}I_C - 3\right) + \frac{1}{2}p(J - 1), \tag{5}$$

where C_{10} is the material parameter which is equivalent with the network density and hence the stiffness. p is the hydrostatic pressure. The kinematic relations are given by

$$I_C = \mathrm{tr}(\mathbf{C}),$$

$$III_C = \det(\mathbf{C}) \quad \text{and} \tag{6}$$

$$J = \sqrt{\det(\mathbf{C})}$$

where $\mathbf{C}$ is the right Cauchy-Green tensor.

By differentiating the potential from equation (5) with respect to the right Cauchy-Green tensor we get the second Piola-Kirchhoff tensor which reads

$$\mathbf{T}_1 = 2C_{ab}(t)\left(_1III_C^{-\frac{1}{3}}\mathbf{I} - \frac{1}{3}\,_1III_C^{-\frac{1}{3}}I_{C1}\mathbf{C}^{-1}\right)$$
$$+ _1p_1J_1\mathbf{C}^{-1} \quad . \tag{7}$$

Since the stiffness is a function of time the material parameter C_{10} was changed to the function $C_{ab}(t)$ which represents the decreasing stiffness due to chain scission. The lower left index 1 points to the current time increment whereas the previous time increment is indicated by the index 0.

2.1.2 Cross linking
The process of cross linking can also be described by a simple model with springs which was developed by Hossain et al. (2009) (Fig. 4).

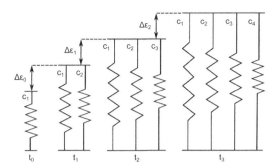

Figure 4. Linear spring model for cross linking.

At the time t_0 the stiffness of the body is defined by the spring c_1. By applying a strain increment $\Delta\varepsilon_0$ the stress is defined as

$$\sigma(t_0) = c_1\Delta\varepsilon_0. \tag{8}$$

At time t_1 a new spring is formed with stiffness c_2 which is under no stress and does not influence the current state of stress. This assumption that cross linking occurs under a stress-free state was made by Tobolsky and Eyring (1943). A new strain increment $\Delta\varepsilon_1$ then leads to the following stress formulation:

$$\sigma(t_1) = c_1\Delta\varepsilon_0 + (c_1 + c_2)\Delta\varepsilon_1 \tag{9}$$

If a new spring is formed in this state and a strain increment is applied, the stress at time t_3 is

$$\sigma(t_1) = c_1\Delta\varepsilon_0 + (c_1 + c_2)\Delta\varepsilon_1 + (c_1 + c_2 + c_3)\Delta\varepsilon_2. \tag{10}$$

For n springs and n strain increments the discrete form reads:

$$\sigma(t_n) = \sum_{i=0}^{n} c_i \sum_{j=i}^{n} \Delta\varepsilon_j \tag{11}$$

Equation (11) can be transformed into a continiuous formulation, namely

$$\dot{\sigma}(t) = c(t)\dot{\varepsilon}(t). \tag{12}$$

The detailed derivation of (12) can be found in (Hossain et al. 2009). This formulation can be used to describe the process of cross linking based on a hyperelastic material. The second Piola-Kirchhoff stress rate reads:

$$\dot{\mathbf{T}}_2(t) = C_{auf}(t)\left(_1III_C^{-\frac{1}{3}}\mathbf{I} \right.$$
$$\left. -\frac{1}{3}_1III_C^{-\frac{1}{3}}{}_1I_{C1}\mathbf{C}^{-1}\right)^{\bullet}, \tag{13}$$

where $C_{auf}(t)$ is the cross linking function. Applying Euler-backward the second Piola-Kirchhoff stress at the current time increment reads:

$$\mathbf{T}_2(t_{n+1}) = \mathbf{T}_2(t_n) + \Delta t C_{auf}(t)\left(_1III_C^{-\frac{1}{3}}\mathbf{I}\right.$$
$$\left. -\frac{1}{3}_1III_C^{-\frac{1}{3}}{}_1I_{C1}\mathbf{C}^{-1}\right)^{\bullet} \tag{14}$$

Evaluation of the time derivative leads to

$$\mathbf{T}_2(t_{n+1}) = \mathbf{T}_2(t_n) + \Delta t \frac{2}{3}C_{auf}(t)_1III_C^{-\frac{1}{3}}$$
$$\cdot\left(\mathbf{1}\otimes{}_1\mathbf{C}^{-1} + {}_1\mathbf{C}^{-1}\otimes\mathbf{1}\right.$$
$$-\frac{1}{3}{}_1I_{C1}\mathbf{C}^{-1}\otimes{}_1\mathbf{C}^{-1}$$
$$\left. - {}_1I_{C1}\overset{4}{\mathbb{H}}\right)\dot{\mathbf{C}} \tag{15}$$

with

$$\overset{4}{\mathbb{H}} = \frac{\partial_1\mathbf{C}^{-1}}{\partial_1\mathbf{C}}. \tag{16}$$

By applying

$$_1\dot{\mathbf{C}} = \frac{d\mathbf{C}}{dt} = \frac{_1\mathbf{C} - {}_0\mathbf{C}}{\Delta t} \tag{17}$$

to equation (15) the second Piola-Kirchhoff stress due to cross linking reads:

$$\mathbf{T}_2(t_{n+1}) = \mathbf{T}_2(t_n) + \Delta t \frac{2}{3}C_{auf}(t)_1III_C^{-\frac{1}{3}}$$
$$\cdot\left(\mathbf{1}\otimes{}_1\mathbf{C}^{-1} + {}_1\mathbf{C}^{-1}\otimes\mathbf{1}\right.$$
$$-\frac{1}{3}{}_1I_{C1}\mathbf{C}^{-1}\otimes{}_1\mathbf{C}^{-1}$$
$$\left. - {}_1I_{C1}\overset{4}{\mathbb{H}}\right)\frac{_1\mathbf{C} - {}_0\mathbf{C}}{\Delta t} \tag{18}$$

Superposition of equation (7) and (18) defines the second Piola-Kirchhoff stress due to both cross linking and chain scission.

2.2 Chemical model

The diffusion-reaction of oxygen with the elastomer can be described by Fick's second law with an additional chemical reaction. The equation for one dimension can be written as

$$\frac{\partial c(x,t)}{\partial t} = D\frac{\partial c(x,t)}{\partial x} - kc(x,t), \tag{19}$$

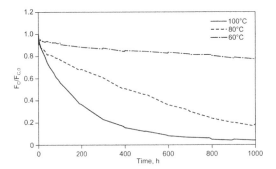

Figure 5. Continious stress relaxation.

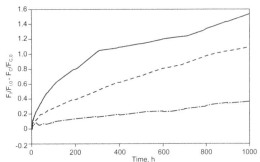

Figure 7. Superposed relaxation data.

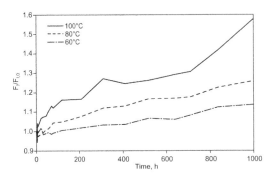

Figure 6. Intermediate stress relaxation.

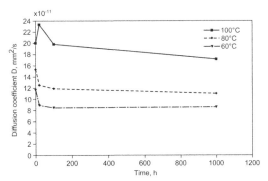

Figure 8. Diffusion coefficient.

where $c(x,t)$ is the oxygen concentration, D is the diffusion coefficient of the elastomer and k is the rate constant of the chemical reaction which represents a reaction of first-order kinetics. Most materials are expected to undergo a dominant chemical reaction which can be characterized by a first-order oder pseudo first-order reaction.

The important information from the chemical model is not the solution of equation (19) but the consumed oxygen $P(x,t)$ which can be calculated by integrating the chemical reaction over time

$$P(x,t) = \int_0^t kc(x,t). \qquad (20)$$

3 PARAMETERISATION OF THE DEGRADATION MODEL

3.1 Mechanical model

To estimate the chain scission function $C_{ab}(t)$ and the cross linking function $C_{auf}(t)$ the stress relaxation method can be used. This technique was developed by Tobolsky et al. (Tobolsky and Eyring 1943, Tobolsky et al. 1944). The process of chain scission can be observed with the so called continuous stress relaxation where a sample is hold under constant elongation at constant temperature and the force is measured over time (see Figure 5).

The intermediate stress relaxation measures both cross linking and chain scission at the same time. The sample is therefore aged at constant temperature and without stress. At discrete time intervals the sample is stressed to a specific elongation under room temperature. The force is then measured at each time interval (see Figure 6). To get information about the cross linking process both data have to be superposed (Fig. 7). From the continuous and superposed relaxation data we can estimate the functions for chain scission $C_{ab}(t)$ and cross linking $C_{auf}(t)$ by fitting mathematical functions to the data.

3.2 Chemical model

The diffusion coefficient was measured by standard permeation measurements with the induction time technique (Beck et al. 2003) for temperatures of 60°C, 80°C and 100°C. In addition different aged samples where measured to take the influence of a changed network on the diffusion coefficient into account. The data is shown in figure 8. The data shows that the diffusion coefficient is increasing with increasing temperature but is constant at constant temperature over time for 60°C and 80°C. At 100°C the diffusion coefficient is decreasing which could result from the reaction of oxygen with the sample due to the high temperature.

The rate constant k from the chemical reaction can be estimated from the oxygen consumption rate r which can be measured with a respirometer (Assink

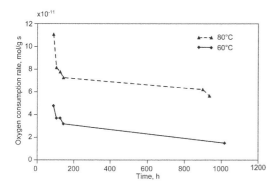

Figure 9. Oxygen consumption rate.

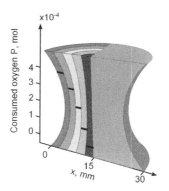

Figure 10. Solution of the chemical model at t = 70 h.

et al. 2005). For the measurement a specific amount of polymer, small enough to neglect the diffusion process of oxygen, was sealed in a chamber with a known amount of air and thus oxygen. The chamber was put in an oven for a certain time. To avoid that solubility and thermal equilibrium effects are influencing the measurement, the chamber was opened after a short time to equilibrate the pressure. After sufficient thermal exposure the air inside the chamber was measured with the respirometer and referenced to standard air conditions. The oxygen consumption rate can be calculated with the known amount of polymer, the ageing time and the measured oxygen difference. The rate constant k can be estimated from the oxygen consumption rate r with

$$k = \frac{r}{C_s} \qquad (21)$$

where C_s is the surface oxygen concentration which can be calculated with Henry's law:

$$C_s = pS \qquad (22)$$

p is the partial pressure of oxygen and S is the solubility of the elastomer. The data shows that the oxygen consumption rate is decreasing over time which is caused by the saturation of the elastomer sample (see Figure 9). Due to the marginal data, an exact conclusion when the saturation of the sample is occurring can't be made. The data at 80°C only shows that this is going to happen at the end of the measurement. Therefore the mean value of the consumption rate is taken to calculate the rate constant.

3.3 Degradation model

To calculate the amount of chain scission and cross linking in dependence of the consumed oxygen the chemical model is first solved for a specific time (see Figure 10). The consumed oxygen P within the sample is then transferred to an equivalent ageing time which can be calculated with

$$t_{eq} = \frac{P}{r}. \qquad (23)$$

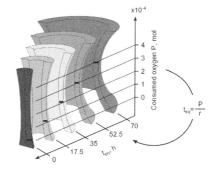

Figure 11. Equivalent ageing time at t = 70 h.

The equivalent ageing time is then applied to the cross linking and chain scission function within the mechanical model. Hence it is possible to calculate the amount of chain scission and cross linking in dependence of the consumed oxygen of the elastomer.

4 RESULTS AND DISCUSSION

To validate the degradation model relaxation experiments were done. Therefore, two samples were compressed and put into an heat oven. One sample was aged at 80°C up to 500 h and the other was aged at 80°C up to 1000 h. The samples were then unloaded and the compression set was measured (Fig. 12). The compression set at 500 h is approximately 4 mm and at 1000 h 6 mm. The degradation model was parameterised with the data at 80°C and an FE-calculation was done. From the calculation the compression set at 500 h resulted in 4.28 mm and at 1000 h 6.78 mm (see Figure 13).

The calculated compression set shows a good correlation with the experimental data, although a few simplifications concerning the chemical reaction and the parameterisation were made. The oxygen consumption rate for example is changing over time which was neglected and would lead if taken into account to a lesser compression set. In addition the oxygen consumption rate will depend on the effective oxygen partial pressure condition within the material, which

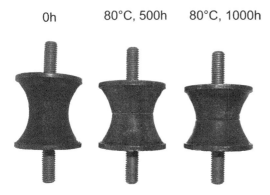

0h   **80°C, 500h**   **80°C, 1000h**

Figure 12. Compression set.

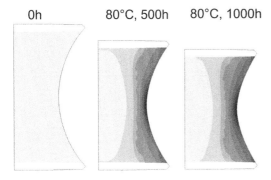

0h   **80°C, 500h**   **80°C, 1000h**

Figure 13. Calculated Compression set.

was neglected here but should be investigated. This will likely result in a slower oxidation rate within the material where lower partial pressures apply, and thus a lower amount of oxidised material. The relaxation behaviour was interpreted as pure chemical process which does not coincide with the reality. At the beginning of stress relaxation physical relaxation is dominant which is associated with the reorientation of the molecular network and with the disengagement and rearrangement of chain entanglements (Gent 2001). The physical relaxation was not taken into account which influences the stress relaxation experiments and therefore the set behaviour of the sample. At the beginning of the FE-calculation a physical relaxation model must be applied to simulate the physical relaxation behaviour whereas at longer ageing time the chemical model must dominante the stress relaxation. The parameterisation of the degradation model was carried out with sulphur cured NR samples.

5 CONCLUSION

An ageing model is presented which links the DLO effect with the changes of the molecular network by chain scission and cross linking functions. Chain scission and cross linking are based on a simple linear spring model and transfered to a hyperelastic formulation. The simulation model parameters were determined by stress relaxation, permeation and oxygen consumption experiments. The first results of the model fits the compression set data reasonably well.

REFERENCES

Achenbach, M. (2000). Service life of seals – numerical simulation in sealing technology enhances prognoses. *Commat 962*, 1–10.

Achenbach, M. (2007). Methodische Ansaetze zur Einbeziehung der Alterung in numerische Berechnungen. Technical report, Parker Hannifin GmbH & Co. KG.

Assink, R., M. Celina, J. Skutnik, & D. Harris (2005). Use of respirometer to measure oxidation rates of polymeric materials at ambient temperatures. *Polymer 46*, 11648–11654.

Beck, K., R. Kreiselmaier, V. Peterseim, & E. Osen (2003). Permeation durch elastomere Dichtungswerkstoffe Grundlagen – Werkstoffiegenschaften – Entwicklungstrends. *Kautschuk Gummi Kunststoffe 56*, 657–660.

Colin, X., L. Audouin, & J. Verdu (2007). Kinetic modelling of the thermal oxidation of polyisoprene elastomers. Part 3: Oxidation induced changes of elastic properties. *Polymer Degradation and Stability 82*, 906–914.

Colin, X., L. Audouin, J. Verdu, & M. Huy (2007). Kinetic modelling of the thermal oxidation of polyisoprene elastomers. Part 2: Effect of sulfur vulcanization on mass changes and thickness distribution of oxidation products during thermal oxiadtion. *Polymer Degradation and Stability 92*, 898–905.

Gent, A. (2001). *Engineering with with Rubber*. Hanser.

Hossain, M., G. Possart, & P. Steinmann (2009). A small-strain model to simulate the curing of thermosets. *Comuptational Mechanics 43*, 769–779.

Tobolsky, A. & H. Eyring (1943). Mechanical properties of polymeric materials. *Journal of Chemical Physics 11*, 125–134.

Tobolsky, A., I. Prettyman, & J. Dillon (1944). Stress relaxation of natural rubber and synthetic rubber stocks. *Journal of Applied Physics 15*, 380–395.

Wise, J., K. Gillen, & R. Clough (1997). Quantitative model for the time development of diffusion-limited oxidation profiles. *Polymer 38*, 1929–1944.

Constitutive Models for Rubber VII – Jerrams & Murphy (eds)
© 2012 Taylor & Francis Group, London, ISBN 978-0-415-68389-0

Proposal of an orthotropic hyperelastic model for fiber-reinforced rubber in the electric generator

A. Matsuda & K. Nakahara

Graduate School of Systems and Information Engineering, University of Tsukuba, Tsukuba, Ibaraki, Japan

ABSTRACT: In this paper, we propose of an orthotropic hyperelastic model for the fiber-reinforced rubber material used as rubber seals in electric generators. The fiber-reinforced rubber is composed of rubber matrix material that is reinforced by two families of fibers. The mechanical properties is anisotropic that depend on directions of two fiber families. We improve an anisotropic model by adding terms to the strain energy function to consider its mechanical properties in small deformation region. The energy functions of the fiber families was represented by power series of invariants of stretch and shear deformation in fiber families. Moreover, proposed strain energy function was polyconvex in totality. In order to identify material parameter of the proposed model, biaxial tensile test result of the rubber matrix and uniaxial tensile test result of the fiber-reinforced rubber were approximated by the nonlinear least squares method. Finally, applicability of the proposed model to the fiber-reinforced rubber was demonstrated by FEM simulation of tensile and compression deformation.

1   INTRODUCTION

The fiber reinforced rubber (FRR) is applied to sealing materials in the electric generators to seal water and oil from mechanical section. FRR is composite rubber material reinforced by orthotropic two families of fibers[1][2][3]. Its mechanical properties are complicated more than isotropic rubber material because of anisotropic reinforcing fibers[4]. Therefore, FEM simulation and constitutive model of the FRR were required for evaluation of mechanical properties and prediction of the lifetime of the sealing material.

In this paper, the constitutive model of a FRR used as sealing gasket in the electric generator was investigated. The matrix material of the FRR was nitrile rubber and reinforcing fibers was plain weaved cotton fibers. We propose an orthotropic anisotropic hyperelastic model which is possible to coordinate strength in warp and weft directions, individually.

2   ORTHOTROPIC HYPERELASTIC MODEL

The second Piola-Kirchhoff stress of hyperelasticity is given by the partial differentiation of the stored energy function W with respect to the right Cauchy-Green deformation tensor C as follows:

$$S = 2\frac{\partial \mathrm{W}(C)}{\partial C}, \tag{1}$$

where, the right Cauchy-Green deformation tensor C is calculated by

$$C = F^T F . \tag{2}$$

A hyperelastic body is characterized by the scalar potential $W(C)$. In the case of orthotropic material behavior, $W(C)$ reduces to an isotropic function of C and following structural tensors $M^{(1)}$ and $M^{(2)}$.

$$\begin{aligned} M^{(1)} &= n^{(1)} \otimes n^{(1)} \\ M^{(2)} &= n^{(2)} \otimes n^{(2)} \end{aligned}, \tag{3}$$

where $\mathbf{n}^{(1)}$ and $\mathbf{n}^{(1)}$ is unit vector which is oriented parallel to the each fibers. The stored energy function W is represented in dependence of the three invariants of C,

$$\begin{aligned} I_1(C) &= C : I, \\ I_2(C) &= Cof(C) : I, \\ I_3(C) &= \det(C), \end{aligned} \tag{4}$$

and the first invariants of $C \cdot M^{(1)}$, $C \cdot M^{(2)}$, $C^2 \cdot M^{(1)}$ and $C^2 \cdot M^{(2)}$, respectively[1]:

$$\begin{aligned} J_4^{(a)}(C, M^{(a)}) &= C : M^{(a)} \quad (a = 1,2), \\ J_5^{(a)}(C, M^{(a)}) &= C^2 : M^{(a)} \quad (a = 1,2), \end{aligned} \tag{5}$$

where $Cof(C)$ is given by $Cof(C) = A^{-T} det(\mathbf{A})$.

The anisotropic invariants $J_4^{(1)}$ and $J_4^{(2)}$ are correspond to the second power of each fiber length. Here we introduce other invariants K_1 and K_2 to satisfy the polyconvex condition as follows[5][7]:

$$K_1^{(a)}(C, M^{(a)}) = J_5^{(a)} - I_1 J_4^{(a)} + I_2 = Cof(C) : M^{(a)} \tag{6}$$

In this study, the stored energy function of orthotropic hyperelastic body was represented as addition of the isotropic term for matrix rubber and orthotropic term for reinforcing-fibers as follows[4][6][8]:

$$W = W_{ani}(\boldsymbol{C}, \boldsymbol{M}^{(a)}) + W_{iso}(\boldsymbol{C}) \tag{7}$$

where $\bar{I}_1$ and $\bar{I}_2$ are the first and second invariants of the volume preserved right Cauchy-Green deformation tensor $\bar{\boldsymbol{C}}$,

$$\begin{aligned} \bar{I}_1(\boldsymbol{C}) &= \bar{\boldsymbol{C}} : \boldsymbol{I} = I_3^{-1/3} \boldsymbol{C} : \boldsymbol{I} \\ \bar{I}_2(\boldsymbol{C}) &= Cof(\bar{\boldsymbol{C}}) : \boldsymbol{I} = Cof(I_3^{-1/3}\boldsymbol{C}) : \boldsymbol{I} \end{aligned} \tag{8}$$

Following Mooney-Rivlin model was introduced to the isotropic part,

$$W_{iso} = C_1(\bar{I}_1 - 3) + C_2(\bar{I}_2 - 3) + p(J-1) \tag{9}$$

where C_1 and C_2 are material coefficients and p is hydrostatic pressure, respectively. J is calculated as $J = det(\boldsymbol{F})$. For anisotropic part, following functions were introduced.

$$\begin{aligned} W_{ani} &= W_{Asai} + W_{lts} + W_{add} \\ W_{Asai} &= \frac{1}{4}\sum_{a=1}^{2}\left[\frac{C_J^{(a)}}{d_J^{(a)}}\left\{ \left(J_4^{(a)}\right)^{d_J^{(a)}} - 1\right\} + \frac{C_K^{(a)}}{d_K^{(a)}}\left\{ \left(K_1^{(a)}\right)^{d_K^{(a)}} - 1\right\}\right] \\ &\quad - \frac{1}{4}\sum_{a=1}^{2}\left\{ \left(C_J^{(a)} - C_K^{(a)}\right)\left(J_4^{(a)} - 1\right) + C_K^{(a)}\ln\left(J^2\right)\right\} \\ W_{lts} &= \frac{1}{4}\sum_{a=1}^{2}C_I^{(a)}\left\{ \left(J_4^{(a)} - 1\right) + \left(K_1^{(a)} - 1\right) - \ln\left(J^2\right)\right\} \\ W_{add} &= \frac{1}{4}\sum_{a=1}^{2}C_{Ja}^{(a)}\left(J_4^{(a)} - 1 + \frac{\exp\left\{d_{Ja}^{(a)}\left(1 - J_4^{(a)}\right)\right\} - 1}{d_{Ja}^{(a)}}\right) \end{aligned} \tag{10}$$

Here, $C_J^{(a)}, C_K^{(a)}, C_{Ja}^{(a)}, C_I^{(a)} \geq 0$ and $d_J^{(a)}, d_K^{(a)} \geq 1$ and $d_{Ja}^{(a)} > 0$. We use stored energy function $W_{Asai}^{(2)(3)}$ and $W_{lts}^{(6)(9)}$ and an additional energy function W_{add} for better representation in small strain region.

3 MATERIAL TESTS AND FEM SIMULATION

In this section, biaxial loading test results of isotropic rubber and uniaxial tensile loading test of FRR were shown.

3.1 Biaxial tensile loading test of matrix rubber

The bi-axial tensile loading tests (in Figure 1) were conducted to evaluate isotropic material behavior of matrix rubber. Rubber sheet without reinforced fibers was 120 mm in width and 2 mm in width. 100% stretch was applied to the tensile deformation and horizontal stretch was constrained. Material coefficient C_1 and C_2 were approximated by test results and relationships between stretch and nominal stresses were shown in Figure 2. Isotropic behavior of matrix rubber was confirmed to show good approximation by the Mooney-Rivlin model.

3.2 Tensile loading test of FRR

Tensile loading test specimens are 120 mm in length, 20 mm in width and 2 mm in thick (in Figure 3). The angle of directions between the warp fiber and tensile deformation is assumed to be the fiber orientation angle θ. The test specimens by which the fiber orientation angle θ became 0°, 15°, 30°, 45°, 60°, 75°, and 90° were produced. 20 mm at both ends were fixed by

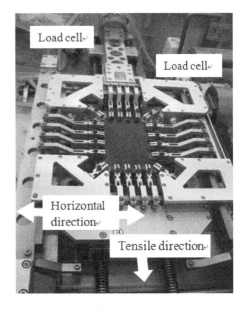

Figure 1. Biaxial tensile loading test.

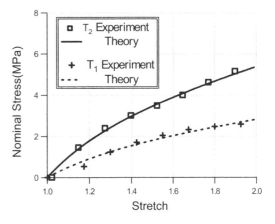

Figure 2. Relationships between nominal stress and stretch of biaxial tensile test.

Figure 3. Specimen of uniaxial tensile test.

jigs and 1.0%/sec of tensile deformation was applied to the specimens.

Material coefficients for the anisotropic stored energy functions were identified by the nominal stress and stretch relationships which were observed by experiments and calculated theoretically using equations 1 and 10. The Levenberg–Marquardt algorithm was applied to the least-square method for identification of material coefficients. The error function is shown as follows:

$$
E = \sum_{i=1}^{n} \left[\left\{ \overline{T}_2^{(i)}(\theta = 0) - T_2^{(i)}\left(\overline{\lambda}_1^{(i)}, \overline{\lambda}_2^{(i)}, \theta = 0\right) \right\}^2 \right.
$$
$$
+ \left\{ T_1^{(i)}\left(\overline{\lambda}_1^{(i)}, \overline{\lambda}_2^{(i)}, \theta = 0\right) \right\}^2
$$
$$
+ \left\{ \beta \overline{T}_2^{(i)}(\theta = 45) - T_2^{(i)}\left(\overline{\lambda}_1^{(i)}, \overline{\lambda}_2^{(i)}, \theta = 45\right) \right\}^2
$$
$$
+ \left\{ T_1^{(i)}\left(\overline{\lambda}_1^{(i)}, \overline{\lambda}_2^{(i)}, \theta = 45\right) \right\}^2
$$
$$
+ \left\{ \overline{T}_2^{(i)}(\theta = 90) - T_2^{(i)}\left(\overline{\lambda}_1^{(i)}, \overline{\lambda}_2^{(i)}, \theta = 90\right) \right\}^2
$$
$$
\left. + \left\{ T_1^{(i)}\left(\overline{\lambda}_1^{(i)}, \overline{\lambda}_2^{(i)}, \theta = 90\right) \right\}^2 \right]
$$

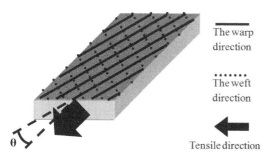

The warp direction

The weft direction

Tensile direction

Figure 4. Fiber orientation angle θ.

Figure 5. Fixture and specimen of tensile loading test.

where $\overline{\lambda}_2^{(i)}$ and $\overline{\lambda}_1^{(i)}$ are stretches of tensile direction and orthogonal direction of tensile deformation, respectively. Notation n is number of measured data.

Moreover, tensile loading test results of which fiber orientation angles are 0°, 45° and 90° were applied to the least-square method because deformation of tensile specimens were concentrated to the weak part of stiffness in the case of 15°, 30°, 60°, 75°. Material coefficients identified by the nonlinear least-square method are shown in Table 1.

3.3 FEM simulation of tensile loading test

3-dimentional FEM simulations are conducted to show the applicability and stability of proposed orthotropic model and simulation code. The FEM simulation code was developed using the total-Lagrange method and kinematic-nonlinearly in large deformation was considered.

To avoid the instability of incompressible behavior, displacement/pressure mixed method was applied. 8-displacement and 1-pressure node solid element was applied to the FEM model. The Newton-Raphson method was introduced for iteration calculation.

FEM model for representation of tensile loading test is shown in Figure 6. The top and bottom surface

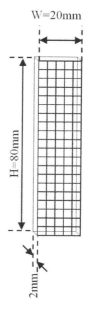

W=20mm

H=80mm

2mm

Figure 6. FEM model for tensile loading test.

Table 1. Isotropic and anisotropic parameters identified by the nonlinear least-square method

C_1	C_2	a	$C_J^{(a)\prime}$	$d_J^{(a)}$	$C_K^{(a)}$	$d_K^{(a)}$	$C_e^{(a)\prime}$	$d_e^{(a)}$	$C_I^{(a)}$
1.287	0.1483	1	296.0	2.3	2.0	1.0	1.0	2.0	0.0
		2	2.575	7.1	2.049	1.0	8.0	15.0	3.0

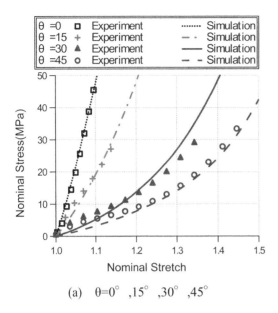

(a) θ=0° ,15° ,30° ,45°

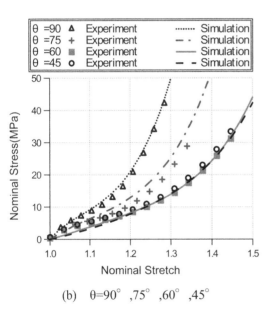

(b) θ=90° ,75° ,60° ,45°

Figure 7. Comparison with uniaxial tensile test and FEM simulation by nominal stress.

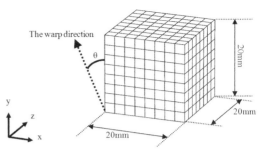

Figure 8. FEM simulation model and fiber orientation angle θ.

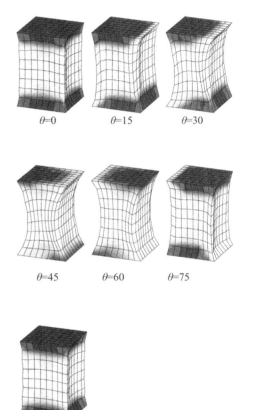

θ=0   θ=15   θ=30

θ=45   θ=60   θ=75

θ=90

0.00MPa   40.00MPa

Figure 9. Mean stress in 50% of tensile deformation.

were constrained and tensile deformation was applied to top surface in the simulation.

Relationships between nominal stress and stretch of which the fiber orientation angles θ are 0°, 15°, 30° and 45° are shown in Figure 7(a) with experimental results. Results of which angle θ are 15°, 45°, 70° and 90° are shown in Figure 7(b), respectively.

Calculated relationship between nominal stress and stretch shows good agreement with tensile test results. Also, different strength in warp and weft directions is considerable by proposed model.

3.4 FEM simulation of rectangle block

In this section, tensile and compress deformation of rectangle block were simulated.

The FEM model is rectangle shape of which all edges are 20 mm. The FEM model and fiber orientation angle is shown in Figure 8.

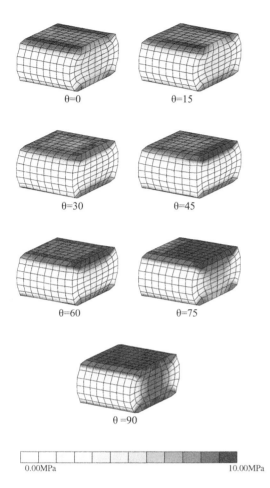

$\theta=0$ $\theta=15$

$\theta=30$ $\theta=45$

$\theta=60$ $\theta=75$

$\theta=90$

0.00MPa   10.00MPa

Figure 10.   Mean stress in 30% of compress deformation.

For the simulation, all nodes in the top and bottom surfaces are constrained and vertical displacement was applied to the top surface.

In Figure 9, the mean stress distributions in 10 mm of vertical displacement are shown. Mean stress distribution and deformation of FEM simulations were distorted by reinforcing fiber. Tensile displacement that corresponds to 50% of initial length is longer than breaking stretch in tensile loading tests.

In Figure 10, the mean stress distributions in 30% of compress deformation are shown. As well as the tensile deformation, stress distributions shows effects of reinforcing fibers.

4   CONCLUSION

In this paper, an orthogonal hyperelastic model was proposed. The anisotropic model is possible to consider different strength in warp and weft direction.

Numerical results calculated by this model show good agreement with experimental results and FEM code shows enough stability for evaluation of mechanical behavior of FRR.

REFERENCES

Reese, S., Raible, T., and Wriggers, P., Finite element modelling of orthotropic material behaviour in pneumatic membranes, *International Journal of Solids and Structures*, Vol. 38 (2001), pp. 9525–9544.

Asai, M., Kimura, Y., Sonoda, Y., Nishimoto, Y. and Nishino, Y., Numerical solution of reinforced rubber based by using the anisotropic hyperelasticity, *Journal of Applied Mechanics JSCE*, Vol. 11 (2008), pp. 467–474

Asai, M., Kimura, Y., Sonoda, Y., Nishimoto, Y. and Nishino, Y., Constitutive modeling for texture reinforced rubber by using an anisotropic visco-hyperelastic model, Journal of Structural Mechanics and Earthquake Engineering A (2010.3), Vol. 66 (2010) , No. 2 pp. 194–205

Holzapfel, G. A., Gasser, T. C., A viscoelastic model for fiber-reinforced composites at finite strains: Continuum basis, computational aspects and applications, *Computer Methods in Applied Mechanics and Engineering*, Vol. 190 (2001), pp. 4379–4403.

Balzani, D., Neff, P., Schroder, J. and Holzapfel, G. A., A polyconvex framework for soft biological tissue. Adjustment to experimental data, *International Journal of Solids and Structures*, Vol. 43 (2006), pp. 6052–6070.

Ehret, A. E., Itskov, M., Modeling of anisotropic softening phenomena: Application to soft biological tissues, *International Journal of Plasticity*, Vol. 25 (2009), pp. 901–919.

Schroder, J., and Neff, P., Invariant formulation of hyperelastic transverse isotropy based on polyconvex free energy functions, *International Journal of Solids and Structures*, Vol. 40 (2003), pp. 401–445.

Boyd, S., and Vandenberghe, L., Convex Optimization(2004), Cambridge University Press.

Itskov, M., and Aksel, N., A class of orthotropic and transversely isotropic hyperelastic constitutive models based on a polyconvex strain energy function, *International Journal of Solids and Structures*, Vol. 41 (2004), pp. 3833–3848.

Constitutive Models for Rubber VII – Jerrams & Murphy (eds)
© 2012 Taylor & Francis Group, London, ISBN 978-0-415-68389-0

A thermo-chemo-mechanical coupled formulation, application to filled rubber

T.A. Nguyen Van & S. Lejeunes
LMA, CNRS UPR7051, France

D. Eyheramendy
ECM, LMA CNRS UPR7051, France

A. Boukamel
EHTP, Casablanca, Maroc

ABSTRACT: Filled rubber exhibits a complex behavior involving strong coupling of thermo-chemo-mechanical phenomena. In this paper, we introduce a nearly incompressible rheological model in finite strains which takes into account the thermo-chemo-mechanical couplings. For the mechanical part, this model is based on a visco-elasticity law. The thermal part takes into account the thermal expansion. For the chemical one, an idealized vulcanization evolution law is considered. The formulation of this model is presented in the Eulerian configuration within a thermodynamic framework. A numerical example is proposed to exhibit the thermo-chemo-mechanical coupling.

1 INTRODUCTION

Filled rubber materials are used in a large number of industrial applications. Rubber parts are often submitted to severe thermo-mechanical loadings (large strain, dynamic loading, self-heating …). Furthermore, the micro-structure of filled rubber is very complex (multi-phase materials: rubber matrix and aggregates/agglomerates of fillers, complex interactions between different components …). This leads to couplings of several physical-chemical phenomena at different scales.

In order to model the phenomenon of self-heating due to dissipative behavior, the coupling of thermo-mechanics was taken into account in the modeling of the behavior of these materials. A phenomenological macroscopic approach for modeling the thermo-mechanical behavior can be found for instance in Boukamel et al (2001), Reese & Govindjee (1997). An alternative microscopic approach to model the thermo-mechanical behavior can be found in Reese (2003).

The thermo-chemo-mechanical couplings present a strong interest for constitutive models for rubber, as it can be applied to different problems: rubber material processing, re-curing phenomena observed during utilization, aging mechanisms …

From the point of view of the chemistry, there is a large number of models for describing the vulcanization and the production of rubber parts in the literature. *E.g.*, one can find the mechanistic approaches based on the simplified schemes of kinetic reactions between the components of elastomer (see Ding & Leonov (1996); Xinhua et al (2007); Han et al (1998)) and the phenomenological approaches based on the kinetic equations (see Labban et al (2009)).

Andre & Wriggers (2005) have proposed an approach of thermo-chemo-mechanical coupling, in small strains, which is based on the introduction of internal variables. This model is able to cover the behavior of material in the state of non-crosslink and crosslink. In Lion & Hofer (2007), a phenomenological approach of thermo-chemo-mechanical couplings in finite strains for polymer material is presented. Other interesting works can be found in Gigliotti & Grandidier (2010); Kannan & Rajagopal (2010).

In this paper, we propose a thermo-chemo-mechanical model that aims at representing the coupled behavior of rubber. This model is based on the principles of thermodynamics of irreversible processes. A nearly incompressible rheological model is established through a decomposition of the deformation into three parts: a mechanical contribution, a contribution of thermal expansion and a contribution of chemical shrinkage. The thermodynamic framework takes into account these three contributions. The evolution of cross-linking is represented by a simple kinetic reaction.

In the first section of this paper, we justify the present work by showing experimental results. In the second section, we present the formulations within the thermodynamic framework in Eulerian configuration. In the last section, one example of thermo-chemo-mechanical couplings is shown to illustrate the coupled phenomena taken into account by the present model.

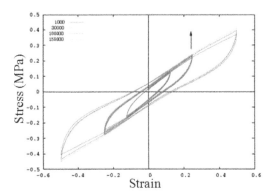

Figure 1. Rigidification phenomenon of rubber part in fatigue shear test.

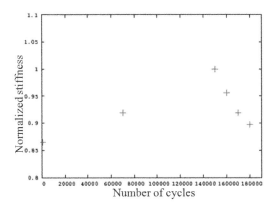

Figure 2. Evolution of the stiffness versus the number of cycles in fatigue shear test.

2 MOTIVATIONS FROM EXPERIMENTAL RESULTS

An experimental campaign of fatigue test on double shearing specimens of silicone rubber filled with silica was realized with a specific protocol: (a) the specimen is submitted to a fixed number of fatigue cycles, (b) the specimen is cooled down by letting itself alone, (c) it is again submitted to a characterization test that consists of cyclic tests with various amplitudes (see Grandcoin (2008)). Figure 1 shows a superposition of some characterization tests at various fatigue level. It can be observed a phenomenon of stiffening of the behavior during the first hundred of thousand cycles of the fatigue test. This phenomenon has the opposite effect than the one of damage.

This phenomenon of stiffening could be explained by the phenomenon of re-curing, which is thermally activated by self-heating process and vulcanizing agents. The new crosslinks created between the chains by these reactions lead to a rigidification of rubber material.

Figure 2 shows the normalized stiffness of double shearing specimen versus the number of fatigue cycles. It is clear that in the range [0, 150.000] cycles, a stiffening behavior occurs and then a damage behavior follows.

3 A RHEOLOGICAL MODEL FOR THERMO-CHEMO-MECHANICAL COUPLING

3.1 Chemical model

The vulcanization process consists in several reactions in three phases of vulcanization: induction, curing, and over-cure phase. For the sake of simplicity, in this paper, we propose a vulcanization process with one reaction:

$$A \rightarrow Vu \tag{1}$$

where A is the non-cured product, Vu is the cured product.

The equation of mass conservation for reaction (1) is written as:

$$m_A + m_{Vu} = m = const$$

$$\xi = \frac{m_{Vu}}{m}; \qquad 0 \leq \xi \leq 1 \tag{2}$$

In a closed system, a chemical reaction is an internal process, the degree of cure ξ measures the progress of the reaction considered (Atkin (1994)).

In the literature, one can find the popular phenomenological kinetic model of Kamal-Sourour that represents the degree of cure as a function of time and temperature in rubber:

$$\dot{\xi} = \left(k_1 + k_2 \xi^m\right)\left(1 - \xi^n\right) \quad \text{with} \quad k_i = k_{i0} e^{-E_i/R\theta} \tag{3}$$

where $R = 8.314$ J/molK is the universal gas constant; E_i is the activation energy of chemical reaction; k_{i0} is the pre-exponential factor.

To model the thermo-chemo-mechanical coupling, we suppose that the degree of cure is a function of temperature and volumetric variation:

$$\dot{\xi} = f\left(\theta, \xi, J_m\right) \quad \text{with} \quad J_m = \det F_m \tag{4}$$

where J_m is the mechanical volume variation.

3.2 Thermodynamic framework in finite strain

In the context of thermodynamics of irreversible processes, the constitutive equation must fulfill the Clausius-Duhem inequality, which takes the following form in the Eulerian configuration:

$$\Phi = \sigma : D - \rho_0 J^{-1} \dot{\Psi} - \rho_0 J^{-1} s \dot{\theta} - \frac{1}{\theta} q \cdot grad\left(\theta\right) \geq 0 \tag{5}$$

where σ is the Cauchy stress, D is the Eulerian rate of deformation, $J = det F$ is the volume variation

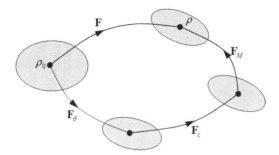

Figure 3. Decomposition of deformation gradient.

(and F is the deformation gradient), ρ_0 is the volumetric mass in initial configuration, s is the entropy, q is the heat flux, Φ is the internal dissipation and Ψ is the specific free energy.

As we know, such a material exhibits chemical shrinkage when curing degree increases and thermal expansion when temperature increases. Following Lion & Hofer (2007), we suppose that the total deformation gradient is multiplicatively decomposed into a mechanical part, a chemical shrinkage part and a thermal expansion part as follows:

$$F = F_m F_\theta F_c = \bar{F}_m\left(J_m^{1/3} J_\theta^{1/3} J_c^{1/3}\mathbf{1}\right) = \bar{F}\left(J^{1/3}\mathbf{1}\right)$$

$$J_\theta^{1/3} = 1 + \alpha\left(\theta - \theta_0\right); \qquad J_c^{1/3} = 1 + \beta g\left(\xi, \xi_0\right) \tag{6}$$

where $\bar{F}_m$ and $\bar{F}$ are the isochoric parts of the mechanical deformation and the total deformation; $\mathbf{1}$ is the identity tensor; α is the coefficient of thermal expansion; β is the coefficient of chemical shrinkage; $g(\xi,\xi_0)$ is the shrinkage function depending on the degree of cure. We suppose the existence of a natural state, identical to initial configuration, $(Fm=1, \theta=\theta_0, \xi=\xi_0)$ which is stress-free.

The specific free energy is given by:

$$\Psi = \Psi\left(\bar{B}_m, J_m, \theta, \xi\right) \tag{7}$$

The time derivative of the free energy is therefore:

$$\dot{\Psi} = \frac{\partial \Psi}{\partial \bar{B}_m} : \dot{\bar{B}}_m + \frac{\partial \Psi}{\partial J_m} \cdot \dot{J}_m + \frac{\partial \Psi}{\partial \theta} \cdot \dot{\theta} + \frac{\partial \Psi}{\partial \xi} \cdot \dot{\xi} \tag{8}$$

The time derivative of J_m is

$$\dot{J}_m = J_m\left[\left(1:L\right) - J_\theta^{-1}\dot{J}_\theta - J_c^{-1}\dot{J}_c\right] \tag{9}$$

and the time derivative of the left Cauchy-Green strain tensor is

$$\dot{\bar{B}}_m = L\bar{B}_m + \bar{B}_m L^T - \frac{2}{3}\left(1:L\right)\bar{B}_m \tag{10}$$

The equation of mass conservation is

$$\rho = \rho_0 J^{-1} \tag{11}$$

The equation of energy conservation is

$$\rho_0 J^{-1}\dot{e} = \sigma : D - div\left(q\right) + \rho_0 J^{-1} r \tag{12}$$

where e is the specific internal energy; r is a radiant term.

By substituting the equations (8), (9) and (10) in (5), we obtain the expression of the specific entropy as follows:

$$s = -\frac{\partial \Psi}{\partial \theta} + 3\alpha J_\theta^{-1/3} J_m \frac{\partial \Psi}{\partial J_m} \tag{13}$$

The remaining internal dissipation (5) is then decomposed into an intrinsic part, a thermal part, and a chemical part.

$$\Phi = \Phi_{int} + \Phi_{thermo} + \Phi_{chemo} \geq 0 \tag{14}$$

$$\Phi_{int} = \left[\sigma - \rho_0 J^{-1}\left(2\bar{B}_m\frac{\partial \Psi}{\partial \bar{B}_m}\right)^D - J_m \cdot \frac{\partial \Psi}{\partial J_m}\mathbf{1}\right] : D \tag{15}$$

$$\Phi_{ther} = -\frac{1}{\theta}q\cdot grad\left(\theta\right) \tag{16}$$

$$\Phi_{chemo} = \rho_0 J^{-1}\left(\frac{\partial \Psi}{\partial \xi} - 3\beta g'\left(\xi\right)J_c^{-1/3}J_m\frac{\partial \Psi}{\partial J_m}\right)\cdot \dot{\xi} \tag{17}$$

where ^{D} stands for the deviatoric operator.

In order to ensure the non-negativeness of internal dissipation for all admissible processes, we suppose, first, the positiveness of each dissipation terms independently of the others, and second, the existence of pseudo-potential of dissipation which is a convex positive function of flux variables. The normality principle is then applied to ensure the positiveness of internal dissipation for all admissible processes.

$$\varphi\left(\bar{D}, q, \dot{\xi}\right) = \varphi_{int}\left(\bar{D}\right) + \varphi_{ther}\left(q\right) + \varphi_{chemo}\left(\dot{\xi}\right) \geq 0 \tag{18}$$

where $\bar{D} = \frac{1}{2}(\dot{\bar{F}}\bar{F}^{-1} + \bar{F}^{-T}\dot{\bar{F}}^T)$

The stress is defined as:

$$\sigma = \left[\frac{\partial\varphi_{int}\left(\bar{D}\right)}{\partial\bar{D}}\bar{D} + \rho_0 J^{-1}\left(2\bar{B}_m\frac{\partial \Psi}{\partial\bar{B}_m}\right)\right]^D + J_m \cdot \frac{\partial \Psi}{\partial J_m}\mathbf{1} \tag{19}$$

The rate of curing is written as:

$$\frac{\partial\varphi_{chemo}\left(\dot{\xi}\right)}{\partial\dot{\xi}} = \rho_0 J^{-1}\left(\frac{\partial\Psi}{\partial\xi} - 3\beta g'\left(\xi\right)J_c^{-1/3}J_m\frac{\partial\Psi}{\partial J_m}\right) \tag{20}$$

The Fourier law is written as:

$$q = -k\cdot grad\left(\theta\right) \tag{21}$$

where k is the material conductivity.

169

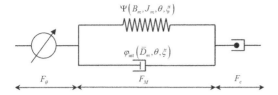

Figure 4. A coupled thermo-chemo-mechanical model.

3.3 Rheological coupled model

In this paper, we propose a rheological model for the coupled thermo-chemo-mechanical behavior of rubber material (see Figure 4). This model consists in an element of thermal expansion, an element of chemical shrinkage and a Kelvin-Voigt element for the visco-elasticity.

The Helmholtz free energy of the model is split into a mechanical contribution, a thermal contribution and a chemical contribution. Then, the mechanical part is split into a volumetric part and an isochoric part.

$$\Psi = \Psi_m^D\left(\bar{B}_m, \theta, \xi\right) + \Psi_m^V\left(J_m\right) + \Psi_\theta\left(\theta\right) + \Psi_c\left(\theta, \xi\right) \quad (22)$$

$$\Psi_m^D\left(\bar{B}_m, \theta, \xi\right) = A\left(\theta, \xi\right)\left(I_1\left(\bar{B}_m\right) - 3\right) + B\left(\theta, \xi\right)\left(I_1\left(\bar{B}_m\right) - 3\right)^3 + C\left(\theta, \xi\right)\left(I_2\left(\bar{B}_m\right) - 3\right) \quad (23)$$

$$\Psi_m^V\left(J_m\right) = K_v\left(J_m - 1\right)^2 \quad (24)$$

$$\Psi_\theta\left(\theta\right) = C_0\left(\theta - \theta_0 - \theta Ln\left(\frac{\theta}{\theta_0}\right)\right) \quad (25)$$

where I_1, I_2 are the first and second invariants; $A(\theta, \xi)$, $B(\theta, \xi)$, $C(\theta, \xi)$ are decreasing functions of the temperature and increasing functions of the curing degree; K_v is the bulk modulus; C_0 is a material coefficient linked to the heat capacity.

$$\Psi_c\left(\theta, \xi\right) = \Gamma\left(\theta\right) \cdot \frac{\left(1 - \xi\right)^{n+1}}{n+1} \quad (26)$$

where $\Gamma(\theta) = P \cdot Exp(-E_a/R\theta)$ is the Arrhenius law which describes the variation of the rate of chemical reaction; E_a is the activation energy of chemical reaction; P and n are the chemical parameters.

The pseudo-potential dissipations of the model are defined as follows:

$$\varphi_{int}\left(\bar{D}; \bar{B}_m, J_m, \theta, \xi\right) = \frac{1}{2}\eta\left(\theta, \xi\right)\bar{D}:\bar{D} \geq 0$$

$$\varphi_{ther}\left(q; \bar{B}_m, J_m, \theta, \xi\right) = -\frac{k}{2\theta}q \cdot q \geq 0 \quad (27)$$

$$\varphi_{chemo}\left(\dot{\xi}; \bar{B}_m, J_m, \theta, \xi\right) = \frac{1}{2}h\left(\theta, \xi\right)\dot{\xi} \cdot \dot{\xi} \geq 0$$

where $\eta(\theta, \xi)$ is viscosity parameter; $h(\theta, \xi)$ is a chemical reaction parameter.

The stress is defined as

$$\sigma = \eta\left(\theta, \xi\right)\bar{D} + \rho_0 J^{-1}\left[2\bar{B}_m\left(A\left(\theta, \xi\right) + 3B\left(\theta, \xi\right)\left(I_1 - 3\right)^2 + C\left(\theta, \xi\right)\right) - 2\bar{B}_m^2 C\left(\theta, \xi\right)\right]^D + 2K_v J_m \cdot \left(J_m - 1\right)\mathbf{1} \quad (28)$$

The evolution of degree of cure is deduced:

$$\dot{\xi} = h\left(\theta, \xi\right)^{-1}\rho_0 J^{-1}\left[\frac{\partial \Psi\left(\bar{B}_m, J_m, \theta, \xi\right)}{\partial \xi} - 3\beta g'\left(\xi\right)J_c^{-\frac{1}{3}}J_m \frac{\partial \Psi\left(\bar{B}_m, J_m, \theta, \xi\right)}{\partial J_m}\right] \quad (29)$$

From the first principle of energy conservation (12), we obtain the heat equation as follows:

$$\rho_0 J^{-1}C\dot{\theta} = \overbrace{\frac{\partial \varphi_{int}}{\partial \bar{D}}:\bar{D}}^{①} + \overbrace{\frac{\partial \varphi_c}{\partial \xi}\dot{\xi}}^{②} + \overbrace{k\Delta\theta}^{③} + \overbrace{\rho_0 J^{-1}r}^{④} + \underbrace{\rho_0 J^{-1}\theta Y_1 \dot{\xi}}_{⑤} + \underbrace{\rho_0 J^{-1}\theta Y_2 : D}_{⑥} \quad (30)$$

$$Y_1 = \frac{\partial^2\Psi}{\partial\theta\partial\xi} + 18\alpha\beta K_v J_m^2 J_\theta^{-\frac{1}{3}}J_\xi^{-\frac{1}{3}}g'\left(\xi\right) + 18\alpha\beta K_v J_m J_\theta^{-\frac{1}{3}}J_\xi^{-\frac{1}{3}}g'\left(\xi\right)\left(J_m - 1\right) \quad (31)$$

$$Y_2 = \left[\frac{\partial^2\Psi_m^D}{\partial\theta\partial\bar{B}_m}\bar{B}_m\right]^D - 12\alpha K_v J_m^2 J_\theta^{-\frac{1}{3}}J_\xi^{-\frac{1}{3}}\mathbf{1} \quad (32)$$

$$C = \theta\frac{\partial s}{\partial\theta} \quad (33)$$

where C is the heat capacity which depends on mechanical volumetric deformation and degree of cure.

In equation (30), the terms ⑤, ⑥ are respectively the coupled thermo-chemical term and the coupled thermo-mechanical term. The terms ① and ② are the mechanical dissipation and the chemical dissipation. The terms ③ and ④ are the heat conduction and the external heat.

Finally, the model proposed is governed by the resolution of the coupled equations (28), (29) and (30) as well as the definition of 16 material parameters.

4 RESULTS AND DISCUSSION

We consider the coupled thermo-chemo-mechanical problem in a homogenous and adiabatic case. The numerical problem is summarized in resolving the differential equations, in which the space dependence is omitted. These equations are resolved with Mathematica.

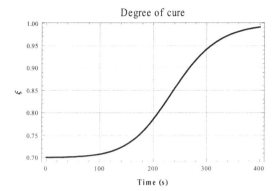

Figure 5.   Time evolution of degree of cure.

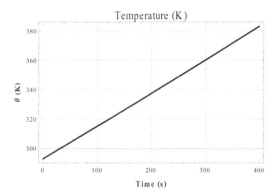

Figure 6.   Time evolution of the temperature.

In this example, a rubber block is submitted to a cyclic shear. The total deformation gradient and the mechanical deformation gradient are given by:

$$F = \begin{bmatrix} 1 & \gamma & 0 \\ 0 & 1 & 0 \\ 0 & 0 & 1 \end{bmatrix}, \ \gamma(t) = \gamma_0 \sin(2\pi ft);$$

$$F_m = \begin{bmatrix} 1+\lambda & \gamma & 0 \\ 0 & 1+\lambda & 0 \\ 0 & 0 & 1+\lambda \end{bmatrix} \quad (34)$$

The stress tensor is given by:

$$\sigma = \begin{bmatrix} \sigma_{11} & \sigma_{12} & 0 \\ 0 & \sigma_{22} & 0 \\ 0 & 0 & \sigma_{33} \end{bmatrix} \quad (35)$$

The resolution of the coupled equations (28), (29) and (30) is done with a numerical scheme based on a predictor-corrector method to reach the equilibrium with a desire precision.

The results of the present model are shown in Figures 5–11. The dissipative behavior of the material induces an increase of temperature from 293°K to

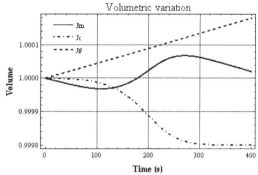

Figure 7.   Time evolution of mechanical, chemical and thermal volumetric deformation.

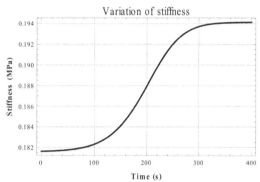

Figure 8.   Time evolution of stiffness of rubber material.

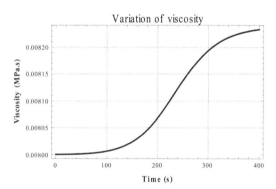

Figure 9.   Time evolution of material viscous coefficient.

380°K (see Fig. 6) during the shearing test. This temperature increase stimulates the chemical reactions and leads to an increase of the curing degree from 0.7 to 1 (see Fig. 5). The variation of each volumetric deformation types are shown in Figure 7, as global incompressibility is enforced, we observe an equilibrium between the mechanical, the chemical and the thermal volumetric terms. The present model takes into account the rigidification phenomenon (see Fig. 8). In figure 10, we observe a small decrease (0.006%) of the heat capacity which seems to be in contrast with the literature (see Ghoreishy & Naderi (2005)). This is due to

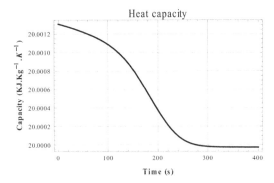

Figure 10. Time evolution of heat capacity.

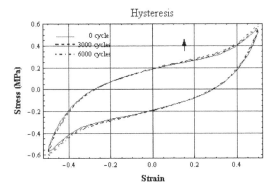

Figure 11. Hysteresis of visco-elasticity material at different levels of loading cycles.

influence of the temperature which has not been taken into account in the present model for the heat capacity as it is chosen constant upon temperature. However, in the experimental results, temperature dependence seems more important than chemical dependence. In the figure 11, the hysteresis loops at 3 different numbers of cycles are shown. This result is in accordance with the experimental result of figure 1.

5 CONCLUSION

We have developed a coupled thermo-chemo-mechanical model which consists in a mechanical part of visco-elasticity, a thermal expansion part and a simplified law of vulcanization. Through a simple homogenous and adiabatic example, we have shown that the present model is able to describe the phenomena of re-curing activated by the mechanical

dissipation. In the future, we aim at, on one hand, enriching the chemical model in order to interpret the complex phenomena of vulcanization, and on other hand, developing a finite element implementation to resolve the coupled problem in 3D.

REFERENCES

Andre, M. & Wriggers, P. 2005. Thermo-mechanical behaviour of rubber materials during vulcanization. *International journal of solids and structures.* 42, 4758–4778.

Atkins, P.W. 1994. *Physical chemistry 5th ed.* Oxford university press.

Boukamel, A., Meo, S., Debordes, O. & Jaeger, 2001. A thermo-viscoelastic model for elastomeric behaviour and its numerical application. *Archive of Applied Mechanics.* 71, 785–801.

Ding, R. & Leonov, A.I. 1996. A kinetic model for sulfur accelerated vulcanization of a natural rubber compound. *Journal of Applied Polymer Science.* 6, 455–463.

Germain, P. 1973. *Cours de mécanique des milieux continus: Théorie générale.* Masson.

Ghoreishy, M.H.R. & Naderi, G. 2005. Three-dimensional finite element modeling of rubber curing process. *Journal of elastomers and plastics.* 37

Gigliotti, M. & Grandidier, J.C. 2010. Chemo-mechanics couplings in polymer matrix materials exposed to thermo-oxidative environments. *Compte Rendu Mecanique.* 338, 164–175.

Grandcoin, J. 2008. Contribution à la modélisation du comportement dissipatif des élastomères chargés: D'une modélisation micro-physiquement motivée vers la caractérisation de la fatigue. Phd Thesis. *Aix-Marseille II University.*

Han, I., Chung, C., Kim, S. & Jung, S. 1998. A Kinetic Model of reversion Type Cure for Rubber Componds. *Polymer korea.* 22, 233–239.

Kannan, K. & Rajagopal, K.R. 2010. A thermodynamical framework for chemically reacting systems. *Zeitschrift für angewandte Mathematik und Physik.* 62, 331–363

Labban, A.E., Mousseau, P., Deterre, R., Bailleul, J. & Sarda, A. 2009. Temperature measurement and control within moulded rubber during vulcanization process. *Measurement.* 42, 916–926.

Lion, A. & Hofer, P. 2007. On the phenomenological representation of curing phenomena in continuum mechanics. *Archives of Mechanics.* 59.

Reese, S. & Govindjee, S. 1997. Theoretical and numerical aspects in the thermo-viscoelastic material behaviour of rubber-like polymers, *Mechanics of Time-Dependent Materials.* 1, 357–396.

Reese, S. 2003. A micromechanically motivated material model for the thermo-viscoelastic material behaviour of rubber-like polymers. *International Journal of Plasticity.* 19, 909–940.

Constitutive Models for Rubber VII – Jerrams & Murphy (eds)
© 2012 Taylor & Francis Group, London, ISBN 978-0-415-68389-0

Microstructural analysis of carbon black filled rubbers by atomic force microcopy and computer simulation techniques

I.A. Morozov
Institute of continuous media mechanics UB RAS, Russia
Leibniz-Institut für Polymerforschung Dresden e.V, Germany

B. Lauke
Leibniz-Institut für Polymerforschung Dresden e.V, Germany

G. Heinrich
Leibniz-Institut für Polymerforschung Dresden e.V, Germany
Technische Universität Dresden, Lehrstuhl Polymerwerkstoffe und Elastomertechnik, Germany

ABSTRACT: Work concerns the 3D modeling of a filler network microstructure in rubber compounds. The model represents the carbon black filler in three states: primary fractal aggregates consisting of spherical overlapped particles, secondary structures or agglomerates and partially broken fragments of micropellets. Information about the structure of the filler and its distribution in the matrix is obtained from the analysis of the AFM-images of the material surface.

1 INTRODUCTION

Accumulation of reliable information on filled rubber morphology, filler distribution and dispersion at micron and submicron levels is currently a topical problem. Recent 3D-TEM investigations of Japanese researchers (Ikeda et al 2007) made it possible to visualize the spatial structure and determine the distances between the closest aggregates in the carbon black networks of rubber compounds; yet some important morphological characteristics of rubber compounds were disregarded in these works. Moreover, the size of the analyzed images ($1 \times 1 \, \mu m^2$) gave no quantitative information concerning the state of filler distribution and microdispersion.

This paper introduces the concept of computer reconstruction of a three-dimensional filler (carbon black) network using the AFM-based data.

Figure 1. AFM images ($20 \times 20 \, \mu m$) of the surface topography of the rubbers with carbon black mass fractions 10 (left) and 30 (right) phr. The scale bars show height differences (in nm).

2 EXPERIMENTS AND DATA PROCESSING

The surfaces of fresh cuts of the materials prepared using the method proposed by Coran et al (1992) were studied. For each specimen, several high quality AFM-images $20 \times 20 \, \mu m$ were scanned for further analysis. The minimal distance between two points in the xy-plane was $\sim 13 \, nm$. The experiments were carried out on Dimension Icon AFM (by Brooker). Custom algorithms developed in Matlab were employed for image processing and further analysis. In this work

we present the results of microstructural analysis and simulation of three SBRs filled with carbon ISAF (N220) with different mass fractions: 10, 30 and 50 phr (Fig. 1). Rubber samples were prepared by Sumitomo Rubber Industries LTD (Japan).

For a weakly filled material, we have observed a lot of large, dense objects (Fig. 1) – partially broken microgranules of carbon black which are almost disappeared in compound with 50 phr. As a result of applied algorithms (Morozov et al 2011), the continuous relief is replaced by the field of discrete structures representing the fragments of the filler network in the material (Fig. 2).

Figure 2. Results of fragmentation of AFM images (Fig. 1). For better visualization only cross-sections of 3D-relief structures are shown. Gray-scale variation corresponds to different height level of structures.

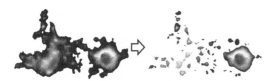

Figure 3. Separation of the agglomerate into primary structures.

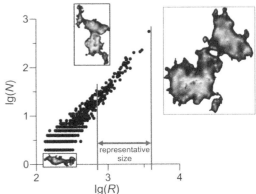

Figure 4. Estimation of representative cluster size.

Table 1. Microstructural parameters of materials under study.

ϕ, phr	ϕ_p	α, μm^{-2}	N	Δ, nm	R, nm	D_{pc}	r, nm	D_{pa}
10	0.51	0.07	1...5	4.3	597	1.02	138	1.00
30	0.38	0.06	20	4.1	1068	1.43	107	1.01
50	0.04	0.05	64	4.0	1296	1.45	101	1.02

Here, ϕ_p is the fraction of microgranules in the total observed filler volume; α is the number of clusters per $1\,\mu m^2$; N is the number of aggregates in a cluster; Δ – distance between nearest aggregates in cluster; R is the average cluster size (mean diameter), D_{pc} and D_{pa} is the perimeter fractal dimension of clusters and aggregates respectively; r is the average aggregate size (mean diameter).

3 RESULTS AND DISCUSSION

Three morphologically distinct classes of objects have been considered:

1. Microgranules – dense and compact, structurally indivisible fragments of the relief (size > 300 nm).
2. Clusters (agglomerates) – structures obtained by breaking the relief into components, which can be further divided into two or even more smaller structures (microgranules or aggregates).
3. Aggregates – structurally indivisible objects whose size is smaller and shape is less compact than the form of microgranules.

Figure 3 shows one of the agglomerates separated from the relief structure, as well as the result of the automatic computer breaking of this agglomerate into number of primary aggregates and a microgranule.

Numerous experimental and theoretical studies have highlighted the fact that the formation of secondary agglomerates is governed by fractal relations: dependence between the size of clusters, R, and the number of primary structures, N: $N \sim R^{Dn}$.

The surface structure allows us to observe only the pieces of real three dimensional filler clusters. The size of these fragments varies over a wide range. The objects for the analysis of the characteristics of clusters must be sufficiently representative. If the relation $R(N)$ is constructed in double logarithmic coordinates, we can see that, for small values of R and N, there is a significant scatter in the experimental data, which causes difficulties for their approximation with the line having a slope equal to D_n (Fig. 4).

In the analysis of the characteristics of the materials under study, we use a minimum cluster size of 700 nm.

Table 1 lists the structural parameters of the examined materials obtained from the analysis of AFM-images.

Filler (aggregates and microgranules) size distribution in the materials under study can be fairly well approximated by the log-normal probability distribution density law (Fig. 5).

The primary aggregates of carbon black are the fractal aggregates of caked spherical particles. The number of particles in aggregates (its mass) n and its size r connected in fractal dependence:

$$n(r) = C \cdot \left(r/r_p\right)^D,$$

where C is a constant, D is the fractal dimension of mass, r is the aggregate size, and r_p is the particle radius as given in the literature (for N220 CB $r_p \approx 10.5\,nm$). Ehrl et al (2009) give next relations between perimeter fractal dimension D_p, mass fractal dimension D and constant C:

$$D = -1.5D_p + 4.4 \text{ and } C = 4.46D^{-2.08}.$$

Using the available data on the geometry of aggregates from AFM and computer analysis, we can restore their spatial structure. More details of the algorithm

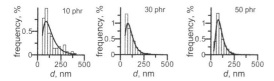

Figure 5. Filler (aggregates and microgranules) size distribution.

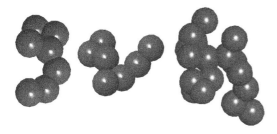

Figure 6. Examples of simulated aggregates.

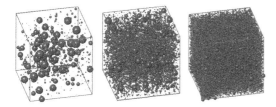

Figure 7. Reconstructed filler network of the materials under study (filler fraction is 10, 30 and 50 phr).

of constructing fractal aggregates of overlapping particles are described in the work by Morozov et al (2010). Examples of reconstructed aggregates are presented in Figure 6.

Average calculated specific surface area of obtained aggregates is $\sim$113 m^2/g which is close to the known CTAB value for given CB grade (109 m^2/g) and independent of filler mass fraction.

Let us describe briefly the main stages of recovery of the three-dimensional structure of the filler in the volume $L_x L_y L_z$. The filler (aggregates and microgranules) particles are presented now as spheres whose size is given in the distribution shown in Fig. 5, and the mass fraction corresponds to the prescribed one (10, 30, 50 phr). The $\alpha^{3/2}$ spheres (centers of the future agglomerates) are randomly arranged in the volume.

The fractal clusters of filler spheres are generated around these centers until the mean size R reaches. Fractal dimension of clusters is calculated in the same way as for aggregates (using D_{pc}). Upon the completion of cluster construction, the remaining part of the free filler is randomly inserted into the total volume of the material (Fig. 7).

4 CONCLUSION AND OUTLOOK

A concept for reconstruction of a three-dimensional filler network from AFM-data was introduced. For illustration three carbon black filled rubber compounds were analyzed. Further research will include replacement of spheres with real-shape aggregates (Fig. 6) and the structural-mechanical modeling of materials with allowance for the interaction forces between aggregates during tension and compression of model volumes.

ACKNOWLEDGEMENT

This work was supported by the RF president grant for state support of young Russian scientists No MK-3914.2011.8 and the RFBR grant No 11-08-00178-a.

REFERENCES

Coran A.Y., Donnet J.B. 1992. The Dispersion of Carbon Black in Rubber Part I. Rapid Method for Assessing Quality of Dispersion. *Rubber Chemistry and Technology*. 65: 973–996.

Ehrl L., Soon M., Lattuada M. 2009. Genaration and geometrical analysis of dense clusters with variable fractal dimension. *The journal of physical chemistry. B*. 113: 10587–10599.

Ikeda Y., Kato A., Shimanuki J., Kohjiya S., Tosaka M., Poompradub S. 2007. Nano-structural elucidation in carbon black loaded NR vulcanizate by 3D-TEM and in situ WAXD measurements. *Rubber Chemistry and Technology* 90: 251–264.

Morozov I. A., Lauke B., Heinrich G. 2010. A new structural model of carbon black framework in rubbers. *Computational materials science*. 47:817–825.

Morozov I. A., Lauke B., Heinrich G. 2011. A novel method of quantitative characterisation of filed rubber structures by AFM. *Kautschuk Gummi Kunststoffe*. 64: 24–27.

Constitutive Models for Rubber VII – Jerrams & Murphy (eds)
© 2012 Taylor & Francis Group, London, ISBN 978-0-415-68389-0

Identification of local constitutive model from micro-indentation testing

Y. Marco
ENSTA Bretagne, Laboratoire LBMS (EA 4325), Brest Cedex, France

V. Le Saux
UBS, LIMATB – Equipe ECoMatH, rue de Saint-Maudé, Lorient Cedex, France

G. Bles & S. Calloch
ENSTA Bretagne, Laboratoire LBMS (EA 4325), Brest Cedex, France

P. Charrier
Modyn Trelleborg, Zone ind. de Carquefou, Carquefou Cedex, France

ABSTRACT: The use of micro-hardness for polymers is commonly restricted to the use of a scalar value (hardness or modulus), that can clearly not be used to identify local constitutive models. In this study load/displacement curves obtained at a micro-scale are used to identify the parameters of an Edwards-Vilgis hyper-elastic model. The proposed protocol is coupling FE simulations achieved with Abaqus® with optimization procedures using specific software. The materials studied are unfilled Natural Rubber and filled PolyChloroprene. Several numerical parameters (indent geometry, friction, thickness …) as well as experimental protocols were tested in order to check the protocol reliability. The identified parameters are compared to the ones obtained from macroscopic measurements (tensile, compression and pure shear tests). A strong scale effect is illustrated preventing from an accurate identification for filled materials. The agreement is very good for unfilled natural rubber which validates several numerical assumptions.

1 INTRODUCTION

1.1 *Industrial motivation*

Mastering the ageing consequences is crucial to certify the long term durability of a new product and to reduce the frequency of maintenance operations and, thus, their cost. But ageing of rubbers is a complex phenomenon, involving numerous factors (material compounding, environmental conditions, temperature and time) and several scales, ranging from chemical kinetics to finite elements simulations. Whatever the material and environment considered, the first need is to identify the physical and chemical mechanisms involved (oxidation, migration of chemical components, hydrolysis, morphological variations, post-curing …). The second step is to define and achieve accelerated ageing tests, consistent with the formerly identified mechanisms, in order to keep reasonable tests and product development durations. The last step is usually to combine the results of the accelerated tests with an Arrhenius-like extrapolation to predict the behaviour at lower temperatures or longer times. Still, the reliability of these accelerated tests is to be carefully validated because, even if the involved mechanisms are the same for ageing in accelerated and service conditions, the properties obtained are driven by competing mechanisms involving several

kinetics. The ageing gradients between accelerated aged samples and naturally aged parts are therefore dependant on their geometry, which restrains the validity of the ageing measurements to the test specific conditions (environment and specimen geometry). Moreover, from a designer point of view the observed criterion is often a macroscopic one (strain or stress at break, life duration …) but the aged material exhibits generally a non uniform ageing. An easy way to evaluate these gradients and to provide data able to link the physical measurements to the mechanics consequences is micro-hardness (Celina *et al.* 2000). Nevertheless, to measure a scalar value (hardness or modulus) is clearly not enough to perform relevant finite element simulations that include the evolution of the local constitutive models, in order to simulate massive aged parts. Instrumented Micro-Hardness is giving a chance to identify this local constitutive law and the aim of this paper is to explore this opportunity.

1.2 *Scientific background*

The gain in precision since 25 years on displacement and load sensors opened the way to inverse analysis using the accurate load-displacement curves (Oliver and Pharr 1992, Loubet et al. 1993, Bec et al. 1996). The first way to perform this inverse analysis is to use analytical modeling. These approaches, based on the principle of geometric similarity (PGS, Tabor, 1971) gave birth to well mastered development of many approximate solutions of elasto-plastic indentation problems and the determination of the material parameters for classical elasto-plastic materials. Nevertheless, the accurate analysis of indentation curves remains complex (see Cheng and Cheng, 2004 for a recent review). Finite Elements simulations can also be used to determine the constitutive law parameters from the indentation curves by the minimization of an objective function quantifying the difference between the experimental curves and the computed ones (Constantinescu and Tardieu, Hamasaki *et al.* 2005). Elastomeric materials, even from a macroscopic point of view, exhibit a complex thermomechanical behavior, including non-linear elasticity, viscosity, complex cyclic response (Mullins and Payne effects) and a strong dependency on the kind of solicitation. Applied to micro-indentation, these points raise numerous questions on the kind of protocol to be used (loading-unloading curve, creep or relaxation tests, with or without pre-load, after a cyclic accommodation …) and on the geometry of indenter that are not yet resolved. These materials are also nearly incompressible, which leads to numerical difficulties, as described by Jabareen and Rubin (2007). Moreover, indentation of rubber involves several experimental difficulties: high difference of stiffness between the tested material and the support used to encapsulate it, which leads to take great care of the influence of the thickness and of the distance to the walls (Gent and Yeoh, 2006 and Busfield and Thomas, 1999), dependency on temperature, which leads to be cautious on the environment and on the surface preparation protocol, and heterogeneous nature coming from the fact that these materials are always obtained by mixing several components. As a conclusion, the identification of the constitutive parameters for elastomeric materials is involving several difficulties and the more recent studies (Giannakopoulos *et al.* 2009, Rauchs *et al.* 2010) illustrates that the parameters identified at a micro-scale are usually not suitable to describe macroscopic tests.

The aim of this study is to investigate the ability to identify the hyperelastic parameters from load-displacement curves obtained at a micro-scale using a Vickers indenter, then to validate these parameters at a macroscopic scale on tensile, compression and plane strain tests. The studied materials are a silica-filled Polychloroprene and an unfilled natural rubber.

2 EXPERIMENTAL SECTION

2.1 *Materials*

Two materials are investigated in this study. The first one is a silica filled PolyChloroprene (called CR in the following) used for offshore applications. The second one is an unfilled Natural Rubber, otherwise fully formulated. These two materials were chosen in order

Figure 1. View of the indenter tip with the fork and top view of embedded samples.

to illustrate the capability of the approach for two very different ranges of properties (viscosity, microstructural heterogeneities and stiffness). It is important to note that all the samples used for both micro-indentation and macroscopic tests were obtained from the same batch of raw material. Therefore, they should exhibit the same properties.

2.2 *Micro-Hardness testing*

The indentation device used is a Micro-Hardness testing device from CSM with a Vickers tip (angle: $136°$). The resolutions are of 0.3 mN and 0.3 nm for ranges of 30 N and 200 μm. This device is equipped with a specific fork system applied around the indenter tip that allows accounting for the thermal dilatation of the system during the test (see Figure 1 for a view of the samples and of the indeter tip). Microindentation testing requires a well validated experimental protocol. The samples cut out from 2 mm thick sheets are embedded at ambient temperature into an epoxy resin and subsequently grinded with grinding media of decreasing roughness (down to grit size 800). The disc rotation speed is low in order to reduce as much as possible the temperature rise. After grinding, a delay of 2 hours is systematically applied in order to let the samples cool down to ambient temperature. The dimensions of the embedded sample are 4 mm width, 2 mm thick and 10 mm long. These dimensions (thickness and width) were validated by tests and numerical simulations. All tests are load controlled.

We used here a classical testing protocol, which consists in applying a weak preload in order to detect the surface contact (set to 10 mN), then a load controlled loading step (up to 100 mN at a loading rate of 200 mN/min), followed by a creep test of 30 s and then a load controlled unloading step (down to 0 mN at a unloading rate of 200 mN/min).

2.3 *Standard macroscopic tests*

The macroscopic measurements are achieved on a Lloyd LR5K+ testing machine, with a 1 kN load cell. The uniaxial tension tests are displacement controlled with a tensile speed of 10 mm/min. The local strain is measured by a laser extensometer LASERSCAN 200. The H2 tensile samples are cut with a punch die from the sheets used for micro-indentation experiments.

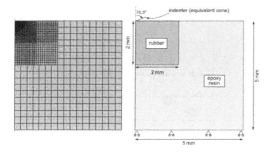

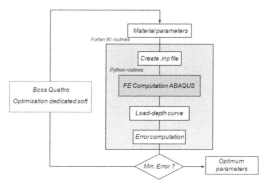

Figure 2. Finite Element model.

The plane strain tensile tests are achieved at the same speed on specific specimens (with the same curing rate). The compression tests are achieved on cylinders with a classical geometry and are displacement controlled with a compression speed of 1 mm/min. Both compression grips are lubricated in order to avoid any barrel effects.

3 NUMERICAL TOOLS

3.1 Finite Elements Model

Among the very numerous models available, we chose the potential proposed by Edwards and Vilgis (1986). This model is able to combine accuracy for several kind of solicitation and numerical stability with only 4 parameters (Nc*, Ns, η and α). Moreover, this model is physically motivated and the parameters can be related to physical interpretations. As our study is part of a larger one dealing with ageing, the choice of this model seemed relevant. This model is implemented into the FE software Abaqus thanks to the UMAT subroutine. For a full description of the numerical implementation, please refer to Le Saux (2010) and Le Saux et al. (2011).

The finite element model must reach a difficult balance between the accuracy of the results and the need for a computation time as low as possible in order to perform the optimization within a reasonable duration (several Finite Element simulations are necessary to insure the identification of the constitutive parameters). To simulate a non-axisymetric indenter like the Vickers tip used in our study is very cost full, both for full 3D and for simulations using 1/8th of the tip. It is possible to reduce this duration by using an equivalent axisymmetric geometry based on the equivalence of the contact projected area. We validate recently (Le Saux et al., 2011) the ability to apply this equivalence for our materials and the Vickers tip is therefore modeled by an equivalent cone whose angle is 140.6°. It is worth noting that the 3D simulations last 30 min on a recent workstation, whereas the axisymmetric simulations last 30 s.

The numerical model is presented on Figure 2. The dimensions of the elastomer part are set according to the experimental ones (height: 2 mm, distance to the edges: 2 mm). A parametric study on these two

Figure 3. Optimization loop.

parameters illustrated that these values were relevant to avoid any effects from the substrate and validated also the used experimental values (Le Saux, 2010). The epoxy resin is also modelled with an elastic linear behaviour (Elastic modulus of 5 GPa and Poisson ratio of 0.25) in order to account for the real stiffness. Because the indenter is much more rigid than the elastomer (the Young modulus of the indenter is 1 141 GPa), the indenter is modeled with a rigid body. The mesh consists in 3132 nodes and 2990 elements (240 CAX4 for the epoxy resin, 2700 CAXH4 for the rubber and 50 RAX2 for the indenter). Far from the indented zone, the mesh is quite coarse. As we get closer to the indented zone, the mesh becomes finer. The elements finally reach a size of 10 µm at the contact with the indenter. At the maximal indenter penetration depth, 30 elements are, at least, in contact with the indenter. At the interfaces between the rubber and the epoxy resin, the nodes displacement variables from the elastomer are set equal to the ones from the resin. The contact law between the indenter and the rubber follows a simple Coulomb friction law with a coefficient of 0.1 which is closed to real measured values for dry contact. There again, a parametric study was achieved. The value of 0.1 was chosen because it minimizes the computational time without degrading the solution. This conclusion is confirmed by a recent study performed by Giannakopoulos and Panagiotopoulos (2009).

3.2 Optimization procedure

The global scheme for the identification of the constitutive parameters of the rubber is presented on Figure 3. The optimization of the parameters search is performed thanks to software developed by Samtech, Boss Quattro v5.0. This software proposes several algorithms developed to converge quickly. In the following, we have used the GCMMA (Globally Convergent Method for Moving Asymptotes) algorithm proposed by Svanberg (1995) to determine material parameters from the indentation curve. Finally, the error computation quoted on Figure 3 is based on the least square method.

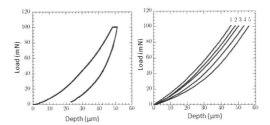

Figure 4. Experimental curves obtained for the silica filled CR (single test on the left, comparison of the loading response at several indentation places).

4 VALIDATION OF THE IDENTIFICATION

4.1 Results for the silica-filled CR

Figure 4 illustrates a typical response obtained by instrumented micro-hardness for this material. On the left, one can see that the creep under constant load is clearly visible and that the unloading curve is definitely different from the loading one. This reveals a visco-elastic behavior. On the right chart, an obvious experimental scattering is illustrated on the loading curves obtained at several places. This is clearly related to the filled nature of the material. Figure 5 illustrates that this scattering is likely coming from the ratio between an intrinsic scale of the material, related to its microstructure and the volume affected by the micro-hardness testing.

It appears from Figure 4 that the identification of constitutive parameters from micro-indentation would probably not be representative of the average volume tested at a macro scale. Nevertheless, we wanted to explore this comparison. The average loading curve (called 3 on Figure 4) was kept and used as the experimental curve feeding the optimization loop described in paragraph 3. The constitutive parameters identified were then used to predict the analytical response of the material for tension, compression and plane strain tensile tests, and compared to experimental values. Figure 6 illustrates that despite the Finite Elements simulation is giving a very satisfactory fit on the micro-hardness results, the obtained constitutive parameters were clearly not suitable at a macroscopic scale!

We also decided to take the problem the reverse way, in order to check if another micro-hardness curves would have given better results. The results are given on Figure 7. The two firsts charts illustrate the classical identification achieved on macroscopic tests achieved using Matlab software optimization toolbox. The third one gives the simulated response of the micro-hardness test using the identified parameters. On this third chart, the stiffest and smoothest experimental curves are plotted. These results shows that once again the parameters identified at one scale are not relevant at the other one, confirming what was suspected and illustrated in Figure 5.

At this stage, this conclusion could also come from the validity of the numerical simulations (equivalence

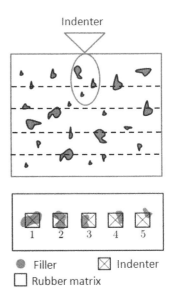

Figure 5. Illustrations (side and top views) of the link between the microstructure typical length and the volume tested by the indenter.

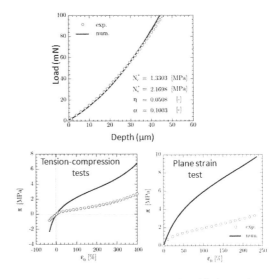

Figure 6. Illustration of the curve identified on micro-hardness results and of the extrapolation obtained for macroscopic tests with these constitutive parameters.

of the indenter, friction coefficient …) and we decided to switch to another material, which would afford well defined features with less coupled difficulties.

4.2 Results for the unfilled NR

A cured unfilled Natural Rubber was chosen because it exhibits very low viscous or residual strain effects. Moreover, the population of inclusions in the material was investigated by micro-structural observations (Dispergrader optical analysis and X-ray microtomography measurements) in order to measure that

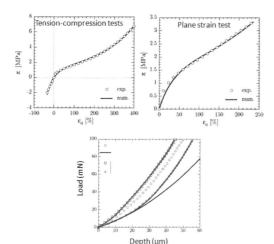

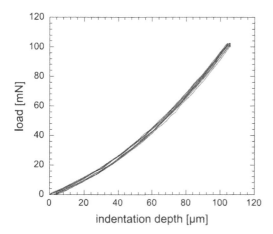

Figure 7. Illustration of the simulated response for micro-hardness test, using the constitutive parameters identified from macroscopic tests.

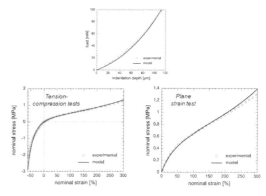

Figure 9. Illustration of the curve identified on micro-hardness results and of the extrapolation obtained for macroscopic tests with these constitutive parameters.

Figure 8. Experimental micro-hardness curve for unfilled natural rubber. Here, the full curves (load-creep-unload) are plotted, for 5 different indentations.

the average radius of the inclusions was around $10\,\mu m$ and that their volumic density was very low.

Figure 8 illustrates a full (load, creep, unload) typical response obtained by instrumented micro-hardness for this material, for 5 different tests. One can see that the creep under constant load is almost not visible and that the unloading curve is very close from the loading one. This confirms its very low viscous behaviour. Moreover 5 curves are plotted and are hardly visible, exhibiting a very low experimental scattering.

We achieved the same procedure as formerly on the silica-filled CR. Figure 9 illustrates the curve identified on the micro-hardness experimental curve. The constitutive parameters identified were then used to predict the analytical response of the material for tension, compression and plane strain tensile tests, and compared to experimental values. Figure 9 illustrates

clearly that the agreement is this time very good. This validates the experimental and numerical proposed approaches and provides, to our knowledge, the first successful transition of constitutive model identification from micro to macro scales for elastomeric materials.

5 CONCLUSION

This study investigated the ability of using microindentation to identify constitutive hyperelastic parameters that could be relevant at the macro-scale. A robust experimental protocol was proposed. An inverse analysis was then performed, coupling FE simulations achieved with Abaqus with optimization procedures using a specific software thanks to fortran90 and Python home made routines. This numerical protocol proved to be very effective and allows investigating several numerical parameters (indent geometry, friction, thickness...). The identified parameters were then used to simulate the curves obtained from macroscopic tests for a wide range of loading (tensile, compression and pure shear tests). The agreement observed is very good, for unfilled natural rubber which was not the case for the studies of the literature trying to perform this change of scales. The same approach applied to a silica filled CR showed on the contrary a strong dependence of the identified parameters to the ratio between a typical length of the microstructure (here, very likely the fillers size) and the tested volume.

In a near future, the tools developed in this study will be used to cover several perspectives, ranging from the identification of the constitutive models at microscopic scales for heterogeneous aged samples, to the characterization of filled materials.

REFERENCES

Bec, S., Tonk, A., Georges, J., Georges, E. & Loubet, J. (1996). Improvements in the indentation method with a surface force apparatus, Philosophical Magazine A 74 (1996) 1061–1072.

Busfield, J. & Thomas, A. (1999). Indentation tests on elastomer blocks, Rubber Chemistry and Technology 75 876–894.

Celina, M., Wise, J., Ottesen, D.K., Gillen, K.T. & Clough, R.L. (2000). Correlation of chemical and mechanical property changes during oxydative degradation of neoprene. Polymer Degradation and Stability, 68 : 171–184.

Cheng, Y. & Cheng, C. Scaling, dimensional analysis, and indentation measurements, Materials Science and Engineering R 44 (2004) 91–149.

Constantinescu, A. & Tardieu, N. On the identification of elastoviscoplastic constitutive laws from indentation tests, Inverse Problems in Engineering 9 (9) 19–44.

Edwards, S. & Vilgis, T. The effect of entanglements in rubber elasticity, Polymer 27 (1986) 483–492.

Gent, A. & Yeoh, O. Small indentations of rubber blocks: effect of size and shape of block and of lateral compression, Rubber Chemistry and Technology 79 (2006) 674–693.

Giannakopoulos, A. & Panagiotopoulos, D. Conical indentation of incompressible rubber-like materials, International Journal of Solids and Structures 46 (2009) 1436–1447.

Hamasaki, H., Toropov, V., Kazuhiro, K. & Yoshida, F. Identification of viscoplastic properties for leadfree solder by microindentation experiments, FE simulation and optimization, J. Japan Soc. Tech. Plast 46 (2005) 397–401.

Jabareen, M. & Rubin, M. (2007), Hyperelasticity and physical shear buckling of a block predicted by the Cosserat point element compared with inelasticity and hourglassing predicted by other element formulation, Computational Mechanics 40 447–459.

Le Saux, V. (2010). Fatigue et vieillissement des élastomères en environnements marin et thermique: de la caractérisation accélérée au calcul de structure. Ph. D. thesis, Université de Bretagne Occidentale.

Le Saux, V., Marco, Y., Blès, G., Calloch, S., Moyne, S., Plessis, S. & Charrier, P. (2011). Identification of constitutive model for rubber elasticity from micro-indentation tests on Natural Rubber and validation by macroscopic tests. To appear in Mechanics of Materials.

Loubet, J., Bauer, M., Tonk, A., Bec, S. & Gauthier-Manuel, B. Nanoindentation with a surface force apparatus. Mechanical Properties and Deformation of Materials having a Ultra-fine Microstructures, Kluwer Academic Publishers (1993) 429447.

Oliver, W. & Pharr, G. A new improved technique for determining hardness and elastic modulus using load and sensing indentation experiments, Journal of Materials Research 7 (1992) 1564–1582.

Rauchs, G., Bardon, J. & Georges, D. Identification of the material parameters of a viscous hyperelastic constitutive law from spherical indentation tests of rubber and validation by tensile tests, Mechanics of Materials 42 (2011) 961–973.

Svanberg, K. A globally convergent version of MMA without linesearch, in: Proceedings of the First World Congress of Structural and Multidisciplinary Optimization, Goslar (Germany), 1995.

Tabor, D. The hardness of solids, Rev. Phys. Technol. 1 (1971) 145–179.

Damage mechanisms in elastomers

Constitutive Models for Rubber VII – Jerrams & Murphy (eds)
© 2012 Taylor & Francis Group, London, ISBN 978-0-415-68389-0

A new approach to characterize the onset tearing in rubber

K. Sakulkaew, A.G. Thomas & J.J.C Busfield
Department of Materials, Queen Mary, University of London, London, UK

ABSTRACT: Since Rivlin and Thomas (1953) the tearing behaviour of rubber has been characterized using fracture mechanics, whereby the rubber has a geometrically independent relationship between the crack growth rate during tearing versus strain energy release rate. This approach works well under conditions of steady tearing when the crack growth rate is easy to measure. However, this approach is much harder to interpret under conditions where the rubber exhibits discontinuous crack growth behaviour such as knotty tearing or stick slip tearing. This type of tearing often arises for filled rubbers and for rubbers that strain crystallize. The measured crack growth rate is now an average of a very rapid tear rate and a zero velocity tear rate. Sakulkaew et al. (2010) developed a new approach that characterizes the rate of increase in the strain energy at the crack tip just immediately before the onset of tearing in a trouser tear test piece. This is measured directly as the time derivative of the strain energy release rate $\dot{T}$. In this work the critical strain energy release rate T_{crit} to propagate the tear relationship with $\dot{T}$ is examined for a range of natural rubber compounds. The aim being to demonstrate the geometric independence of this relationship to verify that the behaviour measured using a trouser tear test piece produces a similar relationship to that measured using a pure shear tear test piece.

1 INTRODUCTION

Tearing in rubber is known to initiate from an inherent flaw present in the rubber (Gent 2005). When the rubber is stretched, the local stress in the vicinity of a flaw is intensified. Once the local stress reaches a critical level, the rubber tears by extension of the crack. It has been widely reported that the rate of crack growth in rubber is determined by a characteristic energy per unit area of the fracture surface created, often known as the tearing energy or the strain energy release rate (Thomas 1994). This is defined as

$$T = -\frac{1}{h}\left(\frac{\partial W}{\partial c}\right)_l \tag{1}$$

where W is the total elastic strain energy in a specimen of thickness h measured in the unstrained state and c is the length of a crack. The suffix l denotes that no external work is done at the system boundaries to create new crack surfaces. In the case of a trouser tear test piece in Figure 1, the tearing energy is given by

$$T = \frac{2F\lambda}{h} - bW \tag{2}$$

where F is the applied tearing force, h is the specimen thickness, λ is the extension ratio in the legs, b is the total width of the specimen and W is the elastic stored energy in the legs of the specimen far removed from tear. W is determined from integration of a tensile stress-strain curve at a strain that corresponds to the extension ratio in the legs of the specimen at the point

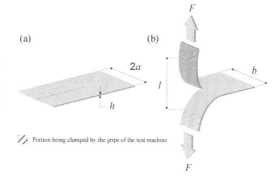

(a)　　　　　(b)

Figure 1. Trouser tear specimen: (a) undeformed; (b) deformed. F denotes the externally applied force to the leg of the specimen.

tearing. The relationship between the rate of tearing and the strain energy release rate is a material characteristic that is independent of test piece geometry (Greensmith and Thomas 1955). In the trouser tear specimen, a tearing force applied to the legs produces tearing at the crack tip. The region of each of the legs is essentially in uniaxial extension with the corresponding extension ratio, λ. The crack length increases by dc which results in an increase in the volume in the region of uniaxial extension in the legs (by $2a \cdot h \cdot dc$). The separation l of the clamps is increased by

$$dl = 2\lambda dc \tag{3}$$

If the tearing is smooth and continuous the average crack growth rate can be determined from the rate of separation of the clamps, S by the relation

$$S = \frac{dl}{dt} = \frac{2\lambda dc}{dt} \tag{4}$$

where l is the separation of the clamps of the test machine, t is time and c is the crack length. Figure 2 shows the four different types of measured force versus time response that are typically observed during tearing of rubber samples. Steady tearing, as depicted in Figure 2(a) and (b) can easily be interpreted using the framework described above as the tearing rate is constant and the relationship between strain energy release rate and the crack growth rate can be evaluated to characterize the tearing behaviour. However, for a lot of rubber materials the behaviour can be either stick slip in nature, as shown in Figure 2(c) or even knotty in behaviour as shown in Figure 2(d).

Under these conditions tearing is not steady and the characteristic rate of crack growth is much harder to determine as it fluctuates between a rapid rate and a zero rate of crack growth. Measurements of the cross head speed do therefore not characterize the rate of crack growth, and the resulting average rate derived from the cross head speed is not very meaningful.

A much more useful measure would be the rate of increase in strain at the tip of the crack which can then be related to the critical strain energy release rate for the material. Thomas (1955) showed that T is approximately the product of the size of the unstrained tip diameter of the tear d and the strain energy density E integrated around the tip of the crack at break,

$$T \approx dE \tag{5}$$

For the purpose of this work it is presumed that the characteristic crack tip diameter, d, does not alter very much during tearing and so a measure of the strain energy release rate is proportional to the strain energy at the tip of the crack (Thomas 1955). It is possible to take the time derivative of Equation 5 to develop the following relationship,

$$\dot{T} \approx d\dot{E} \tag{6}$$

Here $\dot{T}$ (which is dT/dt) is the rate of change in the strain energy release rate with time and $\dot{E}$ (which is dE/dt) is the rate of change of the strain energy with time at the tip of the crack. This equation shows that a measure of the rate of change of strain energy release rate with time can be used as an equivalent measure of the increase in strain energy at the tip of the crack with time. Since Rivlin and Thomas (1953), equation (4) has been used to relate the tearing rate to the applied rate of loading in the sample, without much account being made of the discontinuous nature of the tearing in stick slip and knotty tearing. Sakulkaew et al (2010) showed that a measure of the rate of increase of the strain energy release rate with time was a much

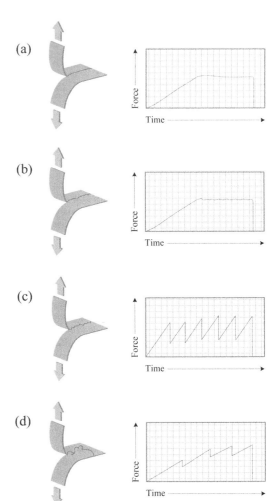

Figure 2. Schematic diagrams showing the tearing force-time relationships and tearing path for different type of crack growth obtained from the trouser tear specimen: (a) and (b) steady tearing, (c) stick-slip tearing and (d) knotty tearing.

more useful measure to determine the characteristic tear strength of a rubber irrespective of the type of tear behaviour observed. This previous work used a trouser tear test specimen to characterize how the critical tearing energy T_{crit} at the onset of the tearing relates to the rate of loading at the crack tip $\dot{T}$ for several different NR and SBR rubber compounds materials. In this work the approach has been extended by making measurements using a pure shear test piece, as shown in Figure 3, and a comparison between these two types of test piece is made. For the pure shear specimen (Thomas 1994; Tsunoda et al. 2000; Papadopoulos et al. 2008), the strain energy released per unit area of crack growth T is given by

$$T = Wh_0 \tag{7}$$

where W is the strain energy density in the central region of the specimen, which is in pure shear, and h_0 is the unstrained height.

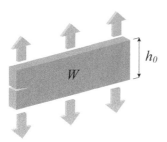

Figure 3. Pure shear specimen.

Table 1. Formulations for the rubber materials.

Ingredients	NR-S1	NR-S3	NR-P1	NR-P3
NR (SMR CV60)	100	100	100	100
Zinc Oxide	3	3	0	0
Stearic acid	1	1	0	0
Antioxidant 6-PPD	1	1	0	0
Accelerator CBS	0.75	2.1	0	0
Dicumyl peroxide (DCP)	0	0	1	3
Sulphur	2.5	7	0	0
Density, g/cm^3	0.96	0.98	0.92	0.93
Crosslink density, $\times 10^{-4}$/mol/cm^3	1.62	3.44	0.84	2.39

2 EXPERIMENTAL

Natural rubber NR (SMR CV60) was supplied by Tun Abdul Razak Research Centre, UK. NR samples employed in this investigation were vulcanized using Dicumyl peroxide (DCP) and sulphur. The compound formulations in parts by weight per hundred parts of rubber (phr) are given in Table 1. Equilibrium swelling in toluene was utilized to estimate the crosslink density of the vulcanizates. The samples of each vulcanizates were swollen in 100 ml toluene at room temperature for 72 hrs. The crosslink density was calculated using the relation (Davies et al. 1994),

$$-\ln(1 - V_r) - V_r - \chi V_r^2 = 2\rho V_0 X V_r^{1/3} \qquad (8)$$

where ρ is the density of the rubber vulcanizate, V_0 the molar volume of the solvent, X is the crosslink density (including chemical and physical interactions) V_r is the volume fraction of rubber in the swollen gel at equilibrium swelling, and χ is the polymer solvent interaction parameter. The polymer-solvent interaction parameter χ for the system NR-toluene is 0.391.

For the trouser specimen, the sample was approximately 24 mm wide, 30 mm long and 2 mm thick. Each leg of the trouser was therefore 12 mm wide and 15 mm long. Approximately 10 mm of the leg of the trouser was clamped in the grip during testing. Pure shear tear test piece dimensions were 60 mm long with a height, h_0 of 15 mm. An edge-cut of depth 10 mm was manually introduced midway and perpendicular to specimen length. The resultant razor cut does not

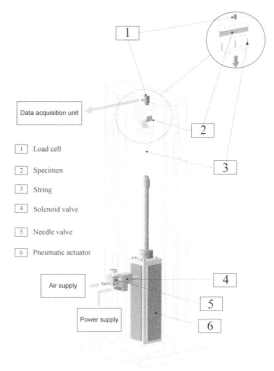

① Load cell
② Specimen
③ String
④ Solenoid valve
⑤ Needle valve
⑥ Pneumatic actuator

Data acquisition unit

Air supply

Power supply

Figure 4. Schematic drawing of the experimental set-up for the pneumatic-driven machine for trouser and pure shear specimens.

represent the geometric profile of the characteristic crack tip (Papadopoulos et al. 2008). Therefore, prior to each experiment, some small scale manual tearing of the tip was done to introduce a more typical crack tip profile. In order to obtain the various strain rates at the tear tip of the specimen, a screw-driven test machine (Instron Model 5564) was used for the relatively slow speed tests where the rate of grip separation was between 8.33×10^{-6} m/s and 0.04 m/s. A different approach was required to investigate higher strain rates. These used either a pneumatic-driven machine, shown in Figure 4 or a drop weight apparatus. The pneumatic-driven machine was used for a higher speed test up to 2 m/s with the loading rate verified by a high speed camera. The speed of a lower clamp attached to a pneumatic actuator was controlled by the amount of the air pressure entering the air cylinder through an adjustable needle valve (Figure 4). The drop weight apparatus was used for the highest rates with a test speed of up 7 m/s. The speed of the lower clamp attached to a string with a drop weight on the other end was controlled by the distance which the drop weight travels before the test piece was loaded. The velocity of the drop weight, v, can be calculated using

$$v = \sqrt{2gh} \qquad (9)$$

where h is the length of the string which corresponds to the distance the drop weight travels without constraint and g is the gravitational acceleration (9.8 m/s^2).

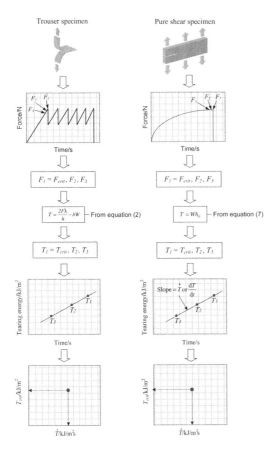

Trouser specimen — Pure shear specimen

Force/N ... Time/s

$F_1 = F_{crit}, F_2, F_3$ — $F_1 = F_{crit}, F_2, F_3$

$T = \dfrac{2F\lambda}{h} - bW$ —From equation (2) $T = Wh_0$ —From equation (7)

$T_1 = T_{crit}, T_2, T_3$ — $T_1 = T_{crit}, T_2, T_3$

Tearing energy/kJ/m² ... T_1, T_2, T_3 ... Time/s

Slope $= \dot{T}$ or $\dfrac{dT}{dt}$

T_{crit}/kJ/m² ... $\dot{T}$/kJ/m²s

Figure 5. Schematic diagram showing the typical calculation to obtain a data point plotted on a graph of T_{crit} against dT/dt or $\dot{T}$ using both trouser specimen and pure shear specimens.

The tearing force for the relatively high speed tests were measured by a piezoelectric force transducer (model ICP® 208C01, PCB Piezotronics, Inc., Depew, NY) used in combination with a signal conditioner (model 480C02, PCB Piezotronics, Inc., Depew, NY). This was chosen due to its rapid response rate of 0.2 ms. The output voltage signal from the load cell was passed through the input channel of a PC-based digital oscilloscope (PicoScope 2203, Pico Technology, UK) for data acquisition and storage. $\dot{T}$ and T_{crit} were determined from the measured force-time curve, a schematic of a typical curve is shown in Figure 5. To obtain $\dot{T}$, the three points just prior to the onset of tearing on the force versus time curve were taken, one at the peak force (critical tearing force) and another two near to but just before the peak force. This is because this is the portion of the curve that corresponds to the maximum strain at the tear tip before the new crack surface is introduced. The three tearing forces ($F_1 = F_{crit}$, F_2 and F_3) are transformed to the strain energy release rate ($T_1 = T_{crit}$, T_2 and T_3) using equation 2 for trouser specimen and equation 7 for pure

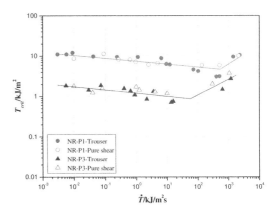

Figure 6. Critical tearing energy T_{crit} as a function of the time derivative of the strain energy release rate dT/dt for peroxide cured NR vulcanizates with different types of test pieces (trouser and pure shear).

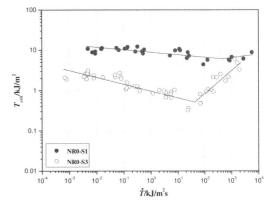

Figure 7. Critical tearing energy T_{crit} as a function of the time derivative of the strain energy release rate dT/dt for NR vulcanizates obtained from the trouser specimen (Kartpan et al. 2010).

shear specimen. The value of $\dot{T}$ was obtained from the slope of the resulting plot of tear energy versus time.

3 RESULTS AND DISCUSSION

The new way of characterizing the critical tearing energy T_{crit} as a function of the rate of change in the energy release rate with time $\dot{T}$ is shown as the data given in Figures 6 & 7.

Figure 6 shows two unfilled peroxide cured samples tested using both the pure shear tear and trouser tear test geometries.

Figure 7 shows the previously reported (Sakulkaew et al. 2010) data measured on two sulphur cured test compounds over a very wide range of loading rates using the trouser tear test specimen. Both figures show the dependence of the critical tearing energy on the rate of testing for the two unfilled NR compounds. It was found that less highly cross linked compounds,

NR-P1 and NR0-S1 were considerably stronger than the more tightly cross linked materials over the entire range of test rates reported. This is consistent with the general behaviour reported previously in the literature (Brown et al. 1987). For these materials the tear strength initially goes down with test speed and then it is seen to increase. The first drop results from a suppression of strain induced crystallization in the more highly cross linked material. The increase in the number of cross links modifies the main chain extensively and also inhibits the movement of the polymer chains. Both of these effects are thought likely to reduce the rate of strain induced crystallization. There is a good correlation between the trouser and pure shear test pieces in terms of the relationship between the critical tearing energy and the time derivative of the strain energy release rate shown in Figure 6.

This indicates that this method of analyzing the rate of tearing is geometrically independent and that test made during stick slip and knotty tearing can now be related from one test piece geometry to another. This clearly suggests that this approach is a significant improvement in the way that knotty or stick slip tearing is characterized.

4 CONCLUSIONS

The geometric independence of the rate of tear and the critical strain energy release rate relationship that can be used to characterize the rate of tearing under knotty or stick slip tearing conditions has been has been demonstrated. Therefore, a much clearer way of characterizing the rate of strain at the tip of the crack has been derived. The rate of loading of the strain at the crack tip is clearly the major factor when determining the strain energy release rate for the extension of a crack in a rubber sample. The method of using the first derivative of the strain energy release rate $\dot{T}$ is easy to measure and can be used to examine all types of non steady tearing such as stick slip or knotty tearing that is observed for a wide range of rubbers. This is a significant improvement over the average rate of crack growth that had been used previously to define the characteristic tear rate.

ACKNOWLEDGEMENTS

One of the authors, K. Sakulkaew, gratefully acknowledges the funding from the Royal Thai Government and thanks Tun Abdul Razak Research Center for the supply of rubber materials.

REFERENCES

Brown, P. S., Porter, M. & Thomas, A.G. 1987. Dependence of strength properties on crosslink structure in vulcanized polyisoprenes, *Kautschuk Gummi Kunststoffe*. Vol. 40: 17–19.

Busfield, J.J.C., Jha, V., Liang, H., Papadopoulos, I.C. & Thomas, A.G. 2005. Prediction of fatigue crack growth using finite element analysis techniques applied to three dimensional elastomeric components, *Plastics Rubbers Composites*, Vol. 34: 349–356.

Davies, C. K. L., De, D. K. & Thomas, A. G. 1994. Characterization of the behavior of rubber for engineering design purposes. 1. Stress-strain relations. *Rubber Chemistry and Technology*. Vol. 67: 716–728.

Gent, A.N. 2005. Strength of elastomers: 455-495. in Mark, J. E., Erman, B. & Eirich, F.R. (eds), *The Science and Technology of Rubber*. New York: Elsevier Academic Press.

Greensmith, H.W. & Thomas, A.G. 1955. Rupture of rubber. III. Determination of tear properties. *Journal of Polymer Science*, Vol. 18: 189–200.

Papadopoulos, I. C., Thomas, A. G. & Busfield, J. J. C. 2008. Rate transitions in the fatigue crack growth of elastomers. *Journal of Applied Polymer Science*. Vol. 109: 1900–1910.

Rivlin, R.S. and Thomas, A.G. 1953 Rupture of rubber. Part 1. Characteristic energy for tearing, *Journal of Polymer Science*, Vol. 10: 291–318.

Sakulkaew, K., Thomas, A.G. & Busfield, J. J. C. 2010. The effect of the rate of strain on tearing in rubber. *Polymer Testing*, Vol. 30: 163–172.

Thomas, A.G. 1955. Rupture of rubber. II. The strain concentration at an incision. *Journal of Polymer Science*, Vol. 18: 177–188.

Thomas, A.G. 1994. The development of fracture mechanics for elastomers, *Rubber Chemistry and Technology*. Charles Goodyear Medal Address: G48–G61.

Tsunoda, K., Busfield, J.J.C., Davies, C.K.L. & Thomas, A.G. 2000, Effect of materials variables on the tear behaviour of a non-crystallizing elastomer. *Journal of Materials Science*, Vol. 35: 5187–5198.

Constitutive Models for Rubber VII – Jerrams & Murphy (eds)
© *2012 Taylor & Francis Group, London, ISBN 978-0-415-68389-0*

Damage variables for the life-time prediction of rubber components

V. Mehling & H. Baaser
Freudenberg Forschungsdienste, KG, Weinheim, Germany

T. Hans
Technical University of Munich, Germany

ABSTRACT: Shortening product life-cycle times, the demand for more intense loading and at the same time for more reliable operation call for generally applicable tools for the evaluation of the life-time of dynamically loaded rubber products. Today, engineers apply more or less sophisticated procedures to estimate the life-time of components. All of these procedures rely on an initial choice of a measure of fatigue loading, the so-called damage variable. This choice may be as simple as the global deformation of a component. More advanced methods require local damage variables, e.g. stress- or strain-measures as well as energy variables or quantities motivated by fracture mechanical approaches. However, it turns out, that some of these variables are only valid for certain applications and fail in general. Damage variables from fracture mechanics yield the most general approach to fatigue evaluation, although their practical computation is challenging. This paper is dedicated to the discussion of local damage variables and to a better and more differentiated understanding of the behavior of small cracks under complex loading conditions.

1 INTRODUCTION

1.1 *Fatigue life-time of elastomer components*

The life-time of rubber products under fluctuating loads is limited by a large number of factors including the composition of the rubber compound, the mechanical and thermal loading history, chemical changes over time by internal processes or external attack as well as reversible and irreversible internal rearrangements (Mars & Fatemi 2004). Here, we want to focus on the mechanical loading only.

The behavior during a typical fatigue experiment can be described as follows. For periodic loading well below the static load limit, the response of a rubber component enters a quasi-stationary state when the initial rearrangement of the polymer chains and the filler network has commenced. After this phase of so-called 'conditioning', further degradation of the material induced by the mechanical loading is attributed to 'fatigue', altering the system response slowly at first, more and more rapidly later. At the end the component fails completely. In this context, the term 'failure' is a matter of definition. While for a tensional test specimen complete fracture may be an adequate and easily detectable criterion, it is often necessary to choose more subtle definitions, e.g. a certain loss of initial dynamic stiffness.

Fatigue data is most commonly visualized by load-number curves, where the fatigue lifetime is given as a function of the level of loading. In the double-logarithmic plotting, this often resembles a linear characteristic. The load-number curves greatly depend on the choice of the quantity that is chosen to represent the loading ('fatigue predictor', 'loading criterion' or 'fatigue load variable') and they can only be considered a representation of material fatigue-properties, if they coincide for any chosen load type and history.

The reason for the above fatigue phenomena is the nucleation and growth of flaws and cracks in the rubber material or on its surface. Today, there is concordance in the rubber community, that there is an abundance of microscopic inhomogeneities such as filler agglomerates, natural impurities of the caoutchouc itself or unevenness of surfaces in all rubbers. They can be reduced by minute control of raw material quality and process parameters, but they cannot be avoided completely. In any case, such initial flaws serve as nuclei for microscopic cracks in the rubber bulk or on its surface. Depending on the loading conditions experienced by these flaws, they grow and have increasing effect on the components behavior. Eventually some of them evolve to macroscopic cracks or local fracture zones which can be observed visually.

It follows from these considerations that physically motivated fatigue load variables should take into account the local loading of the material rather than the overall loading of a component. Methods utilizing such criteria are called local concepts.

1.2 *Local concepts*

In the previous section we have settled for the necessity of local concepts for the description of fatigue. Unfortunately there is a great abundance of local quantities which could be used as load variables, such as stresses, strains, energy densities or derived criteria. Accordingly, a large number of approaches have been published, see Mars & Fatemi (2002). A good choice should be linked to the nature of the materials degradation.

Cracks in rubber usually evolve on critical planes in the material. Similarly, fatigue crack growth is known to depend not only on the amount but also on the direction of local loading. For simple multiaxial loading situations such planes have been found to be oriented normal to the direction of maximum principal strain during a load cycle (Saintier et al. 2006a, Mars & Fatemi, 2006).This indicates that fatigue load variables should incorporate a 'sense of direction'. Newer fatigue load criteria take into account some directional information, as for example the cracking plane approach by Mars (2002) or the critical plane proposed by Saintier et al. (2006b). Such models assume the fatigue damage for a chosen plane to represent the fatigue life independently of fatigue on other planes. Choosing a cracking-plane is an important issue for such models. Other models do not consider specific planes throughout the history, but the currently critically loaded plane at every instance of loading time, e.g. the configurational stress criterion by Andriyana & Verron, (2007) or the tearing energy criterion by Ziegler et al. (2009). In fact, this is similar to approaches working with maximum principal components.

Fracture mechanics based concepts include some directional information per se and they describe the very mechanism of fatigue. Moreover it is known, that the rate of fatigue crack growth is governed by the energy release rate of the defect (Paris et al. 1961). This is why such concepts show the greatest promise to yield solutions for general, multiaxial loading. Specifically, the tearing energy criterion should provide a direct link to life-time prediction by integration of crack propagation relations as the Paris-Erdogan law. However, fracture mechanical quantities are in general quite costly to evaluate and they require certain prerequisites:

- Reliable constitutive law for the description of large deformation rubber material behavior including the characteristic up-turn.
- Knowledge of the location, size, shape and orientation of the considered crack or flaw.

When this information is available and the crack growth characteristics of the material are known, then component life times can be estimated. From the simulation of cracks in complex non-proportional loading situations it is known, that cracks may close partly or completely or they may kink or buckle due to numerical or physical instabilities. Furthermore, crack faces may slide on each other. In such cases information on the roughness of crack surfaces is needed and efficient methods for the identification of fracture mechanical variables in case of contact on the crack-faces are required. In this context a factorization of the tearing energy with respect to the crack-orientation for simple loading situations in 2-d has been proposed by Aït-Bachir et al. (2010). To find a similar relation for general multiaxial loading, including compressed states remains a challenging task. Configurational forces (Gross et al. 2003) may allow for a more detailed understanding of crack behavior in such situations.

In contrast to fracture mechanical predictors, promising sophisticated continuum mechanical concepts like the cracking energy (Mars, 2002), the stress-based predictor by Saintier et al. (2006) circumvent some of these drawbacks by avoiding the simulation of specific cracks. In Ziegler et al. (2009) a method to calculate tearing energies from standard FEA results in a post-processing step without successive micromechanical simulations has been presented. This approach relies on a conservative choice of the cracking plane such that the crack is assumed to be oriented perpendicular to the maximum principal strain at all times. Tearing energies are evaluated for hypothetical penny-shaped cracks of uniform size. Due to symmetry conditions no sliding of crack faces needs to be considered. The location and the orientation of the potentially critical crack can be easily identified and the tearing energy is evaluated without great computational effort. However, because the crack orientation follows the principal axes of the strain tensor, this method requires additional micromechanical simulations if the time history of the tearing energy of specific cracks of fixed orientation is to be generated.

In the first part of this paper the authors take one step back towards the conventional strain-based concepts. The idea of critical planes is adopted in the sense that the stretch of critical fibers of material is taken as fatigue load variable. Such a criterion can be easily incorporated into conventional fatigue analysis tools. The influence of the constitutive law on the result is comparatively small and only strain-information is required from FEA, which is easily accessible. In this context, the fiber orientation can be interpreted as the normal direction of a microscopic crack. Therefore, the second focus of the paper is the behavior of such cracks in micromechanical simulation during non-proportional loading.

2 FIBER STRAIN CONCEPT

2.1 *A strain-based, orientation-dependent predictor*

Although it has been discussed above, that physically sound fatigue-load criteria require a fracture mechanical foundation, many engineers still prefer the use of strain-based criteria. This is due to different reasons. Strains are easily evaluated, often directly measurable and they can be easily interpreted in terms of deformation. Often, insufficient description of the material behavior has little impact on the calculation of strains. In comparison to fracture mechanical quantities, strain-based variables do not require the cumbersome simulation of microscopical cracks. Finally, for many applications, strain criteria have been

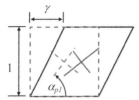

Figure 1. Simple shear deformation.

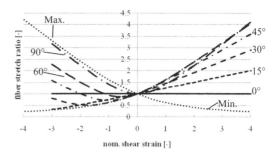

Figure 2. Fiber stretch vs. shear strain for different fiber orientations. Dotted lines indicate upper and lower limits, i.e. maximum and minimum principal stretch.

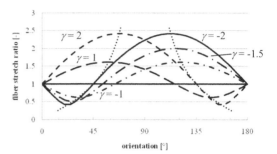

Figure 3. Fiber stretch ratio vs. fiber orientation for different shear strains. Dashed lines indicate positions of curve extrema.

found capable of predicting fatigue life satisfactorily. This is why the authors present a modified strain criterion, regarding the stretching ratio $\lambda_f = dx/dX$ of selected critical material fibers of initial length dX and deformed length dx.

For this purpose it is necessary to select a fiber orientation similarly to the selection of a critical cracking plane. In fact, the fiber orientation can be interpreted as the normal direction of a hypothetical crack. Three choices are possible: The fiber with maximum peak stretch, the fiber with the maximum stretch-amplitude and the fiber with shortest predicted fatigue life. The first one is easily identified while the other two require more elaborate identification methods. All of this is done on the basis of data available at the integration points of any FEA-model in a post-processing analysis. The time history of fiber stretch is analyzed by a counting algorithm. For non-proportional loading, the number of loading events may differ from the number of external load cycles. If fiber strain rather than fiber stretch is used, negative R-ratios will occur. We therefore define $R = \lambda_{f,\min} - 1)/(\lambda_{f,\max} - 1)$ as R-ratio and $A = \lambda_{f,\max} - \lambda_{f,\min}$ as amplitude of such a loading event.

2.2 Example: Simple shear deformation

A cube of rubber is subjected to an alternating simple shear as shown in Figure 1. Here γ denotes the tangent of the shear angle. For very small values of γ, the first principal direction of strain is inclined by 45°. With increasing γ, the principal axes rotate continually and the inclination of principal axes becomes larger when viewed in the undeformed configuration. The maximum and minimum principal stretches are given by equations (1) and (2):

$$\lambda_{p1} = \sqrt{1 + \gamma^2 (1 - a^2)/((1 - a)^2 + \gamma^2)} \qquad (1)$$

$$\lambda_{p2} = 1/\lambda_{p1} \qquad (2)$$

with

$$a = 1 + \frac{1}{2}\gamma^2 + \sqrt{\left(1 + \frac{1}{2}\gamma^2\right)^2 - 1}.$$

The orientation of maximum principal stretch is given by

$$\tan \alpha_{p1} = (a - 1)/\gamma, \qquad (3)$$

And the stretch of a fiber oriented along a given direction α is

$$\lambda_f = \sqrt{(\cos(\alpha) + \gamma \sin(\alpha))^2 + \sin^2(\alpha)}. \qquad (4)$$

The crossed lines in Figure 1 are oriented along the principal directions of strain at the current state. Note, that for positive shear angles, the material lines associated with these directions are perpendicular in the two depicted configurations only. For other values of the shear strain, they are not perpendicular and do not represent principal directions of deformation. In other words, for a loading history, where the orientation of principal axes is not fixed, the history of stretching of any chosen fiber of material is different from the maximum principal stretch. This is illustrated by Figures 2 and 3.

Figure 2 shows stretch ratios for a selection of discrete fiber orientations from the first quadrant of the coordinate system. Here, all orientation angles are given with respect to the horizontal axis which is parallel to the displacement. All other fiber directions are implicitly included due to symmetry considerations. Only the horizontal fiber ($\alpha = 0°$) remains completely undeformed. In case of positive shear strains, fibers with orientation angles larger than 45° experience maximum principal stretch for a specific value of shear strain, while fibers with smaller orientation angles never reach this limit and seem comparatively uncritical with respect to fatigue. For very small amounts of shear strain, the 45°-fiber is stretched most, while

193

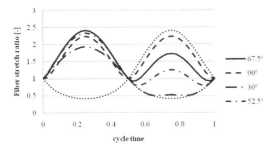

Figure 4. Fiber stretch ratio vs. time for a 1 Hz shear loading with an amplitude of 2.

Table 1. Loading events (nom. fiber strain $\lambda_{f,max} - 1$).

Fiber	30°	52.4°	67.5°	90°
1st loading event				
strain max.	0.93	1.33	**1.41**	1.24
R-ratio	−0.24	−0.16	−0.05	0
strain amplitude A	1.43	**1.54**	1.49	1.24
2nd loading event				
strain maximum	<0	0.26	0.73	1.24
R-ratio	–	−0.81	−0.1	0
strain amplitude A	–	0.46	0.81	1.24

the 90°-fiber remains undeformed. For infinite shear strain, the 90°-fiber approaches the maximum limit line, while all others experience less stretching. In the case of alternating shear loading, the situation becomes more complicated. Clearly the maximum principal strain criterion results in $R = 0$ with two equivalent loading events during one cycle. In Figure 3, the fiber stretches are plotted as functions of the fiber orientation. For very small amounts of shear, half of the fibers are extended, while the other half is compressed. The larger the strain, the less fibers experience compression. With increased loading, the direction of the maximum principal strain shifts from 45° to 90°, as indicated by the dashed line. The peak values of all curves are located on these lines. Here, the fiber with maximum stretch can be easily identified for any chosen load level. In a thee-dimensional FEA one would simply select the maximum principal strain during the loading history and identify the respective principal direction.

The history of fiber stretch for selected fiber orientations during one cycle of shearing is shown in Figure 4. All curves give smaller values than the maximum principal stretch indicated by the dashed line. The 90°-fiber experiences two loading events with $R = 0$ of equal amplitude $A = 1.24$. The 30°-fiber has a large loading event with $A = 1.43$ and $R = −0.54$, and a very small loading event in the range of negative strains, which can be neglected with respect to its impact on fatigue. The fiber with the peak stretch of $\lambda_f = 2.41$ is oriented under $\alpha = 67.5°$. It also shows a large event, $A = 1.49$, $R = −0.05$ and a small event $A = 0.81$, $R = −0.1$. The largest amplitude $A = 1.54$, $R = −0.16$ is found for the 52.4°-fiber which also has

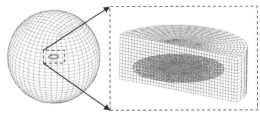

Figure 5. Submodel for crack simulations: Spherical submodel (left) and detail of the crack vicinity (right).

a smaller loading event with $A = 0.463$, $R = −0.812$. Recollecting this data, during cyclic loading, most fibers experience two collectives of stretching with different R-ratios and amplitudes each, as summarized in Table 1. Usually one of the collectives is smaller in amplitude than the other and has smaller impact on fatigue. Neglecting this part of loading results in an error up to a factor of two in the predicted life-time.

When the damage D done by a loading event is described by a simple power law $D \sim A^p$, the fiber sustaining maximum damage depends on the material parameter p. Note, that any superimposed stress does not affect the resulting strain measures in the current example.

3 CRACK SIMULATION

3.1 FEA-model

While all strain-based quantities for the examples in this paper can be evaluated analytically, this is not possible for the quantities characterizing the deformation of a microscopic crack. Here, a crack model is introduced into a homogeneously loaded cube of rubber material by a sub-modeling technique. The spherical sub-model of radius $r_{sm} = 0.5$ mm shown in Figure 5 includes a penny-shaped crack of radius $r_{cr} = 0.05$ mm. The outer surface of the sphere is subjected to boundary conditions conforming the homogeneous macro-field. The crack normal is always oriented in the shearing plane. It is initially parallel to the direction of macroscopic displacement and can be inclined by an angle α. The region around the crack tip is meshed by a structured, axis-symmetric array of C3D8H elements, allowing for a stabile analysis of all considered deformation states. Crack surfaces are modeled as frictionless although it is clear, that friction will have a considerable impact on local deformation and fields in case of closed or partially closed cracks. Furthermore, the pressure inside the crack is assumed to be zero at all times. For the sake of simplicity, the material behavior is assumed to follow Neo-Hooke's law with a shear modulus of $G = 0$ and quasi-incompressible behavior.

Clearly, for real rubber such a description is not sufficient because of the very large deformations found close to the crack tip. Output quantities are the equivalent spherical void-radius r_e (i.e. the radius of a sphere with identical volume as the crack), the relative crack

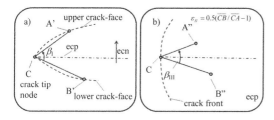

Figure 6. Definition of local crack-opening parameters. A and B are the next neighboring FE-nodes to crack-tip node C.

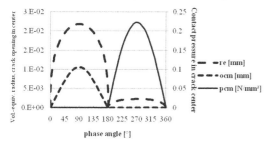

Figure 7. Global crack opening measures for harmonic simple shear, $\alpha = 52°$, $\gamma = 0 \pm 0.5$, $\sigma_y = 0$ MPa.

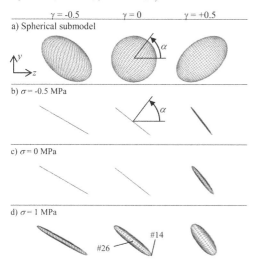

Figure 8. a) Spherical sub-model, psc with normal direction inclined by $\alpha = 52°$ subjected to c) simple shear only and to simple shear with additional b) compressive and d) tensile stress.

opening o_{cm} (given relative to the ecn, see below) as well as the contact pressure p_{cm} in the middle of the crack and local crack opening measures β_I, β_{III} and ε_{II}, where β_I and β_{III} denote the angles of mode-I and mode-III crack opening, respectively and ε_{II} is a crack opening strain attributed to mode-II, see Figure 6. These geometrical local quantities are computed from the deformed coordinates of finite-element nodes A and B which are neighbors to crack tip node C on the upper and lower crack faces. They are therefore dependent on the mesh density. Despite this, they are quite useful to describe and understand the local loading of

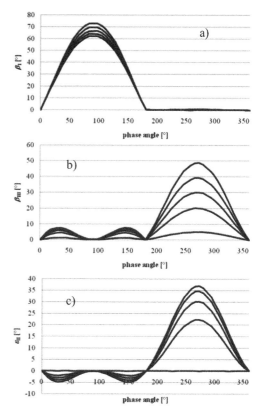

Figure 9. Local crack opening measures: $\alpha = 52°$, $\gamma = 0 \pm 0.5$, $\sigma_y = 0$ MPa a) mode-I b) mode-III c) mode-II. Each line represents a single crack tip node.

the crack. In this context we define two planes of projection: The effective crack plane (ecp) and its normal (ecn) are determined by the deformed coordinates of certain nodes on the crack front. A second plane containing crack tip node C as well as the half distance coordinates between nodes A and B and is chosen to be orthogonal to the ecp. Furthermore, the local J-Integral from ABAQUS/Standard output is given.

3.2 Example: Simple shear and superimposed stress

In the following, we want to focus on the behavior of a small penny shaped crack (psc) in a homogeneous far-field of sinusoidal harmonic simple shear γ with amplitude 0.5 in the zy-plane. In addition, a stress is applied along the y-direction. The deformation of the sub-model with a crack is shown in Figure 8.

The initial inclination of the crack normal of $\alpha = 52°$ is chosen because this is the direction of max. princ. strain for $\gamma = 0.5$. The crack rotates in the direction of shearing and is opened and closed by positive and negative shearing, respectively. The superimposed stress strongly influences the crack configuration. For $\sigma_y = 1$ MPa tensile stress, the crack does not close during the whole cycle of deformation, while it hardly opens for compressive stress.

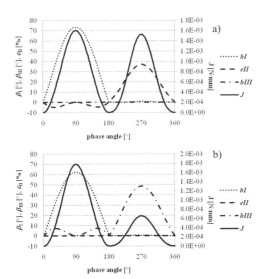

Figure 10. Local crack loading at crack tip nodes a) #14 and b) #26, located along and transverse to direction of shearing, resp. ($\alpha = 52°$, $\gamma = 0 \pm 0.5$, $\sigma_y = 0$ MPa, $\mu = 0$).

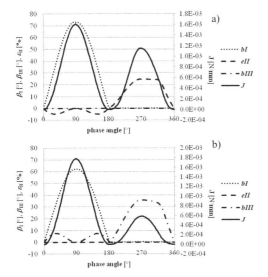

Figure 11. Local crack loading at crack tip nodes a) #14 and b) #26. Influence of crack-face friction, $\alpha = 52°$, $\gamma = 0 \pm 0.5$, $\mu = 0.1$, $\sigma_y = 0$ MPa.

Global measures of opening and closing in the case $\sigma_y = 0$ MPa of are shown in Figure 7. From r_e during negative shear we see that although the crack is in contact in the center, there still is some volume left closer to the crack front.

It is very helpful to learn about the local crack loading. The mode I to III load parameters for selected crack tip nodes are shown in Figure 9. They demonstrate, that the whole crack front opens during positive shear in mode I, while the cracktip is loaded in mode II and III only during negative shear. The local loading varies along the crack front. This is shown in Figures 10 and 11. Node #14 is located in direction of

positive shear loading, while #26 is located transverse to the loading direction, see Figure 8. #14 experiences mode I and mode II loading for positive and negative shear, respectively. Accordingly, #26 experiences mode I and mode II loading. When ABAQUS calculates J-Integrals, the local value at each node has two distinct maxima during one cycle of loading. In our example, the first maximum is due to a mode-I stress/strainfield, while the second one is calculated for the mode II/III stress/strainfield with the crackfaces in contact. This should be interpreted with great caution. The mode II/III loading is reduced by crack face friction (Figure 11).

4 CONCLUSION

In this paper load variables for the description of local fatigue loading are being discussed. Fracture mechanical loading variables seem to be most promising in providing a general fatigue criterion. An orientation-dependent strain-based criterion is proposed and local and global crack loading measures are computed for simple-shear deformation with additionally applied stress.

REFERENCES

Aït-Bachir, M., Verron, E. & Mars, W.V. 2010. Energy release rate of small cracks under finite multiaxial straining. In: Heinrich, G., Kaliske, M. & Lion, A. & Reese, S. (eds), *Constitutive Models for rubber VI*. 313–318.

Andriyana, A. & Verron, E. 2007. Prediction of fatigue life improvement in natural rubber using configurational stress. *Int. J. Sol. Struct.* 44: 2079–2092.

Gross, D., Kolling, S., Müller, R. & Schmidt, I. 2003. Configurational forces and their application in solid mechanics. *Eur. J. Mech. Sol. A/Solids* 22: 669–692.

Mars, W.V. 2002. Cracking Energy Density as a Predictor of Fatigue Life. *Rubber Chem. Technol.* 75: 1–17.

Mars, W.V. & Fatemi, A. 2004. Factors that affect the fatigue life of rubber: a literature survey. *J. Rubber Chem. Technol.* 77: 391–412.

Mars, W.V. & Fatemi, A. 2005a. Multiaxial Fatigue of rubber: Part I: equivalence criteria and theoretical aspects. *Fat. Fract. Engng. Mater. Struct.* 28: 515–522.

Mars, W.V. & Fatemi, A. 2005b. Multiaxial Fatigue of rubber: Part II: Experimental observations and life predictions. *Fat. Fract. Engng. Mater. Struct.* 28: 523–538.

Mars, W.V. & Fatemi, A. 2006. Multiaxial stress effects on fatigue behavior of filled natural rubber, *Int. J. Fat.* 28: 521–529.

Paris, P.C., Gomez, M.P. & Anderson, W.E. 1961. A rational analytic theory of fatigue. *The Trend in Enging.* 13: 9–14.

Saintier, N., Cailletaud, G. & Piques, R. 2006a. Crack initiation and propagation under multiaxial fatigue in a natural rubber. *Int. J. Fat.* 28: 61–72.

Saintier, N., Cailletaud, G. & Piques, R. 2006b. Multiaxial fatigue life prediction for a natural rubber. *Int. J. Fat.* 28: 5300–539.

Verron, E., Le Cam, J.B. & Gornet, L.A. 2006. Multiaxial criterion for crack nucleation. In: *Rubber. Mech. Res. Commun.* 33: 493–498.

Ziegler, C., Mehling, V., Baaser, H. & Häusler, O. 2009. Ermüdung und Risswachstum bei Elastomerbauteilen. *DVM-Bericht* 676: 121–129.

Constitutive Models for Rubber VII – Jerrams & Murphy (eds)
© 2012 Taylor & Francis Group, London, ISBN 978-0-415-68389-0

Formation of crust on natural rubber after ageing

S. Kamaruddin
School of Engineering Sciences, University of Southampton, UK

P.-Y. Le Gac
IFREMER, Brest Centre, France

Y. Marco
ENSTA Bretagne, Laboratoire LBMS (EA 4325), France

A.H. Muhr
Tun Abdul Razak Research Centre, UK

ABSTRACT: A case study of a solid natural rubber tyre, subjected to approximately 80 years natural ageing in woodland in the UK, is presented, showing that at depths greater than 4 mm the rubber is still in good condition, but that a hard crust has developed on the surface. Possible mechanisms accounting for the extreme surface degradation, but modest penetration into the bulk, are discussed, in particular diffusion-limited attack by oxygen and ozone, and the penetration depth of light.

1 INTRODUCTION

Several studies of natural rubber (NR) artefacts that have been exposed to the ambient environment for long periods confirm that a very hard crust is formed. This is not consistent with expectation from some laboratory ageing studies (e.g. Lindley & Teo, 1977). Lindley & Teo's results suggest that only at elevated temperature is the oxidation reaction so rapid that the penetration into the bulk of the rubber is very small.

Lindley & Teo (1977) subjected light-coloured sheets and blocks of NR (protected with the staining antiozonant/antioxidant N-isoproply-N'-phenyl-p-phenylenediamine IPPD), 1, 8 or 25 mm in thickness, to periods of up to 35 weeks in air ovens, at 50 to 200°C, in the dark. They measured uptake of oxygen, loss of weight, and indentation properties of the aged rubber as a function of depth. They also reported time to formation of a hard skin, and ingress h of discolouration from the surface as a function of time and temperature. Oxidation should result in an increase in weight, but loss of accelerator residues and antidegradant resulted in net weight loss of the order of 4% after 5 days at 150°C. In accord with expectation for diffusion-limited oxidation, h appears to be proportional to the square root of time in the first day, at least for ageing at or above 100°C. However, it does not increase as rapidly thereafter. The value of h at one day of ageing, h_1, increases as the temperature is lowered, with an Arrhenius relation, as do t^*, the time to form a hard skin, and the time to absorb 1% by weight

of oxygen (being of similar magnitude). The activation energy is the same for all these fits (85 kJmol^{-1}), and using this to extrapolate to 20°C gives 434 years to form a hard skin. However, the lowest temperature for which a hard skin was observed was 100°C, and the maximum ageing periods at 70°C (11 months) and 50°C fell short of the extrapolated values for hard skin formation at those temperatures. Although the solubility of oxygen in NR falls as the temperature is increased the diffusion rate rises, and the net effect is that the permeability rises (van Amerongen, 1964).

Williams & Neal (1930) investigated the effects of pressure and temperature of oxidation of NR, and reached conclusions not entirely consistent with those of Lindley & Teo (1977):

1. rate of oxygen absorption by NR is independent of concentration/pressure as long as there is sufficient oxygen available
2. at 70°C, pure oxygen at a pressure of 1 bar is needed to give just sufficient dissolved oxygen for oxidation at the maximum rate; in air at 70°C oxidation would be inhomogeneous for thicker sections.
3. at 26°C there may not be enough oxygen dissolved in NR in air for oxidation at the maximum rate especially for grades that are easily oxidised, so oxidation might also be inhomogeneous for thicker sections.

This paper examines a case of rubber that has aged naturally but developed a surface crust, probing the chemical and physical character of the crust and interpreting the results to assess the significance of hypothetical mechanisms leading to its formation:

- oxidation of NR is known to be autocatalytic, and the diffusion coefficient will diminish as the material becomes more glassy with the progress of oxidation, both features usually being ignored in simple theories for diffusion limited oxidation
- UV light, ozone, leaching of antidegradants or fire damage will all impinge predominately on the surface, and may alone or in combination cause the crust formation.

The most useful case-histories for establishing the primary ageing processes and their consequences are those for which the initial properties and the ageing duration and environment are known. However, once a mature understanding has developed, the possibility of deducing the approximate age and environment from the current state of the artefact should emerge.

2 SOLID TYRE FROM BRAMFIELD WOOD

A pair of steel cartwheel rims of diameter 775 mm and width 95 mm, with ~40 mm thick solid rubber tyres attached was noticed in Bramfield Wood (grid reference TL284170) north west of Hertford in April 2004. One was left in-situ, and the other taken to TARRC to investigate the state of the rubber. It is believed that the rims would have originally been fitted to wooden spoked (artillery-type) wheels. The only other readily identifiable items at the site were two steel hubs incorporating brake drums, of a design to accommodate wooden spokes, probably manufactured not later than the 1920s. No wooden parts of the wheels (expected to have been ash) could be identified. A few remnants of ash, in very poor state, were all that remained of the structure of the vehicle, presumably entirely wooden. It is possible that the complete vehicle had not been abandoned at the site, but only a few elements of it, perhaps just those with significant steel content. However, it seems more likely that the steel and rubber parts had simply outlasted the wooden parts, since they comprised a full two-wheel set with no other unrelated artefacts. The Forestry Commission managed the wood from soon after the Second World War until 2009. We found no evidence of fire at the site of the rims.

The exposed surface of the rubber was very hard and cracked (Figure 1), but the interior, once exposed by cutting, was in reasonable, rubbery, condition, having a hardness of 60 to 65 Shore A. Using a pocket hardness gauge on the outer surface, taking care that the indenter is not offered into a crack, gives a value of 90 to 95 Shore A. Results of analysis of the rubber (confirmed as Natural) are given in Table 1.

The total sulfur content was established for the bulk rubber as being 3wt%, giving the basic formulation of NR + 54 pphr black + 3 pphr S + 15 pphr ash + 1 pphr extractables. The ash is likely to include zinc oxide and perhaps some clay; SEM X-ray confirmed presence of S, Zn, Mg, Al, Si and possibly a trace of Ba. The black

Figure 1. Section through rubber showing the hard layer (~0.5 mm thick) and cracks into the transition layer (~4 mm deep).

Table 1. Basic formulation from TGA and PIR.

	Rubber (wt%)	Black (wt%)	Ash (wt%)	Tmax (°C)
surface	47.2	47.7	5.0	445
1–3 mm	55.0	38.3	5.4	446
5–7 mm	59.5	31.5	8.6	418
bulk	59.2	31.9	9.0	413

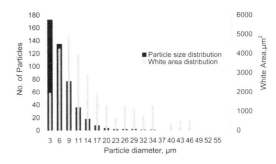

Figure 2. Distribution of particle diameters, and total area contribution of each tranche of particles within the 3 μm ranges of diameter. Area analysed $= 2.1 \times 1.6$ mm $= 3.36 \times 10^6$ μm^2.

type had a mean particle size of 45.5 nm, equivalent to a group 5 black such as FEF (N550). However, it is likely that the carbon black is a channel black, since the furnace manufacturing method was first used in 1922 (from gas, for primary particle sizes above about 50 nm), and it was not until 1942 that the oil furnace method (from which smaller particle size blacks can be made) was developed (Donnet & Voet, 1976). As the first deliberate addition of antioxidants in rubber formulations was in the 1920s (by the US for military gas masks), and antiozonants were first used after the Second World War (eg Huntink, 2003), it seems most likely that the rubber was formulated without antidegradants.

Figure 2 shows the state of dispersion in the bulk rubber, as a distribution of agglomerate diameters, obtained from a Dispergrader at TARRC. This apparatus uses reflection of low angle incident light to detect protrusions and bumps in a fresh surface cut by a sharp

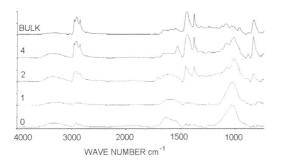

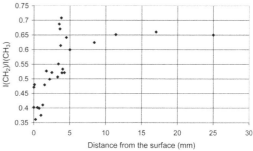

Figure 3. FTIR spectra (relative absorbance vs. wave number) from different depths (labelled in mm) into the rubber.

Figure 5. CH$_2$ intensity, at different distances from the surface (left) to bulk (right).

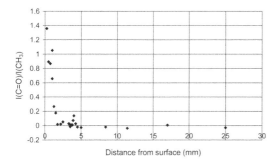

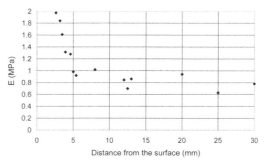

Figure 4. C=O intensity, at different distances from the surface (left) to bulk (right).

Figure 6. Tensile modulus (ratio of nominal stress to strain) at 50% extension at different distances from the surface.

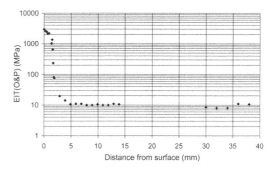

Figure 7. Reduced modulus from micro-indentation according to the analysis of Oliver & Pharr (1992) at different distances from the surface.

razor; such a surface was fairly shiny, qualitatively indicating reasonable dispersion.

Of greatest interest here is the transition from a rubbery material in the bulk, 4 mm or more below the surface, to a much harder material at the surface. It is apparent from Table 1 that there is a loss of polymer relative to black, and possibly decomposition of polymer into carbon, towards the weathered surface, which is strongly marking due to free black, unlike the bulk rubber. The transition layer is creviced with what look like shrinkage cracks, to a depth of about 3 mm. In this transition layer the FTIR spectrum evolves with depth as shown in Figure 3.

The spectra reveal a rise in –OH and C=O groups and a decline in C=CH groups through the transition zone towards the surface, consistent with oxidative attack of the unsaturated backbone of NR. The spatial nature of this change is shown in Figures 4 and 5, using FTIR spectra from IFREMER; normalized with respect to the CH$_3$ band at 2960 cm^{-1}.

The physical properties of the rubber were determined as a function of distance from the surface by two techniques. Figure 6 shows extension modulus from microtensile test strips, cut from 100 micron sections cut with a Leica microtome from samples cooled with liquid nitrogen (IFREMER).

Figure 7 shows reduced modulus obtained from microhardness measurements using an instrument at ENSTA Bretagne, manufactured by CSM. Samples were 8 mm long (in the direction normal to the tyre surface) and 4 mm wide, embedded in epoxy resin ($E = 5$ GPa, $\nu = 0.25$) and ground flat to a thickness of 2 mm, cut from a cross section of the tyre. Once again, special care was taken for surface measurements to avoid the cracks. A 10 mN preload was used to detect contact, followed by loading to 100 mN at 200 mN/min, holding for 30 s and unloading. The retraction stiffness of a Vickers tip (angle: 136°) indenter was used to calculate the "reduced modulus" according to the procedure developed by Oliver & Pharr (1992); for an elastomer this would be 4/3 times the Young's modulus E, taking Poisson's ratio to be ~0.5 and treating the indenter as rigid. Every point in Figure 7 represents the mean of up to 8 separate readings. It is apparent that the reduced modulus is 2 to 3 GPa for the surface crust, more typical of a glassy or highly crystalline polymer, and ~10 MPa for the bulk rubber, rather higher than would be expected from macroscopic measurements. Hyperelastic fits, as a function of position, to the load-indentation plots from this instrument are also possible, using nonlinear FEA and an iterative procedure (Le Saux *et al.* 2011).

3   OZONATION AND OXIDATION

As is well known, the unsaturated double bonds in hydrocarbon polymers are susceptible to attack by ozone and oxygen. Ozone reacts especially rapidly. If the rubber is strained, ozone attack initiates and drives cracks, oriented normal to the maximum tensile strain, into the rubber, since attack at the crack tip causes it to propagate, exposing fresh material. However, for unprotected, unstrained natural rubber the attack is uniform over the surface, and a hard surface layer is formed. It seems probable that the ozone reaction rate depends little on concentration of the gas, but remains very high until falling to zero once all the double bonds are reacted. Thus the rate of thickening of the ozonized skin might be expected to depend on the diffusion coefficient of ozone in the skin, rather than in the bulk rubber, and on the concentration of the ozone in the surrounding air. No penetration of ozone would be expected beyond a sharp front of the fully ozonized material. However, Rodriques *et al.* (2001) found after 60 h attack by 1.5×10^{-4} mol.L^{-1} ozone on uprotected unstretched NR films (10 to 350 microns thick) that below a surface layer, ~0.5 micron thick with intense chemical changes, there was evidence of some ozonation in a transition layer around 250 micron thick.

Similarly, oxygen will diffuse through the rubber, and be absorbed by the chemical reactions, but at a finite rate. For a first order reaction, this rate would be proportional to the oxygen concentration. However, other dependencies might also be anticipated, not least because oxidation is autocatalytic, while on the other hand the reactions must eventually reach completion, and so the rate ceases to depend simply on oxygen concentration. Another factor to be considered is that the diffusion coefficient will depend on the extent of reaction, being much lower in the glassy resin that results from ozonation or oxidation than in the original rubber (van Amerongen, 1964; Clough & Gillen, 1992). Effective antiozonants and antioxidants will greatly influence the reactions and hence the balance between diffusion and reaction. Paraphenylenediamines such as IPPD (used in the rubber investigated by Lindly & Teo, 1977) suppress the autocatalyic oxidative chain reaction, reducing the rate of oxygen consumption and oxidative damage but allowing deeper penetration of oxygen. Most effective antidegradants are themselves mobile and diffuse to replenish regions depleted by such reactions or by leaching or evaporation.

Oxidation can be initiated by UV light, which will penetrate to a depth that depends on scattering and absorption by pigments (e.g. Kubelka, 1948). In the rubber, the loading of black particles is high so they will be the most significant pigment. Most of the light incident on carbon black particles is absorbed, although 5 to 23% is scattered, depending on the black type (Donian & Medalia, 1967). If we take the simplifying assumption that all light incident on the black particles is absorbed, the Kubelka-Munk equations (e.g. Kubelka, 1948) are greatly simplified, to the single differential equation for light intensity I:

$$\frac{dI}{dx} = -KI$$

K, the coefficient of absorption, would be expected to be proportional to the average fraction of cross sectional area occupied by the black, which according to the principle of Delesse will be equal to the volume fraction φ of black. It might also be expected to be proportional to the reciprocal of the effective particle diameter L, since in sections normal to the direction of incidence of the light, separated by ~L, the distribution of opaque particles will become uncorrelated. L would be expected to be a weighted average of primary aggregate and incompletely dispersed agglomerate dimensions, ie to depend on the dispersion. Thus the extent to which UV light might be expected to catalyse oxidative degradation will decline exponentially from the surface, with a decay length of the order of L/φ. From micrographs given by Donnet & Voet (1976) the primary aggregate "diameter" seems to be around $0.2\,\mu$m, but Figure 2 shows there are many agglomerates of a few microns in size; a value for L of ~$0.5\,\mu$m therefore seems reasonable. Using 0.92 gcm^{-3} for the density of the polymer matrix and 2 gcm^{-3} for the density of channel black (Donnet & Voet, 1976), a loading of 54 pphr black is equivalent to $\varphi \sim 0.2$, ignoring the presence of other ingredients. Thus the decay length is estimated as ~$2.5\,\mu$m, so that the depth at which the light intensity falls to 1% of that incident could be estimated as $-\ln(0.01) \times 2.5\,\mu$m $\approx 12\,\mu$m. It thus seems that UV initiation of oxidation could be a significant factor to a depth of no more than a few tens of microns.

4   DISCUSSION

The cracks in the exterior surface of the aged rubber look similar to shrinkage cracks in dried mud. If the rubber had been moulded directly onto the steel rim, the surface would be in a state of biaxial tension. Taking the coefficient of volumetric thermal expansion of the rubber to be $\alpha \approx 5.3 \times 10^{-4}$ K^{-1}, and the vulcanization temperature as ~100°C above ambient implies a linear shrinkage strain of 1.8×10^{-2} in all three orthogonal directions, if the tyre were unconstrained. However, because of the constraint of the steel tyre, no shrinkage is possible in either the hoop or transverse directions at the area of interface between steel and rubber. This will impose tractions on the tyre at the interface, causing tensile stresses in the rubber in planes normal to the interface, which may contribute to the cracking. These stresses will be increased in magnitude by loss of any mobile ingredients in the rubber, such as processing oil, by leaching or evaporation. It is noted that the extractable content is only 1 pphr, less than might be expected to have been used originally.

Gilbert (1962) reported that water or other polar vapour has a substantial effect on ozone attack of rubbers, increasing the ozone uptake for unstrained natural or styrene-butadiene rubber. After an initially high rate of absorption, the rate declines towards zero quite rapidly. The ozonised products initially appear non-uniformly on the surface, and are highly hygroscopic; the reaction products are readily extracted by acetone, in contrast to those generated in dry ozone. He thought that in the presence of a polar solvent, there would be more scission and more production of species soluble in polar solvents. This could obviously have implications for attrition and apparent shrinkage of the rubber surface in the natural environment through a combination of ozone and rain. It could also enable penetration of ozone to greater depths and explain the liberation of carbon black from the highly degraded material, the polymer binder having been rendered soluble and washed out, and, as shown in Figure 1, the shrinkage cracks extend through the severely degraded material, so they cannot be only a consequence of leaching of oxidized or ozonized material.

The depth of the highly degraded material seems to be greater than can be explained by photo-oxidation or ozone attack, leaving oxidation as the primary cause. Extant mathematical models for diffusion-limited oxidation (e.g. Nasdala, 2001; Nasdala *et al.*, 2003; Steinke *et al.* 2011) should provide a way forward to explore this mechanism quantitatively. However, it is anticipated that a significant laboratory effort will be needed to characterise the appropriate material properties, in particular the diffusion coefficient and rate of oxygen uptake as a function of extent of oxidation. The physical consequences of a particular oxygen uptake also require characterisation.

5 CONCLUSIONS

- A crust has formed on a naturally aged unprotected natural rubber, seemingly in conflict with the prediction of Lindley & Teo (1977).
- Ozone attack and photo-oxidation would be expected to lead to degraded layers only a few microns thick, whereas the observation is of a crust several hundredfold thicker.
- Although a complicated interplay of several processes is likely to have contributed to the crust formation, including leaching by rain, it seems most likely that the primary mechanism determining the crust thickness is oxidation.

This suggests that future work to clarify the mechanism should focus on laboratory investigation, separating oxidation from the other mechanisms, to study the effects of degree of oxidation and of temperature on the diffusion coefficient and rate of oxygen consumption, and the theory for diffusion limited oxidation with evolving rates for diffusion and reaction.

ACKNOWLEDGMENTS

Thanks are due to Colin Hull for carrying out analytical work at TARRC, to Ian Stephens for his insight into design history of vehicles and general advice, and John Clarke for sharing his recollections as a forestry worker in Bramfield Wood.

REFERENCES

Clough R L & Gillen K T, 1992 Oxygen diffussion effects in thermally aged elastomers *Polymer Degradation & Solubility* **38**, 47–56

Donet J-B & Voet 1976 A *Carbon Black*, publ. Marcel Dekker, New York.

Donian H C & Medalia A I, 1967 *J Paint Technol.*, **39**, 716

Gilbert J H 1962 Degradation and cracking of elastomers by ozone *Proc 4th Rubber Technology Conf,* London

Huntink N M 2003 *Durability of rubber products*, Thesis, University of Twente

Kubelka P 1948 New contributions to the optics of intensely light-scattering materials. Part I *J Optical Soc America*, **38**, 448–457

Lindley PB & Teo S C 1977 High Temperature ageing of rubber blocks *Plastics & Rubber: Materials & Applications*, **2**, 82–88

Le Saux V., Marco Y., Blès G., Calloch S., Moyne S., Plessis S., P. Charrier (2011). Identification of constitutive model for rubber elasticity from micro-indentation tests on Natural Rubber and validation by macroscopic tests. *Mechanics of Materials* to be published.

Nasdala L, 2001 Oxidative ageing of filled elastomers *Constitutive Models for Rubber II,* ed Besdo D, Schuster R H & Ihlemann J, 205–211

Nasdala L, Wei Y & Rothert H 2003 Homogenisation techniques for the analysis of oxygen diffusion and reaction in fiber-reinforced elastomers *Constitutive Models for Rubber III,* ed Busfield J J C & Muhr A H, p. 85–91

Newton R G 1945 Mechanism of exposure-cracking of rubbers *J Rubber Research ,* **14**, 27–62

Oliver W C & Pharr G M 1992 A new improved technique for determining hardness and elastic modulus using load and sensing indentation experiments. *Journal of Materials Research*, **7**(6): 1564–1582.

Rodriques F H A, Santos E F, Feitosa J P A, Ricardo N M P S & De Paula R C M 2001 Ozonation of unstretched NR Part I: Effect of film thickness *Rubber Chem & Tech*, **74**, 57–68

Steinke L, Spreckels J, Flamm M. & Celina M, 2011, Model for heterogeneous aging of rubber products *Plastics Rubber & Composites*, **40**, 175–179

Van Amerongen G J 1964 Diffusion in elastomers, *Rubber Chem & Tech,* **37**, 1065–1152

Williams I & Neal A M 1930 Solubility of Oxygen in Rubber and Its Effect on Rate of Oxidation *Rubber Chem. Technol.* **3**, 678–688

Constitutive Models for Rubber VII – Jerrams & Murphy (eds)
© *2012 Taylor & Francis Group, London, ISBN 978-0-415-68389-0*

Contribution of accurate thermal measurements to the characterization of the thermo-mechanical properties of rubber-like materials

V. Le Saux
UBS – LIMATB (EA4250), Lorient Cedex, France

Y. Marco & S. Calloch
ENSTA BRETAGNE – LBMS (EA4325), Brest Cedex, France

P. Charrier
TRELLEBORG Modyn, Z.I. de Carquefou, Carquefou Cedex, France

ABSTRACT: Since the pioneer works of Gough (1805) and Joule (1859), the thermal characterization of elastomers under mechanical loading has been of interest to numerous research teams. This is not surprising as the thermal signature of rubber is a very useful data to investigate the dissipation mechanisms as well as the thermodynamics variables and couplings. In former recent studies dealing with fatigue investigations (Le Saux *et al.*, 2010 and 2011), an experimental protocol was developed. This protocol imposes cyclic loading to hourglass shaped samples and takes into account the large displacement and permits a dissociation between the intrinsic dissipation, responsible for the mean temperature variation (called heat build-up in the literature) and thermo-mechanical couplings responsible for the temperature variation around this mean value during one cycle. Up to now, the mean temperature has been investigated in order to feed an energetic fatigue criterion. The aim of the present study is to investigate the thermo-mechanical couplings and the ability of thermal measurements to exhibit some specific thermo-mechanical properties observed for rubberlike materials, *i.e.* detection of thermo-elastic inversion point, evaluation of stress or strain induced crystallization and Mullins effect. The results obtained are correlated to other results from the literature and illustrate that accurate thermal measurements can provide precious information to enhance the thermo-mechanical comprehension of elastomeric materials.

1 INTRODUCTION

The mechanical characterization for engineering purposes is usually limited to a stress and strain analysis, which is clearly not enough to describe the thermodynamics ruling the material behavior. This is especially true for rubbery materials because energetic and entropic contributions are both at work and because the coupling between the volumic change, the temperature and the mechanical variables are strong. The description of the dissipation and coupling terms is required to enhance the understanding and modeling accuracy. A classical way to do so is to measure the temperature changes of a sample submitted to a given mechanical loading or the evolution of the mechanical variables during a not isothermal test. This is not an easy task, whatever the material considered because the temperature evolution is clearly not an intrinsic property of the material, and depends on the geometry and on the mechanical and thermal boundary conditions. Moreover, the temperature changes are usually very low, requiring special experimental care. Once again, elastomeric materials are complex to analyze due to their

specific behavior involving low thermal conductivity (leading to temperature gradients), large changes in shape (large displacements, large change of the boundary conditions), cyclic softening, numerous sources for dissipation (viscosity, cristallinity, damage) and strong dependency of the mechanical parameters on the temperature. . .

Since the pioneer works of Gough (1805) and Joule (1859), the thermal characterization of elastomers under mechanical loading has been of interest to numerous research teams. These studies are focusing on two main issues: the evaluation of the thermomechanical couplings and of the intrinsic dissipation. The first issue has been investigated for long and is funding the understanding of the thermodynamics of rubbery materials, commonly based on the entropy change of a network of chains, with a negligible change of the internal energy between the strained and unstrained states. Various experimental data from the literature and reference works (Treloar, 1975) show that this strong assumption is justified for a wide range of strain, but is very far from the truth at low strain (below 20%) and high strain (if crystallization is involved). Furthermore, other reference works by Joule (1859), Anthony et al. (1942), Caston (1942), Peterson (1942), James and Guth (1943) or Dart and Guth (1945), took advantage of the thermal measurements to illustrate that the volumic change, commonly supposed to be negligible, was to be accounted for in order to describe the thermal behavior at low strains (thermo-elastic inversion). The second issue dealing with thermal measurements is to dissociate the effect the thermo-elastic-couplings from intrinsic dissipation to evaluate their respective contributions to the temperature rise (Honorat, 2006; Boulanger et al., 2004) or to feed an energetic fatigue criterion (Le Saux et al., 2010).

The scope of this study is clearly dealing with the first issue. In order to overcome several specific difficulties (temperature dependence, cyclic softening) a protocol formerly developed is used, based on the cyclic loading of hourglass-shaped specimens. Moreover, as the intensity of the thermal phenomena is very low (evaluated by the literature as less than 0.1°C), a special care is taken to the thermal measurements. The accurate measurements obtained will be compared to the results from the literature to discuss several points: thermo-elastic inversion, Mullins cyclic softening and mechanically induced crystallization.

2 EXPERIMENTAL METHOD

2.1 Materials

In this study, several compounds of natural rubbers filled with different amount of carbon black are considered. Table 1 give some informations on the material compounding. Hourglass shaped samples, called AE2 in the following and detailed elsewhere (see Le Saux et al. 2011 for example), have been used.

2.2 Experimental device

The tests have been conducted on an INSTRON hydraulic testing machine (Model 1342) equipped with a 100 kN cell force. All the experiments have been displacement controlled and conducted under a null R_ε ratio at a frequency of $f_r = 0.5$ Hz. In order to limit the transient effects, the thermal measurements

Table 1. Rubbers compounding. All the quantitites are given in phr (per hundred rubber).

component	NR-a	NR-b	NR-c	NR-d
natural rubber	100	100	100	100
carbon black	0	22	43	56
zinc oxide	9.85	9.85	9.85	9.85
plasticizer	3	3	3	3
sulfur	3	3	3	3
stearic acid	3	3	3	3
antioxydants	5	5	5	5
accelerators	4.3	4.3	4.3	4.3

are performed on the thermal and mechanical stabilized cycle. The main reasons that justify this choice are:

- no crystallization under strain accumulation. At the beginning of every cycle, the material is in an amorphous state;
- with such a low frequency, the well-known heat build-up effect that occurs on rubber-like materials under cyclic loadings (Le Saux et al. 2010) and that can reduce the crystallization (Trabelsi et al. 2003) is minimized;
- the stabilized thermal state is reached after a low number of cycles ($\approx$650 cycles at 0.5 Hz), so that no fatigue damage is affecting the thermal measurements

The temperature variation measured at the skin is lower than 3°C. Moreover, finite element analysis using the model developped by Le Chenadec et al. (2007) and validated by Le Saux et al. (2011) were performed using the previous experimental conditions. The results showed that the temperature rise reached in the sample volume does not exceed 5°C for all the tests, which means that the mechanical and crystallization couplings to the temperature can be neglected.

The temperature measurements have been done with a MWIR 9705 FLIR infrared camera. This device is equipped with a Stirling-cycle cooled Indium-Antimonide (InSb) Focal Plane Array (FPA). The FPA is a 256×320 array of detectors digitized on 14 bits and sensitive in the 3–5 μm spectral band. In order to convert the thermosignal into temperature (in °C), a preliminary calibration operation was performed with a HGH DCN1000 N4 extended black body and a classical 2 points Non Uniformity Correction (NUC) has been applied to the array of detectors. A specific care was taken to minimize the influence of the external environment on the measurements by the use of a "black box" around the sample and the grips of the testing machine. The relative measurement precision is about 20 mK. Emissivity measurements have been performed on our materials according to a protocol already proposed by Poncelet (2007) and based on reflectivity measurements. These tests revealed that the natural emissivity of our materials is very close to 0.98 whatever the material considered (which is also the emissivity of the black body used for the calibration). Consequently, no specific surface preparation of the samples is needed.

2.3 Measurement protocol

The samples are submitted to a sinusoidal loading using the experimental conditions detailed in the section 2.2 and an IR movie is performed once the thermal and mechanical stabilization state is reached. For this movie, the frequency capabilities of the camera are fully exploited (acquisition rate of 50 Hz) and only the central zone (in the thinner section) of the samples is analyzed. To reduce the measurements noise, an average on 15 to 20 pixels is performed for each movie.

A Finite Element calculation using a cyclic behaviour law previously calibrated, allows the conversion of the displacement of the specimen to a local value of the strain measured in the analyzed window.

3 RESULTS

3.1 *Load curves*

Figure 1 presents a comparison of the load curves obtained for the four studied materials submitted to the same mechanical loading. As expected, due to the various mechanical and physical properties, the thermal signature measured are different. It is worth noting that the thermoelastic inversion phenomenon is clearly visible for NR-c and NR-d materials but can hardly be seen for NR-a and NR-b material. This may be due to the weakness of the thermal effect intensity for unfilled materials and materials that contains small amount of fillers. More accurate measurements in the low strain region may lead to a better detection of the phenomenon. Let us focus now in the comparison of the load thermal signature for a same material

but for different maximum loading conditions. These results are presented on figures 2 and 5. Figure 2 illustrates that the higher the maximum load, the higher the thermoelastic inversion point and that the lower the temperature rise reached at the inversion point and the higher the shift of the curves to the right. This phenomenon could be related to the consequences of the Mullins effect. Figure 3 is presenting a schematic view that allows comparing the mechanical and the thermal responses of the material for several loading steps. One can observe that the drop in overall stiffness observed for the mechanical response, can be related to a shift and to an increase of the strain where the "inversion point" is detected. This would therefore lead for a decrease of the entropic contribution at low strain, along with cyclic softening, *i.e.* less entropy change between the strained and unstrained states. The shift of the "inversion point" is clarified on figure 4, which plots the evolution of the thermoelastic

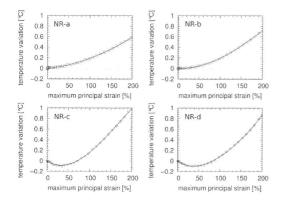

Figure 1. Evolution of the temperature rise (load curves only) for the different materials.

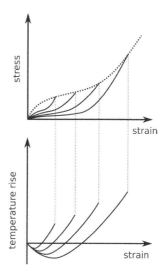

Figure 3. Idealized representation of the thermal signature of the Mullins effect.

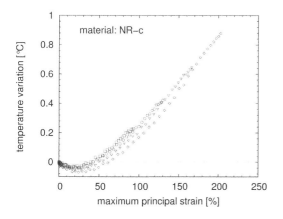

Figure 2. Comparison of the load curves measured for different maximum load. Case of a material sensitive to the Mullins effect.

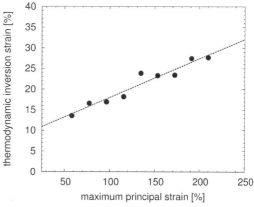

Figure 4. Illustration of the Mullins effect consequences: evolution of the thermodynamic inversion strain (NR-c).

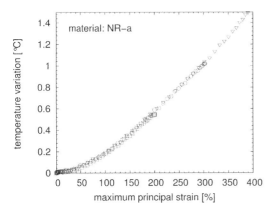

Figure 5. Comparison of the load curves measured for different maximum load. Case of a material that is not sensitive to the Mullins effect.

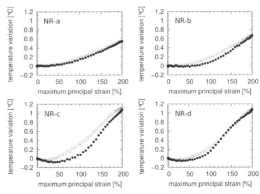

Figure 6. Evolution of the temperature rise for the different materials. The unload curves are filled symbols.

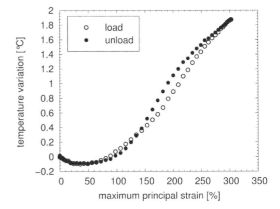

Figure 7. Example of thermal response for a crystallizable rubber (NR-d).

inversion point as a fonction of the maximum strain reached during the cycle. The increase of the inversion strain with the maximum strain is clearly visible and we can notice a linear relationship between the two quantities. It is worth noting that the inversion strain measured are very similar to the values found in the literature (Joule 1859; Honorat 2006). This interpretation of the shift of the thermal curves is motivated by the results given on figure 5 which presents the thermal load signatures for the unfilled rubber that is known to be non sensitive to the Mullins effect. As we can notice, a "master" curve is obtained with clearly no dependance of the temperature rise to the maximum strain reached during the cycle, i.e. there is no shift of the thermal curves. Nevertheless, the conclusion on the evolution of the inversion point along the cyclic softening still needs further investigation as the permanent set of the samples was not mastered. Further testing with a specific driving procedure (imposing the maximum displacement and the minimum load) will be achieved shortly in order to include the influence of this parameter.

3.2 Load-unload curves

This section discuss on both load and unload curves in order to highlight others phenomena. The figure 6 presents some results for the four materials submitted to the same mechanical loading. Several points need to be discussed from these results. The first one is the lack of reversibility between the load and unload curves for all the materials. As a slight hysteresis can be notice for the unfilled rubber for a maximum strain value lower than the minimum recquired to crystallize – evaluated at 300% by Marchal (2006) – one can conclude that this effect can not be attributed solely to the crystallization under strain effect, as suggested by Treloar (1975). Moreover, as the increase of the thermal hysteresis is well correlated with the increase of the carbon black content increase, a link between the thermal and the mechanical hysteresis must exist. It is worth noting that when a certain maximum principal

strain is reached, an inflexion point can be observed on the unloading curve. This effect can be attributed to crystallization under strain that gives rise to the evolution of a "latent heat" (Treloar 1975). When the strain is high enough, i.e. when the amount of crystals is important enough, the unload curve can even get higher than the loading curve as shown on figure 7 for the NR-d material. To analyze further these data, the deformation threshold, i.e. the minimum strain to apply to the sample to observe a partial crystallization of the material,was investigated. To compare our results, we took data coming from the literature, more specifically from Marchal PhD thesis (Marchal 2006). The methodology employed to identify the deformation threshold from her results is presented on figure 8. It is worth noting that both Marchal (2006) and Chenal et al. (2007) showed a dependancy of the crystallization under strain to the number of cycles. The main consequences of this phenomenon is a slight increase of the deformation threshold with the number of cycles until a stabilization is reached (typically 3-5 cycles according to their results) and a reduction of the crystallinity for a given strain, in other words a shift of the crystallinity curve (figure 8) to the right. However, the influence of the number of cycles is low

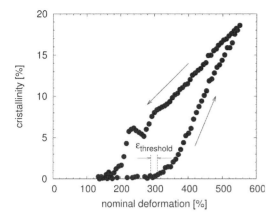

Figure 8. Graphical identification of the deformation threshold from X-ray measurements (data: Marchal (2006)).

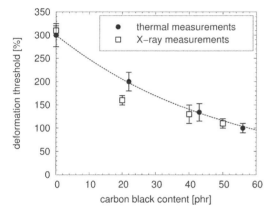

Figure 9. Dependancy of the deformation threshold to the carbon black content. X-ray measurements come from Marchal (2006).

enough to allow the comparison betweenour result (obtained after stabilization of the Mullins effect) and the Marchal's ones (obtained on the first load).

Figure 9 shows a comparison between the deformation threshold identified from our thermal experiments and from Marchal's data. The deformation threshold coming from thermal measurements correspond to the maximum principal strain reached during the cycle for which an inflexion can be observed on the unloading curve (the results of NR-c and NR-d material on the figure 6 are representative results). A fairly good agreement between both approaches is noticed which tends to prove that the crystallization under strain threshold can be easily capture from a careful thermal measurements. The reduction of the crystallization threshold with the rise of the carbon black content is also correctly captures. This reduction is justified by the strain amplification role that fillers play at the microstructure scale.

To the author's opinion, thermal measurements are richer than the analysis presented here, and used with a relevant modelling, it should be possible to get the deformation threshold accurately, but also the evolution of the crystallinity content as a function of the mechanical load.

4 CONCLUSIONS

In this paper, we applied a specific thermal measurements protocols during cyclic loading in order to capture the thermomechanical couplings terms of the heat equation, responsible for the reversible variation of temperature during the cycle. In order to minimize the resolution and the thermal noise, a special care was taken for the calibration of the infrared camera and for the measurements. Several specificities of the thermomechanical properties of rubber-like materials have been highlighted in this paper: the thermoelastic inversion phenomenon, the influence of the Mullins effect on the thermal signature, a thermal hysteresis under cyclic loading and finally, the detection of the crystallization threshold, well correlated with data coming from X-ray measurements. The perspectives of this work are of course numerous. The main point concerns the developement of models in order to better understand these complex experimental data and thus to enhance the comprehension of the thermal signatures of rubbers.

ACKNOWLEDGEMENTS

The authors would like to thank Dr. Cédric Doudard from ENSTA Bretagne for helpful discussions on infrared thermography.

REFERENCES

Anthony, R., R. Caston, and E. Guth (1942). Equations of state for natural and synthetic rubberlike materials. I. Unaccelerated natural soft rubber. *The Journal of Physical Chemistry 46*, 826–840.

Boulanger, T., A. Chrysochoos, C. Mabru, and A. Galtier (2004). Calorimetric analysis of dissipative and thermoelastic effects associated with the fatigue of steels. *International Journal of Fatigue 26*, 221–229.

Caston, R. (1942). *Equation of state of rubber*. Ph. D. thesis, University of Notre Dame.

Chenal, J., C. Gauthier, L. Chazeau, L. Guy, and Y. Bomal (2007). Parameters governing strain induced crystallization in filled natural rubber. *Polymer 48*, 6893–6901.

Dart, S. and E. Guth (1945). Rise of temperature on fast stretching of butyl rubber. *The Journal of Chemical Physics 13*, 28–36.

Gough, J. (1805). A description of a property of caoutchouc or indian rubber. *Memories of the Literaci and Philosophical Society of Manchester 1*, 288–295.

Holzapfel, G. and J. Simo (1996). Entropy elasticity of isotropic rubber-like solids at finite strains. *Computer Methods in Applied Mechanics and Engineering 132*, 17–44.

Honorat, V. (2006). *Analyse thermomécanique par mesures de champs des élastomères*. Ph. D. thesis, Université Montpellier II.

James, H. and E. Guth (1943). Theory of the elastic properties of rubber. *The Journal of Chemical Physics 11*, 455–481.

Joule, J. (1859). On some thermodynamics properties of solids. *Philosophical Transactions of the Royal Society of London 149*, 91–131.

Le Chenadec, Y., C. Stolz, I. Raoult, T. Nguyen, B. Delattre, and P. Charrier (2007, September 4th-7th). A novel approach to the heat build-up problem of rubber. In *Constitutive Models for Rubber V*, Paris (France), pp. 345–350.

Le Saux, V. (2010). *Fatigue et vieillissement des élastomères en environnements marin et thermique: de la caractèrisation accèlérée au calcul de structure*. Ph. D. thesis, Université de Bretagne Occidentale.

Le Saux, V., Y. Marco, S. Calloch, P. Charrier, and D. Taveau (2011). Heat build-up problem of rubbers under cyclic loadings: experimental investigations and numerical predictions. *European Journal of Mechanics – A/Solids (submitted)*.

Le Saux, V., Y. Marco, S. Calloch, C. Doudard, and P. Charrier (2010). Fast evaluation of the fatigue lifetime of rubber-like materials based on a heat build-up protocol and micro-tomography measurements. *International Journal of Fatigue 32*, 1582–1590.

Marchal, J. (2006). *Cristallisation des caoutchoucs chargés et non chargés sous contrainte: effet sur les chaines amorphes*. Ph. D. thesis, Université Paris Sud – Paris XI.

Peterson, L. (1942). *Equation of state of some synthetic rubbers*. Ph. D. thesis, University of Notre Dame.

Poncelet, M. (2007). *Multiaxialité, hétérogénéités intrinsèques et structurales des essais d'auto- échauffement et de fatigue à grand nombre de cycles*. Ph. D. thesis, Ecole Normale Supérieure de Cachan.

Toki, S., T. Fukimaki, and M. Okuyama (2000). Strain-indeuced crystallization of natural rubber as detected real-time by wide-angle X-ray diffraction technique. *Polymer 41*, 5423–5429.

Trabelsi, S., P. Albouy, and J. Rault (2003). Crystallization and melting processes in vulcanized stretched natural rubber. *Macromolecules 36*, 7624–7639.

Treloar, L. (1975). *The physics of rubber elasticity*. Clarendron Press – Oxford.

Nano- to macro-scale modeling of damage in filled elastomers

R. Dargazany & M. Itskov

Department of Continuum Mechanics, RWTH Aachen University, Aachen, Germany

ABSTRACT: Filled elastomers are characterized by their ability to undergo large elastic deformations and also exhibit some inelastic effects such as stress softening, hysteresis and induced anisotropy. Previously, most of these features have been modeled separately. In this contribution, a novel multi-scale study of filled elastomers towards modeling of all these phenomena simultaneously is presented. First, we describe nano-scale interactions between fillers which are the smallest constituents of the rubber matrix. Then, the mechanical behavior of micro-scale aggregates under deformation is modeled. Considering aggregates in the polymer matrix, strain distribution between phases and interactions between polymer and filler network are studied at the meso-scale. Introducing a proper statistical volume element for each network, the macro-scale behavior is finally predicted.

1 INTRODUCTION

Besides the typical hyperelastic behavior and large elastic deformations with non-linear stress-strain response, rubber-like materials may also exhibit notable inelastic effects, like hysteresis and permanent set.

Two important mechanical aspects that can influence elasticity of filled polymer systems are the inelastic behavior of filler aggregates and the non-uniformity of aggregate lengths that leads to localized strain distribution.

So far, many micro-mechanical models have been proposed that can take into account different properties of filled rubber-like materials. Some of these models (see e.g. Dargazany & Itskov 2009a) are based on the decomposition of the rubber matrix into polymer-polymer (CC) and polymer-filler (PP) networks which do not interact with each other. Unfortunately, such models are not able to simulate hysteresis observed in cyclic loading of filled elastomers. Thus, the aim of this contribution is to develop a new micro-mechanical model of this very important phenomenon of filler rubbers.

2 NETWORK DECOMPOSITION

In the context of the rubber matrix the term "polymer chain" is generally referred to as a part of a long polymer molecule which is bonded to other parts of the network just at its both ends. This bond can be a cross-link with another polymer chain or a local adsorption on the aggregate surface. With respect to the bond types, one can categorize polymer chains into three groups: polymer chains with cross-linkage at both ends (CC), with particle adsorption at both ends (PP), and with cross-linkage and particle adsorption at each end (CP).

In comparison to previous works, we take additionally the third network into account which is assumed to consist of the CP chains and the aggregates. Since polymer chains connect the aggregates to the cross-links, deformation of the aggregates plays an important role as well. Accordingly, the energy of the CP network is supposed to consist of the energy of the aggregates and the CP chains.

The movement of the cross-links under deformation is assumed to be affine in the CC network and consequently also in the CP network. The connection of the CP-network to the PP network is due to aggregates. Thus, describing deformation of aggregates in the CP network, enable us to calculate the directional strain localization in the PP network.

The strain energy of the rubber matrix can thus be decomposed by

$$\Psi = \Psi_{CC} + \Psi_{PP} + \Psi_{CP}, \tag{1}$$

where Ψ_{CC}, Ψ_{PP} and Ψ_{CP} represent the total energies of the above mentioned networks relative to the unit reference volume.

In order to calculate Ψ_{CP}, the CP network is represented by a number of subnetworks distributed in different directions. A subnetwork is a uni-directional representation of the CP network with identical properties (see Fig. 3a, b).

Each subnetwork can be divided into a finite number of micro-cells such that each may have one or no aggregate inside. This assumption is justified if the distribution of aggregate inside the rubber matrix is homogeneous, which is not always the case.

However, the above presented concept is valid as far as the filler concentration in rubber compound is below the gelation point. When the filler concentration

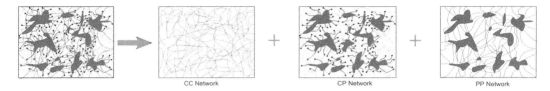

Figure 1. Network decomposition into CC, CP and PP

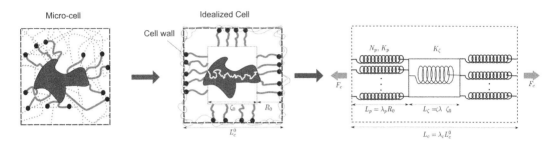

Figure 2. Micro-cell, idealized cell and its representation by a set of nonlinear springs

exceeds the gelation point, a volume filling filler structure is constructed, and the aggregates can no longer be considered as separate bodies.

3 MICRO-CELL

Different lengths of CC and CP chains inside the micro-cell complicate their mathematical modeling. For the sake of simplification, the micro-cells can be idealized at least in the case of the uniaxial loading on the basis of the following assumptions.

- An idealized cell can be considered as a one-dimensional representation of the micro-cell.
- The cross-links placed in each cell are moved and placed on the cell wall. Thus, an idealized cell is bonded to the CC network at its walls.
- Aggregates are assumed to be placed in the middle of the cell.
- The number of CP chains inside a cell depends on the size of the aggregate inside the cell.
- The CP chains inside the cell are replaced by equivalent chains of a constant size.
- At the stress-free state, the equivalent chains are in unperturbed state.

The idealized cell model is illustrated in figure 2. In order to describe the mechanical response of the idealized cell, polymer chains and aggregate are replaced by nonlinear springs as shown in figure 2.

The model parameters used in the following and notation manner are explained in table 1.

3.1 Elasticity of an aggregate

An aggregate under load can be represented by a so-called backbone chain model (Shih et al. 1999). A backbone is a single chain of particles through which the external stress is transmitted. The path and shape

Table 1. Parameter description

Constituents	Description
$\bullet_p$	non-kinematical quantities of polymer chains
$\bullet_\zeta$	non-kinematical quantities of aggregates
$\bullet_c$	non-kinematical quantities of idealized cell

Superscripts	Description
$\bullet^y$	The parameter value at yield point
$\bullet^m$	Maximal value of the parameter in loading history
$\bullet^r$	Maximal value of the parameter in one loading cycle

Variables	Description
$L_\bullet$	Length of a unit of a constituent (u.C.)
$l_\bullet$	Length of an element inside of one u.C.
$n_\bullet$	Number of elements inside of one u.C.
$N_\bullet$	Number of u.C. per unit volume of the sub-network
$V_\bullet$	Volume of one u.C.
$F_\bullet$	Force of one u.C.
$f_\bullet$	force of an element inside of one u.C.
$\psi_\bullet$	Energy of one u.C.
$\lambda_\bullet$	Stretch of one u.C.

of this chain strongly dependents on the aggregation process and load direction.

Generally, colloidal clusters appear in complex geometrical structures and can be characterized by three parameters: correlation length ζ, the fractal dimension d_f and the particle diameter l_ζ. The aggregate correlation length ζ can be considered as the average distance of two arbitrary points on the surface of the cluster (Stauffer and Aharony 1994).

The inter-particle forces within the backbone chain can be decomposed into centro-symmetric and

n	1	3	5	7	9	11	13	15	17	19
C_n	3	$\dfrac{9}{5}$	$\dfrac{297}{175}$	$\dfrac{1539}{875}$	$\dfrac{126117}{67375}$	$\dfrac{43733439}{21896875}$	$\dfrac{231321177}{109484375}$	$\dfrac{20495009043}{9306171875}$	$\dfrac{1073585186448381}{476522530859375}$	$\dfrac{4387445039583}{1944989921875}$

tangential forces and can be represented by two linear elastic elements with an average tensile modulus Q and an averaged bending modulus $\bar{G}$. Thus, the backbone chain is simulated as a combination of elastic beams with tensile spring constant Q and bending spring constant $\bar{G}$. Thus, the energy of a large backbone chain Ψ, is given by

$$\Psi = \frac{\bar{G}}{2} \int \left(\frac{\partial \phi}{\partial s}\right)^2 ds + \frac{Q}{2} \int \left(\frac{\partial \eta}{\partial s}\right)^2 ds, \qquad (2)$$

where s denotes the arc length along the undeformed backbone chain, $\eta(s)$ is the local extension or contraction of the bond along its undeformed contour, and ϕ represents the angle between two successive bonds.

The force F_ζ applied to the aggregate per unit undeformed volume can be expressed in terms of its overall elastic modulus $K_\zeta\left(\hat{\lambda}_\zeta\right)$ by (Dargazany and Itskov 2009b)

$$F_\zeta\left(\zeta, \hat{\lambda}_\zeta\right) = K_\zeta\left(\hat{\lambda}_\zeta\right) \zeta \left(\hat{\lambda}_\zeta - 1\right), \qquad (3)$$

where $\hat{\lambda}_\zeta$ represents the stretch of the aggregate relative to its stress-free state.

3.2 Elasticity of a polymer chain

In order to model polymer macromolecules, we apply the concept of freely jointed chain (FJC) where the orientation of a segment is completely independent of the orientations and positions of its adjacent segments. Consider further a FJC with n_p segments each of length l_p. Let R and r be vectors connecting two ends of this chain at the reference and deformed configuration, respectively. Thus,

$$r = \mathbf{F}_p R, \qquad (4)$$

where $\mathbf{F}_p$ is the deformation gradient tensor at the micro-scale. Accordingly, $\mathbf{F}_p$ describes deformation of a polymer chain. The lengths $R = |R|$ and $r = |r|$ are referred to as the end-to-end distances of the chain in the reference and deformed configuration, respectively. Let us further introduce R_0 as the mean end-to-end distance of a polymer chain in the reference configuration. Thus, one can write

$$R_0 = \sqrt{n_p}l_p, \quad \mathcal{R}_p = n_p l_p, \qquad (5)$$

where $\mathcal{R}_p$ represents the contour length of a polymer chain.

The entropic free energy of a single polymer chain ψ_p per unit volume of rubber is obtained on the basis of the non-Gaussian statistics by

$$\psi_p(n_p, \bar{r}) = n_p KT \left(\frac{\bar{r}}{n_p}\beta + \ln\frac{\beta}{\sinh\beta}\right), \qquad (6)$$

where T stands for the temperature (isothermal condition is assumed) and K is Boltzmann's constant. The overbar indicates a normalized value with respect to the length of its element, e.g. $\bar{r} = r/l_p$. Further,

$$\beta = \mathcal{L}^{-1}\left(\frac{\bar{r}}{n_p}\right), \qquad (7)$$

where $\mathcal{L}^{-1}$ denotes the inverse Langevin function which cannot be expressed in a explicit form. Except of the area of asymptotic behavior close to $\bar{r}/n_p = 1$ the inverse Langevin function can accurately be represented by the Taylor series

$$\mathcal{L}^{-1}(x) = \sum_{i=0}^{\infty} C_i x^i \qquad (8)$$

truncated after the the first 20 terms. The corresponding coefficients of this series are given in table 2 (Itskov et al. 2010).

Assuming the end-to-end distance of the chain to be equal to R_0, the force developed by a single FJC per unit reference volume results from (6) as

$$F_p(\lambda_p, n_p) = \frac{KT}{l_p}\mathcal{L}^{-1}\left(\frac{\lambda_p}{\sqrt{n_p}}\right), \qquad (9)$$

where $\lambda_p = \frac{r}{R_0}$ denotes the stretch of the chain with the end-to-end distance R_0 relative to the reference configuration.

3.3 Governing equations

The variables of the CP network can be categorized according to their application range into two groups: subnetwork and cell variables. The first ones $\mathcal{C}_N = \{L_c^0, \lambda_c\}$ are constant within the subnetwork while the second ones $\mathcal{C}_c = \{L_\zeta, \lambda_\zeta, K_\zeta, \bar{r}, K_p, \lambda_p, N_p, F_c, \lambda_c^y\}$ should be defined in each cell, independently.

As shown in figure 3, the polymer chains are assumed to be unperturbed in the stress-free state of the material and consequently exhibit some resistance force deforming the adjacent aggregates. The amount

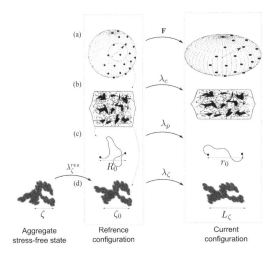

(a)

F

(b)

λ_c

(c)

λ_p

λ_ζ^{res} R_0 r_0

(d) λ_ζ

ζ ζ_0 L_ζ

| Aggregate stress-free state | Refrence configuration | Current configuration |

Figure 3. (a) A continuum of the CP network represented by a micro-sphere, (b) one-dimensional subnetwork, (c) a polymer chain and (d) a filler aggregate

where α_p and β_p are material parameters and l_ζ represents the length of a particle.

4 MECHANICS OF A SUBNETWORK

4.1 Yield of an aggregate

The yield behavior of a filler aggregate is strongly influenced by its geometry and interactions between its particles. Thus, knowing the force and moments applied on different inter-particle bonds of the aggregate, one can locate a critical bond that yields first. The yield force of the critical bond f_ζ^y can then be expressed by

$$f_\zeta^y = f_\zeta^{y-central} \frac{2}{2 + \zeta\left(\frac{1}{l_\zeta} + \frac{2}{3}\frac{Q}{G}\right)}, \qquad (15)$$

where $f_\zeta^{y-central}$ is the maximum central force per unit volume that a bond can withstand in the absence of bending moments. $\frac{1}{l_\zeta} \gg \frac{Q}{G}$ due to the small size of fillers used in elastomers as $l_\zeta \approx 8 - 500\,[\text{nm}]$. Thus 15 is reduced to

$$f_\zeta^y = f_\zeta^{y-central} \frac{2}{2 + \frac{\zeta}{l_\zeta}}. \qquad (16)$$

The yield force of an aggregate F_ζ^y can then be obtained by assuming that the force applied to the aggregate is equal to the one transmitted through the critical bond. Accordingly,

$$F_\zeta^y = f_\zeta^y \frac{V_b}{V_\zeta} = f_\zeta^{y-central} \frac{V_b}{V_\zeta} \frac{2}{2 + \frac{\zeta}{l_\zeta}}, \qquad (17)$$

where the volume of the critical bond V_b is considered to be identical for all bonds. According to Foffi et al. 2005, Manley et al. 2005, the value of $f_\zeta^{y-central}$ does not depend on the correlation length of the aggregate, although several other parameters (e.g., particle type, aggregation mechanism, etc.) effect it. Although the value of $f_\zeta^{y-central}$ is measured and reported for different solutions, V_b should be obtained implicitly by fitting the CP network behavior to experimental results. For this reason, $f_\zeta^y V_b$ can and will be first considered as a material parameter.

Further we assume that the cell of a failed aggregate does no more contribute to the mechanical behavior of the subnetwork and is deactivated. The deactivation stretch of the cell λ_c^y can then be calculated by inserting (17) into (12) which yields

of the aggregates deformation from their stress-free state to the equilibrium state is denoted by λ_ζ^{res}.

The force balance in one cell can further be written for the stress-free state by

$$\frac{1}{2}V_p N_p F_p(1, n_p) = V_\zeta F_\zeta(\zeta, \lambda_\zeta^{res}),$$

$$L_c^0 = \zeta_0 + 2R_0, \qquad (10)$$

where N_p denotes the number of polymer chains on each side of the aggregate in the direction of loading. The lengths of the aggregate in the equilibrium ζ_0 and the current state L_ζ are expressed by

$$\zeta_0 = \lambda_\zeta^{res}\zeta,$$

$$L_\zeta = \hat{\lambda}_\zeta\zeta = \lambda_\zeta\lambda_\zeta^{res}\zeta, \qquad (11)$$

respectively. Thus, (10) is written for the deformed configuration as

$$\frac{1}{2}V_p N_p F_p(\lambda_p, n_p) = V_\zeta F_\zeta\left(\zeta, \hat{\lambda}_\zeta\right)$$

$$\lambda_c L_c^0 = \lambda_\zeta\zeta_0 + 2\lambda_p R_0. \qquad (12)$$

The total force transmitted by the sell can further be written by

$$V_\zeta F_c = V_\zeta F_\zeta(\zeta, \hat{\lambda}_\zeta). \qquad (13)$$

Since an analytical approximation of the number of polymer chains N_p bonded to the aggregate surface is not possible, the following scaling approach is proposed

$$N_p = \alpha_p\left(\frac{\zeta}{l_\zeta}\right)^{\beta_p}, \qquad (14)$$

$$F_\zeta^y = F_\zeta(\zeta, \lambda_\zeta\lambda_\zeta^{res}) \quad \rightarrow \quad \lambda_\zeta$$

$$F_\zeta^y V_\zeta = \frac{1}{2}V_p N_p F_p(\lambda_p, n_p), \quad \rightarrow \quad \lambda_p$$

$$\Rightarrow \quad \lambda_c^y L_c^0 = \lambda_\zeta\zeta_0 + 2\lambda_p R_0 \quad \rightarrow \quad \lambda_c^y. \qquad (18)$$

From (18), one can further conclude that even when the material is in the virgin state, aggregates larger than a certain size

$$\zeta = \left\{ \zeta | F_\zeta^y = F_\zeta(\zeta, \lambda_\zeta^{res}) \right\} \tag{19}$$

cannot exist in the subnetwork.

4.2 Reaggregation and creation of soft bonds

Experimental results on rubber-like materials reveal a considerable softening taking place between first loading cycle and subsequent cycles. This is due to the partial recovery of the filler aggregate during unloading, where the debonded parts of the aggregates come together and form new bonds which are slightly weaker than the original bonds. Although the geometry of the recreated aggregate is not identical with the one of the original aggregate, its geometrical parameters such as the fractal dimension, correlation length, etc. can be assumed to be the same.

In fact, the reaggregation process continuously takes place during unloading. However, for the sake of simplicity we assume it happens just when the material is in stress-free state (see Klüeppel 2003). Due to the lower yield force of the recreated aggregates, the debonded aggregates with $F_\zeta^{ry} < F_\zeta \left(\zeta, \lambda_\zeta^{res} \right)$ do not recreated anymore.

4.3 Aggregate size distribution

The size distribution of aggregates leads to a strongly inhomogeneous stress distribution pattern in the rubber matrix. This distribution influences the mechanical response of the subnetwork.

The aggregate size distribution can be formulated by means of Smoluchowski's equation for the kinetics of irreversible cluster-cluster aggregation of colloids (see e.g. Jullien 1990). Empirically, the distribution is expressed as a logarithmic normal function (Klüeppel 2003) with the following expression

$$\Phi(\zeta) = \frac{4}{g} \frac{l_\zeta}{\breve{\zeta}} \left(\frac{\zeta}{\breve{\zeta}} \right)^{-2\Omega} \exp\left(-(1 - 2\Omega) \frac{\zeta}{\breve{\zeta}} \right), \tag{20}$$

where $\Omega < 0$ is related to the aggregation mechanism and $\breve{\zeta}$ denotes the refrence aggregate size. The parameter g denotes a normalization constant in the reference configuration.

Let further

$$\zeta_{max} = \max\left\{ \zeta^r, \zeta^m \right\}, \tag{21}$$

be the size of the biggest available aggregate in the current configuration, where ζ^m and ζ^r denote the size of the biggest original and recreated aggregate available in the subnetwork, respectively. Since the yielding of the original aggregates takes place only under the primary loading, one can write

$$\zeta^m = \tilde{\zeta}^m(\lambda_\zeta^m), \tag{22}$$

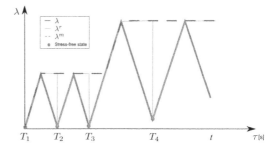

Figure 4. Graphic illustration of the evolution of stretch λ, maximum stretch λ_m, and cyclic maximum stretch λ^r in the case of cyclic loading with increasing amplitude. The stress-free state of the material is highlighted with circles, since due to the residual stretches; the stress-free state is not coinciding with stretch-free state.

where λ_ζ^m is the maximal micro-stretch of the aggregate previously reached in the history of λ_ζ. This parameter is crucial for the description of the initial aggregate yielding and is formulated in a subnetwork as

$$\lambda_\zeta^m = \max_{\tau \in (-\infty, t]} \lambda_\zeta(\tau). \tag{23}$$

The evolution of λ_ζ^m with deformation is shown in figure 4. Substituting (22) into (19) yields

$$\tilde{\zeta}^m(\lambda_\zeta^m) = \left\{ \zeta | F_\zeta^y = F_\zeta(\zeta, \lambda_\zeta^m \lambda_\zeta^{res}) \right\}. \tag{24}$$

In the case of cyclic loading, the recreated aggregates unite and break by turns in the course of reloading and unloading. The value of ζ^r does not change under unloading. Due to the reaggregation, it increases, however, at the stress-free state of the CP network. This means that

$$\zeta^r = \tilde{\zeta}^r(\lambda_\zeta^r), \tag{25}$$

where λ_ζ^r is the maximal micro-stretch of the aggregate over the current load cycle of λ_ζ. In the case of an ideal cyclic loading (unloading down to stress-free state), λ_ζ^r can be given by

$$\lambda_\zeta^r = \max_{\tau \in (T, t]} \lambda_\zeta(\tau) \quad \Leftarrow \quad \max_{T \in (-\infty, t]} \dot{\lambda}_\zeta(T) = 0, \tag{26}$$

which is also illustrated in figure 4. Similar to (24), one can write for recreated aggregates

$$\tilde{\zeta}^r(\lambda_\zeta^r) = \left\{ \zeta | F'^y_\zeta = F_\zeta(\zeta, \lambda_\zeta^r \lambda_\zeta^{res}) \right\}. \tag{27}$$

Finally, (21) can be rewritten by

$$\zeta_{max} = \max\left\{ \tilde{\zeta}^r(\lambda_\zeta^r), \tilde{\zeta}^m(\lambda_\zeta^m) \right\}$$

$$= \tilde{\zeta}_{max}\left(\lambda_\zeta^m, \lambda_\zeta^r \right). \tag{28}$$

The mean force and the energy of cells in a subnetwork are defined in terms of the probability density function $\Phi(\zeta)$ by

$$\langle F_c \rangle = \int_0^{\zeta_{max}(\lambda_\zeta^r, \lambda_\zeta^m)} F_c \Phi(\zeta)\, d\zeta,$$

$$\psi_c(x) = \int_1^x \langle F_c \rangle L_c^0\, d\lambda. \tag{29}$$

Finally, the energy of a subnetwork in the direction $\boldsymbol{D}$ is obtained by

$$\overset{D}{\Psi}{}_{cp}^{sub} = N_c \psi_c\left(\overset{D}{\lambda}\right), \tag{30}$$

where N_c considered as a material parameter denotes the mean number of active cells in a subnetwork.

5 TRANSITION TO THE MACRO-SCALE

Finally, we proceed to define the macroscopic energy of a three-dimensional network. It is generally assumed that the virgin rubber network is initially homogeneous and isotropic and the macroscopic strain energy represents the sum of microscopic strain energies of subnetworks in all directions which can be calculated by means of the integral over the unit sphere. Applying the isotropic space distribution (subnetworks are spread equally in all directions), we can write

$$\Psi_{cp} = \frac{1}{A_s} \int_S \overset{D}{\Psi}{}_{cp}^{sub}\, d\overset{D}{u}, \tag{31}$$

where A_s represents the surface area of the unit sphere. Taking the incompressibility condition

$$\det \mathbf{F} = 1 \tag{32}$$

into account, the constitutive equation for the first-Piola Kirchhoff stress tensor $\mathbf{T}$ can be written by

$$\mathbf{T} = \frac{\partial \Psi_{cp}}{\partial \mathbf{F}} - p\mathbf{F}^{-T}, \tag{33}$$

where $\mathbf{F}$ denotes the macro-scale deformation gradient and p is a Lagrange multiplier which is determined from the equilibrium and the boundary conditions. Using (31) one obtains

$$\frac{\partial \Psi_{cp}}{\partial \mathbf{F}} = \frac{1}{A_s} \int_S w_i \frac{\partial \overset{D}{\Psi}}{\partial \overset{D}{\lambda}} \frac{\partial \overset{D}{\lambda}}{\partial \overset{D}{\chi}} \frac{\partial \overset{D}{\chi}}{\partial \mathbf{F}}\, d\overset{D}{u}, \tag{34}$$

where $\overset{D_i}{\lambda}$ and $\overset{D_i}{\chi}$ denote the micro- and macro-stretches in direction $\boldsymbol{D}_i$, respectively. Since the strain decomposition takes place in the subnetworks, the micro and macro-stretches are considered to be equal $\overset{D_i}{\lambda} = \overset{D_i}{\chi}$ in the CP network. Furthermore, one gets

$$\overset{D_i}{\chi} = \sqrt{\boldsymbol{D}_i \mathbf{C} \boldsymbol{D}_i}$$

$$\Rightarrow \quad \frac{\partial \overset{D_i}{\chi}}{\partial \mathbf{F}} = \frac{1}{2\overset{D_i}{\chi}} \frac{\partial \boldsymbol{D}_i \mathbf{C} \boldsymbol{D}_i}{\partial \mathbf{F}} = \frac{1}{\overset{D_i}{\chi}} \mathbf{F}\left(\boldsymbol{D} \otimes \boldsymbol{D}\right), \tag{35}$$

where $\mathbf{C} = \mathbf{F}^T \mathbf{F}$ denotes the right Cauchy-Green tensor.

6 CONCLUSION

In this contribution, the concept of network decomposition has been modified and a new network (CP) has been introduced. Then, the mechanical behavior of this network has been derived by decomposing it into a number of subnetworks each elongated in one direction. The subnetworks can simulate hysteresis as a result of aggregates yielding under loading and re-aggregation under unloading. The so-obtained model of the CP network also exhibits induced anisotropic behavior in addition to the stress softening in each loading cycle. In the future steps, the developed CP network concept will be implemented into the network evolution model and compared to experimental data.

REFERENCES

Dargazany, R. & M. Itskov (2009a). A network evolution model for the anisotropic mullins effect in carbon black filled rubbers. *International Journal of Solids and Structures 46*, 2967.

Dargazany, R. & M. Itskov (2009b). Non-linear elastic behavior of carbon black filler aggregates in rubber-like elastomers. In *Constitutive Models for Rubber VI*, p. 489.

Foffi, G., C. De Michele, F. Sciortino, & P. Tartaglia (2005). Scaling of dynamics with the range of interaction in shortrange attractive colloids. *Physical Review Letters 94*, 7831.

Itskov, M., A. E. Ehret, & R. Dargazany (2010). A fullnetwork rubber elasticity model based on analytical integration. *Mathematics and Mechanics of Solids 15*, 655–671.

Jullien, R. (1990). The application of fractals to investigations of colloidal aggregation and random deposition. *New Journal of Chemistry 14*, 239.

Küppel, M. (2003). The role of disorder in filler reinforcement of elastomers on various length scales. *Advances in Polymer Science 164*, 1.

Manley, S., H. M. Wyss, K. Miyazaki, J. C. Conrad, V. Trappe, L. J. Kaufman, D. R. Reichman, & D. A. Weitz (2005). Glasslike arrest in spinodal decomposition as a route to colloidal gelation. *Physical Review Letters 95*, 238302.

Shih, W. Y., W. H. Shih, & I. A. Aksay (1999). Elastic and yield behavior of strongly flocculated colloids. *Journal of the American Ceramic Society 82*, 616.

Stauffer, D. & A. Aharony (1994). *Introduction to percolation theory* (2 ed.). London: Taylor & Francis Ltd.

Constitutive Models for Rubber VII – Jerrams & Murphy (eds)
© 2012 Taylor & Francis Group, London, ISBN 978-0-415-68389-0

Influence of strain induced crystallization on the mechanical behavior of natural rubbers

E.A. Poshtan, R. Dargazany & M. Itskov
Department of Continuum Mechanics, RWTH Aachen University, Germany

ABSTRACT: Strain-Induced Crystallization (SIC) is a unique type of crystallization that occurs solely in polymers. Crystallization significantly enhances the stiffness and yield point of natural rubber since crystallines act as a reinforcement. In the present contribution, a constitutive model for strain-induced crystallization in natural rubber is proposed. The crystallization leads to a change of the total strain energy calculated on the basis of the network evolution approach. The stretch distribution in different directions governs the crystalline nucleation procedure. The contribution to the partial immobilization of polymer chains during deformation is accounted for by assuming the crystalline phase and the amorphous phase to be two separate networks. Finally, the proposed model is compared with experimental results of uniaxial tension tests.

1 INTRODUCTION

Elastomer materials are commonly used in everyday life. In order to successfully manipulate their applicability, it is essential to completely understand the relationship between the structure and the properties of the polymer. Its mechanical behavior is closely related to the morphology and consequently depends on the deformation histories experienced by the polymer.

The crystallization in polymers can be triggered both mechanically and thermally. The formation of crystallines can have both a beneficial and a negative impact on the mechanical behavior of polymers. For example, in injection molding, the formation of a highly crystalline network can result in articles that are easily cleaved (Rao & Rajagopal 2002).

The SIC is a mechanical type of crystallization which is unique in polymers. The rapid upturn in stress-strain response of elastomer material under SIC is a phenomenon that repeatedly was observed in experiments. It results from the reinforcing effects of the crystallines generated by the entropy decrease caused by strain. Natural rubber is a polymer that undergoes SIC, and this phenomenon can be well observed in this material.

In literature, different polymer crystallization models have been proposed as for example the folded chain model (Keller & Machin 1967), fringed-micelle type model (Mandelkern 1964) or the extended-chain crystalline model. Recent experiments show that both folded and extended chain types are involved in the crystallization process.

Previously (Dargazany & Itskov 2009) we have proposed a constitutive model describing the main three phenomena observed in elastomer material namely nonlinear elasticity, permanent set and

anisotropic Mullins effect. In the present contribution this model is extended in order to take additionally SIC effect into account. To this end, the influence of SIC on the mechanical behavior is described by a change of the entropic strain energy. The distribution of the crystallines is formulated based on a statistical approach. The contribution of the crystallines to the partial immobilization of polymer chains is accounted for by assuming the crystalline phase and the amorphous phase to be two separate networks. In addition, the model is compared with available experimental results of uniaxial tension tests.

2 STATE OF THE ART

The crystallization rate of polymers is usually observed to be zero at the melting temperature and at the glass transition temperature. The maximum is reached at a temperature in between these two temperatures (Saidan & Motasem 2005). Polymers such as polyethylene in solid form are always semi-crystalline as their crystallization rates at temperatures below the melting point are very high. It is not possible to cool down the polymer rapidly enough to a temperature below the glass transition temperature without substantial crystallization. For polymers such as polyethylene terephthalate (PET), which crystallize slowly, the melt has to be cooled down slowly for the substantial crystallization. If these polymers are quenched to a temperature below their glass transition temperature, they remain in an amorphous state. When amorphous PET is subsequently deformed at temperatures just above the glass transition temperature, strain induced crystallization takes place (Rao & Rajagopal 2002).

It is experimentally observed that the strain induced crystallines appear along the stretching direction (Murakami, Senoo, & Toki 2002, Trabelsi, Albouy, & Rault 2003). The size of the crystalline is independent of stretch value, but their number increases in the direction of the maximal stretch (Gehman & Field 1939, Luch & Yeh 1972, Murakami, Senoo, & Toki 2002). In natural rubber the size of crystallines is about 10 nm in the stretch direction (Yeh 1976). Recent experimental observations on SIC indicate that only about 20 percent of chains are crystallized (Toki et al. 2003, Tosaka et al. 2004).

Nikolov & Doghri (2000) and Nikolov et al. (2002) presented a micro-mechanical study of high-density polyethylenes under small deformations using the lamellar structure model. However they did not take into account the nonlinear response of these materials in unloading. (Kroon 2010) also proposed a constitutive model and predicted the mechanical behavior of natural rubber under SIC although the involvement of folded chains in crystallization process was not considered.

3 STATISTICAL MECHANICS OF POLYMER CHAINS

Experimental observations confirm that polymer molecules inside filled rubbers are usually cross-linked and bonded to aggregate surfaces at many points along their length. Thus, a chain is defined as an ordered polymer molecule between two fixed segments. Each chain consists of a high number of monomers called segments. Each segment might be either bonded to the aggregate surface or cross-linked to other chains.

3.1 Statistical mechanics of a single polymer chain

The probability function of a polymer chain with n segments spanned between two aggregate surfaces at a relative distance $\bar{r}$ from each other is expressed as (for more detail see Dargazany & Itskov 2009)

$$P(n, \bar{r}) = \sqrt{\frac{3\alpha\kappa^2}{2\pi n}} e^B, \tag{1}$$

where

$$B = -G - \frac{\kappa}{\sqrt{\pi}} [\sqrt{6\alpha n} e^{-G} + 3\alpha\bar{r}\sqrt{\pi n} erf(\sqrt{G})$$

$$-\sqrt{6\alpha} e^{-Gn} - 3\alpha\bar{r}\sqrt{\pi} erf(\sqrt{Gn})], \quad G = \frac{3\alpha\bar{r}^2}{2n}. \tag{2}$$

Herein, the relative angle between segments α as well as the active adsorption area of each aggregate κ are considered as the material parameters.

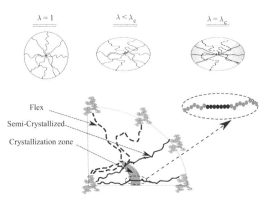

Figure 1. Micro-mechanical modeling of SIC. The above figures illustrate deformation of the unit sphere under uniaxial tension. In the shadowed area chains are extended. A quarter of deformed unit sphere is magnified. The stretched mid-segments in crystallization zone are shown there. Black circles illustrate the crystallized segments.

The non-Gaussian distribution of chains rearranged due to the polymer-filler debonding can further be given by (Dargazany & Itskov 2009)

$$\overset{d}{P}(n, \lambda_m) = \Phi(\overset{d}{\lambda_m}) P(n, r), \tag{3}$$

where $\Phi(\lambda_m)$ denotes a transition function taking into account the number of active chains after the network rearrangement.

3.2 Strain energy of a single polymer chain before SIC

In order to predict crystallization a statistical theory is applied. Accordingly, a polymer is treated as a network of freely-jointed chains cross-linked at junction points, whose motions are affine with the macroscopic deformation (Milchev et al. 1994). The entropic energy of a single chain with the normalized end-to-end distance $\bar{r}$ and n segments of length l is written by (Treloar 1975)

$$\psi(n, \bar{r}) = nKT \left(\frac{\bar{r}}{n}\beta + \ln \frac{\beta}{\sinh \beta} \right), \tag{4}$$

where

$$\beta = \mathcal{L}^{-1} \left(\frac{r}{nl} \right) \tag{5}$$

and $\mathcal{L}^{-1}$ denotes the inverse Langevin function. T stands for the absolute temperature (isothermal condition is assumed) and K is the Boltzmann constant.

3.3 Strain energy of chains after SIC

The crystalline nucleation is a phenomenon mainly influenced by the applied stretch. It is experimentally observed that highly elongated polymer chains are crystallized first. Accordingly, during crystallization polymer chains of the rubber matrix are decomposed into amorphous chains and the crystallized chains. The contribution of the latter ones into the strain energy of the network is relatively small and can be neglected. The volume of crystallized chains relative to the unit volume of the rubber matrix represents the crystallization degree and is denoted in the following by ζ.

As observed by the wide-angle X-ray diffraction experiments, the number of crystallines increases during loading and decreases during unloading. Under loading, the crystallines form above a specific stretch λ_c called crystallization stretch. Under unloading they partly remain until the stress-free state is reached. Above the crystallization stretch, the elongated chains can be divided into Semi-Crystallized (SC) and Flexible (FL) chains (see figure 1). The SC chains (indicated by a solid line in figure 1) are highly stretched and involved in the nucleation of crystallization while the FL chains (indicated by a dashed line in figure 1) are more flexible and effect the crystallization growth. However in this study we only discussed the SC chains while a thorough mathematical description of FL chains will be presented elsewhere. It is also assumed that the crystallines are mostly formed in the middle of SC chains and in their direction.

The stretch applied on a chain in the direction $\boldsymbol{d}$ in the reference configuration is written by

$$\lambda^d = \sqrt{\boldsymbol{d}\boldsymbol{C}\boldsymbol{d}}, \tag{6}$$

where $\boldsymbol{C} = \boldsymbol{F}^T\boldsymbol{F}$ denotes the right Cauchy-Green tensor while $\boldsymbol{F}$ is the deformation gradient.

It is assumed that in the onset of crystallization every chain stretched nearly its extensibility limit ($r = a_c r_0 \lambda$ where $a_c > 1$) is divided into two SC chains. Thus according to (4) the free energy of a single SC chain is given by

$$\psi_{sc}(n, \lambda^d \bar{r}_0) = 2\psi\left(\frac{n(1 - I_c)}{2}, \lambda^d \frac{\bar{r}_0}{2}\right), \tag{7}$$

where $\bar{r}_0$ is the average inter-particle distance in the virgin matrix and I_c is the crystallization ratio of each chain.

The crystallization process starts at λ_c while decrystallization (melting) ends at $\lambda = 1$ (Kroon 2010). Thus, denoting the crystallization and decrystallization rate as a_c and d_c, respectively, in a way that $d_c < a_c$ the length range of SC chains during loading and unloading is calculated as follows

$$D_{sc}^d = \begin{cases} A_c\left(\lambda_m^d\right)\bar{r}_0 & \text{(loading)}, \\ \min\left\{A_c\left(\lambda_m^d\right), d_c\left(\lambda^d\right)\right\}\bar{r}_0, & \text{(unloading)} \end{cases} \tag{8}$$

where the upper range of SC chains is expressed by

$$A_c\left(\lambda_m^d\right) = \lambda_{cr}\left(1 + a_c\left(\frac{\lambda_m^d}{\lambda_{cr}} - 1\right)\right) \tag{9}$$

and $\lambda_{cr} = \lambda/\lambda_c$. The following boundary conditions take into account that the last crystal melts at $\lambda = 1$, the decrystallization process stops at the end of each loading and the crystals should start to melt under unloading.

$$\begin{aligned} &d_c(1) = v\lambda_m, \\ &d_c(\lambda_m) \leq A_c(\lambda_m), \\ &v\lambda_m < d_c(\lambda) \leq A_c(\lambda_m), \end{aligned} \tag{10}$$

where v denotes a sliding ratio. Furthermore, assuming d_c to be a linear function of stretch, one can derive

$$d_c = \frac{A_c(\lambda_m) - v\lambda_m}{\lambda_m - 1}\lambda - \frac{A_c(\lambda_m) - v\lambda_m^2}{\lambda_m - 1}. \tag{11}$$

Using (8) and (3) the total number of current SC chains is calculated as a function of deformation gradient by

$$N_{sc}(\boldsymbol{F}) = \frac{N_0}{4\pi}\int_\Omega\int_{B_c\lambda_m^d\bar{r}_0}^{D_{sc}(\lambda^d)}\Phi(\lambda_m^d)P(n)dnd\Omega, \tag{12}$$

where $B_c = \frac{v}{1-I_c}$, while Ω represents the surface of a unit sphere and N_0 is the initial number of active chains. The direction vector $\boldsymbol{d}$ is expressed in terms of the spherical coordinates.

Thus the energy of total available SC chains can be given by

$$\Psi_{sc}(\boldsymbol{F}) = \frac{N_{sc}(\boldsymbol{F})}{4\pi}\int_\Omega(\psi_{sc}^p + \psi_m^p)d\Omega, \tag{13}$$

where

$$\psi_{sc}^p = \int_{B_c\lambda_m^d\bar{r}_0}^{D_{sc}(\lambda^d)}P(n, \lambda_m^d)\psi_{sc}(n, \lambda^d\bar{r}_0)dn \tag{14}$$

is the energy of the existing SC chains and

$$\psi_m^p = \int_{D_{sc}(\lambda^d)}^{A_c(\lambda_m^d)}P(n, \lambda_m^d)\psi(n, \lambda^d\bar{r}_0)dn \tag{15}$$

is the energy of the melted SC chains.

4 NETWORK DECOMPOSITION

According to Govindjee & Simo (1991) the rubber matrix can be decomposed into a pure rubber network and a polymer-filler network which act parallel to each other with no interaction between them. Thus the free energy of the rubber matrix is expressed by

$$\Psi_r = \Psi_{cc} + \Psi_{pp}, \tag{16}$$

where Ψ_{cc} and ψ_{pp} denote the free energies of chains in pure rubber network and chains distributed between filler aggregates, respectively.

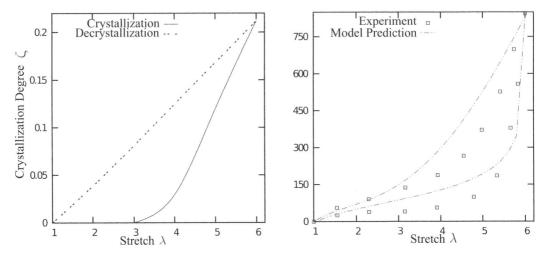

Figure 2. The crystallization degree (left). Comparison of the model prediction and experiments by Toki et al. (2003) (right).

The contribution of the crystallization into the strain energy results from the changes of the entropic energy of the chains involved in crystallines during the crystallization and decrystallization process. Since some of PP chains are involved in crystallines, the PP network is additively decomposed into the amorphous and crystallized network according to the crystallization degree. The energy of the crystallized network is obtained by summing up the energies of SC chains and FL chains, respectively Ψ_{sc} and Ψ_{fl}. Thus,

$$\Psi = \Psi_{cc} + (1 - \varsigma)\Psi_{pp} + \varsigma(\Psi_{sc} + \Psi_{fl}). \qquad (17)$$

5 RESULTS

We consider sulfur vulcanized natural rubber with $\lambda_c = 3$ at $0°C$. The crystallization degree predicted by the above model in one loading-unloading cycle with the stretch amplitude of 6 is plotted versus stretched in figure 2 (left). Accordingly, crystallization starts at $\lambda = \lambda_c$ under loading. During unloading the crystallines decreases linearly until the material is fully unloaded.

The comparison of the model with experimental data of the uniaxial tension test (Toki et al. 2003) is illustrated in figure 2 (right). The results show a substantial influence of SIC on the mechanical behavior of natural rubber. This influence is mainly observed as the upturn in the stress-strain curve above the crystallization stretch. During unloading, this stress hardening still remains even at below the crystallization stretch.

6 CONCLUSIONS

In this paper the Strain-Induced-Crystallization and its influence on the mechanical behavior of the natural rubber have been discussed. In order to take this phenomenon into account in the constitutive modeling

our previous network evolution concept (Dargazany & Itskov 2009) has been extended and adopted accordingly. In the onset of crystallization, elongated chains are subdivided into Semi-Crystallized (SC) and Flexible (FL) chains. The SC chains are highly stretched and involved into the nucleation of crystallization while the FL chains are more flexible and effect the crystallization growth. The influence of the crystallization is described by taking into account entropic strain energies of all evolving subnetworks. The so-obtained model demonstrates fair agreement with experimental data available in literature. Further, the model will be improved by considering the healing effect.

REFERENCES

Dargazany, R. & M. Itskov (2009). A network evolution model for the anisotropic mullins effect in carbon black filled rubbers. *Int. J. Solid. Struct. 46*, 2967.

Gehman, S. & J. Field (1939, Aug.). X-ray investigation of crystallinity in rubber. *Journal of Applied Physics 10*, 564–572.

Govindjee, S. & J. Simo (1991). A micro-mechanically based continuum damage model for carbon black-filled rubbers incorporating mullins' effect. *J. Mech. Phys. Solids 39*(01-04-04), 87.

Keller, A. & M. Machin (1967). Oriented crystallization in polymers. *J. Macromol. Sci., Part B: Phys. 1*(05-01-01), 41.

Kroon, M. (2010, SEP). A constitutive model for straincrystallising rubber-like materials. *Mechanics of Materials 42*(9), 873.

Luch, D. & G. S. Y. Yeh (1972, November). Morphology of straininduced crystallization of natural rubber. i. electron microscopy on uncrosslinked thin film. *Journal of Applied Physics 43*(11), 4326.

Mandelkern, L. (1964). The crystallization kinetics of polymerdiluent mixtures: The temperature coefficient of the process. *Polymer 5*, 637.

Milchev, A., W. Paul, & K. Binder (1994). Polymer chains confined into tubes with attractive walls: A monte carlo simulation. *Macromol. Theory Simul. 3*(02-04-01), 305.

Murakami, S., K. Senoo, & S. Toki, S. and Kohjiya (2002). Structural development of natural rubber during uniaxial stretching by in situ wide angle x-ray diffraction using a synchrotron radiation. *Polymer 43*(7), 2117.

Nikolov, S. & I. Doghri (2000). A micro/macro constitutive model for the small-deformation behavior of polyethylene. *Polymer 41*(5), 1883.

Nikolov, S., I. Doghri, O. Pierard, L. Zealouk, & A. Goldberg (2002). Multi-scale constitutive modeling of the small deformations of semi-crystalline polymers. *Journal of the Mechanics and Physics of Solids 50*(11), 2275.

Rao, I.J. & K.R. Rajagopal (2002). A thermodynamic framework for the study of crystallization in polymers. *Zeitschrift für Angewandte Mathematik und Physik (ZAMP) 53*, 365. 10.1007/s00033-002-8161-8.

Saidan & Motasem (2005, February). *Deformation-Induced Crystallization In Rubber-Like Materials*. Ph. D. thesis, TU Darmstadt, Darmstadt.

Toki, S., I. Sics, S. Ran, L. Liu, & B. S. Hsiao (2003). Molecular orientation and structural development in vulcanized polyisoprene rubbers during uniaxial deformation by in situ synchrotron x-ray diffraction. *Polymer 44*(19), 6003. In Honour of Ian Ward's 75th Birthday.

Tosaka, M., S. Murakami, S. Poompradub, & S. Kohjiya (2004). Orientation and crystallization of natural rubber network as revealed by waxd using synchrotron radiation. *Macromolecules 37*(06-02-05), 3299.

Trabelsi, S., P. Albouy, & J. Rault (2003). Crystallization and melting processes in vulcanized stretched natural rubber. *Macromolecules 36*(06-02-04), 7624.

Treloar, L. (1975). *The Physics of Rubber Elasticity Oxford University*.

Yeh, G. (1976). Strain-induced crystallization. 1. limiting extent of strain-induced nuclei. *Polymer Engineering and sceince 16*(3), 138.

Constitutive Models for Rubber VII – Jerrams & Murphy (eds)
© 2012 Taylor & Francis Group, London, ISBN 978-0-415-68389-0

Tear rotation in reinforced natural rubber

B. Gabrielle, A. Vieyres, O. Sanseau, L. Vanel, D. Long & P. Sotta
Laboratoire Polymères et Matériaux Avancés, CNRS/Rhodia, Saint-Fons, France

P.-A. Albouy
Laboratoire de Physique des Solides, CNRS/Université Paris-Sud, Orsay, France

ABSTRACT: We analyze the impact of tear rotation, that is, an abrupt instability in the direction of propagation of a notch, on the tensile strength of natural rubber elastomers reinforced with carbon black or precipitated silica, in single edge notched samples stretched at constant velocity. As a consequence of tear rotation, the energy at break increases by a factor of 6 to 8 in some cases. We show how the tensile strength of a test sample is related to the presence of tear rotations and analyze semi-quantitatively this increase in tensile strength, based on energetic arguments, without entering into a detailed description of the elastic strain field in the vicinity of the tear tip. The proposed interpretation is based on the idea that tear rotations creates a macroscopic tip radius, which relaxes the local strain (or stress) at the tear tip. Materials reinforced with carbon black or precipitated silica aggregates show similar behavior.

1 INTRODUCTION

In single-edge-notched samples under tensile loading (SENT) at constant velocity, cracks tend to rotate in the direction parallel to the applied tensile stress (Hamed 1996). This is the so-called 'tear rotation' phenomenon (Lake 1991). Associated to this phenomenon, a much higher resistance to cut growth is observed, particularly in filled natural rubber (NR) samples (Hamed 1996, Kim 2000, Hamed 1999a, Hamed 1999b, Hamed 1998a, Hamed 1998b).

In the presence of tear rotation, the relationship between tensile strength (or stress at break) and pre-cut notch length is complex and exhibits different regimes according to the cut length (Hamed 2002, Hamed 2003). It has been suggested that strain-induced crystallization may block mode I notch propagation (i.e. perpendicular to the load direction) in NR (Andrews 1961, Lee 1987). The effects of various material parameters, such as the crosslink density (Hamed 2002), the volume fraction and grade of fillers, were studied.

Describing crack propagation in rubbery materials is difficult for several reasons: rubber materials can undergo extremely large elastic (or quasi elastic) deformations (up to several 100%), and the viscoelastic contribution to tearing energy (dissipation) may be large, as it is evidenced by the large variation of the tearing energy with the tear propagation rate (Persson 2005, Lake 2000).

The first, macroscopic, approach which was developed to analyze fracture toughness of a material is the Griffith approach, based on energetic balance arguments that we recall briefly here (Rivlin 1953). The driving force for propagation is the elastic energy released when a crack propagates at constant overall deformation of the test sample (energy release rate), denoted G in the following:

$$G = -\frac{1}{e}\left(\frac{\partial U_{el}}{\partial c}\right)_\lambda \qquad (1)$$

where e is the sample thickness, c the crack length, U_{el} the elastic energy stored in the sample, W the elastic energy density within the uniformly deformed part of the material (far away from the crack) and λ is the elongation ratio far away from the crack (or more precisely, the elongation ratio in an un-notched sample of same overall elongation).

In single-edge-notched tensile (SENT) samples with a crack (precut notch) of length c, for normal propagation (without instability in the direction of propagation), it was shown that the energy release rate G is (Lake 1970, Lake 1991):

$$G \cong 2\pi \frac{Wc}{\sqrt{\lambda}} \qquad (2)$$

A crack of length c becomes unstable when the energy release rate (equation 1) becomes equal to the tear energy Γ, which may be considered as a characteristic property of the material. Identifying this instability threshold with sample failure gives the following relationship between the energy at break W_b and the precut notch length c is obtained:

$$\frac{W_b}{\sqrt{\lambda_b}} = \frac{\Gamma}{2\pi c} \qquad (3)$$

In this paper, we show measurements of the tensile strength of SENT samples stretched at constant

Table 1. Sample formulations.

Formulation (phr)	NR-s	NR-c
Matrix	100	100
Silica*	50	0
Carbon Black**	0	45
TESPT	4	0

*: precipitated silica, surface $160 \, m^2 g^{-1}$. **: N234.

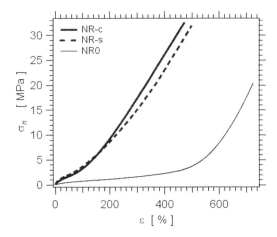

Figure 1. Nominal stress vs strain in un-notched samples stretched at $1.4 \times 10^{-2} \, s^{-1}$, plotted up to failure.

velocity up to failure. Samples with different rubber matrices (SBR, NR), and different reinforcing systems (carbon black, silica with covalent coupling to the rubber matrix) have been compared. We find that tear rotation occurs only in reinforced NR materials and have a large impact on the sample tensile strength. We relate the energy at break measured in pre-notched tensile test samples to the overall crack contour length (which includes tear rotation length). We show that it is possible to draw certain conclusions on tear rotations and on their effect on the measured tensile strength of samples, based on energy balance arguments.

2 MATERIALS

Natural Rubber (NR) matrices have been used with two types of reinforcing fillers (carbon black and precipitated silica). Samples are labeled as NR0 (pure NR matrix), NR-c (carbon black filled) or NR-s (silica filled) (see Table 1). All samples are crosslinked with the same standard sulfur vulcanization system. In silica-filled samples, bis triethoxysilylpropyl tetrasulfur (TESPT) was used to covalently bond the NR matrix to the silica surface.

Both reinforcing systems (carbon black or silica plus coupling agent) provide the same hardness (60 Shore A) and similar reinforcement properties at the same volume fraction (about 20 vol%) (see Figure 1).

Single edge notched samples ($12 \times 70 \times 2 \, mm^3$) are precut to a length varying between roughly 0.3 and 4 mm. The stretching velocity varies between 5 mm/min (tensile strain rate $1.4 \times 10^{-3} s^{-1}$) and 500 mm/min (tensile strain rate $0.14 \, s^{-1}$). Experiments were performed with as prepared samples, without precycling. The elastic energy density W is obtained by integrating the force-extension curve and dividing by the volume of the stretched sample. In precut samples, this only approximates the actual energy density far from the tear, since in that case the strain is not uniform throughout the sample. To describe the tensile strength of samples, we use the elastic energy density at the breaking point W_b measured for SENT samples precut at various cut lengths. Using notched samples ensures that failure will occur at the tip of the precut notch. Thus each sample can be characterized by a relationship between the precut notch length c and the energy density at break $W_b(c)$.

The nominal stress vs strain curves obtained in the unnotched samples up to failure are shown in Figure 1. Even though carbon black or silica filled samples show significant differences, both fillers roughly provide the same reinforcement, at same volume fraction. The energy density at failure W_f can be measured on uncut samples ($W_f = W_b(c = 0)$) from the stress-strain curves in Figure 1. It is of the order 50 MPa in both reinforced samples.

3 RESULTS

We have measured the elastic energy density at break W_b of single edge notched tensile test (SENT) samples as a function of the precut notch length c, at room temperature. It is observed that slow forward propagation of the crack, if any, occurs on a very small length scale, just before fast, catastrophic failure. This slow propagation step therefore contributes only negligibly to the overall energy at break. In Figure 2, the quantity $W_b / \lambda_b^{1/2}$ is plotted as a function of c, according to equation 3 The results obtained in reinforced NR samples are compared to pure (unfilled) NR samples. Energy densities at break in reinforced NR are 1.5 to 8 times larger than in pure NR samples. When fitted with equation 3 (though in a rather limited range of precut notch length c, typically larger than 0.6 mm), a tearing energy $\Gamma_{NR} = 31.4 \, kJ.m^{-2}$ is obtained.

Based on the results in Figure 2, it cannot be concluded that the increased tensile strength (or average resistance to tear propagation) observed in reinforced NR as compared to pure NR may be attributed to an intrinsically higher tear energy Γ, essentially because of the large scattering of data points. It is very difficult to discriminate the samples according to the nature of the reinforcing system (silica or carbon black). Both give widely overlapping clouds of representative points. The scattering of data points is analyzed in the next section in relation to tear rotation.

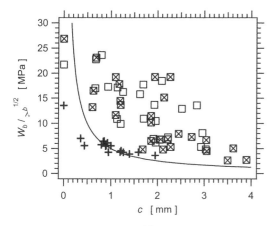

Figure 2. The quantity $W_b/\lambda_b^{1/2}$, where W_b is the energy density at break and λ_b the elongation ratio at break, as a function of the precut length c: $\square$: NR-s; $\boxtimes$: NR-c; $+$: NR0.

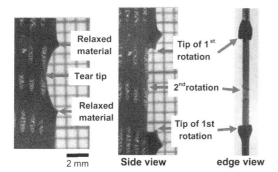

Figure 3. Successive tear rotations in a precut notched reinforced NR tensile strip (some marks have been drawn of the sample surface). Several rotations may occur before final failure occurs.

The tear (initiated at the pre-cut notch) propagates in different ways in the various samples. In pure NR samples, tear propagates perpendicular to the drawing direction, as expected. Propagation is fast. In both CB- and silica-filled NR samples, tear propagation is more complex, with a well-developed tear rotation phenomenon illustrated in Figures 3 and 4. The pre-cut notch first opens widely without significant propagation. A tear then initiates in the central region of the widely open tip and rotates abruptly on both sides along the tensile direction after propagating over a small distance in the usual direction (a few tens to a few 100 s of µm). This results in two folds of relaxed material curling along the crack edges in the tensile direction. Propagation along the tensile direction is slow and stops at some point. Then, as the applied load increases, a second tear eventually initiates and eventually rotates again.

Successive rotations have increasing lengths and as a result, the two relaxed parts of the sample move apart from each other along the sample edge. Up to 5 rotations on each side may occur successively. Sample

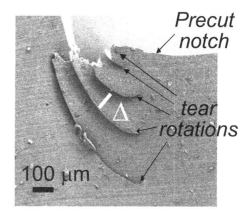

Figure 4. MEB observation of a broken silica filled NR test sample back in the relaxed state, showing 5 successive tear rotations.

failure finally occurs in a catastrophic way starting from the last rotation. Such behavior was already described in filled natural rubber materials (Hamed 1996).

To analyze the effect of tear rotations on the ultimate strength of the samples, we propose a simple approach, essentially based on energetic arguments.

A correlation between the measured energy at break and the overall length of the tear rotations is observed: the longer the rotations, the higher the energy at break. The scattering of results observed in reinforced NR samples (Figure 3) is related to various values of the overall tear rotation length observed for a given value of c.

We propose the following analysis for the relationship between tear rotation and ultimate strength of reinforced natural rubber. When back in the relaxed state (after failure of the sample), rotations take the form of paraboloid curves with the convexity directed towards the direction of the tear tip (see Figure 4). The sample part located inside the concavity is relaxed. This is similar to having a cut with a finite, macroscopic radius ρ in the relaxed state (not to be confused with the opening radius at an infinitely sharp tip), as described by (Thomas 1955) and by (Glucklich 1976).

The radius ρ is directly related to the measured rotation length l_R by $l_R = \kappa\rho$ (Figure 5) with κ of the order π (see Figure 4). Thomas has shown the following approximate relationship between the energy release rate G and the radius ρ (Thomas 1955) for a macroscopic radius ρ:

$$G \approx W_{loc}\rho \tag{4}$$

where W_{loc} is the elastic energy density *at the tear tip*, which is amplified with respect to the energy density far from the tip (stress concentration). The smaller the tip radius, the higher the stress amplification. We can assume that failure occurs (locally) at a critical energy density W_f. Thus, failure at the tip will occur when $W_{loc} \cong W_f$, and equation 4 effectively corresponds to an increase of the measured value of the (macroscopic)

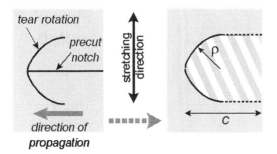

tear rotation

precut notch

stretching direction

direction of propagation

ρ

c

Figure 5. Scheme of tear rotation, which effectively results in a notch of finite radius ρ. The curvilinear length l_R of the rotation (measured in the relaxed state) is $l_R \cong \pi\rho$.

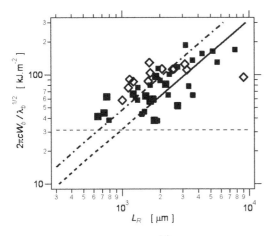

Figure 6. The quantity $2\pi cW_b/\lambda_b^{1/2}$ as a function of the rotation length L_R for various values of the traction speed, temperature and precut notch length. The horizontal dashed line indicates the value 31.4 kJ.m^{-2} obtained from the data in pure NR samples (see Figure 2). Black squares: carbon black-filled NR; white diamonds: silica-filled NR. Full curve: equation 5 with $W_f = 100$ MPa (matches carbon black data); dash-dot line: equation 5 with $W_f = 150$ MPa (matches silica data).

elastic energy density at break W_b as ρ increases. Then, substituting the energy release rate G given by equation 2 in equation 4 gives:

$$2\pi \frac{W_b c}{\sqrt{\lambda_b}} = W_f \frac{l_R}{\pi} \tag{5}$$

This equation relates the ultimate property of a sample, through the quantity $2\pi cW_b/\lambda_b^{1/2}$ (homogeneous to a tearing energy, expressed in kJ.m^{-2}), to the length of the last rotation l_R.

The quantity $2\pi cW_b/\lambda_b^{1/2}$ is plotted in Figure 6 as a function of the total rotation length L_R (not very different from the length l_R of the last rotation) for experiments done on notch tensile strips with various precut notch length values, stretched at different traction speeds and temperatures, in samples reinforced

with carbon black and silica. The horizontal line in Figure 6 corresponds to the value of the tearing energy 31.4 kJ.m^{-2}. The linear curves correspond to equation 5 with W_f values adjusted to match semi-quantitatively either carbon black- or silica-filled sample data (100 and 150 MPa respectively). As expected, these values are larger than the energy at break measured macroscopically in the uncut samples (50 MPa).

4 DISCUSSION

The first rotation occurs below the crystallization threshold. Nevertheless, due to local strain amplification, the material crystallizes in the vicinity of the tear tip quite early (Lee 1987, Trabelsi 2002). The extension of the crystallized region, related to strain concentration at the tear tip, depends on the overall extension ratio (far from the tear tip). The presence of this region may prevent normal propagation of the tear because it induces a strong elastic anisotropy in the material, which would be the main mechanism responsible for tear rotation (Gent 1954, Chenal 2007).

However, strain induced crystallization by itself is not sufficient to induce tear rotation, as pure NR samples (which do crystallize) do not exhibit tear rotation. A combination of strain-induced crystallization and of reinforcement provided by nanometric fillers, which both shift the onset of crystallization to lower overall strain and considerably increase the modulus, is thus needed to induce this phenomenon in our case. Note also that tear rotation does not appear in the absence of a precut notch (of millimeter size).

We have proposed a way to analyze the effect of tear rotations on the tensile strength of a reinforced natural rubber (Gabrielle 2011). When a test sample is stretched at a given extension rate, a precut notch do not propagate as long as the elastic energy density at the tear tip W_{loc} remains below a critical value W_f for initiation of propagation. Propagation initiates when $W_{loc} = W_f$. but is blocked by strain induced crystallization at the tear tip. This induces tear rotation, which effectively creates a *macroscopic* tear tip radius ρ (comparable in size to the rotation length). This provides an elastic mechanism for tip blunting, distinct from the visco-plastic one described by Dugdale & Barenblatt (Dugdale 1960, Barenblatt 1962). In the rotated tear configuration, the relationship between the macroscopic energy density W_0 (far away from the tear) and the local energy density at the tear tip W_{loc} is given by equation 5. Then the test sample can be further stretched, until the local energy density at the tear tip reaches again the critical value W_f. Thus, each successive rotation effectively results in a larger tip radius, which allows reaching larger macroscopic values of the elastic energy density in the bulk of the sample.

We have checked that strain induced crystallization takes place in the whole temperature range investigated (i.e. up to 80°C), with ultimate crystallinity and critical strain at onset little affected by temperature in this range. On the other hand, it is likely that crystallization kinetics is fast enough to remain unaffected or only little affected by the traction speed, in the investigated range of values. A detailed study of strain induced crystallization in our materials will be published separately.

It is observed that the material exhibits well developed fibrilar morphology at very high extension ratios. Then it is likely that such fibrils have a very high modulus (that of a semi-crystalline polymer) in the presence of strain-induced crystallization, thus providing the strong anisotropy in the elastic properties needed to observe tear rotation. Indeed, such anisotropy could favor propagation along the tensile direction, as compared to normal propagation (perpendicular to tensile direction).

5 CONCLUSION

For the first time, a mechanism is proposed to relate the presence of tear rotations observed in reinforced natural rubber to the tensile strength of the material. The measured apparent tensile strength (in terms of energy at break) may be increased by a factor of 6 to 8 in some cases. This large increase in tensile strength associated to the presence of tear rotations is analyzed semi-quantitatively, based on energetic arguments, without entering into a detailed description of the elastic strain field in the vicinity of the tear tip. We show a correlation between the extent of the phenomenon of tear rotation and the energy density at failure of single edge notched tensile test samples. Materials reinforced with either carbon black or silica (with TESPT as coupling agent), formulated to have close mechanical behavior, show similar behavior.

REFERENCES

Andrews, E. H. 1961. *Journal of Applied Physics* 32: 542–548.
Barenblatt, G. I. 1962. *Adv. Appl. Mech* 7: 55.
Chenal, J.; Gauthier, C.; Chazeau, L.; Guy, L.; Bomal, Y. 2007. *Polymer* 48: 6893–6901.
Dugdale, S. D. 1960. *J. Mech. Phys. Solids* 8: 100.
Gabrielle, B., Guy, L., Vanel, L., Long, D. R., Sotta, P. 2011, Macromolecules, to appear.
Gent, A. N. 1954. *Trans. Faraday Soc* 50: 521.
Hamed, G. R.; Kim, H. J.; Gent, A. N. 1996. *Rubber Chemistry and Technology* 69: 807.
Hamed, G.; Zhao, J. 1998. *Rubber Chemistry and Technology* 71: 157.
Hamed, G.; Huang, M. Y. 1998. *Rubber Chemistry and Technology* 71: 846.
Hamed, G.; Kim, H. J. 1999. *Rubber Chemistry and Technology* 72: 895.
Hamed, G.; Park, B. H. 1999. *Rubber Chemistry and Technology* 72: 946.
Hamed, G., N. Rattanasom 2002. *Rubber Chemistry and Technology* 75: 935–941.
Hamed, G., Al-Sheneper, A. A. 2003. *Rubber Chemistry and Technology* 76: 436.
Kim, H. J.; Hamed, G. 2000. *Rubber Chemistry and Technology* 73: 743.
Lake, G. J. 1970. In *Proc. Int. Conf. on Yield, Deformation and Fracture of Polymers, Cambridge*; p. 5.3/1.
Lake, G. J.; Samsuri, A.; Teo, S. C.; Vaja, J. 1991. *Polymer* 32: 2963–2975.
Lake, G. J.; Lawrence, C. C.; Thomas, A. G. 2000. *Rubber Chemistry and technology* 73, 801–817.
Lee, D. J., Donovan, J. A. 1987. *Rubber Chemistry and Technology* 60: 910–923.
Persson, B. N.; Albohr, O.; Heinrich, G.; Uebas, H. 2005. *Journal of Physics: Condensed Matter* 17: R1071–R1142.
Rivlin, R. S.; Thomas, A. G. 1953. *J. Polym. Sci.* 10: 291.
Thomas, A. G. 1955. *Journal of Polymer Science* 18: 177–188.
Trabelsi, S.; Albouy, P.; Rault, J. 2002. *Macromolecules* 35: 10054–10061.

Stress softening and related phenomena

Constitutive Models for Rubber VII – Jerrams & Murphy (eds)
© 2012 Taylor & Francis Group, London, ISBN 978-0-415-68389-0

Modelling Mullins and cyclic stress-softening in filled rubbers

Julie Diani
Laboratoire PIMM, CNRS, Arts et Métiers ParisTech, 151 bd de l'Hôpital, Paris, France

Yannick Merckel & Mathias Brieu
Laboratoire LML, Ecole Centrale de Lille, bd Paul Langevin, Villeneuve d'Ascq, France

ABSTRACT: The study of the stress-strain responses of a carbon-black filled rubber submitted to cyclic uniaxial tension loadings evidenced two specific features common to filled rubbers. First, as originally shown by (Mullins and Tobin 1965), the Mullins softened stress-strain responses collapse into a single master curve by introducing strain amplification factors. Second, the cyclic softened stress-stretch responses collapse into a single stress-stretch response by applying stretch intensity factors. Several Mullins softening models from the literature are confronted to the experimental evidence of the existence of a master curve. Then, a model framework is proposed in order to reproduce the cyclic softening and successfully tested in uniaxial tension.

1 INTRODUCTION

Filled rubbers are often submitted to cyclic loadings. The presence of fillers increase stiffness and lifetime but at the same favor the Mullins softening. On top of that, when submitted to a large number of cycles, filled rubbers experience some cyclic softening that cannot be neglected. Both softenings should be taken into account in an effort to predict the mechanical properties of rubber structures submitted to fatigue. Therefore, we looked at the particular properties of the mechanical behavior of softened filled rubbers. For this purpose we submitted a carbon-black filled styrene butadiene to cyclic loadings at various maximum stretches and up to 1000 cycles. We took a close look at the Mullins effect induced stress-strain responses and at the stress-stretch cyclic responses. Then, several models reported in the literature and designed to reproduce the Mullins softening were tested with more or less success. Finally, a simple model framework is proposed to account for the Mullins induced residual stretch and the cyclic softening. This framework is shown to adapt to any of the Mullins softening model of the literature and prove to be supported by the experimental evidences.

2 EXPERIMENTAL EVIDENCES

In this section, specific features of the cyclic stress-strain responses for non-crystallizing filled rubbers are presented. Figure 1 shows the response of a 40 phr carbon-black filled styrene butadiene rubber (SBR) submitted to a cyclic loading with increasing maximum stretch. The material undergoes a large softening

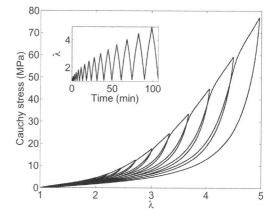

Figure 1. Evidence of Mullins softening during a cyclic uniaxial tension with increasing maximum stretch.

increasing with the maximum stretch. This softening known as the Mullins effect comes with some residual stretch (Figure 2) only partially recoverable. When submitted to a large number of cycles, the filled rubber softening (Figure 3) and residual stretch (Figure 4) increase slowly but steadily with the number of cycles. One notes in Figures 1 and 3 that once the Mullins effect evacuated, the hysteresis evidenced during a cycle is limited. In what follows, we neglect this hysteresis and focus on the material softening only. The softened responses is then characterized by the loading responses. Studying the experimental responses of various filled rubbers, we noticed that the Mullins softening and the Mullins induced residual stretch are not easy to relate. It is likely that both quantities are not completely correlated and it

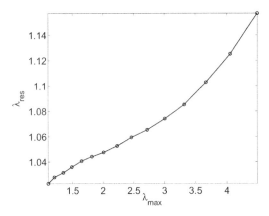

Figure 2. Residual stretch measured after one cycle up to λ_{max}.

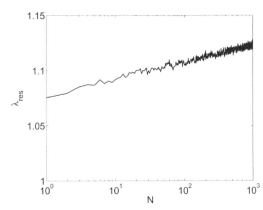

Figure 4. Residual stretch measured after one cycle up to $\lambda_{max} = 3$.

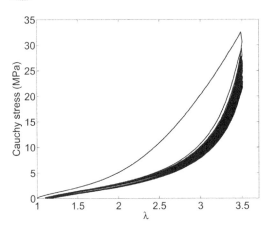

Figure 3. Evidence of cyclic softening during cyclic uniaxial tension at constant maximum stretch.

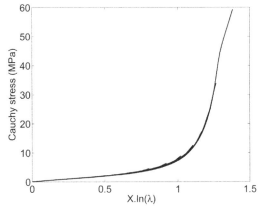

Figure 5. Mullins softened responses superimposition using relation (1).

seems relevant to part the softening from the permanent stretch in the modeling. Therefore for studying the Mullins softening, we assumed that it was unrelated to the residual stretch and we neglected the latter one. This is a path often followed in the literature for the modelling of the Mullins softening. In a procedure proposed by (Mullins and Tobin 1965), revisited by (Kluppel and Schramm 2000) and later by (Merckel et al. 2011), the authors showed that the various maximum stretch-dependent softened responses (or Mullins softened responses) $\mathcal{S}_{\lambda_{max}}(\epsilon, \sigma)$ may collapse into a single master curve $\mathcal{M}$ by writting,

$$\forall \lambda_{max}, \exists D(\lambda_{max}), \mathcal{M}(\epsilon, \sigma) = \mathcal{S}_{\lambda_{max}}((1 - D)\epsilon, \sigma) \quad (1)$$

with σ the Cauchy stress, ϵ a strain measure and D a damage parameter dependent of λ_{max} only. When choosing $\epsilon = \ln(\lambda)$ for the strain measure and when applying the collapse procedure (1) to the material responses from Figure 1, one obtains Figure 5. One notes the satisfying superimposition of all stress-strain responses except for the top parts exhibiting the material viscoelasticity. For the cyclic softening, we

noticed that a procedure of material response superimposition applies also. The N-cycle stress-stretch responses $\mathcal{S}_N(\lambda, \sigma)$ may all collapse into the first cycle stress-stretch curve $\mathcal{S}_1$ according to,

$$\forall N, \exists \lambda_N, \mathcal{S}_1(\lambda, \sigma) = \mathcal{S}_N(\lambda/\lambda_N, \sigma) \quad (2)$$

with λ_N being constant for each cycle N and $\lambda_N \geq 1$. Figure 6 presents the stress-stretch response superimposition and the evolution of λ_N according to the number of cycle N, that correspond to the material stress-stretch responses showed in Figure 3. The specific properties of the material responses exhibited in this section were observed on a large number of non-crystallizing filled rubbers. They seem to be general features of the softened responses of filled rubbers. Therefore relevant modelling of the softened responses of filled rubbers requires the ability of the theoretical responses to superimpose. In the next section, several models are confronted to the experimental evidences presented above.

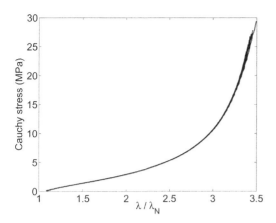

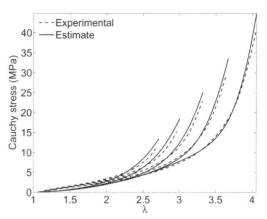

Figure 6. Cyclic loading responses superimposition using relation (2). The inset figure shows the evolution of λ_N vs. N.

Figure 7. Here comparison of the Ogden and Roxburgh model.

3 MULLINS DAMAGE MODELLING

Very few models exist on rubber cyclic softening (usually 1D model) and most of the damage models for filled rubbers focus on the Mullins softening. Therefore, we tested several Mullins softening models on their ability to build softened responses that may superimpose as experimental responses do (Figure 5).

3.1 Strain energy density penalization

One of the first models accounting for Mullins softening was proposed by (Simo 1987) and was based on a phenomenological definition of the strain energy density of the damaged material,

$$\mathcal{W} = (1 - D)\mathcal{W}_0 \tag{3}$$

with $\mathcal{W}_0$ the strain energy density of the virgin material and $(1 - D)$ the reduction factor inspired by Kachanov (1958). Such a strain energy density definition drives to the stress relation $\sigma = (1 - D)\sigma_0$. When assuming D as a function of the maximum stretch or maximum energy only and therefore constant for each stress-strain response, the superimposition showed in Figure 5 is impossible to obtain due to the nonlinearity of the stress-strain responses.

3.2 Ogden and Roxburgh pseudo-elastic model

(Ogden and Roxburgh 1999) defined a pseudo elastic model for the Mullins effect, which writes the uniaxial tension Cauchy stress as:

$$\sigma(\lambda) = \eta(\lambda_{max}, \lambda)\sigma_0(\lambda) \tag{4}$$

Readers are invited to refer to the original paper for details. We tested the model on our experimental data accounting for 8 parameters. Figure 7 shows the difficulties encountered to fit accurately our data. Moreover the superimposition test was performed on these theoretical stress-strain curves and failed.

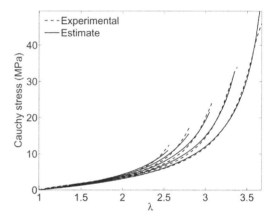

Figure 8. Comparison between the Marckmann et al. (2002) model and the experimental data from Figure 1.

3.3 Strain energy density parameter alteration

Another way to account for Mullins softening is to alter the strain energy density parameters. When considering strain energy density with many parameters this method is likely to generate stress-strain responses presenting the right features, nonetheless the number of parameters often used to fit hyperelasticity may be an obstacle to the proposal of relevant parameter evolutions with softening. In the authors' opinion, using the strain energy parameter alteration presents some interests when few parameters are used only. In (Marckmann et al. 2002) parameter alteration model, the strain energy depends on two parameters only and the alterations of both parameters are linked. Therefore, we tested the model on our experimental data. Constitutive equations can be found in (Marckmann et al. 2002). Figure 8 shows an estimate of the experimental data from Figure 1 by this model. The model proves to fit well the experimental data. The observed discrepancy between the model and the experiments at the small values of stretch is inherent to the fact that two parameters are used only. The superimposition of

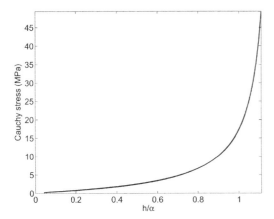

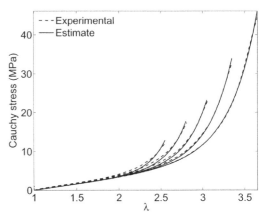

Figure 9. Superimposition of the theoretical stress-strain responses from the Marckmann et al. (2002) model.

Figure 10. Comparison between an hyperelastic model with a strain amplification factor and the experimental data from Figure 1.

the theoretical stress-strain responses apply very well as displayed in Figure 9.

3.4 Hyperelasticity with strain amplification factor

Another modelling approach is to look at the softening as an amplification of the strain at a given stress. The material may be assumed as hyperelastic and the constitutive equations write as,

$$\sigma(\boldsymbol{F}) = \boldsymbol{F} \frac{\partial \mathcal{W}}{\partial \boldsymbol{C}}\bigg|_{\boldsymbol{C}} \boldsymbol{F}^t - p\boldsymbol{I} \tag{5}$$

where $\boldsymbol{C} = \boldsymbol{F}^t \boldsymbol{F}$ and p is an additional pressure resulting from the incompressibility assumption. The strain undergone by the material $\boldsymbol{C}_a$, being amplified compare to $\boldsymbol{C}$. Therefore for a damaged material, on may write,

$$\frac{\partial \mathcal{W}}{\partial \boldsymbol{C}}\bigg|_{\boldsymbol{C}} = \frac{\partial \mathcal{W}_0}{\partial \boldsymbol{C}}\bigg|_{\boldsymbol{C}_a} \tag{6}$$

with $\mathcal{W}_0$ characterizing the virgin material strain energy density. We tested this approach for a strain energy density known for its ability to fit well the uniaxial tension response of rubbers (Lambert-Diani and Rey 1999),

$$\frac{\partial \mathcal{W}_0}{\partial \lambda_i} = 2\lambda_i \exp\left\{ \sum_{j=0}^{3} a_j (I_1 - 3)^j \right\}. \tag{7}$$

The strain amplification factor in uniaxial tension writes as,

$$\lambda_a = X.\lambda \tag{8}$$

with $X(\lambda_{max}) \geq 1$. Figure 10 shows the comparison between the model and the experiments. The model

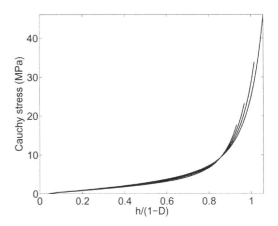

Figure 11. Superimposition of the theoretical stress-strain responses from the hyperelastic model with a strain amplification factor.

is not very successful. Moreover the theoretical stress-strain responses demonstrate a poor superposition in Figure 11. Let us note that (Qi and Boyce 2005) model is based on the same constitutive equations (5) and (6), though their model is based on the physical interpretation of the Mullins effect provided by (Mullins and Tobin 1965) and presented in the next section.

3.5 Composite approach

Following the approach of (Mullins and Tobin 1965), filled rubbers may be recognized as composites made of soft and rigid phases. Considering a two-phase heterogeneous material one may write,

$$\begin{cases} \boldsymbol{\epsilon} = (1 - \Phi)\boldsymbol{\epsilon}_s + \Phi\boldsymbol{\epsilon}_r \\ \sigma(\boldsymbol{\epsilon}) = (1 - \Phi)\sigma_s(\boldsymbol{\epsilon}_s) + \Phi\sigma_r(\boldsymbol{\epsilon}_r) \end{cases} \tag{9}$$

with s and r standing for soft and rigid respectively. The soft phase is made out of gum while the rigid

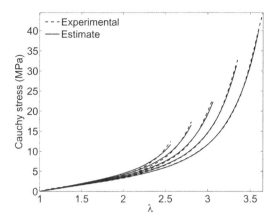

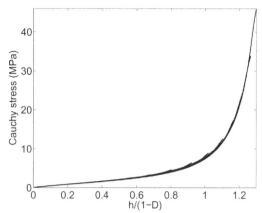

Figure 12. Comparison of the model with the experimental data from Figure 1.

Figure 13. Model stress-strain responses superimposition.

phase is mostly constituted of fillers. The parameter Φ represents the volume fraction of the rigid phase. Since part of the gum is trapped in fillers aggregates, Φ is more likely to be above the volume fraction of fillers. By definition $\epsilon_r = 0$ and we choose to make a simple Reuss assumption ($\boldsymbol{\sigma}_s = \boldsymbol{\sigma}_r$) in the composites and Eq. 9) transforms into:

$$\sigma(\epsilon) = \sigma_s(\epsilon/(1 - \Phi)) \tag{10}$$

Due to the Mullins softening the volume fraction of rigid phase decreases with the maximum stretch ever applied to the material. Let us note that (Kluppel and Schramm 2000) model drives to a similar definition of the stress-strain relation. In order to test the composite approach, we used the strain energy density Eq. (7),

$$\sigma = 2\left(\lambda_a^2 - \lambda_a^{-1}\right) \exp\left\{ \sum_{j=0}^{3} a_j(\lambda_a^2 + 2\lambda_a^{-1} - 3)^j \right\} \tag{11}$$

A satisfying comparison of the model with the experiments is presented in Figure 12. In Figure 13, the superimposition of the theoretical stress-strain responses is validated. As a partial conclusion, we were able to find two modelling approaches capable to fit well the Mullins softened stress-strain responses of the tested material and to satisfy to the stress-strain response superimposition evidenced by the experimental data in section 2. Nonetheless, none of these models have been designed to reproduce the cyclic softening. In the next section, we extend the modelling to account for both the residual stretch induced by the Mullins softening and the cyclic softening.

4 CYCLIC SOFTENING MODELLING

4.1 *Constitutive equations*

In order to account for the residual stretch induced by the Mullins softening (Figure 2) and the strain shift

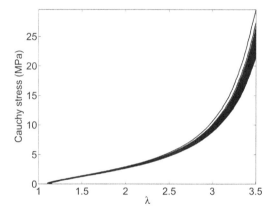

Figure 14. Estimates of the cyclic softened stress-stretch responses.

evidenced during cyclic loadings (Figure 6), one may simply introduce a decomposition of the deformation gradient $\boldsymbol{F}$ into a reversible elastic part $\boldsymbol{F}_e$ and a non reversible part $\boldsymbol{F}_p$,

$$\boldsymbol{F} = \boldsymbol{F}_e \boldsymbol{F}_p \tag{12}$$

According to the experimental observations presented in section 2, the permanent non reversible deformation gradient $\boldsymbol{F}_p$ depends on the strain history. Typically for proportional cyclic loadings, $\boldsymbol{F}_p$ is dependent of the maximum loadings and of the number of cycles. In uniaxial tension, we may compute the non reversible stretch as,

$$\lambda_p = \lambda_{res}(\lambda_{max}).\lambda_N \tag{13}$$

where $\lambda_{res}(\lambda_{max})$ is the residual stretch measured after the first load. Once the decomposition Eq. (13) made, the stress $\sigma(\lambda_e)$ is computed using one of the Mullins softening model exposed in the previous section.

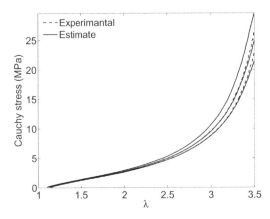

Figure 15. Comparison of the model and the experimental data at the 1000th cycle.

4.2 *Results*

Figure 14 shows the evolution of the theoretical stress-stretch responses when accounting for the cyclic softening with Eq. (13). The comparison of the model estimate with the experimental data appears in Figure 15 and support the model strategy of taking into account the cyclic softening by introducing a irreversible stretch λ_p.

5 CONCLUSIONS

In order to better understand the softening of filled rubbers during cyclic loadings, the stress-strain responses of a carbon-black filled styrene butadiene rubber were examined. On one hand, as (Mullins and Tobin 1965) revealed, the uniaxial tension stress-strain responses of a filled rubber submitted to various stretch intensities collapse into a single curve by the use of a strain amplification factor. On the other hand, we noticed that the cyclic softening stress-stretch responses collapse into a single curve by applying a stretch amplification factor. Both properties seem to be specific features of the mechanical behavior of filled rubbers. Little work may be found on the modelling of the cyclic softening of filled rubbers but many models were proposed for the Mullins softening. We challenged these models to the stress-strain responses superimposition property. Two modelling approaches were shown to work well. Then, we introduced a model framework to account for both Mullins residual stretch and cyclic softening. The model applies easily in uniaxial tension and proves to work well.

ACKNOWLEDGEMENT

This work was supported by grant MATETPRO 08-320101 from the French Agence Nationale de la Recherche. The authors acknowledge their collaboration with J. Caillard, C. Creton, J. de Crevoizier, F. Hild, M. Portigliatti, S. Roux, F. Vion-loisel and H. Zhang.

REFERENCES

Kluppel, M. and M. Schramm (2000). A generalized tube model of rubber elasticity and stress softening of filler reinforced elastomer systems. *Macromo.l Theory Simul. 9*, 74254.

Lambert-Diani, J. and C. Rey (1999). New phenomenological behavior laws for rubbers and thermoplastic elastomers. *Eur. J. Mech. A Solids 18*, 1027–1043.

Marckmann, G., E. Verron, L. Gornet, G. Chagnon, P. Charrier, and P. Fort (2002). A theory of network alteration for the mullins effect. *J. Mech. Phys. Solids 50*, 2011–2028.

Merckel, Y., J. Diani, M. Brieu, and J. Caillard (2011). Characterization of the mullins effect of carbon-black filled rubbers. *Rubber Chem. Technol. 84*(3).

Mullins, L. and N. R. Tobin (1965). Stress softening in natural rubber vulcanizates. Part 1. Use of a strain amplification factor to describe elastic behavior of filler-reinforced vulcanized rubber. *J. Appl. Polym. Sci. 9*, 2993–3009.

Ogden, R. W. and D. G. Roxburgh (1999). A pseudo-elastic model for the Mullins effect in filled rubber. *Proc. Roy. Soc. Lond.*, 2861–2877.

Qi, H. and M. Boyce (2005). Constitutive model for stretch-induced softening of the stressstretch behavior of elastomeric materials. *J. Mech. Phys. Solids 52*, 2187205.

Simo, J. C. (1987). On a fully three-dimensional finite-strain viscoelastic damage model: formulation and computational aspects. *Comput. Methods Appl. Mech. Eng. 60*, 153–173.

Modelling of the Payne effect using a 3-d generalization technique for the finite element method

M. Freund & J. Ihlemann
Professorship of Solid Mechanics, Chemnitz University of Technology, Chemnitz, Germany

M. Rabkin
Vibracoustic GmbH & Co. KG, Hamburg, Germany

ABSTRACT: In order to simulate the Payne effect of elastomeric components as well as the dependence on the static preload a uniaxial visco-elastoplastic model is proposed. The generalization to a fully three-dimensional constitutive model is accomplished by using the concept of representative directions. The generalized model shows a good correspondence with experimental data of a cyclic loaded rubber buffer concerning the storage and loss modulus. The implementation of the model into the finite element program MSC.MARC is verified by a simulation of an elastomeric engine mount.

1 INTRODUCTION

The dynamic stiffness of filled elastomers subjected to cyclic loadings with small strain amplitudes decreases with increasing amount of amplitude. This phenomenon is called Payne effect (Payne 1962) or Fletcher-Gent effect (Fletcher & Gent 1954). It is relevant for optimizing the damping behaviour of elastomeric components in vehicles with respect to noise and vibrations. A significant feature of the Payne effect is also the dependence of the dynamic stiffness on the static preload (Ouyang 2006, Rendek & Lion 2010). To cover these effects a uniaxial visco-elastoplastic model is proposed which represents an advancement of previous approaches (Rabkin & Brüger 2001, Rabkin et al. 2003, Rabkin 2007). The generalization to a fully three-dimensional constitutive model is accomplished by using the concept of representative directions. This enables finite element simulations of inhomogeneous stress conditions even though the original input model describes the material behaviour for uniaxial tension and compression only.

2 THE UNIAXIAL VISCO-ELASTOPLASTIC MATERIAL MODEL

The uniaxial model for describing the Payne effect is based on a micro-elastoplastic approach of Palmov (1998) which is extended by an additional viscoelastic and hyperelastic part according to figure 1.

To describe the hyperelastic behaviour of the parallel spring the non-affine tube model with non-Gaussian extension is used (Heinrich et al. 1988). The

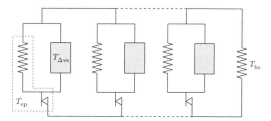

Figure 1. Rheological representation of the uniaxial visco-elastoplastic material model. T_{he}: Hyperelastic part, T_{ep}: Elastoplastic part, $T_{\Delta ve}$: Viscoelastic part.

corresponding uniaxial 1st Piola-Kirchhoff stress T_R in case of ideal incompressibility is given by

$$T_R = G_c \left(\lambda - \frac{1}{\lambda^2} \right) \left[\frac{1 - \frac{1}{n}}{\left(1 - \frac{1}{n} \left(\lambda^2 + \frac{2}{\lambda} - 3 \right) \right)^2} \right.$$

$$\left. - \frac{\frac{1}{n}}{1 - \frac{1}{n} \left(\lambda^2 + \frac{2}{\lambda} - 3 \right)} \right] + 2 G_e \left(\frac{1}{\sqrt{\lambda}} - \frac{1}{\lambda^2} \right). \tag{1}$$

To increase the flexibility of the tube model in representative directions it is extended by an energy-elastic part with two additional material parameters κ and ρ (comp. equation 4).

$$T_{he} = (1 - \kappa) T_R + \kappa T'_{he,0} s_k \tag{2}$$

In this context the quantity $T'_{he,0}$ denotes the derivative of T_{he} with respect to the stretch λ (uniaxial material tangent) referring to the undeformed configuration.

$$T'_{he,0} = 3\left[G_e + G_c\left(1 - \frac{2}{n}\right)\right] \qquad (3)$$

Furthermore the function s_k is given by

$$s_k = \frac{\rho}{|\lambda - 1|}\left[1 - e^{-\frac{|\lambda-1|}{\rho}}\right](\lambda - 1). \qquad (4)$$

The elastoplastic part of the rheological model represents an infinite number of Prandtl elements whose dimensionless yield strains h are distributed according to the following probability density function $p(h)$ (Rabkin et al. 2003):

$$p(h) = \frac{\frac{2\mu}{\nu}\left(\frac{|h|}{\nu}\right)^{2\mu-1}\left[1 + 2\mu + (1 - 2\mu)\left(\frac{|h|}{\nu}\right)^{2\mu}\right]}{\left[1 + \left(\frac{|h|}{\nu}\right)^{2\mu}\right]^3}. \qquad (5)$$

This yields the secant stiffness Λ_{ep} which describes the amplitude dependence or rather the Payne effect (with ν and μ as material parameters).

$$\Lambda_{ep}(x) = 1 - \int_0^{|x|}\left(1 - \frac{h}{|x|}\,p(h)\,dh\right) = \frac{1}{1 + \left(\frac{|x|}{\nu}\right)^{2\mu}} \qquad (6)$$

The elastoplastic stress contribution T_{ep} describes a hysteresis loop consisting of three segments. The corresponding stress responses T_{ep}^0 (first loading), T_{ep}^- (unloading) and T_{ep}^+ (loading) are modelled separately. For this, the strain ε_u and the corresponding stress $T_{ep}(\varepsilon_u)$ at the reverse point of one curve segment are to be stored as internal variables (with the material parameter E_Δ as Youngs modulus of the linear spring within the Prandtl element and $\varepsilon = \lambda - 1$ as engineering strain).

$$T_{ep}^0 = E_\Delta \Lambda_{ep}(\varepsilon)\,\varepsilon \qquad (7)$$

$$T_{ep}^- = T_{ep}^0(\varepsilon_u) + E_\Delta \Lambda_{ep}\left(\frac{\varepsilon - \varepsilon_u}{2}\right)(\varepsilon - \varepsilon_u) \qquad (8)$$

$$T_{ep}^+ = T_{ep}^-(\varepsilon_u) + E_\Delta \Lambda_{ep}\left(\frac{\varepsilon - \varepsilon_u}{2}\right)(\varepsilon - \varepsilon_u) \qquad (9)$$

To account for the dependence on the static preload the correction factor β_{he} as the ratio of the current material tangent T'_{he} to the initial tangent $T'_{he,0}$ of the tube model is introduced.

$$\beta_{he} = \frac{T'_{he}}{T'_{he,0}} = \frac{(1 - \kappa)T'_R + \kappa\,T'_{he,0}\,s'_k}{T'_{he,0}} \qquad (10)$$

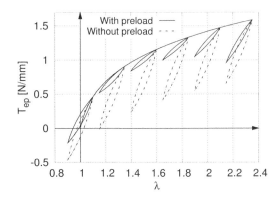

Figure 2. Elastoplastic stress T_{ep} for uniaxial cyclic loading with and without considering the dependence on the static preload.

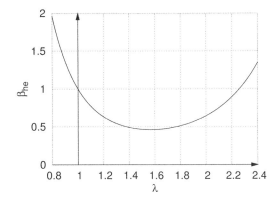

Figure 3. Correction factor β_{he} as a function of the stretch λ.

Thus, by multiplying β_{he} with the parameter E_Δ the slope of the hysteresis loop can be adjusted according to the respective preload strain. This yields a modified formula for computing the elastoplastic stresses of the loading and unloading curve (with $\varepsilon_{u,0}$ as the strain at the reverse point of the first loading curve).

$$T_{ep}^{-(+)} = T_{ep}^{0(-)}(\varepsilon_u) \qquad (11)$$
$$+ \beta_{he}(\varepsilon_{u,0})\,E_\Delta \Lambda_{ep}\left(\frac{\varepsilon - \varepsilon_u}{2(\varepsilon_{u,0} + 1)}\right)\frac{\varepsilon - \varepsilon_u}{\varepsilon_{u,0} + 1}$$

The influence of the correction factor can be demonstrated by simulating a cyclic uniaxial tension test evaluating the elastoplastic stress response T_{ep} only. For this purpose, the hysteresis loops of the original model (equation 7 to 9) as well as the loops of the modified version (equation 11) are depicted in figure 2. Furthermore figure 3 shows the characteristic of β_{he} as a function of the stretch λ.

First of all, the factor β_{he} decreases for small deformations, which causes the hysteresis loops to flatten. After having reached a maximum the correction factor increases so that the loops also exhibit a higher slope. In contrast to this the hysteresis loops of the original model are always identical and do not depend on the static preload.

To account for the time dependent material properties the elastoplastic part of the model is extended by an additional viscoelastic part. The viscoelastic stress contribution $T_{\Delta ve}$ depends on the history of the elastoplastic stress T_{ep} so that both stress responses are directly coupled.

$$T_{\Delta ve} = (q_{ve} - 1) T_{ep} - \frac{(q_{ve} - 1)}{\tau_e} \int\limits_0^t T_{ep}(\tau) e^{-\frac{(t-\tau)}{\tau_e}} d\tau \tag{12}$$

In this particular case, the exponential relaxation function is used to describe the viscoelastic behaviour of a Maxwell model, but in general it can be replaced with other relaxation functions as well. In this context, the viscoelastic parameter q_{ve} links the frequency f of the strain signal with the relaxation time τ_e and the loss factor $\eta = \tan \delta$ whereas τ_e and η denote further material parameters.

$$q_{ve} = \frac{\eta + 2\pi f \tau_e}{2\pi f \tau_e (1 - 2\pi f \tau_e \eta)} \tag{13}$$

The summation of the elastoplastic and the viscoelastic part yields the visco-elastoplastic stress contribution T_{vep}. Finally, the total stress T of the uniaxial model is obtained by incorporating the hyperelastic stress response T_{he}.

$$T = T_{he} + T_{vep} = T_{he} + T_{ep} + T_{\Delta ve} \tag{14}$$

3 THE CONCEPT OF REPRESENTATIVE DIRECTIONS

The concept of representative directions is based on an approach of Pawelski (1998) and intends to generalize one-dimensional material models to fully three-dimensional constitutive models for the finite element method (Freund & Ihlemann 2010). Unlike other approaches, for example the micro-sphere model (Miehe et al. 2004), it does not refer to the physical structure of a polymer but is rather meant as a mere continuum mechanical generalization technique. Furthermore the concept is based on the equivalence of stress power and does not require a free energy of the uniaxial model but only a stress-strain relation.

Starting with the isochoric right Cauchy-Green tensor $\overset{G}{\underline{\underline{C}}}$ as the state of strain within a material point the stretches $\overset{\alpha}{\lambda}$ along discrete directions in space $\overset{\alpha}{\underline{e}}$ (so-called representative directions) are determined by

$$\overset{\alpha}{\lambda} = \sqrt{\overset{\alpha}{\underline{e}} \cdot \overset{G}{\underline{\underline{C}}} \cdot \overset{\alpha}{\underline{e}}} \quad \text{with:} \quad \overset{G}{\underline{\underline{C}}} = J_3^{-\frac{2}{3}} \underline{\underline{C}} \quad . \tag{15}$$

For each direction the respective stretch is used to compute a 2nd Piola-Kirchhoff stress response $\overset{\alpha}{\widetilde{T}}$ by

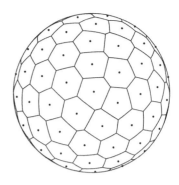

Figure 4. Distribution of 2×50 points on the surface of the unit sphere to use as representative directions with Voronoi cells as corresponding weighting factors.

the one-dimensional material model which in this case is represented by the proposed visco-elastoplastic model.

$$\overset{\alpha}{T} = \overset{\alpha}{T} \left(\overset{\alpha}{\lambda}, \overset{\dot{\alpha}}{\lambda}, \dots \right) \quad \longrightarrow \quad \overset{\alpha}{\widetilde{T}} = \frac{\overset{\alpha}{T}}{\overset{\alpha}{\lambda}} \tag{16}$$

It is postulated that the sum of stress powers of all uniaxial loading processes equals the stress power of the overall deformation process concerning the whole material point.

$$dP = \frac{1}{2} \overset{\sim}{\underline{\underline{T}}}^* \cdot\cdot \overset{\overset{\triangle}{G}}{\underline{\underline{C}}} dV \overset{!}{=} \sum_{\alpha=1}^n \overset{\alpha}{\widetilde{T}} \overset{\alpha}{\lambda} \overset{\dot{\alpha}}{\lambda} d\widetilde{V} \tag{17}$$

This yields the auxiliary stress tensor $\overset{\sim}{\underline{\underline{T}}}^*$ as the response to the deformation $\overset{G}{\underline{\underline{C}}}$.

$$\overset{\sim}{\underline{\underline{T}}}^* = \sum_{\alpha=1}^n \overset{\alpha}{w} \overset{\sim}{\widetilde{T}} \overset{\alpha}{\underline{e}} \circ \overset{\alpha}{\underline{e}} \quad \text{with:} \quad \overset{\alpha}{w} = \frac{\overset{\alpha}{\Delta A}}{4\pi} > 0 \tag{18}$$

This equation describes a numerical integration on the surface of the unit sphere whereas $\overset{\alpha}{w}$ denotes the weighting factors. For an efficient integration the representative directions should be distributed as uniformly as possible. This is accomplished by simulating the repulsion of electric charges on the unit sphere while considering the symmetry relative to the center of the sphere. During the process of iteration the minimization of the total force of the whole system finally leads to an equilibrium where the position vectors of the charges can be considered as evenly distributed (s. figure 4). The associated weighting factors are defined by the surface areas of so-called Voronoi cells which are constructed around the points using a numerical algorithm by Sugihara (2002).

To obtain the total stress tensor $\overset{\sim}{\underline{\underline{T}}}$ as the response to the complete deformation $\underline{\underline{C}}$ the nearly incompressible behaviour concerning hydrostatic loads has to be taken

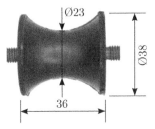

Figure 5. Rubber buffer for displacement controlled uniaxial harmonic tension tests.

Table 1. Identified material parameters of the visco-elastoplastic model in representative directions.

G_c [MPa]	1.55
G_e [MPa]	0.0099
n [–]	50
κ [–]	0.631
ρ [–]	0.085
ν [–]	0.022
μ [–]	0.343
E_Δ [MPa]	3.65
τ_e [s]	0.059
η [–]	0.218

into account, which yields the following relation (with p as the hydrostatic pressure):

$$\underline{\widetilde{T}} = \mathrm{dev}\left(\underline{\widetilde{T}}^* \cdot \overset{G}{\underline{C}}\right) \cdot \underline{C}^{-1} - p\, J_3\, \underline{C}^{-1} . \tag{19}$$

4 MEASUREMENTS AND PARAMETER IDENTIFICATION

The experimental validation of the generalized visco-elastoplastic model was accomplished by means of a rubber sample in form of a buffer (s. figure 5). The sample was subjected to a displacement controlled uniaxial harmonic excitation with a frequency of 3 Hz at five different static preload levels: 0 mm (no-preloaded state), 2 mm, 4 mm, 8 mm and 11 mm. The reaction force was measured for different displacement amplitudes in the range from 0.15 mm to 5.6 mm for each preloaded state.

In order to compare the experimental data with simulation results the given displacement u and the measured force F are converted into engineering strain ε and 1st Piola-Kirchhoff stress T. This is done by means of two conversion factors which allow to map the inhomogeneous stress-strain condition within the buffer to the state of pure uniaxial tension.

$$\varepsilon = \frac{u}{27.6\,\mathrm{mm}} \quad ; \quad T = \frac{F}{\pi\,(12.5\,\mathrm{mm})^2} \tag{20}$$

Unlike the given strain signal the stress response is no longer harmonic due to the nonlinear inelastic material behaviour. Thus, a discrete Fourier transformation is applied in order to define an amplitude $\hat{T}$ for the stress signal. The ratio of the stress amplitude to the strain amplitude is defined as the dynamic modulus E_d which can be interpreted as the absolute value of the complex modulus E_d^*.

$$E_d = \frac{\hat{T}}{\hat{\varepsilon}} = |E_d^*| \tag{21}$$

The complex modulus consists of the storage modulus E' as the real part and the loss modulus E'' as the imaginary part. The storage modulus describes the elastic material characteristics under dynamic loading while

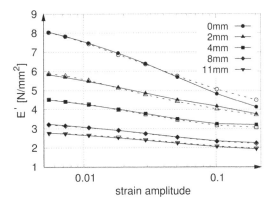

Figure 6. Comparison between the experimental data and the simulation concerning the storage modulus.

the loss modulus represents the damping behaviour. These two quantities can also be formulated by means of the loss angle δ as the angular phase shift between the given strain signal and the resulting stress signal.

$$E_d^* = E' + iE'' \quad \text{with:} \quad \begin{cases} E' = E_d \cos\delta \\ E'' = E_d \sin\delta \end{cases} \tag{22}$$

The storage and loss modulus was evaluated for each measured hysteresis loop concerning the different amplitude values and static preload levels. Then, the experiment was reproduced by a simulation of a homogeneous uniaxial tension test with the visco-elastoplastic model in representative directions. In order to identify the free parameters of the material model the difference between the measured and the simulated storage and loss moduli was minimized by using a nonlinear simplex method. The resulting material parameters are listed in table 1.

Figure 6 and 7 illustrate the quality of the parameter identification by comparing the experimental data (dashed lines) with the simulation results (solid lines) with respect to the storage and loss modulus. For this purpose, the storage and loss modulus is plotted against the logarithmic strain amplitude. The static preload strains can be calculated from the given displacement values by means of equation 20.

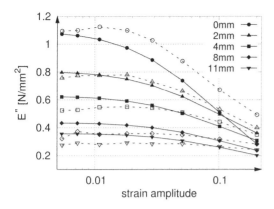

Figure 7. Comparison between the experimental data and the simulation concerning the loss modulus.

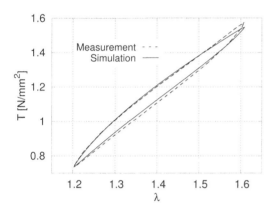

Figure 8. Corresponding hysteresis loops for a displacement amplitude of 5.6 mm and a static preload of 11 mm.

Concerning the storage modulus the simulation shows a remarkable correspondence with the experimental data. On the other hand the loss modulus cannot be reproduced in such a quality, but nevertheless the characteristics of the measured curves are still preserved. Altogether this example shows that the material model is able to describe the Payne effect as well as the dependence on the static preload in a realistic manner. In this context it should be mentioned that the direction of the vertical shift strongly depends on the amount of static preload. In this case the rubber buffer is subjected to comparatively small preload strains so that the dynamic modulus decreases with increasing preload. Other experiments show the opposite behaviour due to much larger preload levels (Ouyang 2006, Rendek & Lion 2010). Within the proposed material model this characteristic is taken into account by the correction factor β_{he} as already stated above.

With regard to a complete validation of the model it was also secured that the corresponding hysteresis loops still show a reasonable material behaviour and resemble the measured stress-strain curves qualitatively. As an example figure 8 shows the hysteresis

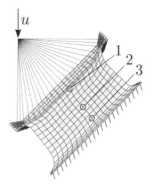

Figure 9. Finite element model of an elastomeric engine mount subjected to cyclic loadings. The numbered nodes refer to the results in figure 11.

loops for a displacement amplitude of 5.6 mm and a static preload of 11 mm.

5 FINITE ELEMENT SIMULATION

The visco-elastoplastic model in representative directions has been implemented into the finite element programm MSC.MARC via the user interface HYPELA2. The FE-implementation is validated by simulating an elastomeric engine mount under cyclic excitation (figure 9). For this purpose, the displacement u is applied to a reference node whose degrees of freedom are coupled with those of the upper surface of the engine mount. The three static preload levels are chosen to 0 mm, 1.4 mm and 5.6 mm whereas the displacement amplitudes are lying in the range from 0.0385 mm to 1.0 mm. The simulation was performed using the material parameters listed in table 1.

In the following the resulting hysteresis loops of the force-displacement curves are used to compute the dynamic stiffness K_d as the ratio of the force amplitude to the displacement amplitude.

$$K_d = \frac{\hat{F}}{\hat{u}} \tag{23}$$

The dynamic stiffness is evaluated for each displacement amplitude and static preload level as depicted in figure 10.

The simulation confirms the ability of the three-dimensional material model to reproduce the Payne effect as well as the dependence on the static preload. In order to illustrate the inhomogeneity of the Payne effect three different nodes of the finite element model (s. figure 9) were evaluated with respect to the corresponding stress-strain behaviour. In this case, the equivalent Cauchy stress σ_{eq} and the equivalent elastic strain ε_{eq} as scalar postprocessing results of MSC.MARC are chosen for the illustration. Figure 11 shows the three hysteresis loops for a displacement amplitude of 1 mm and a static preload of 5.6 mm.

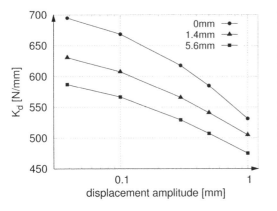

Figure 10. Simulation results concerning the dynamic stiffness of the engine mount.

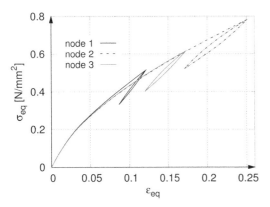

Figure 11. Stress-strain curves of three different nodes of the finite element model for visualizing the inhomogeneity of the Payne effect.

Due to the inhomogeneous loading conditions within the engine mount the three nodes are subjected to different preload strains and strain amplitudes. Thus, the decreasing slope of the hysteresis loops result from the superposition of both effects. This example clearly shows the dependence of the Payne effect on the location within the component, which is of great importance with regard to the material behaviour of complex structures under dynamic loading.

6 CONCLUSIONS

In order to describe the Payne effect of elastomeric components a visco-elastoplastic model for uniaxial tension has been proposed. The model also takes the dependence of the dynamic stiffness on the static preload into account. The generalization of this one-dimensional material model to a fully three-dimensional constitutive model is accomplished by using the concept of representative directions. The generalized model has been compared to experimental data resulting from a cyclic tension test with a rubber buffer. The ability of the model to reproduce the storage and loss modulus for different loading amplitudes and static preload levels is quite satisfying. The model has also been implemented into the finite element program MSC.MARC which enables simulations of complex structures even though the original model predicts the material behaviour for uniaxial tension and compression only. The FE-implementation is tested by simulating the cyclic loading of an elastomeric engine mount. The simulation clearly shows the Payne effect as well as the dependence on the static preload, which confirms the reliability of the proposed model.

REFERENCES

Fletcher, W.P., Gent, A.N. (1954). Nonlinearity in the Dynamic Properties of Vulcanized Rubber Compounds. *Rubber Chem. Tech. 27*, 209–222.

Freund, M., Ihlemann, J. (2010). Generalization of onedimensional material models for the finite element method. *Z. Angew. Math. Mech. 90*, 399–417.

Heinrich, G., Straube, E., Helmis, G. (1988). Rubber elasticity of polymer networks: Theories. *Adv. Polym. Sci. 85*, 33–87.

Miehe, C., Göktepe, S., Lulei, F. (2004). A micro-macro approach to rubber-like materials. Part I: the non-affine microsphere model of rubber elasticity. *J. Mech. Phys. Solids 52*, 2617–2660.

Ouyang, G.B. (2006). Modulus, Hysteresis and the Payne Effect – Network Junction Model for Carbon Black Reinforcement. *Kautschuk Gummi Kunststoffe 6*, 332–343.

Palmov, V. (1998). Vibrations of Elasto-Plastic Bodies. Berlin, Heidelberg, New York: Springer.

Pawelski, H. (1998). Eigenschaften von Elastomerwerkstoffen mit Methoden der statistischen Physik. Aachen: Shaker.

Payne, A.R. (1962). The dynamic properties of carbon blackloaded natural rubber vulcanizates. Part I. *J. Appl. Polymer Sci. 6*, 57–63.

Rabkin, M. (2007). Simulation of the Fletcher-Gent effect by using the subroutine UPHI MSC/MARC. *Constitutive Models for Rubber V*, 263–268.

Rabkin, M., Brüger, T., Hinsch, P. (2003). Material model and experimental testing of rubber components under cyclic deformation. *Constitutive Models for Rubber III*, 319–324.

Rabkin, M., Brüger, T. (2001). A Constitutive Model of Elastomers in the case of Cyclic Load with amplitude-dependent internal damping. *Constitutive Models for Rubber II*, 73–78.

Rendek, M., Lion, A. (2010). Amplitude dependence of filler-reinforced rubber: Experiments, constitutive modelling and FEM-Implementation. *Int. J. Solids Struct. 47*, 2918–2936.

Sugihara, K. (2002). Laguerre Voronoi diagram on the sphere. *Journal for Geometry and Graphics 1*, 69–81.

Constitutive Models for Rubber VII – Jerrams & Murphy (eds)
© 2012 Taylor & Francis Group, London, ISBN 978-0-415-68389-0

Electromechanical hysteresis in filled elastomers

D.S.A. De Focatiis, D. Hull & A. Sánchez-Valencia
Division of Materials, Mechanics and Structures, University of Nottingham, UK

ABSTRACT: This study presents simultaneous experimental measurements of stress, strain and electrical resistivity on three types of conductive elastomers: a carbon-black filled EPDM rubber, a carbon-black filled silicone elastomer, and two grades of thermoplastic polyurethane elastomers filled with small weight fractions of multi-walled carbon nanotubes. The materials are exposed to complex cyclic strain histories. It is found that the resistivity-strain relationships are considerably hysteretic, in a similar fashion to the Mullins effect on the stress-strain relationships. During cyclic loading to a strain of 1, the EPDM rubber exhibits a lower resistivity during the loading part of the cycles than during the unloading part. The opposite effect is seen in the polyurethane elastomers. It is possible that this effect results from bending and buckling of the high aspect ratio nanotubes during unloading. Unexpectedly, in the nanotube-filled elastomers, a minimum in resistivity was observed within both loading and unloading cycles, for a wide variety of strain ranges.

1 INTRODUCTION

The hysteretic phenomenon whereby the mechanical response of an elastomer depends upon the prior stretching history, first identified by Mullins (1947), is widely studied for many types of elastomers, and is generally well known in the rubber community. The similar hysteretic phenomenon whereby the Direct Current (DC) electrical response varies with both strain and strain history was first reported by Bulgin (1945), and also alluded to by Mullins in his 1947 paper. Bulgin was the first to systematically explore the effects of strain, time and temperature on the electrical resistivity of conductive rubbers filled with carbon black. Although much research on carbon black filled rubbers followed (see for example Norman 1970), this phenomenon has yet to be studied in detail on elastomers filled with Carbon Nanotubes (CNTs). CNTs are highly electrically conductive rod-like fillers with large aspect ratios, and their appeal lies in rendering insulating polymers conductive when dispersed at relatively low loadings.

The aim of this paper is to exploit the benefits of modern testing machines and datalogging instruments to revisit electromechanical hysteresis. Electrical resistivity, stress and strain were simultaneously measured during cyclic saw-tooth deformations on two carbon black filled cross-linked rubbers and two CNT-filled thermoplastic elastomers. Possible applications of these conductive elastomers are in the areas of sensors and flexible electronics.

2 EXPERIMENTAL METHOD

2.1 *Materials*

The materials studied in this work consist of two permanently cross-linked elastomers and two thermoplastic elastomers. The thermosetting elastomers are: an accelerated sulphur cross-linked carbon black filled (50 phr) oil extended ethylene-propylene-diene (EPDM) rubber; and a commercial conductive grade of Wacker Chemie Elastosil LR 3162 silicone rubber, also filled with carbon black. The thermoplastic elastomers are Noveon Inc. Estane 58311 ether-based polyurethane elastomers, melt-compounded with 4% and 5% wt/wt% Nanocyl NC7000 Multi-Walled Carbon Nanotubes (MWCNTs) using a twin-screw extruder (Lew et al. 2009).

Sheets of EPDM rubber approximately 0.5 mm in thickness were compression-moulded in a heated press for 13 minutes at 160°C. The two-part Elastosil silicone elastomer was mixed, cast into rectangular moulds, and cured at room temperature overnight following the manufacturer's instructions. The pre-compounded extruded Estane-CNT granules were dried and compression-moulded for 10 minutes at 170°C, followed by cooling to room temperature at a rate of $\sim$0.16° Cs^{-1}. Rectangular test specimens approximately 100 mm in length and 10 mm in width were carefully cut from all the elastomeric sheets using a template and a sharp blade cutter.

2.2 Electromechanical tests

Uniaxial tensile testing was performed using an Instron 5569 tensile testing machine at room temperature, at constant crosshead displacement rate. The reported strain ε was calculated from the grip separation distance l as

$$\varepsilon = l / l_0 - 1 \qquad (1)$$

where l_0 is the initial grip separation distance between the tensile grips. The stretch ratio λ is given by $\lambda = l / l_0$. The true stress σ was computed from the measured force F and the initial cross-sectional area of the specimen A_0 assuming isochoric deformation, as

$$\sigma = F \lambda / A_0 \qquad (2)$$

The electrical response to an applied DC voltage was measured using a Keithley 6517B electrometer with in-built voltage source. A voltage V was applied directly across a central portion l_e of the tensile specimens through spherical metal contacts approximately 3 mm in diameter pressed on to both sides of the specimen using spring clips, as shown in Figure 1. The current I through the circuit was measured using the very sensitive electrometer. The instantaneous resistivity ρ was computed from the measured resistance and the sample dimensions and stretch ratio, assuming isochoric deformation, as

$$\rho = \frac{V A_0}{I \lambda^2 l_e} \qquad (3)$$

Force, cross-head displacement and current signals were simultaneously acquired during deformation using LabView software, at a rate of 1 Hz. The resulting data are expressed below in terms of the nominal tensile strain ε, the true tensile stress σ, and the true resistivity (calculated based on the deformed dimensions) ρ.

The experimental programme was divided into three phases, following three experimental protocols. The first phase recorded the response to continuous deformation at constant nominal strain rate through to failure. The second phase consisted of pseudo-cyclic load-unload deformations at constant nominal strain rate, loading to successively larger strains in each cycle. The third phase consisted of the application of four load-unload cycles of deformation to a fixed strain level (typically 100% strain), followed by further load-unload cycles between different strain levels *within* this envelope, all at constant nominal strain rate, referred to as cyclic tests. Typical imposed strains during the three phases are illustrated in Figure 2 as a function of time.

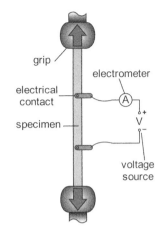

Figure 1. Schematic representation of the electrical and mechanical equipment used in the electromechanical tests.

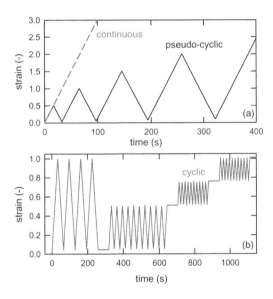

Figure 2. Deformation protocols used in this study. (a) Continuous and pseudo-cyclic strain as a function of time; (b) cyclic strain as a function of time.

3 RESULTS

3.1 Continuous and pseudo-cyclic experiments

Figures 3–4 illustrate the continuous and pseudo-cyclic stress-strain response of a selection of materials studied. All the materials exhibit the Mullins stress-softening effect, and the pseudo-cyclic curves approximately rejoin the continuous loading curves when the applied strain exceeds the previously reached maximum strain. The mechanical response of the Estane-4% CNTs (not shown) was very similar to the Estane-5% CNT material. An observable difference between the EPDM and Elastosil elastomers and the Estane CNT-filled elastomers is the degree of permanent deformation – whereas the permanently

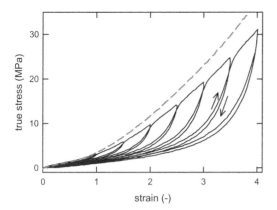

Figure 3. Continuous (dashed line) and pseudo-cyclic (solid line) stress-strain response of EPDM rubber filled with carbon black, deformed at a strain rate of $0.03\,\mathrm{s}^{-1}$.

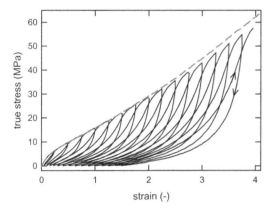

Figure 4. Continuous (dashed line) and pseudo-cyclic (solid line) stress-strain response of Estane filled with 5% MWCNTs, deformed at a strain rate of $0.03\,\mathrm{s}^{-1}$.

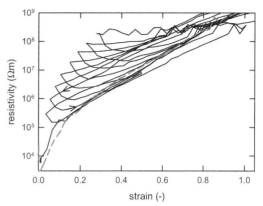

Figure 5. Continuous (dashed line) and pseudo-cyclic (solid line) resistivity-strain response of EPDM rubber filled with carbon black, deformed at a strain rate of $0.03\,\mathrm{s}^{-1}$ and measured at an applied voltage of 100 V.

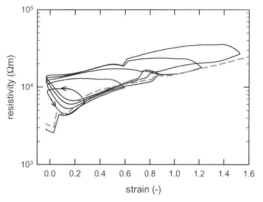

Figure 6. Continuous (dashed line) and pseudo-cyclic (solid line) resistivity-strain response of Elastosil silicone rubber filled with carbon black, deformed at a strain rate of $0.02\,\mathrm{s}^{-1}$ and measured at an applied voltage of 10 V.

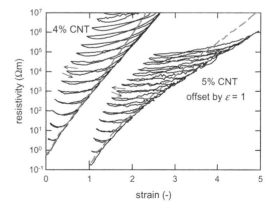

Figure 7. Continuous (dashed line) and pseudo-cyclic (solid line) resistivity-strain response of Estane filled with 4% and 5% (offset) MWCNTs, deformed at a strain rate of $0.03\,\mathrm{s}^{-1}$ and measured at an applied voltage of 30 V.

cross-linked thermosetting elastomers exhibit almost complete recovery upon unloading, there is as as much as 150% permanent strain after unloading the thermoplastics. The same grade of Estane without CNTs (not shown) exhibited only slightly smaller degrees of permanent deformation.

Figures 5–7 illustrate the continuous and pseudo-cyclic resistivity-strain response of the same materials. The electromechanical response of the EPDM rubber is hysteretic in that all of the unloading segments exhibit a larger resistivity than both the first loading and the subsequent segments. The precise values depend on the maximum strain the material has previously been exposed to. Differently to the stress-strain response, the reloading segments are relatively unaffected by the previously reached maximum strain, i.e. the resistivity is almost the same as in the continuous loading curve apart from an initial approach towards it. Very similar behaviour was found with the Elastosil silicone elastomer. Here the subsequent reloading segments rejoin the continuous loading curve very

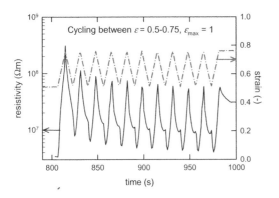

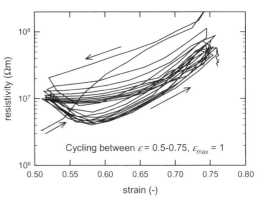

Figure 8. Resistivity measured during cyclic loading between strains of 0.5–0.75 on EPDM rubber at 100 V, after an initial maximum strain of 1, solid line; strain history, dot-dash line.

Figure 9. Cyclic resistivity-strain response measured during cyclic loading between strains of 0.5–0.75 on EPDM rubber at 100 V, after an initial maximum strain of 1.

quickly, at ~20% strain for all cycles. Both of these elastomers are filled with particulate carbon black.

The electromechanical response of the CNT-filled Estane thermoplastic elastomers is visibly different. At low strains, the resistivity during the unloading and reloading segments depends primarily on the maximum strain the material has previously been exposed to. When the strain exceeds the previously reached maximum, the resistivity follows the continuous loading curve, through to large strains. The increased filler content of the 5% CNT grade results in a reduction of resistivity across all strain levels, and the reduction is greatest at the largest strains.

3.2 Cyclic experiments

In this section we examine in greater detail the electromechanical response during repeated cyclic loading. Materials are first exposed to preconditioning consisting of four cycles of loading and unloading to $\varepsilon = 1$. Subsequently, they are exposed to several strain cycles, all within this envelope of maximum strain, as shown in Figure 2(b). Figure 8 illustrates the strain history and resistivity (measured at 100 V) as a function of time during cyclic loading between strains of 0.5–0.75 on EPDM rubber. The resistivity is considerably time-dependent, taking several cycles before settling to a repeatable loop, as can be seen in Figure 9. Once settled, the resistivity is lower in the loading part of the cycle than in the unloading part.

Figures 10–11 illustrate the same strain and resistivity history and resistivity-strain relationship during cyclic loading, now between no load and a strain of 1. Here the time dependence of the electrical response is much smaller (in relative terms), and the loops are more repeatable. The resistivity is again lower on the loading part of the cycle, dipping just after the resumption of loading. Similar findings were observed from cyclic experiments on the Elastosil material (not shown).

Figures 12–13 illustrate the response of Estane filled with 5% MWCNTs to the same smaller strain

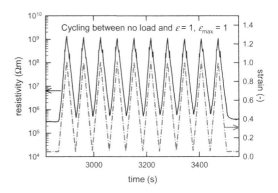

Figure 10. Resistivity measured during cyclic loading between no load and a strain of 1 on EPDM rubber at 100 V, after an initial maximum strain of 1, solid line; strain history, dot-dash line.

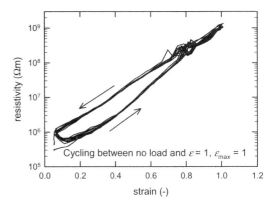

Figure 11. Cyclic resistivity-strain response measured during cyclic loading between no load and a strain of 1 on EPDM rubber at 100 V, after an initial maximum strain of 1, solid line; strain history, dot-dash line.

history as in Figures 8–9. Here the time dependence of the response is again substantial, but once the response settles (after a few cycles), the loading and unloading parts of the response are hard to distinguish. A very

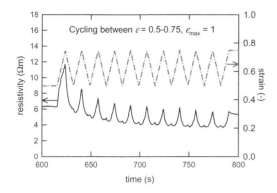

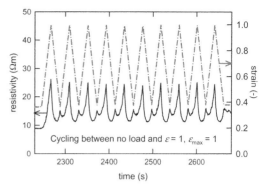

Figure 12. Resistivity measured during cyclic loading between strains of 0.5–0.75 on Estane filled with 5% CNTs at 100 V, after an initial maximum strain of 1, solid line; strain history, dot-dash line.

Figure 14. Resistivity measured during cyclic loading between no load and a strain of 1 on Estane filled with 5% CNTs at 100 V, after an initial maximum strain of 1, solid line; strain history, dot-dash line.

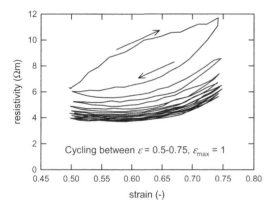

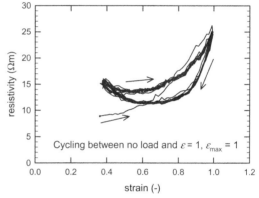

Figure 13. Cyclic resistivity-strain response measured during cyclic loading between strains of 0.5–0.75 on Estane filled with 5% CNTs at 100 V, after an initial maximum strain of 1.

Figure 15. Cyclic resistivity-strain response measured during cyclic loading between no load and a strain of 1 on Estane filled with 5% CNTs at 100 V, after an initial maximum strain of 1, solid line; strain history, dot-dash line.

similar response was observed in the Estane filled with 4% MWCNTs.

Figures 14–15 illustrate the response of Estane filled with 5% MWCNTs to the same larger strain history as in Figures 10–11. Here again the time dependence of the electrical response is much smaller, and the loops are highly repeatable after the first loading. However, the resistivity is now *higher* during the loading part of the cycle than during the unloading part. Both parts of the cycle exhibit a minimum in resistivity at an intermediate strain level within the cycle.

4 DISCUSSION

Although the pseudo-cyclic experiments shown in Figures 6–8 provide a fingerprint of the electrical response of the materials across a wide range of strain, they are not sufficient for an accurate description of the response following an arbitrary strain history. There are several differences in the precise shape of these responses, probably related to a range of microstructural features. The most striking difference is in the

path taken by the resistivity in the *reloading* part of the pseudo-cyclyc loading. In both the EPDM and the Elastosil, this path approximately follows the original continuous loading curve; in the CNT-filled Estanes, it closely follows the previous unloading curve, and only rejoins the continuous loading curve when the strain exceeds the previously reached maximum. All materials exhibited a small reduction in resistivity with strain at small strains upon reloading, across virtually the full range of pre-strains explored. At present, the precise reason for this is unknown.

The cyclic experiments reported in Figures 8–15 were initially intended to explore the applicability of these materials as strain sensors, by observing whether the electrical response to an applied strain was (1) repeatable, and (2) monotonic. The electrical response settles after a few cycles, in a manner similar to the mechanical response. From the data shown it appears that the response to a *smaller* strain amplitude requires a *greater* number of cycles to settle than that of a larger strain amplitude. In none of the cases explored was the response monotonic during the loading part of the cycle, although it was monotonic during the unloading part only in the EPDM rubber.

A striking difference between the materials is most clearly visible in the large amplitude cycles (Figures 11 and 15). In the EPDM (and Elastosil, not shown) the resistivity is lower during the loading part of the cycle than the unloading part. This can (less clearly) be seen also in the pseudo-cyclic experiments. However, the reverse is true in the 5% CNT-filled Estane. Here the resistivity is *higher* during the loading part of the cycle than during the unloading part. This same behavior was also observed in the Estane-4% CNT, and in strain cycles between strains of 0.5 and 1. One possible explanation for the unexpected form of this hysteresis lies in the geometry of the filler particles. The carbon-black fillers are globular, whereas the carbon nanotubes have aspect ratios of around 100. Whereas the nanotubes will behave as rigid rods when subjected to tension, they are prone to bending and buckling when compressed. Although the macroscopic strains imposed on the materials are not compressive, it is possible that, locally, some areas of the nanocomposite experience small degrees of compressive stress. These bent, or buckled nanotubes may be constrained by the presence of other nanotubes, and hence encouraged to form a conductive network during the *unloading* part of the cycle. But they straighten again during loading due to the tensile stresses imposed by the surrounding elastomer, and lose these contacts.

It was also observed that in the cycles performed on the Estane-5% CNT material the resistivity-strain curve has a minimum. Several other strain ranges were explored: from no load to 0.5, no load to 0.75, 0.5 to 1 and 0.75 to 1. The same observations were made on the Estane-4% CNT material on the same set of strain ranges: in all of these cycles, the resistivity-strain curve exhibits a minimum somewhere within the cycle. This is unexpected, and suggests that the configuration of the filler network may be adapting to the strain range that is imposed upon the material in such a way as to lower the average resistivity, or resistance.

5 CONCLUSIONS

This study has presented simultaneous experimental measurements of stress, strain and electrical resistivity during continuous, pseudo-cyclic, and cyclic loading on four elastomers: a carbon-black filled EPDM rubber, a carbon-black filled Elastosil silicone elastomer, and two grades of Estane thermoplastic polyurethane elastomer, melt-compounded with 4% and 5% MWCNTs.

In all materials studied, the resistivity-strain relationship exhibits considerable hysteresis, with the response strongly influenced by the maximum strain previously reached, in a manner similar to the Mullins effect within the stress-strain relationship. When the materials are cycled bween no load and a strain of 1, the response settles quickly with time, and is repeatable. However, whereas in the carbon black filled EPDM rubber the resistivity is lower during loading than during unloading, the reverse is true for the CNT-filled Estanes. A possible explanation for this unusual behavior has been suggested to be related to the very different aspect ratio of the filler particles.

In the CNT-filled materials, the resistivity was observed to reach a minimum *within* the cycle, for a wide variety of strain ranges. This may suggest that the CNT network is adapting to the imposed deformation in such a way as to reduce its resistance to the flow of current.

ACKNOWLEDGEMENTS

The authors wish to acknowledge the contributions of Dr. T. Alshuth of the German Institute of Rubber Technology (DIK) in supplying the EPDM rubber material; of Dr C.Y. Lew of Nanocyl in supplying and compounding the Estane and nanotube materials; and of Dr A. Tognetti of the University of Pisa in supplying the Elastosil material.

REFERENCES

Bulgin 1945. Electrically conductive rubber. *Transactions of the Institution of the Rubber Industry,* 21 (3): 188–218.

Lew, C. Y., Xia, H., Mcnally, T., Fei, G., Vargas, J., Milar, B., Douglas, P., Claes, M. & Luizi, F. 2009. A unified strategy to incorporating nanotubes in twin-screw extrusion processing. In Mitsoulis, E. (ed.) *Proceedings of the Europe/Africa Regional Meeting of the Polymer Processing Society, Larnaca, Cyprus,* 18–21 October 2009.

Mullins, L. 1947. Effect of stretching on the properties of rubber. *Journal of Rubber Research,* 16 (12): 275–289.

Norman, R. H. 1970. *Conductive rubbers and plastics,* Elsevier.

Constitutive Models for Rubber VII – Jerrams & Murphy (eds)
© 2012 Taylor & Francis Group, London, ISBN 978-0-415-68389-0

Effect of the strain amplitude and the temperature on the viscoelastic properties of rubbers under fatigue loading

Pierre Garnier

Clermont Université, Institut Français de mécanique Avancée, Laboratoire de mécanique et Ingénieries,
Clermont-Ferrand, France
PCM, Champtocé sur Loire, France

Jean-Benoît Le Cam

Clermont Université, Institut Français de mécanique Avancée, Laboratoire de mécanique et Ingénieries,
Clermont-Ferrand, France

Michel Grédiac

Clermont Université, Université Blaise Pascal, Laboratoire de mécanique et Ingénieries, Clermont-Ferrand, France

ABSTRACT: This study deals with the effect of fatigue loading on the viscoelastic properties of filled nitrile rubber. Classic strain amplitude sweeps were first carried out at two temperatures (ambient and 80°C) on both a filled and an unfilled nitrile rubber in order to discuss the influence of the fillers and the temperature on the Payne effect. Then, fatigue loading tests were performed to observe the evolution of the viscoelastic properties and the sensitivity of the Payne effect to fatigue.

1 INTRODUCTION

Rubber components are very widely used in various fields of engineering. Designing at best such components must take into account the mechanical properties of rubber, among which their response under fatigue loading. In this context, a major issue is to employ a criterion which reliably models the fatigue limit of the constitutive material, and consequently leads to a correct end-of-life prediction of the structural components. In the literature, the end-of-life prediction is based on the assessment of a mechanical quantity (energy, strain, stress) during a stabilized cycle (Mars and Fatemi 2004). It is assumed that the material properties remain unchanged during the major part of its lifetime, since the maximal loading is also stabilized. To the best knowledge of the authors, the microstructure evolution is not taken into account in the classic procedures used in the mechanical characterization of rubber under fatigue loading (Mars 2001; Brunac, Gerardin, and Leblond 2009; Andriyana, Saintier, and Verron 2010). This can be a decisive matter since it *a priori* impacts the viscoelastic properties of rubber and therefore the value of the mechanical quantity considered. Classically, the viscoelastic properties of filled rubber are studied using Dynamic Mechanical Analysis (DMA) or Dynamic Mechanical and Thermal Analysis (DMTA). These types of analyses provide the storage modulus E', the viscous modulus E'', and the loss factor $\tan(\delta)$ for a given strain amplitude (SA). For rubber materials, a decrease of the storage modulus is observed if the amplitude of the cyclic strain increases. This phenomenon, called the Payne effect (Payne 1962) or the Fletcher-Gent effect (Fletcher and Gent 1953), has been widely investigated. Many theories have been proposed in the literature even though the involved physical mechanisms are not clearly described. These mechanisms, at the microstructure scale, are linked to the different morphological components of rubber, in other words matrix and fillers as well as their interactions. They potentially induce several phenomena such as:

1. the formation of a filler network constituted by filler-filler and filler-matrix interactions (Wang 1999);
2. the existence of occluded polymer located within the filler aggregates. This increases the effective filler concentration (Morozov, Lauke, and Heinrich 2010);
3. the formation of a layer of bound polymer around the filler particle surface (Morozov, Lauke, and Heinrich 2010);
4. the interactions between fillers and matrix along the interface, which leads to adsorption and desorption of polymer chains with several mobility levels (Maier and Goritz 1996);
5. the hydrodynamic effect (Einstein 1906).

However, the influence of the evolution of the matrix itself and the fillers on the rubber viscoelastic properties under fatigue loading has never been investigated till now, to the best knowledge of the authors.

This type of evolution of the rubber microstructure is rather studied by chemists that tend to characterize, using DMA or DMTA, the evolution of the microstructure when the material is subjected to thermo-oxidation (Feller 1994) for instance. Even though these studies have improved the knowledge on the mechanisms involved, to the best knowledge of the authors, none of them investigates the effect of the mechanical cycles on the microstructure itself, and therefore on the evolution of the viscoelastic properties of rubber.

The aim of the present paper is to characterize the evolution of the viscoelastic properties of filled nitrile rubber under fatigue loading, typically up to 10^6 cycles. For this purpose, several strain amplitudes were applied in order to observe and discuss the evolution of the Payne effect with the number of cycles. Moreover, these experiments were carried out at two different temperatures: ambient temperature and 80°C. In this paper, the material, the specimen geometry and the loading conditions are presented first. The obtained results are then discussed. Finally, some perspectives are given for future work.

2 EXPERIMENTAL SETUP

The dynamic properties of both filled and unfilled rubber were measured by means of a Metravib VA2000 viscoanalyzer, using specimens featuring a height of 10 mm and a square cross-section of 5×5 mm^2. Classic DMTA was first carried out with filled and unfilled nitrile rubbers. It consisted in applying first an increasing double strain amplitude (denoted DSA in the following) sweep from 0.1 to 15%, and then a decreasing DSA sweep from 15 to 0.1%. The DSA sampling satisfied a logarithmic distribution, as in Refs. (Wang 1999; Rendek and Lion 2010). This choice is justified by the fact the Payne effect is the most significant for the lowest DSA. The cycle defined by an increasing followed by a decreasing DSA sweep was repeated ten times. The tests were performed at a constant frequency of 10 Hz. The influence of the temperature on the mechanical response was assessed by performing tests at ambient temperature (20°C in the present case) and 80°C. Two types of loading conditions were applied: tensile and compressive. In both cases, the mechanical properties were evaluated in terms of E', E'' and $\tan(\delta)$, and their evolutions along the tests were recorded.

3 RESULTS

3.1 *Classic DMTA*

In light of the DMA experiments previously performed, filled nitrile exhibits a mechanical response in good agreement with those reported in the literature vis-à-vis the Payne effect (Wang 1999; Rendek and Lion 2010; Medalia 1978): the increase of temperature causes the Payne effect to vanish. Indeed, the

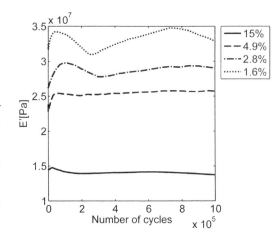

Figure 1. E' versus the number of cycles at ambient temperature, for the four DSA considered.

thermal activation allows the macromolecules to slip easily between each other and in the vicinity and within the filler agglomerates. It also impairs the initial stiffness. Finally, the type of loading conditions changes the relative position between successive sweeps and the values of E', but the Payne effect itself is not really affected.

3.2 *Fatigue tests at ambient temperature*

The effect of the mechanical cycles on the visco-elastic properties is now considered. The phenomena that are observed will be interpreted in light of the chemical and physical evolutions that likely appeared in the tested specimens. Since the previous paragraph shows that fillers influence the viscoelastic properties of the material, only the filled nitrile is examined here. Figure 1 shows the evolution of E' at ambient temperature during the tests. Each DSA previously applied was applied again, but for the sake of simplicity, only four typical curves are plotted in the diagram.

The first remark is that whatever the number of cycles, the higher the strain amplitude, the lower the storage modulus. This is a first analogy with the Payne effect discussed previously. The second observation is that apart from the highest DSA level (15 %), the initial level of E' is lower than the final one. This is somewhat unusual since fatigue damage generally induces a decrease of E'. Another interesting feature is that all curves obtained can be seen as a succession of five elementary steps, which are reported in Figure 2.

The first step, called step 1 in the following, only occurs during the very early stages of the tests. It corresponds to a decrease of E', which can be observed in Figure 3. This first step can be considered as a material softening.

The combined variations of both E' (Figure 1) and $\tan(\delta)$ (Figure 4) during the four following steps are more clearly visible by normalizing the curves. This normalization is performed by dividing E' ($\tan(\delta)$

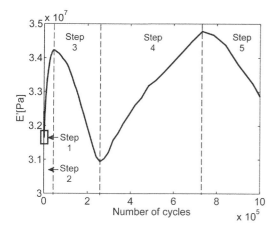

Figure 2. Elementary steps of a fatigue test illustrated at 2.8% DSA at ambient temperature.

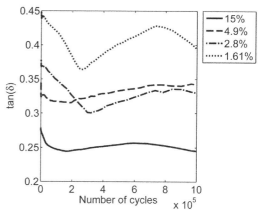

Figure 4. tan(δ) *vs.* the number of cycles at ambient temperature for the four DSA considered.

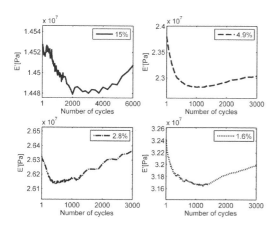

Figure 3. Magnified view of E' on the first thousands of cycles realised at ambient temperature for the four DSA considered.

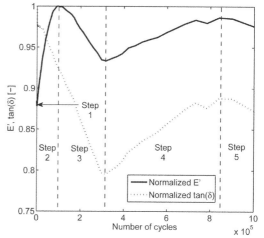

Figure 5. Superimposition of normalized E' and tan(δ) for 2.8% DSA at ambient temperature.

respectively) by its maximum value. The corresponding diagram that illustrates the normalization is given in Figure 5 for the particular case of 2.8% DSA.

After a decrease of E' during step 1, step 2 corresponds to a strong increase up to a maximum value reached at 4.5×10^4, 9.7×10^4, 5.6×10^4, 2.3×10^4 cycles for 1.61, 2.8, 4.9 and 15% DSA, respectively. It is worth noting that the maximum value of E' reached at the end of step 2 is higher than that observed at the beginning of the test, thus meaning that the material becomes stiffer than in its initial state. Step 3 corresponds to a decrease of E'. It is longer than step 2. Step 4 corresponds to an increase of E'. It is longer than step 3. Figure 2 shows that the material hardening rate during step 4 is lower than both the softening rate during step 3 and the hardening rate during step 2. Finally, step 5 corresponds to a decrease of E'. The limits of the steps in terms of number of cycles depend on the DSA.

These results clearly illustrate the complexity of the mechanical response of filled nitrile under fatigue loading. This is probably due to the fact that the material is subjected to both physical and chemical evolutions at the same time. With classic DMTA, the viscoelastic response of rubber depends on the cyclic disruption and reformation of the filler network, as well as on the possible release of occluded rubber (Wang 1999). Under fatigue, some additional phenomena occur such as fatigue of the filler network and chemical evolutions of the rubber matrix (for instance vulcanization, chains scissions (Feller 1994)).

Figure 5 is now used in order to discuss the mechanisms involved during the five steps which are clearly bounded. For steps 1, 3, 4 and 5, the variation of the loss factor is similar to that of E', *i.e.* the viscosity increases as E' increases, and then decreases as E' decreases.

Step 2 is now considered again for comparison purposes with step 4. Similarly to step 4, step 2 corresponds to material hardening, but while the loss factor increases during step 4, either it begins to increase

249

before decreasing during step 2 (for 1.61 and 4.9% DSA) or it first slowly decreases, and then decreases more quickly (for 2.8 and 15% DSA). This is clearly visible in Figure 5 where both the normalized values of E' and $\tan(\delta)$ are represented: during steps 2 and 3, the loss factor regularly decreases while E' increases and then decreases, respectively. On the contrary, during step 4, both E' and the loss factor increase. This exemplifies that the hardening mechanism is different for steps 2 and 4.

As mentioned above, E' significantly increases during step 2. This can be explained by several likely scenarios. The first scenario is that this increase of E' can be attributed to the vulcanization of the rubber matrix. However, since a decrease of E' takes place while the loss factor still decreases (Figure 5, step 3), it can be said that there is another phenomenon which competes with vulcanization. Otherwise, both E' and the loss factor would be strictly monotonic at the same time. Vulcanization would have a stronger influence at the beginning of the test during step 2, whereas the second phenomenon would then be more intense during step 3. Consequently, the authors believe that this scenario is not the most likely. Another explanation is the influence of the reinforcing fillers on the mechanical response of the material. To come to a decision about the cause of the phenomenon, one of the experiments described above was performed on an unfilled rubber specimen. A higher DSA was used (15% DSA), to try to reflect the fact that macromolecules sustain a local stress level which is certainly significantly higher than that sustained by their counterparts constituting the unfilled material. The reason for this is that fillers act as local stress concentrators in the filled material (Suzuki, Masayoshi, and Ono 2005). Figure 6 compares the responses of a filled and an unfilled rubber specimens. It clearly appears that the response of the unfilled rubber specimen is not the same as that of the filled one, since E' continuously decreases up to 1.7×10^5 cycles for the unfilled material only. Thereby, it is possible to conclude that the fillers are the cause of the phenomenon discussed above. To the best knowledge of the authors, this phenomenon concerning fatigue loading with small DSA has not been reported in the literature till now. Finally, similar results can be observed concerning the evolution in the normalized value of E' at higher numbers of cycles. The fact that the normalized value of E' increases, and then decreases, may be due to the macromolecular network, not to the filler network, as this feature can be observed the curves of E' for both filled and unfilled rubber. This will be more detailed in (Garnier, Le Cam, and Grédiac 2011).

It must be emphasized that a stabilized response is generally assumed when the end of life of rubber components is predicted (Brunac, Gerardin, and Leblond 2009; Andriyana, Saintier, and Verron 2010). The current study clearly shows that this assumption is too strong because it does not take into account the microstructural evolution of rubber, at least for the material currently under study.

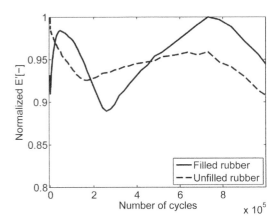

Figure 6. Superimposition of normalized E' of filled and unfilled nitrile rubber at ambient temperature at 2.8 and 15% DSA respectively.

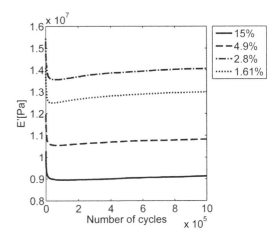

Figure 7. E' vs. the number of cycles at 80°C.

3.3 Fatigue tests at 80°C

The effect of the temperature on the viscoelastic response of filled nitrile rubber is investigated in the current section. The previous tests were performed once again at 80°C for this purpose.

Figure 7 gives the variation of E' during the tests. The curves observed in this case are simpler than these obtained at ambient temperature. First, each curve exhibits the same shape. Second, the viscoelastic response can be split into two steps. Step 1 corresponds to a strong decrease of E' during the 5×10^4 first cycles. This is due to the combination of the well-known material softening and the viscosity. Then, during step 2, E' slightly and continuously increases.

Figure 8 shows a superimposition of the normalized value of E' and $\tan(\delta)$ vs. the number of cycles for 1.61% DSA. $\tan(\delta)$ continuously decreases, first significantly during step 1 and then slightly during step 2.

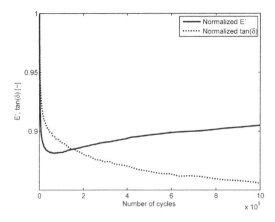

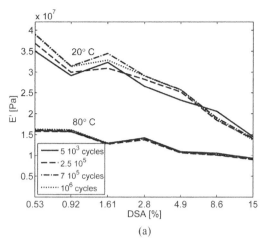

Figure 8. Normalized tan(δ) and E' for 4.9% DSA at 80°C.

(a)

The fact that E' increases and that tan(δ) decreases during step 2 tends to show that the material is subjected to vulcanization.

4 INFLUENCE OF THE NUMBER OF CYCLES ON THE PAYNE EFFECT

In the previous sections, the evolution of some mechanical quantities, *viz.* E' and tan(δ), was given *vs.* the number of cycles. Considering now the Payne effect, it is classically represented in the E'-DSA diagram. Various curves showing the evolution of E' and tan(δ) *vs.* the DSA are plotted at different stages of the tests to characterize and to discuss the influence of the number of cycles on the Payne effect. These curves are plotted for both the ambient temperature and 80°C.

Figure 9 shows some E'-DSA curves obtained for various numbers of cycles, at both the ambient temperature and 80°C. Four values of the number of cycles have been chosen: 5×10^3, 2.5×10^5, 7×10^5 and 10^6 cycles. The following comments can be drawn. First, for both temperatures, the number of cycles does not change the global curve shape. This indicates that the Payne effect is not altered by the number of cycles.

Due to the fact that the step duration is different for each DSA level, it is difficult to compare quantitatively the relative position of the curves. Nevertheless, it clearly appears that at ambient temperature, the gap between the curves observed for low DSA levels is greater than that observed for higher DSA levels. Theses curves tend to the same value at 15% DSA. This means that the higher the DSA level, the lower the influence of the number of cycles. This could be explained by the fact that for low DSA levels, the contribution of the filler network to the initial stiffness of the material is not altered by the DSA. At 80°C, this phenomenon is not observed anymore and the curves are superimposed. As observed during classic DMTA, the DSA has only a small effect on E'. This is certainly the reason why no difference is observed between the

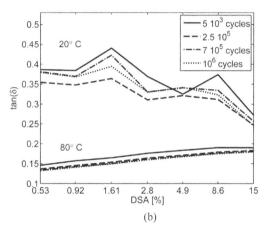

(b)

Figure 9. E' (a) and tan(δ) (b) of filled nitrile rubber *vs.* DSA for different number of cycles at both ambient temperature and 80°C using compressive loading.

curves for different numbers of cycles. Concerning the evolution of tan(δ) *vs.* the DSA at both temperatures, it can be said that it decreases with the number of cycles.

5 CONCLUSION

The influence of fatigue loading on the viscoelastic properties of filled nitrile rubbers is examined in this study. Classic DMTA was first performed to show the influence on the Payne effect of various parameters such as temperature, volume filler fraction and the number of DSA-sweeps applied to the specimens. Fatigue tests were then carried out for several DSAs at both ambient temperature and 80°C. The main conclusion is that at ambient temperature, E' can not be considered as constant. Results obtained at 80°C tend to show that vulcanization initiates in this case. In conclusion, it is clear that the evolution of the viscoelastic properties of rubber must be taken into account in the end-of-life prediction of structural components of rubber. Future work may consist in investigating the microstructure by experimental techniques (for instance IR spectroscopy, Raman) to precisely establish the evolution of the macroscopic mechanical behaviour of rubber related to the microstructural changes.

REFERENCES

Andriyana, A., N. Saintier, and E. Verron (2010). Configurational Mechanics and Critical Plane Approach: Concept and application to fatigue failure analysis of rubber-like materials. *International journal of fatigue 32*(10), 1627–1638.

Brunac, J., O. Gerardin, and J. Leblond (2009). On the heuristic extension of haigh's diagram for the fatigue of elastomers to arbitrary loadings. *International journal of fatigue 31*(5), 859–867.

Einstein, A. (1906). Zur theorie der bownschen bewegung. *Annal of Physics 17*, 549.

Feller, R. (1994). Accelerated aging, photochemical and thermal aspects. In *The Getty Conservation Institute*.

Fletcher, W. P. and A. N. Gent (1953). Nonlinearity in the dynamic properties of vulcanised rubber compounds. *Trans. Inst. Rubber Ind. 29*, 266–280.

Garnier, P., J.-B. Le Cam, and M. Grédiac (2011). On the influence of the number of cycles on the viscoelastic properties of rubber under fatigue loading: another dimension for the Payne effect? *Submitted*.

Maier, P. and D. Goritz (1996). Molecular interpretation of the payne effect. *Kautschuk gummi kunstoffe 49*(1), 18–21.

Mars, W. V. (2001). *Multiaxial fatigue of rubber*. Ph. D. thesis, University of Toledo.

Mars, W. V. and A. Fatemi (2004). Factors that affect the fatigue life of rubber: A literature survey. *Rubber Chemistry and Technology 77*(3), 391–412. Spring Meeting of the ACS-Rubber-Division, Savannah, GA, APR 29-MAY 01, 2002.

Medalia, A. I. (1978). Effect of carbon black on dynamic properties of rubber vulcanizates. *Rubber Chemistry and Technology 51*(3), 437–523.

Morozov, I., B. Lauke, and G. Heinrich (2010). A new structural model of carbon black framework in rubbers. *Computational Materials Science 47*(3), 817–825.

Payne, A. (1962). The dynamic properties of carbon black-loaded natural rubber vulcanizates. part i. *Journal of Applied Physics 6*(19), 57–63.

Rendek, M. and A. Lion (2010). Strain induced transient effects of filler reinforced elastomers with respect to the payne-effect: experiments and constitutive modelling. *ZAMMZeitschrift fur Angewandte Mathematik und Mechanik 90*(5), 436–458.

Suzuki, N., M. I. Masayoshi, and S. Ono (2005). Effects of rubber/filler interactions on the structural development and mechanical properties of nbr/silica composites. *Journal of Applied Polymer Science 95*(1), 74–81.

Wang, M. (1999). The role of filler networking in dynamic properties of filled rubber. *Rubber Chemistry and Technology 72*(2), 430–448. Spring ACS Rubber Division Meeting, Indianapolis, Indiana, may 05–08, 1998.

Constitutive Models for Rubber VII – Jerrams & Murphy (eds)
© 2012 Taylor & Francis Group, London, ISBN 978-0-415-68389-0

Effect of material and mechanical parameters on the stress-softening of carbon-black filled rubbers submitted to cyclic loadings

Yannick Merckel & Mathias Brieu
LML, CNRS, Ecole Centrale de Lille, bd Paul Langevin, Villeneuve d'Ascq, France

Julie Diani
PIMM, CNRS, Arts et Métiers ParisTech, bd de l'Hôpital, Paris, France

Daniel Berghezan
Manufacture Française des Pneumatiques Michelin, CERL, Ladoux, Clermont-Ferrand, France

ABSTRACT: Several carbon-black filled Styrene Butadiene Rubbers (SBR) were tested in cyclic uniaxial tension in order to investigate their cyclic stress-softening. A classic representation using the maximum stress with respect to the number of cycles is first applied to present the experimental data. Nonetheless, the limiting interest of such a representation drove us to propose an original method of cyclic softening characterization for filled rubbers. The method is based on the use of stretch amplification factors. The characterization method provides an interesting tool for the study of the effect of material parameters and loading parameters on the cyclic softening of filled rubbers, as reveals the current study conducted on several SBRs with various amount of fillers and different crosslink densities.

1 INTRODUCTION

Rubber-like materials are commonly used into structures undergoing cyclic loadings, such as tires for instance. In order to increase their stiffness and lifetime the use of reinforcements such as carbon-black fillers is frequent. The main drawback of such reinforcements is to induce some cyclic stress-softening. The larger part of the stress-softening occurs during the first loading and is known as the Mullins effect (Mullins 1969; Diani et al. 2009). During the subsequent cycles, the mechanical behavior evolves slowly becoming undetectable during two successive cycles. However, when considering a large number of cycles, this evolution cannot be neglected anymore. In this study, we focus our attention on the second part of the cyclic stress-softening which occurs once the Mullins effect has been dissipated. This softening is first studied from a classic point of view by following the maximum stress over the number of cycles. Then, an original characterization method is proposed, based on the existence of a stretch amplification factor. Both methods are applied to several materials having various amount of fillers and crosslink densities when stretched up to several maximum stretches for a reasonably large number of cycles (1000). In the following section, materials, experimental protocol and experimental results are presented. Next, the original method of stress-softening characterization is introduced. In a fourth section, the material stress-softening is analyzed in the light of the original characterization. Concluding remarks close the paper.

2 EXPERIMENTAL OBSERVATIONS

2.1 Materials and experiments

For this study several carbon-black filled rubbers (SBR) were supplied by Michelin. In order to study the effect of the material composition on the cyclic stress-softening, materials with various amounts of fillers and crosslink densities were processed. Figure 1 presents the material strategy. Using material B3 as reference, materials B1, B2, B4 and B5 are obtained by varying the amount of fillers from 40 to 5, 30, 50 and 60 phr respectively. Materials A3, C3 and D3 contain the same amount of fillers, but their crosslink densities vary from 3.6 to $10.5 \cdot 10^{-5} \mathrm{mol/cm^3}$. Mechanical tests are performed at room temperature on a conventional INSTRON 5882 uniaxial testing machine. The force is measured by a 2 kN load cell while local strains are measured by a video extensometer. The sample shape is normalized with an initial cross section $S_0 = 2.5 \times 4 \, \mathrm{mm^2}$ and a 30 mm length. Samples are submitted to a thousand cycles from zero stress up to maximum stretch of $\lambda_{max} = 3$. Each test is performed at a constant crosshead speed of 3 mm/min, which corresponds to an approximate average strain rate of $10^{-1} \, \mathrm{s^{-1}}$. Figure 2 presents the stress-stretch response of material B2 when submitted to the cyclic loading. The Cauchy stress $\sigma = F/S$ with F the force and S the current cross section is reached by assuming incompressibility: $S = \lambda^{-1} S_0$. During the first cycle, the material undergoes a strong stress-softening known as Mullins effect. Then, the material softens slowly

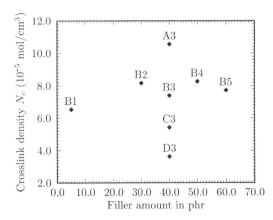

Figure 1. Material strategy.

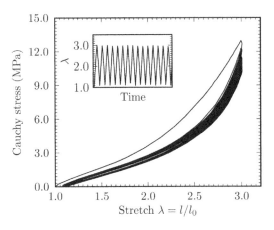

Figure 2. Material B2 stress-stretch response to a uniaxial tensile cyclic test of 1000 cycles at a constant maximum stretch $\lambda_{max} = 3$.

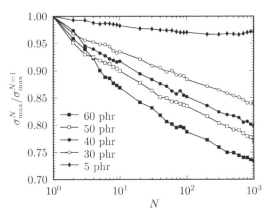

Figure 3. Evolution of the normalized peak stress with respect to the number of cycles N for materials with similar crosslink densities $N_c \approx 7 \cdot 10^{-5}$ mol/cm^3 and different amounts of fillers.

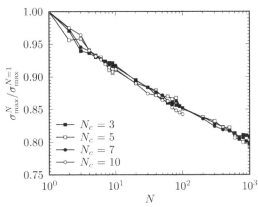

Figure 4. Evolution of the normalized maximum stress with respect to the number of cycles N for 40 phr filled SBRs characterized by differenc crosslink densities (N_c).

with the number of cycles. The softening between two consecutive cycles is extremely low, but its cumulative effect after a large number of cycles cannot be neglected. In the current study, we are interested in the cyclic stress-softening post Mullins effect therefore in what follows the first cycle is not considered and the second cycle stands for $N = 1$. Let us note that a complete study on the effect of the material parameters on the Mullins softening may be found in (Merckel et al. 2011). It concluded on a strong effect of the amount of fillers and a insignificant effect the crosslink density on the Mullins softening.

2.2 Evolution of the maximum of stress

In order to characterize the softening during cyclic loadings, a solution commonly used in the literature, is to follow the evolution of the peak stress or force with respect to the number of cycles (Shen et al. 2001; Gentot et al. 2004; Brieu et al. 2010; Mars and Fatemi 2004; Asare et al. 2009; Berrehili et al. 2010; Yan et al. 2010). The stress value may be normalized by the peak stress at the first cycle. Figures 3 and 4 show the evolution of the normalized peak stress with the number of cycles for every material. One may note that plots become linear with respect to the logarithm of the number of cycles after a relatively low number of cycles ($\simeq 40$). A similar trend has been reported in the literature by (Mars and Fatemi 2004; Gentot et al. 2004; Brieu et al. 2010).

Figure 3 presents the change of the normalized peak stress for materials with similar crosslink densities, $N_c \approx 7.10^{-5}$ mol/cm^3 and different amounts of fillers. This figure shows a significant impact of the amount of fillers on the stress-softening. Adding fillers increase the material softening during cyclic loadings.

Figure 4 shows the change of the normalized peak stress for several 40 phr of carbon-black filled SBR gums characterized by their crosslink densities. One notes that materials A3, B3, C3 and D3, the change of the peak stress is independent of the gum crosslink density.

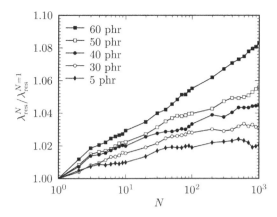

Figure 5. Evolution of the residual stretch with respect to the number of cycles N for materials with similar crosslink densities $N_c \approx 7 \cdot 10^{-5}$ mol/cm^3 and different amounts of fillers.

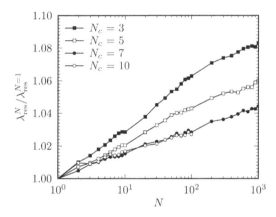

Figure 6. Evolution of the residual stretch with respect to the number of cycles N for 40 phr filled SBRs characterized by differenc crosslink densities (N_c).

2.3 Evolution of the residual stretch

As the stress-softening evolves, the residual stretch (stretch at zero stress) increases with the number of cycles. Figures 5 presents the residual stretch evolution with respect to the number of cycles for materials with amount of fillers. One notes that the residual stretch is larger and increase faster with the increase of the amount of fillers. Figure 6 shows the evolution of the residual stretch with respect to the number of cycles according to the material crosslink density. The residual stretch increases when the crosslink density, N_c, decreases. Therefore while the amount of fillers plays a similar role on the stress-softening and the residual stretch, the crosslink-density shows different effects. It does not affect the stress-softening but affects the residual stretch. As a consequence, characterizing the mechanical behaviour evolution during cyclic loadings by the peak stress evolution only is not satisfying. In the next section, we propose an original characterization

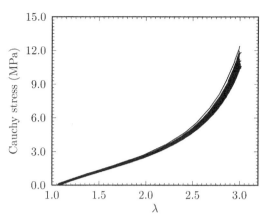

Figure 7. Loading part of the stress-stretch responses of material B2 submitted to a cyclic test of 1000 cycles at a constant maximum stretch $\lambda_{max} = 3$.

of the mechanical behaviour during cyclic loadings in order to better account for the material composition.

3 ORIGINAL CHARACTERIZATION OF CYCLIC SOFTENING

In the literature, the cyclic softening is mostly characterized by the decrease of the peak stress at a given stretch (Shen et al. 2001; Gentot et al. 2004; Brieu et al. 2010; Mars and Fatemi 2004; Asare et al. 2009; Berrehili et al. 2010; Yan et al. 2010). An alternative approach may be based on noticing that the softening may be equivalently characterized by an increase of stretch at a given stress. The interest of such an approach is that it may be applied on the entire stress-stretch response as it is shown below.

3.1 Method

In Figure 2, one notes that material B2 shows some hysteresis during the cycles. Nonetheless, it appears that this hysteresis is not substantial after the first cycle. Therefore the material stress-stretch response may be represented by either the loading or the unloading stress-stretch responses. Figure 7 presents the loading stress-stretch responses extracted from Figure 2, which capture well the material cyclic softening.

From data plotted in Figure 7, one may notice that each cycle N stress-stretch response $\mathcal{S}_N(\lambda, \sigma)$ may compare to the cycle 1 stress-stretch response $\mathcal{S}_1(\lambda, \sigma)$ by introducing a constant parameter $\alpha(N)$ such as:

$$\mathcal{S}_N(\lambda, \sigma) = \mathcal{S}_1(\alpha\lambda, \sigma) \text{ with } \alpha \geqslant 1 \qquad (1)$$

Values of α were calculated by testing the superimposition of the stress-stretch responses $\mathcal{S}_N(\lambda/\alpha, \sigma)$ onto the reference stress-stretch response $\mathcal{S}_1(\lambda, \sigma)$. Values of α are computed with a least squares minimization on the entire stress-stretch responses. Figure 8 show that

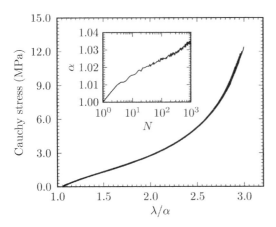

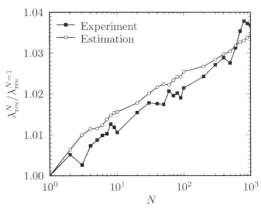

Figure 8. Superposition of the material B2 stress-stretch loading responses. The inset graph presents the values of $\alpha(N)$ providing the superposition.

Figure 10. Comparison of the normalized residual stretch provided by the parameter α (Figure 8) with the measured experimental data for material B2.

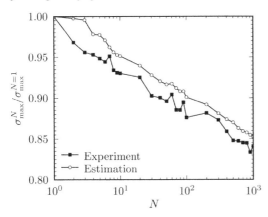

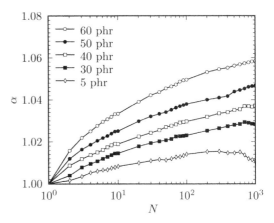

Figure 9. Comparison of the normalized peak stress provided by the parameter α (Figure 8) with the measured experimental data for material B2.

Figure 11. Parameter α read of the impact of the amount of fillers on the cyclic softening.

all cycle loading curves superimpose well. The experimental evidences support the concept of the existence of an stretch amplification factor $\alpha(N)$ as defined in (1), this parameter is now characterizing the cyclic softening.

3.2 *Interest of the proposed characterization*

The main interest of parameter α stands in its definition on the entire stress-stretch response. Using the $N = 1$ loading response and the evolution of α with respect to the number of cycles, it is possible to estimate the complete stress-stretch response at any cycle. It is therefore possible to evaluate both extrema, the peak stress σ_{max} and the residual stretch λ_{res}, for each cycle. Equation (1) provides direct access to $\lambda_{res}^N = \alpha \lambda_{res}^{N=1}$ and the stress ratio is determined once the stress at $\alpha \lambda^{N=1} = \lambda_{max}$ has been computed. Figures 9 and 10 presents a comparison between the normalized peak stress and the normalized residual stretch estimated by α and the measured experimental data. Both quantities are in good agreement with the experiments, which

proves that parameter of α applies well on the entire stress-stretch response emphasizing its interest.

4 RESULTS AND ANALYSIS

In this section, we investigate the impact of the material composition on the cyclic softening, which is now characterized by α.

We first compare computed values of α for materials B1, B2, B3, B4 and B5. Let us remind that these materials have a crosslink density of approximately $7 \cdot 10^{-5}$ mol/cm^3 and various amount of fillers from 5 to 60 phr. Results are shown in Figure 11. The amount of fillers is shown to have a strong impact of the material softening and results presented here are in good agreement with the conventional stress-softening characterization presented in Figure 3. Next, we compare the evolution of the stress-softening parameter α obtained for materials A3, B3, C3 and D3 in order to study the impact of the crosslink density on the cyclic softening. Results are shown in Figure 12. For the

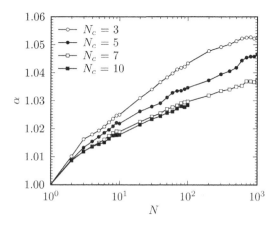

Figure 12.   Effect of the crosslink density on the cyclic softening parameter α for materials with various N_c and a similar filler volume fraction $\Phi \approx 0.166$ (40 phr).

highest value of N_c value, the sample failure occurred before the end of the test. The parameter α appear to depend on the crosslink density, increasing with the decrease of the latter one. This result leads to different conclusion from what was obtained in Figure 4 when focusing on the peak stress only. Actually the stretch amplification factor α contain information on both the peak stress evolution and the residual stretch evolution at the same time, and therefore should be privileged for the characterization of filled rubber cyclic softening. Moreover, the parameter is a relevant to predict the material cyclic softening since it provides access to the entire stress-stretch response.

5   CONCLUSIONS

In this work, we studied the stress-softening of filled rubbers once the Mullins softening evacuated. For this purpose, we used several materials with different amount of fillers and crosslink densities. An original parameter for the cyclic softening characterization was introduced. Unlike the conventional approach using the evolution of the peak stress, the proposed method consists in studying the evolution of the stretch amplification induced by the softening. The stretch amplification parameter is computed in such a way that the stress-stretch response of the first cycle superimposes with the stress-stretch response of cycle N when its stretch measures are amplified by a constant value $\alpha(N)$. The parameter α is estimated on the entire

stress-stretch response and provides good estimate of the evolutions of its extrema: the peak stress and the residual stretch. The parameter was shown to be relevant to characterize the effect of material parameters on the cyclic softening. Actually, both the amount of fillers and the crosslink density were shown to affect the material cyclic softening. By increasing the amount of carbon-black fillers or decreasing the crosslink density, one favors the material softening during cyclic loadings.

ACKNOWLEDGMENT

This work was supported by the French "Agence Nationale de la Recherche" through project AMUFISE (MATETPRO 08-320101). The authors acknowledge useful discussions with J. Caillard, C. Creton, J. de Crevoisier, F. Hild, C. Moriceau, M. Portigliatti, S. Roux, F. Vion-Loisel, and H. Zhang.

REFERENCES

Asare, S., A. G. Thomas, and J. J. C. Busfield (2009). Cyclic stress relaxation (csr) of filled rubber and rubber components. *Rubber Chem. Technol. 82*, 104–112.

Berrehili, A., S. Castagnet, and Y. Nadot (2010). Multiaxial fatigue criterion for a high-density polyethylene thermoplastic. *Fatigue Fract. Engng. Mater. Struct. 33* (6), 345–357.

Brieu, M., J. Diani, C. Mignot, and C. Moriceau (2010). Response of a carbon-black filled SBR under large strain cyclic uniaxial tension. *Int. J. Fatigue 32* (12), 1921–1927.

Diani, J., B. Fayolle, and P. Gilormini (2009). A 5 review on the mullins effect. *Eur. Polym. J. 45*, 601–612.

Gentot, L., M. Brieu, and G. Mesmacque (2004). Modelling of stress-softening for elastomeric materials. *Rubber Chem. Technol. 77*, 758–774.

Mars, W. V. and A. Fatemi (2004). Factors that affect the fatigue life of rubber: a literature survey. *Rubber Chem. Technol. 77*, 391–408.

Merckel, Y., J. Diani, M. Brieu, and J. Caillard (2011). Effect of the microstructure parameters on the mullins softening in carbon-black filled rubbers. *J. Appl. Polym. Sci.*.

Mullins, L. (1969). Softening of rubber by deformation. *Rubber Chem. Technol. 42*, 339–362.

Shen, Y., F. Golnaraghi, and A. Plumtree (2001). Modelling compressive cyclic stress-strain behaviour of structural foam. *Int. J. Fatigue 23*, 491–497.

Yan, L., D. A. Dillard, R. L.West, L. D. Lower, and G. V. Gordon (2010). Mullins Effect Recovery of a Nanoparticle-Filled Polymer. *J. Polym. Sci. Part B: Polym. Phys. 48*, 2207–2214.

Constitutive Models for Rubber VII – Jerrams & Murphy (eds)
© *2012 Taylor & Francis Group, London, ISBN 978-0-415-68389-0*

From the experimental determination of stress-strain full fields during a bulge test thanks to 3D-DIC technique to the characterization of anisotropic Mullins effect

G. Machado, D. Favier & G. Chagnon

Laboratoire Sols, Solides, Structures, Risques (3SR), Université de Grenoble/CNRS, Grenoble, France

ABSTRACT: The bulge test is usually mostly used to analyze equibiaxial tensile stress state at the pole of inflated isotropic membranes. Three-dimensional digital image correlation (3D-DIC) technique allows the determination of three-dimensional surface displacements and strain fields. A method is proposed to calculate from these experimental data the membrane curvature tensor at each surface point of the bulge specimen. Curvature tensor fields are then used to investigate axisymmetry of the test; in the axisymmetric case, membrane stress tensor fields are determined from meridional and circumferential curvatures combined with the measurement of the inflating pressure. Stress strain state is then known at any surface point which enriches greatly experimental data deduced from bulge tests. This method is then used to treat an experimental bulge test on a filled silicone rubber membrane. The results highlight that a global membrane with a very heterogeneous strain history is obtained, from equibiaxial behavior at the center of the membrane until a planar (pure) shear state at the periphery of the bulge. Next, different small tensile specimens are cut from the pre-stretched silicone membrane. Identical cyclic tensile tests are realized on all these specimens. The curves are compared and highlight the difference of the stress-softening according to the place of the cut specimen and according to its orientation with respect to the circumferential or meridional direction.

1 INTRODUCTION

Rubber-like materials exhibit a significant stress softening after a first loading. This phenomenon called Mullins effect has been observed in different paths of deformation during the last six decades. A detailed review for experimental references in different deformations states like uniaxial tension, uniaxial compression, hydrostatic tension, simple shear and equibiaxial tension is given in Diani *et al.* (2009).

More recently the evidence that stress softening is an inherently anisotropic phenomenon can be found for different rubber-like materials under different load cases. However, these investigations are almost restricted to simple deformation histories. Tension tests performed successively in two orthogonal directions have been the most usual method to point out the strain-induced anisotropy. Results can be seen in Laraba-Abbes *et al.* (2003) for a carbon-black filled natural rubber as material, Diani *et al.* (2006) for filled black ethylene propylene diene (EPDM), Itskov *et al.* (2006) for carbon-black filled acrylate rubber (ACM), Hanson *et al.* (2005) for a silica-filled polydimethylsiloxane (PDMS) and Park and Hamed (2000) for different compositions of black filled styrene-butadiene rubber (SBR) and sulfur-cured carbon black filled natural rubber (NR). Simple shear experiments were performed by Muhr *et al.* (1999), they noticed that strain softening of polysiloxane

polymer investigated is smaller in directions orthogonal to the strain cycle causing the softening. Similarly under simple shear, Besdo *et al.* (2003) showed that a prior loading, generated by a cyclic one-side shear, lead to a material anisotropy verified during a symmetric loading, i.e., both-sided shear. Pawelski (2001) carried out two tests, first, a homogeneous plane-strain compression using a cubic sample of an elastomeric material (VLGQ-rubber) rotated and rotated back by 90° with respect to the first preloaded configuration. Here a special procedure was required to correct the original measurements since compression is accompanied by friction which spoils the experimental data. Second, for a polyurethane material, a modified biaxial tension with tensile tests on secondary specimens afterward, was used in a way to find an evidence for more complex memory behavior of the material.

Some models aiming to describe the anisotropy of the stress softening are proposed in the literature see for example, Diani *et al.* (2004), Göktepe and Miehe (2005), Shariff (2006), Diani *et al.* (2006), Ehret and Itskov (2009). All these models provide a solid theoretical basis for mathematical description of the stress softening, like an anisotropic phenomenon, in rubber-like materials. However, the predictive capabilities of models describing the anisotropic Mullins effect are mainly restricted to the uniaxial load case. Horgan *et al.* (2004) reported that the anisotropic model cannot be validated since suitable experimental data are not available. Pawelski (2001) advised that further experimental works are necessary before one can decide how complex a material constitutive equation for rubber-like material should be. Itskov *et al.* (2006) pointed out that additional experiments regarding other load cases are needed. This allows to study further the influence of loading cycles with complex deformation states on the appearance and evolution of the Mullins effect in the case of changing principal stretch directions.

Within this context, the aim of this work is to quantify clearly the induced anisotropy according to the applied strain history. The objective is to focus on the strain-induced anisotropy study by verifying the influence of loading cycles in distinct directions using the bulge test to generate different biaxial deformation histories, and cut specimens from the original plate are thus submitted to tensile tests.

2 EXPERIMENTAL STUDY AND ANALYSIS

The induced anisotropy by stress softening was experimentally studied. For this purpose, samples were made of a filled silicone rubber called Rhodorsil RTV3428. Supplied as two liquid components, the uncured silicone and the curing agent, this liquid mixture is molded by injection to obtain a sheet with constant thickness Meunier *et al.* (2008). The final samples are produced by a polyaddition, curing at 70° C for 4 h in order to accelerate the curing process and assure a sufficient cross-linking density. This material exhibits a pronounced stress softening under different load cases, as reported in Machado *et al.* (2010). In the same work, the manufacture procedure of specimens without any pre-existent anisotropy and testing methods details using three-dimensional image correlation (3D-DIC) are also described. With the 3D-DIC technique, it is possible to determine the 3D contour and the in-plane strain fields of the object surface. This process is carried out by correlation of the images, taken by two cameras in the deformed state with their original reference images. The bulge test were used to precondition the samples, i.e., to induce some primary stress-softening. Thus, subsequent uniaxial tensile tests were conducted on preconditioned specimens.

The objective is to propose an experimental test that permits to analyze second tensile load curves in a material that experienced a more complex first load path. A biaxial tensile pre-stretch can be used to generate a complex in-plane deformation history, under incompressibility assumption. The complexity of this deformation history is given by the biaxiality ratio (μ), that can be expressed as a relation between principal in-plane stretches ($\lambda_{max}, \lambda_{min}$) in the form

$$\mu = \frac{\ln(\lambda_{min})}{\ln(\lambda_{max})}. \tag{1}$$

Thus, biaxiality ratio is very useful to characterize the first history imprinted in the material at the first

loading. Note that in the uniaxial tensile pre-stretch (Machado *et al.* (2009)), the contractive stretches are the minimal principal directions, then the biaxiality ratio is fixed in $\mu = -0.5$.

In this work the bulge test is proposed as an original way to yield very different biaxial strain-histories for first load path. As pointed in Machado *et al.* (2011) bulge test is able to provide more information than the traditional application to determine the material equibiaxial response. It can be used to generate different biaxial states along a radial path of specimen from the equibiaxial state ($\mu = 1$) at the pole until the planar (pure) shear ($\mu = 0$), i.e., a large biaxiality ratio interval ($\mu \in [0, 1]$).

Using the 3D-DIC technique, it is possible to determine the three-dimensional surface displacements and the principal stretches (λ_m, λ_c) in the meridional and circumferential directions, as schematized in Fig. 1. From this point forward, these directions will be denoted by the subscripts m and c respectively. Nevertheless, in the axisymmetric case, deformed membrane principal stresses (σ_m, σ_c) are determined from principal curvatures (κ_m, κ_c) combined with the measurement of the inflating pressure (p). These fields are related as follow

$$\sigma_m = \frac{p}{2h\,\kappa_c} \tag{2}$$

$$\sigma_c = \frac{p}{2h\,\kappa_c}\left(2 - \frac{\kappa_m}{\kappa_c}\right), \tag{3}$$

where h is the current thickness calculated using the incompressibility assumption. κ_m and κ_c are deduced from 3D-DIC measurements. Thus, the stress-strain state is then known at any surface point which enriches greatly experimental data deduced from bulge tests. For a comprehensive explanation see Machado *et al.* (2011).

The specimen geometry consists in thin circular plate, of initial radius $R = 90$ mm with a thickness $h_0 = 2$ mm. The circular plate is simply supported and subjected to an uniform inflating pressure acting perpendicular to the current configuration. Using the 3D-DIC strain field measures, a set of interest points are specified along the radius of undeformed membrane permitting to know the stress and strain histories for every point over the deformed membrane surface. This set of points is represented in the Fig. 1, where each point can be regarded as a small specimen in view of 3D-DIC as a full field strain measure method. Point a corresponds to the membrane center and the other points (b to j) are 10 mm equidistant through the meridional direction.

Fig. 2 presents the first loading strain and stress history of the points a and h in function of the load pressure. Note that each inflation state involves a heterogeneous stress-strain state that evolves from an equibiaxial state at the pole ($R = 0$ mm – point a) tending to planar (pure) shear stretching nearest the clamp ($R = 90$ mm – point j). Machado *et al.* (2011)

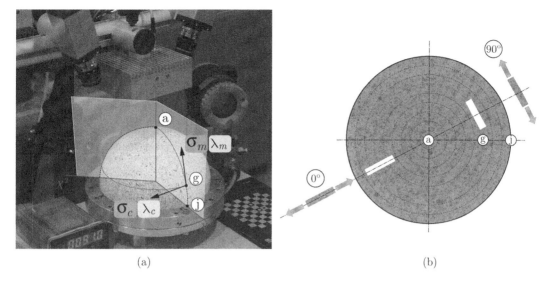

Figure 1. Schematic representation of biaxial preconditioning method and principal stretches: (a) axisymmetric representation of the bulge test with the interest points; (b) Illustration of the cut subsamples over two different directions in the circular membrane.

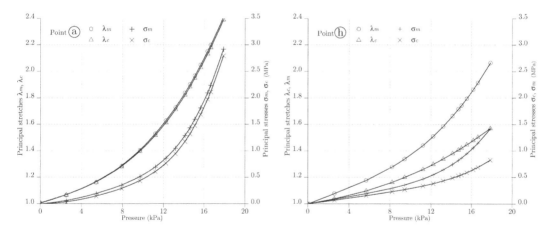

Figure 2. Bulge test deformation history for the points a to h. Principal stretches (λ_m, λ_c) and current principal stresses (σ_m, σ_c) are presented respectively on left and right axis according to the inflating pressure (p).

pointed out the experimental difficulty concerning the measurements on the overall membrane surface and the region near the clamp simultaneously. For materials that experience high displacements a self-shadow effect is observed in this last region.

Fig. 3 presents the mean value and the range of the biaxiality ratio (μ) for all pressure load steps in each point a to i. The measured ratio is almost constant at a given point and variations between the different pressure load steps can be disregarded. Due to the experimental difficulties concerning the measurements near the clamp, the curves in Fig. 3 were extrapolated for the points i and j using polynomial approximation and represented by dotted lines.

After a recovery time of 1 h, a set of tensile test specimens are cut. Two series of samples are defined as follows: samples aligned to the meridional direction denoted by 0° and samples tangent to the circumferential direction (or perpendicular to the meridional direction) denoted by 90°. Both series are illustrated in Fig. 1(b).

At first, the samples aligned to the meridional direction (0°) were submitted to a uniaxial tension test. The loading consists of a simple load unload cycle, where the upper limit overpass the maximum value reached in the biaxial pre-stretch by the major principal stretch (λ_m). Fig. 4 presents the second load curves for the different specimens and a virgin uniaxial tensile

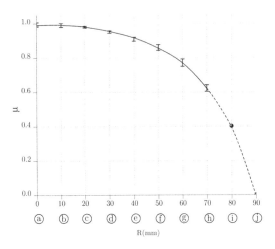

Figure 3. Mean value and range of biaxiality ratio (μ) during all pressure load steps. Points i and j were extrapolated using a polynomial approximation (dotted lines).

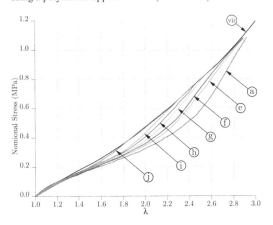

Figure 4. Influence of the biaxial first loading on tensile test response. Uniaxial traction performed at $0°$ with respect to the meridional direction for the subsamples a, e, f, g, h, i, and j.

load curve (*vir*) as reference. The load curves of the points b, c and d are quasi superposed, thus they are not represented in the figure keeping clear the results visualization.

It appears that all the curves are different. The figure highlights that closer the sample is from the bulge pole, more stress-softening is observed. Two aspects are observed, the difference between the reference curve (*vir*) and second load is increasing and the strain-hardening for the return point (*RP*) on the reference curve appears later. These results are in consistency with the first maximal deformation level presented in Fig. 3(a).

Second, the samples cut at the points perpendicularly to the meridional direction ($90°$) are tested. However, the second load–unload curves corresponding to points a and g are presented for both directions ($0°$ and $90°$) in Fig. 5 and compared to the virgin load curve (*vir*). The deformation is evaluated locally, using DIC, in a small zone located in the samples center. It clearly appears that the second load curves, for both directions, come back on the first loading curve at that same maximal deformation but with different levels of stress-softening. A stress softening appears in the orthogonal directions as the material was also stretched in this direction during the preconditioning path. This is due to the biaxial pre-loading where the directional intensity of the stress softening depends on the biaxiality ratio (μ). This emphasizes that subsequent directional behavior is strongly influenced by how the maximal pre-loading is applied.

3 CONCLUSIONS

The use of bulge test, as a biaxial preconditioning test, was originally proposed to find an evidence for an even more complex memory behavior of the material. The measurements made along the meridian of the inflated membrane provide unprecedented information about the history of principal stress and principal

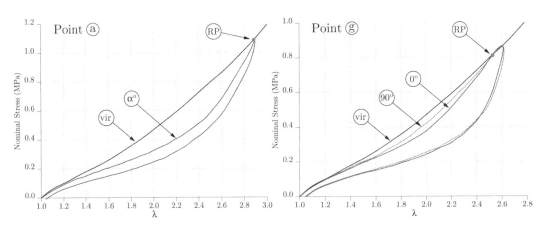

Figure 5. Second load (and unload) curves of specimens a and g cut from the bulge plate at $0°$ and $90°$ with respect to the meridional direction. Note that, for point a, only one curve is presented since the all directions are equivalent.

strain directions without an explicit dependency of any constitutive equation. Results showed that the induced anisotropy is dependent on the pre-stretch biaxiality ratio. The return on the first load reference curve of all second load curves occurs at the same point, independently of the second load direction. Note also that the form and intensity of the anisotropy of the Mullins effect could depend on the elastomer type, the properties of filler particles, and their mass and volume proportion in the polymeric matrix.

ACKNOWLEDGEMENT

We would like to thank the French ANR for supporting this work through the project RAAMO ("Robot Anguille Autonome pour Milieux Opaques").

REFERENCES

Besdo, D., Ihlemann, J., Kingston, J., and Muhr, A. (2003). Modelling inelastic stress-strain phenomena and a scheme for efficient experimental characterization. *In: Busfield, Muhr (eds) Constitutive models for Rubber III. Swets & Zeitlinger, Lisse.*, pages 309–317.

Diani, J., Brieu, M., Vacherand, J. M., and Rezgui, A. (2004). Directional model isotropic and anisotropic hyperelastic rubber-like materials. *Mech. Mater.*, **36**, 313–321.

Diani, J., Brieu, M., and Gilormini, P. (2006). Observation and modeling of the anisotropic visco-hyperelastic behavior of a rubberlike material. *Int. J. Solids Struct.*, **43**, 3044–3056.

Diani, J., Fayolle, B., and Gilormini, P. (2009). A review on the Mullins effect. *Eur. Polym. Journal*, **45**, 601–612.

Ehret, A. E. and Itskov, M. (2009). Modeling of anisotropic softening phenomena: Aplication to soft biological tissues. *Int. J. Plast.*, **25**, 901–919.

Göktepe, S. and Miehe, C. (2005). A micro-macro approach to rubber-like materials. Part III: The micro-sphere model of anisotropic Mullins-type damage. *J. Mech. Phys. Solids*, **53**, 2259–2283.

Hanson, D. E., Hawley, M., Houlton, R., Chitanvis, K., Rae, P., Orler, E. B., and Wrobleski, D. A. (2005). Stress softening experiments in silica-filled polydimethylsiloxane provide insight into a mechanism for the Mullins effect. *Polymer*, **46**(24), 10989–10995.

Horgan, C. O., Ogden, R. W., and Saccomandi, G. (2004). A theory of stress softening of elastomers based on finite chain extensibility. *Proc. R. Soc. London A*, **460**, 1737–1754.

Itskov, M., Haberstroh, E., Ehret, A. E., and Vohringer, M. C. (2006). Experimental observation of the deformation induced anisotropy of the Mullins effect in rubber. *KGK-Kautschuk Gummi Kunststoffe*, **59**(3), 93–96.

Laraba-Abbes, F., Ienny, P., and Piques, R. (2003). A new tailor-made methodology for the mechanical behaviour analysis of rubber-like materials: II. Application to the hyperelastic behaviour characterization of a carbon-black filled natural rubber vulcanizate. *Polymer*, **44**(3), 821–840.

Machado, G., Chagnon, G., and Favier, D. (2009). Experimental observation of induced anisotropy of the mullins effect in particle-reinforced silicone rubber. In *Constitutive Models for Rubber VI*, pages 511–515. CRC Press.

Machado, G., Chagnon, G., and Favier, D. (2010). Analysis of the isotropic models of the Mullins effect based on filled silicone rubber experimental results. *Mech. Mater.*, **42**(9), 841–851.

Machado, G., Favier, D., and Chagnon, G. (2011). Membrane curvatures and stress-strain full fields of bulge tests from 3D-DIC measurements. Theory, validation and experimental results on a silicone elastomer. *Submitted to Exp. Mech.*

Meunier, L., Chagnon, G., Favier, D., Orgéas, L., and Vacher, P. (2008). Mechanical experimental characterisation and numerical modelling of an unfilled silicone rubber. *Polym. Test.*, **27**, 765–777.

Muhr, A. H., Gough, J., and Gregory, I. H. (1999). Experimental determination of model for liquid silicone rubber: Hyperelasticity and Mullins effect. In *Proceedings of the First European Conference on Constitutive Models for Rubber*, pages 181–187. Dorfmann A. Muhr A.

Park, B. H. and Hamed, G. R. (2000). Anisotropy in gum and black filled sbr and nr vulcanizates due to large deformation. *Korea Polymer Journal*, **8**, 268–275.

Pawelski, H. (2001). Softening behaviour of elastomeric media after loading in changing directions. *Constitutive models for rubber, Besdo, Schuster & Ihleman (eds)*, pages 27–34.

Shariff, M. H. B. M. (2006). An anisotropic model of the Mullins effect. *J. Eng. Math.*, **56**(4), 415–435.

Constitutive Models for Rubber VII – Jerrams & Murphy (eds)
© 2012 Taylor & Francis Group, London, ISBN 978-0-415-68389-0

A new isotropic hyperelastic strain energy function in terms of invariants and its derivation into a pseudo-elastic model for Mullins effect: Application to finite element analysis

L. Gornet & G. Marckmann
GeM, UMR CNRS 6183, Ecole Centrale Nantes, Nantes, Cedex, France

R. Desmorat
LMT–Cachan ENS Cachan/CNRS/UPMC/PRES UniverSud Paris, Cachan Cedex, France

P. Charrier
Modyn Trelleborg, Zone ind. de Carquefou, Carquefou Cedex, France

ABSTRACT: The present paper focuses on static stiffness modelling of rubber materials for multiaxial loadings with a minimal number of material parameters in order to ensure robustness of both identification and Finite Element analysis. A physically motivated isotropic constitutive model is proposed for multiaxial large stretch rubber deformation. This model is expressed in terms of classical independent strain invariants. Only three parameters are needed to successfully represent both Treloar and Kawabata experiments (Kawabata *et al.* 1981) with a response equivalent to the one obtained by Ogden six parameters model (Ogden 1972). Mullins effect modelling is finally derived according to a damage mechanics approach (Chagnon *et al.* 2004).

1 INTRODUCTION

The aim of the present paper is to develop new constitutive models for the multiaxial loadings of elastomers which are both physically motivated, well adapted for numerical problems, and accurate for rubber materials. Hyperelastic models dedicated to rubber materials can be classified into two types of strain energy formulations. The first kind of models is issued from mathematical developments such the well-known Rivlin series or as the Ogden model (Ogden 1972). The second kind of models is the one developed from physical motivations. Such models are based on both physics of polymer chains networks and statistical methods. In this work, we propose a strain energy function expressed in terms of independent strain invariants. The proposed model (GD) is dedicated to multiaxial loadings. This model is successfully identified on both Treloar and Kawabata experiments (Treloar 1944, Kawabata *et al.* 1981). We present a bridge between the phenomenological strain energy formulation and the physical motivation of the proposed (GD) model. In the Finite Element context, the proposed model can easily be implemented because of its strain invariants formulation. This has been done in computer codes like Cast3M-CEA (French Atomic Agency) and ABAQUS. Finite Element predictions of a roll restrictor developed by Trelleborg automotive have been carried out with ABAQUS.

The continuous damage approach is applied to soften by damage the material parameters of this initial model in order to describe Mullins effect. A new model for Mullins effect (GDM model) is therefore derived. The identification of the GDM model is made thanks to experimental data from (Marckmann G. & E. Verron 2006).

2 STRAIN ENERGY FUNCTIONS

In this section, the strain energy functions that define the proposed model (GD) are briefly recalled. Links between this model and the eight-chains model are highlighted. A phenomenological energy function expressed in term of the second invariant I_2 is also introduced based on physical motivations.

2.1 *Incompressible GD strain energy density*

Assuming that rubber materials are both isotropic and incompressible, the proposed strain energy function W only depends on the two first invariants of the left Cauchy-Green stretch tensor B:

$$W_{GD}(I_1, I_2) = W_1(I_1) + W_2(I_2)$$

$$= h_1 \int e^{h_3(I_1-3)^2} dI_1 + 3h_2 \int \frac{1}{\sqrt{I_2}} dI_2 \tag{1}$$

where h_1, h_2, h_3 are the material parameters. In this strain energy, the I_1 part of W describes the global response of the material and is equivalent to the Hart-Smith model (Hart-Smith 1966). The second term that involves I_2 improves the accuracy of the model for multiaxial loading conditions (a general power I_2-term has been introduced by Lambert-Diani & Rey, 1999). The true stress tensor is defined by the differentiation of the proposed strain energy with respect to B:

$$\sigma = -pI + 2\left(\frac{\partial W}{\partial I_1} + I_1 \frac{\partial W}{\partial I_2}\right)B - 2\frac{\partial W}{\partial I_2}B^2 \qquad (2)$$

$$I_1 = Tr(B), \quad I_2 = \frac{1}{2}\left[Tr(B)^2 - Tr(B^2)\right] \qquad (3)$$

2.2 Links between GD and eight-chains models

The present part is devoted to the comparison of the GD model to the eight-chains one (Arruda and Boyce 1993) in order to highlight the physical motivations of this proposed phenomenological model. As these constitutive equations are supposed to be qualitatively efficient for the entire range of strains, both small and large strain responses are compared. Comparison of these models is established using their polynomial expansion in terms of I_1 respectively (equations 4 and 5) for small strain. High order terms are then eliminated:

$$W_1(I_1) = C_1 \sum_{i=0}^{\infty} \frac{C_3^i}{(2i+1)i!}(I_1-3)^{2i+1} \qquad (4)$$

The first five terms of the development of the eight-chains model are:

$$W_{8ch} = C_R\left[\frac{1}{2}(I_1-3) + \frac{1}{20N}(I_1^2-3)\right]$$
$$+ C_R\left[\frac{1}{1050N^2}(I_1^3-27) + \frac{19}{7000N^3}(I_1^4-81)\right] \qquad (5)$$
$$+ C_R\left[\frac{519}{673750N^4}(I_1^5-243) + ...\right]$$

The small strain stiffness of the models are respectively defined by first terms of Equations (4) and (5). Then, both the strain energy functions reduce to the neo-Hookean expression. The both models are characterized by their ability to describe the strain-hardening of the material that takes place under large strains. This strain-hardening phenomenon is mainly due to the extensibility limit of polymer chains. Chagnon et al. (2004) established that the first part of the Hart-Smith strain energy $W_1(I_1)$, which is identical to the first part of the GD model, is equivalent to the eight-chains and Gent models for the entire range of strains.

$$\frac{\partial W_1(I_1)}{\partial I_1} = h_1 e^{h_3(I_1-3)^2} \qquad (6)$$

We proposed now to highlight the physical motivation of the second part of strain energy $W_2(I_2)$ of the GD

Figure 1. Entanglement of the eight-chains rubber network. Eight-chains model completed by surrounding chains (in bold) is the physical motivations of the GD model.

model. As mentioned by Treloar (Treloar 1975), term as function of I_2 in constitutive equations can be seen as corrections of the phantom network theory (terms as function I_1). We propose to constrain the eight-chains model by a new network of chains on the surface of the cube (figure 1). The confinement of the eight-chains model is governed by a strain energy potential. This potential constrains the eight-chains cube surface. Let us recall that the surface of the eight-chains model (a cube) is $I_2^{1/2}$ and that its increase under deformation is $I_2^{1/2} - 3^{1/2}$. We therefore define a pressure constrain of the eight-chains rubber network. This phenomenon is modeled by the second invariant energy part:

$$W_2(I_2) = 3h_2 \int \frac{1}{\sqrt{I_2}} dI_2 \quad \text{or} \quad \frac{\partial W_2(I_2)}{\partial I_2} = \frac{3h_2}{\sqrt{I_2}} \qquad (7)$$

where $3h_2$ stands for the pressure constrain of entanglement of the eight-chains rubber network.

2.3 Compressible GD strain energy density

In this part we introduce the proposed strain energy function for compressible isotropic hyperelastic materials in terms of strain invariants. Based on kinematic assumption, we use a decoupled representation of the strain energy function originally proposed by (Flory 1961, Ogden 1984).

$$W(\overline{I}_1, \overline{I}_2, J) = W_{iso}(\overline{I}_1, \overline{I}_2) + W_{vol}(J) \qquad (8)$$

$$J = \det(F), \quad \overline{F} = J^{-\frac{1}{3}}F, \quad \overline{B} = \overline{F}\,\overline{F}^T \qquad (9)$$

$$\overline{I}_1 = Tr(\overline{B}), \quad \overline{I}_2 = \frac{1}{2}\left[Tr(\overline{B})^2 - Tr(\overline{B}^2)\right] \qquad (10)$$

where $W_{iso}(I_1, I_2)$ and $W_{vol}(J)$ stand for the isochoric and volumetric elastic strain energy of the material, respectively. The isochoric part is equivalent to the equation (1) with modified invariants. The volumetric part is chosen to enforce a nearly quasi incompressible behaviour.

$$W_{iso}(\overline{I}_1, \overline{I}_2) = h_1 \int e^{h_3(\overline{I}_1-3)^2} d\overline{I}_1 + h_2 \int \frac{1}{\sqrt{\overline{I}_2}} d\overline{I}_2 \qquad (11)$$

$$W_{vol} = \frac{1}{D}(J-1)^2 \qquad (12)$$

The volumetric properties part is presented in (Doll and Schweizerhof 2000). The true stress tensor is defined by the differentiation of the proposed strain energy with respect to B:

$$\sigma_{iso} = 2J^{-\frac{5}{3}}\left(\frac{\partial W_{iso}}{\partial \bar{I}_1} + \bar{I}_1\frac{\partial W_{iso}}{\partial \bar{I}_2}\right)B - 2J^{-\frac{7}{3}}\frac{\partial W_{iso}}{\partial \bar{I}_2}B^2 \quad (13)$$

$$\sigma_{vol} = \frac{\partial W_{vol}}{\partial J}I, \ \sigma = \sigma_{vol} + \sigma_{iso} \quad (14)$$

3 EXPERIMENTAL DATA AND IDENTIFICATION OF PARAMETERS

In order to compare the efficiency of the models, we choose two complementary data sets issued from classical references (Marckmann & Verron 2006). The first set is due to Treloar (1944). In the current study, data from Treloar for unfilled natural rubber (cross-linked with 8 parts of S phr) were used. This material exhibits highly reversible elastic response and no stretch-induced crystallization up to 400%. Thus it is well-modeled by hyperelastic constitutive equations. Experimental measures were performed for four different loading conditions: equibiaxial extension of a sheet (EQE), uniaxial tensile extension (UE), pure shear (PS) and biaxial extension (BE). The second data set is due to Kawabata et al. (1981). It was obtained using an experimental apparatus for general biaxial extension testing. In terms of stretch ratios, unfilled polyisoprene specimens were stretched from 1.04 to 3.7 in the first direction (λ_1) and from 0.52 to 3.1 in the perpendicular direction (λ_2). These values correspond to moderate strain but lead to deformation conditions from uniaxial extension to equibiaxial extension. Here, both experimental data sets are simultaneously considered to compare the models because the two materials are quite similar. Thus, for a given model, a unique set of material parameters must be able to reproduce these data with a good agreement. The parameter identification is performed using genetic algorithms as presented in (Marckmann and Verron 2006). The GD model includes entanglement of the eight-chains rubber network and identified responses are depicted on figures 2–4. The GD model responses on figures 2 to 4 are almost equivalent to Ogden six parameters results for bi-axial extensions presented on figures 5 to 7. Harts-Smith identified responses are depicted on figures 8–10.

The GD parameters are: $h_1 = 0.142$ MPa, $h_2 = 1.585 \times 10^{-2}$ MPa and $h_3 = 3.495 \times 10^{-4}$. The proposed model GD is able to accurately reproduce the whole "S" shaped response of the material. The model behaves satisfactory under all the presented loadings (figures 2–4).

The Ogden six parameter strain energy density is classically

$$W = \sum_{n=1}^{3}\frac{\mu_n}{\alpha_n}\left(\lambda_1^{\alpha_n} + \lambda_2^{\alpha_n} + \lambda_3^{\alpha_n} - 3\right) \quad (15)$$

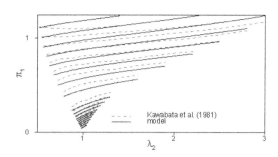

Figure 2. Experimental data (- -) and GD model identification for biaxial tensile tests. Piola-Kirchhoff function of extension for several transverse extensions. Incompressible GD model response.

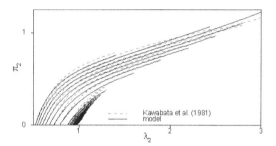

Figure 3. Experimental data (- -) and GD model identification for biaxial tensile tests. Piola-Kirchhoff function of extension for several transverse extensions. Incompressible GD model response.

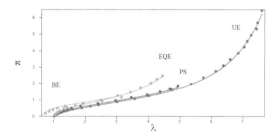

Figure 4. Experimental data (•) and GD model identification for biaxial extension (BE), equibiaxial extension (EQE), pure shear (PS), uniaxial tensile extension (UE). Incompressible GD model response.

Parameters are: $\mu_1 = 0.63$ MPa, $\mu_2 = 1.2 \times 10^{-3}$ MPa, $\mu_3 = -1 \times 10^{-2}$ MPa, $\alpha_1 = 1.3$, $\alpha_2 = 5$ and $\alpha_3 = -2$. The conditions $\mu_i\alpha_i > 0$ ensures the positive definite character of the strain energy. In order to recall the influence of $W_2(I_2)$, let us recall the classical strain energy of Hart-Smith (equation 16).

$$W_{HS}(I_1, I_2) = W_1(I_1) + W_2(I_2)$$
$$= a_1\int e^{a_3(I_1-3)^2}dI_1 + a_1 a_3 \ln(I_2/3) \quad (16)$$

The Hart-Smith parameters are: $a_1 = 0.140$ MPa, $a_2 = 5.254 10^{-4}$ and $a_3 = 1.290$.

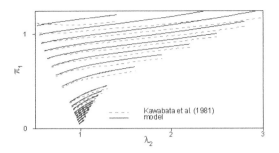

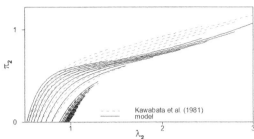

Figure 5. Experimental data (- -) and Ogden model identification for biaxial tensile tests. Piola-Kirchhoff function of extension for several transverse extensions. Incompressible Ogden six model response.

Figure 8. Experimental data (- -) and Hart-Smith model identification for biaxial tensile tests. Piola-Kirchhoff function of extension for several transverse extensions. Incompressible Hart-Smith model response.

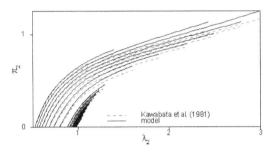

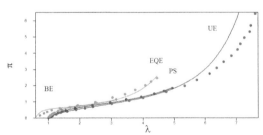

Figure 6. Experimental data (- -) and Ogden model identification for biaxial tensile tests. Piola-Kirchhoff function of extension for several transverse extensions. Incompressible Ogden six model response.

Figure 9. Experimental data (- -) and Hart-Smith model identification for biaxial tensile tests. Piola-Kirchhoff function of extension for several transverse extensions. Incompressible Hart-Smith model response.

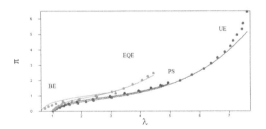

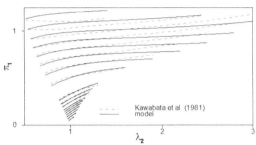

Figure 7. Experimental data (•) and Ogden model identification for biaxial extension (BE), equibiaxial extension (EQE), pure shear (PS), uniaxial tensile extension (UE). Incompressible Ogden six model response.

Figure 10. Experimental data (•) and Hart-Smith model identification for biaxial extension (BE), equibiaxial extension (EQE), pure shear (PS), uniaxial tensile extension (UE). Incompressible Hart-Smith model response.

4 MULLINS EFFECT

Elastomers present a loss of stiffness after the first loading cycle of a fatigue experiment (Mullins, 1969). It has been observed that this phenomenon is only dependent on the maximum deformation previously reached in the history of the material. It is quite important to model it because the mechanical behaviour of rubber products is highly modified by this softening phenomenon. Moreover, as the Mullins effect depends on the maximum deformation endured previously, material REV of the product are not identically affected. As a consequence, it is not acceptable to

determine experimentally an accommodated hyperelastic constitutive equation for the material. The stress-softening should be explicitly included in the model. The continuum damage mechanics has often been used to model the Mullins effect even if phenomenon undergoing Mullins effect is not a strictly speaking an irreversible damage phenomenon and can even be modelled without damage (Cantournet *et al.*, 2009). For example, it can be recovered with time and annealing accelerates this recovery. A thermodynamic variable d is introduced to represent the loss of stiffness and the corresponding stress-softening. The general theory of Continuum Damage Mechanics is detailed in Lemaitre and Chaboche (1990) book. When applied to GD hyperelastic density, it yields to GDM model. This new strain energy function for hyperelastic model with damage variables can be written (eq 17):

$$W_{GDM}\left(I_1, I_2\right) = \tilde{h}_1 \int e^{\tilde{h}_3(I_1-3)^2} dI_1 + 3\tilde{h}_2 \int \frac{1}{\sqrt{I_2}} dI_2$$

$$\tilde{h}_1 = h_1\left(1-d_1\right), \tilde{h}_2 = h_2\left(1-d_2\right), \tilde{h}_3 = h_3\left(1-d_3\right) \qquad (17)$$

The incompressible state laws (eq. 2) associated with this model are classically obtained by equation 18.

$$\sigma = \left.\frac{\partial W_{GDM}}{\partial B}\right|_{I_3=1} \qquad (18)$$

We define the thermodynamic forces Y_{d_i} associated with damage internals variables d_i by:

$$-Y_{d_i} = \frac{\partial W_{GDM}}{\partial d_i} \qquad (19)$$

A non standard damage model is build here as the damage thermodynamics forces are not used to describe damage evolution. We consider instead that the quantity governing the damage evolution laws is the maximum of the first invariant. This assumption is achieved according to the physical motivations of maximum strain state endured during the history of the deformation (Marckmann *et al.* 2002). In order to do this we can introduce damage criterion functions

$$f_i = I_1 - k_i(d_i) \qquad (20)$$

such as $f_i < 0$ implies no damage evolution and as damage evolves at $f_i = 0$ with then $d_i - k_i^{-1}(I_1)$. Choosing particular expressions for k_i-functions allows to derive the damage evolution laws (equations 19–20) and make the model complete. The proposed GDM model is based on an improved method already proposed in (Chagnon *et al.* 2004). The evolution equation of the damage variable is expressed thanks to the first strain invariant and presents an exponential form (equations 21). This model is able to represented unloadings Mullins effect responses (figures 11 and 12).

$$d_1 = d_{1\infty}\left(1 - \exp\left(-\frac{I_1^{\max}}{\eta_1}\right)\right)$$

$$\qquad (21)$$

$$d_2 = d_{2\infty}\left(1 - \exp\left(-\frac{I_1^{\max}}{\eta_2}\right)\right)$$

where $d_{1\infty}$, η_1, $d_{2\infty}$, η_2 and b are material parameters. $I_1^{\max}$ represents the maximum value of the first strain invariant obtained during loadings. The coupling with damages d_1 and d_2 is similar to the discontinuous damage part of the constitutive equations proposed by Miehe (1995). Considering the mass conservation of the polymer network, which implies that the number of monomer segments per unit volume Nn must

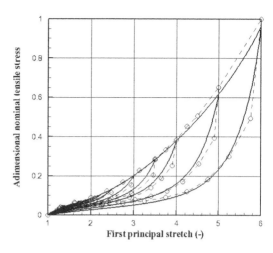

Figure 11. Tensile behaviour of a hyperelastic incompressible GDM model with stress-softening. Identification is performed for a material used on a roll restrictor.

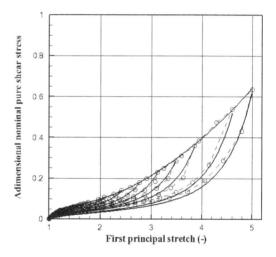

Figure 12. Pure Shear behaviour of a hyperelastic incompressible GDM model with stress-softening. Identification is performed for a material used on a roll restrictor.

remain constant (Marckmann *et al.* 2002), where N is the mean number of monomer per chain and n the number of chain per unit of volume. According to relationship between parameters derived in (Chagnon *et al.* 2004), this leads to link the mechanical properties $\tilde{h}_1$ and $\tilde{h}_3$ and shows that d_3 is not an independent thermodynamics damage variable,

$$d_3 = 1 - F\left(d_1\right), \quad F\left(d_1\right) = \frac{1}{\left(3\left(b\,\tilde{h}_1-1\right)\right)^2} \qquad (22)$$

The GDM is identified on idealized experimental data as presented in (Chagnon *et al.* 2004). Behaviour is considered time-independent. Figure 11 represents the response of the model for a tensile cyclic test. Pure Shear response is presented by Figure 12. These

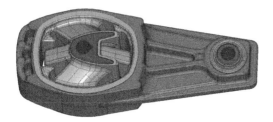

Figure 13. Engine roll restrictor made of steel and rubber parts.

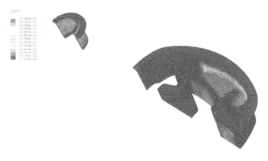

Figure 15. White parts represent the damage level d_2 (Mullins effect) pattern in the roll restrictor during the first "first gear full torque" engine's acceleration.

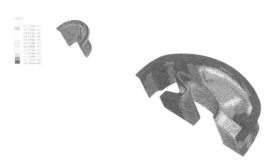

Figure 14. White parts represent the damage level d_1 (Mullins effect) pattern in the roll restrictor during the first "first gear full torque" engine's acceleration.

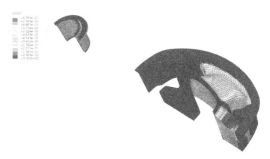

Figure 16. White parts represent the damage level d_3 (Mullins effect) pattern in the roll restrictor during the first "first gear full torque" engine's acceleration.

figures shows how the GDM model is able to accurately reproduce the response of successive loadings for different maximal stretches which characterizes the Mullins effect.

The GDM parameters are: $h_1 = 0.0157$ MPa, $h_2 = 0.0098$ MPa, $h_3 = 0.000561$, $d_{1\infty} = 0.5525$, $\eta_1^{-1} = 0.0119$, $d_{2\infty} = 1.0$, $\eta_2^{-1} = 0.3645$ and $b = 62.69$.

5 FINITE ELEMENT PREDICTIONS

In the finite element context, the proposed GD and GDM models can easily be implemented because of their strain invariants formulations. This has been done in computer codes Cast3M-CEA (French Atomic Agency) and ABAQUS. The two previous constitutive models were implemented in the finite element context, thanks to the UMAT and UHYPER facilities.

Finite element predictions of a roll restrictor (figure 13) developed by Trelleborg automotive have been performed with ABAQUS and the GDM model. This behavior law is able to describe the local loss of stiffness of the material and the non-homogeneity of the structure after the first "first gear full torque" engine acceleration. The material properties used for finite element simulations are not the ones used for real structures. Predictions are performed with the GDM model discussed in previous sections. The 3D model is composed of C3D8H elements. It corresponds to an incompressible finite element hybrid formulation. It appears that the stress softening level is heterogeneous in the roll restrictor rubber parts (figures 14–16).

Figure 17. Isotropic damage model ($d = d_1 = d_2$, $d_3 = 0$). White parts represent the damage area "d" (Mullins effect) during the first "first gear full torque" engine's acceleration.

Damage is located on the surface for most part of the structure excepted in the both snubbers.

Figure 17 illustrates the above Mullins observations in respect of a GDM model with isotropic damage evolution law (equation 23). Mullins effect is located on the surface for most part of the structure excepted in the both snubbers.

$$d = d_1 = d_2, \, d_2 = 0$$

$$d = d_\infty \left(1 - \exp\left(-\frac{I_1^{\max}}{\eta} \right) \right) \tag{23}$$

6 CONCLUSIONS

We propose here a simple isotropic hyperelastic model (GD) expressed in terms of classical independent strain invariants of the symmetric Cauchy-Green tensor. The strain-energy part as function of I_1 is taken identical to the Hart-Smith one. This part is equivalent to the Eight chain model. Concerning the function of the second invariant I_2 a squareroot part is proposed. The corresponding energy density contribution is connected to the non-affine deformation of the entanglement Eight-chains network. The proposed model is successfully identified on both Treloar and Kawabata experiments. Only 3 parameters are needed to describe the experimental results. The response quality is equivalent to the one of the Ogden six parameter model.

The Mullins effect is finally taken into account by coupling with damage the GD model. The GDM model is thus derived stating that the loss of stiffness depends on the maximum value of the first invariant I_1. In the finite element context the proposed models GD and GDM are implemented in the finite element codes Cast3M and ABAQUS. Damage predictions observed on the FE simulations of the Engine roll are in good agreement with experimental data.

REFERENCES

Arruda E. & Boyce M.C. 1993. A three dimensional constitutive model for the large stretch behavior of rubber elastic materials. *J. Mech. Phys. Solids*, 41, 2, 389–412.

Chagnon G. Marckmann G. & Verron E. 2004. A comparison of the hart-smith model with arruda-boyce and gent formulations for rubber elasticity, *Rubber Chemistry and Technology*, 77, 724–735.

Chagnon, G. Verron, E. Gornet, L. Marckmann, G. & Charrier, P. 2004. On the relevance of continuum damage mechanics as applied to the Mullins effect in elastomers. *J. Mech. Phys. Solids*, 52, 1627–1650.

Cantournet S., Desmorat R. & Besson J., 2009. Müllins effect and cyclic stress softening of filled elastomers by internal sliding and friction thermodynamics model, *International Journal of Solids and Structures*, 46, pp. 2255–2264.

Doll S. and Schweizerhof K. 2000. On the developpement of volumetric strain energy function, *J. Appl Mech T ASME* 67, 17–21.

Flory, P. J. 1961. Thermodynamic relations for highly elastic materials, Transactions of the Faraday Society 57 (6,7), 829–838.

Hart-Smith, L. J. 1966. Elasticity parameters for finite deformations of rubber-like materials. *Z. angew. Math. Phys.* 17, 608–626.

Kawabata, S., Matsuda, M., Tei, K., & Kawai, H. 1981, *Macromolecules* , 154–162.

Lambert-Diani J. & Rey C., 1999. New phenomenological behavior laws for rubbers and thermoplastic elastomers, *Eur. J. Mech. A/Solids*, 18, 1027–1043, 1999.

Lemaitre, J. & Chaboche, J. L. 1990. *Mechanics of solid materials*, Cambridge University Press.

Marckmann G. & E. Verron 2006. Comparison of hyperelastic models for rubberlike materials, *Rubber Chemistry and Technology*. 5, 835–858.

Marckmann, E. Verron, L. Gornet, G. Chagnon, P. Charrier & P. Fort, A theory of network alteration for the Mullins effect, *J. Mech. Phys. Solids* **50** (2002), pp. 2011–2028.

Miehe, C. 1995. Discontinuous and continuous damage evolution in Ogden type large strain elastic materials. *Eur. J. Mech., A/Solids* 14, 697–720.

Mullins, L. 1969. Softening of rubber by deformation. *Rubber Chem. Technol.* 42, 339–362.

Ogden, R. W., 1984. Recent advances in the phenomenological theory of rubber elasticity. Rubber Chemistry and Technology. 59, 361–383.

Ogden, R. W., 1972. Large deformation isotropic elasticity – on the correlation of theory and experiment for incompressible rubberlike solids. Proc. R. Soc. Lon. A. 326, 565–584.

Treloar, L. R. G. 1975, The Physics of Rubber Elasticity, Oxford Classic Texts.

Treloar, L. R. G. 1944, *Trans. Faraday Soc.*, 59–70.

Cast3M, *www-cast3m.cea.fr*

ABAQUS, *www.simulia.com*

Constitutive Models for Rubber VII – Jerrams & Murphy (eds)
© 2012 Taylor & Francis Group, London, ISBN 978-0-415-68389-0

The Mullins effect

Stephen R. Rickaby & Nigel H. Scott
School of Mathematics, University of East Anglia, Norwich, UK

ABSTRACT: In this paper we consider the inelastic features associated with stress relaxation, residual strain and hysteresis. These features are then combined with the (Arruda and Boyce 1993) eight-chain model, to develop a constitutive equation that is capable of predicting the Mullins Effect for uniaxial, equibiaxial and pure shear extension subject to cyclic stress-softening for an anisotropic, incompressible elastic material.

1 INTRODUCTION

When a rubber specimen is loaded, unloaded and then reloaded, the subsequent load required to produce the same deformation is smaller than that required during primary loading. This phenomenon is known as stress-softening, which can be described as a decay of elastic stiffness. Stress-softening is particularly evident in specimens of filled rubber vulcanizates.

Figure 1 represents the idealised stress-softening behaviour of a rubber specimen under simple tension. The process starts from an unstressed virgin state at 0 and follows path A, being the primary loading path. If unloading occurs at 1 on this path, then the sample of rubber will follow path B and will return to the unstressed state at 0. If the material is then reloaded the stress-strain behaviour will follow path B. If the rubber is now strained beyond point 1 then path D is activated, being a continuation of the primary loading path. If further unloading occurs at 2, the rubber will retract along path C to the unstressed state at 0. The shape of this second stress-strain cycle differs significantly from the first. If the material is now reloaded the stress-strain behaviour follows path C to the primary loading path at 2.

This stress-softening phenomenon is known as the Mullins effect, which has been named after (Mullins 1947) following an extensive study that he conducted into carbon-filled rubber vulcanizates. (Diani *et al.* 2009) have conducted a current detailed review of this effect.

Many authors have modelled the Mullins effect, including (Mullins 1947), though it is prevalent for authors to model a simplified version of this phenomenon where the following inelastic features are neglected:

- Stress relaxation
- Residual strain
- Hysteresis

The aim of this study is to develop a usable model that captures the main features of stress-softening for

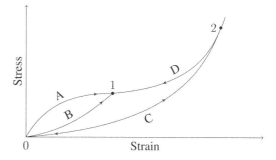

Figure 1. The idealised behaviour of stress-softening in rubber.

raw experimental data. Not all softening features may be relevant for a particular application, thus in order to develop a functional model it is required that specific parameters could be set to zero to exclude any particular inelastic feature, and still maintain the integrity of the model.

2 CONCEPTUAL MODEL

Following the work of (Merodio and Ogden 2005), for an incompressible fibre reinforced material a strain energy function can be defined as,

$$W = W_{\mathrm{i}}(I_1, I_2) + W_{\mathrm{oi}}(I_4, I_5, I_6, I_7),$$

where W_{i} is related to the isotropic base material and W_{oi} is related to the orthotropic character of the material, with I_1, I_2, I_4, I_5, I_6 and I_7 being material invariants defined by,

$$I_1 = \mathrm{tr}\,\mathbf{C}, \; I_2 = \frac{1}{2}\left\{(\mathrm{tr}\,\mathbf{C})^2 - \mathrm{tr}\,\mathbf{C}^2\right\}, \; I_4 = \mathbf{u} \cdot (\mathbf{Cu}),$$

$$I_5 = \mathbf{u} \cdot (\mathbf{C}^2\mathbf{u}), \; I_6 = \mathbf{v} \cdot (\mathbf{Cv}), \; I_7 = \mathbf{v} \cdot (\mathbf{C}^2\mathbf{v}),$$

where $\mathbf{C} = \mathbf{F}^{\mathrm{T}}\mathbf{F}$ is the right Cauchy-Green strain tensor and $\mathbf{F}$ is the deformation gradient, with $\mathbf{u}$ and $\mathbf{v}$

being preferred unit material directions. The invariant $I_3 = \det \mathbf{C}$ is absent because of incompressibility. The Green strain tensor can be written as,

$$\mathbf{E} = \frac{1}{2}\left[\mathbf{C} - \mathbf{I}\right].$$

The strain energy function W may be expressed as $W(I_1, I_2, I_4, I_5, I_6, I_7, \eta_1, \eta_2)$, assuming that η_1 is function of $\mathbf{E}$ only and η_2 is function of I_1, I_4 and I_6 only. Then for an incompressible material we obtain,

$$\mathbf{T}_0 = -p_0\mathbf{I} + 2\left(\frac{\partial W}{\partial I_1} + I_1\frac{\partial W}{\partial I_2}\right)\mathbf{B} - 2\frac{\partial W}{\partial I_2}\mathbf{B}^2$$

$$+ 2\frac{\partial W}{\partial I_4}\mathbf{Fu} \otimes \mathbf{Fu} + 2\frac{\partial W}{\partial I_6}\mathbf{Fv} \otimes \mathbf{Fv}$$

$$+ 2\frac{\partial W}{\partial I_5}[\mathbf{Fu} \otimes \mathbf{BFu} + \mathbf{BFu} \otimes \mathbf{Fu}]$$

$$+ 2\frac{\partial W}{\partial I_7}[\mathbf{Fv} \otimes \mathbf{BFv} + \mathbf{BFv} \otimes \mathbf{Fv}]$$

$$+ \mathbf{F}\frac{\partial W}{\partial \eta_1}\frac{\partial \eta_1}{\partial \mathbf{E}}\mathbf{F}^{\mathrm{T}} + 2\frac{\partial W}{\partial \eta_2}\frac{\partial \eta_2}{\partial I_1}\mathbf{B}$$

$$+ 2\frac{\partial W}{\partial \eta_2}\frac{\partial \eta_2}{\partial I_4}\mathbf{Fu} \otimes \mathbf{Fu} + 2\frac{\partial W}{\partial \eta_2}\frac{\partial \eta_2}{\partial I_6}\mathbf{Fv} \otimes \mathbf{Fv},$$

where $\mathbf{B} = \mathbf{FF}^{\mathrm{T}}$ is the left Cauchy-Green strain tensor, with η_1 being the softening variable and η_2 the residual strain variable of (Dorfmann and Ogden 2004).

Now, upon performing the replacements

$$\mathscr{G}^t_{\tau=-\infty}\{\mathbf{E}(\tau)\} = \frac{\partial W}{\partial \eta_1}\frac{\partial \eta_1}{\partial \mathbf{E}}, \quad \mathscr{R}^t_{\tau=-\infty} = 2\frac{\partial W}{\partial \eta_2}\frac{\partial \eta_2}{\partial I_1},$$

$$\mathscr{T}^t_{\tau=-\infty} = 2\frac{\partial W}{\partial \eta_2}\frac{\partial \eta_2}{\partial I_4}, \quad \mathscr{S}^t_{\tau=-\infty} = 2\frac{\partial W}{\partial \eta_2}\frac{\partial \eta_2}{\partial I_6},$$

we obtain the following general orthotropic model for an equibiaxial and pure shear extension, which can easily be reduced to a transversely isotropic model to represent uniaxial extension:

$$\mathbf{T}_0 = -p_0\mathbf{I} + \aleph_1\mathbf{B} + I_1\aleph_2\mathbf{B} - \aleph_2\mathbf{B}^2$$

$$+ \aleph_4\mathbf{Fu} \otimes \mathbf{Fu} + \aleph_5[\mathbf{Fu} \otimes \mathbf{BFu} + \mathbf{BFu} \otimes \mathbf{Fu}]$$

$$+ \aleph_6\mathbf{Fv} \otimes \mathbf{Fv} + \aleph_7[\mathbf{Fv} \otimes \mathbf{BFv} + \mathbf{BFv} \otimes \mathbf{Fv}]$$

$$+ \mathbf{F}\left[\mathscr{G}^t_{\tau=-\infty}\{\mathbf{E}(\tau)\}\right]\mathbf{F}^{\mathrm{T}} + \left[\mathscr{R}^t_{\tau=-\infty}\right]\mathbf{B}$$

$$+ \left[\mathscr{T}^t_{\tau=-\infty}\right]\mathbf{Fu} \otimes \mathbf{Fu} + \left[\mathscr{S}^t_{\tau=-\infty}\right]\mathbf{Fv} \otimes \mathbf{Fv}. \tag{1}$$

The tensor quantity $\mathscr{G}^t_{\tau=-\infty}\{\mathbf{E}(\tau)\}$ is the proportion of stress relaxation, the scalar $\mathscr{R}^t_{\tau=-\infty}$ is the proportion

of isotropic residual strain, and the scalars $\mathscr{T}^t_{\tau=-\infty}$ and $\mathscr{S}^t_{\tau=-\infty}$ are the proportion of anisotropic residual strain relating to the preferred directions involved in I_4 and I_6, respectively. The scalars $\aleph_i$ are defined by,

$$\aleph_1 = 2\frac{\partial W}{\partial I_1}, \quad \aleph_2 = 2\frac{\partial W}{\partial I_2}, \quad \aleph_4 = 2\frac{\partial W}{\partial I_4},$$

$$\aleph_5 = 2\frac{\partial W}{\partial I_5}, \quad \aleph_6 = 2\frac{\partial W}{\partial I_6} \quad \text{and} \quad \aleph_7 = 2\frac{\partial W}{\partial I_7}.$$

The I_5 term has been described by (Merodio and Ogden 2005) as being related to the fibre stretch and the behaviour of the reinforcement under shearing deformation. For uniaxial, equibixial, and pure shear extension there are no shear stresses, and then it is possible to neglect the $\aleph_5$ and $\aleph_7$ terms.

By employing the (Arruda and Boyce 1993) eight-chain model, which is given below, we may additionally neglect the $\aleph_2$ term.

$$W_{\mathrm{i}} = \kappa\Theta nN_8\left\{\left[\sqrt{\frac{I_1}{3N_8}}\right]\mathscr{L}^{-1}\left(\sqrt{\frac{I_1}{3N_8}}\right)\right.$$

$$\left.+ \log\left\{\frac{\mathscr{L}^{-1}\left(\sqrt{\frac{I_1}{3N_8}}\right)}{\sinh\left(\mathscr{L}^{-1}\left(\sqrt{\frac{I_1}{3N_8}}\right)\right)}\right\}\right\} - h_8,$$

where κ is the Boltzmann constant, Θ is the absolute temperature, n is the chain density, N_8 is the number of links, $\mathscr{L}^{-1}(\,\cdot\,)$ is the inverse Langevin function and h_8 is a constant such that the strain energy vanishes in the natural undeformed state.

Equation (1) may now be reduced to

$$\mathbf{T}_0 = -p_0\mathbf{I} + \aleph_1\mathbf{B} + \aleph_4\mathbf{Fu} \otimes \mathbf{Fu} + \aleph_6\mathbf{Fv} \otimes \mathbf{Fv}$$

$$+ \mathbf{F}\left[\mathscr{G}^t_{\tau=-\infty}\{\mathbf{E}(\tau)\}\right]\mathbf{F}^{\mathrm{T}} + \left[\mathscr{R}^t_{\tau=-\infty}\right]\mathbf{B}$$

$$+ \left[\mathscr{T}^t_{\tau=-\infty}\right]\mathbf{Fu} \otimes \mathbf{Fu} + \left[\mathscr{S}^t_{\tau=-\infty}\right]\mathbf{Fv} \otimes \mathbf{Fv}. \tag{2}$$

The non-linear stress relaxation function $\mathscr{G}^t_{\tau=-\infty}\{\mathbf{E}(\tau)\}$ may be represented by the (Bernstein et al. 1963) model as defined by,

$$\mathscr{G}^t_{\tau=-\infty}\{\mathbf{E}(\tau)\} = A_0(t)\mathbf{I} + \mathbf{I}A_1(t)\mathrm{tr}\,\mathbf{E}(t) + 2A_2(t)\mathbf{E}(t),$$

where A_0, A_1 and A_2 are functions of time. Within the developed stress-softening model, under cyclic loading conditions these function will further depend upon whether the material is being loaded or unloaded.

Following the work of (Dorfmann and Ogden 2003) a softening function can be defined by

$$\mathbf{T} = \left\{1 - \frac{1}{r}\tanh\left(\frac{b}{\mu}\left[W_{\max} - \hat{W}(\lambda_1, \lambda_2)\right]^{\frac{1}{\vartheta}}\right)\right\}\mathbf{T}_0, \tag{3}$$

with r, b and ϑ being dimensionless material constants and μ the ground state shear modulus.

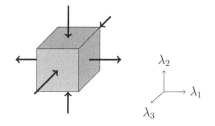

Figure 2. Lateral expansion during uniaxial extension.

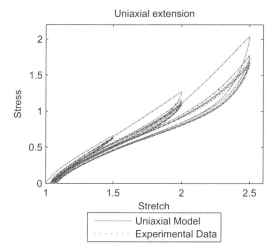

Figure 3. Comparison with experimental data of Dorfmann and Ogden.

3 ANALYSIS

3.1 *Uniaxial extension*

Uniaxial extension is obtained by applying a deformation in the λ_1 direction, such that $\lambda_1 = \lambda > 1$ as shown in Figure 2. If the material is incompressible, then it must contract in the lateral λ_2 and λ_3 directions in the ratio $1/\sqrt{\lambda}$, thus generating a single preferred direction, which is in the direction of the extension. The material can be modelled as transversely isotropic.

A typical graphical representation of equation (2) when mapped to raw experimental data for uniaxial extension is as shown in Figures 3 and 4. The experimental data came courtesy of (Dorfmann and Ogden 2004) and was presented in their paper.

These mappings have been obtained by applying the Padé approximation derived by (Cohen 1991) for the inverse Langevin function, and using suitable approximations for the remaining constants and functions.

3.2 *Equibiaxial extension*

Equibiaxial extension is achieved by applying an equal deformation in the λ_1 and λ_2 directions, such that $\lambda_1 = \lambda_2 > 1$, as shown in Figure 5. If the material is incompressible, then it must contract in the lateral λ_3

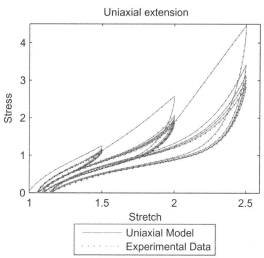

Figure 4. Comparison with experimental data of Dorfmann and Ogden.

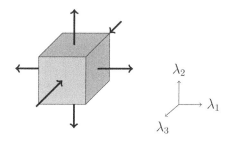

Figure 5. Lateral expansion during equibiaxial extension.

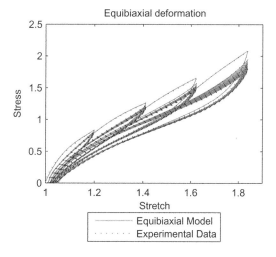

Figure 6. Comparison between the model and the experimental data of Németh *et al.*

direction in the ratio $1/\lambda^2$ where $\lambda_1 = \lambda_2 = \lambda$, thus generating two preferred directions **u** and **v**, being in the directions of λ_1 and λ_2 respectively. The material can be modelled as orthotropic.

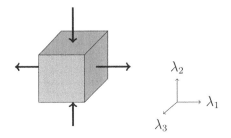

Figure 7.   Lateral expansion during pure shear extension.

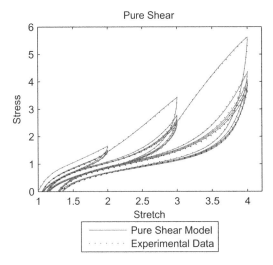

Figure 8.   Comparison with experimental data of Raoult *et al.*

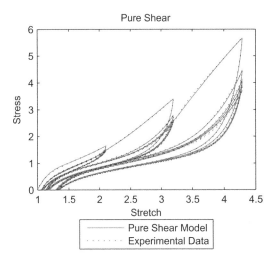

Figure 9.   Comparison with experimental data of Raoult *et al.*

A typical graphical representation of equation (2) when mapped to raw experimental data for equibiaxial extension is as shown in Figure 6. The experimental data came courtesy of (Németh *et al.* 2005) and was presented in their paper.

3.3   *Pure shear*

Pure shear extension is achieved by applying a deformation in the λ_1 direction, such that $\lambda_1 = \lambda > 1$, and keeping $\lambda_3 = 1$ fixed, as shown in Figure 7. If the material is incompressible, then it must contract in the lateral λ_2 direction in the ratio $1/\lambda$, thus generating two preferred directions, those in the directions of the extension and compression. The material can be modelled as orthotropic.

A typical graphical representation of equation (2) when mapped to raw experimental data for pure shear extension is as shown in Figures 8 and 9. The experimental data came courtesy of Trelleborg and PSA Peugeot Citroën, and was partly presented in the paper (Raoult *et al.* 2005).

4   CONCLUSIONS

The model developed here provides an accurate representation of the Mullins effect for uniaxial, equibiaxial and pure shear extension. The model has been developed in such a way that any of the salient inelastic features could be excluded and the integrity of the model would still be maintained.

The details of the stress relaxation, residual strain and anisotropic models will be discussed in detail in a separate paper.

REFERENCES

Arruda, E. M. and M. C. Boyce (1993). A three-dimensional constitutive model for the large stretch behavior of rubber elastic materials. *J. Mech. Phys. Solids 41*, 389–412.

Bernstein, B., E. A. Kearsley, and L. J. Zapas (1963). A Study of Stress Relaxation with Finite Strain. *Trans. Soc. Reheology VII 71*, 391–410.

Cohen, A. (1991). A Padé approximation to the inverse Langevin function. *Rheol. Acta 30*, 270–273.

Diani, J., B. Fayolle, and P. Gilormini (2009). A review on the Mullins effect. *Eur. Polym. J. 45*, 601–612.

Dorfmann, A. and R. W. Ogden (2003). A pseudo-elastic model for loading, partial unloading and reloading of particle-reinforced rubber. *Int. J. Solids Structures 40*, 2699–2714.

Dorfmann, A. and R. W. Ogden (2004). A constitutive model for the Mullins effect with permanent set in particle-reinforced rubber. *Int. J. Solids Structures 41*, 1855–1878.

Merodio, J. and R. W. Ogden (2005). Mechanical response of fiber-reinforced incompressible non-linear elastic solids. *Int. J. Non-Linear Mech. 40*, 213–227.

Mullins, L. (1947). Effect of stretching on the properties of rubber. *J. Rubber Research 16*(12), 275–289.

Németh, I., G. Schleinzer, R. W. Ogden, and G. A. Holzapfel (2005). On the modelling of amplitude and frequency-dependent properties in rubberlike solids. *Constitutive Models for Rubber 4*, 285–298.

Raoult, I., C. Stolz, and M. Bourgeois (2005). A Constitutive model for the fatigue life predictions of rubber. *Constitutive Models for Rubber 4*, 129–134.

Design and applications

Constitutive Models for Rubber VII – Jerrams & Murphy (eds)
© 2012 Taylor & Francis Group, London, ISBN 978-0-415-68389-0

Evaluation of magneto-rheological elastomers for spacecrafts

G. Aridon

EADS-ASTRIUM Satellites, Avenue des Cosmonaute, Toulouse Cedex, France

T. Lindroos & J. Keinänen

VTT Technical Research Centre of Finland, Sinitaival, Tampere, Finland

ABSTRACT: This study focuses on the stiffness controllability of magneto-rheological elastomer (MRE) but also on the damping dependency with the magnetic field. It is well known that elastomer visco-elastic behavior is highlighted by elliptic hysteresis loops. Force-deflection loops of MRE were measured for different amplitudes and frequencies conditions. It results rectangular shape hysteresis loops. Hence, it has to be wondered which dissipation mechanisms are present in addition to the viscous in the MRE. The aim of this study is to have a better understanding of the working principles of MRE in order to evaluate their ability to be used for innovative solutions in damping and stiffness variation devices. This knowledge will confirm or re-orientate, based on the outputs, the applicability of MRE's to Space applications.

1 INTRODUCTION

1.1 Spacecraft requirements

The design of spacecraft structures, instruments, and components is conditioned by the constant need to withstand a combination of all vibration environments. Three types of vibration events are present:

– Shocks due to the spacecraft separation from the launcher or the release of solar arrays, antennas, radiators and inducing working anomaly on equipments.
– Sine & acoustic vibrations due to the launcher and generating problem of structure/equipment strength.
– In-orbit micro-vibrations due to reaction wheels decreasing observation instrument performances.

Thus, the need to minimize vibration across a broad frequency range is the new driver of the structural design.

Astrium Satellites is used to employ elastomer stiffness and damping properties to provide solutions of passive damping or local isolation. Nevertheless, it would be easier is the non-linear behaviour of such visco-elastic material would be controllable. This is a priori the capability of MR technology with a flexible control.

1.2 Background on magneto-rheological elastomers

Different magneto-rheological forms exist. The most usual are MR fluids but they present an important disadvantage such as sedimentation. This drawback can be eliminated by changing the matrix for foams or

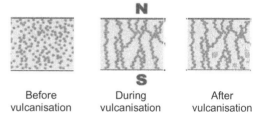

Figure 1. Aligning particles in the elastomer matrix.

elastomers. Typically, MRE's composition consists of approximately 30 volumic % of ferromagnetic particles in soft elastomer matrix like natural and synthetic rubbers, silicones or polyurethanes.

Eugene Guth (1945) characterized the behaviour of such doped composites. Zero-field modulus E0 increases with particle loading (approximately +65% compared to a classical plain elastomer) but is not dependent to the particle orientation.

However, by applying magnetic field, highest performance (i.e. highest stiffness increase) is obtained by aligning the magnetic particles during matrix cure. Insertion of aligned particles also increases the material stiffness when no magnetic field is applied.

Following figure (Ruddy *et al.* (2007)) highlights the particle fill rate dependency under quasi-static loading. The best increase of modulus induced by magnetic field occurs when the particles become magnetically saturated. It can reach as much as 50% stiffness increase with an iron content of 27% by volume (Davis, 1999). Nevertheless, this result has to be completed by the fact that the MR effect depends on the initial elastomer modulus: i.e. the lower the initial

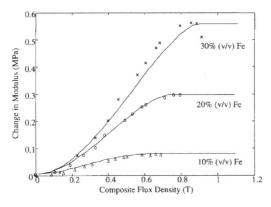

Figure 2. Elastic modulus response to applied field for 10, 20% and 30% iron by volume (Ruddy *et al.* (2007)).

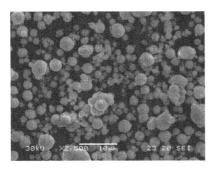

Figure 3. Scanning electron microscopy (SEM).

stiffness is (elastomeric gels for example), the higher is the stiffness MR effects.

2 UNDERSTANDING INTERDEPENDENCY OF THE PARAMETERS

MRE properties typically depend on the particles concentration, particle size, properties of elastomer matrix, applied field, preload, amplitude and frequency of the loading and other non investigated factors (temperature, ageing, ...). The interdependency of these factors is very complex; therefore it is important to carry on an investigation in order to identify dedicated applications depending on the performances.

2.1 *Elaboration of MRE samples*

The first task is the evaluation of different polyurethane and silicone grades for the selection of the unfilled matrix material. In the manufacturing point of view, low viscosity is a favorable property because of an easier dispersion of filling particles and degassing of compound. A dynamic Mechanical Thermal Analysis (DMTA) is utilized to characterise the unfilled matrix.

Then, the selection of magnetic filling particles is performed with the evaluation of different carbonyl

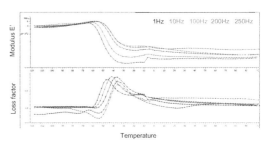

Figure 4. Dynamic properties of aligned MRE samples.

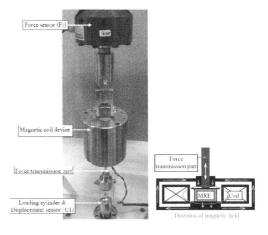

Figure 5. Experimental set-up.

iron grades through the determination of particle size distribution and morphology. Figure 3 shows the particles with scanning electron microscopy (SEM).

Finally, aligned MRE samples were manufactured with different iron powder contents (26% to 36%). Dynamic properties without any magnetic field have been evaluated via DTMA.

Following results hereafter are presented for Silicone MR-elastomer with the best percentage of iron particles.

2.2 *Experimental set-up*

The used experimental set-up used to evaluate MRE efficiency is illustrated on the following figures.

Magnetic fields were generated by combining permanent and electrical magnetic circuits. The magnetic flux of the circuit was determined as the function of the coil current.

The direct method was used in the dynamic test (detailed in standard ISO-10846). In direct method measurement force is measured from the force output side (F2) of the part (see Figure 5) and the displacement is measured from the force input side (U1). Dynamic stiffness was calculated from time displacement-force plots using trend line of the graph. Loss factor was determined by the basis of dissipated energy related to maximum of potential energy.

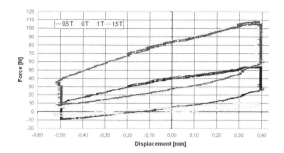

Figure 6.   Quasi-static hysteresis loops.

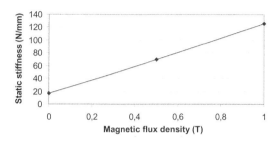

Figure 7.   Quasi-static stiffness versus magnetic flux density.

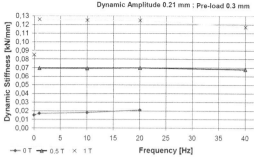

Figure 8.   Dynamic stiffness versus frequency for different magnetic flux densities.

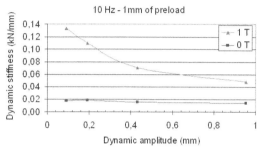

Figure 9.   Quasi-static stiffness dependency versus magnetic flux density with and without magnetic field.

2.3   Evaluation of quasi-static performances

Static test parameters are saw-tooth waveform, 0.1 Hz, amplitude ±0.5 mm = peak to peak 1 mm, room temperature of 20°C. Magnetic flux density is varied from 0 to 1.5 Tesla.

The saturation effect of the particles is highlighted on the graph: A magnetic flux density higher than 1T is no more efficient to increase the MR effect. The magnetic flux density behaves as an asymptotic law. Usually, rectangular shape hysteresis loops are typical of dry friction phenomenon (in opposite with classical visco-elastic behaviour of elastomers). Hence, behaviour of MRE is different from classical elastomers. Indeed, in addition to the viscous damping, other phenomena have to be taken into account.

Thus, four main dissipation mechanisms can be distinguished:

– Viscous damping (more effective on dynamic tests)
– Stick-slip occurring between the particles and the matrix.
– Dissipation due to the magnetic device (permanent magnet exhibits rectangular hysteresis loops)
– Dissipation due to the magnetic forces acting in the particles chains.

Slopes of the hysteresis loops represent static stiffness values. They are reported in the figure 7 in order to show the linear dependency with the applied field (before magnetic saturation of particles).

At least, it has to be noticed that static stiffness can be increased by a factor 6 with 1T of magnetic flux density (MR effect).

2.4   Evaluation of dynamic performances

Force-deflection loops have been achieved in order to highlight frequency and amplitude influences.

The low dynamic rigidification seems not to be affected by the magnetization. This phenomenon depends of the elastomer matrix only.

Dynamic stiffness plotted hereafter decreases with the dynamic amplitude. This behaviour is an elastomer typical behaviour but measures show that the Payne effect is amplified with the application of a magnetic field. At low deformation amplitude, stiffness can be increased by almost a factor of 7 with magnetic field, while, at higher amplitudes, this factor drops to 2. MR effect depends on the strain amplitude (due to magnetic forces dependence on the distance between the dipoles).

Loss factor is based on the dissipated energy related to maximum of potential energy by using the following equation:

$$\eta = \frac{Wd}{2\pi\left(\frac{1}{2}kX^{2}\right)} \qquad (1)$$

where η is loss factor, k is stiffness, X is amplitude and Wd is dissipated energy (area of the hysteresis loop).

The variation of the loss factor with the dynamic amplitude is representative from classical viscous behaviours. Very high values of damping are measured for low amplitudes.

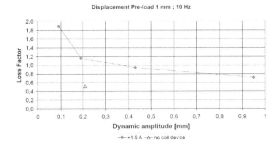

Figure 10.   Variation of loss factor with dynamic amplitude.

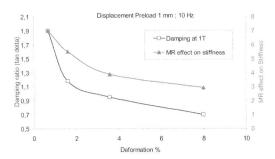

Figure 11.   Variation of the damping ratio (loss factor) and the MR effect with dynamic amplitude.

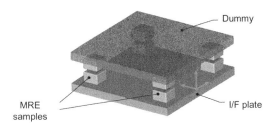

Figure 12.   Configuration for the dynamic test.

Loss factor is multiplied by a factor 2.3 with magnetic field (1T).

3   CONCLUSIONS

This research work produces large amount of data about dynamic properties of MRE's. Investigated sample configurations were polyurethane and silicone matrices filled with different percentage of particles.

The obtained optimum change on stiffness is higher than expected according to bibliography:

– Stiffness can be increased until a factor 6 with 1 Tesla of magnetic flux density
– Loss factor can be increased until a factor 2.3 with 1 Tesla of magnetic flux density
– The stiffening due to the magnetization (MR effect) depends on the initial modulus and on the strain amplitude.

Following applications could be considered for semi-active vibration control devices:

– Flexible ability for on orbit reconfiguration.
– Stiffener configuration during launch events.
– A Tuned Mass Damper with an adjustable resonant frequency.
– An isolator with adaptive properties able to correct the stiffness deviation under thermal solicitations.
– Adaptive tuned shocks absorbers.

Nevertheless, the design of the device would have to take into account the dynamic performances dependency with the amplitude (sum-up in the following figure).

After this study, some points have been highlighted as points to be performed before considering a use on a spacecraft such as the mass of magnetizing equipment, the MRE out-gazing for vacuum use and the mechanical strength to avoid creeping and cracking.

The next step will consist in performing an elementary dynamic test in order to evaluate stiffness and damping properties of such magneto-rheological suspension for representative vibrations events.

BIBLIOGRAPHY

Davis, L. C., 1999, Model of magnetorheological elastomers, *Journal of Applied Physics*, vol. 85, no. 6, pp. 3348–3351.
Guth, E., 1945, Theory of filler reinforcement, *Journal of Applied Physics*, vol. 16, no. 20, pp. 20–25.
ISO 10846-1, 1997, Acoustic and vibration – Laboratory measurement of vibro-acoustic transfer properties of resilient elements-part 1: Principles and guidelines, pages 20.
Keinänen, J., Lindroos, T., Liedes, T., Vessonen, T., Klinge, P., 2008, Adaptive Tuned Mass Damper Concept, Actuator 08, pp. 412–419.
Ruddy, C., Ahearne, E. and Byrne, G., 2007, A review of magneto-rheological elastomers: Properties and applications, University College Dublin, Ireland.

Constitutive Models for Rubber VII – Jerrams & Murphy (eds)
© 2012 Taylor & Francis Group, London, ISBN 978-0-415-68389-0

Coupling between diffusion of biodiesel and large deformation in rubber: Effect on the mechanical response under cyclic loading conditions

A. Andriyana & A.B. Chai
Department of Mechanical Engineering, University of Malaya, Kuala Lumpur, Malaysia

E. Verron
Institut de Recherche en Génie Civil et Mécanique, GeM UMR CNRS, École Centrale de Nantes, Nantes, France

M.R. Johan & A.S.M.A. Haseeb
Department of Mechanical Engineering, University of Malaya, Kuala Lumpur, Malaysia

ABSTRACT: Motivated by a variety of environment, political and economic concerns over the use of conventional energy sources, the development of biodiesel as alternative source of energy has been extensively explored during the last decade. In the context of industrial rubber components, the use of biodiesel, e.g. palm biodiesel, introduces additional requirement on the material compatibility since it often creates many problems in rubber seals, pipes, gaskets and o-rings in the fuel system. Hence, durability in service of rubber components in this aggressive environment becomes a critical issue. The present work can be regarded as a first step toward an integrated durability analysis of industrial rubber components exposed to aggressive environments, such as oil environment in biofuel systems, during their service. More precisely, the emphasis of this research is to investigate the diffusion of palm biodiesel into rubber undergoing concurrently large deformation. To this end, a specially-designed coupled diffusion-large deformation device is developed. Different types of rubber are considered. Particular attention is given on the effect of the corresponding coupled phenomenon on the cyclic behavior of rubber.

1 INTRODUCTION

Due to energy insecurity, increase in energy consumption and environmental concern, Asian countries, in particularly Malaysia and Indonesia, are actively promoting the use of biofuels as a partial substitution of petroleum fuels (Jayed *et al.* 2011). The corresponding policy is in line with the Kyoto Protocol which emphasizes the introduction of biofuels produced from plants absorbing CO_2 in order to reduce global warming. The use of biofuels such as palm biodiesel as alternative fuel for diesel engine appears to be attractive since it does not require extensive engine modification. Nevertheless some issues remain to be addressed, e.g. compatibility with engine components in particularly with those made of elastomers.

During the service, the elastomeric components are subjected to fluctuating multiaxial mechanical loading conditions which can lead to fatigue failure. Moreover, additional material degradation due to the presence of hostile environment such as oil rich environment is expected. One main form of degradation in rubber exposed to liquid is swelling which can be described in terms of mass or volume change (Haseeb *et al.* 2010). Thus, the need to study the interaction between fluctuating mechanical loading and diffusion of liquids into rubber materials and its resulting effect on the durability becomes a critical issue.

It is well known that under cyclic loading conditions, rubber shows some inelastic responses such as mechanical hysteresis, stress softening and permanent set (Diani *et al.* 2009). The hysteresis corresponds to the amount of energy loss during a cycle and can be related to either viscoelasticity (Bergström and Boyce 1998), viscoplasticity (Lion 1997) or more recently to strain-induced crystallization (Trabelsi *et al.* 2003). The stress softening was firstly observed by Mullins (1948) and often referred to as Mullins effect. In fact, during the first loading cycles, for a given strain level in the uploading, the stress decreases with the number of cycle before stabilizes after couple of cycles depending on the type of rubber. Up to this date, no general agreement has been found either on the physical source or on the mechanical modeling of this softening at the microscopic or mesoscopic scales (Diani *et al.* 2009).

In this study, the interaction between diffusion of palm biodiesel and large deformation in rubber is considered. In particularly, our focus is on the effect of the above interaction on the mechanical response under cyclic loading conditions with the main emphasis on the stress softening. Two types of rubbers are investigated: Nitrile Butadiene Rubber (NBR) and Polychloroprene Rubber (CR).

The present paper is organized as follows. In Section 2, experimental works including materials, specimen geometry and the types of test conducted in this study are described. The experimental results are presented and discussed in Section 3. Concluding remarks are given in Section 4.

2 EXPERIMENTAL PROGRAM

2.1 *Materials*

Rubber specimens used in this research are provided by MAKA Engineering Sdn. Bhd., Malaysia. The materials investigated are commercial grade of Nitrile Butadiene Rubber (NBR) and Chloroprene Rubber (CR) with 60 shore hardness. Biodiesel is prepared by blending palm biodiesel (provided by Am Biofuels Sdn. Bhd., Malaysia) with diesel. The analysis report of the palm biodiesel investigated is shown in Table 1. The immersion tests conducted are immersion in B0 (100% diesel), B25 (blend of 25% of biodiesel and 75% of diesel), B75 (blend of 75% of biodiesel and 25% of diesel) and B100 (100% biodiesel).

2.2 *Specimen geometry and compression device*

In order to investigate the interaction between diffusion and large deformation in rubbers and its resulting mechanical response under cyclic loading condition,

Table 1. Properties of B100 palm biodiesel

Test	Unit	Methods	Results
Ester content	% (m/m)	EN 14103	96.9
Density at 15°C	kg/m^2	EN ISO 12185	875.9
Viscosity at 40°C	mm^2/s	EN ISO 3104	4.667
Flash point	°C	EN ISO 3679	168
Cetane number	–	EN ISO 5165	69.7
Water content	mg/kg	EN ISO 12937	155
Acid value	mgKOH/g	EN ISO 3679	0.38
Methanol content	% (m/m)	EN 14110	<0.01
Monoglyceride content	% (m/m)	EN 14105	0.67
Diglyceride content	% (m/m)	EN 14105	0.2
Triglyceride content	% (m/m)	EN 14105	0.2
Total glycerine	% (m/m)	EN 14105	0.25

an annular rubber specimen having height, outer diameter and wall thickness of 10 mm, 50 mm and 6 mm respectively is used. The specimens are placed in a specially designed compression device as illustrated in the Figure 1. As depicted in the Figure 1, four stainless steel plates are arranged successively and separated by three spacer bar of different heights: 9.8 mm, 9 mm and 8 mm. Four rubber specimens are placed in each level. The plates are tightened using bolts and nuts until they are in contact with the respective spacer bar. Since the height of the rubber specimen is 10 mm, the above arrangement allows the introduction of different compressive strains to the specimen: 2%, 10% and 20%. The device containing rubber specimens are subsequently immersed into different palm biodiesel blends for durations of 30 and 90 days. The detail of the immersion tests is given in the Table 2. It is to note that in practice, the 2% compressive strain is so small that its effect on the macroscopic mechanical response is negligible. Nevertheless, this level of strain is retained to represent stress-free condition while ensuring that the diffusion occurs only along radial direction. Detail description of the compression device is given in another paper (Chai *et al.* 2011).

2.3 *Measurement of volume change*

The test procedure for the measurement of volume change can be summarized as followed:

1. Before the immersion, the weight of the rubber specimen is measured in air and in distilled water. The specimen is then quickly dipped into alcohol and blotted dry with filter paper.

Table 2. Immersion tests

Biodiesel Blend	Level of Compressive Strain (%)	Immersion Duration (days)
B0	2/10/20	30/90
B25	2/10/20	30/90
B75	2/10/20	30/90
B100	2/10/20	30/90

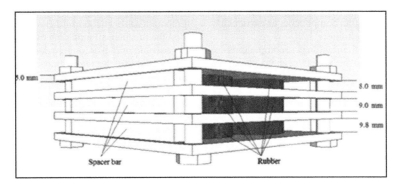

Figure 1. Specially-designed compression device for the observation of the coupling between diffusion and large deformation.

2. After weight measurement, the rubber specimens are placed in sequence on the compression plates. Grease is applied on the surface of the specimens that are in contact with the compression plate to avoid bulging of the specimens. Thereby ensuring the specimens to be in a simple uniaxial compressive stress state.
3. Bolts and nuts are used to tighten the compression device until the compression plates are uniformly in contact with the spacers. The device containing rubber specimens is subsequently immersed completely into different biodiesel blends for 30 or 90 days.
4. At the end of the immersion period, the specimens are removed from the compression device and quickly dipped into acetone; it is then clean with filter paper to remove the excess oil. The specimens are left for 30 minutes to allow for recovery before any measurement is made after immersion.
5. Step 1 is repeated to measure the weight of rubber specimen after immersion.

The percentage of volume change is calculated using the following relations (Trakarnpruk and Porntangjitlikit 2008):

$$\% \, Volume \, Change = \frac{(M_2 - M_4) - (M_1 - M_3)}{(M_1 - M_3)} \times 100$$

(1)

where M_1 and M_2 are the mass in air (gram) before and after immersion while M_3 and M_4 are mass in water (gram) before and after immersion.

2.4 Mechanical testing

After each period of immersion, the mechanical response of the material under cyclic loading condition is investigated. For this purpose, cyclic tests using Instron 5500 uniaxial test machine equipped with 10 kN load cell at room temperature are conducted. To ensure uniform displacement control on the specimens, circular compression plates are attached to the machine. The experimental setup is connected to a computer to record the experimental data. All tests are conducted at a strain rate of $0.01 \, s^{-1}$ to avoid excessive increase in the temperature of the specimens, i.e. thermal effect is not considered in the present study. The specimens are subjected to cyclic compressive loading at two different maximum compressive strains: 30% and 40% of 6 cycles each.

3 RESULTS AND DISCUSSION

Figures 2 and 3 presents the volume changes experienced by NBR and CR respectively in different contents of biodiesel, different durations of immersion and different pre-compressive strains. The graphs clearly show the increase in volume of both materials after immersion. In general, the volume increases with biodiesel content and duration of immersion. As indicated in this figure, CR shows significantly higher change in volume than NBR. Indeed, it is observed that the volume change ranges between 18–100% and 3–18% for CR and NBR respectively.

Except for CR immersed in B100 for 3 months, the pre-compressive strain appears to give restriction to the diffusion of liquid into rubber. Indeed, as the pre-compressive strain increases, the resulting volume change decreases. The corresponding results can be explained by the fact that as higher pre-compressive strain is introduced, the initial effective area for diffusion of liquid to occur becomes smaller. Moreover, as evoked by Treloar (1975), the presence of compressive stress will reduce diffusion since the hydrostatic part of this stress state is positive. According to the author, positive hydrostatic stress will restrict the diffusion. In the case of CR immersed in B100 for 3 months, the above trend is not followed. It might be attributed to the fact that the swelling level is so high (more than 80%) which could generate strong interaction

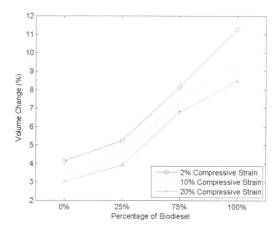

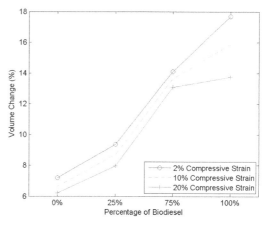

Figure 2. Volume changes of NBR at different compressive strains after 1 month (top) and 3 months (bottom) of immersion in different biodiesel blends.

285

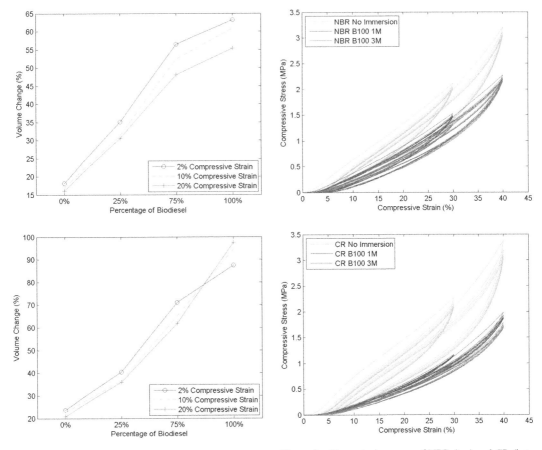

Figure 3. Volume changes of CR at different compressive strains after 1 month (top) and 3 months (bottom) of immersion in different biodiesel blends.

Figure 4. Stress-strain curves of NBR (top) and CR (bottom) under cyclic loading conditions at dry states (without immersion) and after 1 month (1M) and 3 months (3M) of immersion in B100. Results correspond to pre-compressive strain of 2%. Note that for immersed rubbers, the stress is expressed with respect to *unswollen-unstrained* configuration.

between liquid-rubber modifying significantly the effects of initial effective area and hydrostatic stress.

The stress-strain responses under cyclic compressive loading conditions at two different maximum compressive strains of previously non-immersed (dry) and immersed (swollen) NBR and CR are depicted in Figure 4. For each of maximum compressive strains, the specimen experiences six cycles of loading. It is important to highlight that the stress given in this figure is expressed with respect to *unswollen-unstrained* configuration (dry cross section). As indicated in Figure 4, no significant difference in the nature of stress-strain behavior between dry and swollen rubbers is observed. However, for given strain, lower stresses are recorded for swollen rubbers with CR appears to exhibit larger stress drop than NBR. Whether previously experiencing immersion or not, it appears that NBR and CR exhibit mechanical hysteresis and stress softening. The former, characterized by the difference between uploading and unloading paths during one cycle, decreases with loading cycle and stabilizes after 5 cycles. Moreover, the mechanical hysteresis

exhibited by dry CR appears to be larger than that of dry NBR. For both materials, the presence of liquids appear to decrease the size of hysteresis loop. In the latter, i.e. stress softening which is characterized by the difference in stress during the first few uploading, disappears after 5 cycles. It is to note that all aforementioned inelastic phenomena under cyclic loading conditions decrease significantly with the presence of liquids.

Before proceeding further with the nature of stress softening in the presence of biodiesels, it is important to recall the difference between the stress decrease associated with the decrease in strength of rubber with that associated with the stress softening. In the former, the idea is to compare the stress level during the first uploading between dry rubber and swollen rubber. The latter compares the stress level during the first and the second uploadings in one rubber (either dry or swollen). To characterize the former and the

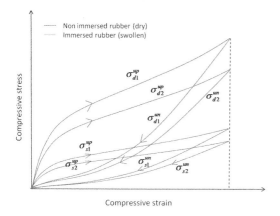

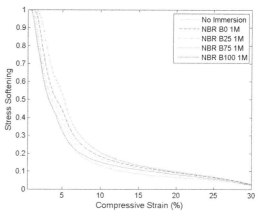

Figure 5. Illustration of two first cycles stress-strain curve of previously non immersed (dry) and immersed (swollen) rubbers under cyclic loading.

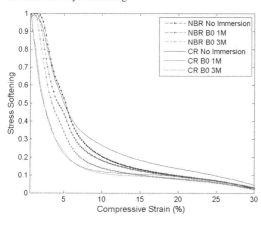

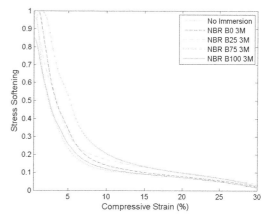

Figure 7. Stress softening in NBR previously immersed in various biodiesels for two different durations of immersion: 1 month (1M) (top) and 3 months (3M) (bottom). Results correspond to pre-compressive strain of 2%.

$$Stress\ softening\ (dry) = \frac{\sigma_{d1}^{up} - \sigma_{d2}^{up}}{\sigma_{d1}^{up}}$$

$$Stress\ softening\ (swollen) = \frac{\sigma_{s1}^{up} - \sigma_{s2}^{up}}{\sigma_{s1}^{up}}$$
(3)

where σ_{d1}^{up} is the stress in dry rubber during uploading of the first cycle, σ_{d2}^{up} is the stress in dry rubber during uploading of the second cycle, σ_{s1}^{up} is the stress in swollen rubber during uploading of the first cycle and σ_{s2}^{up} is the stress in swollen rubber during uploading of the second cycle. The use of Equation (3) for the definition of stress softening implies that we only focus on the softening which occurs between the first and the second uploading, i.e. further stress softening after the second uploading is not considered. In the next paragraph, the evolution of stress softening in the presence of biodiesel is discussed.

It is observed that in general the stress softening decreases as the strain level approaches the maximum strain previously endured by the material. Furthermore, the presence of liquids decreases the stress softening as can be consulted in Figures 6–8.

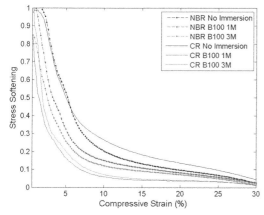

Figure 6. Stress softening in NBR and CR previously immersed in B0 (top) and B100 (bottom) for two different durations of immersion: 1 month (1M) and 3 months (3M). Results correspond to pre-compressive strain of 2%.

latter, we use the terms *stress drop* and *stress softening* respectively defined by (see Figure 5):

$$Stress\ drop = \frac{\sigma_{d1}^{up} - \sigma_{s1}^{up}}{\sigma_{d1}^{up}}$$
(2)

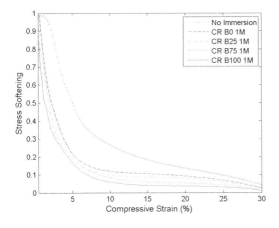

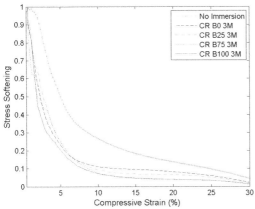

Figure 8. Stress softening in CR previously immersed in various biodiesels for two different durations of immersion: 1 month (1M) (top) and 3 months (3M) (bottom). Results correspond to pre-compressive strain of 2%.

Increasing the immersion time from 1 month to 3 months, i.e. increasing the swelling level, appears to further decrease the stress softening. Initially, at dry state, CR shows slightly higher level of stress softening than NBR as depicted in Figure 6.

The effect of biodiesel content on the stress softening in NBR and CR is presented in Figures 7 and 8. In this figure, it is found that for both materials, the increase in the content of biodiesel appears to decrease the level of stress softening.

4 CONCLUSIONS

In this work, the interaction between diffusion of liquids and large strain in rubber and its resulting mechanical response under cyclic loading conditions

was investigated. Two types of rubber were considered: NBR and CR. For given biodiesel content and given duration of exposure, it was found that CR experienced higher change in volume than NBR. Pre-compressive strain appeared to give restriction for diffusion of liquids. Moreover, for given duration of immersion, higher content of biodiesel yields to higher level of swelling in both rubbers.

Under cyclic loading conditions, inelastic phenomena in swollen rubbers appeared to be significantly smaller than that of dry rubber, i.e. smaller stress softening and mechanical hysteresis were recorded. The increase in the content of biodiesel decreased further the aforementioned phenomena. More extensive studies on the physical mechanism by which the presence of liquids modify the mechanical response under cyclic loading condition are needed in the near future.

REFERENCES

Bergström, J. S. & M. C. Boyce (1998). Constitutive modeling of the large strain time-dependent behavior of elastomers. *J. Mech. Phys. Solids 46*(5), 931–954.

Chai, A. B., A. Andriyana, E. Verron, M. Johan, & A. S. M. A. Haseeb (2011). Development of an experimental device to investigate mechanical response of rubber under simultaneous diffusion and large strain compression. In *7th European Conference on Constitutive Models for Rubber (ECCMR)*. Dublin, Ireland.

Diani, J., B. Fayolle, & P. Gilormini (2009). A review on the Mullins effect. *Eur. Polym. J.l 45*(3), 601–612.

Haseeb, A. S. M. A., H. H. Masjuki, C. T. Siang, & M. A. Fazal (2010). Compatibility of elastomers in palm biodiesel. *Renew. Energ. 35*(10), 2356–2361.

Jayed, M. H., H. H. Masjuki, M. A. Kalam, T. M. I. Mahlia, M. Husnawan, & A. M. Liaquat (2011). Prospects of dedicated biodiesel engine vehicles in malaysia and indonesia. *Renew. Sust. Energ. Rev. 15*, 220–235.

Lion, A. (1997). On the large deformation behaviour of reinforced rubber at different temperatures. *J. Mech. Phys. Solids 45*(11–12), 1805–1834.

Mullins, L. (1948). Effect of stretching on the properties of rubber. *Rubber Chem. Technol. 21*, 281–300.

Trabelsi, S., P. Albouy, & J. Rault (2003). Crystallization and melting processes in vulcanized stretched natural rubber. *Macromolecules 36*(20), 7624–7639.

Trakarnpruk, W. & S. Porntangjitlikit (2008). Palm oil biodiesel synthesized with potassium loaded calcined hydrotalcite and effect of biodiesel blend on elastomer properties. *Renew. Energ. 33*(7), 1558–1563.

Treloar, L. R. G. (1975). *The Physics of Rubber Elasticity*. London: Oxford University Press.

Constitutive Models for Rubber VII – Jerrams & Murphy (eds)
© 2012 Taylor & Francis Group, London, ISBN 978-0-415-68389-0

Characterization and numerical study of rubber under fast depressurization

J. Jaravel, S. Castagnet, J.C. Grandidier, G. Benoit & M. Gueguen
Institut Prime, CNRS – ENSMA – Chasseneuil du Poitou, France

ABSTRACT: Elastomers exposed to high-pressure gas suffer from internal fracture when the high-pressure gas is rapidly decompressed. In some cases, this kind of solicitation – referred to as explosive decompression – leads to cavity growth and blister fracture. To understand this phenomenon, a tensile machine is fitted with a pressure cell so that mechanical tests can be performed in gaseous environment (hydrogen or carbon dioxide). In order to understand explosive decompression failure, a simulation of the response of a hyperelastic hollow sphere under a coupled mechanical-diffusion load is proposed.

1 INTRODUCTION

Elastomers are used in industrial applications with a wide range of gas pressure and temperature. For instance, rubbers are used as seals in pumps and pipelines for gas extraction and transport and in the tanks of hydrogen vehicles. Hydrogen embrittlement in metals has been brought to light by means of high-pressure gas tests (Viswanadham, Green & Montague 1976), and it is necessary to study the behavior of elastomers in gaseous environments in order to predict their response under high-pressure gas. Specifically some elastomers suffer damage under fast depressurization after gas saturation. This type of solicitation is sometimes referred to as explosive decompression. This paper presents experimental tests of rubber undergoing explosive decompression, as well as a numerical simulation of a hyperelastic hollow sphere subjected to explosive decompression. The phenomenon of cavitation in rubber was discovered by Gent (Gent & Lindley 1959) with the "Poker chip test": a tension loading is applied to a rubber cylinder placed between two metal plates, thus inducing hydrostatic stress in its center. In specific conditions, cavities appear in the center of the rubber cylinder. This problem has been studied and modeled (Dollhofer, Chiche, Muralidharan, Creton & Hui 2004, Gent & Tompkins 1969b, Ball 1982, Gent 2005, Bucknall 2007) but few studies have been conducted to understand the effect of gas pressure on elastomers (Gent & Tompkins 1969a, Briscoe & Liatsis 1992, Briscoe, Savvas & Kelly 1994, Stevenson & Morgan 1995, Embury 2004, Li, Mayau & Song 2007, Yamabe & Nishimura 2009). In order to explain damage mechanisms in rubber, Gent & Tompkins (1969a) assume the existence of an initial defect in the material, which will grow for a pressure higher than $5E/6$, where E stands for the Young's modulus of the rubber. Other researchers postulate that the observed damage in rubber is cracks (Stevenson & Morgan 1995).

The aim of the experimental campaign is to verify experimentally the damage criterion for rubbers subjected to explosive decompression, and to check whether the criterion depends only on pressure level during saturation and does not depend on the rate of decompression. A coupled diffusion-mechanics simulation of a cavity in a hyperelastic material will give us some insight on the influence of mechanical and diffusion properties on the development of a cavity during explosive decompression.

2 EXPERIMENTAL STUDY

2.1 Material

A commercial silicone (vinyltrimethoxysilane) is used. Its transparency allows a spatial follow-up of damage in the sample. Its Young's modulus is 1.6 MPa at room temperature.

2.2 Experimental conditions

Tests are conducted on an Instron 8802 tensile machine fitted with a pressure chamber that allows traction tests under high gas pressure. Tests can be conducted with nitrogen, hydrogen, and carbon dioxide at a pressure ranging from 0 to 40 MPa. Decompression under CO_2 must be slow enough to avoid solid CO_2 plug in the purge system, as the triple point for CO_2 is at $-56.6°C$ and 0.51 MPa. The chamber has a diameter of 150 mm, a depth of 100 mm and a volume of 1.77 l to avoid any risk of explosion in case of an automatic purge. Consequently, tensile tests have a limited stroke.

As shown in Figure 1, chamber is fitted with one of the two doors that can rotate around the columns of the traction machine to allow easy installation of the samples. One door has a 40 mm central window and is suited for gas pressure up to 4 MPa, while the other

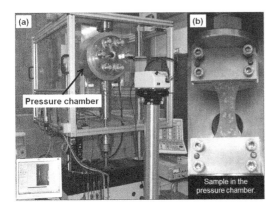

Figure 1. Tensile machine fitted with a pressure chamber (a) and sample clamped in the gas cell (b).

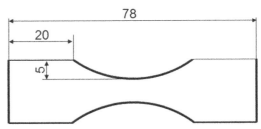

Figure 2. Size of the samples (thickness: 2 mm).

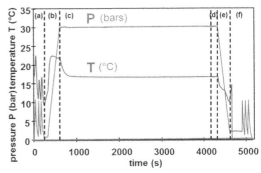

Figure 3. Pressure and temperature in the chamber during a test.

one can resist up to 40 MPa of gas pressure and has an off-centered 25 mm sapphire window that allows the follow-up of cracks on notched samples. The back side of the chamber is equipped with a centered sapphire window, which allows the spatial and temporal follow-up of damage using a Sony XCD SX90 CCD camera fitted with an objective. The mobile piston go trough the bottom of the chamber, chamber fitted with a water-cooled load cell that allows a measurement without seal friction. A system has been introduced under the mobile piston to equal gas pressure exerted on its top. The system is partially corrected because the surface on which off-setting pressure – that comes directly from the chamber through a pipe – is slightly different from the surface of the piston in contact with the gas in the chamber. The value of this difference is known and depends only on the gas pressure in the chamber and the contact surfaces of the gas. The chamber is fitted with a heating band, allowing tests at temperatures ranging from room temperature to 150°C. The thermal inertia of the system is important, therefore it is impossible to do experiments with complex thermal cycles. The precision of temperature measures is ±0.1°C when thermal equilibrium is reached. Furthermore, the tensile machine is located in a ventilated room for safety reasons, therefore room temperature varies from one day to another.

As for classical tensile tests, it is possible to apply displacement or stress when the chamber is filled with high-pressure gas. Stress measurement is done using a 20 kN ± 10 N internal load cell. Also, it is possible for specimens to undergo complex gas loadings (ramp, cycle...) synchronized with mechanical loadings with a precision of ±0.01 MPa during the stabilized stage.

2.3 Experimental protocol

Test into account the reduced size of the chamber, they are designed as depicted in Figure 2. Its thickness allows the specimen to be saturated by hydrogen after a few hours at a constant pressure. This is estimated by a one-dimensional calculation of the gas saturation

time in an infinite sheet of constant thickness e submitted to a gas concentration $C_{\infty H2}$ at both ends. The following equation allows time calculating of gas concentration in the sheet; it is based on Fick's theory and was proposed by Crank (1956).

$$\frac{C(t)}{C_{\infty H2}} = 1 - \frac{8}{\pi^2} \sum_{n=1}^{\infty} \frac{1}{(2n-1)} \exp\left(-(2n-1)^2 \pi^2 \frac{Dt}{e^2}\right) \quad (1)$$

where D is the diffusion coefficient of hydrogen in rubber.

For further investigation a slight stress concentration is introduced at the center of test specimens.

In order to avoid the mixture of oxygen and hydrogen in the pressure chamber, three successive pressurization/depressurization cycles are performed – before starting the test – by introducing nitrogen up to 1 MPa (Fig. 3.a). Hydrogen or carbon dioxide saturation of the sample (Fig. 3.c) is realized after pressurization of the chamber (Fig. 3.b). A mechanical loading can be applied to the sample (Fig. 3.d) before starting the decompression phase, which occurs at a controlled rate up to 90 MPa/min (Fig. 3.e). For safety reasons, three nitrogen purges must be performed before opening the chamber (Fig. 3.f).

2.4 Hydrogen effect on rubber damage

A test campaign is currently conducted to study damage in rubber under high hydrogen pressure. The first study consists in determining at what time and under which conditions damage appears under explosive decompression. Tests with various saturation times,

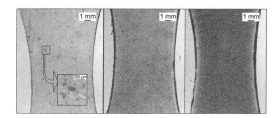

Figure 4. Effect of decompression rate on damage of a commercial rubber. Decompression at 0.9 MPa/min (left), 9 MPa/min (center), and 90 MPa/min (right) after hydrogen saturation at 9 MPa during 1 hour.

decompression rates, saturation levels and stress levels are currently performed on samples described on Figure 2.

The first results show the effect of the decompression rate after pressurization at 1 MPa/min and saturation at 9 MPa.

Figure 4 shows the apparition of damage in the sample after decompression. The density of cavities increases with the decompression rate. Observe the development of cavities in the sample from primary cavities. These first cavities lead to the development of satellite bubbles. This phenomenon is described by Gent et al (Gent & Tompkins 1969a).

3 NUMERICAL STUDY

3.1 Model description

The aim of the model is to describe the behaviour of a micrometric cavity in a hyperelastic environment submitted to a coupled diffusion-mechanical loading. The goal is to simulate an experimental test with pressurization, saturation, and explosive decompression using a two-dimensional calculation with Abaqus®. This calculation is done with coupled temperature-displacement elements, thanks to the similarities between mass diffusion and thermal diffusion equations. A hollow sphere is modeled and hyperelasticity of the material is implemented using a subroutine UHYPER and a Mooney-Rivlin density of strain energy, as indicated in Eq. 2. C_{ij} and D_1 are material constants, and I_1, I_2, and J stand for the three invariants of the strain tensor.

$$w = C_{10}(I_1 - 3) + C_{01}(I_2 - 3) + C_{20}(I_1 - 3)^2 + \frac{1}{D_1}(I_3 - 1)^2 \tag{2}$$

Subroutine UMATHT is used in order to define the user diffusion of gas in the material. The components of the flux are defined as follows (Eq. 3).

$$flux(i) = -\alpha \left(\frac{\partial C}{\partial x_i} \right) \tag{3}$$

Furthermore, subroutine DFLUX allows us to implement the diffusion of gas between the material

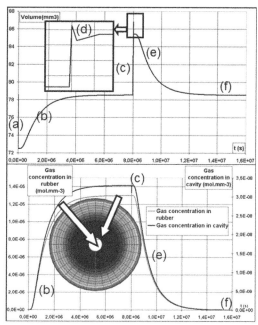

Figure 5. Follow-up of the volume (up) and gas concentration (down) of the cavity during a cycle of pressurization, saturation, and decompression.

and the cavity, in which a ideal gas law is used (Eq. 4.). This flux is expressed as a function of the pressure difference between the cavity and the material. In the rubber, Henry's law is used (Eq. 5).

$$flux = \beta(\Delta P) \tag{4}$$

$$Pressure = \frac{Concentration}{solubility} \tag{5}$$

Pressure in the cavity apply a mechanical pressure on the cavity. A mechanical loading and temperature are applied on the exterior boundary of the model they are linked with one another by Henry's law (Eq. 5).

3.2 First results

First simulations have been done to model the response of a cavity submitted to decompression after saturation.

Figure 5 shows a decrease of the size of the cavity during pressurization (Fig. 5.a), followed by its progressive growth as the gas diffused in the sample (Fig. 5.b). At the start of decompression, the mechanical pressure is abruptly removed, inducing a short swelling phase (Fig. 5.c), as well as an increase of cavity volume. Thus, the concentration inside the cavity drops, and gas starts filling the cavity (Fig. 5.d). The last phase consists in a decrease of gas concentration (Fig. 5e.), then the cavity goes back to its initial state (Fig. 5.f).

The proposed model did not lead to a catastrophic development of the cavity, which could be caused by an instability due to material parameters or coupling between mechanics and diffusion. The addition of a rupture criterion is necessary to model the apparition of macroscopic cavities from initial bubbles.

4 CONCLUSION

Experiments are currently conducted in order to understand the damage mechanisms in action during fast depressurization at high pressure level (hydrogen, carbon dioxide ...). These tests aim at developing a damage criterion able to predict the behavior of elastomers under explosive decompression. To help us understand these phenomena, a numerical model of a hollow sphere was built using Abaqus®.

REFERENCES

Ball, J.M. 1982. Discontinuous equilibrium solutions and cavitation in nonlinear elasticity. *Phil. Trans. R. Soc. Lond.*(A 306): 557–611.

Briscoe, B.J. & Liatsis, D. 1992. Internal crack symmetry phenomena during gas-induced rupture of Elastomers. *Rubber Chemistry And Technology*(65): 350–73.

Briscoe, B.J., Savvas, T. & Kelly, C.T. 1994. Explosive decompression failure of rubber: a review of the origins of pneumatic stress induced rupture in elastomer. *Rubber Chemistry And Technology*(67): 384–416.

Bucknall, C.B. 2007. New criterion for craze initiation. *Polymer*(48): 1030–1041.

Crank, J. 1956. *The Mathematics of diffusion*, Oxford: Clarendon.

Dollhofer, J., Chiche, A., Muralidharan, V., Creton, C. & Hui, C.Y. 2004. Surface energy effects for cavity growth and nucleation in an incompressible neo-Hookean material – modeling and experiment. *Int. J. of Solids and Structures*(41) 22–23: 6111–6127.

Embury, P. 2004. High pressure gas testing of elastomer seals and a practical approach to designing for explosive decompression service. *Sealing Technology*: 6–11.

Gent, A.N. 2005. Elastic instabilities in rubber. *Int. J. of Non-Linear Mechanics*(40): 165–175.

Gent, A.N. & Lindley, P.B. 1959. Internal rupture of bonded rubber cylinders in tension. *Proceedings Of The Royal Society Of London Series A-Mathematical And Physical Sciences*(249) 1257: 195–205.

Gent, A.N. & Tompkins, D.A.1969a. Nucleation and growth of gas bubbles in Elastomers. *J. Of Applied Physics*(40): 2520–5.

Gent, A.N. & Tompkins, D.A. 1969b. Surface energy effects for small holes or particles in Elastomers. *J. Polym. Sci. Part A2*(7): 1483–1488

Stevenson, A. & Morgan, G. 1995. Fracture of Elastomers by gas decompression. *Rubber Chemistry And Technology*(68): 197–211.

Li, J., Mayau, D. & Song, F. 2007. A constitutive model for cavitation and cavity growth in rubber-like materials under arbitrary triaxial loading. *Int. J. of Solids and Structures*(44): 6080–6100.

Viswanadham, R.k., Green, J.A.S. & Montague, W.G. 1976. Hydrogen embrittlement of an fe-based amorphous metal. *Scripta Metallurgica*(10) 3: 229–230.

Yamabe, J. & Nishimura, S. 2009. Influence of fillers on hydrogen penetration properties and blister fracture of rubber composites for O-ring exposed to high-pressure hydrogen gas. *Int. J. Of Hydrogen Energy*(34) 4: 1977–1989.

Constitutive Models for Rubber VII – Jerrams & Murphy (eds)
© 2012 Taylor & Francis Group, London, ISBN 978-0-415-68389-0

Fatigue peeling of rubber

T.L.M. Baumard & A.G. Thomas
Department of Materials, Queen Mary University of London, London, UK

W. Ding
Dunlop Aircraft Tyres Limited, Erdington, Birmingham, UK

J.J.C. Busfield
Department of Materials, Queen Mary University of London, London, UK

ABSTRACT: A fatigue peeling test has been developed to evaluate the failure of rubber to rubber interfaces under cyclic loading. The test specimen has legs wider than the contact width between the two layers being unpeeled. Results obtained through this method have been compared to those of a typical fatigue crack growth experiment. Both measurements are shown over a wide range of peeling or tearing energies and the results shown that the trends in behavior between these two failure modes are similar with the peeling necessary to drive the crack being slightly higher than the tearing energy at the same crack growth rate. This is probably due to higher energy dissipation as a result of crack tip blunting due to a wider peel interface being present in the peel specimen. Cyclic and time dependent contributions to the fatigue crack growth behavior have been calculated using this test for an SBR compound and the results appear to be consistent with previous work.

1 INTRODUCTION

Tyres are made of several layers bonded together either by physical or chemical interactions. The interfaces formed are of potential concern when it comes to predicting the lifetime of the whole structure with cracks likely to initiate or to be driven in the interface regions. In the worst case, catastrophic failure can result from poor bonding properties or adhesion between two layers.

Peeling tests have been widely used before to evaluate the adhesion between two materials or the tear resistance of a thin layer of adhesive. Depending on the material tested and its behavior at the test temperature, several peel configurations have been adopted previously.

For strong interfaces with a wide contact width such as that shown in Figure 1 results in the legs tearing through. Several methods have previously been proposed to overcome this such as introducing a cloth backing to the legs of the test piece or by reinforcing the legs by sticking them to a stiffer material. The technique used in this study is to reduce the contact width between the two layers, making it less than the width of the legs. The stress is thus concentrated at the interface and this forces the crack to be driven along the interface. In order to validate the procedure, a conventional pure shear fatigue test was conducted and the results were compared to those obtained from this peeling experiment.

Crack growth in elastomers has been successfully described using a fracture mechanics approach based

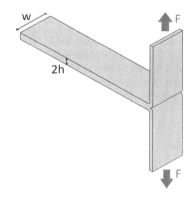

Figure 1. Peel Specimen.

on the tear energy or energy release rate concept. The tear energy is calculated using the relation

$$T = -\left(\frac{\partial U}{\partial A}\right)_l \tag{1}$$

where U is the elastically stored energy in the test piece, A is the surface area created as the crack propagates and l indicates that the differentiation is made at constant elongation so that the external forces do not work. For very brittle materials such as glass this energy is equal to the surface energy of the two new created surfaces. However for elastomers, the energy required to drive the crack greatly exceed this surface energy, resulting from the visco-elastic energy dissipation during tearing.

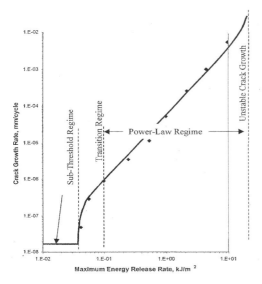

Figure 2. Four regimes of fatigue behavior of elastomers (taken from Lindley (1973)).

Table 1. Formulation of the compounds given in parts per hundred rubber (phr).

	SBR50	NR50
SBR 1500	100	–
NR (SMR CV60)	–	100
Carbon black		
N330 HAF	50	50
Stearic acid	2	2
Zinc oxide	5	5
6PPD[a]	3	3
CBS[b]	–	1.5
DPG[c]	1.3	–
MBTS[d]	1	–
Sulphur	1.5	1.5
Curing temperature /°C	160	150
Cure time/min	30	12.5

[a] N-(1,3-Dimethylbutyl)-N0-phenyl-p-phenylenediamine
[b] N-cyclohexyl-2-benzothiazolsulphenamide
[c] 1,3-diphenylguanidine
[d] Dibenzothiazyl disulfide

The energy release rate concept was first applied to the analysis of rubber specimens under static loading but it was quickly realized that the concept could also be applied to cyclic loading, with the maximum tear energy during a cycle determining the crack growth rate. Lake & Lindley (1965) showed that the fatigue crack growth behavior was divided into four regions (Fig. 2). At low tear energies (sub-threshold regime), crack growth only depends on the environmental attack. After a transition, the crack growth rate per cycle follows a power law relationship of the form

$$\frac{dc}{dn} = B\left(\frac{T}{T_u}\right)^\beta \qquad (2)$$

where B and β are both constants characteristic of the material tested and $T_u = 1\ \text{J/m}^2$ is introduced to make T/T_u dimensionless. For natural rubber (NR) and styrene-butadiene rubber (SBR) based compound, β is respectively about 2 and 4. In this regime, the behavior is purely related to the mechanical loading applied to the test piece.

Strain crystallizing rubber materials such as NR compounds essentially show only a very modest change in the fatigue behavior with test frequency but SBR materials have been shown by Busfield et al. (2002) to have a frequency dependence on the rate of tearing even when the maximum tearing energy in the cycle is maintained. In this case it has been proposed that the fatigue crack growth results from a combination of two effects. The first is a time dependent contribution that only depends on the length of time of a cycle and the second is a cyclic component that reflects only the crack growth the sample is

submitted to because of the numerous loading cycles independently of the length of time of each cycle.

$$\left(\frac{dc}{dn}\right)_{total} = \left(\frac{dc}{dn}\right)_{time} + \left(\frac{dc}{dn}\right)_{cycle} \qquad (3)$$

Busfield et al. (2002) deducted the magnitude of both contributions for unfilled SBR using a pure shear test piece. They observed that at low frequencies and for high tear energies, the fatigue behavior was largely dominated by the time dependant contribution, the cyclic contribution being almost negligible.

Assuming this observation was valid for peeling, the time dependent contribution on fatigue behavior on a peeling test was evaluated by extrapolating the data at the lowest frequencies and at high tear energies for a styrene butadiene rubber filled with 50 parts of carbon black. From this the static crack growth rate as a function of the tear energy was deduced.

$$\frac{dc}{dt} = B_s\left(\frac{T}{T_u}\right)^{\beta_s} \qquad (4)$$

Integrating this relationship during a loading cycle for a range of different test frequencies gives the time dependent contribution (Busfield 2002). This allows the cyclic contribution per cycle to be calculated using equation 3.

2 EXPERIMENTAL METHODS

2.1 Materials used

The formulation of the materials used to check the validity of the test in terms of fracture mechanics are shown in Table 1. They have been chosen to reflect different fatigue behavior in terms of the strain-crystallizability of the polymer.

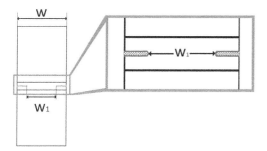

Figure 3. Reduction of the surface of contact. The Teflon film is shown as hatched.

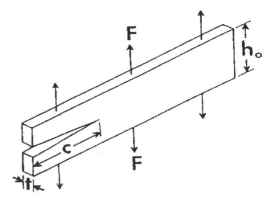

Figure 4. Pure shear test specimen.

2.2 Peeling specimen

The specimens used in this study were approximately 120 mm long, 25 mm wide and 2 mm thick for each leg. The reduced contact width was obtained by pressing together two rubber layers separated by a 25 μm thick Teflon film that has been cut beforehand to allow contact between the two layers only on a 15 mm by 80 mm surface centered on the width of the sample as can be seen on Figure 3.

Uncured rubber sheets of the correct dimensions were prepared using different procedures depending on the material used. For compounds with low adhesion to metal, the layer has been prepared by heating up the rubber to 70°C in order to mold it to the right dimension in a hot press. Materials that adhered too strongly to were instead carefully cut from accurately calendared sheets prepared on a two roll mill.

Once the rubber sheets were brought into contact, they were left to allow interdiffusion of the molecules through the interface for 3 hours at a pressure of 2 Bar. The peel test sample was then cured in a hot press under a pressure of about 4 Bar. After removal of the Teflon film, the samples present a reduced bonded surface of width equal to w_1.

2.3 Peeling test

For a peel specimen, the variation of energy is equal to the difference between the work applied on the legs and the energy dissipated through the deformation of those legs when the crack propagates distance dc. The peel energy is then given by

$$P = \frac{2F\lambda}{w_1} - \frac{2hwW}{w_1} \tag{5}$$

where λ is the extension of the legs, F is the force applied and W is the strain energy density calculated from a tensile test experiment at the same uniaxial stress as the peel test. In order to minimize the influence of the Mullins effect, the tensile data have been evaluated after 1000 cycles as suggested by Asare et al. (2009).

The crack growth rate per cycle has been deducted from the peak displacement during the fatigue peel test which was done to a constant maximum force in each cycle. The crack growth rate has been obtained by taking the slope of the displacement against number of cycles curve divided by twice the extension ratio, λ, at the maximum force after the sample has reached the steady crack growth rate (Papadopoulos et al. 2008).

$$\frac{dc}{dn} = \frac{1}{2\lambda}\frac{dl}{dn} \tag{6}$$

2.4 Pure shear experiments

To validate the design of the peeling test, pure shear fatigue tests were conducted on identical compounds to compare with the results obtained with peel experiments based on the assumption that the crack growth rate per cycle does not depend on the specimen geometry. The pure shear fatigue test piece has been widely used to evaluate the fatigue properties of rubber compounds in different aspects of rubber physics such as wear (Liang et al. 2009) or fatigue crack growth prediction (Papadopoulos et al. 2003, Busfield et al. 1999). A typical pure shear specimen is shown in Figure 4. As long as the width of the sample is about 8 times its height and as the two ends of the sample are kept parallel by the clamps, a region of the sample is in pure shear (Thomas 1994). If a crack is progressing through this region then the tear energy is independent of the length of the crack tip and equal to

$$T = Wh_0 \tag{7}$$

where W is the strain energy density in pure shear and h_0 is the height of the sample.

Papadopoulos et al. (2008) showed that for a pure shear test piece, the tear energy could be calculated from the equation below

$$T = \frac{V}{t(l_0 - c + x)} \tag{8}$$

where l_0 is the length of the test piece, t is its thickness, c is the size of the crack and U is the elastically stored

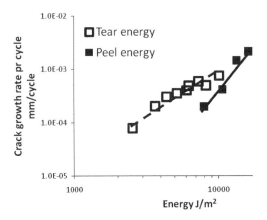

Figure 5a. Comparison between tear and peel fatigue behavior for NR50.

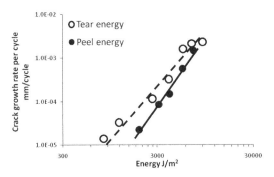

Figure 5b. Comparison between tear and peel fatigue behavior for SBR50.

energy in pure shear that can be deduced from the force displacement relationship in pure shear for the material tested. x is the length of a strip in the uncracked region which is not energy free and contributes to the release of energy as the crack is driven and has been found to be 28% of the length l_0 by finite element analysis (Asare et al. 2011).

3 RESULTS AND DISCUSSION

3.1 Comparison between pure shear and peeling

The relationship between the peel energy and peel rate differs slighty from the tear energy versus fatigue crack growth rate for both SBR and NR compounds. It appears that both materials have slightly better fatigue peel properties as is seen in Figures 5a and 5b. The difference tends to decrease at the higher energy regions. This is thought to be due to the addition of two geometric effects. The first being due to the contact width being significantly wider in the peel test than the sample thickness in the pure shear fatigue test. The second due to energy being possibly dissipated due to bending of the test piece during the peel test.

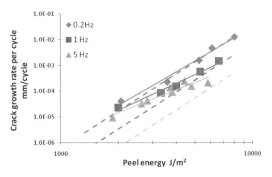

Figure 6. Frequency effect on fatigue peel behavior for SBR50.

Kadir & Thomas (1981) evaluated the effects of pure shear specimen thickness on the fatigue or fracture properties of materials. They showed that crack growth dependence on thickness is related to the development of crack tip roughness. They observed that for gum SBR, the change in crack growth rate could be of an order of magnitude when the thickness increased from 0.5mm to 5mm at constant energy release rate. For thicker specimen the crack growth rate is significantly higher than for thinner test pieces. This effect has also been reported by Tsunoda et al. (2000). They hypothesised that cavitation could happen at the crack tip under hydrostatic pressure and result in the roughness around the tip. They explained the change in crack growth rate by the change in crack tip roughness which depended significantly on the sample thickness.

Peeling samples, due to their wider crack tip (15 mm) compared with pure shear test pieces (2 mm thickness) develop rougher profiles at the same energy release rate resulting in a lower crack growth rate per cycle. Moreover the data tends to converge at high energy release rates.

Energy dissipation via bending of the legs has been neglected for the calculation of the peel energy. However, depending on the force applied to the legs (and thus the peel energy), the radius of curvature of the legs can vary. When the crack grows at a constant elongation, the radius of curvature changes especially at low energies where the contribution is proportionally higher. This modification of the radius causes a release in the bending energy stored in the material that is not taken into account in the calculations. The calculated peel energy is therefore overestimated. The precise contribution of this bending term can be estimated using a finite element model and this is the subject of an ongoing project.

3.2 Time dependent contribution for peeling

The effect of the frequency on the fatigue behavior of a styrene butadiene compound filled with 50 parts of carbon black can be seen on Figure 6. It is obvious that the higher the frequency the lower the crack growth rate per cycle especially at higher frequencies. At relatively low peel energies the crack growth behavior of SBR50 at different frequencies converge and the frequency effect becomes less pronounced.

The dashed lines in Figure 6 represent the calculated values of the time dependent contribution to fatigue crack growth $(dc/dn)_{time}$ obtained from the static crack growth coefficients. It appears that at high peel energies the total crack growth rate per cycle equals the time dependent contribution. The fatigue behavior is totally time dependant at those peel energies. Conversely, it appears that at low peel energies the fatigue behavior of the compound is dominated by a cyclic contribution especially at high frequency. Why the fatigue behavior of the styrene butadiene rubber is dominated by time dependant contribution at high peel energies and by cyclic component at low ones is still unclear and needs further investigations. However the results obtained with peeling are consistent with those obtained from a pure shear fatigue test by Busfield et al. (2002) even though the peel and the tear energies involved in those processes are different. Further investigations on the magnitude of both contributions have to be done to explain the dependence of the fatigue behavior on both contributions.

4 CONCLUSION

A fatigue type peeling experiment has been successfully developed and both strain and non strain crystallizing rubber compounds have been tested by reducing the contact width between the two legs of the specimen. Compared with a pure shear fatigue test, the fatigue behavior in peeling appears to give lower crack growth at the same energy release rate. This difference is thought to relate to the test width of the front of the crack resulting in higher energy dissipation via cavitation and crack tip blunting in a peeling design. This is a topic of ongoing research whereby the width of the peel region is being systematically varied to investigate the effect of contact width on the measured behavior.

The time dependent component has been calculated at different frequencies for a non strain-crystallizing compound and shows similar results measured in fatigue by Busfield et al. (2002). The fatigue peel behavior is again largely dominated by time dependent contribution at high peel energies and low frequencies while it is mainly affected by a cyclic component at low energy release rates and high frequencies.

One interesting investigation would be to evaluate both contributions for the same material in pure shear and see if the difference between the tear and the peel energies is similar. The extent of the difference could lead us to draw conclusions on whether the cavitation at the crack tip blunting is a purely time-dependant effect or not for non-strain crystallizing compounds.

ACKNOWLEDGEMENTS

One of the authors, Thomas Baumard, would like to acknowledge the sponsor, Dunlop Aircraft Tyres Ltd. for their generous financial support.

REFERENCES

Asare, S. & Busfield, J.J.C. 2011. Fatigue life prediction of bonded rubber components at an elevated temperature. *Plastics Rubber and Composites* 40: 192–198.

Asare, S., Thomas, A.G. & Busfield, J.J.C. 2009. Cyclic stress relaxation (CSR) of filled rubber and rubber components. *Rubber Chemistry and Technology* 82: 104–112.

Busfield, J.J.C., Tsunoda, K., Davies, C.K.L. & Thomas, A.G. 2002. Contributions of time dependent and cyclic crack growth to the crack growth behaviour of non strain crystallising elastomers. *Rubber Chemistry and Technology* 75: 643–656.

Busfield, J.J.C., Thomas, A.G. & Ngah, M.F. 1999. Application of fracture mechanics for the fatigue life prediction of carbon black filled elastomers. In Dorfmann, A. & Muhr, A. (eds), *Constitutive Models for Rubber*: 249–256.

Kadir, A. & Thomas A.G. 1981. Tear behavior of rubbers over a wide range of rates, *Rubber Chemistry and Technology* 54: 15–23.

Lake, G.J. & Lindley, P.B. 1965. The mechanical fatigue limit for rubber, *Journal of Applied Polymer Science* 9:1233–51.

Liang, H., Fukahori, Y., Thomas, A.G. & Busfield, J.J.C. 2009. Rubber abrasion at steady state. *Wear* 266: 288–296

Lindley, P.B. 1973. Relation between hysteresis and dynamic crack growth resistance of natural rubber, *International Journal of Fracture* 9: 449–462.

Papadopoulos, I.C. Thomas, A.G. & Busfield, J.J.C. 2008. Rate transitions in fatigue crack growth of elastomers, *Journal of Applied Polymer Science* 109: 1900–1910.

Papadopoulos, I.C., Liang, H., Busfield, J.J.C. & Thomas, A.G. 2003. Predicting cyclic fatigue crack growth using finite element analysis techniques applied to three-dimensional elastomeric components. In Busfield, J.J.C. & Muhr, A. (eds), *Constitutive Models for Rubber III*: 33–40

Thomas, A.G. 1994. The development of fracture mechanics for elastomers, *Rubber Chemistry and Technology* 67: G50–G60.

Tsunoda, K., Busfield, J.J.C., Davies, C.K.L. & Thomas, A.G. 2000. Effect of materials variables on the tear behaviour of a non-crystallising elastomer. *Journal of Materials Science* 35: 5187–5198.

Constitutive Models for Rubber VII – Jerrams & Murphy (eds)
© 2012 Taylor & Francis Group, London, ISBN 978-0-415-68389-0

Elastomer prediction method for space applications

P. Camarasa

ASTRIUM Satellites, Avenue des Cosmonautes, Toulouse Cedex, France

ABSTRACT: Satellites are subject to various types of vibrations ranging from high amplitude vibrations during launch to small vibrations (so-called micro-vibrations) during in-orbit operations. Such vibrations are key design drivers for structures, payload and equipment items. The level of residual vibrations is a major issue in the design of satellites as it directly impacts the structural strength and/or the functionality of payload and equipment items during launch, and the pointing performance during in orbit operations.

This is why, ASTRIUM develops, for many years, isolators and dampers of in-orbit micro-vibrations and launch vibrations.

Historically the design of such products was mainly based on engineering experience obtained in particular through prototyping phases. In order to reduce development cost and schedule, a methodology for predicting the elastomer characteristics and behaviour is under development. The goal is to improve the performance prediction of elastomer-based isolation and damping devices.

1 SATELLITE AND LAUNCHER NEEDS

For space applications three kinds of vibrations shall be considered:

- Shock induced on the satellite. Several events can be dimensioning for the satellite, shock events are due to launcher stages or satellite separations but also due to reflector or solar arrays release on the satellite.
- Static and transient vibrations transmitted by the launcher. These events are translated in steady state vibrations (sine and random) for satellite qualification.
- Microvibration in orbital configuration for high pointing accuracy satellite. All disturbances coming from moving parts in equipment, like reaction wheels or cryocoolers can lead to pointing performance degradation.

In order to reduce vibration levels, three possible implementations of isolators or dampers can be considered on the satellite: the first one is at the interface of the disturbing source, the second one is at the interface of the sensible instrument and the third one is an implementation along the vibration propagation path.

2 ELASTOMER CHARACTERISATION

2.1 *Introduction*

Elastomer devices are commonly applied for the reduction of vibrations levels. Elastomer formulation depends on applicable requirements. The main parameters are stiffness and damping performances, thermal and mechanical environment. From these end-level

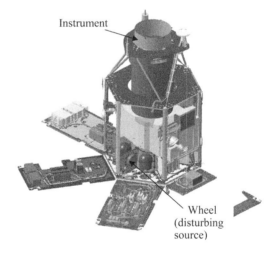

Instrument

Wheel (disturbing source)

Figure 1.   Observation satellite example.

specification, a need exists to predict the behaviour of the elastomer in terms of superposed response under static and vibratory loads. Components characteristics such as static force-deflection behaviour and dynamic stiffness at various frequencies and pre-loads are also important design conditions.

Therefore, in order to obtain elastomer laws for prediction by finite element, a complete static and dynamic characterization is performed in the relevant mechanical environment.

However,   elastomers   present   visco-elastic behaviour. At the present time, still few industrial dimensioning calculations are done by considering the non-linear visco-elastic phenomena. Therefore, one achieves finite elements calculations with the

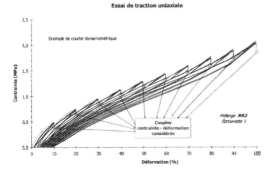

Figure 2. Unidirectional traction/compression example.

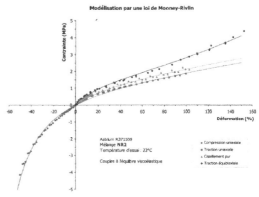

Figure 3. Mooney Rivlin law for modeling.

assumption that compact elastomers present elastic behaviour in large strains. Objective for the future is to implement visco-elastic behaviour in dynamics.

2.2 Static characterization and validation

For that, it is essential to establish a data base sufficiently complete including experimental results obtained from unidirectional and multidirectional tests.

This data base is created by applying static experimental protocols, making it possible to avoid the main part of the dissipation (viscoelasticity). The experimental results from the various strains allow calculating the coefficients of hyperelastic laws (stress/strain relation in large strain).

Before calculation on final device, it is necessary to validate the hyperelastic models. For that, representative samples in term of elastomer confinement (loaded/free surfaces ratio) are tested in traction/compression and shearing. A test/prediction correlation up to large strain is analyzed. A selection among Mooney and Ogden laws for different strain ranges is performed; the straightforward Neo-Hooke law with a single coefficient is often used.

2.3 Elastomer dynamic properties and characterization

Elastomer viscoelasticity leads to dynamic stiffness increasing with frequency and decreasing with stain amplitude (Payne effect).

In order to complete the dynamic behaviour, one considers a superposition of hyperelastic law with linear viscoelasticity in small strain. Both dynamic stiffness and damping behaviour are determined by so-called phi-functions. Coefficients of these functions are identified from viscoelastic modulus E', E".

These modulus are obtained by sine sweep in frequency and amplitude for different temperatures. First, the samples are tested in small strain without preload.

Then, for large dynamic strains, previous samples with metallic interfaces for strength validation are tested. In this case, the complex stiffness K* or K', K" are analyzed.

Figure 4. Traction/compression and shearing tests.

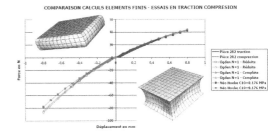

Figure 5. Hyperelastic law validation.

3 STATIC AND HARMONIC ANALYSIS METHOD

3.1 Analysis and tests methodology

For both static and dynamic loads, the method consists in comparing analytical analysis with tests results and MARC FEM.

In statics, analytical approach and FEM assess the non-linear stiffness with respect to displacement. These results are correlated with stiffness measurement after relaxation steps.

In dynamics, it is more difficult: analytical approach assesses the frequency mode taking into account the stiffness increasing with frequency. An iterative algorithm is used. Then the final transfer function of isolation is assessed with MARC and correlated with test results.

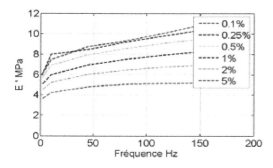

Figure 6. Example of Elastic modulus E' with respect to frequency and strain.

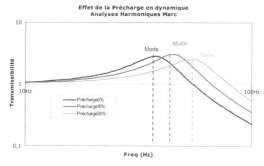

Figure 8. Static pre-load impact for transmissibility assessment.

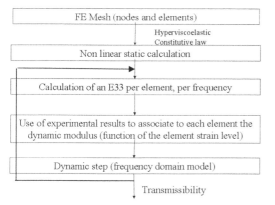

Figure 7. Iterative procedure for Payne effect.

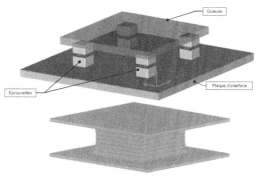

Figure 9. Mock-up with four elastomer block – Block with elastomer base plate in traction.

3.2 Modeling technique with MARC FEM

In statics, it is possible to use a wide range of Mooney-Rivlin law with several coefficients. In dynamics, it is necessary to use a Mooney-Rivlin with a single coefficient (Neo-Hooke law). Therefore, the analysis is organized with the following two steps:

- Static pre-load (gravity, thermo-elastic if necessary)
- Small linear vibrations around a non-linear pre-stressed static equilibrium

The two non-linear effects in dynamics are assessed with the phi-functions for the stiffness increasing with frequency, and with an iterative approach for the stiffness decreasing with amplitude.

Basically, industrial finite element codes such MARC or ABAQUS are not able to take into account large strain; nevertheless the suggested iterative approach assesses small non-linearity with modulus variation versus strain level.

Furthermore, in general, it is not possible to calculate the dynamic modulus (E', E") dependence with the static preload (linear viscoelasticity assumption). A replacement solution consists to take into account measurement of these modulus with static preload. Therefore a dynamic transmissibility can be assessed with two load cases in MARC. The first one is a static load case with a hyperelastic law for the preload, the

second is a harmonic load case with the dynamic modulus. Transmissibility curve shows an increasing of eigen frequency with preload.

4 STATIC AND HARMONIC ANALYSES EXAMPLES

Following examples of application are presented to illustrate the analytical method for static and harmonic environments:

- A mock-up in order to give the simplest example to illustrate the analytical methods.
- An equipment isolator.
- A disturbance isolator
- A shock attenuator

4.1 Mock-up with four simple devices

The mock-up is made of four parallelepiped blocks and a simple mass (metallic plate). In statics, a Mooney-Rivlin with four coefficients is required, but it was also necessary to model the details like base plate of the elastomer blocks. Analytical analysis without elastomer base plate assesses the stiffness.

In dynamics, iterative analytical approach provides a good assessment of the eigen frequency.

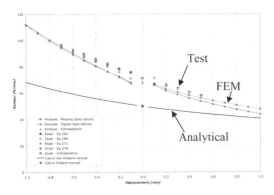

Figure 10. Test/prediction correlation in traction/compression (static). Analytical and Marc analyses for stiffness assessment.

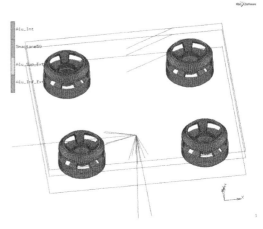

Figure 13. Equipment isolator with four devices.

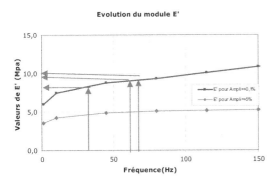

Figure 11. Dynamic modulus characterisation (dynamic stiffness coefficient).

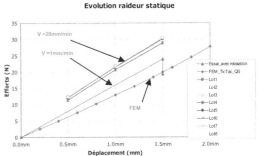

Figure 14. Static force/displacement – Test/prediction correlation (Equilibrium state after relaxation).

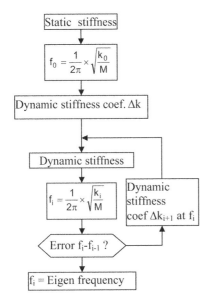

Figure 12. Analytical eigen frequency assessment.

Initial step for frequency assessment uses the static stiffness; then modulus frequency characterisation provides data for frequency assessment iterative loop. Analytical prediction provides an eigen frequency in traction/compression at 38 Hz for a measurement at 35 Hz.

4.2 Equipment isolator

The real equipment isolator is represented by four devices with elastomer beam. Static prediction with MARC finite element model provides a good test/prediction correlation (error of 3%). The stiffness prediction in statics is compared with test at equilibrium state after sufficient relaxation. Quasi-static test results with traction speed above 1 mm/min show stiffness increasing due to viscoelastic effect of the butyl base of the used elastomer.

This butyl elastomer with high dissipation needs iterative algorithm to take into account the dynamic modulus (E', E") dependence with the frequency and the strain rate. Even for strain lower than 5%, the dynamic transmissibility is sensible with the strain rate. Curves plotted here after show a sine response for

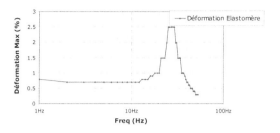

Evolution du taux de déformation Max Elastomère

Figure 15. Strain rate assessment with Marc and iterative algorithm.

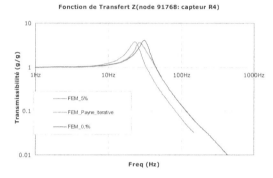

Figure 16. Transmissibility for strain of 0.1% (red), iterative algorithm (green) and strain of 5%.

0.2 g, the strain rain assessed by Marc and the iterative algorithm is 2.5%. The transmissibility is well with constant strain of 0.1% and 5%.

4.3 Disturbance isolator

The most efficient microvibration reduction is to use an isolator at the disturbing source interface. To assess the performance, the transmissibility force is measured with a dynamometric table.

The following curve highlights a good correlation between test results and Marc analyses.

4.4 Shock attenuator for spacecraft

SASSA (Shock Attenuator System for Spacecraft and Adaptor) is an ESA R&D. This study focuses on the attenuation of shock due to the launcher stages separations. In order to withstand high load due to the spacecraft mass, it is necessary to design the device with sufficient strength and stiffness. To increase the stiffness in traction/compression, elastomer is confined with a large ratio between loaded and free surfaces. Nevertheless, such pieces are sensitive to the hydrostatic depression. Therefore, in order to meet the requirements, detailed non-linear analyses with Marc

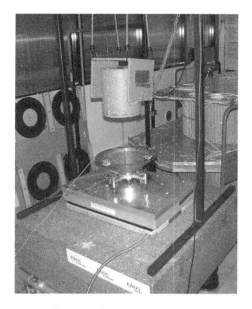

Figure 17. Test set-up for transmissibility assessment.

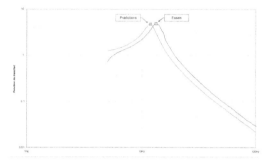

Figure 18. Test (blue)/prediction (pink) correlation.

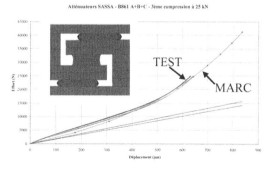

Figure 19. Test/prediction correlation in statics for high load.

are performed for elastomer sizing. Then, some linear models are derived for linear analysis at system level for spacecraft or launcher verification. The filtering performance obtained with Nastran linear model is well correlated with test.

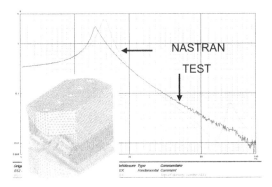

Figure 20. From detailed model to Nastran linear model.

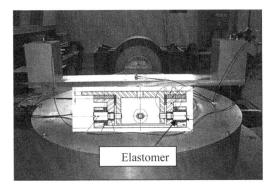

Figure 21. Mode at 40 Hz and shock attenuator.

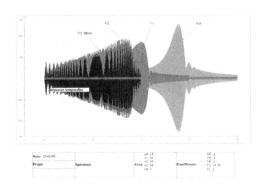

Figure 22. Time response without shock attenuator (green) and for two shock attenuator design (black and red).

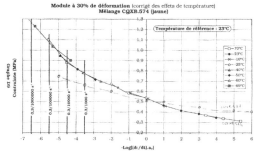

Figure 23. Example of time/temperature master curve.

5 TRANSIENT ANALYSES EXAMPLES

Two examples of shock attenuators are presented to illustrate the analytical method for transient environments.

5.1 Shock attenuator for reflector release

This device is used to reduce shocks due to pyro-mecanism for antenna reflector release on spacecraft. Shock attenuator and isolator for microvibrations are based on a non-linear stiffness. The low stiffness is used for low displacements involved in microvibration or shock and the high stiffness is necessary to withstand high loads. In the case of the reflector and shock attenuator design, it was required to identify the impact of the stiffness non-linearity on the reflector frequency mode at 40 Hz. Thus, a sine test with the shock attenuator equipped of a mock-up with a plate and two masses was defined to obtain a mode at 40 Hz.

Transient analyses with Marc allow assessing the impact of elastomer non-linearity with large displacement on the mode frequency decreasing. Accelerations responses with typical wave forms due to stiffness increasing are plotted here below. These analyses assess the shocks created by the non linearity and the associated forces.

The method is also a very powerful design tool as it allows tuning the stiffness non-linearity with physical parameters (elastomer modulus, thickness, gap . . .)

5.2 Shock attenuator for solar array release

This device is an improvement of the previous shock attenuator for reflector release. Therefore, the instantaneous elastomer stiffness increasing due to shock high speed has to be taken into account to assess attenuation performance with a sufficient accuracy. Nevertheless, test with high strain velocity involved in shock, is not possible with classical facilities.

Thus, the idea is to use time/temperature reciprocity for elastomer. That means that there is equivalence between the stiffness increasing with high velocity or with cold temperature. For that, elastomer master curve time/temperature is used.

Thus, method for shock analysis is the following:

- Mooney-Rivlin hyperelastic law identification
- High velocities involved in shock identification (1/1000 s to 1/100000 s) for different temperatures.
- Elastomer time/temperature master curve
- From master curve, identification of a stiffness increasing coefficient to apply to Mooney-Rivlin hyperelastic law

Then, to obtain accurate temporal response, a test set-up and analysis with Marc were defined. The shock

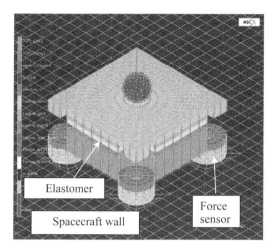

Elastomer

Force sensor

Spacecraft wall

Figure 24.   Marc FEM for ball drop (contact analysis).

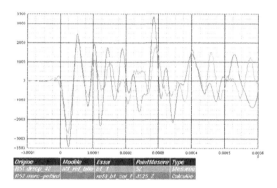

Figure 25.   Test (green)/prediction (red) correlation.

is obtained with the drop of a ball and the contact on the shock attenuator. A complete instrumentation with accelerometers and force sensors at attenuator interface is defined.

To predict accurate temporal response with Marc model, it is recommended to:

- Use 3D element for wave shock propagation
- Use flexible element for the ball, in order to obtain the contact mode frequency
- Take into account the force sensors in the model

6   SYNTHESIS AND PERSPECTIVE

The methodology developed at Astrium for elastomeric device performance prediction is based on the combination of characterisations by tests and software simulations. Static analyses in large deformation and dynamic analysis in small strain are easily accessible. Improvements have been developed in dynamics to take into account the dependence with small strain level (Payne effect).

The methodology shall be generalized to multiaxial large strain in dynamics occurring for sine qualification test with elastomer devices. It will then be necessary to take into account the elastomer nonlinear viscoelasticity. The ultimate step will consist in modeling the non-linearities of contact isolator end stops that withstand the load during launch.

Constitutive Models for Rubber VII – Jerrams & Murphy (eds)
© 2012 Taylor & Francis Group, London, ISBN 978-0-415-68389-0

Structural optimization of a rubber bushing for automotive suspension

G. Previati
Politecnico di Milano, Department of Mechanical Engeneering, Milan, Italy

M. Kaliske
Technische Universität Dresden, Institute for Structural Analysis, Dresden, Germany

M. Gobbi & G. Mastinu
Politecnico di Milano, Department of Mechanical Engeneering, Milan, Italy

ABSTRACT: In this paper, the problem of the design of a rubber component for automotive application is discussed. The design of such components is complex due to the highly nonlinear response of the material and to the difficulty in estimating its fatigue life. Many different material models can be found in the literature and employed for the description of the stress strain relationship. Referring to the static stress-strain characteristic, the designer has to choose among different models and to define the model parameters. Moreover, the fatigue life of the rubber is very difficult to be predicted and many different predictors and approaches have been proposed in the literature. The influence of different material models and of different life predictors on the design of a rubber component are investigated in the contribution.

A simple case study related to the optimal design of a rubber bushing for the lower control arm of an automotive suspension system is presented. The design of the component is kept as simple as possible, thus, only the static stress/strain characteristic of the material is considered. A series of experimental tests is performed, both to characterize the material and to define the loading acting on the component. To objectively evaluate the influence of the different material models and life predictors, the design is performed by applying optimal design theory. The Pareto set, including the best design solutions is computed. By comparing the Pareto optimal sets, the effect of different material models and life predictors is discussed.

1 INTRODUCTION

Design of rubber components is usually quite complicated. Rubber material has a very complex behavior both statically and dynamically. Just referring to the static stress/strain characteristic, many different models are present in the literature (Marckmann & Verron 2006), but only a few of them are available in commercial software. The designer has to choose a model and to define its parameters. Moreover, the fatigue life of the rubber components is very difficult to analyze and many different life predictors and approaches have been proposed (Mars & Fatemi 2002).

In such a situation, it is quite difficult to understand what is the influence of different material models or of different life predictors on the fatigue life estimation of a rubber component. In this paper, the design of a rubber bushing for automotive applications is selected as a case study to analyze the influence of different material models (chosen among the most diffused ones and available in commercial applications) and of different life predictors.

The design of the component is kept as simple as possible, thus only the static characteristics of the material are considered. Experimental tests have been realized to estimate the parameters of the different material models. The load acting on the bushing has been experimentally measured. The design is performed by applying the concepts of optimal design (Mastinu et al. 2006; Matusov 1995). The Pareto set of the best compromise solutions is computed for different materials by considering different life predictors. By this approach, the definition of the parameters of the system is objective and the influence of material models and life predictors can be evaluated by comparing the different Pareto optimal sets.

The paper is organized as follows. In the first section, the design problem is formulated and the loads acting on the component are discussed. Then, the considered material models and the life predictors are briefly introduced by using a continuum mechanical formulation. In the fourth section, the material models are calibrated on basis of experimental tests. Then, the optimality theory from Pareto is introduced and the system model described. Finally, the results of the optimization process are presented and discussed.

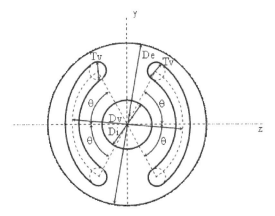

Figure 1. Schematic representation of a section of the bushing for automotive applications.

Table 1. Constraints on the design of the bushing.

Constraint	Value	Tolerance
Longitudinal stiffness (x direction)	900 N/m	±180 N/m
Lateral stiffness (y direction)	5000 N/m	±1000 N/m
Vertical stiffness (z direction)	2500 N/m	±500 N/m
Difference between vertical and lateral stiffness	2500 N/m	±250 N/m
Sum of vertical and lateral stiffness	7500 N/m	±7500 N/m
Maximum Cauchy stress (static)	7 MPa	–

2 DESIGN PROBLEM FORMULATION

Rubber bushings are introduced in automotive suspension systems in order to adjust the compliance of the system and to add a filtering stage to the road unevenness. In the design of the bushing some given predefined values of the stiffness have to be obtained, while the mass has to be kept as small as possible.

In Figure 1, the section of a bushing for automotive applications is shown. The bushing has two grooves, in this way two different stiffness can be obtained in the x and y direction. The rubber core of the bushing is comprised between two steel tubular elements (the outer is 4 mm thick and the inner one 5 mm). The geometry of the bushing (see Fig. 1) is given by the length L, the inner diameter Di and the outer diameter De. The grooves are defined by the width Tv and the angular length θ. The mean diameter of the grooves Dv is the mean of the outer and inner diameters of the bushing. The width S of the rubber part of the bushing is given by half of the difference between the outer and the inner diameter. By changing the values of the five quantities θ, L, Di, S and Tv, the geometry of the bushing can be varied. These five parameters are the design variables of the optimal design problem.

The optimization of the bushing is performed with respect to 2 objectives: mass reduction (taking into account the mass of the rubber and the mass of the steel) and the life of the component (the fatigue damage predictor has to be minimized).

The constraints listed in Table 1 are considered during the optimization process. These constraints represent the required behavior of the bushing in term of stiffness and maximum Cauchy stress allowed for the material.

2.1 Load cases

The suspensions system of a vehicle, and therefore its components, is subjected to different load conditions. For the current optimization problem, we consider a

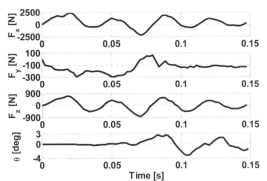

Figure 2. Measured time histories of the forces acting on the bushing and of the rotation angle of the lower arm of the suspension.

passenger car with a mass of 2350 kg, a wheel base of 3050 mm and a track width of 1580 mm. The bushing to be optimized belongs to the front suspension (double wish-bone layout, lower arm, rear bushing). From the vehicle data and the geometry of the suspension, a static load of −2250 N in the y direction and −650 N in the z direction has been computed (forces conventions in Figure 1).

For the design of the bushing, two load conditions are considered (starting from the static load on the bushing). The first condition is the maximum braking. In this situation, the maximum load on the bushing is reached. The load has been estimated in 7000 N in the x direction, −18000 N in the y direction and 5500 N in the z direction. This type of load is reached few times in the life of the vehicle, thus it is considered as a static load. From this load case the maximum Cauchy stress is computed and it must be less than the allowed stress of Table 1.

The second load condition is a passing over a cleat. The load is estimated on the basis of experimental tests as shown in Gobbi et al. (2010). The measured time histories of the forces acting on the bushing and of the rotation angle θ of the arm of the suspension are shown in Figure 2. This load case is representative of the fatigue load on the suspension. From this second load case, life predictors are computed.

For each design variables configuration, starting from the static pre-load, the three linearized stiffness in the three directions are computed by imposing a displacement to the bushing and calculating the corresponding force via the FEM model described in Section 7.

3 MATERIAL MODELS

Considering the usual notation of the continuum mechanics, the following material models can be defined (ABAQUS 2007, Marckmann & Verron 2006).

Neo-Hooke model: this model is the simplest form of strain energy function, with just a linear dependence on the first invariant of the deformation gradient ($I_1 = \lambda_1^2 + \lambda_2^2 + \lambda_3^2$, where λ_i are the principal stretches). The strain energy density function is

$$W = C_{10}(I_1 - 3) \tag{1}$$

Yeoh model: this model follows a phenomenological approach in which the material response is considered as a polynomial function of the first invariant only. The strain energy function takes the form

$$W = \sum_{i=1}^{3} C_{i0}(I_1 - 3)^i \tag{2}$$

Arruda-Boyce model: This model is also known as *8-chain model* and it takes into account only the first invariant, with strain energy function

$$W = \mu \sum_{i=1}^{5} \frac{C_i}{\lambda_m^{2i-2}}(I_1^i - 3^i) \tag{3}$$

The five parameters C_i are computed from the first five terms of the series expansion of the inverse Langevin function ($C_1 = 1/2$, $C_2 = 1/20$, $C_3 = 11/1050$, $C_4 = 19/7000$, $C_5 = 519/673750$).

Ogden model: this model has a phenomenological foundation and derives the strain energy function in terms of principal stretches

$$W = \mu \sum_{i=1}^{3} \frac{C_i}{\lambda_m^{2i-2}}(I_1^i - 3^i) \tag{4}$$

4 FAILURE CRITERIA

The definition of a unique failure criterion for rubber materials is still a discussed open problem (for an extensive references list see Mars & Fatemi 2002). In this paper, the fatigue life is predicted by considering different continuum mechanics quantities (predictors) that can be related to the life span of the component. This approach is suitable for the design of components when different component shapes can be considered and it is very difficult to have some knowledge about the initial location and characteristics of the crack.

In the literature, many different mechanical quantities have been proposed as life predictors. In this paper, the maximum principal stretch, the maximum principal Cauchy stress, the strain energy density and a predictor based on configurational mechanics will be considered.

The latter predictor has been proposed in Verron & Andriyana 2008. Being based on configurational mechanics, in this paper it will referred to as configurational predictor. A short description of this predictor follows.

The predictor is derived by considering the damaging of the material during one stabilized load cycle. The damaging of the material is computed by considering the damage in each time instant of the cycle and by integration over the cycle. For the formulation of the predictor, Verron considers that the Eshelby configurational tensor (Eshelby 1975; Kienzler & Herrmann, 2000) is the driving force of the material damaging. The Eshelby tensor Σ can be computed from continuum mechanics quantities

$$\Sigma = W\mathbf{I} - J\mathbf{F}^T\boldsymbol{\sigma}\mathbf{F}^{-T} \tag{5}$$

From the consideration that a crack can grow only if the material stress state tends to open the defect, Verron and Andriyana consider that the load cycle can damage the material only in the instants in which the defect is opening. Thus, the predictor Σ^* has a quite complex definition

$$\Sigma^* = \left|\min\left((\Sigma_i^d)_{i=1,2,3}, 0\right)\right| \tag{6}$$

where $(\Sigma_i^d)_{i=1,2,3}$ are the eigenvalues of the damage part of the configurational stress tensor Σ^d. This tensor is obtained by the integration over the cycle of

$$\mathbf{d\Sigma^d} = \sum_{i=1}^{3} d\Sigma_i^d V_i \otimes V_i \tag{7}$$

with

$$d\Sigma_i^d = \begin{cases} d\Sigma_i & \text{if } d\Sigma_i < 0 \text{ and } V_i \cdot \Sigma V_i < 0 \\ 0 & \text{otherwise} \end{cases} \tag{8}$$

$d\Sigma_i$ and V_i being the eigenvalues and eigenvectors of the configurational stress tensor increment.

5 MATERIAL PARAMETER IDENTIFICATION

For the identification of the material parameters, a uniaxial traction test and a pure shear test have been completed. The uniaxial traction test has been realized according to the ASTM D412 C norm (ASTM 2006). For the pure shear test, a hollow cylinder of material has been used. The cylinder has an inner diameter of 116 mm, an outer diameter of 136 mm and a length of 85 mm. The rubber material has been vulcanized to two steel rings at the outer and inner part of the cylinder. The outer steel ring has been fixed and the inner

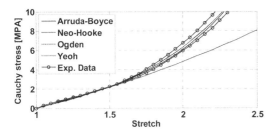

Figure 3. Uniaxial traction test, experimental data and computed material characteristics.

Table 2. Identified material parameters

Material model	Parameters
Arruda-Boyce	$\mu = 0.9687, \lambda_m = 1.6572$
Ogden	$\mu_1 = 0.33214, \lambda_1 = 4.6889,$
	$\mu_2 = 0.00069038, \lambda_2 = 4.6888,$
	$\mu_3 = 0.9435, \lambda_3 = 1.3216$
Neo-Hooke	$C_{10} = 0.6882$
Yeoh	$C_{10} = 0.62617, C_{20} = 0.060984,$
	$C_{30} = 0.0023121$

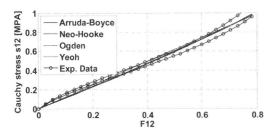

Figure 4. Pure shear test, experimental data and computed material characteristics.

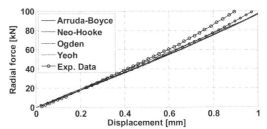

Figure 5. Generic state of deformation, experimental data and computed material characteristics.

cylinder has been displaced in the axial direction. The resulting force has been measured. The resulting state of deformation of the rubber material is very close to pure shear. In Figures 3 and 4, the results of the material tests and the identified material responses are reported. As expected, the Arruda-Boyce, Ogden and Yeoh models can follow all the traction curve of the material, while the Neo-Hooke curve can follow only the very first part of the curve. For the pure shear test, all of the material models have shown a less non-linear response of the material. The parameters of the material models are reported in Table 2.

In Figure 5, the response of the same rubber cylinder used for the pure shear test while subject to a radial deformation is shown. In this case, the resulting state of deformation is quite complex. This load case has been used to verify the behaviour of the material models calibrated on the traction and pure shear tests. All of the material models show a satisfactory agreement with the experimental data.

6 OPTIMIZATION PROCESS

The design of the bushing has been formulated as a multi-objective optimisation problem and solved by applying the Pareto-optimal theory (Mastinu et al. 2006; Matusov 1995). The solution of an optimisation problem is a set composed by infinite solutions, the so called Pareto-optimal set. The multi-objective programming problem reads

$$Find \min_{x \in \Re^n}(g_1(x), g_2(x), ..., g_k(x)) \quad subject \ to$$
$$h_i(x) \le 0, \qquad i = 1, ..., q \qquad x \in X \tag{9}$$

where $g_i(x)$ are the k objective functions, x is the vector of n design variables, X is the domain of variation of the design variables and $h_i(x)$ are the q inequality constraints.

The definition of the optimal solutions in a multi-objective framework is called Pareto-optimal solution $x*$ and it is that for which there does not exist another solution such that

$$g_r(x) \le g_r(x*) \ r = 1, 2, ... k$$
$$\exists l : g_l(x) \le g_l(x*) \tag{10}$$

The set of all Pareto-optimal solutions defines the Pareto-optimal set. For the computation of the Pareto-optimal set, a huge number of numerical simulations of the system is usually required. In this paper, to have an accurate representation of the Pareto-optimal set and an affordable computation time, the physical models of the system (described in the following section) will be approximated by using an "approximate" mathematical model.

7 SYSTEM MODEL

The finite element model of the bushing has been realized by using commercial software. Figure 6 shows the mesh of the model. The inner and outer surfaces of the rubber element have a rigid body constraint to simulate the presence of the steel rings. The outer surface is connected to the ground. The load is applied to the inner surface. The mesh is realized by linear bricks, hybrid formulation. The configurational predictor is

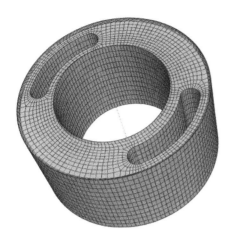

Figure 6.   FEM model of the bushing.

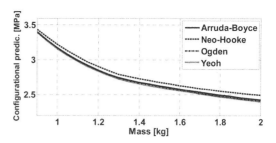

Figure 7.   Pareto-optimal set, objective functions.

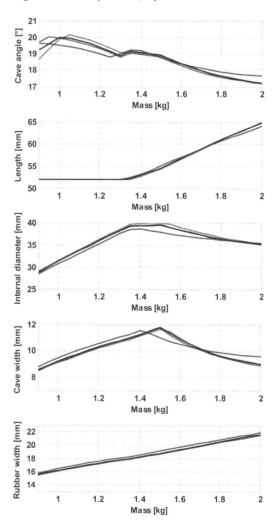

Figure 8.   Pareto-optimal set, design variables. Continuous line: Arruda-Boyce. Dashed line: Neo-Hooke. Dash-dot line: Ogden. Dotted line: Yeoh.

computed by post processing the results of the analysis (see Section 4).

The physical model of the bushing has been used to simulate about 200 different combinations of parameters for each material model. These simulations have been used to calibrate the parameters of a series of interpolating functions. The interpolating functions are quadratic functions of the design variables of the system.

8   RESULTS

In this section, the results of the analysis are reported. Firstly, the optimal parameters computed by varying the material models are reported by considering the configurational predictor as a measure of the damaging of the material. Then, by using the Yeoh model, the optimal parameters are shown as function of the different life predictors.

8.1   Comparison of different material models

Figure 7 shows the objective functions computed for the Pareto-optimal sets of the four different material models. It can be noticed that the Arruda-Boyce, Yeoh and Ogden models have very close solutions. Only the Neo-Hooke model shows some differences. In Figure 8, the computed sets of optimal design variables are reported. Also for the design variables, only the Neo-Hooke model shows some differences with respect to the other models.

In this application, where not excessively large deformations are considered, the three more complex models behave almost in the same way. It has to be considered that few experimental data have been used for the calibration of the material models. In this situation, the calibration of the Ogden model is quite complex because six parameters have to be defined. The Yeoh and Arruda-Boyce models are simpler to be calibrated and, at least in this case, behave as the Ogden model.

8.2   Comparison of different life predictors

In this section, the results of the optimization process for the four different life predictors are presented.

Figure 9 shows the sets of the optimal parameters computed for the different life predictors by

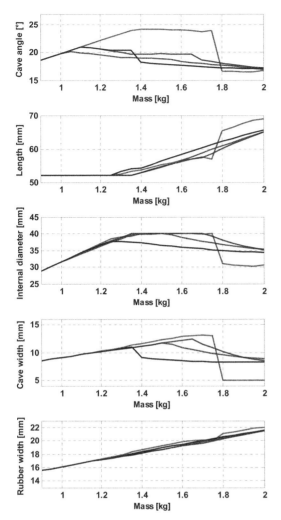

Figure 9. Pareto-optimal set, design variables. Continuous line: Maximum stretch. Dashed line: Maximum stress. Dash-dot line: Configurational predictor. Dotted line: Strain energy density.

using the same material model (Yeoh model). It can be seen that different results are obtained for different life predictors. In particular, the maximum stress and the configurational predictor have quite close solutions. This can be due to the strong dependence of the configurational predictor on the stress tensor. Slightly different results are obtained for the maximum strain. The maximum strain is related to the maximum stress, however, the nonlinearity of their relationship allows for the differences in the solutions. Completely different solutions are obtained for the strain energy density. In fact, this index has a physical foundation quite different compared to the other ones. The four indices have shown the same results for low values of mass. This is probably due to the reaching of the lower limit of the allowed length of the bushing.

9 CONCLUSION

In this paper, the design of a rubber bushing to be employed in a car suspension system has been analyzed in order to evaluate different mechanical models for rubber materials and different life predictors while varying the shape of the component. The concepts of the optimal design have been used to define, in an objective way, the design of the component.

Four different material models, available in most commercial software, and four different life predictors have been compared. Referring to the material models, the Yeoh, Arruda-Boyce and Ogden models have given almost the same results on the design of the component. Small differences in the design can be found by using the Neo-Hooke model. This effect is due to the very simple formulation of the Neo-Hooke model that can be used only for very small deformations of the material.

Quite different results on the sets of the optimal parameters have been found while comparing the four life predictors. Only the maximum stress and the configurational predictor have shown similar results. The maximum stretch has shown some differences with respect to these two indices but a similar tendency. The strain energy density, being based on a different physical motivation, has shown results quite different from all the others indices.

REFERENCES

Abaqus Analysis Theory Manual. 2008. Dassult Systems, USA.

ASTM D 412. 2006. *Standard Test Methods for Vulcanized Rubber and Thermoplastic Elastomers-Tension.* ASTM International.

Eshelby, J.D. 1975. The elastic energy-momentum tensor. *Journal of Elasticity* 5: 321–335.

Gobbi, M., Mastinu, G. & Previati, G. 2010. A method for the indoor testing of a road vehicle suspension system. AVEC conference, Loughborough.

Kienzler, R. & Herrmann, G. 2000. *Mechanics in material space with applications to defect and fracture mechanics,* Berlin: Spinger-Verlag.

Marckmann, G. & Verron, E. 2006. Comparison of hyperelastic models for rubberlike materials. *Rubber Chemistry and Technology,* 79: 835–858.

Mars, W.V. & Fatemi, A. 2002 A literature survey on fatigue analysis approaches for rubber. *International Journal of Fatigue* 24: 949–961.

Mastinu, G., Gobbi, M. & Miano, C. 2006. *Optimal Design of Complex Mechanical Systems With Applications to Vehicle Engineering.* Berlin: Springer Verlag.

Matusov, J. 1995. *Multicriteria Optimization and Engineering.* New York: Chapman & Hall.

Verron, E. & Andriyana, A. 2008. Definition of a new predictor for multiaxial fatigue crack nucleation in rubber. *Journal of Mechanics and Physics of Solids* 56: 417–443.

Constitutive Models for Rubber VII – Jerrams & Murphy (eds)
© 2012 Taylor & Francis Group, London, ISBN 978-0-415-68389-0

Optimisation and characterisation of magnetorheological elastomers

Jennifer McIntyre & Stephen Jerrams
Dublin Institute of Technology, Ireland

Timo Steinke, Aleksandra Maslak, Piotr Wagner, Markus Möwes,
Thomas Alshuth & Robert Schuster
Deutsches Institut für Kautschuk Technologie e.V. (German Institute for Rubber Technology), Germany

ABSTRACT: Two series of samples were fabricated to investigate the factors affecting the performance of magnetorheological elastomers (MREs), particularly the increase in shear storage modulus that can be achieved when an MRE is subjected to shear in the presence of an external magnetic field. The first series consisted of samples with varying viscosities; the second were prepared with a range of ferromagnetic particle loadings. Half of the samples in each series were vulcanised in a magnetic field of 0.6 T. Results show that the increase in shear storage modulus observed when an MRE is sheared in the presence of an external magnetic field is higher for specimens cured in a magnetic field. This increase was observed to be higher in materials with a low viscosity matrix.

1 INTRODUCTION

Magnetorheological elastomers (MREs) are smart materials consisting of ferromagnetic particles in an elastomer matrix. The magnetic particles are added to the matrix mixture prior to curing, and a magnetic field is applied as the composite cures. The particles are aligned into chains along the magnetic field lines. When the composite MRE is cured, the particles are fixed in their aligned positions within the chains, parallel to the direction of the field (see Fig. 1). The cured composite MRE is intended to be used in the presence of a magnetic field.

In the presence of an external magnetic field, more work is required to shear an MRE than to shear an unfilled elastomer. This extra work rises with increases in the applied magnetic flux density, so shear modulus is dependent on the magnetic flux density and can be controlled by it (Boczkowska & Awietjan (2009), Jolly et al. (1996)).

MREs will be useful in sensing and damping applications. The shear storage modulus, G', represents the ability of the material to store the energy of deformation, which contributes to the material stiffness. The loss modulus, G", represents the material's ability to dissipate the energy of deformation (Boczkowska & Awietjan (2009)).

2 THEORY

Jolly et al. (1996) developed a model to calculate the interaction energy between the particles in a chain and the energy per unit volume of the composite. The

Figure 1. Chains of ferromagnetic particles in a matrix which was cured in the presence of a magnetic field.

version presented in this text was derived by Alshuth et al. (2007) from Jolly's model. The new model highlights the role of particle size: an increase in particle diameter leads to an increase in shear modulus of the composite when an external magnetic field is applied.

Consider two adjacent particles in a chain in the presence of an external magnetic field (fig. 2).

The interaction energy, E_{12}, between two dipoles of the same strength and direction, m_1 and m_2, is given by:

$$E_{12} = \frac{|m|^2 \left(1 - 3\dfrac{h^2}{h^2 + \Delta x^2}\right)}{4\pi\mu_0\mu_r \left(h^2 + \Delta x^2\right)^{3/2}} \quad (1)$$

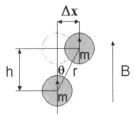

Figure 2. Diagram of adjacent particles in a chain (Alshuth et al. (2009)).

where μ_r = relative permeability of the particle; μ_0 = relative permeability of the free space

Assuming the particles are spherical and multiplying by the total number of particles in the composite, the interaction energy for the sample can be calculated. From this, the energy density (energy per unit volume) is given by:

$$U_D = \frac{3\phi|m|^2(\gamma^2 - 2)}{2\pi\mu_0\mu_r d_p^3 h^3(\gamma^2 + 1)^{5/2}}$$ (2)

where ϕ = volume faction of particles; $\gamma = (\Delta x/h)$; d = particle diameter

The shear strength can be obtained from the first derivative of the energy density with respect to strain:

$$\tau = \frac{\partial U_D}{\partial \gamma} = \frac{9\phi\gamma|m|^2(4 - \gamma^2)}{8\mu_0\mu_r d_p^3 h^3(\gamma^2 + 1)^{7/2}}$$ (3)

The induced magnetic polarisation of the particles is given by:

$$J_p = \frac{|m|}{V_p}$$ (4)

where V_p = volume of one particle

Subsituting this into equation (3) gives:

$$\tau = \frac{\phi(4 - \gamma^2)d_p^3 J_p^3}{8\mu_0\mu_r h^3(\gamma^2 + 1)^{7/2}}$$ (5)

Equation (5) shows that both particle size and magnetisation saturation are important factors in the change in shear modulus that can be achieved.

3 MATERIALS

Experiments to identify suitable particles were conducted using magnetorheological fluids (MRF) comprising particles in an oil matrix. This was quicker and hence more economical than beginning experimentation with a rubber matrix. The particles were free to align into chains easily when a magnetic field

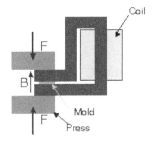

Figure 3. Diagram of the electromagnetic press.

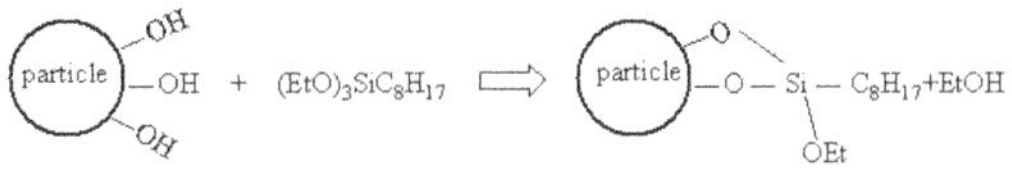

Figure 4. The chemical reaction due to silanisation.

was applied. The potential increase in shear storage modulus, G' was observed for each particle type tested, as well as the quantity of particles required. Various particle types and concentrations were investigated. Carbonyl iron particles exhibited a much higher change in G' than any of the other particle types tested.

3.1 Series A

A series of samples consisting of a natural rubber matrix and micro-sized carbonyl iron particles (BASF Grade SM) was fabricated. This series of four samples contained two different concentrations of carbon black (5 phr and 30 phr) some of which had 20 phr of softener added. An electromagnetic coil induced a current to flow through a C-shaped clamp, which held the material mixture, so that the rubber material was in a magnetic field of 0.6 T while it was vulcanised in a press under pressure (see Fig. 3).

3.2 Series B

Series B contained manganese ferrite ($MnFe_2O_4$) nano particles that were synthesised at DIK from their metallic salts using a bottom-up process; from atoms and molecules of materials, rather than by breaking down materials into smaller parts (top-down process). The particles were ground to reduce the size of any agglomerates that had formed.

These particles were modified with silane (Evonic Si208) in order to change their surface properties. The silanisation changed the polarity of the particles allowing them to bond with the rubber matrix.

Many rubbers are non-polar and hydrophobic. The manganese ferrite particles used in Series B were highly polar. The surface of an unmodified particle contained OH groups. Functionalisation of the particle with silane caused a chemical reaction that changed the surface of the particle interfacing with the rubber matrix (see Fig. 4), and caused it to become non-polar. Without this surface treatment, the polar

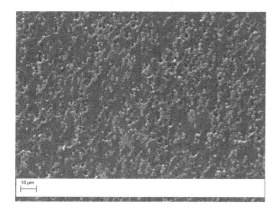

Figure 5.   Series A, Sample 2 SEM 1.

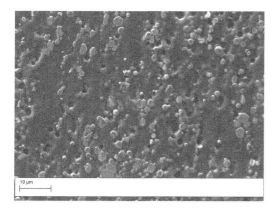

Figure 6.   Series A, Sample 2 SEM 2.

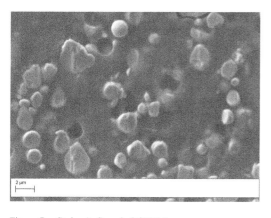

Figure 7.   Series A, Sample 2 SEM 3.

particles would not have mixed well with non-polar natural rubber matrices and a homogenous dispersion would not have been achieved.

Series BC consisted of seven samples: six samples of three different particle loadings (400, 200 and 100 phr), three of which were vulcanised in a magnetic field and one 'control' sample which contained no particles. The details and results are shown in Table 2.

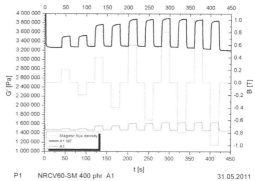

Figure 8.   Series A, Sample 1 Rheometric Test.

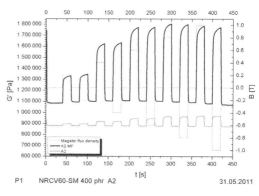

Figure 9.   Series A, Sample 2 Rheometric Test.

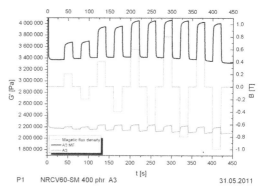

Figure 10.   Series A, Sample 3 Rheometric Test.

4   RHEOMETRIC EXPERIMENTS

The four Series A samples were subjected to an oscillatory shear strain of 0.5% deformation, at a frequency of 10 Hz, in a plate-plate rheometer (Anton Paar Physica MCR501).

The current controlling the magnetic field was switched on and off at intervals, and was increased incrementally. The current direction was changed to investigate the responses of hard and soft magnetic materials. Plots of the change in shear modulus as the magnetic flux density was raised are shown in Figs. 7 to

315

Table 1. Series A vulcanised in a magnetic field (0.6 T).

Sample	CB (phr)	Softener (phr)	G_0 (MPa)	G_1 (MPa)	ΔG (MPa)	$\Delta G/G_0$ (%)	Mooney viscosity (MU)
A1 (M)	5	–	3.28	3.88	0.60	18.1	42.5
A1	5	–	1.47	1.65	0.18	12.24	42.5
A2 (M)	5	20	1.10	1.78	0.68	61.82	27.6
A2	5	20	0.88	0.98	0.10	10.80	27.6
A3 (M)	30	20	3.41	4.06	0.65	19.09	29.0
A3	30	20	2.11	2.20	0.09	4.27	29.0
A4 (M)	30	–	2.88	3.21	0.33	11.48	57.3
A4	30	–	1.53	1.58	0.05	3.28	57.3

Table 2. Series C.

Sample	Particle loading (phr)	Magnetic field (0.6 T)	G_0 (MPa)	G_1 (MPa)	ΔG (MPa)	$\Delta G/G_0$ (%)
B1	400	✓	3.60	3.71	0.10	2.89
B2	200	✓	1.00	1.13	1.13	13.00
B3	100	✓	0.65	0.68	0.03	5.10
B4	400	–	4.54	4.60	0.06	1.38
B5	200	–	2.67	2.73	0.06	2.06
B6	100	–	0.96	0.96	0.01	0.65
B7	–	–	0.60	0.60	0.00	0.00

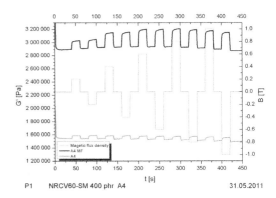

P1   NRCV60-SM 400 phr A4   31.05.2011

Figure 11. Series A, Sample 2 Rheometric Test.

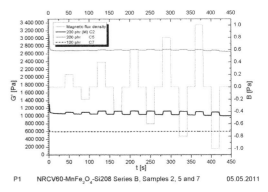

P1   NRCV60-MnFe$_2$O$_4$-Si208 Series B, Samples 2, 5 and 7   05.05.2011

Figure 12. Series B, Sample 2, 5 and 7 Rheometric Test.

11 in the results section. The variations in composition and results are shown in Table 1.

5 RESULTS

5.1 Microscopy

Scanning Electron Microscope (SEM) (model EVO® MA10) imaging was carried out in order to check dispersion of particles and filler and alignment of the particle chains. Figures 1, 2 and 3 show Series A, Sample 2 with increasing magnification.

The results show that the greatest increase in relative shear storage modulus occurs in sample A2, which contained the lower amount of carbon black with added softener (see Table 1). The next greatest increase was exhibited by the other sample containing softener, which had a higher concentration of carbon black. The smallest increase in shear storage modulus, G', was observed in the sample without softener and the highest carbon black loading.

Results from the Series B experiments are given in Table 2. All of the samples vulcanised in the magnetic field exhibited a greater increase in shear storage modulus, G' than those vulcanised without the magnetic field, but the greatest relative increase in G' was observed in the 200 phr sample (B2) (see Fig. 12).

The 200 phr specimens, cured with and without an external magnetic field (Samples B2 and B5 respectively) were compared with the control sample (B7), containing no magnetic particles (refer to fig. 11).

6 DISCUSSION

The first task was to identify particles of an appropriate size with a relatively high magnetic saturation. According to the theory originally derived by Jolly et al. (1996) and developed by Alshuth et al. (2007), large particles should exhibit a greater MR effect than small particles (Equation. no. 5). The experiments using MRFs confirm this theory, as the greatest increase in shear modulus was observed in the micro-sized carbonyl iron particles. The smallest nano-sized (average diameter of 15 nm) manganese ferrite particles exhibited a lesser change in shear modulus. However, the same equation also states that the higher the magnetisation saturation, the greater the potential increase in shear modulus.

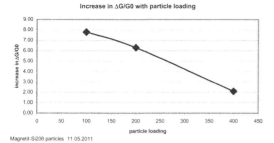

Increase in ΔG/G0 with particle loading

Magnetit-Si208 particles 11.05.2011

Figure 13.   Series B, Sample 2, 5 and 7 Rheometric Test.

Particle alignment in sample A2 can be seen from the SEM images (Fig. 4 to 6), although the particle loading was quite high. A lower concentration of particle would have allowed the chains to be more easily distinguished.

The particle chains are not neatly ordered like a string of pearls, but consist of uneven columns that are two or three particles wide in places. The alignment is resisted by the viscosity of the elastomer matrix. This resistance leads to an irregular structure.

With greater magnification, holes were seen where particles had been pulled out of the matrix when the sample was cut (Fig. 5 and 6). The image with the greatest magnification (Fig. 6) shows most of the particles were a few microns in diameter, but some larger particles existed and some were agglomerated.

The results show that the greatest increase in relative modulus occurs in the sample with the lowest amount of carbon black and with the softener. This indicates that the increase in modulus depends on the viscosity of the sample. Tests were conducted with a Mooney viscometer (Mooney MV 200 E) in order to confirm this finding. The highest relative shear modulus occurred in the sample with the lowest viscosity. This confirms the findings of Boczkowska and Awietjan (2007), whose work has shown that the MR effect achieved is highly dependent on the particle structure formed by the aligned chains, which in turn depends largely on the matrix viscosity.

The plots shown in Fig. 7 to 10 show that magnetic saturation of the particles was almost achieved with a magnetic flux of 0.6 T. By 0.8 T, saturation was fully achieved, and no further increase in shear modulus was observed for further increases in magnetic field strength.

Saturation did not occur at exactly the same time or level of magnetic flux density for each sample. The sample with the lowest viscosity, sample A2, took longer to achieve its maximum value of saturation at a higher magnetic flux density than sample A4, which had the highest viscosity. This suggests that the mobility of the particles was more limited in the harder more viscous samples, which achieved magnetic saturation earlier. The orientation of the particles and the gain in shear modulus was restricted in the more viscous samples.

Observing Series B, if the samples cured in the magnetic field (directionalised samples) are compared with the samples of the same concentration cured without the magnetic field, it is clear that the directionalised samples displayed a greater increase in the relative shear modulus: the particle orientation due to the magnetic field improved the relative shear storage modulus, ($\Delta G'/G_0$).

Table 2 shows the size of the increase for each particle loading. The increase was greatest at the lower particle loadings: for the 100 phr sample, the relative modulus was seven times greater for the directionalised sample than for the one cured without the magnetic field. However, as the particle concentration was increased, the rise in relative shear storage modulus due to orientation dropped. The value for the relative shear modulus of the directionalised 400 phr sample was twice the magnitude of the sample cured in the absence of a magnetic field (see Table 2 and Fig. 11).

7   CONCLUSIONS

The following conclusions were drawn from the experiments:

1. MR performance (increase in shear storage modulus, G') is largely dependent on the viscosity of the MRE matrix. A low viscosity is required to enable the magnetic particles to align into chains at the time of curing.
2. MR effect is largely dependent on particle size. Particles must be large enough to exhibit a high magnetic interaction between them, but small enough not to become stress-raisers, from which cracks can form and propagate in the matrix.
3. Vulcanisation in a magnetic field causes the magnetic particles to align into chains running in the direction of the applied field. Later when these MREs are sheared in a magnetic field, the chains reinforce the material, which leads to a much higher MR effect than that seen in composites with homogenously dispersed particles.

The optimum recipe for an MRE is a low viscosity matrix, containing micro-sized particles (perhaps 40 μm diameter), which have a high saturation magnetisation. The particle concentration depends on the density of the particle type and the matrix they are to be mixed with.

REFERENCES

Alshuth, T., Ramspeck, M., Schuster, R.H., Halbdel, B., and Zschunke, F. 2007. Magnetorheologische Elastomere: Einfluss der Partikelausrichtung auf die Schaltbarkeit. *Kautschuk Gummi Kunstoff* 60 (9): 448–455.

Boczkowska, A. and Awietjan, S.F. 2009. Smart composites of urethane elastomers with carbonyl iron. *Journal of Materials Science* 44 (15): 4104–4111.

Jolly, M.R., Carlson, C., Muñoz, B.C. and Bullions, T. 1996. The Magnetoviscoelastice Response of Elastomer Composites Consisting of Ferrous Particles Embedded in a Polymer Matrix. *Journal of Intelligent Material systems and Structures* 7 (10): 613–622.

Rosenweig, R.E. (1st ed.) 1985. *Ferrohydrodynamics.* Cambridge: Cambridge University Press.

Constitutive Models for Rubber VII – Jerrams & Murphy (eds)
© 2012 Taylor & Francis Group, London, ISBN 978-0-415-68389-0

Marine ageing of polychloroprene rubber: Validation of accelerated protocols and static failure criteria by comparison to a 23 years old offshore export line

V. Le Saux
UBS – LIMATB (EA4250), Rue de Saint-Maudé, Lorient Cedex, France

Y. Marco & S. Calloch
ENSTA BRETAGNE – LBMS (EA4325), Brest Cedex, France

P.-Y. Le Gac
IFREMER, Service Matériaux et Structures, Plouzané, France

ABSTRACT: Polymers are widely used in marine environment due to their good insulation properties and weathering resistance. Despite this extensive use, their long term behavior in such an aggressive environment is still not well known. This study focuses on the influence of marine ageing on the mechanical properties of a fully formulated Chloroprene Rubber (CR) used for offshore applications. The ageing is characterized at several scales: the microscopic scale (degradation mechanisms) by the use of physical measurements, the mesoscopic scale (hardness profiles) by the use of micro-indentation tests and macroscopic scale (mechanical testing). The analysis is based on monotonic tension tests on samples cut from sheets and on instrumented micro-indentation in order to describe accurately the gradients induced by ageing. One specific accelerated ageing protocol in renewed natural seawater is investigated for temperatures ranging from 20 to 80°C. In order to dissociate the effect of water absorption from the ageing consequences, a specific drying protocol is also used. The relevancy of a Time-Temperature superposition based on the Arrhenius principle is evaluated on several indicators (strain and stress at break, secant modulus for 100% strain) by the comparison to the values obtained on samples cut from an offshore export line aged under service conditions for 23 years.

1 INTRODUCTION

The assessment of the durability of elastomeric products submitted to marine environment is a major issue for several industrial fields (offshore oil and gas industries, harbour and naval applications, renewable marine energy, etc.) and is needed, for example, for new products certification, better evaluation of maintenance periods and lifetime prediction.

It is striking to observe that if the thermal ageing of rubbers is widely investigated in the literature (Celina *et al.* 2000; Ha-Anh and Vu-Khanh 2005; Gillen *et al.* 2005), it is clearly not the case for marine ageing (Stevenson 1984; Ab-Malek and Stevenson 1986; Pegram and Andrady 1989; Davies and Evrard 2007). Despite many marine applications (wetsuits, harbour drydock seals, etc.), sometimes very sensitive such as offshore export line, very few articles investigate the marine ageing of elastomers. We focus in this article on the marine ageing of a polychloroprene rubber used as coating layer of an offshore floating export line.

2 METHODOLOGY

Polymer lifetime prediction in an aggressive media such as seawater is ideally a three steps process. First, the ageing mechanisms observed for natural ageing must be identified. Then, accelerated ageing tests are performed (in order to keep reasonable characterization durations) and the kinetics of the ageing reactions are investigated through mechanical indicators (strain at break, etc.). The main difficulty is to set-up relevant accelerated ageing tests (these tests have to accelerate the degradation previously identified without modifying the involved mechanisms). Finally, a time-temperature superposition rule is applied in order to evaluate the ageing consequences under service conditions. The comparison to a naturally aged material is here a key point in order to validate the master curves and the associated extrapolation.

The literature shows that most of the scientific prodution deals with either the identification of the degradation mechanisms (ageing causes) or the evaluation

319

Table 1. Polychloroprene compound.

compound	phr
rubber	100
zinc oxyde	6
plasticizer	<10
fillers (CB/Si)	6/46
stearic acid	2
antioxidants	2
accelerators	1
magnesie	4

of the ageing consequences on the mechanical properties, but very few studies try to combine these approaches. This could be explained by the huge gap between the physical and the mechanical scales. Nevertheless, the improvement of the design methods needs the set-up of a dialogue between these scales. In the last ECCMR (Le Saux *et al.* 2009), we presented the first results of this project. Here, we propose to go further and discuss the following points:

1. accelerated ageing tests are very often performed without questionning about their relevancy. A first point is to design appropriate accelerated ageing tests, *i.e.* tests that reproduce the mechanisms that occur during natural ageing;
2. rubber degradation appears as a non uniform process resulting from a competition between diffusion and consumption of chemical species. A second requirement is to evaluate these ageing gradients in order to check if the measurements achieved on thin samples in the laboratories can be easily extended to massive industrial structures;
3. to evaluate the quality of the time/temperature superposition rule and the extrapolation to service conditions by comparison to a naturally aged sample and the relevance of the indicators followed.

3 EXPERIMENTAL METHOD

3.1 Material

The material studied is a silica-filled polychloroprene used for the external layer of offshore floating flowlines. Its main components are given in table 1. Square plates of 2 mm thickness and 250 mm width were manufactured by our industrial partner from the same material batch, in order to ensure the reliability of mixing and moulding conditions. Specimen are obtained using 3 specific methods (microtome, razor blade, punch die) depending on the measurement performed.

3.2 Accelerated ageing

In order to be as close as possible from the service conditions, specimens were immersed in several tanks with renewed natural seawater coming directly from the Brest estuary and maintained at different temperatures: 20, 40, 60 and 80°C. Water was continuously renewed using a peristaltic pump leading to the replacement of the vessel volume (60 L) every 24 hours without any modification of the temperature. The plates of 2 mm were removed periodically from the tanks and subsequently dried at 40°C under N_2 inert atmosphere until a constant weight was reached. This drying protocol was optimized and validated by a specific study detailed elsewhere (Le Saux 2010). It must be underlined that the durations mentionned in the following do not take this drying time into account and therefore represent the time spent in the ageing vessels. To compare this specific ageing protocol to more classical ones, we have also performed thermal ageing and marine ageing in non renewed artificial seawater according to a standard protocol currently performed by industrials.

3.3 Natural ageing

The natural aged sample comes from the coating layer of an offshore floating export line. This flow line was used in the Atlantic Ocean (near the Cameroun coasts) for 23 years at 10 meters depth and at mean temperature of 20°C. The flow line was still meeting the structural mechanical requirements after its withdrawal. The 2.3 mm thick topcoat neither presented any damage after its service life. The compound used has exactly the same recipe as the samples used for the accelerated ageing tests.

3.4 Experimental devices

Tensile tests The tensile tests were achieved on a Lloyd LR5K+ testing machine equipped with a 1 kN load cell. The tests were displacement controlled with a grip speed of 10 mm/min. A laser extensometer LASER-SCAN200 was used to measure the local elongation. Measurements were performed on classical H_2 samples. For each ageing condition, at least three samples were tested and the results averaged.

Micro-indentation measurements The hardness profiles were obtained using a CSM-instruments micro-hardness tester with a Vickers tip. The samples cut out from 2 mm thick sheets were embedded at ambiant temperature into epoxy resin and subsequently grinded with grinding media of decreasing granulometry (down to grit size 800). The disc rotation speed was low in order to reduce as much as possible the rise of temperature. After grinding, a delay of 2 hours was systematically applied before performing a new measurement in order to let the samples cool down to ambient temperature. All tests were load controlled. We used here a classical testing protocol, which consists in applying a weak preload in order to detect the surface contact (set to 10 mN), then a load controlled loading step (up to 100 mN at a loading rate of 200 mN/min), followed by a creep test of 30 s and then a load controlled unloading step (down to 0 mN at a unloading rate of 200 mN/min). Each point is an average of 5 measurements.

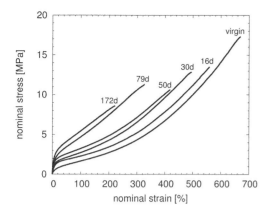

Figure 1. Stress-elongation curves obtained for an ageing temperature of 80°C on H_2 samples.

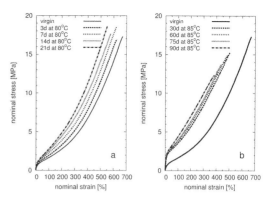

Figure 2. Stress-elongation curves obtained for the thermal ageing protocol (a) and synthetic marine environment (b) on H_2 samples.

Physical measurements FTIR characterization was performed with a Thermo Nicolet spectrometer using a $4\,cm^{-1}$ resolution and 32 scans. When ATR mode is used, spectra are normalized using the C-Cl bond situated at $825\,cm^{-1}$ (Socrates 2005). The stability of this bond along ageing was confirmed with NMR measurements (see Le Gac *et al.* (2011) for more details). For the chemical profiles along the samples thickness, transmission measurements were performed on 50 microns films. These films were cut out from the 2 mm sheets with a Leica microtome after being cooled by liquid nitrogen.

4 RESULTS

4.1 *Tensile tests*

Figure 1 shows the evolution of the tensile behaviour of the material after accelerated ageing at 80°C for different durations. Ageing leads to a significant increase of the initial modulus (up to 200 MPa after 6 months) and a considerable decrease of ultimate strain and stress values. It can also be observed that the longer the ageing duration, the more significant the changes of the tensile properties. These evolutions are observed for all ageing temperatures and a strong dependance of the degradation kinetics to the temperature is noticed.

Figure 2 shows the evolution of the tensile behaviour of aged materials submitted to the other ageing protocols, *i.e.* thermal ageing and in synthetic non renewed seawater. For thermal ageing (figure 2-a), the evolutions are very different from those observed on figure 1 and very similar to the ones classicaly found for thermal ageing (see Hamed and Zhao (1999) for example), which tends to prove that the ageing mechanisms that occur with our specific ageing protocol are different from oxydation ones. The results will be confirmed in section 4.3 dealing with physical measurements. Dealing with the synthetic marine ageing (figure 2-b), the same observations as the one highlighted previously (increase of the initial stiffness and reduction of the ultimate properties) can first be

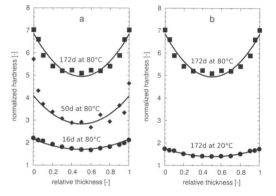

Figure 3. Micro-hardness profiling for various ageing duration and a given temperature (a) or various ageing temperature for a given duration (b).

done, but the degradation kinetics seem to slow down in a second time with a saturation of the initial stiffness rise, not observed on figure 1. To the author's opinion, this phenomenon is linked to the fact that the water is not renewed (more details are given in Le Saux (2010)). Even if only based on a simple mechanical analysis, these results show the importance of the environments in which the accelerated ageing tests are performed.

4.2 *Micro-hardness profiles*

Even if tensile tests can bring useful informations regarding the consequences of ageing on the mechanical properties, they do not permit to quantify the gradients induced by ageing. These gradients are directly related to the ageing mechanisms characterized by a competition between the diffusion and the reaction of chemical species. One simple way to evaluate these gradients from a mechanical point of view is to use modulus profiling (Gillen *et al.* 1987) as shown on figure 3 which presents the influence of the ageing duration for a given temperature (figure 3-a) and the influence of the ageing temperature for a given

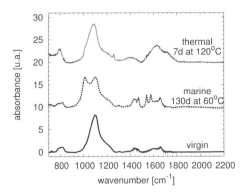

Figure 4. Comparison between the IR spectrum of various materials.

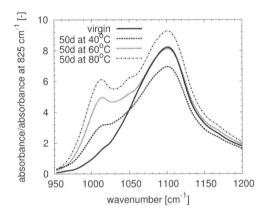

Figure 5. Apparition of a spectral band located at 1014 cm^{-1} and modification of the 1100 cm^{-1} one.

duration (figure 3-b) on the modulus profilings. These results illustrate a clear gradient between the skin and the core of the sample. The ratio between the skin and the core values is increasing as the severity of the ageing is increasing, *i.e.* higher temperatures and/or durations. The presence of gradients and their strong dependance on the ageing temperatures and durations implies that the correlation between the materials submitted to accelerated tests and the materials submitted to naturally ageing is complex and that special care need to be brought to the extrapolation.

4.3 *Physical measurements*

We present in this section the main results concerning the identification of the ageing analysis. The idea here is not to propose a full analysis but rather give the main results needed to validate the accelerated ageing tests and to propose a correlation with the modulus profilings results. For a deeper analysis, please refer to the forthcoming paper of Le Gac *et al.* (2011). Figure 4 shows some FTIR measurements performed on 3 materials: a virgin one, a thermally aged one (thermo-oxydation) and a marine aged one. As we can notice, the classical bands related to oxydation (3400 cm^{-1}, 1720 cm^{-1} and 1175 cm^{-1}) are not observed on the

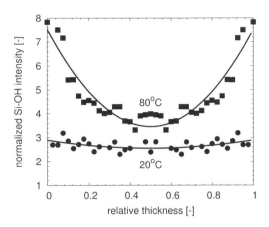

Figure 6. Si-OH concentration gradients through the thickness of the samples for an ageing duration of 172 d.

specimen aged in marine environment but can clearly be seen on the thermal aged sample, which proves that the ageing mechanisms that occur in marine environment are differents from oxydation ones, which confirms the analysis based on mechanical results presented previously. One can conclude that performing thermal ageing to characterize the marine ageing of the CR is clearly not relevant, which differs from the results presented in one recent paper dealing with NR (Mott and Roland 2001). For the sample aged in marine environment, we can notice a slight reduction of the 1100 cm^{-1} band attributed to the Si-O species (Socrates 2005) and the apparition of a band at 1014 cm^{-1} which can be correlated to the formation of C-C, Si-OH (Socrates 2005) or C-OH bounds (Davies and Evrard 2007). The intensity of these phenomena is increasing with the severity of the ageing, as shown on figure 5. Le Gac *et al.* (2011) performed solid NRM measurements and showed that the apparition of the 1014 cm^{-1} spectral band can be attributed to the silica hydrolysis, *i.e.* Si-OH bounds formations. Figure 6, which shows the ratio between the 1014 cm^{-1} and the 825 cm^{-1} spectral band, clearly indicates that the degradation gradients are related to a gradient of water concentration into the specimen. Therefore, it seems that the Si-OH FTIR peak is relevant as a local indicator of marine ageing for our material. This conclusion is motivated by the very good correlation between the Si-OH peak gradients (figure 6) and the hardness profilings (figure 3-b).

5 LIFETIME PREDICTION AND COMPARAISON TO A NATURALLY AGED MATERIAL

5.1 *Master curves and activation energies*

We propose in that section to analyze all the data to apply the Arrhenius lifetime prediction methodology. The first step consists in building a master curve related to a given indicator and for a given temperature, as shown on figure 7-a for the nominal strain at

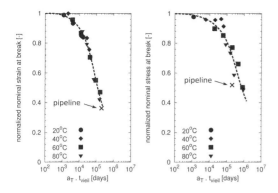

Figure 7. Master curve obtained for the nominal strain at break for a reference temperature of 20°C.

Table 2. Identification of the different activation energies.

indicator	E_a [kJ.mol^{-1}]
ε_{rupt}	52
π_{rupt}	79
M_{100}	42

break and 7-b for the nominal stress at break. These master curves are obtained for a reference temperature set to 20°C. These shift factors are then plotted versus the inverse of the temperature in order to evaluate the activation energy related to the property followed during the ageing. Table 1 presents the activation energies evaluated from the master curve related to classical indicators, *i.e.* nominal strain at break ε_{rupt}, nominal stress at break π_{rupt} and secant modulus for a nominal deformation of 100% M_{100}.

Table 2: Identification of the different activation energies.

5.2 *Correlation with a naturally aged sample*

To feed the discussion and bring additionnal results, we propose to correlate the informations coming from accelerated ageing tests to data coming from a naturally aged material. The tensile curves obtained for an unaged sample, a sample aged naturally for 23 years and a sample that has been previously submitted to an accelerated ageing (172 d at 80°C) are plotted on figure 8. Natural ageing leads to a significant rise of the initial stiffness and a strong decrease of the elongation at break. These consequences are indeed very similar to the ones observed for accelerated ageing. However, due to the hardness profiling which differs from accelerated ageing ones (figure 9), a slight modification of the form of the curve is to be noticed. It is worth noting that for the results plot on the figure 9, only the external face (left) was exposed to water, which explains the non-symmetrical profile and the shape of the gradients. We should also remark that the

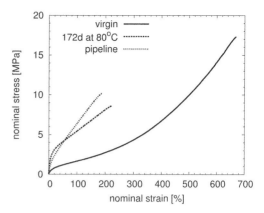

Figure 8. Strain and stress curve obtained for the naturally aged sample.

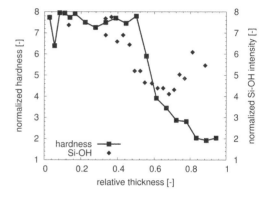

Figure 9. Correlation between the modulus and Si-OH profilings for the naturally aged sample.

hardness value measured on the exposed face is very similiar to the one measured on accelerated aged samples (172 d at 80°C for example). Let us focus on the correlation between the master curves (figure 7) and the data coming from the naturally aged sample. As shown on figure 7, a fairly good agreement is observed for the strain at break, but not for the stress at break. This result could be justified regarding the differences in the hardness profiling. Due to ageing, samples can be modelled as multi-layered materials. Each layer is submitted to the same strain, but not to the same stress. Therefore, the arrhenius methodology works with a strain value, but not with a stress value, illustrating the necessity to take into account the gradients effects to correctly analyze ageing data. Finally, FTIR measurements performed on the naturally aged sample revealed that the 1100 cm^{-1} peak almost disappeared after ageing, whereas the 1014 cm^{-1} peak is obviously generated while no oxydation products are to be noticed. Here again, a good correlation between the mechanical value and Si-OH concentration is observed (figure 9) as previously highlighted for accelerated aged samples.

6 CONCLUSION

The ageing in seawater of a silica-filled Polychloroprene rubber (CR) has been investigated in this paper. The CR degradation has been analysed in order to highlight the main features of the chemical mechanisms and the mechanical consequences for accelerated and long term natural ageing, which is rarely found in the literature. The main achivements of this study are:

- the proposal of a relevant accelerated ageing protocol to speed up the polymer degradation in marine environments;
- the use of harness profiling as a tool to mechanically quantify the gradients induced by ageing needed for a better interpretation of ageing results;
- the good correlation between the Si-OH gradients and modulus ones, illustrating the relevance of the hardness profiling as a bridge between physical and mechanical measurements;
- the limits of the Arrhenius methodology due to the gradient effects that are not taken into account.

ACKNOWLEDGEMENTS

The authors would like to thank the Britanny region for its financial support, all the actors of the FEMEM project, espacially V. Laguarrigue from Trelleborg Engineered Systems for providing the materials.

REFERENCES

Ab-Malek, K. and A. Stevenson (1986). The effect of 42 years immersion in seawater on natural rubber. *Journal of Materials Science 21*, 147–154.

Celina, M., J. Wise, D. Ottesen, K. Gillen, and R. Clough (2000). Correlation of chemical and mechanical property changes during oxidative degradation of neoprene. *Polymer Degradation and Stability 68*, 171–184.

Davies, P. and G. Evrard (2007). Accelerated ageing of polyurethanes for marine applications. *Polymer Degradation and Stability 92*, 1455–1464.

Gillen, K., R. Bernstein, and D. Derzon (2005). Evidence of non-arrhenius behaviour from laboratory ageing and 24-year field ageing of polychloroprene. *Polymer Degradation and Stability 87*, 57–67.

Gillen, K., R. Clough, and C. Quintana (1987). Modulus profiling of polymers. *Polymer Degradation and Stability 17*, 31–47.

Ha-Anh, T. and T. Vu-Khanh (2005). Effects of thermal ageing on fracture performance of polychloroprene. *Journal of Materials Science 40*, 5243–5248.

Hamed, G. and J. Zhao (1999). Tensile behaviour after oxydative ageing of gum and black-filled vulcanizates of SBR and NR. *Rubber Chemistry and Technology 72*, 721–730.

Le Gac, P., V. Le Saux, Y. Marco, and M. Paris (2011). Investigation of ageing mechanisms and mechanical consequences of marine environment on Polychloroprene rubber: validation of accelerated protocols by comparison to a 23 years old offshore export line. *Polymer Degradation and Stability (submitted)*.

Le Saux, V. (2010). *Fatigue et vieillissement des élastomères en environnements marin et thermique: de la caractérisation accélérée au calcul de structure*. Ph. D. thesis, Université de Bretagne Occidentale.

Le Saux, V., Y. Marco, S. Calloch, P. Le Gac, and N. Ait Hocine (2009, September 7th-10th). Accelerated ageing of polychloroprene for marine applciations. In *Constitutive Models for Rubber VI*, Dresden (Germany), pp. 3–9.

Mott, P. and M. Roland (2001). Ageing of natural rubber in air and seawater. *Rubber Chemistry and Technology 74*, 79–88.

Pegram, J. and A. Andrady (1989). Outdoor weathering of selected polymeric materials under marine exposure conditions. *Polymer Degradation and Stability 26*, 333–345.

Socrates, G. (2005). *Infrared and raman characteristic group frequencies: tables and charts (third edition)*. Wiley.

Stevenson, A. (1984). *Rubber in offshore engineering*. Agam Hilger Ltd.

Constitutive Models for Rubber VII – Jerrams & Murphy (eds)
© 2012 Taylor & Francis Group, London, ISBN 978-0-415-68389-0

Nanomechanics of rubber for fuel efficient tyres

Keizo Akutagawa, Satoshi Hamatani & Hiroshi Kadowaki
Bridgestone corporation, Ogawahigashi-cho, Kodaira-shi, Tokyo, Japan

ABSTRACT: From global warming concern a tyre rolling resistance is one of the important performances in the tyre development. Approximately 80% of tyre rolling resistance can be attributed to energy loss of the rubber component. The rubber used in tyre applications is filled with carbon-black and silica, but the mechanical behaviours in nano-scopic scale have not been fully understood due to the limitation of the mechanical analysis in such a small scale. Combination of transmission electron micro tomography and 3 dimensional finite element analysis enables us to investigate the deformation of the filled elastomer in nano-scopic scale. The rubber phase between filler aggregates showed a large strain concentration over 200%, even if the overall strain was only 15%. This is due to the fact that the trapped rubber inside the filler network shell can behave as a hard domain and induce the strain amplification, which produces the highly strained area[1]. It was found that the amount of the trapped rubber decreases with increasing overall strain, as the filler network shell is broken up. This may be associated the non-linear stress-strain behaviour of filled rubber system. To control this behavior is important for designing rubbers with lower energy loss for the fuel efficient tyres.

1 INTRODUCTION

For tyre compound design the energy loss properties around 60°C are often predictor in reducing tyre rolling resistance. The agglomeration of the filler aggregates in the rubber matrix is major concern in this temperature range, but the general principle governing the energy loss of filled compounds has not been fully understood. Recently, Transmission Electron Micro Tomography (TEMT) of filled rubber by 3-D electron microscope has been carried out by Jinnai et al.[2]. From this method, the reconstruction of 3-D image can be constructed by computerized tomography (CT). This work consists of the combination techniques of the reconstruction of 3-D image by TEMT and Voxel 3-D FEM calculation which enable us to predict strain distribution of filled rubber in nanoscopic scale[1]. The purpose of this study is aiming at understanding the nano-scopic mechanical behavi-ous of the filled rubber, which governs the energy loss properties.

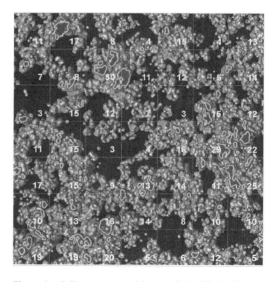

Figure 1. 3-D reconstructed image of the filled rubber in nano-scopic scale.

2 EXPERIMENTAL

The filled rubber used in this study is N234 carbon black filled SBR#1500. The 3-D TEMT experiments were carried out using TEMT operated at 200 kV with 1024 × 1024 pixel elements. The 3-D tomography image of the filled rubber in nano-scale taken by TEMT is shown in Figure 1. The size of captured 3-dimensional image was 1000 × 1000 × 150 nm. The image was divided into 49 cubic clusters and the volume fraction was calculated for each cluster. The numbers superimposed in Figure 1 represent the values of volume fraction of rubber. The finite element model was constructed from one of the clusters which has the same value of volume fraction of rubber to the overall value of 14%. The voxel method was applied to transfer the 3-diminutional image to the finite element model. Each voxel size is 2 nm × 2 nm × 2 nm, which is equivalent to that of one pixel size of 3-dimentional image. The size of voxel is also corresponding to the average length of the cross links.

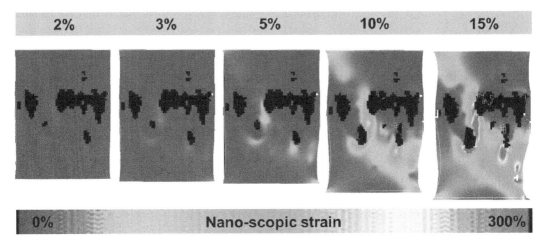

Figure 2. The macro-scopic strain dependence of nano-scopic strain distribution calculated at the strain of 2%, 3%, 5%, 10% and 15%.

3 RESULTS AND DISCUSSION

The 3-D finite element analysis was carried out with deformations applied with tensile strain of 1%, 3%, 5%, 10% and 15% respectively. The interaction between filler and polymer is assumed to be contact without sliding. The constitutive equations used for this model are the nonlinear equations for rubber part and the linear equations for filler part with a magnitude of modulus 3 GPa. The non-linear equation was derived from the stress-strain curve experimentally measured on the unfilled cross linking rubber, which has the same cross linking formulation as the filled rubber used. The further discussion of nonlinear constitutive equations is reported elsewhere[3]. The boundary condition to neighboring unit cell is assumed to be free surface condition[4]. The black part of these results represents filler and the rest of the part is rubber. The details of the methods of the transmission electron microscopic tomography and the 3-dimensional finite element analysis are shown in the elsewhere[1,2]. The overall stress-strain relation calculated in this condition is shown in Figure 3 together with those derived experimentally on the specimen used.

It shows a quite good agreement between calculations and experimental results. From this results it is assumed that the mechanical behaviours calculated in nano-scale are able to represent those appeared in the actual situation.

As seen in Figure 2 at overall strain of 15% there are highly deformed regions over 200% strain and less deformed region less than 5% strain. This is due to the fact that the trapped rubber inside the filler network shell can behave as an immobilized domain and induce the strain amplification, which produces the highly strained area. The immobilized domain start to move as the filler network shell is broken up with increasing overall strain. It was also reported that the strain distribution in nano-scale was classified into three strain

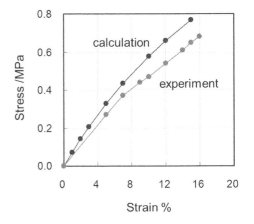

Figure 3. The comparison between calculated and experimental results of macroscopic stress-strain relation.

groups and the distribution of each strain group shows a log normal distribution[2].

The strain dependences of them were calculated with overall strain of 5%, 10% and 15% and shown in Figure 4. The magnitude of each peak decreases with broadening its strain range. It was also found that the peak magnitude of the trapped rubber decreases with increasing overall strain and the strain in nano-scale at peak shifts to higher strain region. From this shifts it is assumed that the trapped rubber flows into more strained region, as the filler network shell is broken up. This can be also seen in Figure 2 and it can be described as if the eggshell is hatched with increasing external stress and stuff inside comes outside. The unstrained rubber inside the shell flows into the strained rubber outside, which makes overall stress relaxed. This may be associated with the non-linear stress-strain behaviors of filled rubber system. To control this mechanics between filler network and rubber is also important

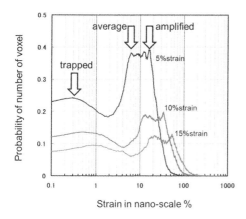

Figure 4.   The strain distribution in nano-scale calculated by 3D-FEA at strain of 5, 10 and 15%.

for designing the filled rubbers for fuel efficient tyres. The further investigation on this is still underway.

4   CONCLUSIONS

The nonlinear relationship between stress and strain is historically described as Fletcher-Gent effect at lower strain region and Mullins effect at higher strain region. The egg-hatch effect predicted in nano scale finite elemental analysis may contribute to explain some part of these effects. It can be expected to contribute the design of the compounds for fuel efficient tyres.

ACKNOWLEDGEMENTS

This work was carried out with partly supported by the New Energy and Industrial Technology Development Organization.

REFERENCES

[1] K. Akutagawa, K. Yamaguchi, A. Yamamoto, H. Heguri, H. Jinnai, Y. Shinbori, *Rubber Chem. Techno*l., 2008, 81, 182.
[2] H. Jinnai, Y. Shinbori, T. Kitaoka, K. Akutagawa, N. Mashita, T. Nishi, *Macromolecules*, 2007, 40, 6758.
[3] G. Heinrich, M.Kaliske, A. Lion, S. Reese, *Constitutive Models for Rubber VI: Proceedings of the 6th ECCMR*, Taylor & Frances, 2009, 409.
[4] V. Jha, A. Hon, A.G. Thomas, J.J.C. Busfield, *J. Appl. Polym. Sci*., 2007, 13, 2573.

Fatigue and time dependent behaviour

Constitutive Models for Rubber VII – Jerrams & Murphy (eds)
© 2012 Taylor & Francis Group, London, ISBN 978-0-415-68389-0

Definition and use of an effective flaw size for the simulation of elastomer fatigue

F. Abraham

Department of Mechanical & Design Engineering, University of Portsmouth, Portsmouth, UK

ABSTRACT: Previous research has shown the importance of flaws for the fatigue life of elastomers. However, there are other factors that may limit fatigue life apart from flaw size such as their numbers per volume, their distribution and the flaw density. This research aims to establish the use of an effective flaw size to bring together the fracture mechanics approach and fatigue life simulation for complex filled industrial style elastomers. Fatigue data and fracture mechanics are used to calculate a representative Effective Flaw Size (EFS) which allows the calculation of fatigue properties of commercial filled elastomers that are inhomogeneous by nature. The Effective Flaw Size in combination with a stress concentration factor and dynamic strain energy can as well be used to simulate the fatigue behaviour under non-relaxing conditions. Until now the fracture mechanics approach did not work for the calculation of fatigue life when minimum loads, especially in tension were applied.

1 INTRODUCTION

The fatigue life behaviour of elastomers has been studied extensively ((Lake & Lindley, 1965), (Cadwell, Merrill, Sloman, & Yost, 1940), (Schöpfel, Idelberger, Schütz, & Flade, 1996)). The same holds true for the fracture mechanics of elastomers (Selden, 1995). Rarely have experiments on fatigue life been carried out on the same elastomers that were used for the fracture mechanics, especially when they are not under non-relaxing condition. The reason is that they could show inconsistencies or even contradictions between calculations and experimental results, especially if different materials are compared (Abraham F., 2002). Another reason is that fatigue life testing is extremely time consuming and companies are not always willing to publish their research to protect their advantage in fatigue simulation knowledge.

The general fatigue life behaviour is usually described by plotting maximum stress over cycles to failure. Such tests are done under fully relaxing conditions and are called Wöhler curves (s-n curves) as shown in 1. A comparison of unfilled and filled EPDM and SBR has shown that a direct comparison of experimental fatigue life and fracture mechanics is not possible as a ranking reverse occurred. For this simulation the maximum flaw sizes and the dynamic crack growth rate were used, besides other measured values. The maximum flaw size was determined by cutting thin slices out of the rubber test specimen and examining them with a half automated system using a light microscope (Abraham, Alshuth, & Clauß, 2005). The dynamic crack growth rates were measured with a Coesfeld Tear Analyzer System Bayer.

Micro Computed Tomography (Micro CT) imaging has given further insight into the fatigue behaviour of commercial elastomers. Micro CT imaging carried out before testing, has shown the shape, sizes and distribution of naturally occurring flaws in the test specimen (Le Gorju Jago, 2008). Micro CT imaging during testing described the growth of cracks starting from these natural flaws. These micro cracks are growing at the beginning independently from each other. Only if flaws are in close proximity or the crack has grown close enough to another flaw or crack, only then they are interacting and combining to a big crack. These results demonstrate the need for a new definition of the initial flaw size. It would be wrong to use the biggest flaw when it is in a not highly stressed volume element when there are a few smaller flaws in very close proximity that could act like a big flaw. Other important factors for the fatigue life are the geometry/shape, material/density/stiffness, the distribution and the numbers of these initial flaws. Previous research has shown that even small flaws weaken the material and reduce the fatigue resistance extremely when there are enough of them (Figure 1 and (Abraham, Alshuth, & Clauß, 2005)). The conclusion of these individual results is that a direct fracture mechanical simulation of a commercial rubber component is virtually impossible. First, it would be necessary to know the precise 3 dimensional description of all flaws and their position on a micrometre level in the part which would take a very long time. Second, these data would be the basis for an extremely time intensive Finite Element Simulation including fracture mechanics. This method would be very expensive and impractical.

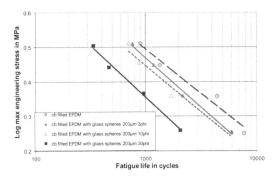

Figure 1. Influence of glass sphere flaw density on fatigue life.

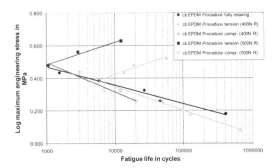

Figure 2. Fatigue life behaviour of carbon black filled EPDM, including minimum load variations.

All these reasons indicated the necessity of a simple quick simulation method for the fatigue behaviour of industry standard filled elastomers.

2 METHODOLOGY

This research uses carbon black filled EPDM dynamic crack growth data to simulate the fatigue life behaviour of this elastomer (Figure 2 (Abraham F., 2002)). A direct calculation was shown not to be possible as the two used elastomers displayed a ranking reverse in their fatigue life and dynamic crack growth behaviour. Even the inclusion of the maximum flaw size determined by light microscopy did not enable a simulation of the fatigue behaviour (Abraham, Alshuth, & Clauß, 2003). The inclusion of the minimum load effect into the simulation makes the calculation even more problematic as the standard fracture mechanics approach does not accommodate such testing conditions.

The following fracture mechanics equations (Selden, 1995) (Abraham, Alshuth, & Clauß, 2005) are used for the simulation of the measured fatigue behaviour and iteration of the initial crack length:

$$\frac{dc}{dn} = B \cdot T^{\beta} \tag{1}$$

$$T = 2 \cdot k \cdot w \cdot c \tag{2}$$

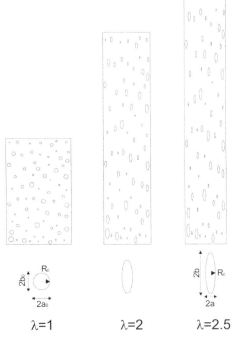

Figure 3. Schematic depiction of flaw geometry changes due to strain.

$$k = \frac{\pi}{\sqrt{\lambda}} \tag{3}$$

$$n = \frac{1}{\beta - 1} \cdot \frac{1}{B \cdot (2 \cdot k \cdot w)^{\beta}} \cdot \left[\frac{1}{c_0^{\beta - 1}} - \frac{1}{c^{\beta - 1}} \right] \tag{4}$$

T - tearing energy
B - crack growth factor
β - crack growth exponent
w - strain energy density
c - crack length
k - constant
λ - strain
n - number of cycles
c_0 - initial crack length

The results of the standard approach are not satisfactory, see chapter 3. RESULTS. The introduction of an additional factor to accommodate for the high strains applied to the specimen and flaws appears to be necessary. Otherwise it would be not possible to simulate the fatigue behaviour of a complex industrial style filled elastomer. This holds especially true for non-relaxing testing condition with minimum loads in tension. A schematic depiction of the specimen and flaw geometry changes during tension is shown in the next Figure 3.

Figure 3 shows how the radius of the flaw increases with higher strains, which causes the crack tip to blunt independently of the original flaw size of geometry. Such behaviour has been observed by Micro CT

Table 1. Simulation results of initial crack sizes

				dyn Energy	$k_t = 2/\lambda$ dyn Energy	$k_t = 2/\lambda^{1.5}$ dyn Energy	$k_t = 2/\lambda^2$ dyn Energy
min. Load N	max. Load N	c_{0-1} μm	c_{0-2} μm	c_{0-3} μm	c_{0-4} μm	c_{0-5} μm	
0	250	90.3	90.3	114.0	76.3	51.1	
0	300	106.4	106.4	123.0	78.8	50.5	
0	350	130.7	130.7	123.5	71.6	41.5	
−200	200	55.3	55.3	96.1	75.5	59.3	
−150	250	79.1	79.1	113.7	81.2	58.0	
−100	300	93.2	93.2	114.1	75.2	49.6	
−50	350	145.0	145.0	154.1	94.7	58.2	
0	400	153.3	153.3	140.5	80.2	45.7	
50	450	78.7	133.3	111.2	60.5	32.9	
100	500	62.3	168.9	149.8	84.0	47.1	
150	550	36.7	136.2	139.2	83.9	50.5	
0	450	243.9	243.9	210.0	116.2	64.2	
−200	300	129.0	129.0	171.1	117.4	80.6	
−150	350	165.4	165.4	186.3	117.8	74.5	
−50	450	164.4	164.4	142.0	78.6	43.5	
0	500	243.6	243.6	194.9	103.9	55.4	
100	600	109.6	271.6	213.8	113.0	59.7	
200	700	59.4	231.2	249.1	154.1	95.4	
	standard dev.	62.7	62.7	43.6	23.8	15.3	
	c_0 average	**119.2**	**152.3**	**152.6**	**92.4**	**56.5**	

measurements (Le Gorju Jago, 2008). This effect justifies the inclusion of the stress concentration factor commonly used in fracture mechanics to accommodate diameter changes in components geometries (Callister, 1994).

$$k_t = 2 \cdot (\frac{a}{R_c})^{\frac{1}{2}} \tag{5}$$

The relevant crack radius R_c (5) can be calculated using the standard equation for an ellipse (6).

$$\frac{x^2}{a^2} + \frac{y^2}{b^2} = 1 \tag{6}$$

$$R_c = \frac{b^2}{a} \tag{7}$$

Defining an initially spherical flaw $2a_0 = 2b_0 = c_0$ and assuming a linear change just depending on the overall strain.

$$2a = \frac{c_0}{\lambda} \tag{8}$$

$$2b = c_0 \cdot \lambda \tag{9}$$

Entering (8) and (9) in (7) and (5) creates a very simple relationship for the stress intensity factor k_t under this simplifications (10).

$$k_t = \frac{2}{\lambda^2} \tag{10}$$

This stress concentration factor (10) is used to scale the strain energy density w of equation (4).

$$n = \frac{1}{\beta - 1} \cdot \frac{1}{B \cdot (2 \cdot k \cdot w / k_t)^\beta} \cdot \left[\frac{1}{c_0^{\beta-1}} - \frac{1}{c^{\beta-1}} \right] \tag{11}$$

Note using a constant flaw width $2a_0 = 2a = c_0$ to represent an incompressible flaw which can only be elongated according to (9) results in a $K_t = 2/\lambda$. The same factor holds true for the assumption that the flaws behave like the overall diameter and height of the test specimen under constant volume.

3 RESULTS

The calculated initial crack lengths using equation (4) results in a wide range of sizes from 37 μm to 244 μm depending on testing condition and an average size of 119 μm with a high standard deviation of 63 (c_{0-1} in Table 1) simulation using this average initial crack length cannot replicate the tested fatigue behaviour even when the minimum load variation were ignored, as shown as thin solid lines in Figure 4. The different K_t factors in Table 1 represent variations corresponding to different simplifications concerning the flaw behaviour under straining.

The introduction of the "dynamic" strain energy density (Abraham F., 2002) and (Abraham, Alshuth, & Jerrams, 2005) into equation (4), instead of the normal

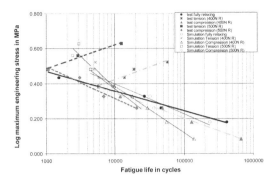

Figure 4. Simulation result without correction.

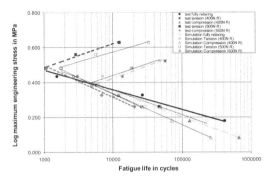

Figure 5. Comparison of fatigue tests with simulation using $K_t = 2/\lambda$.

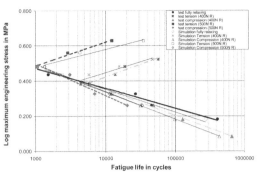

Figure 6. Comparison of fatigue tests with simulation using $K_t = 2/\lambda^{1.5}$.

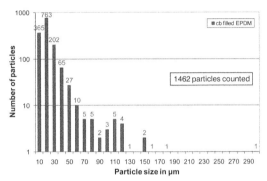

Figure 7. Particle size distribution of the carbon black agglomerates in the rubber matrix.

strain energy density, improves the simulation results. This calculated average initial crack length (c_{0-2} in Table 1), using the dynamic energy, could roughly reproduce the fatigue behaviour under non-relaxing conditions. Despite this, the simulation is still not of an acceptable quality.

The use of the stress concentration factor $K_t = 2/\lambda$ results in a relatively high initial average crack length of 153 μm and a standard deviation of 44 (c_{0-3} in Table 1) and could be used as the effective initial flaw size for the fatigue life simulation as shown in Figure 5.

The use of the stress concentration factor $K_t = 2/\lambda^{1.5}$. results in a very plausible initial average crack length of 92 μm and a standard deviation of 24 (c_{0-4} in Table 1) and could be safely used as the effective initial flaw size for the fatigue life simulation as shown in Figure 6. The measured particle size distribution of the carbon black agglomerates in the rubber matrix is shown Figure 7 (Abraham, Alshuth, & Clauß, 2005).

The calculated effective initial flaw sizes using the stress concentration factor $K_t = 2/\lambda^2$ (10) are much closer together and more realistic (c_{0-5} in Table 1) compared to the original simulation without any correction. This initial average crack length of 57 μm has a relative low standard deviation of 15 and can be used as the effective initial flaw size for the fatigue life simulation. The effective flaw size value itself is very realistic

as can be seen in Figure 7, only a few flaws a larger than this EFS. 8 displays the simulation results using the dynamic stored energy and the stress concentration factor $K_t = 2/\lambda^2$.

4 DISCUSSION

The simulation of individual initial crack lengths for the different testing conditions results in very different initial crack sizes, if the standard fracture mechanics approach is used. The average crack length cannot be used to simulate the genuine fatigue behaviour properly, especially in the case of minimum load variations. These tests under non-relaxing conditions, mainly in pure tension, still result in much too small initial crack sizes. A way to overcome this problem is the introduction of the dynamic strain energy density instead of the normal strain energy. A dynamic energy criterion has previously been defined as the "normal" energy (stored or total) minus any existing positive static part of this energy ((Abraham F., 2002), (Abraham, Alshuth, & Jerrams, 2005)). This means that the energy available for the testing conditions purely in tensions is reduced according to the minimum load. The energies under fully relaxing conditions or with a minimum load in compression are not affected by this energy correction.

The results for the simulations using the dynamic strain energy in conjunction with the stress concentration factor for the flaws are shown in 8. Here is shown a very good slightly conservative calculation of the fatigue results. Only the very high amplitude test with 500 N range in tension and compression predict a longer than measured fatigue life. The reduced life could be explained by the high temperatures developed in the specimen during testing and/or the not directly measured maximum strains due to test equipment limitations at these strains. Overall, the current simulation demonstrates a very good representation of the fatigue behaviour especially when the high scatter naturally occurring in fatigue test is taken into account (factor 6 for this material).

The calculated initial effective flaw size using the stress concentration factor is working very well. Even a variation of the stress concentration factor exponent between 1, 1.5 and 2 gives stable results. The different exponents are result of the different boundary conditions and simplification concerning the flaw shape changes or their mean value. The fatigue simulations are still generally very good as only the initial flaw sizes varies as they decrease with increasing exponent (Table 1). They stay in the region of the measured real flaws existing in this type of rubber, but they do not reach the level of the largest found flaw Figure 7 (Abraham, Alshuth, & Clauß, 2005) which is more than plausible as the largest possible flaw might not be in a highly stressed region of the specimen.

The proposed calculation method explains the minimum load influence of elastomers containing active fillers but the same geometrical changes apply to unfilled elastomers as well. Unfilled elastomers do not show such a minimum load effect. This might be explained by the far lower strains applied to unfilled non strain-crystallising elastomers in combination with smaller flaw sizes which do not change their shape so dramatically. If this is the case, then this would explain why unfilled strain-crystallising elastomers like Natural Rubber show the same minimum load effect (Andre, Cailletaud, & Piques, 1999), as they are capable of even higher maximum strains.

5 CONCLUSION

The new proposed simulation method using the stress concentration factor k_t in combination with the dynamic strain energy density is very successful in the calculation of the effective initial flaw size. This effective initial flaw size can be used to calculate the fatigue life under relaxing test conditions and under non-relaxing conditions. Figure 8 shows the best simulation method with a $K_t = 2/\lambda^2$.

The next step would be to establish a method to directly measure/calculate the initial flaw size for standard industrial filled elastomers, using Micro CT data for example. This is important for practical use, as

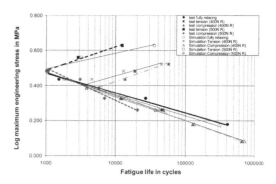

Figure 8. Comparison of fatigue tests with simulation using $K_t = 2/\lambda^2$.

the suggested simulation method is a reverse calculation of the initial flaw size using fatigue and dynamic crack growth data and the whole testing and simulation process is very time extensive.

REFERENCES

Abraham, F. (2002 December). PhD-Thesis The Influence of Minimum Stress on the Fatigue Life of non Strain-Crystallising elastomers. UK: Coventry University.

Abraham, F., Alshuth, T., & Clauß, G. (2003 15–17 September). Poster: Comparison of Fatigue Life and Dynamic Crack Growth Measurements on Elastomers. ECCMR III, Third European Conference on Constitutive Models for Rubber. London, UK.

Abraham, F., Alshuth, T., & Clauß, G. (2005). Testing and simulation of the influence of glass spheres on fatigue life and dynamic crack propagation of elastomers. In A. &. (eds, Constitutive Models for Rubber IV (pp. 71–76). London: Balkema.

Abraham, F., Alshuth, T., & Jerrams, S. (2005). The effect of minimum stress and stress amplitude on the fatigue life of non strain crystallising elastomers. Materials and Design, 26, 239–245.

Andre, N., Cailletaud, G., & Piques, R. (1999). Haigh diagram for fatigue crack initiation prediction of natural rubber components. KGK Kautschuk Gummi Kunststoffe, 52. Jahrgang, Nr 2/99, 120.

Cadwell, S. M., Merrill, R. A., Sloman, C. M., & Yost, F. L. (1940). Dynamic fatigue life in rubber. Industrial and Engineering Chemistry, Analytical Edition. 12, 19.

Callister, W. D. (1994). Materials Science and Engineering: An Introduction – 3rd Edition. New York: John Wiley & Sons, Inc.

Lake, G. L., & Lindley, P. B. (1965). The Mechanical Fatigue Limit for Rubber. Journal of Applied Polymer Science Vol. 9, 1233.

Le Gorju Jago, K. (2008). Fatigue life of rubber components: 3D damage evolution from X-ray computed microtomography. In L. M. Boukamel, Constitutive Models for Rubber V (p. 173). London: Taylor & Francis/Balkema.

Schöpfel, A., Idelberger, H., Schütz, D., & Flade, D. (1996). Betriebsfestigkeit von Elastomerbauteilen DVM-Tag 1996, Bauteil'96, Elastomerbauteile, (p. 103).

Selden, R. (1995). Fracture Mechanics Analysis of Fatigue of Rubber – A Review. Progress in Rubber and Plastics Technology, 11, 56–83.

Constitutive Models for Rubber VII – Jerrams & Murphy (eds)
© 2012 Taylor & Francis Group, London, ISBN 978-0-415-68389-0

Determination of effective flaw size for fatigue life prediction

J.G.R. Kingston & A.H. Muhr
Tun Abdul Razak Research Centre, UK

ABSTRACT: Effective flaw sizes (c_i) for initiation of mechanical failure have been determined for unfilled and filled natural rubber and filled styrene-butadiene rubber by comparing experimental results for failure scenarios with predictions made using fracture mechanics, treating effective flaw size as a fitting parameter. Two independent techniques are used; the different failure scenarios being cyclic fatigue life of tensile strips in one case, and ozone cracking in the other. For the filled rubbers there was a surprisingly large dependence of apparent effective flaw size on the strain amplitude used in the fatigue tests. The fatigue method also called for the crack growth parameters to be determined, and these are compared to values reported in the literature for similar materials.

1 INTRODUCTION

According to the theory of Gent, Lindley & Thomas (1964) the fatigue life of rubber is determined by the number of cycles required for a pre-existing flaw to grow from its initial effective size to such a size (eg 1mm) that mechanical failure is imminent. To enable the calculation of fatigue life to be made, several simplifying assumptions are necessary: (1) there is no interaction between flaws, so that fatigue life may be calculated on the basis of a single flaw (2) the growth of flaw size is controlled by the energy release rate according to the same characteristic relationship as determined from test pieces incorporating artificially introduced cracks (3) the effective flaws have a population density sufficiently high to ensure that there is one with the worst-case location and orientation to minimise fatigue life.

In this work we seek to characterise three rubbers in terms of crack growth behaviour and effective flaw size. The most commonly used testpieces for determining crack growth characteristics in rubber are tensile (with a single small edge crack), pure shear and trouser tear (Rivlin & Thomas, 1953), cut from moulded sheet. If the resulting crack growth data is going to be useful, these different techniques must be equivalent so that it is possible to apply the results from any of them to a different situation. Attention is restricted here to a through-thickness crack in the edge of a thin strip in simple extension, for which the energy release rate (or tearing energy) T is given by (eg Timbrell et al. 2003)

$$T \equiv \frac{d\Pi}{dA} \cong \frac{2\pi}{\sqrt{\lambda}} Wc \qquad (1)$$

In Equation 1 Π is the total energy of the system, A is the area of one half of the fracture surface, W is the retraction strain energy density in the region of

Table 1. Composition (parts per hundred rubber by weight) and basic properties of the materials. Static and dynamic stress relaxation (at 100 or 250% strain) rates are expressed as % change per tenfold increase in time or cycles, respectively.

Material	NR+0	NR+45N330	SBR+77N339
SMR CV60	100	100	0
SBR1712	0	0	137.5
N330	0	45	0
N339	0	0	77
Strukthene	0	4.5	0
ZnO	5	5	2.5
Stearic acid	2	2	1
HPPD	3	2	2
CBS	0.6	0.6	0
TBBS	0	0	1.6
TMTD	0	0	0.2
Sulfur	2.5	2.5	1.6
Cure, min@ 150°C	20	18	30
Hardness, Shore A	36	60	65
Static relaxn (100%)	3.4%	5.1%	7.0%
Dyn relaxn (100%)	0.8%	4.1%	–
Dyn relaxn (250%)	–	10.8%	7.8%

uniform strain far from the crack tip, c is the crack length in the strain-free state and λ is the extension ratio of the rubber.

2 MATERIALS

Table 1 shows the compositions, moulding conditions, hardness and stress relaxation rates of the three materials used in this study. For SBR+77N339 a grade of Styrene-butadiene Rubber (SBR1712) incorporating 37.5 pphr oil was used, so that the hardness was not very different to the filled Natural Rubber

(NR+45N330) despite the higher loading, and higher structure, of carbon black (N339).

3 CRACK GROWTH RATE MEASUREMENTS

Parallel-sided tensile testpieces, nominally 25×2 mm in cross section and at least 125 mm in length, were used. They provide a simple and quick way of measuring the characteristics of the rubber, with the principal advantage that the energy release rate depends on the size of the crack. Therefore it is possible to quickly sample a wide range of energy release rates including very small values, for which the strain is low and the crack is small. Not only is it easy to make the crack small, the length of the testpiece makes it easier to set up low strains accurately.

A dedicated fatigue machine was used, having an eccentric drive to deliver a fixed stroke to the clamp at one end of each testpiece, while the other clamp was mounted on a load cell. The lengths of the cracks in the rubber were measured with a travelling microscope fitted with a graticule eyepiece. The strain energy density was interpolated from previously obtained 10th cycle stress-strain retraction curves using the trapezium method. Load-deflection behaviour of the test pieces was not monitored during the crack growth tests.

The results are given in Figures 1 to 3 as log-log plots of the crack growth rate versus the maximum energy release rate T_{max} in the fully relaxing cycles. According to the literature (eg Lake & Lindley 1964) the cyclic crack growth characteristic may be described by three regimes:

$$T_z < T_{max} \le T_0 \qquad dc/dn = r \equiv t_c \rho_z q \qquad (2)$$

$$T_0 < T_{max} \le T_t \qquad dc/dn = \oint \rho(T)dt + r_A\left(\frac{T_{max} - T_0}{T_{ref}}\right) + r \quad (3)$$

$$T_t \le T_{max} < T_c \qquad dc/dn = \oint \rho(T)dt + r_B\left(\frac{T_{max}}{T_{ref}}\right)^\beta + r \quad (4)$$

In Equation 2, t_c is the time during each cycle that the rubber is strained such that $T > T_z$, q is the ozone concentration and ρ_z is the susceptibility of the rubber to ozone attack. For $T_{max} < T_z$ there is no crack growth. Estimated values are $q \sim 0.003$ ppm by volume (the test laboratory varied between 0.002 and 0.003 ppm), $\rho_z \sim 6.8$ nm $\cdot$ s^{-1} ppm^{-1} (Braden & Gent, 1960, I), and $t_c \sim 0.17$ s (for the test frequency of 4.44 Hz and typical time fraction of full relaxation), so that $r \sim 3$ μm/Mcycle. In Equations (3) and (4) $\rho(T)$ is the rate of time-dependent crack growth, assumed in the present work to make negligible contribution to dc/dn. This is known to be justified for NR except at values of T_{max} approaching T_c(Gent et al., 1964; Lake et al., 1991), but whether or not it is adequate for SBR + 77N339 depends on t_c.

In Equations 3 and 4 a reference energy release rate T_{ref}has been introduced, so that the parameters

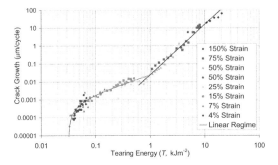

Figure 1.   Crack growth characteristics of NR + 0.

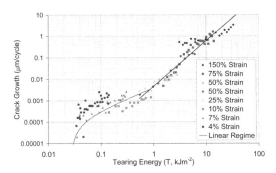

Figure 2.   Crack growth characteristics of NR+45N330.

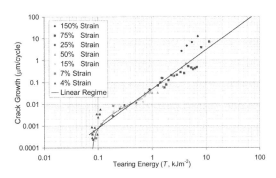

Figure 3.   Crack growth characteristics of SBR+77N339.

r_A and r_B have simple dimensions of length increment per cycle. We shall take T_{ref} to be 1 kJm^{-2} for convenience, since this is in the middle of the crack growth characteristics and close to the transition value T_t for most rubbers. This approach is thought more convenient than replacing T_{ref} by the catastrophic tearing energy T_c (eg Mars & Fatemi, 2003) which is only evaluated in determinations of the complete crack growth characteristics, and, being material-dependent, makes comparison of values of r_A and r_B for different materials more complicated to interpret. The tearing energy T_t at the transition between rate laws is not here regarded as a parameter, but can be related to the other parameters by equating dc/dn according to Equations 3 and 4.

Table 2. Comparison of crack-growth characteristics parameters from Sections 3 and 5 and (a) Braden & Gent (1960II) (b) Lake & Lindley (1964)

Material	T_z	T_0	r_A	r_B	β
	Jm^{-2}		$\mu m/kcycle$		–
This work					
NR+0	0.16	32	25	25	2.8
NR+45N330	0.87	28	4.0	4.7	2.1
SBR+77N339	4.2	77	32	49	1.8
Literature:					
NR+0	0.12[a]	40[b]	24[b]	58[b]	2.1[b]
NR+50N330	–	–	–	5.3[b]	1.3[b]
SBR+50N330	–	–	–	121[b]	4.2[b]

A comparison of our results with those reported by Lake & Lindley (1964) is given in Table 2. The compositions were fairly similar except that of the filled SBR, which in Lake & Lindley's work had a loading of 50 pphr N330 and no oil.

4 ESTIMATION OF EFFECTIVE FLAW SIZES FROM FATIGUE RESULTS

Fatigue life of tensile dumbbells can be estimated from an effective initiating flaw size c_0, the appropriate value of T_{max} as a function of crack length, and the crack growth characteristics by integration of Equations 2 to 4, neglecting the final term of Equation 4 and using Equation 1 for the relationship for T_{max} and the current flaw size:

$$N = \int_{c_i}^{1mm} \frac{dc}{dc/dn} \approx \int_{c_i}^{\infty} \frac{dc}{dc/dn}$$

$$= \frac{1}{r}\left(c_0 - c_i\right)$$

$$+ \frac{T_{ref}}{2kWr_A} \ln\left(\frac{(2kWc_t - T_0)r_A + T_{ref}r}{(2kWc_0 - T_0)r_A + T_{ref}r}\right) \qquad (5)$$

$$+ \frac{1}{r_B(\beta - 1)\left(2kW/T_{ref}\right)^\beta c_t^{\beta-1}}$$

In Equation 5, c_0 and c_t are, respectively, effective flaw sizes corresponding to $T = T_0$ and $T = T_t$. As the fatigue specimens were dumbbells the strain is uniform in the parallel-sided section where failure is anticipated.

Often estimates of effective flaw size in a material use only the last term in Equation 5, which only considers the power law regime (eg Gent et al. 1964, Choi & Roland 1996). At flaw sizes lower than c_0, the fatigue life of the rubber would become infinite without ozone attack. Figures 4 to 6 show the full fatigue life prediction of Equation 5 for the materials (solid lines), and that assuming the power law dependence (last term of Equation 5 only; dotted line).

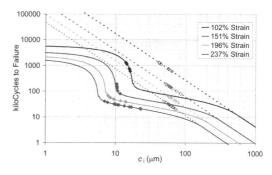

Figure 4. — Predicted fatigue lives for NR+0 ---- extrapolated power laws; points, fatigue tests and deduced effective flaw sizes.

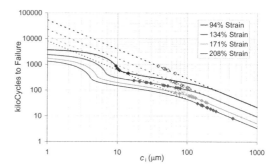

Figure 5. — Predicted fatigue lives for NR+45N330 ---- extrapolated power laws; points, fatigue tests and deduced effective flaw sizes.

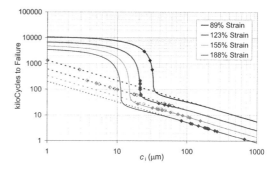

Figure 6. — Predicted fatigue lives for SBR+77N339 ---- extrapolated power laws; points, fatigue tests vs deduced effective flaw sizes.

The value (3 μm/Mcycle) of r used to generate the predictions in Figures 4 to 6 is similar in magnitude to that found appropriate to the conditions and materials of Lake & Lindley (1965). Varying r has only a small effect on the shape of the predicted curve for $c_0 < c_i < c_t$, imperceptible as c_i approaches c_t, justifying the mathematical simplification of neglecting the contribution of r when integrating Equation 4.

Fatigue tests were carried out at a frequency of 5 Hz, on fully relaxing cycles such that $t_c \sim 0.1\,s$ at four nominal strains $e = 100, 150, 200$ or 250% – based on $e = (38 + x)/38$ where x is the extension and 38 mm

Table 3. Estimated mean effective flaw sizes ± SD (μm); italics indicate extrapolations beyond the power law region

Method	NR+0	NR+45N330	SBR+77N339
Power ~ 100%	51 ± 10	48 ± 11	1.1 ± 0.92
Power ~ 150%	63 ± 9.2	79 ± 17	22 ± 21
Power ~ 200%	64 ± 7.3	134 ± 34	70 ± 38
Power ~ 250%	61 ± 7.0	209 ± 85	219 ± 153
Full ~ 100%	15 ± 0.9	12 ± 3	11 ± 1.8
Full ~ 150%	11 ± 0.3	42 ± 22	28 ± 12
Full ~ 200%	10 ± 2.3	129 ± 39	67 ± 40
Full ~ 250%	13 ± 5.3	209 ± 85	219 ± 153
Ozone	31	26	45

is the "effective length" of the standard dumbbells, originally determined for an unfilled reference rubber. The actual strains, determined from gauge marks on the parallel-sided portion of the dumbbells, were not the same as the nominal, since the stress-strain constitutive relation for each material differed from that of the reference rubber. Fatigue results (six or twelve replicates for each condition) are included in Figures 4 to 6, wherein the actual strains are also reported, and compared to the predictions. Estimates of apparent flaw size were made by identifying the abscissa values corresponding to the measured fatigue lives (see points mapped onto the predictions). Mean values are given in Table 3.

As most of the measured fatigue lives do not correspond to c_i values much lower than c_0, the predicted effective flaw sizes will not be unduly sensitive to the value of r, with the possible exception of NR+0 and SBR+77N339 for the lowest fatigue strains.

5 ESTIMATION OF EFFECTIVE FLAW SIZE FROM OZONE ATTACK

Rubber can be significantly degraded by the attack of ozone on exposed surfaces. In unstrained rubber, ozonolysis occurs uniformly on the rubber surface. Unlike sunlight crazing, cracks only develop on stretched rubber, perpendicular to the direction of strain. From experiments with varying strains and incised cuts in the rubber, it is known that cracks initiate and grow from pre-existing flaws on the rubber surface (Braden & Gent, 1960 II), but only if T exceeds a critical energy release rate T_z (Braden & Gent, 1960 I).

As the strain is increased, there will be more flaws on the surface large enough for their associated energy release rates, calculated according to a relation such as Equation 1, to be higher than T_z. Counter-intuitively, this is less damaging for the rubber than when only a few flaws are growing (Braden & Gent, 1960, II). Firstly, the crack growth rate due to ozone attack is a constant, so the flaws do not grow any faster under greater strain. Secondly, each flaw creates a zone around it where the rubber is relaxed by the crack growth (Braden & Gent, 1960, II) and the

ozone concentration is reduced by consumption at active crack tips (Lake & Thomas, 1967). When other flaws enter this zone, their growth rate will diminish. Thus, higher strain results in much smaller, but more numerous, cracks.

If T_z and the strain energy in the rubber are known, it is possible to calculate how big a flaw would need to be for it to grow under ozone attack. Hence, by exposing rubber to ozone at different extensions and measuring the density of cracks that develop from the flaws in the rubber surface, it is possible to build up a profile of the flaw size distribution in the material.

5.1 Method

All the ozone work was carried out in an ozone cabinet at 40°C, with an ozone concentration of 50 parts per hundred million.

It would be expected that flaw size depends on the individual batch of rubber, not just on the formulation. Since the formulations given in Table 1 include antiozonants, it was considered necessary to extract them prior to use of the ozone method of flaw size determination, not just remix the rubbers without anitozonant. Strips were cut from each material and extracted in a cold Soxhlet apparatus using an azeotrope of acetone, chloroform & methanol for 48 hours. The solvents were then removed by drying the rubber in an unheated vacuum oven for at least five days. This process also removed much of the process oils and many other volatile materials from the rubber. The flaw population was not expected to have been affected, but it was necessary to remeasure the stress strain properties of the material, in order to calculate the strain energy densities.

5.2 Measuring flaw density

After extraction, the strips were strained to set extensions (10%, 20% & 30%) in jigs. The ends were wrapped in PVC tape before clamping and they and the cut edges were covered in wax to prevent ozone cracking from occurring at these points. Consequently, only the moulded rubber surface in uniform simple extension was exposed to the ozone. The rubber strips were then allowed to relax for at least 64 hours before being exposed to ozone.

The rubber strips were exposed to ozone for 72 hours. This was enough time for the ozone to "develop" surface flaws in the rubber strips into visible cracks.

The rubber strips were then photographed at moderate magnifications (5x – 20x) using a Nikon Macrophot low magnification microscope. Rulers were also photographed under the same conditions to determine the magnification accurately. These photographs were used to quantify the density of cracks in the rubber surface.

5.3 Determination of T_z

In order to measure T_z, a number of slits of varying length were cut in the edges of tensile specimens of the extracted rubbers with a razor blade.

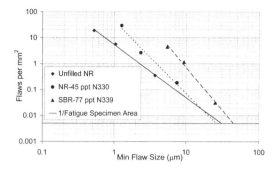

Figure 7. Number of flaws exceeding given minimum effective size

The strips, with ends protected with adhesive tape, were hung from clips and strained by hanging weights from them, chosen to give extensions of approximately 1%. After allowing at least 48 hours for creep to occur, the extensions were measured accurately using a cathetometer. The strain energy density was calculated from the load and strain, making an assumption of linear elastic behaviour at such small strains. The strips were then placed in the ozone cabinet. After 48 hours, the strips were removed from the ozone cabinet and the cuts examined to see which, if any, had grown. If none had, the process was repeated with heavier weights. T_z was calculated from the cut off length for the cracks to grow and the associated strain energy density in the material, using Equation 1; values are reported in Table 2.

5.4 Deduced values for c_i

With T_z and W known, it is possible to calculate the minimum effective size for a flaw to grow under a given strain. Thus we deduce that the number of cracks formed at that strain is equal to the number of flaws in the exposed surface exceeding this minimum size. The assumption was again made that flaws in the rubber surface behaved like through-thickness edge cracks and thus T calculated from Equation 1. In fact the cracks develop in the major surface and are much shorter than the width. This may have implications for the energy release rate, in particular for the numerical prefactor in Equation 1; any change in this factor caused by the different geometry will be absorbed into the qualification "effective" for the flaw size. Figure 7 shows how the number of flaws in the rubber surface exceeding a given minimum size depends on the given minimum size. On a log-log chart, the distribution was found to be approximately linear. It was extrapolated to give the effective flaw size such that the expected number greater or equal to it in a dumbbell testpiece is one, reported in Table 3.

6 DISCUSSION

Table 2 reveals some differences between our results and those from the literature for some crack growth

parameters, particularly for the exponent β. Contributions to this could come from the intrinsic scatter in the data, the fact that the power law region prevails for little more than one tenfold change in T_{max}, and the absence in our work of a measurement of the upper limit T_c or account being taken of time dependent crack growth. Even for natural rubber the latter becomes significant as T approaches T_c (Gent et al. 1964, Lake et al, 1991). Gent et al. (1964) reported T_c to be about $10\,\mathrm{kJm^{-2}}$ for a material differing from NR+0 only in grade of NR and type of antidegradant, but Figure 2 shows no sign of upturn for T_{max} up to $20\,\mathrm{kJm^{-2}}$. According to Lindley & Thomas (1962), the time-dependent crack growth contribution ($\rho(T)$ in Equations 3 and 4) is not very significant for fully relaxing tension cycles of unfilled SBR at frequencies above $\sim$2 Hz.

Had the determination of effective flaw size from fatigue life only been done on NR+0, it would have been deemed to work well, the effective flaw size being independent of fatigue strain (see Table 3). It is interesting to note that most of the literature addresses only unfilled rubbers (Gent et al. 1964; Lake & Lindley, 1964 and 1965; Choi & Roland, 1996), and reached similar conclusions, albeit effective flaw sizes tend to be larger ($\sim$25μm).

The filled materials have greater scatter in their crack growth characteristics and fatigue life. Arguably, they also have greater effective flaw sizes, in accord with the expectation that filler clusters or agglomerates could serve as crack initiators. The fact that the apparent T_z is substantially larger than for NR+0 is not expected from the work of Braden & Gent (1960II). The data in Table 3 showing a dependence on fatigue strain amplitude of the apparent flaw size calls for a thorough investigation, since it would appear to undermine the primary motivation for determining the effective flaw size – to use it as a parameter in prediction of fatigue life of products. There is relatively little literature on application of fracture mechanics to filled rubbers, although the first paper expounding the theory for fatigue (Lindley & Thomas, 1962) did include results for several. They pointed out the necessity of allowing for stress-softening and set in determination of W in Equation 1, but we have met their recommendation: to calculate W from at least the sixth cycle. Moreover, the fatigue life for the different fatigue strains changed by a maximum factor of $\sim$1000, for SBR+77N339; from the dynamic stress relaxation rate given in Table 1 this would lead to a change in W of only 26%, much smaller than the 20-fold change in apparent energy release rate.

Lindley & Thomas also found shorter fatigue lives at strains $>\sim$150% for filled vulcanizates than anticipated from the linear dependence of log (cycles to failure) versus log (W) established for smaller amplitudes (strains $>\sim$150%). This they attributed to the anomalously fast crack growth in the first few hundred cycles, the crack tip "sharpness" apparently being initially greater and reaching steady state roughness only after such an amount of natural growth. This phenomenon occurs also for artificial razor cuts. Gent et al. (1964) used the same explanation for anomalously short fatigue lives of unfilled NR for strains $> \sim 400\%$, corresponding to fatigue life of ~ 1 kcyle or less, and showed that if cut growth is "smooth" with a rate constant (r_B) about 20 times higher than for steady-state cracks this mechanism can explain the fatigue behaviour at very high strains. It is noted that the apparent effective flaw sizes in Table 3 for the filled materials are about 20 times higher than those at low strain, in accord with the explanation.

This suggests that it is not sufficient, for our purpose of a predictive methodology for fatigue life, to characterize the material only by steady-state crack growth characteristics (according to assumption (2)) and effective flaw size. For high-strain fatigue prediction we need also to determine the transient behaviour from rapid crack growth at the start of cycling to slower, steady-state crack growth. For many engineering applications this may not be necessary, since the anomaly only becomes manifest for strains greater than $\sim 400\%$ for unfilled materials or $\sim 150\%$ for highly filled materials. It appears then that the effective flaw size in the filled materials is best taken from the lower strain fatigue fits, making it much the same in magnitude as that of NR+0, rather surprisingly.

It is apparent that further work is needed to improve the predictive methodology for fatigue life, based on measured behaviour of macroscopic cracks and deduced characteristics of material flaws.

7 CONCLUSIONS

The work highlights several considerations for methods of determining effective flaw sizes c_i:

- set and relaxation are especially significant for filled materials. Ideally retraction energy should be monitored during fatigue and for a reference testpiece undergoing the same deformation cycles as for the crack-growth testpieces. At high frequencies and amplitudes, hysteretic heating could also be an issue.
- For severe fatigue conditions, when the lifetime is little more than one kcycle, non-steady state crack growth could predominate, leading to anomalously high apparent effective flaw sizes, or shorter than expected lifetimes on the basis of a fixed size.

- If $c_i < c_0$, the fatigue life will depend on ozone concentration, the polymer, and the effectiveness of the antiozonant system, which would all need to be quantified if low strain fatigue or spontaneous ozone cracking is used to determine effective flaw size.

ACKNOWLEDGMENTS

Thanks are due to the Parties to a Joint Industry Project on Modelling Rubber, which helped to fund the work, and to our colleague Dr Julia Gough for drawing our attention to several flaws in the manuscript.

REFERENCES

Braden, M. & Gent, A.N. 1960. The attack of ozone on stretched rubber vulcanizates *J Appl Polym Sci* **3** (I)90–99 (II)100–106

Choi, I. S. & Roland, C. M. 1996. Intrinsic defects and the failure properties of cis-1,4-polyisoprenes. *Rubber Chem & Tech*, **69**, 591–9

Gent, A .N., Lindley, P. B. & Thomas, A. G. 1964. Cut growth & fatigue of rubber I The relationship between cut growth and fatigue *J Appl Polym Sci*, **8**, 455–466

Greensmith, H. W. 1964 Rupture of rubber XI. *J Polym Sci*, **8**, 1113–1128

Lake, G.J. & Lindley, P.B. 1964. Ozone cracking, flex cracking and fatigue of rubber. *Rubber Journal*, **146** (1964); pp. 10:24–30, 11:30–36

Lake, G.J. & Lindley, P.B. 1965. The mechanical fatigue limit for rubber. *J Appl Polym Sci* **9**, 1233–1251

Lake, G. J., Samsuri, A., Teo, S. C. & Vaja, J. 1991. Time-dependent fracture in vulcanized elastomers, *Polymer*, **32**, 2963–2975

Lake, G. J. & Thomas, A. G. 1967. Ozone attack on strained rubber *Proc International rubber Conf*, Brighton, publ Maclaren

Lindley, P. B. & Thomas, A. G. 1962. Fundamental study of the fatigue of rubbers, *Proc 4th Rubber Technology Conf*, London

Mars, W. V. & Fatemi, A. 2003. A phenomenological model for the effect of R ratio on fatigue of strain crystallizing rubbers. *Rubber Chem & Tech*, **76**, 1241–1258

Rivlin, R. S. & Thomas, A. G. 1953. Rupture of rubber Part 1 Characteristic energy for tearing *J Polym Sci*, **10**, 291

TARRC 1979. Engineering Data Sheets; EDS19 & 16 are available on www.rubber.demon.co.uk

Timbrell, C., Wiehahn, M., Cook, G. & Muhr, A. H. 2003. Simulation of crack propagation in rubber *Constitutive Models for Rubber III*, publ Swets & Zeitlinger, Lisse

Constitutive Models for Rubber VII – Jerrams & Murphy (eds)
© *2012 Taylor & Francis Group, London, ISBN 978-0-415-68389-0*

Fatigue crack growth dynamics in filled natural rubber

L. Vanel
LPMCN, Université de Lyon, Université Lyon 1 and CNRS, UMR, France

L. Munoz, O. Sanseau, P. Sotta, D. Long & L. Odoni
LPMA, UMR CNRS/Rhodia CRTL, Saint Fons, France

L. Guy
Rhodia Operations – CRTA, Collonges au Mont d'Or, France

ABSTRACT: We present fatigue experiments performed on filled natural rubber and study the correlations between crack growth dynamics and fracture morphologies imprinted by an irregular crack path. Slow crack growth dynamics is obtained by cyclic fatigue in a pure shear test. We will show that an unstable crack growth regime exists for high loads. We will also discuss the appearance of sawtooth striations which follow a scenario that significantly differs from previous results reported in the literature.

1 INTRODUCTION

Although there is a good knowledge of factors that can affect the fatigue life of rubbers (Mars 2004), crack propagation in rubbers is still not well understood (Persson 2005). In reinforced elastomers, rupture dynamics is a much more complex process than in pure elastomers due to the intrinsic heterogeneous mixture of a rubber matrix with filler particles at submicronic scale. In the case of natural rubber, an additional source of heterogeneity is the strain-crystallization effect. How rupture dynamics and crack path are affected by filler particles and strain-crystallization is still a matter of debate. Actually, understanding how rupture dynamics and crack path are correlated to each other is probably an important key in order to improve long time resistance of reinforced rubbers.

We present fatigue experiments performed on filled natural rubber and study the correlations between crack growth dynamics and fracture morphologies imprinted by an irregular crack path. Slow crack growth dynamics is obtained by cyclic fatigue. In order to control properly the crack growth dynamics, we use a pure shear test. The goal of this study is also to understand how rupture dynamics and morphologies depend on the control parameters of a fatigue experiment (frequency, stress or strain, strain rate, temperature). We will discuss in particular the appearance of sawtooth striations and will show that their formation follows a scenario that significantly differs from previous results reported in the literature.

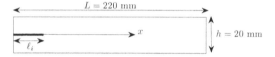

Figure 1. Sketch of pure shear sample geometry.

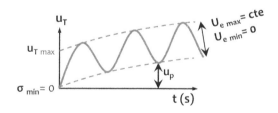

Figure 2. Sketch of trimodal fatigue procedure. Total displacement u_T increases progressively as plastic elongation u_p occurs so as to keep constant the applied displacement u_e.

2 EXPERIMENTAL PROTOCOL

The material considered is a 50 phr Silica-filled natural rubber. The material is molded into a pure shear test piece of height $h = 20$ mm, length $L = 220$ mm and thickness $e = 1.4$ mm (figure 3). An initial crack of length $\ell_i = 30$ mm is precut with a razor blade at one end of the sample. In mode I loading, fracture mechanics predicts that the elastic energy release rate is independent of the crack length at constant applied strain. In this case, crack growth is expected to occur at constant velocity.

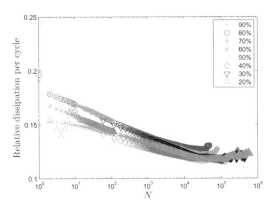

Figure 3. Relative dissipation per cycle is decreasing as a function of cycle number N.

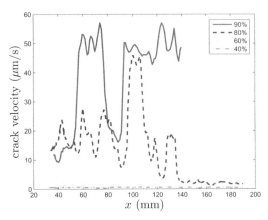

Figure 4. Instantaneous crack velocity as a function of crack tip position along the x axis. Crack growth at high applied strain is unstable (amplitude of fluctuations 100%) and rather stable at low strain (amplitude of fluctuations 20%).

Uniaxial fatigue tests are performed using an Instron 8800 servohydraulic test machine. Fatigue cycles are applied using a trimodal control with the applied force returning to zero at the minimum position of the cycle and a constant applied displacement amplitude u_e with respect to this minimum position (figure 2). In that way, we take into account plastic creep elongation u_p and avoid any compression of the sample at later stages. In contrast with a standard fatigue loading procedure where the minimum and maximum displacement are fixed, the trimodal control allows us to obtain a rather stable elastic strain u_e/h during the entire experiment while the total displacement $u_T = u_e + u_p$ applied to the sample increases progressively.

Video tracking of the crack growth is done with an Imperx 4872×3248 pixels camera allowing a spatial resolution about $50\,\mu m$. Image acquisition is phase-locked to the fatigue cycles. Image analysis allows us to extract the crack tip position and measure the instantaneous crack velocity. In this paper, only the component of the velocity along the x axis will be considered.

SEM fractography measurements have been used to characterize post-mortem the crack path at spatial resolutions of $1\,\mu m$ or $2\,\mu m$.

3 RELATIVE DISSIPATION

The relative dissipation per fatigue cycle $\mathcal{D}$ is computed from the force F vs displacement u relation during a fatigue cycle as:

$$\mathcal{D} = \frac{\int_{\text{cycle}} F.\mathrm{d}u}{\int_{\frac{1}{2}\text{cycle}} F.\mathrm{d}u} \qquad (1)$$

There is no systematic and overall little dependence of the relative dissipation on the amplitude of the applied strain. Also, the relative dissipation is observed to gradually decrease as the experience proceeds. This effect is not understood yet but could be a signature of

the interaction properties between silica particles and rubber.

4 CRACK GROWTH DYNAMICS

Here, we present results obtained at room temperature for fatigue cycles at a frequency of 1Hz. In figure 4, we plot as a function of the crack tip x-coordinate the instantaneous crack velocity (actually its x-component). In contrast with the expected behavior, the crack velocity is not always stationary. At large strains, it fluctuates a lot between two main distinct velocity levels. This multistable behavior is reminiscent of the well-known intermittent crack growth behavior observed in the trouser test geometry for SBR samples for a given range of mean velocity (Tsunoda 2000). When decreasing the maximum applied strain, we observe that the multistable crack growth regime eventually disappears although relative fluctuations in velocity of about 20% are still observed.

5 FRACTURE MORPHOLOGY

Fracture morphologies observed by SEM depend on the crack velocity and can be split into four main types (figure 5):

- at very low velocity, a rough disordered interface is observed (A)
- at high velocity are observed very regular and mainly straight sawtooth striations oriented perpendicularly to the crack propagation direction and extending across the whole sample thickness (B)
- increasing the velocity starting from type A is observed a mixture of a rough disordered interface with small, slightly irregular, sawtooth striations (A')

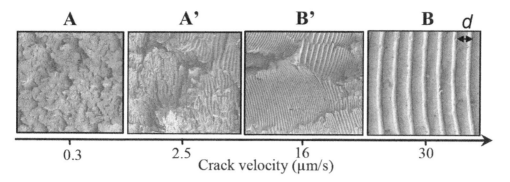

Figure 5. SEM micrographs of rupture interface showing the effect of crack velocity on morphology. A: Rough disordered surface. A': Mixture of sawtooth striations and rough disordered surfaces. B': Several zones of smaller sawtooth striations. B: Large sawtooth striations across the sample.

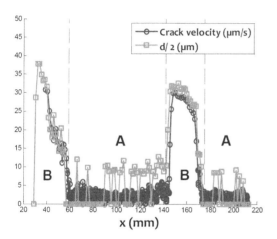

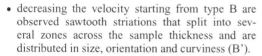

Figure 6. Instantaneous crack velocity and half-width d/2 of the sawtooth striations. A and B refers to the morphologies shown in figure 5. The excellent correlation between the two sets of data in A shows that exactly two fatigue cycles are needed to form a sawtooth striation.

- decreasing the velocity starting from type B are observed sawtooth striations that split into several zones across the sample thickness and are distributed in size, orientation and curviness (B').

From the SEM images, we can extract the striation width d (Figure 5) as a function of the crack tip x-position. In figure 6, we plot the half striation width $d/2$. An excellent correlation is observed between the instantaneous crack velocity v and the striation half-width so that we can write: $vT = d/2$, where T is the duration of one fatigue cycle (here $T = 1$ s). Note that the correlation is very good in the zones noted B of figure 6 where morphologies B and B' are observed. In the zones noted A where morphologies A and A' are observed, a measurement of striation size is sometimes possible at a local scale but does not correlate well with the macroscopic crack velocity. It is likely that the crack front velocity is not uniform and can locally be high enough to trigger striations locally without affecting much the overall crack velocity.

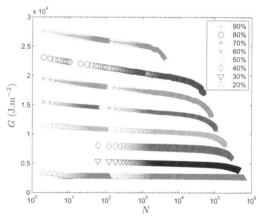

Figure 7. Estimated tear energy G as a function of cycle number N for various applied strains.

6 TEAR ENERGY

The tear energy G (or elastic release rate) in pure shear geometry is defined as: $G = Wh$ where W is the elastic energy density and h the sample height. It can be estimated for each fatigue cycle as:

$$G = \int_{\frac{1}{2}\text{cycle}} \sigma.\mathrm{d}u \qquad (2)$$

where σ is the uniform stress applied away from the crack tip and u is the sample elongation. The stress is approximated using the knowledge of the crack length as: $\sigma = F/(L - \ell_x)$ where F is the measured force, L the pure shear sample length and ℓ_x the x-component of the crack tip position or equivalently the crack length projected along the x direction.

As shown in figure 7, the tear energy computed according to equation (2) has a stable value during most of the experiment, whatever is the amplitude of the applied strain. Deviations are found during the first cycles, probably due to an initially strong Mullins effect, and also at the end of the experiment when

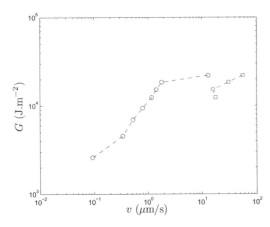

Figure 8. Tear energy G as a function of crack velocity. For high values of G, the minimum (circles) and maximum (squares) velocities observed during crack growth have been reported.

the crack length is so large that the geometry can not be considered pure shear anymore. Note that for each curve in figure 7 the data have been plotted up to the cycle number for which the crack length has reached 190 mm.

7 TEAR ENERGY VS CRACK VELOCITY

In figure 8, we report the tear energy G as a function of crack velocity. More precisely, the tear energy G corresponds here to the average over an experiment of the values reported in figure 7. For the crack velocity, we report the mean crack velocity value when crack growth is stable, and we report the two extreme velocity values in the case of unstable crack growth. In the unstable regime, the high velocity branch (square symbols) also correspond to a fracture morphology with saw-tooth shaped striations as described in section 5. Thus, there is a strong correlation between the apparition of a faster crack growth dynamics and a change in the crack path morphology. The origin

of this change in behavior is at the moment still unexplained. We are currently checking how the existence of the unstable regime depends on the loading characteristics (frequency) and the material properties of the samples.

8 CONCLUSIONS

We have described fatigue experiments in filled natural rubber and analyzed both the crack growth dynamics and morphology. We have shown that there exist a range of tear energy above which the crack growth is unstable and statiscally evolve between two different regimes. We have shown that fracture morphology and crack growth velocity are strongly correlated, especially in the unstable regime. In contrast with (Bathias 1997) and (Le 2010) claiming sawtooth striations occur in one fatigue cycle, we clearly show that it takes two cycles to form a striation. Our results suggest that striations result from the alternative selection of two symmetric inclined rupture planes. More work is in progress to validate the generality of these observations.

REFERENCES

Bathias, C., K. Le Gorju, C. Lu, & L. Menabeuf (1997). Fatigue crack growth damage in elastomeric materials. In *STP 1296. Fatigue and Fracture Mechanics*, Volume 27, pp. 505–513.
Le Cam, J.-B. & E. Toussaint (2010, May). The mechanism of fatigue crack growth in rubbers under severe loading: the effect of stress-induced crystallization. *Macromolecules* 43(10), 4708–4714.
Mars, W. V. & A. Fatemi (2004). Factors that affect the fatigue life of rubber: A literature survey. *Rubber Chem. Tech.* 77(3), 391–412.
Persson, B. N. J., O. Albohr, G. Heinrich, & H. Ueba (2005). Crack propagation in rubber-like materials. *J. Phys. Condens. Matter 17*, R1071.
Tsunoda, K., J. J. C. Busfield, C. K. L. Davies, & A. G. Thomas (2000, October). Effect of materials variables on the tear behaviour of a non-crystallising elastomer. *J. Mater. Sci.* 35(20), 5187–5198.

Constitutive Models for Rubber VII – Jerrams & Murphy (eds)
© *2012 Taylor & Francis Group, London, ISBN 978-0-415-68389-0*

Heat build-up and micro-tomography measurements used to describe the fatigue mechanisms and to evaluate the fatigue lifetime of elastomers

Y. Marco
ENSTA Bretagne, Laboratoire LBMS (EA 4325), Brest Cedex, France

V. Le Saux
UBS, LIMATB – Equipe ECoMatH, Lorient Cedex, France

S. Calloch
ENSTA Bretagne, Laboratoire LBMS (EA 4325), Brest Cedex, France

P. Charrier
Modyn Trelleborg, Zone ind. de Carquefou, Carquefou Cedex, France

ABSTRACT: The aim of this paper is to investigate the ability of a recently developed protocol to identify the fatigue lifetime of elastomers. This protocol is based on thermal and micro-tomography measurements that are coupled throughout an energetic criterion. Interrupted fatigue tests are performed for several global displacements and micro-tomography measurements are used to investigate the fatigue mechanisms involved (initiation and propagation) along the fatigue tests. Moreover, we consequently use micro-structural information obtained from a X-ray computed micro-tomography investigation to evaluate the ratio of the global dissipated energy to the one related to fatigue damage. This global dissipated energy can be related to the skin thermal measurements thanks to finite element simulations using a model developed recently (Le Chenadec *et al.*, 2007). A critical energy criterion is then applied and the results are compared to a classical Wöhler curve.

1 INTRODUCTION

The failure of industrial structures is mostly related to ageing and/or fatigue. Both phenomena are difficult to master because the characteristic times are long, the influent parameters are numerous and involve probabilistic aspects. Dealing with fatigue, the classical way to identify a design criterion is to build a so called "Wöhler curve", linking the chosen parameter to the number of cycles leading to failure (break or initiation of a crack). To be reliable, this classical approach presents at least two main disadvantages: it requires long duration tests and a large number of specimens (usually a minimum of 25 specimens is needed) in order to have a good estimation of the fatigue intrinsic dispersion. These two disadvantages obviously limit the material studied and restrained the scientific study of the influence of the numerous parameters involved. To accelerate the identification of the fatigue properties is therefore a main concern for both industrial and academic partners.

The aim of this paper is to investigate the opportunity to use thermal measurements as a quick indicator of the fatigue behavior of elastomeric materials. Based on a specific protocol, this pragmatic approach is giving interesting results but suffers from two flaws. The first one is that temperature is not an intrinsic data of the material. It is therefore mandatory to link the temperature at skin to the dissipation sources. This link is presented here, using a recent dissipation model from Y. Le Chenadec *et al.* (2009). The second one is that energetic measurements like heat sources measurements makes difficult to split the sources involved or not in the fatigue mechanisms, like viscous dissipation, for example. To get closer from the fatigue mechanisms, a microstructural study using X-ray Computed tomography is achieved. Not only the basic initiation mechanisms but also the scenario of the fatigue damage evolution can be identified by this technique. These informations are very relevant but are still based on geometry, giving no hints on the energy dissipated. In a last part, we consequently use the microstructural information on the cavities population to evaluate the ratio of the global dissipated energy to the one related to fatigue damage. The global dissipated energy is evaluated from the thermal measurements. A critical energy criterion is then applied and the results are compared to a classical Wöhler curve. This comparison exhibits a very good correlation, opening a new field of investigation.

Figure 1. Experimental set-up.

2 EXPERIMENTAL SECTION

2.1 Materials and samples

More than 15 different materials were tested during this study, in order to cover a wide range of industrial standard recipes. The analysis based on the thermal measurements will include all of them, while the specific study including tomography measurements will be restrained to a silica-filled Polychloroprene (CR). For all these compounds, hourglass shaped specimens (called AE2 in the following) were manufactured from a single batch in order to ensure the reliability of mixing and moulding conditions. The geometry of the specimen are classical for fatigue investigations and are presented elsewhere (Le Saux et al., 2010). This geometry of specimen was chosen for several reasons:

- it is classically used to obtain Wöhler curves;
- the initiation zone is well mastered and located in the thinner section;
- the central section is thin enough to prevent a high temperature gradient between the skin and the core due to the low thermal conductibility of rubber, and also limits the rise of temperature under cyclic loading.

2.2 Experimental campaigns

Three experimental campaigns have been achieved in this study: classic fatigue testing to build the Wöhler curves of the materials studied; heat-build up specific tests; interrupted fatigue tests for the tomography investigation. The fatigue tests were performed at Trelleborg Modyn laboratory on servo-hydraulic machines described elsewhere (Ostoja-Kuczynski et al., 2005). The experimental determination of the initiation Wöhler curve was achieved using an end-of-life criterion based on the variation of the effective stiffness (Ostoja-Kuczynski et al., 2005). It has been shown that the criterion is equivalent to the apparition of crack of 2 mm on the surface of the specimen (Ostoja-Kuczynski et al., 2003). The heat build-up tests were performed on a servo-hydraulic test machine (Instron 1342) at ENSTA Bretagne, at a frequency of

2 Hz and were displacement controlled. This frequency was chosen to be the same than the one used to perform the fatigue tests, in order to be as representative as possible from the fatigue lifetime evaluation. A heat build-up experiment can be defined as a succession of cyclic tests of increasing loading conditions during which the temperature of the specimen is measured. The number of cycles used for each loading condition is the number of cycles needed for the temperature to stabilize (for example, 2000 cycles at 2 Hz are sufficient for rubber-like materials). For the interrupted fatigue tests, performed at ENSTA Bretagne on the same machine as the heat build-up tests, three macroscopic maximum displacements (2, 4, and 6 mm) were chosen, associated with the respective numbers of cycles that lead to initiation (referenced Ni in the following and obtained from the Wöhler curve). For each of these displacements, at least five interrupted tests were achieved. The tests were stopped after five cycles and after 10, 25, 50, and 100% of Ni.

2.3 Thermal measurements

The thermal measurements were performed thanks to an infrared camera, which gives access to a 2D measurement with a high acquisition rate (50 frames/s) and a very good precision (about 30 mK). Even if it provides only a surface measurement, AE2 specimen are thin enough to avoid a too high core to skin temperature gradient. The infrared camera that has been used is a Flir Systems camera (reference Phoenix MWIR 9705) with a Stirling-cycle cooled Indium Antimonide (InSb) Focal Plane Array (FPA). The FPA is a 320 × 256 array of detectors digitized on 14 bits, sensitive in the 3–5 lm spectral band. A careful preliminary calibration operation, detailed elsewhere (Le Saux, 2010) allows the conversion of the thermosignal (proportional to the thermal radiation) into a temperature in degree Celsius (°C).

The temperature measurement raised two major technical problems: to take into account the large displacements of the specimen and to define and measure accurately a given heat build-up temperature. For each loading block, a dedicated protocol allowed us to measure both the mean temperature rise for a un-strained position, and the amplitude of the temperature induced by thermomechanical couplings. This study focuses on the intrinsic dissipation and will therefore use the mean temperature rise. The study of the evolution of the temperature along a cycle is presented in another paper of ECCMR 2011 (Le Saux et al., 2011).

2.4 X-ray tomography measurements

The X-ray CT device is an industrial Phoenix device used (v|tome|x L 240) with a resolution high enough to detect the defect sizes classically measured by SEM measurements (range from 10 to 400 μm). A dedicated commercial software (myVGL 2.0) performing 3D image analysis was then used to study the inclusions and the defects population of the samples. The minimal size of detection for defects was set to 4 voxels (i.e., an equivalent radius of 13 μm) to ensure

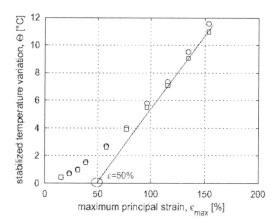

Figure 2. Heat build-up curve and graphical analysis.

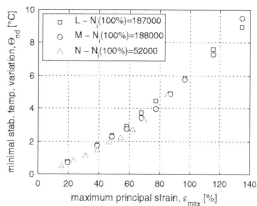

Figure 3. Illustration of the difficulty to directly relate the fatigue lifetime to the rise of temperature.

an effective detection. All the specimens were slightly stretched (relative displacement of 2 mm of the inserts) to open the cavities and to make them easier to detect. For a more detailed description, please refer to Le Saux *et al.*, 2011.

3 ANALYSIS FROM THE THERMAL MEASUREMENTS ONLY

3.1 *Method calibration*

At this stage of the study, we were evaluating if a quick empirical protocol could be used to evaluate quickly the fatigue lifetime (defined here as initiation after 10^6 cycles). The first step was therefore to identify an empirical protocol on a given material, with a known fatigue lifetime.

Fig. 2 presents the temperature measurements obtained for this material. On this curves, one can see the temperatures (for strained and unstrained state along the cyclic loading) stabilized after 2000 cycles. The first striking point is that the evolution of the heat build-up curve is much smoother than for metallic materials. This was expected as no clear fatigue limit is to be seen on the Wöhler curves obtained for elastomers. Moreover, rubbers exhibit many dissipative sources (viscosity, damage, microstructural changes . . .).

The empirical analysis proposed to meet the 10^6 cycles strain identified on a Wöhler curve for this material, is illustrated on figure 2 and consists in considering only the curve related to the intrinsic dissipation (square symbols on figure 2) and to draw a line from the lasts points.

It is worth noting that the repetability of the tests was checked for several materials, tested with small variations of the ambient temperature and for different loading histories (different numbers of loading blocks, with different increasing amplitudes).

3.2 *Results for a wide range of elastomers*

More than 15 materials were tested. The aim was to cover a wide range of materials, crystallizing, non

crystallizing, unfilled, monomer or copolymers matrix with a wide range of mechanical properties (fatigue lifetime, tan delta). The protocol defined in paragraph 3.1 was then applied to analyze the heat build-up curves obtained from other materials. It was observed (Le Saux *et al.*, 2010) from the good agreement with the values provided by the Wöhler curves that the proposed graphical analysis is giving good results.

3.3 *Limits*

This good agreement is surprising as viscosity is obviously a first order dissipation phenomenon that could be dissociated from the fatigue properties. To challenge the protocol, it was consequently applied to specific materials, with either the same fatigue limit but different tan delta, or the same tan delta but different fatigue limit. In the first case, the protocol is still giving satisfactory results, but in the second one, illustrated on figure 3, one can see that the curves are very difficult to dissociate despite very different numbers of cycles to initiation. It can therefore be concluded that even if the rough estimation analysis presented above has proven to be quite effective for several materials, the global dissipated energy could not be directly related to the fatigue damage and that the description of the damage at lower scales was mandatory. A second step is needed as well. In this section, only temperature variation is considered, which is clearly not an intrinsic material variable. It is therefore required to link the temperature measured at the skin to the dissipation sources activated in the volume. These two points are investigated in the next paragraphs.

4 FROM THEMPERATURE MEASUREMENTS TO DISSIPATION SOURCES

4.1 *Heat build-up model*

The first step of the demarch is to solve the mechanical problem in order to compute the dissipated energy. A first way would be to use a dissipative model, giving

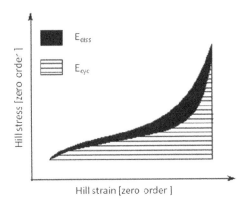

Figure 4. Schematic 1D illustration of the cyclic and dissipated energy.

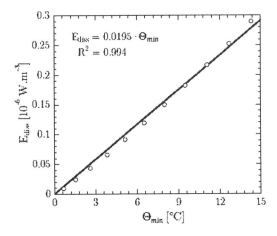

Figure 5. Illustration of the linear relation between the volumic heat sources and the skin temperatures for AE2 samples.

a direct access to the dissipation sources. Nevertheless, this approach requires numerous parameters and requires an excellent reproduction of the hysteresis loop. Moreover, the computation on several cycles is needed to reach the stabilization of the behaviour, which can be time consuming for industrial components. A hyperelastic approach has been used to model the behaviour of our material and because no large temperature variation was expected, an isothermal hyperelastic model was chosen. The hyperelastic potential is identified on the stabilized behaviour of the material, with a classical Mooney-Rivlin potential.

As the hyperelastic modelling leads to no dissipation, a specific approach needs to be developed to evaluate the dissipated energy from the hyperelastic modeling. This approach requires the definition of two energetic quantities (Y. Le Chenadec *et al.*, 2007), which are illustrated on figure 4: a cyclic energy (Ecyc) representative of the cyclic hyperelastic loading and a dissipated energy (Ediss) which corresponds to the fraction of the elastic energy converted into heat. In this study, these energies have been computed according to a model developed recently (Le Chenadec *et al.*, 2009). The cyclic energy writes:

$$E_{cyc} = \int_{E_{min}^{(0)}}^{E_{max}^{(0)}} \left(T^{(0)} - T_{min}^{(0)} \right) : d E^{(0)}$$

where $E^{(0)}$ and $T^{(0)}$ are the order 0 conjugated Hill tensor. Conjugate Hill tensors are used to define the cyclic energy in order to guarantee the objectivity of the quantity (Hill, 1968). Thanks to pure shear experiments where E_{cyc} can be obtained analytically, Le Chenadec *et al.* (2007) showed that the dissipated energy can be related to the cyclic energy through a power law function:

$$E_{diss} = \kappa E_{cyc}^{\gamma}$$

Here, κ and γ will be considered as not temperaturedependant which is justified by the fact that the expected heat build-up level are lower than 20°C. Once

these quantities have been evaluated from the mechanical response of the structure, the power loss quantity, defined as:

$$P_{diss} = F_{frequency} . E_{diss}$$

is then introduced as heat sources in a thermal finite element calculation thanks to user subroutines programmed in fortran 77 (Abaqus USDFLD and HETVAL subroutines). To take advantage of the difference between mechanical and thermal characteristic times (many mechanical cycles are needed to reach the steady-state temperature) and to reduce the computational costs, a strategy based on an uncoupled cyclic algorithm is adopted: the mechanical problem is first solved, a dissipated energy is then computed from the mechanical quantities and put as heat sources in a thermal calculation.

4.2 Parameters identification

Two kinds of parameters are to be evaluated in this step: the first one is related to the mechanical behavior and the other one to the thermal properties. The identification of the constitutive model for rubber elasticity and the material thermal parameters (volumic density, specific heat and thermal conductivity) is very classic. Otherwise, the boundary conditions and the parameters κ and γ needs a careful identification, which is detailed in a forthcoming paper (Le Saux *et al.*, submitted to European Journal of Mechanics – A/Solids).

4.3 Approach validation and relation between temperature and dissipation sources

Using the modelization and numerical tools detailed in the paragraph 5, we are now able to relate the temperature at the skin, measured in paragraph 2 to the dissipated energy in a given volume. Figure 5 plots the

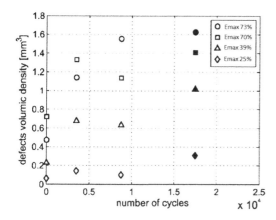

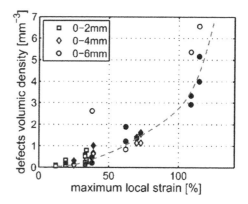

Figure 6. Stabilization along the fatigue cycles of the volumic density of the cavites for several local maximum strain.

Figure 7. Evolution of the stabilized volumic density of the cavites along the local maximum strain.

dissipated energy in the central section of the sample (central section of 1 mm thick evaluated by tomography) and illustrates a linear dependence between these variables. We will take advantage of this relation in section 6 dealing with the proposal of an energetic fatigue life criterion.

5 X-RAY TOMOGRAPHY MEASUREMENTS

X-ray computed micro-tomography was used to investigate the fatigue damage occurring in a polychloroprene rubber. This non destructive technique allowed us to study the initiation and propagation mechanisms (Le Saux *et al.*, 2011). We will focus here on the evolution of the defect population along the fatigue cycles and its dependency on the maximum local strain. This study revealed that the defect volumic density depends both on the number of cycles and on the maximum local strain. Figure 6 illustrates that the defect density evolves quickly during the first 10% of the fatigue test and then increases very slowly. The value obtained after a few thousands cycle is therefore representative of the final value (at the initiation stage).

Figure 7 illustrates the dependency of the volumic density of the cavities on the maximum local strain. This plot was obtained from several samples and for several fatigue test durations and exhibits a remarkable "master curve". We therefore know the dependency of the defect density on the maximum strain but we still do not know how the material dissipates energy.

6 WÖHLER CURVE PREDICTION

In this study, the maximal strain is kept as the fatigue parameter and an initiation criterion based on the cumulated dissipated energy is used. The principle of the approach is therefore very classic and simple: whatever the fatigue parameter may be, the cumulated dissipative energy needed to initiate a crack will

be a constant, called here Critical Dissipated Energy (CDE). This energy based approach is close to the one proposed by Mars (2001) or Grandcoin (2008). In elastomers, the energy dissipated evolves during the fatigue tests but reaches a stabilized value after the number of cycles needed to stabilize the rise of temperature induced by the heat build-up. Once this stabilization step is achieved, the dissipated energy measured along the fatigue test is almost a constant, for a given global displacement. This is confirmed by the stabilization of the heat build-up along a fatigue test, showing that the dissipation sources are constant. It is therefore possible to write that:

$$CDE = N_i \cdot E_{Fatigue,diss/cycle}$$

with CDE the Critical Dissipated Energy considered as an intrinsic constant, N_i the number of fatigue cycles needed to reach the initiation and $E_{fatigue,diss/cycle}$ the energy dissipated per cycle by the fatigue mechanisms, which is a function of the maximum strain. Once the defect population and its evolution with respect to the maximum local strain are described by tomography, a first way to evaluate the energy dissipated by the fatigue mechanisms would be to sum all the energy dissipated by each defect along one fatigue cycle. This is clearly not an easy task, from both modelization and numerical points of view.

Here, we have chosen a much more phenomenological approach: we chose to relate the total dissipated energy (evaluated from the stabilized rise of temperature) to the energy dissipated by the defect population. The first hypothesis assumes that the ratio between the total dissipated energy and the one related to the fatigue mechanisms (whatever the kind of dissipation it leads to) is linearly dependant on the defect density. It would then come:

$$E_{Fatigue,diss/cycle} = A.\varpi_d. E_{diss/cycle}$$

with A being a constant, ϖ_d the defect volumic density in the central zone of the sample, V the considered volume and $E_{diss/cycle}$ the total dissipated energy

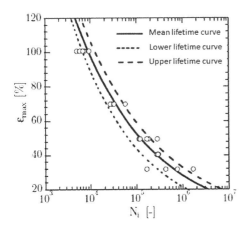

Figure 8. Validation of the Wôhler curve prediction and illustration of the correlation between the dispersion on the population defects and the fatigue dispersion.

during one fatigue cycle in that volume. The second hypothesis is that the total dissipated energy is can be evaluated from the temperature rise measured at the skin, which was proved in paragraph 4. We could consequently write that:

$$E_{diss/cycle} = B. \theta$$

with B a constant parameter and θ the stabilized minimum rise of temperature in the central zone of the sample, which is dependant on the maximum imposed deformation. This global evaluation is also consistent as because the rise of temperature, measured at the skin, is an average value of the dissipation sources that are located in the volume underneath, which is also true for the defect volumic density. From the former equation, it can now be written:

$$CDE = N_i.A.\varpi_d. B. \theta$$

which also writes:

$$N_i.\varpi_d. \theta = \mathbf{Constant}$$

for a given volume. To validate the approach, it is possible to build a lifetime curve from this last equation. The first step is to take one point of the Wôhler curve (here for $\varepsilon_{max} = 30\%$), for a given local strain, to evaluate the constant. Then, it is possible to fit power laws from the curves giving the dependence of ϖ_d on strain (from tomography measurements) and from the curve giving the dependence of θ on strain (from thermal measurements). The curve obtained is plotted on Fig. 8 (called Mean lifetime curve) and exhibits a very good correlation with the other experimental fatigue data.

As the temperature rise is very repeatable from a sample to another, it was also very tempting to try to link the dispersion observed on figure 5 (volumic density of the cavities along strain) with intrinsic fatigue dispersion. We therefore identified two power laws from figure 5, from the lowest values and the highest values, and then plotted the curves obtained on the figure 8 (respectively upper and lower lifetime curves). The evaluation is actually very good, even if more numerous tomography data are required to be confident in the scattering range evaluated.

7 CONCLUSION

The present study needs of course further validation on other materials. Nevertheless, it illustrates how powerful can be the coupling of microscopic description and macroscopic energetic evaluation to understand and model the fatigue behavior of elastomeric materials

REFERENCES

J. Grandcoin (2008). Contribution à la modélisation du comportement dissipatif des élastomères chargés: d'une modélisation micro-physiquement motivée vers la caractérisation de la fatigue. Ph.D. thesis, Université d'Aix-Marseille II.

R. Hill (1968), On constitutive inequalities for simple materials I, Journal of the Mechanics and Physics of Solids 16 229–242.

Y. Le Chenadec, Y., Stolz, C., Raoult, I. Nguyen T., M.-L., Delattre, B. Charrier, P. (2007). *A novel approach to the heat build-up problem of rubber. ECCMR V, Paris.*

Y. Le Chenadec, I. Raoult, C. Stolz, M. Nguyen-Tajan (2009). Cyclic approximation of the heat equation in finite strains for the heat build-up problem of rubber, Journal of Mechanics of Materials and Structures 4 (2) 309–318.

Le Saux, V. (2010). Fatigue et vieillissement des élastomères en environnements marin et thermique: de la caractérisation accélérée au calcul de structure. Ph. D. thesis, Université de Bretagne Occidentale.

Le Saux, V., Y. Marco, S. Calloch, P. Charrier, and D. Taveau (2011). Heat build-up problem of rubbers under cyclic loadings: experimental investigations and numerical predictions. European Journal of Mechanics – A/Solids (submitted).

Le Saux, V., Y.Marco, S. Calloch, C. Doudard, and P. Charrier (2010). Fast evaluation of the fatigue lifetime of rubberlike materials based on a heat build-up protocol and micro-tomography measurements . International Journal of Fatigue 32, 1582–1590.

Le Saux V, Marco Y, Calloch S, Charrier P. (2011) Evaluation of the fatigue defect population in an elastomer using X-ray computed micro-tomography. J Polym Eng Sci, accepted for publication.

Mars W (2001). Multiaxial fatigue of rubber. Ph.D. thesis, University of Toledo.

Ostoja-Kuczynski, E., Charrier, P., Verron, E., Marckmann, G., Gornet, L. & Chagnon, G. 2003. *Crack initiation in filled natural rubber: experimental database and macroscopic observations. ECCMR III.*

Ostoja-Kuczynski, E., Charrier, P., Verron, E., Gornet, L. & Marckmann, G. 2005. *Influence of mean stress and mean strain on fatigue life of carbon black filled natural rubber. ECCMR IV.*

In-situ synchrotron X-ray diffraction study of strain-induced crystallization of natural rubber during fatigue tests

S. Beurrot, B. Huneau, E. Verron & P. Rublon
LUNAM Université, Ecole Centrale de Nantes, GeM, UMR CNRS 6183, Nantes cedex, France

D. Thiaudière, C. Mocuta & A. Zozulya
Synchrotron Soleil, Gif Sur Yvette, France

ABSTRACT: A home made stretching machine has been developed to perform fatigue tests on natural rubber in the synchrotron facility Soleil. Strain-Induced Crystallization (SIC) is investigated by Wide-Angle X-ray Diffraction (WAXD) during in-situ fatigue tests of different minimum and maximum strain levels. The index of crystallinity χ decreases with the number of cycles when the minimum strain level reached during the fatigue test is lower than the critical stretch ratio for melting λ_M. On the contrary, when the stretch ratio is maintained higher than the critical stretch ratios for melting λ_M and crystallization λ_C, χ increases with the number of cycles. The crystallites size and orientation have the same evolution during the different fatigue tests: the crystallites have a constant size but their orientation is enhanced with the number of cycles.

1 INTRODUCTION

Natural Rubber (NR), cis-1,4-polyisoprene, has remarkable fatigue properties (Cadwell et al. 1940, Mars and Fatemi 2004) which are generally explained by Strain-Induced Crystallization (SIC). SIC is commonly investigated by Wide-Angle X-ray Diffraction (WAXD). But studies on the evolution of SIC during fatigue testing are very rare, mainly because the typical frequencies of fatigue tests (1 Hz or more) are not compatible with the long time acquisition required by X-ray diffraction measurements (from a few seconds to an hour). Kawai (1975) succeeded in measuring SIC during fatigue by using a stroboscopic technique to accumulate the weak intensity of the diffracted beam over several hundreds of cycles. However, this technique requires to average the crystallinity over a large number of cycles. Moreover, this work was limited to a large R ratio ($R = \lambda_{min}/\lambda_{max} = 3.5/4.5$) for which crystallites never melt because λ_{min} always remains greater than the critical stretch ratio for melting. Furthermore, Rouvière et al. (2007) recently performed interrupted fatigue tests and WAXD measurements on stretched fatigued samples to obtain the evolution of the crystallinity along fatigue life for different values of R. However, this method does not allow to separate SIC due to fatigue from SIC due to constant elongation during the 45-minute acquisition of the X-ray diffractogram.

In the present study, we use a synchrotron radiation to reduce the exposure time and to perform in-situ fatigue tests for different R ratios. A versatile testing machine allowing uniaxial and biaxial loading conditions as well as large strain was especially designed and built for that purpose.

2 EXPERIMENTAL METHOD

2.1 *Material and sample*

The material used in this study is a carbon black-filled natural rubber, cross-linked with 1.2 phr (per hundred of rubber) of sulphur and CBS accelerator. It also contains ZnO (5 phr) and stearic acid (2 phr) and is filled with 50 phr of N330 carbon black. The samples are classical flat dumbbell specimen with a 10 mm gauge length and a $2 \times 4\,mm^2$ section.

2.2 *Synchrotron*

The synchrotron measurements have been carried out at the beamline in the French national synchrotron facility SOLEIL. The wavelength used is 1.319 Å and the beam size is 0.3 mm in diameter at half-maximum. The 2D WAXD patterns are recorded by a MAR 345 CCD X-ray detector. In order to make an accurate correction of air scattering, a PIN-diode beam stop was used.

2.3 *Fatigue testing machine*

The fatigue tests have been conducted with a home-made stretching machine shown in Figure 1. It is composed of four electrical actuators, but only two opposite ones were used in this study. Their movements are synchronized, in order to keep the center of the

Figure 1. Uniaxial and biaxial stretching machine in DiffAbs.

Table 1. Fatigue loading conditions.

test No.	l_{min} (mm)	l_{max} (mm)	f (Hz)
1	0	20	2.5
2	4	33.2	0.8
3	9.3	33.2	1
4	25	45	1.5

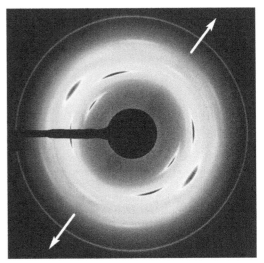

Figure 2. An example of diffraction pattern. The dark arcs stem from the crystalline phase, whereas the large ligth ring is the reflexion of the amorphous phase. The white arrows show the tensile direction.

specimen fixed during the fatigue tests. Their loading capacity is $\pm500\,\text{N}$ and their stroke is 75 mm each.

2.4 Procedure

All the experiments are conducted by prescribing constant displacements of the clamps at each cycle. Just before fatigue testing, specimens are first cycled at a higher deformation (during 55 cycles) in order to lower the remaining elongation of the sample due to Mullins effect and viscoelasticity during the fatigue tests.

The minimum exposure time of the CCD detector to record a workable scattering pattern is 1 s, which is about the duration of a fatigue cycle. It is then not possible to record diffraction patterns during fatigue tests while the actuators are in motion. Hence, to measure the evolution of crystallization during fatigue testing, it has been decided to pause the test at maximum deformation of the sample every 250 cycles to record a complete diffraction pattern. The fatigue machine being triggered by the monitoring system of the X-ray beam, the duration of the pause is less than 1.5 s. The first scattering pattern is recorded during the first cycle of a test.

Four different fatigue tests have been performed; Table 1 shows for each the minimum and maximum displacements of the grips (l_{min} and l_{max}) and the loading frequency (f).

2.5 Scattering pattern analysis

An air scattering pattern (without sample) was first collected and has been used to correct the patterns. Moreover, the change in thickness of the sample under extension and the change of intensity of the incident photons have also been considered. All these corrections are performed by following the well-established method of Ran et al. (2001). Both the determination of the pattern center and the calibration of the diffraction angles were achieved by considering the first diffraction ring of ZnO ((100)-plane, $a = 3.25\,\text{Å}$ (Reeber 1970)). Here, small angles scattering was not investigated; the range of diffraction angles is $2\theta \in [8°, 26.7°]$. An example of diffraction pattern is shown in Figure 2.

The intensity of photons diffracted by the isotropic phases in the material $I_{\text{isotropic}}(2\theta)$ is extracted from the diffraction patterns by considering the minimum intensity along the azimuthal angle β for each Bragg angle 2θ. Then, the intensity of photons diffracted by the anisotropic material $I_{\text{anisotropic}}(2\theta, \beta)$ is calculated as the difference between the total intensity of photons diffracted $I_{\text{total}}(2\theta, \beta)$ and $I_{\text{isotropic}}(2\theta)$. The spectra extracted from $I_{\text{anisotropic}}(2\theta, \beta)$ and $I_{\text{isotropic}}(2\theta)$ are classically fitted by a series of Pearson functions (Chenal et al. 2007, Trabelsi et al. 2003, Rault et al. 2006, Toki et al. 2000). Figures 3a and 3b presents examples of fitting and deconvolution of a $(2\theta, I_{\text{isotropic}})$ spectrum and a $(2\theta, I_{\text{anisotropic}})$ spectrum, respectively.

The index of crystallinity χ is calculated from the simplified Mitchell formula (Mitchell 1984):

$$\chi = \frac{\mathcal{I}_{\text{cryst}}}{\mathcal{I}_{\text{cryst}} + \mathcal{I}_{\text{amorphous}}} \qquad (1)$$

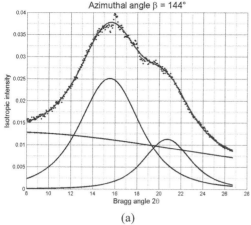

(a)

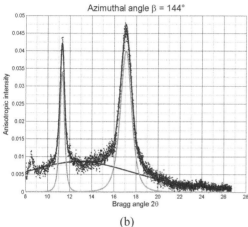

(b)

Figure 3. Examples of deconvolution of (a) a spectrum $(2\theta, I_{\mathrm{isotropic}})$ and (b) a spectrum $(2\theta, I_{\mathrm{anisotropic}})$ fitted by series of Pearson functions.

where $\mathcal{I}_{\mathrm{cryst}}$ is the integrated intensity of the (120) and (200) Bragg reflections of NR and $\mathcal{I}_{\mathrm{amorphous}}$ is the integrated intensity of the amorphous halo, considered equal to the integrated intensity $I_{\mathrm{isotropic}}$.

The crystallites size is deduced from the Scherrer formula (Guinier 1963):

$$l_{hkl} = \frac{K\lambda}{\mathrm{FWHM}_{2\theta}\cos\theta} \qquad (2)$$

where l_{hkl} is the crystallites size in the direction normal to the hkl diffraction plane, K is a scalar which depends on the shape of crystallites (here we adopt 0.78 as Trabelsi et al. (2003)), λ is the radiation wavelength, θ is the Bragg angle and $\mathrm{FWHM}_{2\theta}$ is the full width at half maximum of the peak hkl in 2θ. Finally, the crystallites disorientation is simply given by half the full width at half maximum (FWHM_{β}) of the peaks, measured on the azimuthal profiles of the reflection.

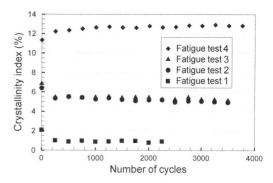

Figure 4. Evolution of the crystallinity index during fatigue tests.

3 RESULTS

The evolution of the index of crystallinity of carbon black-filled NR during fatigue tests for different levels of deformation is given in Figure 4. As expected, larger is the strain, higher is the crystallinity. During the three fatigue tests with the lowest minimum and maximum strain levels (tests No. 1, 2 and 3), the index of crystallinity decreases (from 15% to 50%) during the first 250 cycles, and then continues to decrease at a lower but constant rate until the end of the test. In Fig. 4, the results are shown only for the first 4000 cycles, but fatigue test No. 3 was performed during 40,000 cycles and the index of crystallinity decreases at a constant rate during the whole test, down to about $\chi = 2\%$ (in comparison with $\chi = 7\%$ at the beginning of the test). During fatigue test No. 4, which admits the highest mininum and maximum strain levels, the evolution of the index of crystallinity is very different: χ increases with the number of cycles. The minimum and maximum local stretch ratios reached during the fatigue tests have been measured by an optical technique, as well as the critical stretch ratios for crystallization (λ_C) and melting (λ_M) at room temperature and low strain rates (measured with the same method as Trabelsi et al. (2003)). Those stretch ratios are sketched in Figure 5. It appears that the three fatigue tests for which the index of crystallinity decreases with the number of cycles have been performed with minimum stretch ratios lower than λ_C; whereas the minimum stretch ratio during test No. 4, for which the index of crystallinity increases with the number of cycles, is greatly higher than λ_C.

Figure 6 shows the evolution of the mean size of the crystallites during fatigue tests No. 2, 3 and 4 (χ is too low during fatigue test No. 1 for accurate measurements of the crystallites size). Only the sizes in directions normal to the diffraction planes (200), (120) and (201) are shown because they correspond to the most intense diffraction arcs and hence to less scattered measurements. The crystallites size remains constant with the number of cycles and is similar for the three fatigue tests: $l_{200} = 125$ Å, $l_{201} = 115$ Å and $l_{120} = 50$ Å. Sizes in directions normal to the planes (121),

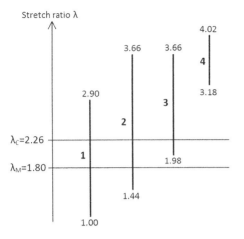

Figure 5. Minimum and maximum stretch ratios reached during fatigue tests compared to the critical stretch ratios for crystallization λ_C and melting λ_M (measured for quasi-static test). Fatigue test numbers are in bold.

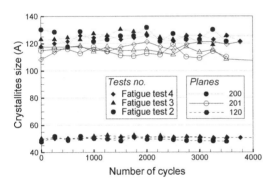

Figure 6. Evolution of the crystallites mean size during fatigue tests.

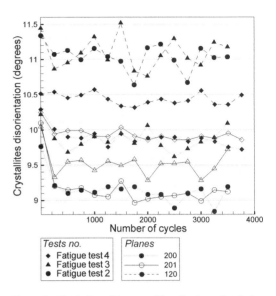

Figure 7. Evolution of the crystallites disorientation during fatigue tests.

(202) and (002), which correspond to less intense diffraction arcs, have been measured as well; they are also constant during fatigue tests and similar for the three tests.

The evolution of disorientation of the diffraction planes (200), (120) and (201) of the crystallites during fatigue tests No. 2, 3 and 4 is shown in Figure 7. Disorientation of the crystallites from the mean orientation varies from $\pm 9°$ to $\pm 11.5°$, depending on the diffraction planes and the fatigue tests considered. Nevertheless, for all fatigue tests and the three planes studied, and despite scattering of data, the same evolution is observed: the disorientation of the crystallites decreases with the number of cycles, particularly during the first 250 cycles of the fatigue tests.

4 DISCUSSION AND CONCLUSION

4.1 Evolution of χ during fatigue tests

Studying fatigue of elastomers, experimenters and mechanicians face a recurring issue: it is extremely difficult to perform fatigue tests controlling any local mechanical quantity (such as stress, strain or energy). Indeed, fatigue tests are almost always performed controlling either the displacement of the machine grips or the force applied to the sample, from which local mechanical quantities can partially be calculated *a posteriori*. In our study, fatigue tests were performed by prescribing constant maximum and minimum displacements of the grips. As the material is cycled, stress-softening and viscous effect modify both stress and strain reached at a given displacement for different numbers of cycles. This phenomenon can be minimized by pre-cycling the material (to avoid the Mullins effect at the beginning of the test), but it cannot be completely suppressed. Thus, the evolution of the crystallinity index, and the size and disorientation of crystallites during fatigue tests presented in the previous section were measured at constant displacement but at decreasing local stress and strain.

It is not yet established whether crystallization of rubber is induced by strain, stress, energy or a combination of them. However, we believe that a mechanical quantity that controls crystallization -even if undefined yet- exists; in the following, it is referred to as the "crystallization driving quantity". As stress, strain and energy are closely related, we can assume that this crystallization driving quantity evolves similarly as stress and strain during fatigue tests. As a consequence, the maximum value of this quantity reached at each cycle decreases too with the number of cycles during the fatigue tests. In the following, we discuss our results in the light of this assumption.

- As shown in Fig. 5, fatigue tests No. 1 and 2 are performed with minimum stretch ratios lower than the critical stretch ratio for melting λ_M. Moreover, even if the initial minimum stretch ratio in fatigue test No. 3 is slightly higher than λ_M, it becomes lower than λ_M after a few cycles. For these three tests, the same phenomenon takes place during each cycle: all the crystallites melt when the minimum displacement is reached, and new crystallites nucleate when the sample is stretched again. Nevertheless, as the maximum value of the crystallization driving quantity reached during one cycle is lower than at the previous cycle, the crystallinity index is lower too. Hence, χ decreases with the number of cycles.

- During fatigue test No. 4, local stress and strain decrease with the number of cycles as well, but the local stretch ratio is always greatly larger than λ_C. In other words, the minimum value of the crystallization driving quantity decreases from one cycle to the next one, but it always remains higher than its threshold of crystallization (the counterpart of λ_C). As a consequence, when the grips are at the minimum displacement, all crystallites do not melt (as λ_M is lower than λ_C, the local stretch ratio is maintained higher than λ_M too). As the sample is stretched, the crystallization driving quantity increases, as well as χ, until the maximum displacement of the grips is reached. Then, when the sample is unstretched again, the crystallization driving quantity decreases and the change in χ results from the balance of two phenomena: on the one hand, χ decreases because the crystallization driving quantity decreases; and on the other hand, the crystallization driving quantity remains higher than the threshold of crystallization, so crystallization is still induced and χ continues to increase. Quantitatively, the latter phenomenon is not as significant as the first one, so χ decreases during the unstretching phase. But this decrease during the unstretching phase is lower than the increase during the stretching phase. Hence, from one cycle to another, χ increases for the same value of crystallization driving quantity. During fatigue test No. 4, this evolution of χ is minimized by the fact that the maximum crystallization driving quantity reached during each cycle decreases during the test.

As stated previously, it would be very difficult to perform fatigue tests controlling the crystallization driving quantity even if it was known. Nevertheless, in order to clarify our theory of crystallization in fatigue, Figure 8 shows our prediction of the evolution of χ with the number of cycles if the tests were performed at constant maximum and minimum values of crystallization driving quantity (as compared to the original data measured during tests performed at constant maximum and minimum displacements of the grips). To summarize, four different situations are possible:

- The maximum crystallization driving quantity is lower than the threshold of crystallization: χ remains equal to 0.
- The maximum crystallization driving quantity is higher than the threshold of crystallization and the minimum crystallization driving quantity is lower than the threshold of melting (equivalent to tests No. 1, 2 and 3): all the crystallites melt at each

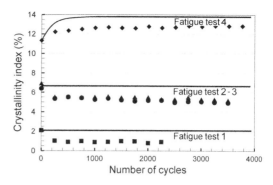

Figure 8. Evolution of the crystallinity index with the number of cycles during fatigue tests under constant crystallization driving quantity as assumed by the authors (lines), compared to original data measured during tests under constant displacement of the grips (symbols).

cycle, and the maximum crystallinity index reached at each cycle is constant.
- The maximum crystallization driving quantity is higher than the threshold of crystallization and the minimum crystallization driving quantity is between the thresholds of melting and crystallization (this is not equivalent to test No. 3, as during this test the minimum stretch ratio becomes lower than λ_M after a few cycles only): all the crystallites do not melt at each cycle and crystallization is induced during the unstretching phase of the cycles only at the beggining of the phase. It is not possible to conclude whether χ at maximum deformation is constant during the fatigue test or if it increases with the number of cycles.
- Both minimum and maximum crystallization driving quantities are higher than the thresholds of melting and crystallization (equivalent to test No. 4): all the crystallites do not melt at each cycle, and crystallization is induced during the unstretching phase. The maximum crystallinity index reached at each cycle increases with the number of cycles.

4.2 Evolution of the crystallites size and disorientation during fatigue tests

While the evolution of χ with the number of cycles depends on the stretch ratios reached during the fatigue tests, the size and orientation of the crystallites evolve similarly during the different tests. As shown in Fig. 6, the size of the crystallites is identical for the three fatigue tests and constant with the number of cycles. It means that the evolution of χ with the number of cycles (decreasing for tests No. 2 and 3, increasing for test No. 4) is only due to a variation of the number of crystallites nucleated at each cycle. On the contrary, the disorientation of the crystallites varies from one test to another as shown in Fig. 7. But no correlation between the stretch ratios reached during fatigue and the level of disorientation is observed. An additional quasi-static test performed on the same material and in the same conditions shows that crystallites have very similar size and orientation during quasi-static tests as during fatigue tests. It suggests that once the crystallites are nucleated, cycling the material has little effect on the crystallites size and orientation.

REFERENCES

Cadwell, S. M., R. A. Merril, C. M. Sloman, & F. L. Yost (1940). Dynamic fatigue life of rubber. *Industrial and Engineering Chemistry (reprinted in Rubber Chem. and Tech. 1940; 13:304–315) 12*, 19–23.

Chenal, J.-M., C. Gauthier, L. Chazeau, L. Guy, & Y. Bomal (2007). Parameters governing strain induced crystallization in filled natural rubber. *Polymer 48*, 6893–6901.

Guinier, A. (1963). *X-ray Diffraction*. W. H. Freeman & Co.

Kawai, H. (1975). Dynamic X-ray diffraction technique for measuring rheo-optical properties of crystalline polymeric materials. *Rheologica Acta 14*, 27–47.

Mars, W. V. & A. Fatemi (2004). Factors that affect the fatigue life of rubber: a literature survey. *Rubber Chemistry and Technology 77*, 391.

Mitchell, G. R. (1984). A wide-angle X-ray study of the development of molecular-orientation in crosslinked natural rubber. *Polymer 25*, 1562–1572.

Ran, S., D. Fang, X. Zong, B. S. Hsiao, B. Chu, & P. F. Cunniff (2001). Structural changes during deformation of kevlar fibers via on-line synchrotron SAXS/WAXD techniques. *Polymer 42*, 1601–1612.

Rault, J., J. Marchal, P. Judeinstein, & P. A. Albouy (2006). Chain orientation in natural rubber, part II: 2H-NRM study. *The European Physical Journal E 21*, 243–261.

Reeber, R. R. (1970). Lattice parameters of ZnO from 4.2 degrees to 296 degrees K. *Journal of Applied Physics 41*, 5063–5066.

Rouvière, J. Y., A. Bennani, D. Pachoutinsky, J. Besson, & S. Cantournet (2007). Influence of mechanical and fatigue loading on crystallization of carbon black-filld natural rubber. In A. Boukamel, L. Laiarinandrasana, S. Méo, and E. Verron, *Constitutive Models for Rubber V*, pp. 323–326.

Toki, S., T. Fujimaki, & M. Okuyama (2000). Strain-induced crystallization of natural rubber as detected real-time by wide-angle X-ray diffraction technique. *Polymer 41*, 5423–5429.

Trabelsi, S., P. A. Albouy, & J. Rault (2003). Crystallization and melting processes in vulcanized stretched natural rubber. *Macromolecules 36*, 7624–7639.

Constitutive Models for Rubber VII – Jerrams & Murphy (eds)
© *2012 Taylor & Francis Group, London, ISBN 978-0-415-68389-0*

The influence of inelasticity on the lifetime of filled elastomers under multiaxial loading conditions

D. Juhre & M. Doniga-Crivat
German Institute for Rubber Technology, Hannover, Germany

J. Ihlemann
University of Chemnitz, Germany

ABSTRACT: In this work we present some important remarks on the lifetime of complex rubber parts under multiaxial loading conditions like simple shear with rotating axis. The results from the experiment are accompanied by FE simulations which show the lack of hyperelastic models used for estimating the inhomogeneous stress distribution in cyclic loaded rubber parts.

1 INTRODUCTION

To predict the service life of rubber parts by means of finite element simulations it is necessary to find a unique relationship between the varying operational loads and the corresponding material behaviour. In general this relationship is experimentally investigated via cyclic uniaxial tension tests with varying load amplitudes and frequencies. The results are then used as background for lifetime simulations of complex rubber parts. The very fact of assuming a correlation between the uniaxial material response and the multiaxial stress state in a real rubber part is daring, since experimental tests show that the service life strongly depends on the applied load type, e.g. tension-dominated loading in comparison to shear-dominated loading. A further drawback comes along with the wrong choice of material models for the present rubber. In the most rubber components highly filled elastomers are used which show a strong inelastic behavior due to material softening, hysteresis and residual strains. All effects are very typical for these kinds of elastomers and they influence the structural behavior as well as the service life of rubber parts most notably in inhomogeneous and cyclic loading conditions. Pure elastic material models (e.g. [1], [2], [3] and [4]) are not able to describe these effects and their usage leads to a strong overestimation of the stress-strain correlation in the material. Improved material models, considering e.g. viscoelasticity [5] or damage-like internal variables [6], fail to represent all of the above mentioned phenomena of filled elastomers, too. A material model, which is able to describe the three mentioned phenomena in filled elastomers, is the so-called MORPH (Model Of Rubber PHenomenology) model [7,8]. In this work, we want to utilize this model to compare it with hyperelastic material models (e.g. Yeoh model) to highlight the drawbacks, if we choose pure elastic models in predicting the lifetime of rubber materials.

2 CONSTITUTIVE MODELS

The MORPH model has been originally proposed by Ihlemann and Besdo [7,8] and it is based on an additional split of the Cauchy stress σ into three parts:

$$\sigma = \frac{2\,\alpha}{J}\,\mathbf{b}'_{iso} + \sigma'_Z - q(J)\,\mathbf{1} \tag{1}$$

wherein q is a pressure function depending on the Jacobian J, $\mathbf{b}'_{iso}$ the deviatoric part of the isochoric left Cauchy-Green tensor, $\alpha = \hat{\alpha}(p_1, p_2, p_3, b_S)$ a function depending as well on material parameters p_1, p_2 and p_3 as on the history function b_S which depends itself on the maximum value of the isochoric part of the equivalent left Cauchy-Green strain b_{iso}^{eq} at each material point during all loading cycles:

$$b_S(t) := \max\left[b_{iso}^{eq}(\tau), \quad 0 \leq \tau \leq t\right] \tag{2}$$

The auxiliary stress σ_Z is computed through its time derivative:

$$\overset{*}{\sigma} = \beta\, \overset{*}{b_{iso}^{eq}}\,(\sigma_H - \sigma_Z) - \frac{5}{3}\,(\mathbf{d}\cdot\mathbf{1})\,\sigma_Z \tag{3}$$

$$+\,(\mathbf{d}\,\sigma'_Z + \sigma'_Z\,\mathbf{d})$$

wherein $\mathbf{d}$ is the symmetric part of the rate of the deformation gradient, $\beta = \hat{\beta}(p_3, p_4, b_S)$ a function depending on the material parameters p_3 and p_4 and the history function b_S and $\boldsymbol{\sigma}_H$ the cladding stress defined by:

$$\boldsymbol{\sigma}_H = \frac{\gamma}{J} \exp\left(p_7 \frac{\overset{*}{\mathbf{b}}_{iso}}{\overset{*}{b}_{iso}^{eq}} \frac{b_{iso}^{eq}}{b_S} \right) + \frac{p_8}{J} \frac{\overset{*}{\mathbf{b}}_{iso}}{\overset{*}{b}_{iso}^{eq}} \quad (4)$$

with $\gamma = \hat{\gamma}(p_5, p_6, b_S)$. Hence, the MORPH model depends on 8 material parameters $(p_1, \ldots, p_8)$. Their influence on the form of the stress-strain curve is sketched in Fig. 1. A detailed derivation of the model can be found in [8].

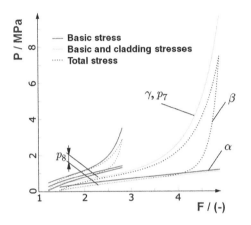

Figure 1. Structure of the MORPH model and influence of its included material parameters.

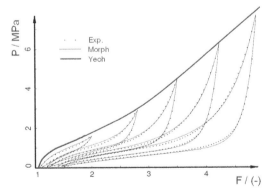

Figure 2. Fitting of Yeoh and MORPH model to experimental data.

To compare the model with a standard hyperelastic material model, we choose the Yeoh model defined by the following strain energy function:

$$W = p_1 (I_1 - 3) + p_2 (I_1 - 3)^2 + p_3 (I_1 - 3)^3 + q(J) \mathbf{1} \quad (5)$$

Both models are fitted to a uniaxial tension test for a filled natural rubber (see Fig. 2). The resulting material parameters are summarized in Tab. 1.

3 SIMPLE SHEAR WITH ROTATING AXES

A conclusive evidence for the inappropriate choice of a hyperelastic material model for predicting the lifetime of rubber material is shown by investigating a new experimental approach considering simple shear with rotating axes. The test rig for realizing the exceptional load type is drafted in Fig. 3. It is based on prior investigations done by Gent [9] and consists of three parts, whereby, during a measurement, the outer parts are fixed in their position. The middle part has a degree of freedom in radial direction. For an experiment, two rotationally symmetric rubber samples are clamped into the rig in a double-sandwich-arrangement, each between one outer and the middle part. The experiment is initiated by the radial displacement of the middle part. Thus, a simple shear deformation is initiated in the samples leading to a radial force F_R. The geometry of the samples is optimized in such a way, that simple shear is the predominating deformation throughout the whole volume. At the same time, the maximum load is located within the sample, far away from surfaces and contact areas, so that a failure is most probable

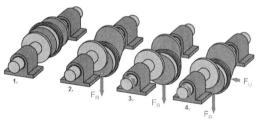

Figure 3. Sketch of the test rig: 1. double-layered sandwich specimen; 2. appliying shear load resulting in the radial force F_R; 3. rotating specimen and deflection of middle part; 4. surpressing deflection leads to circumferential force F_U.

Table 1. Material parameters for Yeoh and MORPH model

Material model	Material parameters							
	p1	p2	p3	p4	p5	p6	p7	p8
Yeoh	0.739	−0.019	0.136					
MORPH	0.062	0.364	0.219	3.09	0.00846	5.92	5.70	0.201

in the interior. After initiating the simple shear deformation, one of the shafts of the outer parts is rotated to start a simple shear deformation with rotating axes. For inelastic materials, the middle part would move

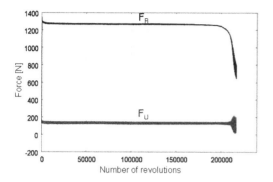

Figure 4. Lifetime test with the new test rig: measurement of reaction forces F_R and F_U until material failure.

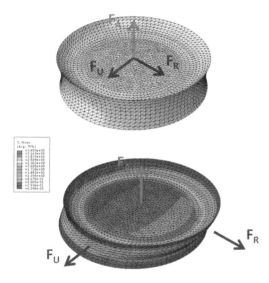

Figure 5. Finite element simulation of simple shear with rotating axes including the directions of the resulting forces.

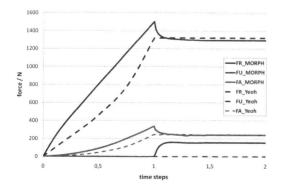

Figure 6. Reaction forces F_R, F_U and F_A from FE simulations using MORPH and Yeoh model.

sideways due to energy-dissipation, as shown in the third image of Fig. 3. If it is restricted (fourth image in Fig. 3), a resulting circumferential force F_U could be measured instead.

Fig. 4 show the results of a long time fatigue experiment with the experimental rig, whereby the radial force F_R and the circumferential force F_U during the experiment are plot against the revolutions with a frequency of 1 Hz. The experiment has been carried out till failure of one of the samples in the double-sandwich-arrangement. During the first 10000 cycles, stress softening effects can be observed in the curve progression of the two measured forces. The material becomes weaker and both forces decrease in the course of the experiment. Due to the work done by the circumferential force F_U, the samples have a finite lifetime accompanied by an abrupt decrease of the resulting forces.

For simulating this loading type we apply at first a shear load on the sample and afterwards we let rotate the lower and upper surface within one revolution. In Fig. 5 the discretized geometry of the specimen and its deformation during the loading process is shown. The resulting forces are plotted in Fig. 6. By using the MORPH model (solid lines) we get the characteristic progress of the resulting forces F_R and F_U. The magnitude of the values as well as the decreasing of the forces during the first revolution reflect very well the real material behaviour as shown in Fig. 4. Since the test rig is fixed in longitudinal direction, the specimen cannot compensate the large shearing deformation thus, an additional positive force F_A in the axial direction is induced. This force cannot be measured with the test rig in its current state, but a modification of the machine concerning this point is in progress. By using the above fitted material models, the magnitude of the axial force is a fifth of the radial force. The ratio generally depends on the type of material. By using a hyperelastic material model like the Yeoh model the resulting forces are captured only partially. The radial force F_R and the axial force F_A can be modelled unless the decrease of both forces during the revolution is not reproduced. Obviously, this is an inelastic effect coming from material softening (Mullins effect). Even worse is the fact that the circumferential force F_U is cannot be displayed at all, since it is a force generated by inelastic phenomena, too. If we now want to predict the lifetime of the specimen utilizing a hyperelastic material model, the missing circumferential force in the system leads to a misleading prediction. In the worst case it results in an infinite lifetime prediction which is definitely unphysical.

REFERENCES

[1] Ogden, R. W. (1972). Large Deformation Isotropic Elasticity – On the Correlation of Theory and Experiment for Incompressible Rubberlike Solids, Proc. of the Royal Society of London. Series A, Mathematical and Physical Sciences 326, pp. 565–584.

[2] Yeoh, O. H. (1993). Some forms of the strain energy function for rubber, Rubber Chemistry and Technology, 66, pp. 754–771.

[3] Gent, A.N. (1996). A new constitutive relation for rubber, Rubber Chemistry and Technology 69, pp. 59–61.

[4] Arruda, E. M. and Boyce, M. C. (1993). A three-dimensional model for the large stretch behavior of rubber elastic materials, Journal of the Mechanics and Physics in Solids 41, pp. 389–412.

[5] Bergstrom, J.S. and Boyce, M.C. (2000). Large strain time-dependent behavior of filled elastomers. Mechanics of Materials 32, pp. 627–644.

[6] Dorfmann, A. and Ogden, R.W. (2004). A constitutive model for the Mullins effect with permanent set in particle-reinforced rubber, International Journal of Solids and Structures 41, pp. 1855–1878.

[7] Besdo, D. and Ihlemann, J. (2003). A phenomenological constitutive model for rubberlike materials and its numerical applications, International Journal of Plasticity 19, pp. 1019–1036.

[8] Ihlemann, J. (2003). Kontinuumsmechanische Nachbildung hochbelasteter technischer Gummiwerkstoffe. Duesseldorf: VDI.

[9] Gent, A. N. (1960). Simple rotary testing machine. British Journal of Applied Physics 11, p. 165.

Constitutive Models for Rubber VII – Jerrams & Murphy (eds)
© 2012 Taylor & Francis Group, London, ISBN 978-0-415-68389-0

Time-dependence of fracture behaviour of carbon black filled natural rubber

M. Boggio, C. Marano & M. Rink
Dipartimento di Chimica, Materiali e Ingegneria Chimica "Giulio Natta" Politecnico di Milano, Milan, Italy

ABSTRACT: Fracture toughness at crack onset rate dependence was investigated performing videorecorded fracture tests on compounds having different carbon black content, under pure shear test configuration at different displacement rates using the J-integral fracture mechanics approach. Filled rubber compounds fracture toughness slightly increases at low displacement rates while, at high displacement rates, the mechanisms which induce time-dependence are different and therefore different trends are observed depending on the carbon black content. Results were analyzed considering the input strain energy as made of two components, one "stored" and released when fracture occurs, and the other "dissipative" related to non-catastrophic fracture events. To separate energy into these components, "loading-unloading" tensile tests were performed at increasing maximum strains on unnotched pure shear specimens. From these analyses, fracture toughness rate-dependence has been interpreted considering mechanisms which involve the "bound rubber", a part of the rubber molecules bonded to the carbon black aggregates' surface, which have restricted mobility, modifying the material orientability.

1 INTRODUCTION AND THEORETICAL BACKGROUND

As the relaxation time of rubbers is very low, generally it would not be expected to observe rate dependence in their mechanical behavior. In fact, unfilled natural rubber exhibits hyperelastic non-linear stress-strain behaviour. Nevertheless, in the case of filled natural rubber, time dependence of mechanical and rheological properties has been widely reported in literature and has been related to phenomena taking place at the filler/polymer interface (e.g. Lake et al. 2001, Leblanc et al. 2002, Fukahori et al. 2005). It is possible to consider the strain energy during a deformation process as made of two components: one "stored" by the molecules, and the other "dissipative" related to non-catastrophic fracture events which take place during deformation of filled rubber compounds. These non-catastrophic fracture events are those which cause the Mullins effect (Diani et al. 2009, Dargazany et al. 2009) and include chain rupture, chain detachment from the filler surface or slippage (adsorption/desorption) at the filler-polymer interface. As proposed by Hamed et al. (1994) for elastomers and in (Webber et al. 2007) for double network hydrogels, on the basis of the Lake – Thomas physical model (Lake et al. 1965) together with junction rupture firther energy dissipation occurs as a consequence of strain energy release by load bearing chains. Indeed, a macromolecule between two junctions in the undeformed material is in a random coil conformation. During deformation, the distance between the junctions increases, the macromolecule

orients and assumes an extended conformation with a consequent entropy decrease. With further increase of deformation, the chain bonds deform and the macromolecule starts to store enthalpic strain energy. When the energy is enough to break a chain bond or to break the junction, the polymer chain retract from the extended towards the coiled conformation and releases the energy previously stored. If this phenomenon is distributed in all the strained volume, the strain energy elastically stored by the relatively short extended chains before the rupture is dissipated, through these non – catastrophic events. Moreover, a part of the rubber molecules are bonded to the carbon black aggregates' surface and have restricted mobility: a gradient of relaxation times decreasing from the aggregate surface to the bulk rubber is expected (Leblanc et al. 2002, Fukahori et al. 2005). Both the stored and the dissipative components of the strain energy are expected to involve the "bound rubber" on which a viscoelastic effects, at times involved by the tests, may be present and can dissipate viscoelastic energy and, on the other hand, shorten the length of the chains with high mobility. As the bound rubber content depends on the total aggregates surface area, the mechanical behaviour time-dependence can be expected to depend on the CB content.

This topic was investigated, in this work, with regard to fracture. Videorecorded fracture tests on unfilled natural rubber and on compounds having different carbon black content were performed, under the pure shear test configuration using a suitable specimen geometry, at different displacement rates. Since in rubbers there are large strains and non linear behaviour

Table 1. Materials composition, nomenclature and compounds density.

	Carbon Black (CB) content			ρ_{comp} [g/cm^3]
	[phr]*	[w%]	[vol %]	
NR0	0	0	0	0.931
NR50	50	33.3	21.1	1.102
NR75	75	42.8	28.5	1.165

* parts in weight per hundred parts of rubber.

at the crack tip, the stress intensity factor criterion cannot be applied, hence J-integral fracture mechanics approach was applied.

2 MATERIALS AND METHODS

2.1 Materials

Natural rubber (NR) based compounds filled with ASTM N330 (ASTM D 3053:2008) carbon black (CB) kindly supplied by Bridgestone TC, Italy were analyzed. Materials composition details are reported in Table 1.

The CB content expressed by volume percent, vol%, was calculated from the weight percent content, w%, as:

$$vol\% = w\% \cdot \frac{\rho_{comp}}{\rho_{CB}}$$

in which ρ_{comp} and ρ_{CB} are the densities of the relevant rubber compound and of the CB respectively. The latter value was determined from the density of the compounds and that of the unfilled rubber and was equal to 1.747 g/cm^3. The compounds density, measured by immersion method in a mixture of ethylic alcohol and distilled water (ISO 1183 1:2004E) are also reported in Table 1.

All the compounds were vulcanized in a compression moulding press at a temperature of $T_{curing} = 160°C$ and a pressure of $P_{curing} = 8$ MPa for 15 min. The curing procedure, as verified by Bridgestone with an oscillating disc curemeter (ASTM D 2084:2007) ensures complete vulcanization.

2.2 Sample preparation and geometry

All the tests were carried out under pure shear deformation state.

2.2.1 Pure shear (PS) fracture specimens
Figure 1 shows the 97 mm × 84 mm PS fracture specimens. In the clamped part of the specimens, reinforced rubber, which consists in a natural rubber containing aligned fibers of polyamide, was cured with the compound during the vulcanization process.

The reinforced rubber ensures that, during testing, deformation of the reinforced part of the specimen can

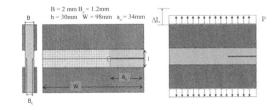

Figure 1. Test configuration and specimen geometry.

be neglected along the fibers alignment direction. This solution permits to avoid the thinning of the specimen under the clamp and the consequent specimen sliding.

A notch of 34 mm length was introduced with a cutter and sharpened with a razor blade just before the test starting. In order to invite crack propagation along the notch plane, side-grooves were introduced on both sides of the specimen. The introduction of side-grooves allowed to study the particular fracture phenomenology of CB filled natural rubber compounds previously described, separating the sideways cracks onset from the forward crack onset so as to analyze the relevant fracture toughness applying fracture mechanics.

2.2.2 PS specimens for loading-unloading tests
The specimens were similar in dimensions to the PS fracture specimens but without grooves and notch.

2.3 Mechanical tests

All tests were performed using a screw driven dynamometer, on the compounds with different CB contents and at displacement rates of 5-50-500 mm/min.

2.3.1 Fracture tests
Fracture tests (Fig. 1) were video recorded to determine cracks onset, to observe fracture phenomenology and the crack tip shape and size during the test.

Fracture toughness was evaluated at cracks onset from the load displacement curves. J integral was determined using the expression:

$$J = \eta \frac{U}{B_c(W - a_0)}$$

in which $\eta = \eta(a_0/W)$ is a dimensionless factor which depends on specimen geometry and U the input energy up to crack onset. For the PS specimen as previously confirmed by Hocine et al. (2003) by experimental and a numerical analysis $\eta(a_0/W) = 1$. This expression is in agreement with a previous work by Kim et al. (1989).

2.3.2 Loading-unloading tests
Loading[7] unloading tests were performed on pure shear ungrooved and unnotched specimens in order to separate the overall energy at a given strain (ε_{max}) in its dissipated and stored components.

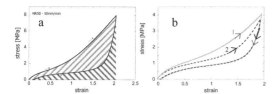

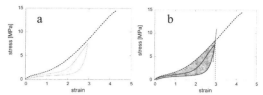

Figure 2. a. Typical loading-unloading test 2.b. Two successive loading-unloading curves up to similar maximum strain.

Figure 3. a. Two successive loading-unloading curves at increasing maximum strain 3.b. Adopted linear approximation.

In figure 2a a typical loading-unloading test on NR50 at 50 mm/min is shown. From the stress-strain curves the dissipated energy, u_{diss}, the stored energy, u_{st}, and the total input energy, u_{tot}, were measured at ε_{max} as shown in figure 2a.

The following testing procedure was adopted:

– one specimen was strained up to fracture
– loading – unloading tests were performed on a single new specimen at increasing maximum strains (form $\varepsilon_{max} = 0.5$ up to a strain the nearest possible to fracture).

In figure 2b an example of to two successive loading-unloading curves up to similar ε_{max} on the same specimen are reported (NR50 50 mm/min).

The reloading curves did not overlap onto the previous unloading curve (ideal Mullins effect), while the unloading curve for the first and the second unloading perfectly overlap.

Another important feature is that $\sigma(\varepsilon_{max})$ in the first loading is higher than in the second loading. In figure 3a are represented two successive loading-unloading curves performed at increasing ε_{max}, on the same specimen of NR50 at 50 mm/min together with the loading curve up to fracture, on a virgin specimen.

As it can be observed in figure 3a, at high values of maximum strain, when the specimen is reloaded, at strains larger than that of the previous cycle the stress does not reach the value observed on the virgin specimen. Therefore, the input energy was determined from the area under the stress strain curve of the virgin specimen up to the maximum strain of the relevant cycle while the elastic energy was obtained from the unloading curve relevant to the last cycle. At high strains when the stress did not reach the level of the virgin sample (Fig. 3.a), a linear extrapolation of the unloading was performed (Fig. 3.b).

3 RESULTS

3.1 *Fracture tests*

It has been frequently reported (e.g. Hamed et al. 1999, Medalia 1987) that for filled rubber compounds, in notched specimens under mode I loading, instead of a crack propagating, as expected, along the initial notch plane, one or more cracks develop parallel to the applied loading direction. The specimen adopted in this work, with the presence of side-gooves which favor the crack to propagate along the notch plane,

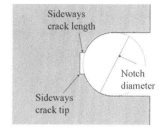

Figure 4. Crack tip at forward crack onset.

makes it possible to better analyze this particular phenomenology (Boggio 2010) For the materials and testing conditions in which this type of fracture phenomenology occurs, with this particular test specimen, two cracks propagated along the loading direction (the sideways cracks) after which a crack, propagates perpendicular to the loading direction (the forward crack), along the notch plane across the whole specimen cross-section.

Figure 4 shows the magnification of the crack tip in the frame just before the forward crack onset and a schematic representation of the crack tip. The two sideways cracks which propagated in the load direction are visible.

This fracture phenomenology is strongly correlated with strain induced molecular orientation and/or crystallization occurring in crosslinked rubbers, during loading, near the crack tip. Indeed, when strained, molecules tend to assume a preferential orientation in the loading direction so that the material becomes anisotropic and the resistance to crack propagation across the aligned polymer chains increases (Gent et al. 2003). Moreover, in the case of stereoregular polymers such as natural rubber, chains alignment can also give rise to strain-induced crystallization (Trabelsi et al. 2004). It is known that both strain induced molecular orientation (Trabelsi et al. 2003, Raoult et al. 2006) and crystallization (Poomadrub et al. 2005) are influenced by the presence of carbon black even if the mechanisms involved are not totally understood. Hamed et al. (1999) proposed that the presence of CB and strain induced crystallization act synergically to induce the necessary strength anisotropy for sideways cracks to occur.

In the present work fracture with the presence of sideways cracks occurred depending on compound composition and displacement rate.

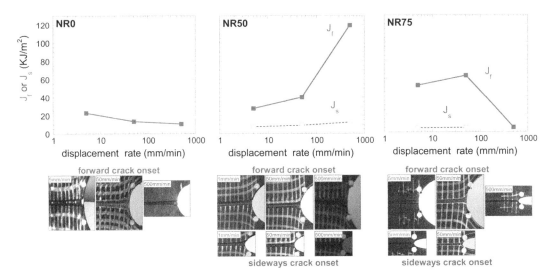

Figure 5. Fracture toughness at forward and sideways cracks onset and the relevant frames at crack onset.

Figure 5 shows the fracture toughness at forward and sideways cracks onset as a function of displacement rate for the three compounds with the relevant frames corresponding to forward and when present sideways cracks onset.

NR0 does not show sideways cracks at any displacement rates. J_f decreases as displacement rate increases and the frames at forward crack onset suggest that also the deformation at the crack tip decreases. This rate dependence was not expected because the relaxation time of rubber is definitely below the times involved during the tests. This dependence could be correlated to rate dependence of crystallization kinetic and/or crystallization degree at the crack tip.

NR50 and NR75 at 5 and 50 mm/min show a similar behaviour:

- sideways cracks are present
- the crack opening and hence local deformations at sideways cracks onset are rate independent
- J_s is rate independent
- both the sideways cracks length and crack opening at forward crack onset seem to be rate-independent
- J_f slightly increases as displacement rate increases

At 500 mm/min the two filled compounds show a very different behaviour. NR50 shows sideways cracks and their depth at forward crack onset is larger than for lower displacement rates. Therefore a higher crack opening and hence higher overall deformations are observed at forward crack onset. As a consequence J_f strongly increases. NR75 at 500 mm/min does not display sideways cracks. The anisotropy at the crack tip is not sufficient so that sideways cracks onset before forward crack propagation. Forward crack initiates at a local deformation similar to that reached at sideways cracks onset at lower displacement rates. Also the corresponding J_f has a value similar to J_s at 5 and 50 mm/min.

3.1.1 Loading-unloading tests

Figure 6 shows the stored, u_{st}, and dissipated, u_{diss}, energies as a function of the rubber deformation, ε_c, which is calculated from the overall deformation, ε, considering that CB particles are undeformable:

$$\varepsilon_c = \frac{\varepsilon}{1 - \phi^{\frac{1}{3}}}$$

where ϕ is the CB volume fraction.

NR0 does not dissipate energy for strains below approximately 3.25 at which strain induced crystallization takes place (Boggio 2010). For the filled compounds for all the considered displacement rates, at a given ε_c, stored energy is the same irrespective of CB content; while dissipated energy increases as CB content increases.

From these results, knowing from the fracture tests J_f and the overall deformation, ε_f, of the fracture specimen at forward crack onset, it is possible to separate the stored and the dissipative components of the fracture toughness (J_{f-st} and J_{f-diss} respectively) considering that:

$$\begin{cases} J_f = J_{f-st} + J_{f-diss} \\ \left. \dfrac{u_{f-st}}{u_{f-diss}} \right|_{\varepsilon_f} = \dfrac{J_{f-st}}{J_{f-diss}} \end{cases}$$

Results are shown in figure 7.

For NR0, being ε_f lower than strain induced crystallization onset, as expected no energy dissipation is observed. The rate dependence of the stored energy is probably correlated to local phenomena occurring at the crack tip.

For both filled compounds at 5 and 50 mm/min the stored components of fracture toughness is rate independent and therefore the rate dependence of J_f is related to the rate dependence of the energy dissipative mechanisms in the overall specimen. These are due

366

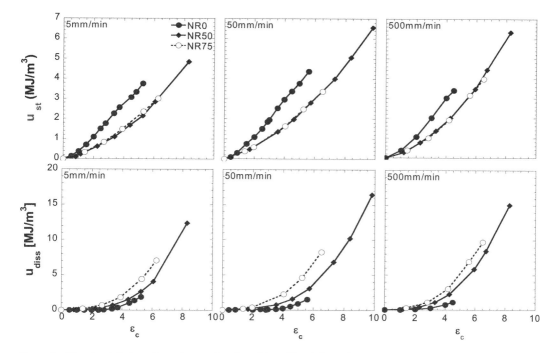

Figure 6. Stored and dissipated energies as a function of the rubber deformation ε_c for all the compounds of Table 1.

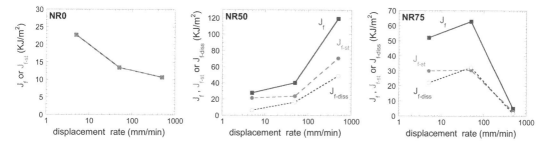

Figure 7. Stored and dissipative components of fracture toughness at forward crack onset for all the compounds of Table 1.

to non-catastrophic events previously described which take place at rubber-filler interface and are related to the presence of the bound rubber around the CB aggregates whose mobility is rate dependent.

For NR50 the large increase in J_f at 500 mm/min is due to an increase in both energy components. This is linked to the fact that sideways cracks propagate up to higher length at this displacement rate and therefore large overall displacements can take place. The reason why sideways cracks propagate further at high rates has still to be investigated and must take into account local phenomena at the crack tip.

For NR75 both stored and dissipated energy strongly decrease since a forward crack occurs at small overall displacement.

4   CONCLUDING REMARKS

The fracture toughness separation method, adopted in this work gives information about the stored and dissipated energy in the whole specimen during the deformation process but does not provide an insight into the local mechanisms occurring at the crack tip. Nevertheless it has shown that in the filled compounds, the rate dependence of J_f between 5 and 50 mm/min, at which fracture phenomenology at the crack tip is practically identical, is linked to the rate dependence of the dissipative component of toughness. The rubber around the CB aggregates (bound rubber) displays a gradient of relaxation times and therefore, since the non-catastrophic events at the rubber particle interface involve the bound rubber, they may be expected to be rate dependent.

The rate dependent restricted mobility of the bound rubber may also influence the molecular orientability and hence the anisotropy at the crack tip. For NR75 at 500 mm/min anisotropy is not enough for sideways cracks initiation.

Further, the fact that for NR50 at the same rate not only sideways cracks are present but also propagate further than at the lower rates may also be related to material anisotropy at the crack tip. To better understand this issue work aimed at measuring local deformations in the process zone at the crack tip is in progress.

ACKNOWLEDGEMENT

M. Boggio gratefully acknowledges for support from Bridgestone Technical Center Roma Italia. The authors thank Bridgestone Technical staff for the helpful discussion.

REFERENCES

Ait Hocine, N. & Nait Abdelaziz, M. 2003. A new alternative method to evaluate the J-integral in the case of elastomers. *Intl. J. Fracture* 124: 79–92.

ASTM D 3053:2008 Standard Terminology Relating to Carbon Black.

ASTM D 2084:2007 Standard Test Method for Rubber Property Vulcanization Using Oscillating Disk Cure Meter.

ISO 1183 1:2004(E) Plastics Methods for determining the density of non cellular plastics Part 1: Immersion method, liquid pyknometer method and titration method .

Boggio, M. Ph.D. Thesis: Fracture behaviour of carbon black filled natural rubber 2010.

Dargazany, R. & Itskov, M. 2009. A network evolution model for the anisotropic Mullins effect in carbon black filled rubbers. *International Journal of Solids and Structures* 46(16): 2967–2977.

Diani, J. & Fayolle, B. & Gilormini, P. 2009. A review on the Mullins effect. *European Polymer Journal* 45(3): 601–612.

Fukahori, Y. 2005. New Progress in the Theory and Model of Carbon Black Reinforcement of Elastomers. *Journal of applied polymer science*. 95(1): 60–67.

Gent, A.N. Razzaghi-Kashani, M. & Hamed G.R. 2003. Why Do Cracks Turn Sideways?. *Rubber chemistry and technology* 76(1) 122–132.

Hamed, G.R. 1994. Molecular Aspects of the Fatigue and Fracture of Rubber. *Rubber Chemical Technology* 67(3): 529–536.

Hamed, G.R. & Park, B.H. 1999. The Mechanism of Carbon Black Reinforcement of SBR and NR Vulcanizates. *Rubber chemistry and technology* 72(5): 946–960.

Kim, B.H. & Joe, C.R. 1989. Single specimen test method for determining fracture energy (Jc) of highly deformable materials. *Eng. Fracture Mechanics* 32(1): 155–161.

Lake, G.J. & Lindley, P.B. 1965. The mechanical fatigue limit for rubber. *Journal of Applied Polymer Science* 9(4): 1233–1251 .

Lake G.J. & Thomas AG. Strength. In Gent A.N. editor. Engineering with rubber, How to design rubber components Munich: Carl Hanser Verlag, 2nd edition 2001.

Leblanc, J.L. 2002. Rubber–filler interactions and rheological properties in filled compounds. *Progress in Polymer Science* 27(4): 627–687.

Medalia, A.I. 1987. Effect of Carbon Black on Ultimate Properties of Rubber Vulcanizates. *Rubber chemistry and technology* 60(1): 45–62

Poompradub, S. Tosaka, M. & Kohjiya, S. 2005. Mechanism of strain-induced crystallization in filled and unfilled natural rubber vulcanizates. *Journal of Applied Physics* 97(10): 103529–103538.

Raoult, J. Marchal, J. Judeinstein, P. & Albouy, PA. 2006. Stress-induced crystallization and reinforcement in filled natural rubber: 2H NMR study. *Macromolecules* 39: 8356–8368 .

Trabelsi, S. Albouy, P.A. & Rault, J. 2003. Effective local deformation in stretched filled rubber. *Macromolecules* 36: 9093–9099.

Trabelsi, S. Albouy, P.A. & Rault, J. 2004. Strain-induced crystallization properties of natural and synthetic cis-polyisoprene. *Rubber Chemical Technology* 77: 303–316.

Webber, R.E. Creton, C. Brown, H.R. & Gong, J.P. 2007. Large Strain Hysteresis and Mullins effect of tough Double-Network Hydrogels. *Macromolecules*. 40: 2919–2927.

Constitutive Models for Rubber VII – Jerrams & Murphy (eds)
© 2012 Taylor & Francis Group, London, ISBN 978-0-415-68389-0

Investigations regarding environmental effects on fatigue life of natural rubber

J. Spreckels & U. Weltin
Institute of Reliability Engineering, Hamburg University of Technology, Hamburg, Germany

M. Flamm & T. Steinweger
Beratende Ingenieure Flamm, Buchholz, Germany

T. Brüger
Vibracoustic GmbH & Co. KG, Hamburg, Germany

ABSTRACT: This paper presents investigations with specimens of sulphur-linked natural rubber concerning effects of ageing, preload and ozone on mechanical fatigue life. In service, rubber parts often are exposed to these influences, therefore fatigue life may be affected. Especially for natural rubber, this aspect is very important because this material is very susceptible to ageing and ozone effects. Thus, impacts of thermal ageing and ozone in interaction with mechanical loads have been investigated to broaden knowledge regarding their effects and interdependencies. Well-known approaches can be confirmed and extended. Additionally, new aspects have been found.

1 INTRODUCTION

Fatigue life of many elastomers is influenced not only by mechanical loads but also by environmental influences, especially oxygen, ozone and temperature causing ageing effects with changes in the material and inducing ozone cracking. In service, these influences often appear at the same time and some have interdependencies to each other.

Tests used to predict lifetime or to secure a minimum of service lifetime often try to accelerate ageing effects artificially e.g. with an elevation of temperature or an exposure to a test atmosphere containing an elevated ozone level (e.g. Brown 1996). With these tests, ageing-affected changes in properties or damages shall be induced in a rather short time to judge service suitability of the material or the part with only a low expenditure of time.

In this paper, effects of the mentioned influences have been investigated in accelerated form. The focus is laid especially on the influences' effects on mechanical fatigue properties.

2 BASIC EFFECTS

2.1 Short-time thermal effects

As known, mechanical properties of elastomers strongly depend on temperature. Often, this is shown in changes of the stiffness and damping of the material, which particularly is of special interest when temperature is near the glass transition region.

Regarding mechanical fatigue, most literary sources report of a decrease in fatigue life when temperature is increased e.g. from room temperature (e.g. Cadwell et al. 1940, Härtel 1985, Platt 1988, Charrier et al. 2009). Lake and Lindley showed in fracture mechanical investigations a slight decrease of fatigue life with increasing temperature for NR material but also a large decrease of fatigue life for SBR material (Lake & Lindley 1964). Young observes also using fracture mechanics that fatigue crack growth is slowest at 25°C and at temperatures of up to 75°C fatigue crack growth increases which results in a decrease in fatigue life. Even at 0°C, fatigue life is lower than at 25°C (Young 1986). Flamm reports about tests on natural rubber specimens with pulsating loads (Flamm 2003). The author observed a slight decrease of fatigue life when temperature is increased from about 30°C to about 60°C and an increase in fatigue life at temperatures above 60°C to 84°C. Especially the increase seems surprising at first sight. It is assumed that it results due to a superposition of mechanical fatigue with relaxation and creep effects respectively, which are accelerated by the higher temperature.

2.2 Long-time thermal ageing

Ageing of elastomers can be described as a proceeding change in properties of the material. In this present section, anaerobic and aerobic ageing are viewed at.

Aerobic ageing occurs by a reaction of the elastomer with oxygen. The oxidation is a series of chemical reactions resulting in hardening and embrittlement by e.g. additional crosslinks or a softening due to degradation. Both, hardening and softening take place simultaneously (e.g. Ahagon 1990, Ngolemasango et al. 2008). For NR it is reported that chain scissing softening processes dominate at high temperatures (Dolezel 1978, Stevenson 2001). Aerobic ageing needs the presence of oxygen, so effects like diffusion limited oxygen may occur especially at high temperatures (e.g. Wise et al. 1997, Celina & Gillen 2009, Le Saux et al. 2009). At these temperatures, the reaction of the oxygen is accelerated to such an extent, that the atmospheric oxygen chemically reacts with the elastomer before it diffuses into the bulk of the material, so the oxidation is located near the surface of the elastomer and the bulk is far less oxidated. This leads to a heterogeneous profile of the material. Steinke et al. present an approach for a calculational estimation of this effect (Steinke et al. 2011a).

Anaerobic ageing occurs without the influence of oxygen. It includes processes like post-curing (increase of sulphur-linkages using residual sulphur) and transformation of polysulfide linkages to di- and monosulfide linkages (which releases sulphur possibly useable for additional linkages). Also intramolecular linkages can be formed. Assuming a homogenous temperature distribution, anaerobic ageing can be expected to appear homogenous as well.

Fatigue life often is reported to be decreased by ageing (e.g. South 2003, Woo et al. 2009).

Both, anaerobic and aerobic ageing are accelerated by elevated temperatures which in many cases is used for short-time tests of long-time behavior. Often, time-temperature superposition (TTS) is used to predict ageing behavior (e.g. Gillen et al. 1996). Based e.g. on the Arrhenius-equation, ageing behavior at rather low temperatures is predicted using accelerated test data at elevated temperatures which then is calculated down to lower temperatures (see equation 1).

$$a_t = e^{-\frac{E_A}{R}\left[\frac{1}{T_1} - \frac{1}{T_2}\right]} \tag{1}$$

In equation (1), a_t is the acceleration factor (ratio of the two times to reach the same change in property), E_A is the activation energy (in kJ/mol), R is the universal gas constant and T_1 and T_2 are absolute temperatures (in K). Often, authors mention the limits of this approach especially for application on ageing because ageing involves a multitude of different chemical reactions. So, the assumption of one constant activation energy for complete ageing does not always lead to accurate results. Especially, results of excessively accelerated ageing tests may lead to large uncertainties for lifetime prediction (Celina & Gillen 2009).

Celina shows the temperature-dependency of the activation energy for PU, EPDM, Butyl rubber and PP and attributes this to two different ageing processes occurring at the same time with two different activation energies each. Considering this, the prediction with the time-temperature superposition can be improved considerably (Celina et al. 2005). Contrary to high-grade unsaturated main chains of natural rubber, these four materials have main chains which are completely or widely saturated. Thus, investigations with these materials may lead to different results compared to NR.

2.3 Ageing in combination with static load

As is known, relaxation and creep phenomena occur when the elastomer undergoes a static load. Relaxation describes the decrease of stresses when the part or specimen is loaded statically with a constant strain and creep (or retardation) refers to the increase in displacement due to a constant load (e.g. constant force). Both processes are accelerated by elevated temperatures which can be used for accelerated tests (Flamm et al. 2005).

Superposition of relaxation and ageing additionally lead to permanent changes of the geometric shape of the elastomer part as similar described by Achenbach for seals (e.g. Achenbach 2000). This results because an additional new network is formed by ageing reactions in the loaded state which therefore is the non-loaded state for the new network. So, the original and the generated network have two different equilibrium positions each. This leads to a new equilibrium position which the specimen takes up after unloading. The position is located between the virgin position and the loaded position, thus the shape undergoes the permanent geometric changes (Achenbach 2000).

2.4 Ozone effects

Ozone is a molecule consisting of three oxygen atoms (O_3). As a trace gas, ozone appears only in very low concentrations of about a few pphm in the ground-level atmosphere compared to the main elements like nitrogen and molecular oxygen (O_2). But even these low concentrations of ozone may damage elastomers.

As is known, ozone reacts with the carbon-carbon double bonds of the main chain and causes ozone cracks when the elastomer is strained above a certain critical strain level (e.g. Braden & Gent 1960b). There is nearly no effect on elastomers with saturated main chains like EPM.

An atmosphere with an increased ozone level during a fatigue life test causes a decrease in lifetime because ozone cracks may act initiating for fatigue crack growth (e.g. Dolezel 1978, Spreckels et al. 2009).

Ozone crack growth rate of statically loaded rubber is reported to be proportional to the ozone concentration (e.g. Braden & Gent 1960a). This approach can be extended for dynamic crack growth rate regarding frequency and time above critical strain and below (Ellul 2001). The time to first visibility of ozone cracks is (for natural rubber) in accordance to the previor correlation reversely proportional to ozone concentration (e.g. Zuev et al. 1962).

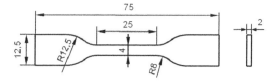

Figure 1. Drawing of tensile bar specimen type "S2" according to DIN 53504.

Figure 2. Rotation-symmetric waisted specimen.

3 EXPERIMENTS

3.1 *Materials and specimens*

The experiments predominantly have been performed with flat tensile specimens of a standardised geometry (figure 1) called "S2" in standard DIN 53504. Additionally, a few rotation-symmetric waisted specimens (figure 2) have been used to investigate compressional preload during ageing, chapter 3.2.3.

The specimens have been made of a natural rubber mixture which is filled with carbon black and equipped with an anti-ageing protection of wax and chemically acting ingredients. The network is sulphur-linked.

3.2 *Effects of thermal ageing*

3.2.1 *Fracture strain of artificially aged specimen*
Often, fracture strain or elongation at break of artificially aged specimens are used to investigate the applicability of time-temperature superposition and to establish activation energies (e.g. Mott & Roland 2001, Ngolemasango 2008). Thus, specimens are exposed to elevated temperatures for certain ageing times, after which fracture strain is determined in a tensional test.

Figure 3 shows results of such a test series with tensile bar specimens. Temperatures of 60°C up to 120°C have been used at ageing times of up to about 100 days. For each ageing condition at least three specimens have been used.

In spite of a certain mean variation the results clearly show the acceleration of the fracture strain development. On average, an activation energy of 95.5 kJ/mol is calculated using acceleration values at a relative fracture strain of 0.7, 0.6, 0.5 and 0.4. This value is in accordance with typical values for the activation energy for natural rubber ageing (Mott & Roland 2001). The dashed lines in figure 3 have been calculated from the curves at a temperature of 10°C higher,

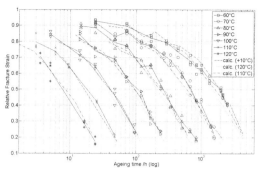

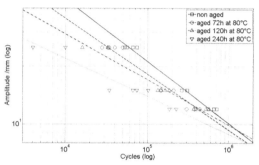

Figure 3. Developing of fracture strain (relative) at different exposition temperatures (solid) and calculated curves with identified activation energy (dashed, dotted and dot-dashed).

Figure 4. S/N curves after artificial ageing.

each using equation (1) and the identified activation energy. The dotted lines have been calculated from values at 120°C to all other temperatures and the dot-dashed lines have been calculated from values at 110°C respectively. It can be seen that the calculations fit the results reasonably well. Additional investigations will show if this can be extended to lower temperatures like 50°C.

3.2.2 *S/N curves of artificially aged specimens*
Fatigue life results of artificially aged tensile bars (all at a temperature of 80°C) have been arranged in S/N curves which are depicted in figure 4. Three different load amplitudes of 25 mm, 15 mm and 12 mm have been used with a load ratio of R = 0 (pulsating load). The tests were run with displacement control.

The results show that fatigue life decreases with increasing ageing exposure time. In spite of the mean variation, the figure clarifies that the mathematical slope of the S/N curve changes with increasing ageing. In this case it becomes flatter. The "turning region" is in the region of more than 1,000,000 cycles, so at the high amplitude, the decrease of fatigue life is very distinct, while at lower amplitude, the influence of ageing on fatigue life is lower but still present.

The extreme case of raising the amplitude would be an increase up to the fracture strain. Fracture strain value is reported to be underneath the S/N curve expanded to one loadcycle (Flamm et al. 2011). Results like in figure 3 show a huge influence of ageing

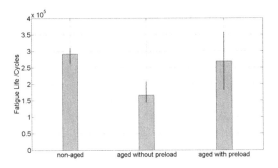

Figure 5. Results of fatigue test of tensile bar after artificial ageing with preload.

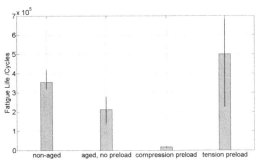

Figure 6. Results of fatigue test of rotation-symmetric waisted specimens after artificial ageing with and without preload.

on fracture strain as it can be expected from expanding the S/N curve.

3.2.3 Effect of pre-load during artificially ageing

When the specimen is exposed to a static mechanical load during the artificial ageing, the ageing process is superposed by a relaxation and retardation process respectively as described above.

To investigate effects of this on fatigue life tensile bars have been artificially aged at 80°C for 72 hours. During ageing, the specimens have been elongated to a displacement of 20 mm in tension. The changed equilibrium point of the geometric shape has been used as the new starting point for the loading that follows in the fatigue test. The tension bars have been exposed to a displacement-controlled mechanical load of 15 mm amplitude and 15 mm mean load with a frequency of 1 Hz at room temperature till fatigue rupture. Results are illustrated in figure 5.

The left bar shows the lifetimes of non-aged specimens. Lifetime of the middle bar (aged without preload) is decreased compared to the left bar which is the effect from ageing without static loading like it can be also seen in figure 4. Lifetime values of the right bar (aged with tensional preload) however, are increased compared to the middle bar. So, the lifetime is significantly influenced by the preload applied during artificial ageing. In this case it leads to an increase in fatigue life.

To investigate effects also of compression loads, a similar test series has been performed using the rotation-symmetric waisted specimens (depicted in figure 2) because the tensile bars start buckling when they are loaded with compression. The specimens were artificially aged for one week at 80°C and simultaneously preloaded 7 mm in compression and 7 mm in tension respectively. The compressed specimens showed after ageing a compression set of in average 4.5 mm and the tensioned specimens showed a tension set of in average 3.2 mm. The following displacement-controlled fatigue test used a load amplitude of 9 mm and a mean load of 9 mm with a frequency of 2 Hz. The new geometric equilibrium position has been used as starting point for the loads. Results are depicted in figure 6.

In the very left bar the fatigue life of non-aged specimens is depicted, the middle left bar shows fatigue life of specimens aged without preload. A decrease in fatigue life is visible. Next bar (middle right) shows lifetime after ageing under compression load and the very right bar shows fatigue life of artificially aged specimens under tensional load. The results clarify that the compressional preload decreases lifetime significantly compared to ageing without preload. The tensional preload during ageing increases lifetime which is in accordance to the tests with tensile bars reported in above. In this case, fatigue life after ageing with tensional preload is even in the same region as without ageing at all, some specimens even have a higher lifetime than the specimens without ageing.

The preload during ageing causes several affects which are assumed to influence fatigue life in the observed way. The geometric shape is changed permanently which results in an increased (compression) and decreased (tension) cross section area. Additional residual stresses are caused by the second network, Steinke reports of induction of tension stresses in an aged specimen (under compression) after unloading (Steinke et al. 2011b). So, the stress situation is changed (additionally possibly the R-ratio) and thus the forces needed for the same displacement. To investigate this aspect, further experiments are in the moment on the test rig which run similar tests but with the use of force-control instead of displacement-control.

3.3 Ozone tests

Figure 7 shows results of a test series of ozone attack on tensile bar specimens that are loaded with a preload of 20 mm and an amplitude of 4 mm with a rather low frequency of 1/3600 Hz. Results of a test series ran with thermally aged specimens for 3 days at 80°C are also included in the figure. Two specimens have been used for each test concentration. The y-axis marks the time to complete rupture of the specimens. The lines in the figure are the application of the mathematical correlation between time and ozone concentration mentioned in chapter 2.

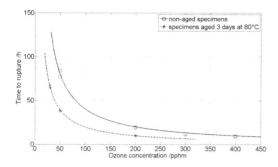

Figure 7. Time to rupture under different ozone concentrations (dynamic mechanical load).

The figure shows that the mathematical correlation of ozone concentration and time to first ozone cracks and ozone crack grow rate are both valid even when time to rupture is used as the damage criterion. The mean variation of the results is very low. The correlation is also valid for the thermally aged specimens. Ozone cracking and ozone induced rupture occur earlier in thermally aged specimens than in non-aged specimens.

The reduced ozone durability caused by thermal ageing seems to be an interesting and important interaction which suggests an additional decrease in fatigue life of thermal aged specimens in an ozone atmosphere. Therefore, this is currently under investigation.

4 CONCLUSIONS

This paper presents several investigations of ageing effects on fatigue life of natural rubber elastomers which on the one hand confirm and expand well-known approaches and on the other hand show some new aspects.

Results of fracture strain after artificial ageing clarify that the time-temperature superposition can be applied very well for at least the investigated temperature regions. An activation energy is identified to use the time-temperature shifting which seems to fit reasonably well to the results.

Artificial ageing leads to a change of fatigue life. Fatigue life at high loads is decreased most while fatigue life at lower amplitudes is decreased less which causes a significant change in the slope of the S/N-curve. "Turning region" of S/N curve is in high-cycle region.

Preload during ageing causes compression or tension sets as well as significant changes in fatigue life especially with displacement-control. The effect can be explained with the changes in the polymer network due to chain scission and cross linking in the loaded state. Decrease and increase of fatigue life are both possible. Because of the importance especially for rubber parts with preloads in service, this should be considered when predicting lifetime.

Ozone tests clarify, that the mathematical relation of time to ozone cracking and ozone concentration

(inversely proportional) is valid for the investigated concentrations even if time till rupture is used as a criterion. The approach is also valid for thermally aged specimens, whose ozone durability is significantly reduced.

REFERENCES

Achenbach, M. 2000: Service life of seals – numerical simulation in sealing technology enhances prognoses. Computational Materials Science 19, pp. 213–222

Ahagon, A., Kida, M. & Kaidou, H. 1990: Aging of tire parts during service. I. Types of aging in heavy-duty tires. Rubber Chemistry and Technology 63, pp. 683–697

Braden, M. & Gent, A. 1960a: The Attack of Ozone on Streched Rubber Vulcanizates. I. The rate of crack growth. In: Journal of Applied Polymer Science III, 7, pp. 90–99

Braden, M. & Gent, A. 1960b: The Attack of Ozone on Streched Rubber Vulcanizates. II. Conditions for Cut Growth. In: Journal of Applied Polymer Science III (1960), 7, pp. 100–106

Brown, R. 2006: Physical Testing of Rubber. Springer Science + Business Media

Cadwell, S. M., Merril, R. A., Sloman, C. M. & Yost, F. L. 1940: Dynamic fatigue life of rubber. In: Industry and Engineering Chemistry 12, pp. 19–23

Celina, M., Gillen, K. & Assink, R. 2005: Accelerated aging and lifetime prediction: Review of non-Arrhenius behaviour due to two competing processes. In: Polymer Degradation and Stability 90, pp. 395–404

Celina, M. & Gillen, K. T. 2009: Advances in Exploring Mechanistic Variations in Thermal Aging of Polymers. In: Martin, J. W., Rynth, R. A., Chon, J. & Dickie, R. A. (ed.): Service Life Prediction of Polymeric Materials. Springer Science + Business Media, pp. 45–56

Charrier, P., Ramade, T., Taveau, D., Marco, Y. & Calloch, S. 2009: Influence of temperature on durability of carbon black filled natural rubber. In: Heinrich, G., Kaliske, M., Lion, A. & Reese, S. (ed.): Constitutive Models for Rubber VI, pp. 179–185

Doležel, B. 1978: Die Beständigkeit von Kunststoffen und Gummi. Hanser

Ellul, M. D. 2001: Chapt. 6: Mechanical Fatigue. In: Gent, A. N. (ed.): Engineering with Rubber. 2nd Edition. Hanser, pp. 137–176

Flamm, M. 2003. Ein Beitrag zur Betriebsfestigkeitsvorhersage mehraxial belasteter Elastomerbauteile VDI Düsseldorf

Flamm, M., Groß, E., Steinweger, T. & Weltin, U. 2005: Creep prognosis for elastomeric parts through accelerated tests. In: Austrell, P.-E. & Kari, L. (ed.): Constitutive Models for Rubber IV, pp. 445–450

Flamm, M., Spreckels, J., Steinweger, T. & Weltin, U. 2011: Effects of Very High Loads on Fatigue Life of NR Elastomer Materials. In: International Journal of Fatigue 33, pp. 1189–1198

Gillen, K. T., Clough, R. L. & Wise, J. 1996: Chapt.: Prediction of Elastomer Lifetimes from Accelerated Thermal-Aging experiments. In: Clough, R. L., Billingham, N. C. & Gillen, K. T. (ed.): Polymer Durability, American Chemical Society, Washington DC, 1996

Härtel, V., Hofmann, M. & Schreiber, F. 1985: Lebensdauerprüfungen an Gummiprüfkörpern und ihre Korrelation zu Fertigbauteilen. In: DVM (ed.): Erprobung von Gummiteilen und Gummi-Metallverbindungen unter dem

Gesichtspunkt der Betriebsfestigkeit. Vorträge der 11. Sitzung des Arbeitskreises Betriebsfestigkeit

Lake, G. J. & Lindley, P. B. 1964: Cut Growth and Fatigue of Rubbers. II. Experiments on a Noncrystallizing Rubber. In: Journal of Applied Polymer Science 8, pp. 707–721

Le Saux, V., Marco, Y., Calloch, S., Le Gav, P. Y. & Ait Hocine, N. 2009: Accelerated ageing of polychloroprene for marine applications. In: Heinrich, G., Kaliske, M., Lion, A. & Reese, S. (ed.): Constitutive Models for Rubber VI, pp. 3–8

Mott, P. H. & Roland, C. M. 2001: Aging of natural rubber in air and seawater. In: Rubber Chemistry and Technology 74, pp. 79–88

Ngolemasango, F. E., Bennett, M. & Clarke, J. 2008: Degradation and Life Prediction of a Natural Rubber Engine Mount Compound. In: Journal of Applied Polymer Science 110, S. 348–355

Platt, W. 1988: Betriebssicherheit von elastomerbestückten Wellenkupplungen unter besonderer Berücksichtigung der Einsatztemperatur, RWTH Aachen, Dissertation

South, J. T., Case, S. W. & Reifsnider, K. L. 2003: Effects of Thermal Aging on the Mechanical Properties of Natural Rubber. In: Rubber Chemistry and Technology 76, 4, pp. 785–802

Spreckels, J., Flamm, M., Steinweger, T. & Weltin, U. 2009: Consideration of Environmental Influences on Fatigue Tests of Elastomer Components. In: Heinrich, G.,

Kaliske, M., Lion, A. & Reese, S. (ed.): Constitutive Models for Rubber VI, pp. 9–14

Steinke, L., Spreckels, J., Flamm, M. & Celina, M. 2011a: Model for heterogenous ageing of rubber products. In: Plastics, Rubber and Composites 40, pp. 175–179

Steinke, L., Weltin, U., Flamm, M., Seufert, B. & Schmid, A. 2011b: Numerische Analyse des thermooxidativen Alterungsverhaltens von Elastomerbauteilen, DVM-Arbeitskreis Elastomerbauteile, 1. Tagung.

Stevenson, A. & Campain, R. 2001: Chapt. 7: Durability. In: Gent, A. N. (ed.): Engineering with Rubber. 2nd Edition. Hanser, pp. 177–221

Wise, J., Gillen, K. & Clough, R. 1997: Quantitative model for the time development of diffusion-limited oxidation profiles. Polymer 38, pp. 1929–1944

Woo, C. S., Kim., W. D., Choi, B. I., Park, H. S., Lee, S. H. & Cho, H. C. 2009: Fatigue life prediction of aged natural rubber material. In: Heinrich, G., Kaliske, M., Lion, A. & Reese, S. (ed.): Constitutive Models for Rubber VI, pp. 15–18

Young, D. G. 1986: Fatigue Crack Propagation in Elastomer Compounds: Effects of Strain Rate, Temperature, Strain Level and Oxidation. In: Rubber Chemistry and Technology 59, pp. 809–825

Zuev, Y. S. & Pravednikova, S. I. 1962: Influence of Concentration of Ozone upon Cracking of Vulcanized Rubber. In: Rubber Chemistry and Technology 35, pp. 411–420

Constitutive Models for Rubber VII – Jerrams & Murphy (eds)
© 2012 Taylor & Francis Group, London, ISBN 978-0-415-68389-0

Potentials of FEA-simulation for elastomer stress softening in engineering practice

J. Präffcke & F. Abraham

Department of Mechanical and Design Engineering, University of Portsmouth, Portsmouth, UK

ABSTRACT: Since L. Mullins in 1969 first reviewed the stress softening effect in rubber materials, it has been subject of continuous research within the field of elastomer materials science. Depending on the material properties, this stress softening, commonly referred to as the *Mullins effect*, can drastically affect the mechanical behaviour. Due to its complex nature, the Mullins effect is still considered as a challenge for modelling and simulation. Several models have been introduced for its prediction, most of which are based on a phenomenological approach and particular assumptions and simplifications. During this project the potentials of the commercial finite elements suite Abaqus for the simulation of the Mullins effect were investigated. This was not only undertaken with contemplation of the implemented material model itself, but instead with particular regard to its practical applicability. Practice relevant techniques for the modelling of rubber components with stress softening behaviour were introduced. Several examples of relevant use cases were indicated, corresponding simulations were carried out, and the results were evaluated. It was found that the stress softening effect can be predicted with a sufficient accuracy for a broad range of engineering applications. It was further shown that, at least for simple applications, this is even possible with a limited amount of calibration test data. However, the investigations also showed that in some cases extensive test data preparation and modelling effort might be necessary. Furthermore, the simulation results were found to follow several simplifications that contrast with the actual material behaviour.

1 INTRODUCTION

Many rubbers and rubber-like materials show a significant decrease in their stiffness upon repeated loading and unloading. This phenomenon was first characterised by Mullins (1969), and is commonly referred to as *Mullins effect* in literature. It has been studied and investigated since, but due to its complexity the Mullins effect is still considered as a challenge within modelling and simulation, as well as in its physical understanding (Diani, Fayolle, & Gilormini 2009).

The aim of this project is to revise and to evaluate the simulation capabilities of the finite elements suite Abaqus for the simulation of the Mullins effect. In doing so, a particular focus is set on the applicability within engineering practice.

2 BACKGROUND

2.1 *Abaqus Mullins effect in practice*

As part of his investigation about the accuracy of elastomer simulations, Bergström (2005) tested the Ogden & Roxburgh based Mullins effect model in Abaqus, as well as an own implementation of the Qi & Boyce model. He found, that the relative error of simulation results versus test data can be improved from around 25% to around 7% by incorporating either one of the two Mullins effect models, instead of using a perfectly hyper-elastic model.

Paige & Mars (2004) also reviewed the Ogden & Roxburgh model implementation in Abaqus. However, they focused on a rather practical approach. By means of the example of a pre-loaded rubber mount, they demonstrated how the Mullins effect can affect the overall response of a rubber component assembly in manners that can be difficult to anticipate.

2.2 *Purpose of this project*

The previous section gives some examples of effort that has been undertaken in order to evaluate the capabilities and limits of Abaqus for the stress softening behaviour simulation in hyper-elastic materials. Several authors have tested the implemented Ogden & Roxburgh model and reviewed the simulation results in comparison to actual material test data. However, these investigations have either been rather theoretical and have covered the assessment of the material model in general (e.g. Bergström 2005), or have focused on one very specific application (e.g. Paige & Mars 2004).

Nevertheless, as Paige & Mars (2004) point out, the stress softening effect is necessary to consider, because its consequences can be rather difficult to predict in certain applications. Work that reflects and reviews the use of the Abaqus Mullins effect implementation in actual engineering applications, however, is rare.

Now the question arises, to what extent the Mullins effect simulation in Abaqus is applicable for general engineering applications and what quality of results can be expected.

This is why this project is undertaken. It covers an investigation about the potentials of the commercial finite elements suite Abaqus 6.7[1] for the simulation of the Mullins effect in hyper-elastic materials. This investigation is not only performed with contemplation of the material model itself, but instead with particular regard to its practical applicability.

3 MODELLING

3.1 *Hyper-elastic material model*

The investigations are carried out on a carbon black filled ethylene propylene diene Monomer (cb filled EPDM) with 110 phr low active carbon black and 70 phr softener. This synthetic elastomer with rubber-like material properties is used as an example material, for which the following test data sets are available:

- tensile test data of a cyclic test, covering five cycles up to a constant stretch and beginning with a virgin material sample (source: Abraham 2002)
- tensile test data hysteresis loops up to five different load amplitudes from a sample that has been preconditioned with approx. 50 loading cycles for each amplitude (source: Abraham, Alshuth, & Jerrams 2005)

This test data was obtained from simple material tests which can be conducted with relatively little expense.

Due to a limited amount of test data, the hyper-elastic material model is set up by using the Marlow strain energy form.

3.2 *Stress softening behaviour*

3.2.1 *The Abaqus Mullins effect model*
The Mullins effect model in Abaqus 6.7 is based on the formulation described by Ogden & Roxburgh (1999). In particular, it is an extension of their model that accounts for material compressibility (Dassault Systèmes 2007, p. 17.6.1–4). It is, however, possible to manually implement other models, such as the Qi & Boyce model for example (Bergström 2005), by using the Abaqus user subroutine UMULLINS.

Diani, Fayolle, & Gilormini (2009, p. 605) characterise the Ogden & Roxburgh model as a form that results in

- coincident unloading and reloading responses
- zero damage when stretching to a previously unreached strain level

- damage as function of a damage variable for previously encountered strain levels

In Abaqus, no hysteresis effects are supported in combination with this model (Dassault Systèmes 2007, p. 17.6.1–1) and, in Abaqus 6.7, also no permanent set effect (Dassault Systèmes 2007, p. 17.6.1–4)[2]. Progressive damage due to cyclic unloading and reloading to the same strain level is not supported either (Dassault Systèmes 2007, p. 17.6.1–8).

This Mullins effect implementation is to be understood as an extension to the hyper-elastic material model. It is based on several parameters, which describe the damaged response curve characteristics and can either be entered directly or be evaluated from provided test data. These parameters are rather based on experience and do not have a specific physical interpretation in general (Dassault Systèmes 2007, p. 17.6.1–6). It is notable that Paige & Mars (2004), nevertheless, reviewed the physical meaning of these parameters by investigating several limiting cases.

3.2.2 *Curve fitting*
As permanent set behaviour is not supported in Abaqus 6.7, any unloading and reloading simulation curves necessarily pass through the origin of zero-stress/zero-strain. Furthermore, no hysteresis effects are supported. Consequently, the curves of unloading from a certain strain and reloading to this strain are coincident.

When using unloading-reloading calibration test data that contains permanent set, hysteresis or progressive damage effects, these effects are ignored by the curve fitting algorithm. The hysteresis and permanent set effects are then simply lost (Dassault Systèmes 2007, p. 17.6.1–8). When the test data contains progressive damage effects due to repeated unloading and reloading to the same strain level, a best fitting curve is determined through all these curves (Dassault Systèmes 2007, p. 17.6.1–8).

It seems that Abaqus' curve fitting algorithm tries to fit the unloading/reloading curve in the position of an arithmetic average curve between the test data's hysteresis affected unloading and reloading curves. Consequently, this exact arithmetic average curve can be considered as a benchmark to evaluate curve fit quality in the following.

As a consequence, the primary loading simulation curve never coincides with the reloading curve when calibration test data that contains hysteresis is used; even if little or no softening occurs. to the unloading/reloading curve being considered as arithmetic average between the hysteresis test data curves.

Fig. 1 illustrates how a perfect stress-strain curve to a corresponding curve of real test data would look like, when considering the discussed simplifications and assumptions.

---

[1] In the course of this project, version 6.7 of Abaqus is used. Where necessary, newer versions of this software suite may be referred to. However, unless stated otherwise, all observations and evaluations are based on version 6.7.

---

[2] Permanent set with Mullins effect is, however, supported in Abaqus version 6.8 and newer (Dassault Systèmes 2008).

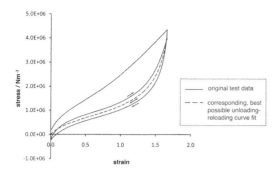

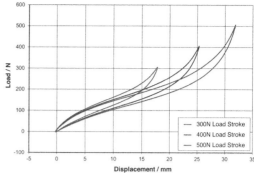

Figure 1. Best possible curve fit for unloading-reloading calibration test data, regarding discussed simplifications and assumptions.

Figure 2. Hysteresis loops test data of pre-loaded material (approx. 50 preceding cycles for each strain level). Source: Abraham et al. (2005).

4 APPLICATIONS IN ENGINEERING

4.1 Homogeneous stress softening

When considering a rubber component that is subjected to constant stress and strain throughout its volume, then the stress softening effect consequently is homogeneous throughout the material as well. This is the simplest possible use case and might only rarely occur in actual engineering applications. However, in many simple applications this approximation of the actual material behaviour might be a satisfactory approach.

4.1.1 Test data usage

Dassault Systèmes (2007, p. 17.6.1–8) recommend to choose the Mullins effect calibration test data depending on the type of loading that the rubber component is prospected to undergo:

- When using several sets of curves that unload from and reload to the same strain level, Abaqus' curve fitting algorithm determines an overall stiffness from all these curves
- When using a single unloading-reloading curve set, Dassault Systèmes (2007, p. 17.6.1–8) suggest to choose the unloading-reloading data of

  - the first cycle, if only monotonic loading with little unloading is prospected
  - a stabilised[3] cycle, if repeated loading and unloading is prospected

In the following example the latter case is considered. Hence, several sets of single test data hysteresis loops, each to a different strain level, are used to calibrate the Mullins effect behaviour. These loops, as shown in Fig. 2, are taken from test data of a pre-loaded material (approx. 50 cycles for each strain level). Permanent set occurrence due to pre-loading was removed by shifting the hysteresis loops along the strain-axis, so that they approximately pass through the origin

of zero-stress/zero-strain. This does not necessarily deteriorate the curve fit quality, as Abaqus 6.7 already ignores any permanent set effects in combination with the Mullins effect (cf. Section 3.2.2).

4.1.2 Curve fitting

When using this prepared test data to calibrate the Mullins effect in Abaqus, a warning concerning the curve fit is output. It states that the Mullins parameter r has been determined to be smaller than $r = 1.0$. It further reads that this is not possible for the Ogden-Roxburgh model, and that r is thus fixed to $r = 1.001$. As r directly controls the shape of the fitted curve, the curve fit is likely to be inaccurate.

To check the curve fit, the calibration test data itself is compared to simulation data up to the same three strain levels. This data is acquired from a simple, single-element model simulation. The comparison confirms the presumption that the fitted curve does not match the actual test data well.

Because the automated fix of the parameter r to 1.001 has obviously caused the curve misfit, a manual correction of this parameter should consequentially help to provide a more accurate result. As the parameter r (just as the other parameters, m and β, as well) does not generally have a direct physical interpretation, this manual correction can mainly be undertaken on the basis of experience or trial and error. After a couple of iterations and coarse simulations, for example, the curve fits the test data significantly better with a manually fixed $r = 2.5$ (Fig. 3).

Because the theoretical, perfect curve fit is considered to be an average curve between the unloading and reloading curve (cf. Section 3.2.2), the simulation error in comparison to the actual material behaviour depends on the amount of hysteresis. For the simulation outcome shown in Fig. 3, the maximum of this error is around 17%. In comparison to the theoretical, perfect curve fit, the simulation outcome shows a maximum relative error of around 6%.

The considered strain levels in Fig. 3 are identically equal to the strain levels which have been provided in the Mullins effect calibration test data. Fig. 4, on the

---

[3] Although the stress softening progresses with every cycle, the response can be considered as nearly stabilised after a couple of cycles (Ciesielsky 2001, p. 126).

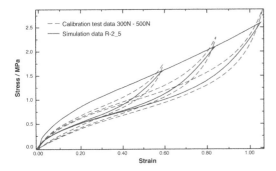

Figure 3. Manual correction of Mullins parameter r to provide a better curve fit.

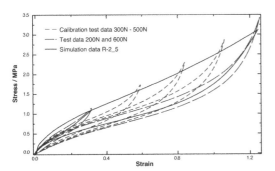

Figure 4. Simulation with strains above and below the provided Mullins effect calibration test data strain range.

other hand, shows the use of the same Mullins effect calibration (with manually fixed $r = 2.5$ as well) for the simulation of strains which are below and above the provided calibration test data strain range.

The simulation curve for the lower strain level fits the test data as well as the curves in Fig. 3 do. The simulation curve of the higher strain level still shows a good fit as well. However, the fit is obviously not as accurately placed as for the other curves. This indicates that in simulations with significantly higher strain levels than provided by the calibration test data, the curve fit might not provide sufficient accuracy. Associated with this issue, Dassault Syst'emes (2007, p. 17.6.1–7) recommend the use of Mullins calibration test data that covers the whole prospected simulation strain range.

4.2 Local stress softening effects

In actual engineering applications, the material is often subjected to significant local stress/strain differences. This inhomogeneous stress/strain distribution throughout the material can, for example, be caused by geometrical complexity as well as concentrated or asymmetric loading.

As the material then softens mainly in local spots which are subjected to high strains, the rest of the material remains stiffer. Taking this into consideration, it is clearly recognisable that local material softening can lead to a more even stress distribution throughout

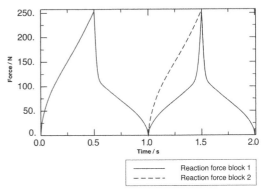

Figure 5. Reaction force curves of both rubber blocks throughout two loading steps.

the component. It can thus be an important aspect to consider when predicting crack growth and fatigue behaviour by means of stress that has been determined by finite element analysis. Accordingly, in order to get accurate simulation results which take the stress softening effect into account, local stress softening effects have to be considered in certain applications.

To investigate whether it is possible to simulate this local softening behaviour in Abaqus, a simple model consisting of two blocks of rubber is used. The blocks are connected by a thin rib, so that they form a single cell with a single section definition within the Abaqus model.

The simulation includes two loading steps. During step 1 only block 1 is stretched, whereas during step 2 both blocks are stretched to the same extension.

Fig. 5 shows the reaction force curves of both rubber blocks throughout the two loading steps.

The different loading curves during the second step indicate a different material response of the two blocks. The significantly softer response of block 1 suggests, that the material has been subjected to local stress softening during the first loading step. A look at the local damage dissipation energy also confirms this. Energy is dissipated due to stress softening damage in block 1 during step 1, whereas a similar amount of energy dissipation occurs in block 2 during step 2.

4.3 Progressive softening due to cyclic loading

When subjected to cyclic loading up to a constant strain level, elastomer materials do not only soften during the first cycle. It rather occurs a progressive damage throughout each cycle. Dassault Systèmes (2007, p. 17.6.1–3) and Ciesielsky (2001, p. 126) describe this behaviour to be stabilising after a few cycles, whereas Sommer & Yeoh (2001, p. 312) characterise the progressive softening to be continuous over many cycles, yet to diminish significantly during the first one.

As mentioned in Section 3.2.1, Abaqus ignores progressive stress softening effects in the provided calibration test data and does not include this effect in the material model.

In the following, approaches to adapt this effect are investigated and their practical applicability is evaluated.

4.3.1 Reduction to overall softening

Dassault Systèmes (2007, p. 17.6.1–8) recommend to cover the total amount of softening which occurs until the (quasi-) stabilised response ensues within the first cycle. To realise this simple approach, they suggest to use the unloading-reloading test data of the (quasi-) stabilised response to calibrate the Mullins effect.

The modelling and calibration for this approach coincide with the techniques for simple softening as described in Section 4.1.

This approach is justified by the fact, that most of the material softening generally occurs during the first loading cycle. Nevertheless, for applications in which a progressing stiffness loss is relevant, it has to be kept in mind that the material response during the directly subsequent cycles is stiffer than depicted by the simulation. To what extent this error affects the result quality clearly depends on the individual material's softening characteristics.

4.3.2 Cycle-wise material re-calibration

This approach is based on the consideration that the material response during each cycle can be considered as the response of an individual, cycle specific material behaviour. The material in a cycle n is thus considered to be the, by means of the Mullins effect, softened material of preceding cycle $n - 1$. By setting up a separate material model for every loading cycle, each cycle can then be simulated separately in Abaqus.

4.3.3 Test data usage

The primary hyper-elastic and the stress softening behaviour are determined separately for each cycle on the basis of its own test data set. The calibration test data for each subsequent cycle $n \geq 2$ thus is

Primary loading:	loading data of test cycle n
Unloading:	unloading data of test cycle n
Reloading:	loading data of test cycle $n + 1$

This approach obviously requires an extensive preparation of the test data, because separate test data sets for each cycle are used.

The displacement simulation results for cycles $n \geq 2$ cannot be considered as absolute displacement values, because of possible permanent set effects. Instead, the total permanent set that results from subsequent cycles needs to be taken into account.

4.3.4 Curve fitting

For the Mullins effect calibration, cyclic test data up to a constant strain level is used.

The curve fit check for each cycle is again carried out on a single-element model. For the first cycle, the Mullins parameter β is manually corrected in order to provide a more accurate curve fit over the whole strain range (cf. Section 4.1.2). For the second and third cycle, the curve fit algorithm provides approximately sufficient curve fit results.

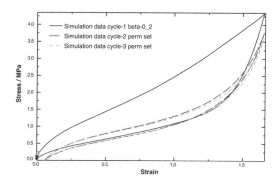

Figure 6. Stress-strain simulation curves of three cycles with progressive stress softening (reconstructed permanent set for cycle 2 and 3 curves).

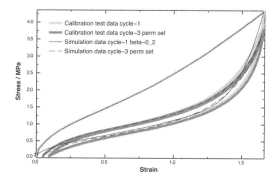

Figure 7. Comparison between progressive stress softening simulation curves and material test data (reconstructed permanent set for cycle 3 curve).

Fig. 6 shows the stress-strain simulation curves of all three cycles, whereas Fig. 7 compares the simulation curves of cycle 1 and 3 to the calibration test data. The curves of cycle 2 and 3 have been shifted along the strain-axis in order to approximately reconstruct the priorly removed permanent set.

The cyclic stress/strain behaviour in Fig. 6 and 7 conforms qualitatively well to the actual material test data. Quantitatively, however, the same considerations about accuracy which have been discussed in Section 4.1.2 apply for each cycle.

The results show that this approach makes it technically possible to simulate a cyclic stress softening behaviour throughout a finite number of cycles. However, the following restrictions apply:

• Because the material model is re-calibrated for each cycle, the maximum strain has to be constant during all cycles and homogeneous throughout the material.
• Furthermore, this maximum strain level has to be equal to the maximum strain level of the calibration test data. This is due to the amount of softening being dependent on the maximum previously encountered strain.

- Because test data is required to calibrate the material for each cycle, only a finite number of cycles can be simulated.

5 CONCLUSION

The investigations showed that it is possible to simulate Mullins stress softening that is homogeneous throughout the material, as well as local softening effects within the same material and even within the same geometric cell and section definition. An assessment of local stresses in a rubber material that has been subjected to prior loading is hence justifiable.

It was further shown that the Mullins effect can be modelled with a limited amount of test data. Such a model was tested in several applications and seems to be able to provide reasonable results, at least when the loading modes (e.g. uni-axial tension) in simulation and calibration test data coincide. It was shown that a good representation of the material behaviour can be achieved by manually correcting the Mullins parameters. It should, however, be taken into account that numerous simplifications might apply, as well during test data preparation as in the constitutive material model itself. It is thus questionable, to what extent the remaining error in such a manually improved curve fit necessarily affects the simulation result quality.

Progressive stress softening damage was found not to be supported by the Abaqus Mullins Effect model. However, an approach was presented to adapt this missing feature. This approach, nevertheless, was proven to be significantly restricted in its versatility. Furthermore, an extensive amount of test data preparation and curve fitting effort were found to be necessary. Due to these aspects, along with the priorly discussed considerations about the general Mullins effect simulation accuracy, a more simplified approach probably brings the better cost-benefit ratio in many practice applications. In this simpler approach, the stress softening solely occurs during the first cycle and covers the complete softening from virgin material to quasi-stabilised response.

It should also be noted that the curve fitting algorithm in Abaqus has been found as quite sensitive regarding test data curve imperfections or unexpected curve shapes. the test data for a smooth and effective curve fit have been introduced as part of this project.

It is further notable that the Mullins effect implementation has been extended since Abaqus version 6.7. These improvements and extensions include

- permanent set with Mullins effect, which is implemented since Abaqus 6.8 (Dassault Systèmes 2008) and makes the *PLASTIC option available for rubber materials with Mullins effect. This model has been reviewed by Bergström (2008).
- a graphical user interface for the Mullins effect calibration in Abaqus/CAE since Abaqus 6.8 (Dassault Systèmes 2008).
- finite-strain viscoelasticity with Mullins effect since Abaqus 6.10 (Dassault Systèmes 2010).

REFERENCES

Abraham, F. (2002, December). *The Influence of Minimum Stress on the Fatigue Life of Non Strain-Crystallising Elastomers*. Ph. D. thesis, Coventry University.

Abraham, F., T. Alshuth, & S. Jerrams (2005). The effect of minimum stress and stress amplitude on the fatigue life of non strain crystallising elastomers. *Materials and Design* 26, 239–245.

Bergström, J. S. (2005). Constitutive Modeling of Elastomers – Accuracy of Predictions and Numerical Efficiency. *PolymerFEM.com*.

Bergström, J. S. (2008). Review of the FeFp Model. *PolymerFEM.com*.

Ciesielsky, A. (2001). *Introduction to Rubber Technology*. Shrewsbury: Smithers Rapra.

Dassault Systèmes (2007). *Abaqus Analysis User's Manual, Volume III: Materials (Version 6.7)*.

Dassault Systèmes (2008). *Abaqus Release Notes (Version 6.8)*.

Dassault Systèmes (2010). *Abaqus Release Notes (Version 6.10)*.

Diani, J., B. Fayolle, & P. Gilormini (2009). A review on the Mullins effect. *European Polymer Journal 45*, 601–612.

Mullins, L. (1969). Softening of rubber by deformation. *Rubber Chem. Technol. 42*, 339–362.

Ogden, R. W. & D. G. Roxburgh (1999). A pseudo-elastic model for the Mullins effect in filled rubber. *Proceedings: Mathematical, Physical and Engineering Sciences 455*, 2861–2877.

Paige, R. E. & W. V. Mars (2004). *Implications of the Mullins Effect on the Stiffness of a Pre-loaded Rubber Component*.

Sommer, J. G. & O. H. Yeoh (2001). Tests and Specifications. In A. N. Gent (Ed.), *Engineering with Rubber* (2nd ed.), pp. 307–355. Munich: Hanser Publishers.

On the influence of heat ageing on filled NR for automotive AVS applications

P. Charrier
Modyn Trelleborg, Zone ind. de Carquefou, Carquefou Cedex, France

Y. Marco
ENSTA Bretagne, Laboratoire LBMS (EA 4325), Brest Cedex, France

V. Le Saux
UBS, LIMATB – Equipe ECoMatH, Lorient Cedex, France

R.K.P.S. Ranaweera
Dept. of Mechanical Engineering, University of Moratuwa, Katubedda, Moratuwa, Sri Lanka

ABSTRACT: Durability of filled Natural Rubber (NR) is still an open issue to a wide range of scientific studies. In this paper, we focus on the effect of heat ageing (thermo-oxidation) on the fatigue life duration for a range of temperatures of [40°C, 120°C] relevant for under hood applications. First, the heterogeneous nature (DLO effect) of the thermo-oxidation process is highlighted on thick samples through micro-Shore hardness measurements and failure surface analysis. These measurements also exhibit the strong dependency of the ageing kinetics on temperature. Second, this database is used to analyze the fatigue tests through a time-temperature approach, which revealed three activation energies for three temperature ranges. In the third part, influences of several factors likely to affect the design of industrial accelerated ageing procedures are investigated (effect of pauses, effect of ageing sequence).

1 INTRODUCTION

1.1 *Industrial motivation*

During the last years, three main evolutions lead to a global increase of the severity of specification for automotive parts. The first one is the worldwide extension of the market. The second one is a constant search for decreasing the cost. As a result carmakers focused on having the same AVS components (and therefore the same compound) for different word regions dealing with very diverse road and environment conditions. Ultimately this lead to the study of best compromise for the compounds used in AVS components. The third one is the COV emission reduction enforced by international institutions which lead to higher engine temperatures and higher optimization in order to decrease the component weight. Accordingly, the new specification requirements demanded higher thermal resistance level and more precisely a very good heat ageing strength. In this article, we focus on the effect of a preliminary heat ageing on the fatigue behavior of an elastomeric material as it is the common method used by Asian carmakers to specify the durability resistance of AVS components.

1.2 *Scientific background*

Thermal ageing of rubbers has been widely investigated (Bolland and Gee, 1946; Grassie and Scott, 1985; Gent, 1992; Celina, 2005) throughout the last decades. Many factors govern the degradation mechanisms in rubber; however it is generally admitted that oxygen is the most critical factor. Thermal ageing may give rise to main-chain scission, crosslink formation and crosslink breakage, which ultimately lead to severe changes in mechanical properties. Depending on whether there is more chain scission or crosslinking, an elastomer will soften or stiffen respectively after oxidation. The main idea here is not to try to highlight these consequences through physical measurements but rather to follow one or several relevant indicators in order to be able to assess lifetime prediction based on the use of the Arrhenius methodology (Celina, 2005). Because degradation of rubber usually exhibit ageing gradients, it is important to describe such heterogeneities in order to correctly interpret the ageing results and thus the reliability of the extrapolation to service conditions (Celina, 2000). Finally, a comparison to naturally aged materials is of course crucial to challenge the activation energies identified with the Arrhenius approach.

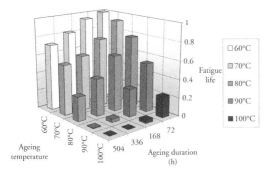

Figure 1. Normalized duration life (cyclic strain level = 100% – room temperature – 5 Hz) after various preliminary heat ageing (from 60°C to 100°C@3 days to 21 days).

2 EXPERIMENTAL CONTEXT

The fatigue samples geometry (Ostoja-Kuczynski et al. 2005) and micro-hardness protocols (Le Saux 2010) have been detailed in former publications and will not be recalled here.

3 PRELIMINARY RESULTS

3.1 AE2 samples

Our first test campaign was targeted to challenge standard ageing post treatment procedure and to investigate the evolution of the fatigue strength at room temperature of samples previously submitted to various thermal ageing conditions. These tests were performed on AE2 samples, and the preliminary ageing durations ranged between 3 to 21 days (72 to 504 hours) for temperatures from 60°C to 100°C. These conditions represent the specifications provided by most of the carmakers. The fatigue tests were displacement controlled with a maximum local strain of 100% at a frequency of 5 Hz to minimize heat build-up. All tests were performed at a controlled temperature of 23°C. As expected, the 3D bar plot given in Figure 1 illustrates the decrease in fatigue strength with increasing ageing temperature or duration. Ageing temperature seems to have a greater influence than the ageing duration.

The XY plot given in Figure 2 shows the evolution of fatigue life with ageing duration and the evolutions seem to follow an exponential law for each temperature. This enables us to perform a time-temperature superposition successfully, which is illustrated on Figure 3 for a reference temperature of 80°C. The black curve follows the ageing kinetic equation given below:

$$N_i = N_{i-0} \times e^{-t \cdot k_0 \cdot e^{-E_0/R.T}} \qquad (1)$$

The deduced activation energy of 94 kJ/mol is very close to other values found in the bibliography for several ageing indicators.

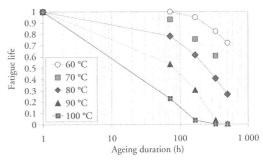

Figure 2. Same results as in Figure 1 – 1 curve for 1 heat ageing temperature.

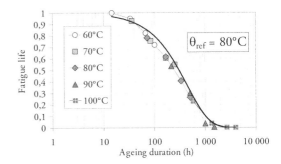

Figure 3. Time-temperature superposition with a reference temperature of 80°C – data coming from Figure 2.

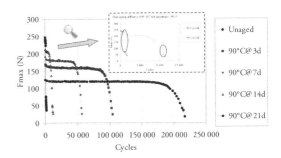

Figure 4. Normalized cyclic force evolutions for various preliminary heat ageing durations at 90°C.

Figure 4 presents the cyclic force evolution during the fatigue tests after ageing at 90°C. The cyclic stiffness of the sample increases with the ageing duration. Moreover, in the zoomed frame, it appears that for severe heat ageing, the stiffness drops suddenly due to macroscopic crack propagation.

Figure 5 shows the fatigue fracture surfaces obtained for all ageing conditions. Depending on the ageing conditions, the surfaces illustrate zones with progressive or brittle propagation. A detailed example is provided on Figure 6 (left fascia is obtained after 14 days at 100°C, right fascia is obtained after 21 days at 100°C). On the left, one can observe (going from the skin to the centre of the sample) first a progressive propagation (rough surface), then a brittle one (smooth surface) and again a progressive propagation. On the

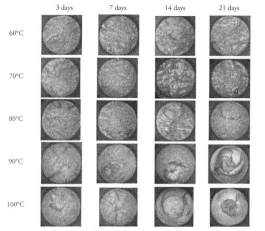

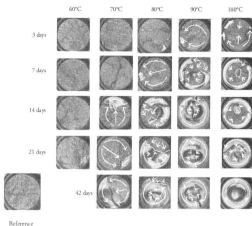

Figure 5. Failed surfaces for fatigue tests (cyclic strain level = 100% – room temperature – 5 Hz) after various preliminary heat ageing (from 60°C to 100°C@3 days to 21 days).

Figure 7. Failed surfaces for fatigue tests (cyclic strain level = 100% – room temperature – 2 Hz) after various preliminary heat ageing (from 60°C to 100° C@3 days to 42 days).

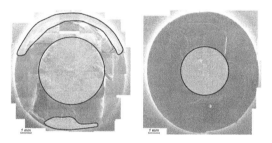

Figure 6. SEM view of the failed surfaces after heat ageing 14 days@100°C (left view) and 21 days@100°C (right view) – orange translucent areas stand for progressive propagation.

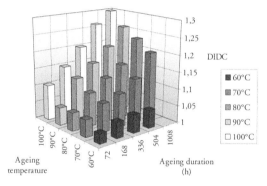

Figure 8. Normalized skin DIDC hardness after various preliminary heat ageing (from 60°C to 100°C@3 days to 42 days).

right, there are only 2 stages: a brittle one (annular external smooth ring – failure happens in 1 cycle) and a mechanical one (the rough disc in the middle – few cycles for the complete failure). This is due to the DLO effect resulting in a sharp ageing profile.

3.2 AE42 samples

A second experimental campaign was carried out with larger samples (AE42) following the same principle and increasing the longest ageing duration up to 42 days. Figure 7 is comparable to Figure 5 except that the brittle failures appear earlier due to the more homogeneous strain/stress fields for AE42 samples. White arrows represent the crack propagation paths deduced from surface analysis. A classical μ-DIDC hardness device was used to study the evolution of the hardness along the sample radius after heat ageing. Figure 8 shows that the μ-DIDC hardness at skin evolves oppositely to the fatigue strength. At the sample's core (plotted on Figure 9), the μ-DIDC hardness increases a little bit when the ageing temperature is higher than 60°C but is not sensible to the temperature beyond 70°C. Moreover, it seems that the hardness only slightly increases with ageing duration.

To provide more accurate resolution, another micro-shore hardness measuring device based on Instrumented Indentation Technique (IIT) was used (Le Saux 2010). This technique allowed measurement of the hardness profile along a radius. Figure 10 presents such measurements after 14 days with various ageing temperatures and exhibits clearly a so called DLO effect, especially for temperatures above 80°C. The depth of this plateau decreases when the ageing temperature increases. The hardness in the middle of the sample increases with temperature but stabilizes at 90°C. This explains the failure surfaces observed on Figure 7. Figure 11, focuses on the affect of ageing duration for an ageing temperature of 90°C. In that case, the depth of the plateau increases along with time and the hardness in the middle of the sample appears to be stabilized.

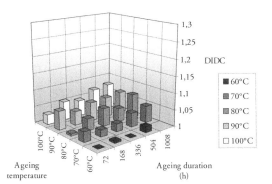

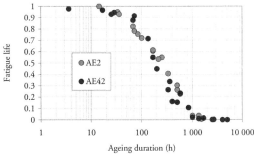

Figure 12. Time-temperature superposition of normalized duration lives for the AE2 and the AE42 samples at the same reference temperature.

Figure 9. Normalized DIDC hardness in the centre of the AE42 sample after various preliminary heat ageing (from 60°C to 100°C@3 days to 42 days).

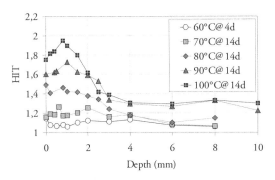

Figure 10. Normalized HIT profiles along a radius of the AE42 sample (0 mm corresponds to the skin and 10 mm to the center of the sample) after 14 days at various ageing temperatures.

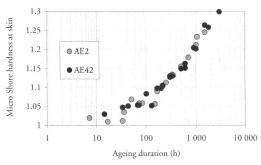

Figure 13. Time-temperature superposition of normalized skin shore hardness for the AE2 and the AE42 samples at the same reference temperature.

3.3 Affect of the sample geometry

Having test results for 2 samples (AE2 with a diameter of 10 mm and AE42 with a diameter of 20 mm), it is appealing to compare both ageing kinetics to see if they are intrinsic. This is the purpose of Figure 12 (fatigue lifetime for initiation) and Figure 13 (µ-DIDC hardness at skin). The conclusion is that these kinetics do not depend on the thickness of the chosen component for the range studied. Both parameters are driven by the skin behavior where oxygen seems to be in excess. Therefore, the evolutions after heat ageing do not depend on the depth.

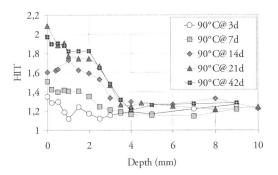

Figure 11. Normalized HIT hardness profiles along a radius of the AE42 sample (0 mm corresponds to the skin and 10 mm to the center of the sample) after various ageing durations at 90°C.

Oxygen diffusion-consumption simulations become mandatory to go further explaining the notable evolution of the size of the plateau and the changes of the shore hardness in the center of the sample (slow oxygen diffusion inside the sample and/or consumption of the remaining oxygen, plasticizers exudation …). This falls in to the scope of another research project that will not be detailed here.

3.4 Application to AVS components

Because of warranty costs issues (after technical expertise of parts, these costs are shared between carmakers and automotive suppliers), it is now possible to get AVS components subjected to severe conditions in different countries. The HIT profile obtained on a torque rod coming back from Turkey after many years of usage is plotted on Figure 14. The profile obtained is similar to the ones obtained formerly in our study. Since the size of the DLO plateau is broad, we can conclude to that the ageing temperature was moderate and the exposure time was very long. Furthermore, the value of the hardness at skin gives us precise information about the ageing severity faced by this component.

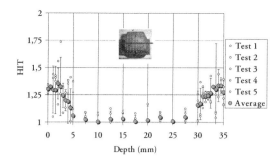

Figure 14. Normalized HIT profiles along the filtering arm of a roll restrictor after many years of use in Turkey.

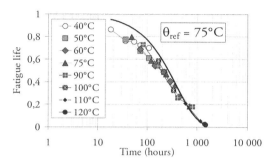

Figure 15. Standard time-temperature superposition with a reference temperature of 75°C.

As a conclusion, HIT profile is a very interesting tool to study heat ageing of AVS components.

4 EXTENDED RANGE OF AGEING

4.1 Test results and post processing

After the discussions with carmakers, it became obvious that the formerly chosen temperature range covers specification conditions but only provides a narrow view of the temperatures really experienced by the AVS components. During the 6.000h of engine life, the average temperature under hood is rather between 50 and 60°C. Most of the time, it hovers below 60°C and during very short periods it can rise above 100°C. As a consequence, we decided to extend the temperature range to [40°C; 120°C] and to increase the maximal ageing duration to 90 days (2.160 h). AE2 samples were used in this test campaign.

The standard time-temperature superposition of the normalized duration life at a reference temperature of 75°C is plotted on Figure 15. It appears that the predicted black curve according to the ageing kinetic given by equation 1 is not successful. To resolve this issue, a new kinetic equation was suggested:

$$N_i = N_{i-0} \times e^{-t^\alpha \cdot k_0 \cdot e^{-E_0/_{R \cdot T}}} \qquad (2)$$

Figure 16 demonstrates the relevance of our choice. Deduced activation energies are plotted on the Figure 17. Clearly, the activation energy decreases when

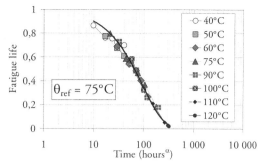

Figure 16. Enhanced time-temperature superposition with a reference temperature of 75°C.

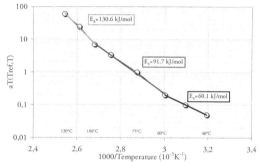

Figure 17. Deduced activation energies from Figure 19.

the temperature decreases making the thermal ageing easier at low temperatures. Moreover, two threshold temperatures could be highlighted at 60°C and 100°C. Using the activation energy evaluated between 60°C and 100°C to extrapolate the rubber behavior at 40°C and at 110°C would clearly lead to an over or under estimation of the ageing severity.

4.2 Oxygen diffusion and consumption

The question that arises is 'how we can explain these three activations energies?' One can refer to the Basic Auto-oxidation Scheme (Wise J. et al. 1997 and Launay A., 2008). Depending on the oxygen diffusion rate at skin (which is temperature dependant) together with oxygen consumption rate due to thermal ageing (also temperature dependant) various ageing profiles could be observed:

- On the left graph of Figure 18, the O_2 concentration is high enough and all the free radicals R° are almost instantaneously transformed into ROO° ones. The termination processes involving R° radicals can be neglected;
- On the right graph of Figure 18, the O_2 concentration is lower than a critical value. A part of the free radicals R° cannot then react with O_2 and will be involved in other reactions, especially termination processes. Ageing is therefore driven by of diffusion kinetics.

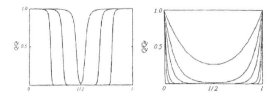

Figure 18. Shape of depth distributions of the oxygen reduced conversion due to ageing – left view: low temperature, oxygen is in excess – right view: very high temperature with a lack of oxygen diffused.

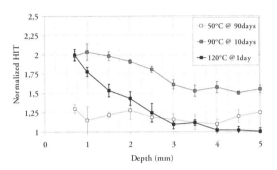

Figure 19. ITT profiles along a radius of AE2 samples (0 mm corresponds to the skin and 5 mm to the center of the sample) for different ageing temperatures.

IIT measurements were performed along the radial axis of AE2 samples aged at three temperatures (50, 90 and 110°C – see Figure 19) and expected profiles were obtained:

– At low temperature, the ageing profile is nearly constant along the radius of the AE2 sample;
– For intermediate temperature, a plateau is clearly visible at the skin;
– At very high temperatures, there is no more plateau at the skin.

5 ON THE DIFFICULTIES TO SET SIMPLE AND RELEVANT SPECIFICATIONS

5.1 Today situation

The carmakers follow two strategies to specify the thermal environment in component specifications:

– Superpose the most severe mechanical profile to the most severe thermal conditions with a fatigue test duration between 5 to 15 days depending on the allowed time editing;
– Evaluate a thermal profile when the engine is turned on and modify it into an equivalent preliminary static heat ageing (using Van't Hoff law) condition. Afterwards perform the fatigue tests at a moderate temperature.

Both approaches seem to be very conservative and need to be challenged. Regarding compound choice, they lead to choices significantly different!

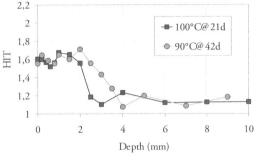

Figure 20. HIT profiles along a radius of the AE42 sample (0 mm corresponds to the skin and 10 mm to the center of the sample) after 2 "equivalent" heat ageing conditions.

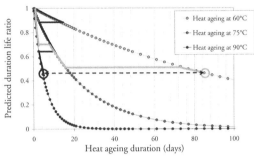

Figure 21. Heat ageing kinetics at 60°C, 75°C and 90°C – black curve: 14d@60°C + 7d@75°C + 2d@90°C – grey curve: 2d@90°C + 7d@75°C + 14d@60°C.

5.2 Simplified Arrhenius law

First, we would like to highlight the restrictions associated to the simple Van't Hoff law (heat + 10°C = duration/2). As shown on Figure 20, "equivalent" ageing conditions do no lead to the same plateau depth. Even if the hardness at the skin is similar, the affected zones are different, leading potentially to different failure scenarios.

5.3 Ageing damage accumulation

In order to guess an equivalent simple ageing condition, severities of ageing blocks are compared and accumulated. Figure 21 illustrates an example of two combinations of the same three ageing blocks using theoretical ageing kinetics. They are assumed to lead to the same material resistance reduction. Using three ageing conditions (14d@60°C, 7d@75°C & 2d@90°C), six combinations are obtained, plus the "equivalent" one, that is 6d@90°C. Experimental results are plotted on Figure 22. The ageing damage accumulation does not depend on the test sequence. Moreover, the "simplified" accumulation rule gives only a rough approximation to the severity of the tests.

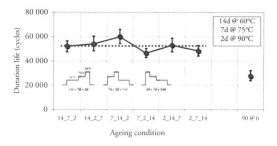

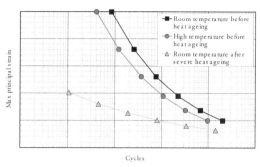

Figure 22. Fatigue duration lives (AE2 samples, cyclic strain level = 100% – room temperature – 5 Hz) after various combinations of heat ageing (14d@60°C, 7d@75°C and 2d@90°C) and after "equivalent" heat ageing (6d@90°C).

Figure 24. Poor heat ageing resistant NR – Wöhler curves for various testing conditions.

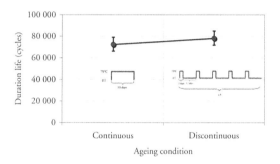

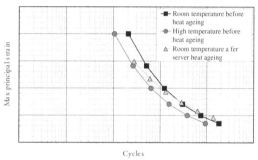

Figure 23. Fatigue duration lives (AE2 samples, cyclic strain level = 100% – room temperature – 5 Hz) after continuous and discontinuous heat ageing (in total: 10d@75°C).

Figure 25. Good heat ageing resistant NR – Wöhler curves for various testing conditions.

5.4 Discontinuous heat ageing

Usually, a customer drives only 10% of the total day duration. As a consequence, the effect of pauses during heat ageing is to be checked and compared to continuous heat ageing. In a similar manner, the use of the vehicle during the remaining 90% of the day is often ignored but should be studied carefully. Comparing the ageing (75°C@10d) performed in continuous and discontinuous ways (5 times: 2 days of ageing + 10 days of rest at room temperature), we observed both conditions are equivalent (see Figure 23). Nevertheless, thermal profiles measured by carmakers on vehicles are mandatory to provide a definitive conclusion on the affect of the unused periods.

5.5 Temperature measured on vehicle and fatigue specification

To conclude about the two kinds of durability specifications, we can compare Figure 24 and Figure 25 which are the Wöhler curves of two materials at room temperature before and after a severe heat ageing or tested at high temperature without preliminary heat ageing. First material (Figure 24) could easily face short durability test at high temperature. However, after a severe heat ageing, it will break quickly after some fatigue cycles. In an opposite manner the second material (Figure 25), although its initial durability resistance is lower, will be able to sustain the fatigue

test even after a severe heat ageing. Two kinds of fatigue specifications therefore lead to two possible material choices.

6 PERSPECTIVES

The present study is clearly not finished. Further developments will deal with ageing during the fatigue tests and finite element analysis of samples taking in to account the gradients in properties induced by thermal ageing.

REFERENCES

Bolland, J.L. & Gee, G 1946. Kinetics studies in the chemistry of rubber and related materials. II. The kinetics of oxidation in unconjuga ted olefins. *Transactions of the Faraday Society*, 42:236–243.

Cadwell, S. M, Merril, R.A, Sloman, C.M & Yost, F.L. 1940. Dynamic Fatigue Life of Rubber. *Industrial and Engineering Chemistry, 12, n° 1: 19–23.*

Celina, M, Wise, J., Ottesen, D.K., Gillen, K.T. & Clough, R.L. 2005. Correlation of chemical and mechanical property changes during oxidative degradation of neoprene; *Journal of Polymer Degradation and Stability*, 68: 171–184.

Celina, M, Gillen, K.T. & Assink, R.A. 2005. Accelerated aging and lifetime prediction: Review of non-Arrhenius behaviour due to two competing processes. *Journal of Polymer Degradation and Stability*, 90:395–404.

Gent, A.N. 1992. *Engineering with rubber*. Hanser.

Grassie, N. & Scott, G. 1985. *Polymer degradation and stabilisation*. Cambridge University Press.

Lake, G. J. & Lindley, P. B. 1964. Cut growth and fatigue of rubbers. II. Experiments on a noncrystallizing rubber. *Journal of Applied Polymer Science*, 8, 707–721.

Launay A., 2008 PSA – TRELLEBORG master research thesis.

Le Chenadec, Y., Stolz, C., Raoult, I. Nguyen T., M.-L., Delattre, B. & Charrier, P. 2007. *A novel approach to the heat build-up problem of rubber. Proceedings 5th European Conference on Constitutive Models for Rubber*, Paris, September-2007.

Le Cam, J.-B., Verron, E., Huneau, B., Gornet, L. & Pérocheau, F. 2005. *Micro-mechanism of fatigue crack growth: comparison between carbon black filled NR and SBR. Proceedings Fourth European Conference on Constitutive Models for Rubber*, Stockholm, June-2005.

Le Saux, V., 2010. Fatigue et vieillissement des élastomères en environnement marin et thermique: de la caractérisation accélérée au calcul de structure, Ph.D. thesis, UBO/UEB

Marchal, J. 2006. *Cristallisation des caoutchoucs chargés et non chargés sous contrainte: Effet sur les chaînes amorphes*. Ph-D Dissertation, University of Paris XI.

Ostoja-Kuczynski, E., Charrier, P., Verron, E., Marckmann, G., Gornet, L. & Chagnon, G. 2003. *Crack initiation in filled natural rubber: experimental database and macroscopic observations. Proceedings Third European Conference on Constitutive Models for Rubber*, London, September 2003.

Ostoja-Kuczynski, E., Charrier, P., Verron, E., Gornet, L. & Marckmann, G. 2005. *Influence of mean stress and mean strain on fatigue life of carbon black filled natural rubber. Proceedings Fourth European Conference on Constitutive Models for Rubber*, Stockholm, June-2005.

Wise, J., Gillen, K.T., & Clough, R.L., 1997. Quantitative model for the time development of diffusion-limited oxydation profiles. Polymer, 38(8):1929–1944.

Test methods and analytical techniques

Development of an experimental device to investigate mechanical response of rubber under simultaneous diffusion and large strain compression

A.B. Chai & A. Andriyana
Department of Mechanical Engineering, University of Malaya, Kuala Lumpur, Malaysia

E. Verron
Institut de Recherche en Génie Civil et Mécanique, GeM UMR CNRS 6183, École Centrale de Nantes, Nantes, France

M.R. Johan & A.S.M.A. Haseeb
Department of Mechanical Engineering, University of Malaya, Kuala Lumpur, Malaysia

ABSTRACT: Rubbers are massively used as seals and gaskets in the automotive industry where they are mainly subjected to compressive loading during their service. The introduction of biodiesel such as palm biodiesel, motivated by the environmental and economic factors, has placed additional demands on these components due to compatibility issue in the fuel system. Hence, it is crucial to investigate the durability of rubber components in this aggressive environment. A number of works on the static immersion test investigating the diffusion of liquids in rubber can be found in the literature. Nevertheless, from the experimental work viewpoint, studies focusing on the coupling between diffusion and large deformation in rubber are less common. In the present work, a compression device for coupled diffusion and large strain in rubber is developed. The apparatus comprises of four stainless steel plates and spacer bars in between which are specifically designed such that compression can be introduced on the rubber specimens while they are immersed into biodiesel simultaneously. Thereby allowing coupled diffusion and large strain to take place. Different immersion durations and pre-compressive strains are considered. At the end of each immersion period, the resulting mechanical response of rubber specimens are investigated. The features of this compression device are discussed and perspectives are drawn.

1 INTRODUCTION

Fossil fuel is depleting rapidly due to its limited reserve and increasing demands from various industry. The corresponding issue, which causes environmental degradation as well as political and economic concern, has encouraged the needs of searching for alternative fuel. One of the solution is the biodiesel which is derived from plant materials or animal fats. The biodiesel has properties similar to that of diesel and it is biodegradable and has low sulfur content. However, the fatty acid ester in biofuel is different from hydrocarbon in diesel and investigation on the material compatibility in the fuel system has been a great interest of many researchers. (Haseeb et al. 2010, Fazal et al. 2011). Indeed, in the case of rubber, changes in fuel composition often create many problems in rubber seals, pipes, gaskets and o-rings in the fuel system (Trakarnpruk & Porntangjitlikit 2008).

In diesel engine, the rubber sealing components are in contact with the fuel which may lead to degradation of rubber due to compatibility issue. One main form of degradation in rubber exposed to liquid is swelling which can be described in terms of mass or volume change (Haseeb et al. 2010, Trakarnpruk

and Porntangjitlikit 2008). In addition to exposure to potentially hostile environments, the rubber sealing components are simultaneously subjected to fluctuating mechanical loading during their service. The durability of rubber components hence becomes a critical issue. A number of static immersion tests investigating the diffusion of liquid in rubber have been extensively studied (see (Treloar 1975) and references herein). However, investigations on more complex problems involving the swelling of polymer network in the presence of a stress (strain), in particularly multiaxial stress state, are less common. The earliest work deal with the problem dated back to the work of Flory & Rehner (1944). Since this pioneering work, more recent accounts on coupling diffusion-deformation can be found in the literature (Soares 2009, Hong et al. 2008, Baek 2004, Nah et al. 2010). It is to note that the aforementioned studies deal with the interaction between diffusion of liquid and large deformation without explicitly relating them to cyclic and fatigue behaviors of rubber.

The present work can be regarded as a first step toward an integrated durability analysis of industrial rubber components exposed to aggressive environments, e.g. oil environment in biofuel systems, during

Table 1. Properties of B100 palm biodiesel.

Test	Unit	Methods	Results
Ester content	% (m/m)	EN 14103	96.9
Density at 15°C	kg/m^2	EN ISO 12185	875.9
Viscosity at 40°C	mm^2/s	EN ISO 3104	4.667
Flash point	°C	EN ISO 3679	168
Cetane number	–	EN ISO 5165	69.7
Water content	mg/kg	EN ISO 12937	155
Acid value	mgKOH/g	EN ISO 3679	0.38
Methanol content	% (m/m)	EN 14110	<0.01
Monoglyceride content	% (m/m)	EN 14105	0.67
Diglyceride content	% (m/m)	EN 14105	0.2
Triglyceride content	% (m/m)	EN 14105	0.2
Total glycerine	% (m/m)	EN 14105	0.25

their service. In this work, a compression device is developed to investigate the interaction between diffusion of liquids and large deformation in rubber and the effect of swelling on the mechanical response under cyclic loading in common rubbers used in sealing application such as Nitrile Butadiene Rubber (NBR) and Polychloroprene Rubber (CR) is investigated.

This paper is organized as follows. In Section 2, experimental works including materials, specimen geometry, development of a compression device and the types of test conducted in this study are detailed. The experimental results are presented and discussed in Section 3. Concluding remarks are given in Section 4.

2 EXPERIMENTAL PROGRAM

2.1 Materials

Rubber specimens used in this research are provided by MAKA Engineering Sdn. Bhd., Malaysia. The material investigated is commercial grade of NBR and CR with 60 shore hardness. Biodiesel is prepared by blending palm biodiesel (provided by Am Biofuels Sdn. Bhd., Malaysia) with diesel. The analysis report of the palm biodiesel investigated is shown in Table 1. The immersion tests conducted are immersion in B0 (100% diesel), B25 (blend of 25% of biodiesel and 75% of diesel), B75 (blend of 75% of biodiesel and 25% of diesel) and B100 (100% biodiesel).

2.2 Specimen geometry

In order to investigate the interaction between diffusion and large deformation in rubber, a specially designed hollow cylindrical rubber specimen is used. Since the specimen will be subjected to compressive loading, its wall thickness should be large enough to avoid buckling. At the same time, it should be thin enough to ensure that equilibrium diffusion (swelling) can be achieved within a reasonable period of time. For this purpose, an annular rubber specimen having

height, outer diameter and wall thickness of 10 mm, 50 mm and 6 mm respectively is used in the present study.

2.3 Compression device

The hollow cylindrical specimens mentioned in the previous subsection are subsequently subjected to different pre-compressive strains by attaching it into a specially designed compression device prior to immersion as shown in Figure 1. The compression device has special features as described below:

1. It consists of four stainless steel plates for corrosion resistance as it is to be immersed into diesel and biodiesel which are deemed to be corrosive.
2. The device is able to accommodate a total of 12 specimens arranged in 3 different levels between two successive plates.
3. Each plate has 4 main holes to allow liquid to flow and diffuse into the inner surface of the rubber specimens. In this way, each rubber specimen is subjected to diffusion of liquid from both inner and outer wall surfaces, i.e. diffusion along radial direction only as illustrated in Figure 2.
4. For each level between two successive plates, a pre-compressive strain is applied to the specimens located at the corresponding level: 20% for the level 1, 10% for the level 2 and 2% for the level 3. Different pre-compressive strains are ensured by using spacers of appropriate height. Bolts and nuts located at each corner of the plates are used to tighten the device until the compression plates are uniformly in contact with the spacers. Additional ring spacers are placed around the bolt in the middle of the plates to prevent bending of the plates.
5. In practice, the 2% strain is so small that its effect on the macroscopic mechanical response is negligible. Nevertheless, this level of strain is retained to represent stress-free condition while ensuring that the diffusion occurs only along radial direction.

In order to investigate the effect of interaction between diffusion of biodiesel and large compressive strain on swelling behavior and mechanical response under cyclic loading condition, the device containing rubber specimens are subsequently immersed into different palm biodiesel blends for the duration of 30 and 90 days. The detail of the immersion tests is given in Table 2.

2.4 Swelling measurement

The swelling of rubber specimens after immersion is described in terms of mass change and volume change and the test procedure for swelling measurement is summarized as followed:

1. Before the immersion, the weight of the rubber specimen is measured in air and in distilled water. The specimen is then quickly dipped into alcohol and blotted dry with filter paper.

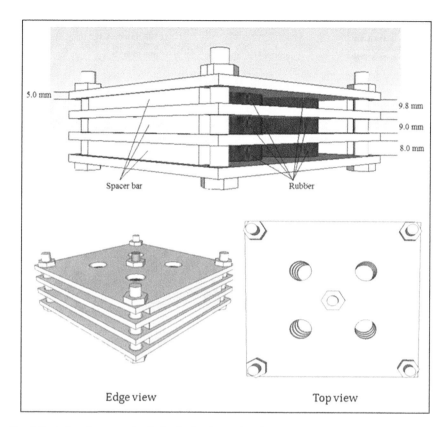

Figure 1. Specially-designed compression device for the observation of the coupling between diffusion and large deformation.

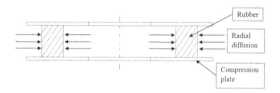

Figure 2. Diagram of radial diffusion for swelling under compressive strain.

Table 2. Immersion tests

Biodiesel Blend	Level of Compressive Strain (%)	Immersion Duration (days)
B0	2/10/20	30/90
B25	2/10/20	30/90
B75	2/10/20	30/90
B100	2/10/20	30/90

2. After weight measurement, the rubber specimens are placed in sequence on the compression plates. Grease is applied on the surface of the speci-mens that are in contact with the compression plate to avoid bulging of the specimens. Thereby ensuring the specimens to be in a simple uniaxial compressive stress state.

3. Bolts and nuts are used to tighten the compression device until the compression plates are uniformly in contact with the spacers. The device containing rubber specimens is subsequently immersed com-pletely into different biodiesel blends for 30 days and 90 days.

4. At the end of the immersion period, the speci-mens are removed from the compression device and quickly dipped into acetone; it is then clean with fil-ter paper to remove the excess oil. The specimens are left for 30 minutes to allow for recovery before any measurement is made after immersion.

5. Step 1 is repeated to measure the weight of rubber specimen after immersion.

The percentage of mass change and volume change are calculated using the following relations (Trakarnpruk & Porntangjitlikit 2008):

$$\% \; Mass \; Change = \frac{M_2 - M_1}{M_1} \times 100 \qquad (1)$$

$$\% \; Volume \; Change = \frac{(M_2 - M_4) - (M_1 - M_3)}{(M_1 - M_3)} \times 100$$

$$(2)$$

where M_1 and M_2 are the mass in air (gram) before and after immersion while M_3 and M_4 are mass in

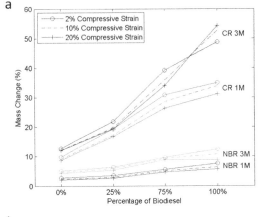

a

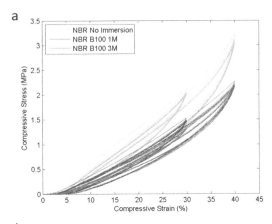

a

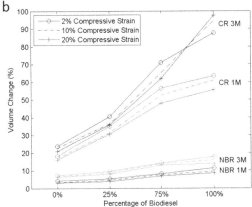

b

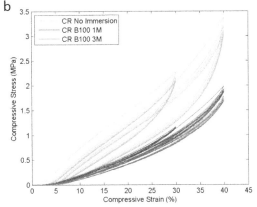

b

Figure 3. (a) Mass change and (b) volume change of NBR and CR at different compressive strains after 1M and 3M of immersion in different biodiesel blends.

Figure 4. Stress-strain curves of a) NBR and b) CR at dry states (without immersion) and after 1 month (1M) and 3 months (3M) of immersion in B100. Results correspond to pre-compressive strain of 2%. For immersed rubbers, the stress is expressed with respect to unswollen-unstrained configuration (dry cross section).

water (gram) before and after immersion. The tests are conducted on four specimens under each compressive strain and for each biodiesel blend.

2.5 Mechanical response measurement

In order to gain insight on the effect of coupled diffusion-large deformation on the mechanical response under cyclic loading conditions, mechanical tests using Instron 5500 uniaxial test machine equipped with 10 kN load cell at room temperature are conducted. To ensure uniform displacement control on the specimens, circular compression plates are attached to the machine. The experimental setup is connected to a computer to record the experimental data. All tests are conducted at a strain rate of $0.01 \, s^{-1}$ to avoid excessive increase in the temperature of the specimens, i.e. thermal effect is not considered in the present study. The specimen is subjected to cyclic compressive loading at two different maximum compressive strains: 30% and 40% of 6 cycles each. To ensure repeatability of the results, at least three specimens are used to performed each test.

3 RESULTS AND DISCUSSION

3.1 Swelling of rubber

Figure 3 shows the percentage of mass change and volume change of NBR and CR with different compressive strains (2%, 10% and 20%) after immersion in different biodiesel blends (B0, B25, B75 and B100) for 30 days (1M) and 90 days (3M) respectively. Both plots show similar patterns. The percentage of fuel uptake is increasing with the increase of palm biodiesel content. It is clear that for low biodiesel content (B0 and B25), no significant fuel uptake is recorded for NBR. At the higher percentage of biodiesel content (B100), both NBR and CR shows significant mass change and volume change. The corresponding trend can be attributed to the segmental mobility of the polymer and free volume of the polymer (George 2001). Several factors influencing the segmental mobility of the polymer chain including unsaturation, crosslinking degree, crystalization and nature of the substituents. The nature of the substituent is particularly critical in

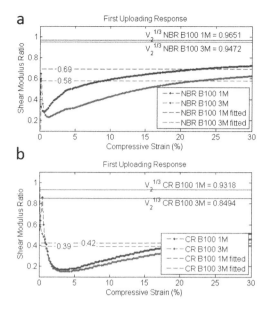

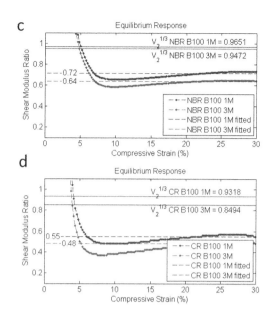

Figure 5. Shear modulus ratio vs compressive strain.

this study. This is because CR is made from emulsion polymerization of 2-chloro-1, 3-butadiene and NBR is emulsion copolymer of acrylonitrile and butadiene (Dick 2001). The polar substituent of acrylonitrile in NBR and chlorine substituent in CR are resistant to oil absorption. However, the swelling of rubber is by the principle of "like dissolve like" – polar solvent are more likely to dissolve polar substances and non-polar substances are more likely to dissolve in non-polar solvent (Zhang & Cloud 2007). The high polarity of ester in palm biodiesel favors the forming of polymer-solvent interaction resulting to the increase of swelling in CR as the ester content in the palm biodiesel blend increases (Pekcan 2002).

In addition, it is observed that the fuel uptake is affected by the level of compression, except for CR after 3 months immersion. The increase of pre-compressive strain has restricted the fuel uptake into the elastomeric materials, i.e. the compressive stress appears to reduce the amount of swelling compared with that for stress-free rubber. Absorption of liquid into rubber occurs when liquid dissolve on the surface (adsorption) and penetrate further into the rubber by diffusion. As the compressive strain increases, the initial effective area for diffusion to occur along radial direction in hollow cylindrical rubber specimens becomes smaller. Hence, the resulting swelling is lower. Furthermore, the reduction in swelling of rubber is affected by the hydrostatic component of the applied stress. A compressive stress, for which the hydrostatic component is positive leads to a decrease in the swelling of rubber (Treloar 1975, Fukumori et al. 1990). The fluctuating trend of CR after 3 months immersion in B100 might be caused by the maximum swelling in the polymer network and strong polymer-solvent interaction which has remedied the effect of compressive strain.

3.2 Influence of swelling on the mechanical response of the rubbers

The mechanical response of NBR and CR corresponding to 2% pre-compressive strain after immersion in B100 under cyclic compressive loading at two different maximum compressive strains are depicted in Figure 4. To compare the mechanical response of the swollen rubber with the dry rubber (unswollen), the compressive stress of the swollen specimen (after immersion) is computed by dividing the measured compressive force with the *unswollen unstrained* cross section. For each maximum compressive strains, the specimen experiences six cycles of loading. It is shown that there is not much difference in the nature of stress-strain behavior after immersion. However, lower stress is recorded for CR after 90 days immersion given the same pre-compressive strain. The corresponding behavior can be related to the swelling in the CR which decreases its strength due to strong interaction of rubber-solvent matrix system (George 1999).

According to Treloar's theory (Treloar 1975), the only effect of the swelling is to reduce the modulus in inverse proportion to the cube root of the swelling ratio, without changing the form of the stress-strain relations, i.e.,

$$\frac{G'}{G} = V_2^{1/3} \qquad (3)$$

where G' and G are respective shear modulus in swollen and dry state and V_2 is volume fraction of rubber in the mixture of rubber and liquid. In Figure 5, three types of shear modulus ratio are plotted as a function of applied compressive strain. The first one is the cube root of the swelling ratio as predicted by Treloar in Equation 3. Secondly, the shear modulus ratio is obtained by fitting the first uploading response of the stress-strain curve by assuming the response can be represented with Neo-Hookean model and the third ratio is obtained from the real experimental data (equivalent with the ratio of stresses at swollen and dry states) as shown in Figure 5 (a) and (b). The same shear modulus ratios are plotted in Figure 5 (c) and (d) by using the equilibrium response. In our case, the equilibrium response is given by the imaginary curve lines between the uploading and unloading of the 6th cycle, after stress-softening effect is removed (Bergström & Boyce 1998). All the plots in Figure 5 shows that the shear modulus ratio deviates from the cube root of the swelling ratio as predicted by Treloar. This might be attributed to the fact that Treloar assumes rubber networks to follow Gaussian statistical model.

4   CONCLUSIONS

In the present work, a simple device for the observation of the interaction between diffusion of liquids and large deformation in rubber was developed. The device consists of four stainless steel plates with spacer bars in between. The presence of spacer bars allow the introduction of pre-compressive strain while exposing simultaneously rubber specimens to biodiesel. It was found that the swelling in rubbers increases with the increase of palm biodiesel content and decreases with the increase of pre-compressive strain.

The effect of the presence of biodiesel on the mechanical response of rubber was also studied. It was observed that the presence of biodiesel and the increase in its content reduce the mechanical strength of the rubber. Furthermore, the shear modulus ratio of swollen and dry rubbers is found to deviate from the one predicted by Treloar.

Finally, it is to note that only uniaxial stress state is observed in the present study. Further investigation on how multiaxial stress state plays role on the diffusion is needed.

REFERENCES

Baek, S. (2004). Diffusion of a fluid through an elastic solid undergoing large deformation. *Int. J. Nonlinear Mech. 39*(2), 201–218.

Bergström, J. S. & M. C. Boyce (1998). Constitutive modeling of the large strain time-dependent behavior of elastomers. *J. Mech. Phys. Solids 46*(5), 931–954.

Dick, J. S. (Ed.) (2001). *Rubber technology: compounding and testing for perfomance.* Hanser.

Fazal, M. A., A. S. M. A. Haseeb, & H. H. Masjuki (2011). Biodiesel feasibility study: An evaluation of material compatibility; performance; emission and engine durability. *Renew. Sust. Energ. Rev. 15*, 1314–1324.

Flory, P. J. & J. Rehner (1944). Statistical Mechanics of Cross-Linked Polymer Networks II. Swelling. *Rubber Chem. Technol. 35*(1), 521–526.

Fukumori, K., T. Kurauchi, & O. Kamigaito (1990). Swelling behaviour of rubber vulcanizates: 2. Effects of tensile strain on swelling. *Polymer 31*(12), 2361–2367.

George, S. (1999). Effect of nature and extent of crosslinking on swelling and mechanical behavior of styrenebutadiene rubber membranes. *J. Membrane Sci. 163*(1), 1–17.

George, S. (2001). Transport phenomena through polymeric systems. *Prog. Polym. Sci. 26*(6), 985–1017.

Haseeb, A. S. M. A., H. H. Masjuki, C. T. Siang, & M. A. Fazal (2010). Compatibility of elastomers in palm biodiesel. *Renew. Energ. 35*(10), 2356–2361.

Hong, W., X. Zhao, J. Zhou, & Z. Suo (2008). A theory of coupled diffusion and large deformation in polymeric gels. *J. Mech. Phys. Solids 56*(5), 1779–1793.

Nah, C., G. B. Lee, J. Y. Lim, Y. H. Kim, R. SenGupta, & A. N. Gent (2010). Problems in determining the elastic strain energy function for rubber. *Int. J. Nonlinear Mech. 45*(3), 232– 235.

Pekcan, O. (2002). Molecular weight effect on polymer dissolution: a steady state fluorescence study. *Polymer 43*(6), 1937– 1941.

Soares, J. S. (2009). Diffusion of a fluid through a spherical elastic solid undergoing large deformations. *Int. J. Eng. Sci. 47*(1), 50–63.

Trakarnpruk, W. & S. Porntangjitlikit (2008). Palm oil biodiesel synthesized with potassium loaded calcined hydrotalcite and effect of biodiesel blend on elastomer properties. *Renew. Energ. 33*(7), 1558–1563.

Treloar, L. R. G. (1975). *The Physics of Rubber Elasticity.* London: Oxford University Press.

Zhang, H. & A. Cloud (2007). Research Progress in Calenderable Fluorosilicone with Excellent Fuel Resistance. *SAMPE.*

Constitutive Models for Rubber VII – Jerrams & Murphy (eds)
© 2012 Taylor & Francis Group, London, ISBN 978-0-415-68389-0

Determination of the behaviour of rubber components under hydrostatic pressure

M. Stommel & J. Zimmermann

Chair for polymeric materials, Saarland University, Saarbruecken, Saarland, Germany

ABSTRACT: Dimensioning of rubber components by using FE-simulations is state of the art and offers a satisfactory result's quality for most applications. In contrast to this especially for mounts or bushings with a confined rubber component problems arise. For example the material stiffness can show a large deviation between test and simulation results. The small free surface and the hydrostatic pressure these components are exposed to can lead to an insufficient simulation quality since the assumption of incompressibility becomes more and more invalid. Therefore, the final products have to be tested under real conditions to determine their behaviour in use which means high effort and costs. To reduce the development costs and optimise the quality of the simulations, an exact knowledge of the material parameters such as the bulk modulus is essential. This study presents a test bench which enables the determination of this parameter for a range of pressure from 1 to 300 bar. First results are recorded for a filled NR. Furthermore the interrelationship of the mechanical behaviour and the hydrostatic pressure is investigated. Tensile tests under hydrostatic pressure (range from 1 to 300 bar) are conducted and the results are presented.

1 INTRODUCTION

Dimensioning of rubber components by using finite elements tools is state of the art in industry. Common simulation tools provide a various number of implemented material models to simulate the hyperelastic material behaviour of these components. These models are based on either a strain energy density function e.g. Mooney-Rivlin Mooney (1940), Ogden (1972) or molecular-statistical approaches e.g. Arruda-Boyce Boyce (1996), Kilian (1981). The bulk modulus or its reciprocal, the compressibility, are important material parameters to optimise the quality of the simulation results achieved by these models.

Simulations are usually carried out by the assumption of incompressible material behaviour or by setting the bulk modulus K to a constant value which is several orders of magnitude higher than the shear modulus. Figure 1 shows a typical dependence of pressure on volume ratio for a rubber component. The bulk modulus corresponds to the gradient of the graph. In practise the dependence of the bulk modulus on the value of pressure is neglected. To cover a wide range of pressure, the bulk modulus is approximated by the secant's gradient, which offers a mean value for K. Particularly for lower and higher pressure values the bulk modulus deviates significantly from the secant's modulus, which degrades the quality of simulation results in these pressure areas. However, this method offers a satisfactory results' quality for most of the technical applications rubber is used in.

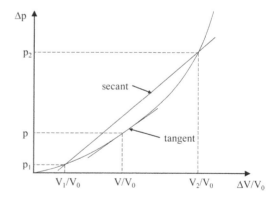

Figure 1. Compression behaviour of a rubber component.

In case of highly confined rubber components such as embedded mounts or bushings (Fig. 2) this assumption is no longer valid. Due to the large fraction of rubber surface bonded to metal a hydrostatic pressure sets up under load. Therefore, a large deviation between simulation results and the components' real life behaviour of characteristics like material stiffness could be obtained. In this case the quality of the simulation results depend on the value approximated for the bulk modulus.

Similar problems arise for rubber components under pressure loads like seals. Figure 3 shows exemplary the dependence of the nominal stress on the bulk

Figure 2. Example of the rubber component of a calibrated and radial deflected confined bushing used in automotive vehicles.

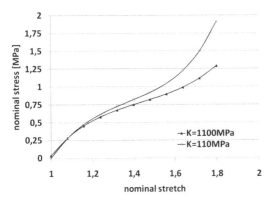

Figure 3. Dependence of simulation result for tensile stress on bulk modulus.

modulus for an NR in a tensile test under a constant pressure of 300 bar.

These examples show the importance of the bulk modulus in simulating rubber components. The exact knowledge of the material's bulk modulus at different hydrostatic pressure values is essential to avoid elaborate tests under real circumstances. This can be obtained in compression tests, where the volume change of a rubber specimen over an increase in pressure is recorded.

Besides the bulk modulus, the tensile behaviour of rubber components under hydrostatic pressure plays an important role in many applications like seals or bearings. In Le Cam (2010) the significance of the change in volume during deformation for characterising rubber components is pointed out. Therefore the change in volume according to the hydrostatic pressure, which influences the material characteristics like material stiffness of these components as well, has to be regarded. In the following, a test bench is introduced, that enables the determination of compression behaviour of rubber components for a pressure range from 1 bar to 300 bar.

Furthermore, the material models for FE-simulations of rubber components assume a split of the

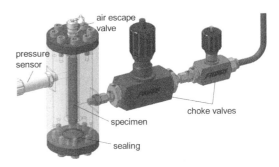

Figure 4. Test bench for determination of bulk modulus.

strain energy function in a deviatoric, incompressible, and a volumetric part. To verify this assumption, the test bench allows tensile tests under various hydrostatic pressure levels as well.

2 TEST BENCH

In this chapter the test bench to conduct compression tests is presented. Furthermore, with small modifications, tensile tests under hydrostatic pressure can be conducted.

2.1 Compression test

The method used to determine the bulk modulus is a direct measurement of change in pressure due to a reduction in volume of a rubber specimen. Therefore, the specimen is placed in a metal cylinder filled with water. The increase in pressure is realised by a reduction of volume due to a screwing in of a spindle, which is realised by a choke valve (Fig. 4).

The bulk modulus can be determined for a pressure range from 1 bar to 300 bar by this test setup. The test specimen is cylindrically shaped with a volume fraction of rubber compared to the total chamber of 0.2.

2.2 Tensile test

In figure 5 the test bench for conducting tensile tests under hydrostatic pressure is shown. The strain of the specimen is applied by another spindle in the head and a nut. The measurement of the force is realised indirectly by a combination of a path sensor and a compression spring. Due to the known spring stiffness, the force and therewith the tensile tension is determined. The tested specimen is shaped like a stadium (see figure 5). So there is no need for a clamping, and the specimen is hold by 2 bolts. With the length of this specimen, tensile tests up to an elongation of 100% are possible. The test bench itself allows much higher elongation dependent on the employed specimen. The strain rate is also variable. The hydrostatic pressure is adjustable between 1 bar and 300 bar.

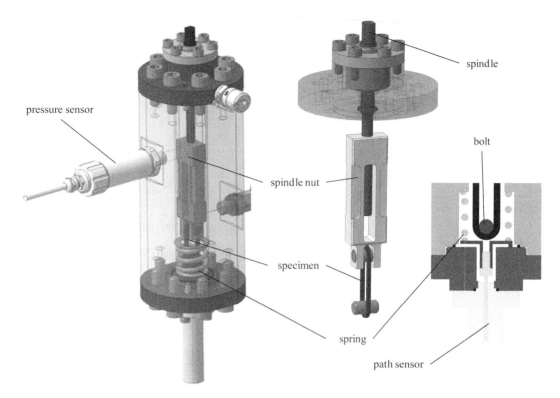

Figure 5. Test bench for conducting tensile tests.

3 RESULTS

First tests are conducted for a filled NR with 50 Shore A. The results achieved in compression tests as well as tensile tests under hydrostatic pressure are presented in the following. Simulated tensile tests under hydrostatic pressure with different values for the compressibility are compared to the achieved results from the test bench. Furthermore simulations of the bushing (Fig. 2) were conducted and the dependence of their results on the choice of the bulk modulus is carried out.

3.1 Compression tests

Before measuring the bulk modulus of a rubber specimen, the compressibility of the system is to determine. Therefore a cylindrical steel specimen is used to detect the compression behaviour of the assembly. Its volume corresponds to the rubber's volume. Based on the well-known bulk modulus of steel, the compression behaviour of the system can be determined and used as a reference for the following experiments with rubber specimens. Figure 6 shows the measured system response. The pressure over a reduction of volume is recorded. This reduction is variable up to 3100 mm³. For a volume ratio of more than 0.005 curve linearity is obtained. The setting of the system's components causes the curvature at the beginning.

The testing of the rubber specimen is carried out under the same conditions. In this case, the gradient of

the system response compared to the reference measuring is significantly lower. However the shape of the curve is similar to the reference curve. The result for the tested rubber specimen after eliminating the influence of the test bench is shown in figure 6.

The bulk modulus of the tested rubber specimen corresponds to the gradient of the displayed graph. In the beginning of compression, for volume ratios up to 0.005, the incline is at a very low level, which means due to the very low pressure values, a bulk modulus of

$$K = \frac{\Delta p}{V / V_0} = 112 \, MPa. \qquad (1)$$

In the field of volume ratio from 0.005 to 0.01, the curvature increases. For volume ratios higher than 0.01, there is an approximately linear relationship with pressure. Therefore, the bulk modulus for greater volume ratios than 0.01 is $K = 1070 \, MPa$.

Several further tests with using other types of rubber and shore hardness show the same phenomena. A linear progression at lower volume ratio is followed by an increase in curvature and another constant level of bulk modulus for higher ratios. There is an increase in bulk modulus of a factor about 10. It is evident, that because of this behaviour, the commonly used approximation of the bulk modulus by considering the secant's gradient is not feasible for wider fields of pressure.

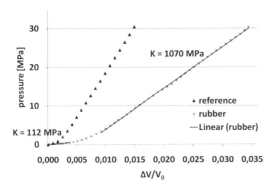

Figure 6. Results of compression testing.

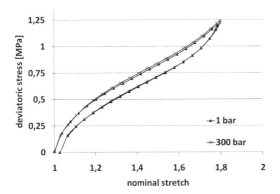

Figure 7. Comparison of nominal stress for two different states of hydrostatic pressure.

3.2 Tensile tests

Along with the compression tests, the tensile behaviour of rubber under hydrostatic pressure is analysed. The used test specimens are of 2 mm thickness. They have two ligaments 4 mm wide (Fig. 5). For this type of specimen, tests were conducted at hydrostatic pressure values of 1 bar, 100 bar, 200 bar and 300 bar.

In figure 7 the deviatoric stress component S_{11} in loading direction of the fourth load cycle for the described specimen is shown. The results for hydrostatic pressure values of 1 bar and 300 bar are exemplary represented. All tests conducted show approximately the same material behaviour. The maximum deviation of the curves is about 1%. This confirms the assumption, that there is no apparent influence of hydrostatic pressure up to 300 bar on the deviatoric behaviour of an NR for lower stretches up to 100%.

The results obtained so far show an increase in bulk modulus by a factor of 10 from lower to higher hydrostatic pressure values. In contrast to this, the influence on deviatoric stress has proved negligible. Therefore, the split of the strain energy function W in a deviatoric and a volumetric part is valid.

3.3 Comparison of test results with FE-simulations

In the following the effects of these results on the quality of FE-simulations are determined. The FE-model

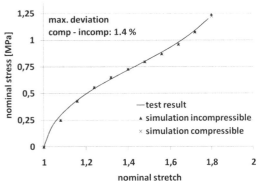

Figure 8. Comparison of FE-simulation results (incompressible and compressible) with tensile test results.

used is based on Kilian (1981) with the strain energy function formulated as:

$$W = W^{dev}(\bar{I}_1, \bar{I}_2) + W^{vol}(J) \qquad (2)$$

$$W = \mu \left\{ -(\lambda^2 - 3)[\ln(1 - \eta) + \eta] - \frac{2}{3}\alpha \left[\frac{\tilde{I}-3}{2}\right]^{\frac{3}{2}} \right\}$$
$$+ \frac{1}{D}\left(\frac{J^2-1}{2} - \ln(J)\right) \qquad (3)$$

with:

$$\eta = \sqrt{\frac{\tilde{I}-3}{\lambda_m{}^2-3}} \qquad (4)$$

$$\tilde{I} = (1 - \beta)\bar{I}_1 + \beta\bar{I}_2 \qquad (5)$$

and

$$D = \frac{2}{K} \qquad (6)$$

where μ = initial shear modulus; λ_m = locking stretch; α = global interaction parameter; β = linear mixture parameter; $\bar{I}_i$ = strain invariants of the left Cauchy-Green strain tensor; and J = total volume change.

3.3.1 Comparison of tensile tests

At first the influence of the assumption of incompressibility on the tensile behaviour of rubber components is pointed out. Therefore, FE-simulations with the incompressible part of the upper strain energy function, the first term, are compared to simulation's results with an assumed bulk modulus of $K = 110$ MPa and the test results for 1 bar.

As shown in figure 8 the maximum deviation between the both simulation results is 1.4% at a stretch of 1.8. This deviation is as well as the difference between these results and the test results negligible. That means, that provided lower hydrostatic pressure values, the bulk modulus shows no significant influence on the tensile behaviour of rubber components.

According to the tests conducted, simulations for a hydrostatic pressure of 300 bar were carried out. As shown in chapter 3.1, the bulk modulus in this

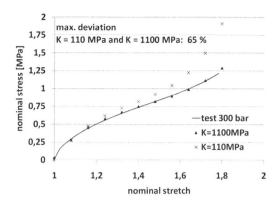

Figure 9. Comparison of FE-simulation results with different value of K and tensile test results under 300 bar.

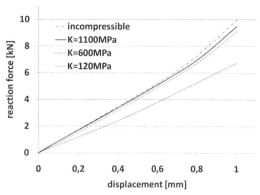

Figure 10. Simulation results of bushing for different values of the bulk modulus.

field of pressure is about 1100 MPa. This is compared to $K = 110$ MPa, which matches the value for lower hydrostatic pressures in figure 6. The results shown in figure 9 emphasize the importance of the knowledge of the exact value of the bulk modulus.

The bulk modulus achieved by the compression tests for 300 bar leads to a satisfying result's quality for the simulation of a tensile test under hydrostatic pressure. The maximum deviation from the test results is 2%. In contrast to this a factor 10 lower bulk modulus shows a different material response in the tensile test. The nominal stress deviates at a nominal stretch of 1.8 about 65% from the experimental results. Based on this, there is no reliable dimensioning possible by assuming the bulk modulus under similar conditions.

3.3.2 Influence on the simulation of confined rubber components

In the following the effect of the represented results on the simulation of practical rubber components is presented for the example of the bushing shown in figure 2. For this kind of bushings used in automotive vehicles, the material stiffness is a crucial parameter to influence significantly its road handling. Due to the confined rubber components, hydrostatic pressure up to 40 MPa occurs. This affects, as shown above, the mechanical behaviour of the rubber component.

The simulation contains the calibration of the bushing which leads to a hydrostatic pressure in the areas of the rubber component bonded to metal. Furthermore a radial deflection of the inner surface shell is applied. The rubber used is the same NR as in the conducted test. Due to a hydrostatic pressure higher than 10 MPa, a bulk modulus $K = 1100$ MPa is used in the FE-simulation. The reaction force occurring at the inner shell is represented in figure 10. Further simulations with $K = 120$ MPa, which relates to the value for lower pressure in figure 6, and $K = 600$ MPa as a mean value which could present the secant's modulus were conducted. Besides, the bushing was simulated with the assumption of an incompressible material behaviour.

A deviation of up to 30% from the simulation with $K = 1100$ MPa occurs for the lower pressure value.

The deviation for the other simulation results from the first one is about 5%. Too high values for the bulk modulus lead to the assumption of a too high material stiffness. For too low values this effect reverses.

4 CONCLUSIONS

The tests conducted with the presented test bench validate the assumption of incompressibility or neglecting the dependence of the bulk modulus on hydrostatic pressure to achieve a satisfactory result's quality in FE-simulations. It is shown that the deviatoric stress is not influenced significantly by the hydrostatic pressure. Therefore the split of the strain energy function in a deviatoric and a volumetric part is valid.

In contrast to this, for confined rubber components the exact knowledge of the bulk modulus' dependence on pressure is important for the result's quality. The compression tests underline this dependence. With the use of this information FE-simulations were conducted. A significant deviation in simulation results was obtained due to a variation of the bulk modulus. It is shown, that the behaviour under hydrostatic pressure and the bulk modulus at different pressure values are crucial for the dimensioning of rubber components used under these conditions.

REFERENCES

Boyce, M.C. 1996. Direct comparison of the Gent and Arruda-Boyce constitutive models of rubber elasticity. *Rubber chemistry and technology* 69: 781–785.

Kilian, H.-G. 1981. Equations of state of real networks. *Polymer* 22: 209–216.

Le Cam, J-B 2010. A review of volume changes in rubbers: the effect of stretching. *Rubber chemistry and technology* 83: 247–269.

Mooney, M. 1940. A theory of large elastic deformation. *Journal of applied physics* 11: 582–592.

Ogden, R.W. 1972. Large deformation isotropic elasticity – on the correlation of theory and experiment for incompressible rubberlike solids. *Proc. Roy. Soc. A, Mathematical and Physical Sciences* 326: 565–584.

Constitutive Models for Rubber VII – Jerrams & Murphy (eds)
© *2012 Taylor & Francis Group, London, ISBN 978-0-415-68389-0*

Creating a uniform magnetic field for the equi-biaxial physical testing of magnetorheological elastomers; electromagnet design, development and testing

D. Gorman, S. Jerrams, R. Ekins & N. Murphy
Centre of Elastomer Research, Dublin Institute of Technology, Republic of Ireland

ABSTRACT: This paper investigates a method to provide the magnetic field requirements for physical testing of magnetorheological elastomers (MREs) subjected to equi-biaxial loading using the bubble inflation method. For accurate physical testing of MREs, detailed knowledge of the properties of the applied magnetic field is required. To obtain reliable data it is essential to determine the strength, uniformity and directionality of flux density. A Halbach cylinder array can produce a magnetic field of approximately uniform flux density in one direction for a reference plane perpendicular to the direction. However, it is limited by the fixed field strength. To overcome this significant limitation, an electromagnetic array based on the geometry of a Halbach cylinder is proposed. This electromagnetic array will be capable of generating a uniform magnetic field, for the reference plane and in the perpendicular direction, that is capable of having the flux density varied to offer a range of field strengths for tests on different elastomer samples. FEA simulations of uniform electromagnetic arrays have been modelled. Ultimately, a model is offered that simulates the behaviour of an electromagnetic array and the capability to generate a uniform magnetic field with different flux densities and directionality over the required volume. The advantages and disadvantages of an electromagnetic array over a fixed strength Halbach cylinder were investigated and a detailed comparison of both was carried out. Preliminary tests have been conducted on prototype electromagnets and the measured magnetic fields have been found to be in agreement with the FEA model. In addition to the magnetic field experiments, tests have been carried out on a compressed air cooling system to allow continuous operation of the electromagnets for the duration of a fatigue test without test samples becoming overheated and chemical degradation occurring. These tests are also necessary to establish conditions where there is minimal drop in field strength due to the increased resistance associated with temperature increases during prolonged dynamic testing. In conclusion a design for an electromagnetic array for the equi-biaxial testing of MREs is presented along with proposals for further testing to fully develop the array and establish standard dynamic test procedures for the material.

1 INTRODUCTION

MRE's are smart elastomers which change physical characteristics in the presence of a magnetic field (A. Boczkowska, 2009). These changes are due to ferromagnetic particles (usually iron) in the elastomer moving to align with the applied magnetic field (G.V. Stepanov, 2007). There are two types of MRE's, Isotropic and Anisotropic (Zsolt Varga, 2006). The difference between the two types is due to alternative curing processes. For Isotropic MRE's the ferromagnetic particles are added during curing and no external magnetic field is applied resulting in a random approximately uniform distribution. In the anisotropic case, an external magnetic field is applied during the curing process resulting in the ferromagnetic particles moving in the gel to form aligned chains and being locked into this arrangement once the curing process is completed and the elastomer is formed.

The majority of experimental evaluation and testing of MRE's has been carried out on uniaxially loaded samples with the magnetic field assumed to be homogeneous in both flux density and directionality over the entire sample volume. Due to the larger sample volume required for the equi-biaxial physical testing of elastomers using bubble inflation (N Murphy, 2007), generating a magnetic field which maintains the necessary uniformity requires a more complex magnetic array.

A magnetic field of uniform strength is required as the MR (magnetorheological) effect is due to the magnetic iron particles in the MRE attempting to align in the direction of the applied magnetic field (Bica, 2009). This alignment is caused by the interactions of magnetic dipoles (G.V. Stepanov, 2007). The force on a magnetic dipole of moment (m) in a magnetic field (B) is given in equation 1.1 (I.S. Grant, 1990).

$$F grad(m.B) \tag{1.1}$$

As the force which causes the alignment of particles is dependent on the magnetic field, the MR effect is dependent on the magnetic field strength. Therefore a non uniform strength field would result in an altered

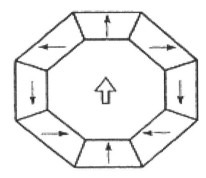

Figure 1. Halbach Cylinder (Coey, 2002).

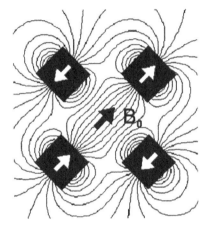

Figure 2. Open access Halbach array (B.P. Hills, 2005).

alignment and change in mechanical properties over the volume of the sample.

Uniform magnetic field direction is required for testing of MREs in both uniaxial and equi-biaxial cases, as the directional alignment of the field lines, applied force, and particle chains (Anisotropic MREs only) influence the material behaviour in a magnetic field. Both (Zsolt Varga, 2006) and (A. Boczkowska) found that a magnetic field applied parallel to the particle chains produces a larger MR effect compared with a field of the same flux density applied perpendicular to the particle chains.

A Halbach cylinder is a permanent magnet array arranged in a cylinder to produce a uniform field inside the cylinder and zero field outside (Coey, 2002). The Halbach cylinder produces a field of uniform strength and field lines in the required direction as shown in figure 1 but the fixed field strength limits its use as an effective device over a full range of test requirements.

Hills (B.P. Hills, 2005) developed an open access Halbach array to allow greater access to the sample while under NMR (nuclear magnetic resonance) imaging. This more open design came at the expense of both field strength and uniformity. A schematic of the Hills design is shown in figure 2.

In order to provide a more comprehensive testing system which allows for a variable field strength, an equivalent arrangement of electromagnets is required. There are two significant problems with using electromagnets in preference to permanent magnets. These are, the electric power requirement and providing adequate cooling of the system (Montgomery, 1963). To generate a static magnetic field, a constant source DC voltage is required as field is proportional to current. It is also necessary to operate the electromagnetic array at a constant temperature as resistance is proportional to temperature and increases in temperature cause a drop in current and magnetic field for a given voltage.

2 EXPERIMENTAL METHODS AND RESULTS

2.1 Modelling the magnetic field

An FEA model was created using Finite Element Method Magnetics, FEMM4.2 (Meeker). This software was used to solve magnetic equations in which the fields are time-invariant. The field intensity (H) and flux density (B) must obey equations 2.1 and 2.2

$$\nabla \times H = J \qquad (2.1)$$

$$\nabla \times B = 0 \qquad (2.2)$$

Each material is modelled based on its magnetization curve which can be generated by plotting values determined from equation 2.3

$$B = \mu H \qquad (2.3)$$

Equation 2.3 is non linear for magnetic materials as permeability (μ) is a function of B

The FEMM software has been use in the design of a magnetic system for the testing of MRFs (magnetorheological fluids) by (S.A. Mazlan, 2009). The results obtained by the simulations where in close agreement with the measured field.

The initial stage of the design method was to create an electromagnetic model based on the open access design presented by Hills (B.P. Hills, 2005) and shown in figure 2. An electromagnetic version of this arrangement was produced using 4 identical iron core electromagnets each with 1500 turns of 1 mm copper wire drawing a current of 15 amps. The field produced by such an arrangement is presented in figure 3 and the area inside the coils is 80 mm × 80 mm. To obtain the same magnetic orientation as Hills' design, current in the two non central coils flows in the opposite direction to the other two coils.

The field generated along the central line in figure 3 is shown graphically in figure 4.

It is clear from the data shown in figures 3–4 that a simple replication of Hills opens access Halbach cylinder will not produce a field of uniform flux density to allow for the testing of MREs under equi-biaxial conditions.

A series of modifications were made to the basic design shown in figure 3 and these are presented in figure 5.

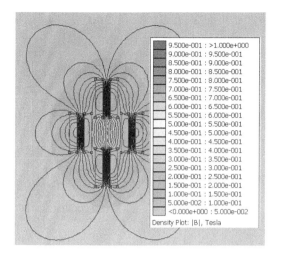

Figure 3. Electromagnetic array based on Hills' design.

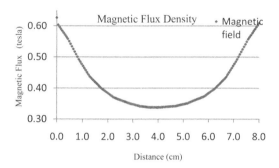

Figure 4. Magnetic flux of array shown in figure 3.

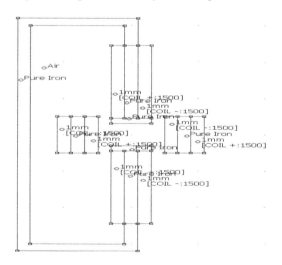

Figure 5. Proposed array.

Figure 5 shows the changes made to the array. These include the connection of the two central electromagnets by an iron structure. This iron structure forms a pathway for the flux line to follow, thus increasing field strength.

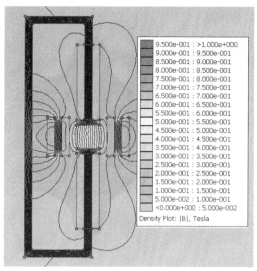

Figure 6. Simulated array field.

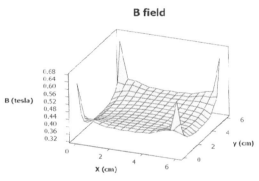

Figure 7. Magnetic Flux density of array shown in figure 6.

The second notable change is the elongation of the central magnets. This allows for the requisite number of turns to be made with less wire, reducing the resistance of the coil and creating a more efficient electromagnet.

This modification was not applied to the other two electromagnets as their position was maintained to provide a field with uniform direction and they have less of an effect on the overall flux density.

The final changes where the addition of iron pole pieces to the central elongated electromagnets. This had the effect of increasing the uniformity of the field but also reduces the area between the coils to 80 mm × 60 mm.

The magnetic field produced by this modified array is shown in figure 6. All electromagnets had a current of 15 amps with 1500 turns.

The field flux density in the 60 mm × 60 mm square highlighted in figure [6] is shown graphically in figure 7.

It can be seen from figures 6–7 that the array shown in figure 5 is capable of producing a field with both

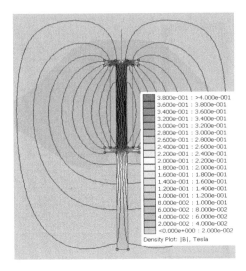

Figure 8. Simulated prototype field.

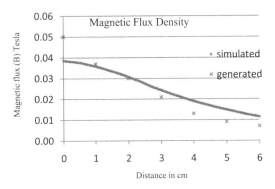

Figure 9. Simulated v generated field from prototype.

the required uniformity of direction and flux density for the testing of MREs using bubble inflation.

2.2 The simulated and generated magnetic fields

To verify that the array presented in figure 5 is capable of producing the magnetic field simulated in figures 6-7, one of the central elongated electromagnets was tested and the actual measured magnetic field was compared with the simulated field produced by such a coil modelled with FEMM software.

The model of the prototype is shown in figure 8.

The field produced along the line v distance from the coil and the actual field measured using a hall probe are shown in figure 9.

2.3 Magnetic array cooling

To ensure that the magnetic array operates with a constant current it must be held at a fixed temperature. This is achieved by cooling the electromagnet with compressed air via intakes and vents incorporated into the pole pieces of the magnets. (Outer radius 60 mm inner, radius 20 mm).These are shown in figure 10.

Figure 10. Pole Pieces.

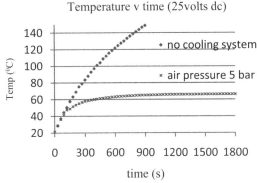

Figure 11. Temperature v time.

The left image in figure 10 shows the vents from which the hot air exits the magnet and is the pole piece facing the sample. The right image in figure 10 shows the air intake system. A hole is drilled through one of the eight grooves in the same plane as the axis of the central bore. The air flow enters through this hole and then flows into the other seven grooves and from them down cooling channels incorporated into the windings and out through the vents.

Figure 11 shows the effectiveness of the cooling system.

When the magnet was powered from a 25 V supply without cooling, a temperature of 140°C was reached after fifteen minutes continuous running. This was accompanied by a reduction in current from 6.84 amps to 4.64 amps. However, when the cooling system was actuated, a steady state was achieved at 66°C degrees with a current of 6.12 amps.

3 CONCLUSIONS

The electromagnetic array presented in figure 5 is capable of producing a suitable field for the equibiaxial physical testing of MREs. The field produced is uniform in both the required flux density and directionality of the field lines similar to that produced by a Halbach cylinder. The advantage of such a design over a Halbach cylinder is that electromagnetic arrays can produce a uniform field over a range of flux densities. For this design, the flux density can be varied in the range 0–420 mT.

By comparison with a permanent magnet design the disadvantages of any electromagnetic design, are the requirements for constant input of energy and the removal of heat energy.

The heating problems can be overcome by the proposed cooling system and a steady state can be maintained for the duration of a fatigue test on a magnetorheological elastomer sample.

4 PROPOSED FURTHER WORK

The prototype will be evaluated at higher currents and once the design is proven, the full electromagnetic array will be manufactured. This will lead to an extensive test programme investigating the dynamic properties of a range of MREs subjected to complex loading. In particular, fatigue resilience, stress softening and set will be determined for MREs based on natural rubber (NR), silicone and ethylene propylene diene monomer (EPDM) matrices with a range of ferromagnetic particle sizes and volume fractions.

REFERENCES

Bica I. Compressibility modulus and principal deformations in magneto-rheological elastomer: The effect of the magnetic field: J. Ind. Eng. Chem., 2009.

Boczkowska A., Awietjan S. Mechanical properties of magnetorheological elastomers

Boczkowska A., Awietjan S. Smart composites of urethane elastomers with carbonyl iron – [s.l.]: Journal of material science, 2009. – 15: Vol. 44.

Coey J.M.D. Permanent magnet applications – [s.l.]: Elsevier, 2002. – 248 (441–456): Journal of magnetism and Magnetic Materials.

Grant I.S., Phillips W.R. Electromagnetism [Book]. – [s.l.]: Wiley, 1990. – Vol. 2nd edition. chapter 4 p. 131.

Hills B.P., Wright K.M., Gillies D.G. A low field, low cost Halbach magnet array for open access NMR [s.l.]: Journal of Magnetic Resinance, 2005. – 175 (336–339).

Mazlan S.A. Issa A., Chowdhury H.A, Olabi A.G. Magnetic circuit design for the squeeze mode experiments on magnetorheological fluids: Materials and Design, 2009. – 30 1985–1993.

Meeker D. [SoftwareOnline]. – http://www.femm.info.

Montgomery D.B. The Generation of High Magnetic Fields Reports on progress in physics 1963. Vol. 26

Murphy N., Hanley J., Ali H., Jerrams S., The effect of specimen geometry on the multiaxial deformation of elastomers: Taylor & Francis, 2007. – ECCMR V: Paris.

Stepanov G.V., Abramchuk S.S., Grishin D.A., Nikitin L.V., Kramarenko E Yu., Khokhlov A.R. Effect of a homogeneous magnetic field on the viscoelastic behaviour of magnetic elastomers. – [s.l.]: Elsevier, 2007. 488–495: Vol. Polymer 48.

Varga Z., Filipcsei G., Zrınyi M. Magnetic field sensitive functional elastomers with tuneable elastic modulus: Elsevier, 2006. 227–233: Vol. Polymer 47.

Constitutive Models for Rubber VII – Jerrams & Murphy (eds)
© 2012 Taylor & Francis Group, London, ISBN 978-0-415-68389-0

A method of real-time bi-axial strain control in fatigue testing of elastomers

N. Murphy

School of Manufacturing and Design Engineering, Dublin Institute of Technology, Dublin, Ireland

J. Hanley & S. Jerrams

Centre for Elastomer Research, Dublin Institute of Technology, Dublin, Ireland

ABSTRACT: Previous publications by the authors (Murphy *et al*, 2009) demonstrated a method of conducting equi-biaxial fatigue tests of elastomers using the bubble inflation method. Each of these tests was carried out at a constant engineering stress amplitude. A question posed by this research was whether a sample could be tested bi-axially under strain amplitude control over multiple cycles and large deformations. A limitation of the testing system used to determine stress and strain values in the aforementioned tests was its ability to process the results from the vision system in real-time. This paper presents an improved real-time strain measurement and control system and the accompanying results for tests carried out using different parameters, including control at constant pressure, volume, stress and strain amplitudes. This is a further development of the Dynamet system first presented at ECCMR 2005.

1 INTRODUCTION

A method of accurately controlling engineering stress during equi-biaxial cyclic testing of elastomers using the validated Dynamet system has previously been reported (Murphy *et al*, 2007, 2009). Using this method, stress amplitudes and peak stresses were maintained at constant values during fatigue tests. By monitoring the change in stretch ratio with accumulation of cycles, the peak engineering stress throughout a fatigue test was controlled, allowing constant engineering stress amplitude equi-biaxial fatigue tests to be carried out for an elastomer. A separate calibration curve was required for each material type or specimen thickness and this time consuming procedure had to be carried out prior to long-term fatigue tests. Changes in values of λ will be more rapid for materials which exhibit greater set during successive cycles and specimen thickness will also affect the rate of change of λ values. Using this method of loading, S-N curves for an elastomer can be generated which exhibit lower levels of scatter than those typically associated with uniaxial results. Hence, tests to establish fatigue lives of elastomers can be conducted in shorter time periods than in uniaxial testing. While the initial system was capable of producing high quality equi-biaxial fatigue data, the level of user intervention in the control and data processing was considerable and further development was necessary to address this.

A more direct method of controlling engineering stress during a dynamic test was required and led to the development of a machine vision system, enabling the three-dimensional profile of the sample surface to be measured in real-time. The improved high speed image acquisition system allows the possibility of continuous data recording during testing and facilitates the employment of higher cycling test frequencies.

Selectable control parameters in the revised system include engineering stress, true stress, volume, pressure and strain (stretch ratio). This paper discusses the development of the revised system that employs these control parameters and some indicative results are presented.

2 THEORY

When values of stress are to be evaluated for a specific test-piece undergoing deformation, it is crucial to specify if the stress values are to be expressed as engineering stress (σ_{eng}) or true stress (σ_{true}). This is of particular importance if comparisons are to be made between different test loading methods. The influence of the use of engineering stress as a measurement standard has pronounced implications when dealing with hyperelastic materials while for other materials which undergo very small elastic deformation in service, the effect is negligible. The relationship of true stress to engineering stress for both uniaxial tension and equi-biaxial tension can be expressed as;

$$\sigma_{true}/\sigma_{eng} = \lambda \tag{1}$$

where λ represents the stretch ratio in the direction of the applied load. This relationship is valid also in the case of equi-biaxial deformation using bubble inflation. For the bubble inflation test method developed in this study, these relationships form the basis of the control system to allow testing to preset values of engineering stress.

Table 1. Stress-strain relationships for uniaxial, bi-axial and bubble inflation.

	Uniaxial tension of dumb-bells	Equi-biaxial tension of in-plane stretched sheets	Equi-biaxial tension of inflated thin shells
Stretch Ratio, λ	l/l_o	l/l_o	l/l_o, r/r_o
Strain Relation, $I_3 = 1$	$A = A_o/\lambda$	$t = t_o/\lambda^2$	$t = t_o/\lambda^2$
$\sigma_{nominal}$ (also called σ_{eng})	F/A_o	$F/A_o \cdot t$	$P \cdot (r/2t_o) \cdot \lambda$
σ_{true}	$\sigma_{nominal} \cdot A_o/A$	$\sigma_{nominal} \cdot (l_o/l) \cdot (t_o/t)$	$\sigma_{nominal} \cdot (t_o/t)$
$\sigma_{true}/\sigma_{nominal}$	λ	λ	λ

Engineering stress for the bubble inflation case is a function of pressure, radius, original thickness and stretch ratio (Murphy *et al*, 2009, Mott *et al*, 2003, Javořik, Dvořák, 2007) and can be expressed as;

$$\sigma_{eng} = P(r/2t_o) \cdot \lambda \qquad (2)$$

Unlike other testing modes where the parameters controlling stress are easily measured or inferred, bubble inflation requires continuous measurement of P, r and λ in real-time during the test procedure. While this is not difficult for simple static tests to failure, where the deformation data can be post processed, it is difficult where repeatable dynamic measurement and control is required. Table 1 illustrates the importance of accurate dynamic deformation data for the bubble inflation case by comparison with biaxial stretch frame and uniaxial tests.

The introduction and validation of real-time strain monitoring was the initial developmental stage in achieving real-time stress control for the Dynamet system. The ability to record three dimensional coordinates of markings on the bubble surface provided a means of calculating r and λ in real-time. Consequently, this resulted in a system with the ability to provide instantaneous engineering stress control, or strain control as required. Markings used for stretch ratio calculations were required to be placed close to the bubble pole within the region where equi-biaxial deformation occurs (Johannknecht *et al*, 2002). Two methods were employed using three or five markings respectively. In the case of the three marker system, the markings were placed within the region of equi-biaxial deformation giving stretch ratio measurement and radius calculations based on a three point curve fit. A flat sheet specimen subjected to bubble inflation will assume an elliptical shape when the bubble height exceeds the value of the radius of the inflation orifice (Treloar, 1944). The five point method obtained stretch ratio measurements from three points within the region exhibiting equi-biaxial tension, while the extra points were used to calculate the radius of curvature of an ellipse using a least squares fit based on determining the constants $P_{(1-6)}$ for the conic representation of an ellipse given in equation 3.

$$P_{(1)}x^2 + P_{(2)}xz + P_{(3)}z^2 + P_{(4)}x + P_{(5)}z + P_{(6)} = 0 \qquad (3)$$

Once resolved using regression analysis, this conic equation can be converted to a conventional equation

of an ellipse. However, when using the five marker method and the elliptical curve fit, it is necessary to calculate an average radius of curvature for the equi-biaxial region to avoid over estimations of the stress values based on the maximum radius value which occurs at the bubble pole.

3 MATERIALS

EPDM rubber of 70 Shore A hardness, cross-linked with sulphur and containing low activity carbon black was chosen for this investigation. The test samples consisted of 50 mm diameter, 2 mm thick EPDM discs inflated through a 38 mm orifice using a non swelling inflation fluid (Jerrams *et al*, 2008).

4 METHODOLOGY

When designing the real time control system, the relationship between the speed of testing and the elimination of overshoot was considered. A camera system was specified with the capability of providing a frame capture rate of 100 Hz, giving 100 data samples for an inflation cycle carried out at a sample fatigue cycling frequency of 1 Hz, with control action executed every 10 milliseconds. As a comparison, the previous system was capable of obtaining 29 data points for every sampled fatigue cycle. Figure 1 demonstrates the operation of the modified system.

Methods of lighting and thresholding for the vision data were developed to address the difficulties posed by the change in orientation and position of the target markers during inflation cycles. Thresholding and centroid extraction, as shown in Figure 2 are essential to filter out noise from the vision data such as that caused by samples with smooth surface finishes which produce varying reflections throughout the inflation cycles.

The stress and stretch ratio were calculated in real-time in the control program, by using equation (2) for engineering stress and computing the stretch ratio from the following equation:

$$1 + l/l_o = \lambda \qquad (4)$$

where l is the measured circumferential deformation on the test sample surface, obtained from the measured movement in the x-axis and corrected for curvature using the calculated value of r.

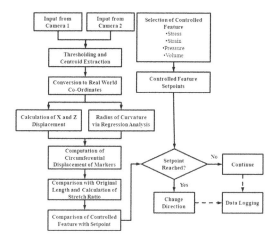

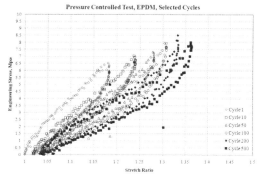

Figure 3. Pressure Control Results, Selected Cycles.

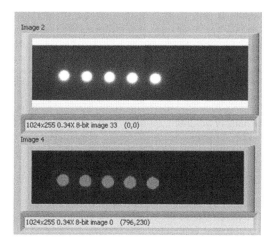

Figure 2. Thresholding and Centroid Extraction.

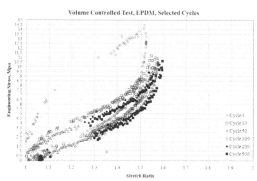

Figure 4. Volume Control Results, Selected Cycles.

Figure 1. Process Flow Diagram for the Dynamet strain control system.

The system stores all critical parameters during tests, including cycle number, pressure, volume, stress and strain. Data at a sampling resolution of 10 milliseconds can be obtained for subsequent analysis from any cycle in the test.

5 TEST RESULTS AND DISCUSSION

The preliminary tests from the new system are presented in this section. Four tests were carried out for two purposes; to validate the system and to demonstrate the behaviour of EPDM under different fatigue regimes, including pressure, volume, stress and strain amplitude controlled cycling regimes.

5.1 Pressure controlled test results

The results of the pressure amplitude controlled tests are presented in Figure 3. This test was carried out

between limits of 0 Barg and 0.8 Barg. It is apparent from this plot that the peak values of engineering stress increased in the sample due to stress softening with the accumulation of cycles. The effect of permanent set is demonstrated by the increases in the values of stretch ratio in the depressurised state as the number of cycles increased. This test regime is of benefit when testing components subjected to repeated pressure cycling over long periods (e.g. diaphragms).

5.2 Volume controlled test results

The results of the volume controlled tests are presented in Figure 4. This test was carried out between limits of 0 cc and 8 cc and demonstrated that there is a high degree of stress softening in early cycles, where the volume delivered to the sample is accompanied by high stressing of the elastomer. In subsequent cycles, the peak values of engineering stress decreased and the permanent set in the sample resulted in no strain data being recorded below stretch ratios of 1.3. Constant volume tests are of benefit when testing components where the same volume is repeatedly delivered to the sample. However, it can also be observed that constant volume inflations of the bubble do not provide constant strain control limits, as the maximum value for the stretch ratio for a given cycle increases throughout the test due to stress softening in the bubble pole region.

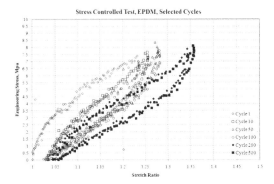

Figure 5. Engineering Stress Control Results, Selected Cycles.

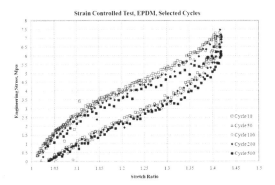

Figure 6. Strain Control Results, Selected Cycles.

5.3 *Engineering stress controlled test results*

The results of the engineering stress controlled tests are presented in Figure 5. This test was carried out between engineering stress limits of 0 MPa and 8 MPa. There was automatic compensation throughout the test to account for changes in bubble radius and increased stretch ratios due to stress softening throughout the test. This self correction of parameters by the control system during the test removed the requirement for user intervention in adjustment of parameters or sample specific calibration constants to ensure correct stress control. This method of control facilitates comparisons with data from load controlled fatigue tests under simple uniaxial and complex bi-axial conditions using stretch frames. The production of true stress control fatigue test data is also possible by the inclusion of the term λ^2 in place of λ in the control equation (2). While the results compare favourably with previously published results for the Dynamet system (Murphy *et al*, 2009), the time taken to produce these results has been greatly reduced.

5.4 *Strain controlled test results*

The results of the strain controlled tests are presented in Figure 6. This test was carried out between stretch ratio limits of 1 and 1.4. Real-time monitoring and calculation of l, the circumferential deformation on the test sample surface and r, the bubble radius, allowed

the stretch ratio of the sample to be maintained within the control limits during cycling. This method of control allows the determination of long-term stress softening of elastomers under complex loading condition at different loading rates. As in the constant volume tests, permanent set of the samples caused difficulties in obtaining data for stretch ratios when approaching a value of unity beyond the initial cycle. This phenomenon would be less pronounced for samples tested under preload (or pre-strained) conditions where the lower control values of stress or stretch ratio are greater than 0 or 1 respectively.

6 CONCLUSIONS

The Dynamet system has been developed from a system capable of providing long-term stress controlled fatigue data requiring a high level of user intervention, into one which provides real time data recording and control with minimal user intervention. This is combined with superior data collection capabilities. The new system allows a number of parallel research themes to be explored in detail. These themes include stress softening under equi-biaxial loading conditions and stress controlled fatigue tests comparing material behaviour under engineering stress control with behaviour under true stress control.

As previously discussed, (Murphy *et al*, 2009, Jerrams *et al*, 2008), an investigation will be carried out on subsequent fatigue tests to determine if there is a limiting value of elastic modulus (E*) at failure for EPDM. This will determine if failure occurs within a material specific complex elastic modulus range regardless of loading methods and will therefore provide a reliable basis for fatigue life prediction. Thereafter, the research programme will be broadened to observe if this failure criterion can be adopted for other non strain-crystallising and strain-crystallising rubbers.

A number of further improvements to the Dynamet system are in progress, particularly around the development of marking methods to accommodate changes in marker contrast due to variations in the colour or transparency of a sample during inflation. This has particular relevance for dynamic analysis of very thin membranes and biomedical materials.

7 NOTATION

l	Strained length (surface length)
l_0	Unstrained length
t	Strained specimen thickness
t_0	Initial specimen thickness
r	Radius of curvature
A	Strained cross-sectional area
A_0	Original cross-sectional area
$I_1; I_2; I_3$	Strain invariants
p	Pressure
λ	Stretch ratio
σ	Stress

ACKNOWLEDGEMENTS

The authors would like to thank the School of Manufacturing and Design and the Centre for Elastomer Research (CER) in the Dublin Institute of Technology.

REFERENCES

Javořik, J., Dvořâk, Z., "Equi-biaxial Test of Elastomers", KGK Kautschuk Gummi Kunststoffe, September 2007, p. 456–459 (2007).

Jerrams, S., Hanley, J., Murphy, N., Ali, H., Equi-Biaxial Fatigue of Elastomers: The Effect of Oil Swelling on Fatigue Life, Rubber Chemistry and Technology Journal, Vol. 81, Issue 4, p. 638–649, September/October 2008.

Johannknecht, R, Jerrams, S. Clauss, G. 2002 "Determination of non-linear, large equal bi-axial stresses and strains in thin elastomeric sheets by bubble inflation". Proceedings of the Institute of Mechanical Engineers, Vol 216 Part L, No. L4, (ISSN 1464-4207) Journal of Materials, Design and Applications.

Mott, P., Roland, C. M., Hassan, S., "Strains in an Inflated Rubber Sheet", Rubber Chemistry and Technology, 76, 326–333 (2003).

Murphy, N., Hanley, J., McCartin, J., Lanigan B., McLoughlin S., Jerrams S., Clauss G., Johannknecht R. Determining multiaxial fatigue in elastomers using bubble inflation. In Constitutive Models for Rubber; Austrell, P.E., Kari, L., Ed.; Balkema, 2005; Vol. 4; p. 65.

Murphy, N., Hanley, J., Jerrams, S. The Effect of Pre-Stressing on the Equi-Biaxial Fatigue Life of EPDM, proceedings of the 6th European Conference on Constitutive Models for Rubber, Dresden, September 2009 and a chapter in the book 'Constitutive models for rubber VI', pp. 269–273, 2009.

Treloar, L.R.G., "Strains In An Inflated Rubber Sheet, and The Mechanism Of Bursting", Rubber Chemistry and Technology, Volume 17, p. 957–967 (1944).

Constitutive Models for Rubber VII – Jerrams & Murphy (eds)
© 2012 Taylor & Francis Group, London, ISBN 978-0-415-68389-0

Measurements and simulation of a jumping rubber ball

Herbert Baaser, Guido Hohmann, Claus Wrana & Jochen Kroll
Freudenberg Forschungsdienste, Weinheim & Lanxess, Leverkusen, Germany

ABSTRACT: In this contribution we compare the results of a jumping rubber ball experiment to FE-simulations. For this purpose we apply two formulations of finite strain viscoelasticity. The experiment is observed by a high speed camera to catch the time range in between two floor contacts. The FE-formulations are based on an integral formulation and on a differential approach for the time evolution, respectively, where the latter is able to describe the PAYNE effect for filled systems, i.e. the amplitude dependence of the stiffness. In order to discuss the performance of high frequency time domain models and computations, both approches are compared to the measured data. Here, the model damping characteristics are found to be overestimated, which we currently attribute to a lack of parameter calibration.

1 INTRODUCTION & EXPERIMENT

We deal with the simple experiment of a jumping rubber ball, observing the frequency and height of the jumps. This test is modeled by the finite element method incorporating time domain viscoelasticity. In a first test we drop a rubber jumping ball ($\varnothing = 39$ mm) from an initial height of $h_0 = 800$ mm onto a rigid ground floor neclecting any friction effects in between the bodies. The impact zone is captured by a high speed camera with 1'000 frames per second (*fps*), see Figure 1. The movie is split at the time points of impact and the time intervals are determined by frame counting. Moreover, the energy theorem can be applied to calculate the height of jumps, see Figure 2. The aim is to describe these observations by a time domain viscoelasticity within the scope of finite elements using the ABAQUS/Standard software package.

The dynamic material behaviour is characterised by a temperature sweep $G^*(\vartheta)$ at fixed frequency of $f = 10$ Hz. The mastercurve is obtained by an extraction of the activation data from the temperature sweep using a technique introduced by (Kroll 2008) and thereby compassing the impact of cristallization at lower temperatures, see Figure 3. This mastercurve is approximated by a PRONY series of 10 terms to provide the base for the FE computation. Additionally,

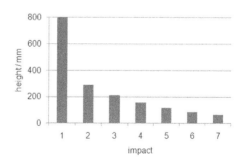

Figure 2. Height of jumping ball (before impact #) vs. impact number #.

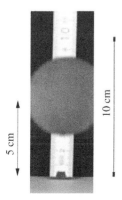

Figure 1. Jumping rubber ball experiment at $h \simeq 50$ mm. $h_0 = 800$ mm.

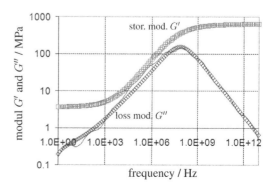

Figure 3. Master curve: stiffness vs. frequency.

amplitude sweeps are performed to catch the strain dependent effects.

To summarise, this simple experiment seems to provide a good validation for high frequency capabilities of the time domain representation of a material model for rubbers.

2 NUMERICAL MODELS

In the framework of modeling (linear) viscoelastic behaviour, the time dependence of stiffness is incorporated by adding damper elements. Regarding the description of stffness this can be done by a generalised MAXWELL model as given in Figure 4, which leads to an overall stress response as

$$\sigma = \sigma_\infty + \sum_{\alpha=1}^{m} \sigma_\alpha \quad \text{with} \quad \sigma_\infty = G_\infty \gamma \qquad (1)$$

and

$$\sigma_\alpha = \eta_\alpha \dot{\gamma}_\alpha^D = G_\alpha \gamma_\alpha^S \qquad (2)$$

$$= G_\alpha (\gamma - \gamma_\alpha^D), \qquad (3)$$

where the *viscosity* $\eta_\alpha = \tau_\alpha G_\alpha$ in each string α is given by its characteristic relaxation time τ_α and string stiffness G_α. The local strains encountered in the spring and the damper are denoted by γ_α^S and γ_α^D, respectively.

Differentiation of (3) and the application of (2) as a constitutive law for the damper yields

$$\dot{\sigma}_\alpha = G_\alpha (\dot{\gamma} - \dot{\gamma}_\alpha^D), \qquad (4)$$

which directly results in the first order differential equation

$$\dot{\sigma}_\alpha + \frac{\sigma_\alpha}{\tau_\alpha} = G_\alpha \dot{\gamma}. \qquad (5)$$

The solution of (5) is given in the form of a *convolution* or *heredity* integral

$$\sigma(t) = \int_{-\infty}^{t} G(t - \tau) \frac{d\gamma}{d\tau} d\tau, \qquad (6)$$

where $G(t)$ denotes the *relaxation function* in time t. As usual in that context, we also proceed by a representation of $G(t)$ in the form of a PRONY series, i.e.

$$G(t) = G_\infty + \sum_{\alpha=1}^{m} G_\alpha e^{-t/\tau_\alpha}, \qquad (7)$$

where G_∞ again denotes the parallel spring in Figure 4 and G_α and τ_α the characteristic properties in the dissipative strings.

2.1 *Viscoelasticity in ABAQUS*

In ABAQUS the transition to the finite strain regime is realized by a representation of (5) in quantities of KIRCHHOFF stresses. As reflected by experimental observations and argued there (see ABAQUS manual and (Holzapfel 2000)), just the isochoric part τ_{iso} of the stress response is formulated in equivalence to the results above. Here, the solution is given in equivalence to (6) in terms of

$$\tau_\alpha = \frac{\bar{g}_\alpha^P}{\tau_\alpha} \int_0^T e^{-s/\tau_\alpha} \left[\bar{\mathbf{F}}_t^{-1} \cdot \tau_0 \cdot \bar{\mathbf{F}}_t \right]_{T-S} ds \qquad (8)$$

with

$$\tau = \tau_0 - \sum_{i=1}^{N} \text{dev} \, \tau_\alpha \qquad (9)$$

and $\bar{\mathbf{F}}$ as *relative deformation gradient* in the ABAQUS typical incrementation setup from a time increment to the next. With respect to the convolution integral in (6) and (8), we call this formulation the *integral type* or *stress concept*. In contrast, models such as those proposed in (Reese and Govindjee 1998), (Hartmann 2002) or (Rendek and Lion 2010), will be called *differential type* or *deformation concept*.

2.2 *Enhanced treatment of viscoelasticity*

In this treatise, we also use the model of (Rendek and Lion 2010), which applies (4) as evolution equation directly and formulates a more general type of viscoelasticity, which do not need expressions like (6) or (8). In addition, this model is also able to capture *amplitude dependent* effects, see (Payne 1960) or (Lion and Kardelky 2004), by introducing an intrinsic time scale, which bulges or stretchs the physical time scale.

The here presented models are based on the one-dimensional rheological assumption as depicted in Figure 4 to capture the long-term behaviour as well as the physical relaxations processes by MAXWELL strings. Furthermore that constitutive modell changes the intrinsic viscosity η_α^0, see Figure 4, by a shifting depending on the actual process status in the form

$$\eta_\alpha = \frac{\eta_\alpha^0}{H_\alpha}, \qquad (10)$$

where $H_\alpha(t) = 1 + d_\alpha q_\alpha(t)$ and the rate equation

$$\dot{q}_\alpha(t) = \frac{1}{\lambda_\alpha} (||\dot{\gamma}(t)|| - q_\alpha) \qquad (11)$$

for the internal variable q_α is assumed. So, two additional material parameters, d_α and λ_α, for each string α are introduced.

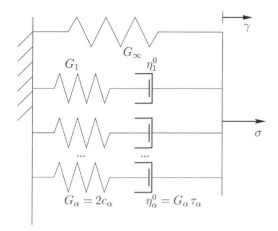

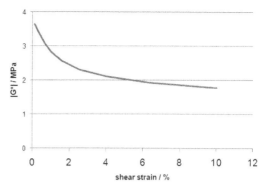

Figure 5. DMA: Amplitude dependence.

Figure 4. Generalized MAXWELL model introducing the material parameters G_∞, G_α and $\eta_\alpha^0 = G_\alpha \tau_\alpha$.

A formulation in tensorial quantities, see again (Rendek and Lion 2010), yields an update for the inelastic strains as

$$\mathbf{e}_{in,\alpha}^{n+1} = \frac{\tau_\alpha}{\tau_\alpha + \Delta t} \mathbf{e}_{in,\alpha}^n + \frac{\Delta t}{\tau_\alpha + \Delta t} \mathbf{e}^{n+1} \quad (12)$$

from increment n to increment $n + 1$ of length Δt. The formulation with respect to the reference configuration based on the PIOLA strains $\mathbf{e} := 1/2(\mathbf{C}^{-1} - \mathbf{I})$ and the right CAUCHY–GREEN tensor $\mathbf{C} = \mathbf{F}^T \cdot \mathbf{F}$ enables an objective time integration. Futhermore, the implementation in ABAQUS needs a transformation of the computed stresses and their derivate w.r.t. the deformation measure into the actual configuration.

The procedure finally leads to a sum of stresses of

$$\sigma_{in} = \sum \sigma_\alpha = - \sum \frac{2c_\alpha}{J} (\mathbf{F} \cdot \mathbf{e} \cdot \mathbf{F}^T - \mathbf{F} \cdot \mathbf{e}_{in,\alpha} \cdot \mathbf{F}^T), \quad (13)$$

where the stiffness for the neo-HOOKEian behaviour in each string is denoted by c_α.

Here, we use the formulation of viscoelasticity in ABAQUS in comparison to an umat implementation of the above model in that software environment.

3 PARAMETER CALIBRATION

For our task the calibration procedure is realized in a two-step algorithm: In a first step the parameters c_α and η_α^0, see Figure 4, are determined by a shear experiment with 1% shear strain following the ideas of (Kaliske and Rothert 1998), where we represent amplitude dependence by the following procedure. The amplitude dependence is captured by a second step which relys on an optimization process provided by the ALTAIR HYPERSTUDY software package. In this regard a standalone material model is applied to realize very fast optimization runs. This small application is represented by the original ABAQUS umat code embedded

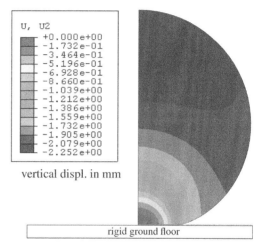

vertical displ. in mm

rigid ground floor

Figure 6. Numerical result of a deformation wave through the ball's cross section (axisymmetric model) at impact.

in an PYTHON- and FORTRAN77-environment to simulate the DMA-shear strain situation. Then the same high strain amplitudes as given in Figure 5 are applied.

The model results for the proposed parameter set are computetd as long as the parameters give a good approximation for the experimental data.

4 NUMERICAL RESULTS

We show exemplary different simulation results obtained by the above discussed types of models. At first, the numerical answer of the jumping height and the computed impact times resulting from the "classical" ABAQUS viscoelastic material model, see Section 2.1, are shown.

To give an idea of the FE solution, we show the deformation of the ball at the first impact in Figure 6, where one can see a deformation wave going vertically through the ball after the first instant of contacting the ground floor. Furthermore the results obtained by the differential model of Section 2.2 (without and with strain dependence) are provided by choosing a

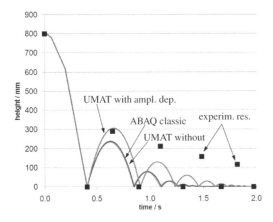

Figure 7. Results: Experimental impact times and simulated jumping heights.

basic parameter set such that integral and differential approach lead to similar results. Finally, the additional parameters of the enhanced model are choosen according to the procedure of Section 3. Figure 7 shows a very good agreement in between the integral and the differential method without strain dependence. One can see the good agreement between the two formulation *without* amplitude dependence inspite of their different type of modeling in time domain. Nevertheless, both models show a too pronounced damping character compared with the experimantal observations. The umat implementation considering amplitude dependence by an adaquate calibration shows a good agreement within the first two impacts. Afterwards also this simulation fails to model the damping character adequate: The damping is again much too pronounced.

5 CONCLUSIONS

The quite simple experiment of a jumping rubber ball seems to be a reasonable example to verify the capabilities of different viscoelastic formulations in comparison. Both, the (high) frequency elastic behaviour (*storage*) as well as the damping factor (*loss modulus*) can be observed in simulation and be easily compared with experimental data by looking on the jumping times and jumping heights between the impacts.

Is is shown, that the formulation of (linear) viscoelasticity – despite in finite strain regime – seems not to be able to give reasonable results in simulation.

Inspite of the capabilties of the enhanced viscoelastic model (Rendek and Lion 2010), the parameter calibration procedure following Section 3 and the lonely information of Figure 5 seems to be not valid enough to obtain a precise parameter fit. Here, further investigatino have to follow.

Both formulations show deficits with respect to the long-term behaviour, especially on the view of dissipative effects: The damping is too large.

REFERENCES

Baaser, H. (2007, Dec). Representation of Elastomers in Industrial Applications. *PAMM 7(1)*, 4060011–4060012. online: Mar 31 2008.

Baaser, H. and G. Hohmann (2009). Representation of dynamic elastomer behaviour with focus on amplitude dependency. In *ECCMR*. ISBN 978-0-415-56327-7.

Hartmann, S. (2002). Computation in finite-strain viscoelasticity: finite elements based on the interpretation as differential-algebraic equations. *Comp. Meth. Appl. Mech. Eng. 191*(13–14), 1439–1470.

Holzapfel, G. (2000). *Nonlinear Solid Mechanics*. Number ISBN 0-471-82304. Wiley.

Kaliske, M. and H. Rothert (1998). Constitutive approach to rate–independent properties of filled elastomers. *International Journal of Solids and Structures 35(17)*, 2057–2071.

Kroll, J. (2008). An analytical method for the determination of the high-frequency behaviour of elastomers. *Rubber Fall Colloquium*.

Lion, A. and C. Kardelky (2004). The payne effect in finite viscoelasticity: Constitutive modelling based on fractional derivatives and intrinsic time scales. *Int. J. Plasticity 20*, 1313–1345.

Payne, A. (1960). A note on the existence of a yield point on the dynamic modulus of loaded vulcanisates. *J. Appl. Polym. Sci. 3*, 127.

Reese, S. and S. Govindjee (1998). A theory of finite viscoelasticity and numerical aspects. *Int. J. Sol. Struct. 35*, 3455–3482.

Rendek, M. and A. Lion (2010). Amplitude dependence of filler-reinforced rubber: Experiments, constitutive modelling and FEM – implementation. *International Journal of Solids and Structures 47*, 2918–2936.

Swelling of bent rubber strips and recovery when the stresses are removed

A.N. Gent & C. Nah
Chonbuk National University, Jeonju, South Korea

ABSTRACT: Swelling of rubber is strongly affected by applied stress. For example, when rubber sheets are bent, they will swell more on the tension side and less on the compression side. When the bending constraint is removed, only partial recovery towards the flat state occurs, followed by slow further recovery as the swelling liquid migrates internally. These effects have been explored using strips of natural rubber swollen by dodecane (Nah et al. 2011). The "set" on release from bending was predicted by simple swelling theory, and the time dependence of later recovery was consistent with the rate of diffusion of dodecane in rubber. Thus, bending and recovery experiments provide a simple way of studying the internal mobility of compatible liquids. Other ways of studying recovery from an imposed bending deformation are also proposed. We infer that liquid migration can make a significant contribution to deformational energy losses.

1 INTRODUCTION

If a rubber sample is deformed unevenly, the degree of swelling will vary with position in accordance with the local stress field. Moreover, if the stress distribution is changed, the swelling liquid will migrate internally. This readjustment will be slowed by the finite rate of internal diffusion of the liquid, and thus can contribute to energy consumption in time-dependent deformation processes. We have studied the recovery of a bent strip of rubber swollen in a slowly-diffusing liquid, dodecane, and then released from constraints (Nah et al. 2011). The swelling liquid was absorbed initially to a greater degree in the outer (stretched) region than in the inner (compressed) region, and when the bending stresses were removed, the liquid migrated internally to reduce the amount of bending. This motion is assumed to take place only in the thickness direction, i.e., under constrained conditions, with the length and width of the bent sheet remaining substantially constant. Similar one-dimensional swelling processes have been examined by Southern & Thomas (1965), Rabin & Samulski (1992), and Kang & Huang (2010). Also, recent work by Hong et al. (2008, 2009) has explored the dependence of the distribution of swelling liquid on stress gradients and the associated time dependence. However, no previous work deals with recovery of shape of swollen samples when the constraints are removed, as far as we are aware. The partial recovery that occurs immediately on release was found to be in accord with Treloar's (1950) theory of swelling of rubber under an imposed strain, and the protracted return towards the unstrained (straight) condition was comparable to the measured rate of diffusion of dodecane in natural rubber (Nah et al. 2011). This agreement suggests that a bending and recovery experiment is a useful way of examining the internal mobility of absorbed liquids. Some other ways of performing such experiments are also described here.

2 THEORETICAL CONSIDERATIONS

2.1 *Amount of swelling*

The basic relation between the volume swelling ratio Q of a crosslinked rubber sample and the applied stress t was developed by Treloar (1950):

$$t_1 = (RT/V_1)\,[F(Q) + A\,L_1^2/Q] \tag{1}$$

with similar relations for stresses t_2 and t_3 in the perpendicular directions. In Equation 1, L_1 denotes the ratio of the length in the 1-direction in the swollen, stretched state relative to that in the unswollen, unstretched state and the function $F(Q)$ represents the entropy and heat of dilution of rubber by the swelling liquid:

$$F(Q) = \ln\,[1 - (1/Q)] + 1/Q + X/Q^2 \tag{2}$$

where X is the polymer/solvent interaction parameter, given a representative value of 0.4 here. The term A $(=\rho V_1/M_c)$ characterizes the elastic resistance of the rubber molecular network to expansion; ρ is the density of the swelling liquid, V_1 is its molar volume, and M_c is the average molecular weight of network strands. A value of A of 0.019 gives the best fit to the swelling data reported by Treloar for a lightly-crosslinked sample of natural rubber swollen by heptane, and is employed here for swelling by dodecane also.

2.2 Swelling in constrained extension

In constrained extension, one dimension is held constant while a tensile strain is imposed in the perpendicular direction. For example, we employ the length ratio L_1 to represent the tensile strain imposed on a swollen sample, while the length L_2 in a perpendicular direction is held constant at the value attained in homogeneous swelling. The stress t_3 in the third direction is assumed to be zero. Thus the new dimensions are: L_1(imposed), $L_2 = Q_o^{1/3}$ (held constant), and $L_3 = Q/Q_o^{1/3}L_1$ (from the relation for the swollen volume: $Q = L_1.L_2.L_3$). From Equation 1, on putting $t_3 = 0$,

$$F(Q) = - A\, L_3^2/Q = - AQ/(Q_o^{2/3}\, L_1^2) \qquad (3)$$

Swelling ratios Q predicted by Equation 3 for a given imposed strain were found to be considerably higher than in simple extension (Nah 2011). In both cases they followed an approximately linear increase with applied strain.

When the applied strain L_1 is released, the stresses in the L_1 and L_3 directions both become zero, and the lengths L_{1R} and L_{3R} after release are therefore also equal, while the length ratio L_{2R} remains unchanged. Thus the strain ratios become $L_{1R} = L_{3R} = (Q/L_{2R})^{1/2}$, $L_{2R} = Q_o^{1/3}$. Hence $L_{1R}^2 = (Q/Q_o^{1/3})$ and the residual strain e_r is given by:

$$e_r = (L_{1R} - L_{1,o})/L_{1,o} = [(Q/Q_o)^{1/2} - 1] \qquad (4)$$

where $L_{1,o}$ is the original swollen length in the 1-direction, $Q_o^{1/3}$.

Calculated values of the residual strain e_r are plotted in Figure 1 as a function of the imposed strain ratio $e(=L_1)$. They are seen to be substantially higher than for samples swollen in simple extension.

3 MIGRATION OF A SWELLING LIQUID IN RUBBER

A rubber strip was first swollen homogeneously and then bent into a circular arc and allowed to reach equilibrium, Figure 2a. During this period the swelling liquid migrated internally so that more was accommodated in the outer (stretched) region and less in the inner (compressed) region. For relatively thin rubber strips, these changes in the degree of swelling can be assumed to take place under constrained conditions, because the length and width of the sample remain virtually unchanged.

We now consider the changes in shape that took place when the bending stresses were removed. The mean radius of the bent strip in the swollen state is denoted R_o. On removing the constraint the strip returned to a less-bent shape, with a larger radius R, Figure 2b, and then, as the swelling liquid diffused internally, the strip slowly straightened. The initial value of the ratio R_o/R gives the amount of "set" due

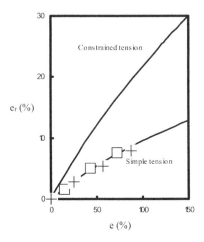

Figure 1. Residual strain e_r vs imposed strain e. Upper curve, calculated from Equation 4 for constrained extension. Lower curve, calculated for simple extension. Points from Treloar's data (1950) for swelling with heptane (crosses) or benzene (squares). (Figure from Nah et al. 2011).

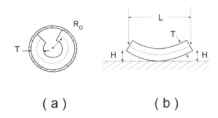

(a)   (b)

Figure 2. Sketch of experimental procedure. (a) Attaining swelling equilibrium in the bent state, at a mean radius R_o. (b) Recovery towards the unstrained state.

to unequal swelling in the bent state, and can be compared with the residual set e_r predicted by Equation 4. The gradual increase in R as the strip straightened gives a measure of the rate of internal migration of the swelling liquid. Experimental details are given in the following section.

4 EXPERIMENTAL DETAILS

Strips of soft vulcanized natural rubber with thicknesses of about 1 and 2 mm were swollen in dodedcane and then inserted into plastic tubes, see Figure 2a, to hold them in a bent configuration. In the fully-swollen state, the volume swelling ratio Q_o was about 4, but a somewhat smaller degree of swelling, 70% or 90% of the equilibrium value, was used to avoid exudation of the swelling liquid on the tension side when the bent strip was released.

After removing the samples from the tube, they were allowed to recover in a heated chamber containing pads moistened with dodecane to minimize evaporation during the slow recovery process. The radius of curvature R after release from the imposed bend

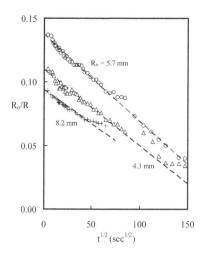

Figure 3. Increase in radius of curvature R with time t for strips subjected to imposed bends of various radii R_o. Swollen thickness $T_s = 2.9$ mm. (From Nah et al. 2011).

was observed to increase slowly with time as the sample straightened. Values of R were calculated from the average height H of the ends of the now partially-bent strip above the horizontal plane on which the sample rested, see Figure 2b, using the geometrical relation:

$$R = (L^2/8H) + (H/2) - (T_s/2) \tag{5}$$

where T_s is the thickness of the swollen sample.

An unswollen sample was also bent and heated for a somewhat longer time to examine the amount of "set" that could arise from other causes. It was found to be small, about 5%, and did not change significantly over periods of several hours. It was concluded that residual deformation as a result, for example, of chemical changes, might cause incomplete recovery but it would not affect the observed time-dependence.

5 EXPERIMENTAL RESULTS

Recovery of partially swollen strips from imposed bending is shown in Figure 3, where values of the ratio R_o/R are plotted against $t^{1/2}$, where t is the recovery time. R_o/R is the ratio of strains remaining in the released strip to the higher strains imposed by bending, and is thus equivalent to the residual strain ratio or set, e_r/e. For constrained extension, the value is therefore expected to be about 20%, see Figure 1. Samples swollen to 90% recovered initially to about 20% of the imposed strain level, in good agreement with the less-swollen samples, Figure 3, showed smaller amounts of initial recovery, $12 \pm 2\%$.

The linear dependence of the amount of recovery upon $t^{1/2}$ in the initial stages of recovery, Figure 3, is consistent with the mechanism of slow recovery being internal diffusion (migration) of swelling liquid towards a homogeneous distribution. Indeed, the slopes of the linear relations in Figure 3 were in

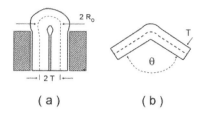

Figure 4. Recovery from severe bending.

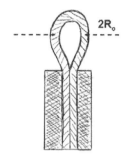

Figure 5. Sketch of a strip bent into the shape of an elastica. The section above the broken line has a radius R_o of $0.11 L$, approximately, where L is the length of the unclamped portion of the strip.

reasonable agreement with the diffusion coefficient D determined directly from the initial rate of uptake of dodecane by a thin rubber sheet, 37×10^{-12} m^2/sec.

6 OTHER TEST ARRANGEMENTS

To increase the range of strains in the bent sample, and thus the possible amount of slow recovery, more severe bending conditions can be applied. For maximum bending, approaching 180° (Figure 4), the strain set up at the outer surface is about 100%.

The mean radius R_o is slightly greater than $T/2$, where T is thickness of the swollen strip, and the length L subjected to a bend of 180° is approximately πR. On release, the angle θ between the straight arms becomes a measure of the residual radius R: $L/R = (\pi - \theta)$, and thus an estimate of the residual strain ratio e_r/e is given by

$$e_r/e \ (= R_o/R) = 1 - (\theta/\pi) \tag{6}$$

Some swollen materials will crack if tensile strains approaching 100% are imposed. In these cases, less severe strains can be created by bending a strip into the shape of an elastica, Figure 5. The upper portion of the bent strip, indicated by broken lines in Figure 5, has a length of $0.35L$ (Gent 1994) where L is the length of the unclamped portion of the strip, and a mean radius R_o of approximately $0.11L$. Thus the peak strain, given by $T/(2R_o)$, where T is the thickness of the swollen strip, is approximately $4.5(T/L)$. However, in employing these calculations, it should be noted that bending becomes unstable when the inner compressive strain is greater than about 35%, and sharp creases appear on the inner surface (Gent and Cho 1999).

Another test mode is suggested by the work of Fukahori et al. (1996). They showed that compression of a thin bonded rubber block will induce migration of an absorbed liquid towards the free surfaces. It follows that recovery when the compressive load is removed will be delayed, as in the bending experiments discussed above.

7 CONCLUSIONS

When a rubber sample is swollen in the bent state, the amount of immediate recovery of shape when the applied stresses are removed can be estimated from Treloar's (1950) theory of swelling of rubber under strain. Further recovery towards the unstrained state was found to follow the characteristic time dependence for diffusion of compatible liquids into rubbery solids, at least initially, and the rates of recovery were similar to those for absorption of the liquid used, dodecane. Thus, recovery from an imposed bend appears to provide a simple method for studying internal migration of absorbed liquids under gradients of stress. Other test configurations are briefly outlined. They employ more severe bending strains and thus a greater amount of recovery is expected, with a concomitant increase in accuracy.

ACKNOWLEDGEMENT

This research was carried out under the World Class University Program of the Korea Science and Engineering Foundation, funded by the Ministry of Education, Science and Technology (R33-2008-000-10016-0).

REFERENCES

Fukahori, Y., Seki, W. & Kubo, T. 1996. Mass transfer in compression creep of bonded rubber blocks. *Rubber Chem. Technol.* 69: 752–768.

Gent, A. N. 1994. Buckles in adhering elastic films and a test method for adhesion based on the elastica. *J. Adhesion Sci. Technol.* 8: 807–819.

Gent, A. N. & Cho, I. S. 1999. Surface instabilities in compressed or bent rubber blocks. *Rubber Chem. Technol.* 72: 253–262.

Hong, W. et al. 2008. A theory of coupled diffusion and large deformation in polymeric gels. *J. Mech. Phys. Solids* 56: 1779–1793.

Hong, W., Liu, Z. & Suo, Z. 2009. Inhomogeneous swelling of a gel in equilibrium with a solvent and mechanical load. *Internatl. J. Solids and Structures* 46: 3282–3289.

Kang, M. K. & Huang, R. 2010. Swell-induced surface instability of confined hydrogel layers on substrates. *J. Mech. Phys. Solids* 58: 1582–1598.

Nah, C., et al., 2011. Swelling of rubber under non-uniform stresses and internal migration of swelling liquid when the stresses are removed. *Macromolecules* 44: 1610–1614.

Rabin, Y. & Samulski, E. T. 1992. Swelling of constrained polymer gels. *Macromolecules* 25: 2985–2987.

Southern, E. & Thomas, A. G. 1965. Effect of constraints on the equilibrium swelling of rubber vulcanizates. *J. Polymer Sci. Part A* 3: 641–646.

Treloar, L. R. G. 1950. The swelling of cross-linked amorphous polymers under strain. *Trans. Faraday Soc.* 46: 783–789.

Constitutive Models for Rubber VII – Jerrams & Murphy (eds)
© 2012 Taylor & Francis Group, London, ISBN 978-0-415-68389-0

Experimental determination of mechanical properties of elastomeric composites reinforced by textiles made of SMA wires and numerical modelling of their behaviour

B. Marvalova, J. Kafka & J. Vlach
Technical university of Liberec, Czech Republic

L. Heller
Institute of Physics AS CR, Prague, Czech Republic

ABSTRACT: The paper presents an experimental and theoretical research of composites with elastomeric matrix reinforced by textile structures namely by fabrics made of SMA fibers. Applications are focused on thin inflatable membranes and tubes and on active thin-walled elastomeric plates embedded with hybrid SMA knitted fabrics. Parameters of material models are determined on a basis of experimental measurements and the material models are verified by the comparison of the experimental measurements with numerical simulations in finite element code.

1 INTRODUCTION

Shape Memory Alloys (SMA) possess both sensing and actuating functions due to their shape memory effect, pseudo-elasticity, high damping capability and other remarkable properties. Because of these unique properties, SMAs are prospective in many applications such as structural vibration control, biomechanics and smart textiles. Combining the SMAs with other materials can create intelligent or smart composites by utilizing the unique properties of SMAs. Mechanical and thermal properties of composites reinforced by fabric structures were investigated in our laboratory (Marvalova 2000, Marvalova & Palan 2001). Recently, the experimental research and the modelling of smart elastomeric composites reinforced by SMA woven or knitted fabrics have been taken up. Preliminary results of the experimental research of the elastomeric composites reinforced by different types of SMA knitted fabrics are presented in this paper.

2 EXPERIMENTAL

2.1 *Preparation of composite specimens reinforced by NiTi fabrics*

Studied NiTi knitted fabrics were made of commercially available drawn NiTi fibres Fort Wayne Metals Ltd. of the chemical composition 55.82 wt. % Ni giving the fibres superelastic behaviour at temperatures above 10°C.

The knitted fabrics of NiTi files were fabricated at the Department of Textile Structures of Technical University of Liberec and their thermal processing was performed at the Institute of Physics AS CR in Prague.

The knitted fabric was made of NiTi fibers of diameter 0.1 and 0.2 mm, which had been previously treated by cold drawing. This technological process, which is based on plastic deformation of fiber to a final diameter, creates the fibers with a highly distorted microstructure with a high density of defects and with an occurrence of amorphous regions (Pilch et al. 2009).

The fibers with such microstructure do not show phenomena such as a shape memory effect and superelasticity which are typical for intermetallic alloys of nickel and titanium. To incite these properties it is necessary to recover microstructure of fibers by a heat treatment with appropriate time and temperature. These two parameters guide the microstructure recovery and recrystallization. The resulting microstructure determines significantly functional features of fibers such as the shape of superelastic stress-strain curve, the length and the stress magnitude of superelastic plateau, the stress necessary for an activation of martensitic transformation, and the stability of functional properties in a cyclic loading (Delville et al. 2010).

A specific geometrical form called parent shape can be also given to the fibers or to the fiber textile structures as yarns and knitted or woven fabrics by a mechanical shape setting in the course of the heat treatment. This fiber structure if loaded and unloaded subsequently it will return to the shape defined by the thermo-mechanical processing after a warming above the characteristic temperature of the end of the martensitic transformation. A quality of shape setting and the

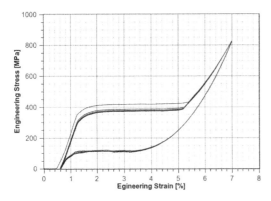

Figure 1. Superelastic stress-strain response of NiTi fiber in tension.

Figure 2. Composite with silicon matrix and reinforcement of NiTi knitted fabric.

shape resistance to mechanical loading depend also on temperature and duration of the thermo-mechanical treatment.

Specimens of knitted fabric 20×20 cm were treated for 30 min at 450°C in a resistance furnace at normal atmosphere. The fabrics were clamped in a device stretching the fabric in two directions. The stress-strain graph of a fiber of diameter 0.1 mm submitted to such thermal processing is shown on Figure 1.

The reinforcement of the silicon caoutchouc composite on Figure 2 is a single jersey weft knitted fabric of the NiTi wire with a diameter of 0.1 mm.

The knits were made by hand on a flat knitting V-bed machine 5E and the wire was fed without passing through a thread tension device. An accurate placement of NiTi wires and their positions are difficult to maintain in course of their embedment into matrix. Therefore hybrid warp knitted fabrics of polyester and polypropylene multifilament yarns were also prepared with NiTi wires inserted in weft and warp directions.

The knitted NiTi fabric was cleaned thoroughly in isopropyl alcohol. A dilute solution of triethoxysilane

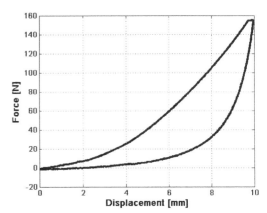

Figure 3. Tensile test of NiTi knitted fabric.

in the isopropyl alcohol has been used to increase an adhesion between the matrix and the knitted NiTi fabric. The VARTM technology (Vacuum Assisted Resin Transfer Molding) with a rigid mold was used for the preparation of samples. The matrix of polyaditive silicone RTV ZA-13 was injected to the mold under 25 kPa vacuum. Due to this process, the resulting composite is devoid of bubbles and its quality is very good.

2.2 Testing of NiTi knitted fabric

The result of unidirectional quasistatic tension of a 200 mm wide specimen with maximum elongation about 5% is on Figure 3. Cruciform samples of knitted fabric were tested also in bidirectional tension.

Although the knitted structure of NiTi wires is highly deformable, higher in-plane tension forces cause often the failure of wire loops still before the phase transition. The failure of one sole loop leads to the significant failure of the fabric.

Cyclic tests of knitted NiTi fabric specimens were conducted on Instron Electropuls at room temperature. The fabric was fixed in a circular frame shown in Figure 4 and loaded perpendicularly to its plane by a spherical punch. Tests were performed in the frequency range of 1–10 Hz with amplitudes varying between 1 and 5 mm. After the loading on the chosen initial force, the process was controlled by deflection.

The record of a loading cycle of the sample is on the Figure 5. The NiTi specimens of knitted fabric failed without any observed phase transition. Similar hysteresis loops were observed for fabrics knitted of steel wires with the same parameters.

2.3 Testing of NiTi composite

The experimental bubble-inflation technique is widely used for measuring the mechanical properties of thin flexible structures.

The tested composite flat membrane reinforced by the knitted fabric of NiTi is shown in Figure 2. Tests were performed at room temperature.

Figure 4. Cyclic loading of knitted NiTi fabric.

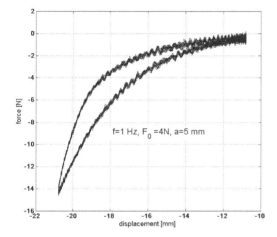

Figure 5. Experimental data – cyclic loading of NiTi sample.

Figure 6. Inflation of composite membrane. Pattern of black and white spots deposited on the membrane surface.

The samples of composite membrane were clamped in a test device between an aluminium vessel and an steel ring of inner diameter 144 mm (Fig. 6). Pressurised air was introduced into the aluminium vessel, resulting in a spherical deformation of the membrane.

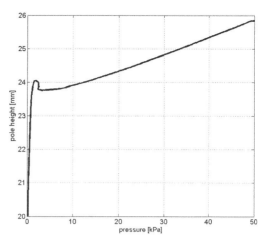

Figure 7. Height of the pole of deformed composite membrane at the different stages of loading.

The pressure varying slowly between 0 and 50 kPa was recorded with a digital pressure sensor and the deformation of the bubble was measured by 3D – digital image correlation system Istra of Dantec Dynamics.

This system allowed measurement of the plane strains on the surface of the bubble. To enable the digital image correlation process, a pattern consisting of chalk and coal dust was spread on the surface of membrane and fixed with hairspray.

Around 200 images were captured by two digital cameras in the course of loading and unloading. The bubble shapes were almost spherical.

The coordinates of the deformed shape were fitted to the spherical surface by least squares and the radius r of sphere was calculated for each stage of loading and the height of pole was approximated (Fig. 7). The radius was used for an approximate calculation of the membrane inner force per unit length $F = pr/2$.

The principal Green Lagrangian strains were calculated from the Istra data as mean values in grid points in the vicinity of the pole. The graphs of the strains in Figure 8 indicate that the shape of membrane is not spherical as assumed. The principal directions were found identical with the directions of course and wale of the knitted NiTi fabric reinforcing the membrane.

It is clear from the Figures 7 and 8 that the membrane deformation depends linearly on the load within the applied pressure. Thus the internal force per unit length F is also a linear function of pressure.

Result demonstrate that phase transition was not reached in the range of loading pressures used.

3 NUMERICAL MODELLING

A simple finite element model of knitted fabric was set up in MSC.Marc. The model represents the characteristic geometrical structure of the fabric. The loop geometry is simplified and it is represented by a

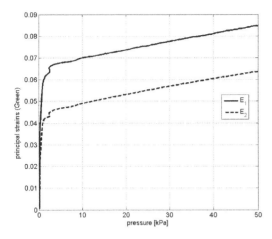

Figure 8. Measured principal Green strains at the pole of membrane.

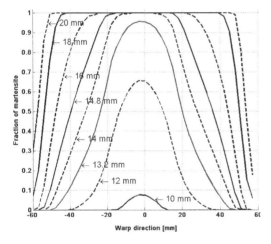

Figure 9. Fractions of martensite in warp direction with the punch vertical displacement as a parameter.

plane hexagonal grid of nonlinear beam elements. The shape of a grid cell captures the different geometry in the direction of course and wale which results to the different stiffness of fabric model in the main directions.

The mechanical model of SMA in MSC.Marc was used which is based on Auricchio's model of superelasticity (Auricchio 2001).

A quasistatic punch test depicted on the Figure 4 was simulated and the fractions of martensite were calculated for different punch positions.

The fraction of martensite on the diameter of specimen in the warp direction of the knitted fabric is shown on the Figure 9.

4 CONCLUSION

Preliminary results of experimental and numerical research of the composites with the matrix of silicone caoutchouc reinforced by knitted NiTi fabric were presented. The experimental measurements have shown that the main problem of the knitted fabric of very thin NiTi wires is a small resistance to the repeated tensile loading. The finite element model of knitted fabric gives realistic results. Further investigations would be required in order to better understand the superelastic and memory behaviour of NiTi knitted fabrics and composites.

ACKNOWLEDGEMENT

This work was supported by the subvention from Czech Grant Agency GACR No P108/10/1296.

REFERENCES

Auricchio, F. 2001. A robust integration algorithm for a finite strain shape memory alloy superelastic model. *Int. J. Plasticity*, Vol. 17, pp. 971–990.

Delville, R., Malard, B., Pilch, J., Sittner, P., Schryvers, D. 2010. Microstructure changes during non-conventional heat treatment of thin Ni-Ti wires by pulsed electric current studied by transmission electron microscopy. *Acta Materialia*, 58, 13, pp. 4503–4515.

Marvalova, B. 2000. Determination of Effective Stiffness Tensor of Textile Reinforced Composites. in Proc. *7th Int. Conf. Comp. Eng. ICCE/7*, Denver, pp. 593–4.

Marvalova,B., Palan, M. 2001. Determination of the effective thermal conductivity of textile reinforced carbon-carbon composites. in Proc. *8th Int. Conf. Comp. Eng. ICCE/8*, Tenerife, pp. 621–2.

Pilch, J., Heller, L., Sittner, P. 2009. Final thermomechanical treatment of thin NiTi filaments for textile applications by electric current. *ESOMAT 2009, 05024, www.esomat.org*, DOI: 10.1051/esomat/200905024.

Constitutive Models for Rubber VII – Jerrams & Murphy (eds)
© *2012 Taylor & Francis Group, London, ISBN 978-0-415-68389-0*

Simple shearing of soft biomaterials

C.O. Horgan
University of Virginia, Charlottesville, VA, US

J.G. Murphy
Dublin City University, Dublin, Ireland

ABSTRACT: Shearing is induced in soft biomaterials in numerous settings. The limited experimental data available suggest that a severe *strain-stiffening* effect occurs in the shear stress when soft biomaterials are subjected to simple shear in certain directions. This occurs at relatively small amounts of shear (when compared to the simple shear of rubbers). This effect is modelled here within the framework of nonlinear elasticity by consideration of a class of incompressible *anisotropic* materials. Due to the large stresses generated for relatively small amounts of shear, particular care must be exercised in order to maintain a homogeneous deformation state in the bulk of the specimen. The results obtained are relevant to the development of accurate shear test protocols for the determination of constitutive properties of soft biomaterials.

1 INTRODUCTION

In the mechanics of rubber-like solids modelled by nonlinear elasticity theory, the classical problem of *simple shear* has played an important role as a basic canonical problem that is rich enough to illustrate several key features of the nonlinear theory. However, even for this basic problem for isotropic materials, there are some fundamental issues regarding boundary conditions that that remain to be resolved (see, e.g., Horgan & Murphy 2010 for a discussion). Here we examine simple shear for *anisotropic* materials that model the mechanical behaviour of wide classes of soft biomaterials. The results are relevant to shear test protocols for determination of constitutive properties of fibrous soft tissues. Although shearing is induced in soft tissues in numerous physiological settings, the study of shear deformation has received relatively little attention in the biomechanics literature compared to extension or compression perhaps due to the fact that testing in shear is more difficult to implement. Simple shear was the basis for a shear test device proposed by Dokos et al. (2000) that was subsequently used by Dokos et al. (2002) to measure shear properties of passive ventricular myocardium. Other examples of applications of *simple shear* to the biomechanics of soft tissues are the works of Schmid *et al.* (2006) on myocardial material parameter estimation and that of Gardiner & Weiss (2001) on human medial collateral ligaments.

Simple shear of transversely isotropic fibrous biomaterials will be discussed here. One would intuitively expect that because of the fibres, there will be some directions in which severe resistance to shearing will be encountered. This is manifested in the simple shear experimental data of Dokos *et al.* (2002). A simple constitutive model within the framework of nonlinear elasticity will be used here to reflect this feature. It will be shown that there is a narrow band of angles of orientation of the fibres for which unlimited shear is possible (shearing in the approximate direction of the fibres) while, for *all* other angles, there is essentially the *same* limit as to the amount of shear allowed. We anticipate that this model will have wide applicability in biological systems where shearing is a dominant mode of deformation. The constitutive model used reflects the stretch induced stiffening of collagen fibres with increased shear loading. The corresponding strain-energy densities are of logarithmic form in the anisotropic invariant and are a viable alternative to the exponential forms usually used in biomechanical modelling of soft tissues.

Even for rubbers, accurate experimental simple shear data is difficult to obtain and some protocols have been developed to deal with these difficulties. No such protocols exist for biomaterials, as far as the authors are aware. The need for such guidance is demonstrated by our results, which indicate that large stresses are generated in sheared blocks of soft biological tissue for relatively small amounts of shear when compared to the shearing of rubber blocks and consequently large tractions must be applied on the inclined faces of the block to maintain homogeneity of deformation. These tractions are never applied in practice and thus the question arises as to the conditions under which the assumption of simple shear is appropriate when one face of a block of biological tissue is moved relative to the parallel face. Some guidance will be provided here based on shear test methods for *rubbers* and it is hoped that the results obtained will be relevant to shear test protocols for the determination of constitutive properties of soft biomaterials.

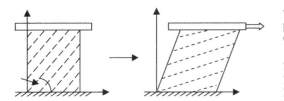

Figure 1. Simple shear of a block with one family of parallel fibres.

2 SIMPLE SHEAR

The deformation known as simple shear has the mathematical representation

$$x_1 = X_1 + \kappa X_2, \; x_2 = X_2, \; x_3 = X_3, \qquad (1)$$

where (X_1, X_2, X_3) and (x_1, x_2, x_3) denote the Cartesian coordinates of a typical particle before and after deformation respectively and $\kappa > 0$ is an arbitrary dimensionless constant called the amount of shear. The *angle of shear* is $\tan^{-1} \kappa$. The usual interpretation of simple shear is *two-dimensional* where a rectangular specimen, whose dimensions are all of the same order, is deformed into a parallelogram. It is assumed that the block is composed of an incompressible transversely isotropic fibre-reinforced nonlinear hyperelastic material. For the simple shear deformation (1), the deformation gradient tensor $\boldsymbol{F}$, the left Cauchy-Green strain tensor $\boldsymbol{B} = FF^T$ and its inverse are readily found. The three principal invariants of $\boldsymbol{B}$ are defined as

$$I_1 = tr \, \boldsymbol{B}, \; I_2 = \frac{1}{2}\left[(tr\boldsymbol{B})^2 - tr\boldsymbol{B}^2 \right], \; I_3 = \det \boldsymbol{B}, \qquad (2)$$

which in the case of simple shear are

$$I_1 = I_2 = 3 + \kappa^2, \; I_3 = 1. \qquad (3)$$

We confine attention here to materials reinforced with one family of parallel fibres aligned at an angle θ to the X_1 axis in the undeformed state (see Figure 1).

The undeformed unit vector $A = (\cos\theta, \sin\theta, 0)$ transforms to $a = (\cos\theta + \kappa \sin\theta, \sin\theta, 0)$ in the current configuration. We introduce the usual anisotropic invariant $I_4 = a \cdot a$ which, for simple shear, has the form

$$I_4 \equiv \lambda^2 = (c + \kappa s)^2 + s^2 = \kappa^2 s^2 + 2\kappa cs + 1, \qquad (4)$$

Here we have introduced the convenient notation $c = \cos\theta, s = \sin\theta$ and the symbol λ for the stretch in the fiber direction. We will only be concerned with the range $0 \le \theta \le \pi/2$ so that $I_4 \ge 1$ and so the fibres are always in extension. It is sufficient for our purposes to consider the constitutive law for the Cauchy stress T for incompressible transversely isotropic hyperelastic materials with $W = W(I_1, I_4)$ so that

$$\boldsymbol{T} = -p\boldsymbol{I} + 2W_1\boldsymbol{B} + 2W_4 \, \boldsymbol{a} \otimes \boldsymbol{a}, \qquad (5)$$

where $W_i = \partial W / \partial I_i \; (i = 1, 4)$ and $\otimes$ denotes the tensor product with Cartesian components $a_i a_j$. The in-plane Cauchy stresses are therefore given by

$$\begin{aligned} T_{11} &= -p + 2\left(1 + \kappa^2\right) \, W_1 + 2\left(c + \kappa s\right)^2 W_4, \\ T_{22} &= -p + 2W_1 + 2s^2 W_4, \\ T_{12} &= 2\kappa W_1 + 2s\left(c + \kappa s\right) \, W_4, \end{aligned} \qquad (6)$$

while the out-of-plane stress is

$$T_{33} = -p + 2W_1, \qquad (7)$$

where the derivatives in (6), (7) are evaluated at values of the invariants given by (3) and (4) and the quantity p is the arbitrary hydrostatic pressure arising due to the incompressibility constraint. Since the deformation (1) is *homogeneous*, the equilibrium equations in the absence of body forces are satisfied if and only if p is a constant.

3 SHEAR STRESS RESPONSE

We consider the class of strain-energy densities

$$W = \frac{\mu}{2}(I_1 - 3) + F(I_4), \qquad (8)$$

where $F(1) = 0$ and $F'(1) = 0$ so that $W = 0$ in the undeformed state and the stress is purely hydrostatic there. We also assume that $F'(I_4) > 0$ for $I_4 > 1$ which ensures that the axial stress is tensile for simple extension along the fibre direction. For the model (8), the shear stress has the form

$$T_{12} = \mu\kappa + 2s\left(c + \kappa s\right)F'(I_4) > 0. \qquad (9)$$

The general class of models (8) has been widely used in continuum mechanics modelling of soft tissues. Perhaps the most well-known specific form of (8) is the *standard reinforcing* model

$$W = \frac{\mu}{2}(I_1 - 3) + \frac{E(I_4 - 1)^2}{4}, \qquad (10)$$

where E is a nonnegative material modulus that measures the *degree of anisotropy*. The model (10) with quadratic nonlinearity in I_4 was proposed initially by Polignone & Horgan (1993a,b) in the context of cavitation problems for rubber-like materials (see, e.g., Horgan & Polignone 1995 for a review). It was shown in Polignone & Horgan (1993a) that this quadratic nonlinearity represents the *simplest polynomial form* for $F(I_4)$ in (8) that satisfies the conditions on F listed after (8). The form (10) has subsequently been adopted by many authors (see e.g., Merodio & Ogden 2005, Merodio *et al.* 2006 and references cited therein). The quadratic nonlinearity in (10) is used to reflect the presence of oriented collagen fibres in an elastin matrix. The fibres are stiffer than the matrix but are extensible. A number of more general choices for $F(I_4)$

428

were considered by Horgan & Saccomandi (2005) that reflect the *stretch induced stiffening of collagen fibres* as they are loaded. One of the simplest of these models is given by

$$W = \frac{\mu}{2}(I_1 - 3) - \frac{E}{4}J\ln\left[1 - \frac{(I_4 - 1)^2}{J}\right],\tag{11}$$

where the *additional* parameter J is a positive dimensionless parameter that measures the rapidly increasing stiffness of the fibres with increasing stretch. In order for the logarithm function in (11) to be well defined, it is required that the deformation satisfy the constraint $I_4 < \sqrt{J} + 1$ or equivalently

$$\lambda^2 < \sqrt{J} + 1,\tag{12}$$

where the fibre stretch λ is defined in (4). As is discussed by Horgan and Saccomandi (2005), the model (11) was motivated by the Gent model (Gent 1996) of rubber elasticity that reflects *limiting chain extensibility* of the molecular chains. See, e.g., Horgan & Saccomandi (2006) for a review of such models and their applications to strain-stiffening rubber-like and biological tissues. Application of the Gent model to the mechanics of arterial walls is discussed by Horgan & Saccomandi (2003), Holzapfel (2005) and Ogden & Saccomandi (2007). *The Gent model has been shown to be a viable alternative to the classical Fung exponential model that has had wide-spread application in the biomechanics of soft tissues.* In the limit as $J \to \infty$ in (11), we recover the *standard reinforcing* model (10). The new model (11) has the *additional* feature of measuring the increased stiffness of the collagen fibres with deformation. The constraint (12) provides a value for the maximum fibre stretch (or *locking stretch*) as $\lambda^2 = \lambda_m^2 = \sqrt{J} + 1$. A generalization of (11) in which the quadratic dependence on I_4 is replaced by an arbitrary power-law was also proposed by Horgan & Saccomandi (2005) and was shown there to reflect the gradual *engagement* of crimped collagen fibres on loading. An application of this generalized model to soft-cuticle biomechanics is given in Lin *et al.* (2009). Another model proposed by Horgan & Saccomandi (2005) similar to (11) was used by Sadovsky *et al.* (2007) to study the plant mechanics of angiosperm roots. The *isotropic* part of (11) can also be generalized to reflect increased stiffening of the elastin matrix (see, e.g., Ogden & Saccomandi 2007). For exposition purposes, however, we shall confine attention to (11), for which

$$W_4 = \frac{E(I_4 - 1)/2}{1 - (I_4 - 1)^2/J} = \frac{EJ(\lambda^2 - 1)/2}{J - (\lambda^2 - 1)^2},\tag{13}$$

where λ is defined in (4). Thus, regardless of the choice of the hydrostatic pressure, we see from (13) and (6) that the in-plane stresses have a singularity as

$\lambda^2 \to \sqrt{J} + 1$, reflecting the ultimate stiffness of the fibres. On using (4), we write the constraint (12) as

$$\kappa\left(\kappa s^2 + 2cs\right) < \sqrt{J}.\tag{14}$$

Thus, for a given material parameter J and a given fibre angle θ, the constraint (14) restricts the amount of shear that the specimen can undergo. Specifically κ must satisfy the constraint

$$\kappa < \kappa_m = \frac{-c + \sqrt{c^2 + \sqrt{J}}}{s},\tag{15}$$

where $0 < \theta \le \pi/2$. The limiting shear κ_m will be called the *locking shear.* If $\theta = 0$ the constraint (14) is automatically satisfied and it is easily verified that, in this case, the results coincide with those for an isotropic neo-Hookean material. The other limiting case of $\theta = \pi/2$ yields $\kappa_m = J^{1/4}$ and so if one has accurate experimental data on shearing the material perpendicular to the direction of the fibres, the crucial parameter J can be determined.

4 EFFECTS OF FIBRE ORIENTATION

Horgan & Saccomandi (2003) proposed values of the limiting chain parameter in the isotropic Gent model based on experimental data of other authors for human thoracic aorta segments. In the absence of experimental data corresponding to the new model (11), here we simply use the *same* values for the parameter J as were proposed by Horgan & Saccomandi (2003) for the Gent model. We anticipate that this will simply provide a crude estimate for realistic values of this parameter for soft tissues. Thus, for a 21-year old male, we adopt the value $J = 2.3$ while for the stiffer aorta of a 70-year old male, we take the value $J = 0.4$, on rounding off the values proposed by Horgan & Saccomandi (2003) to one decimal place. Plots of the locking shear versus the angle of orientation of the fibres for these two values of J are given in Figure 2. The striking feature of both curves in Figure 2 is their 'L'-shaped character, which is especially notable for the smaller values of J. These plots reveal that the simple model (11) has a complex response in shearing. Specifically, this model allows virtually unrestricted shear when shearing almost parallel to the direction of the fibres and when shearing in *any* other direction there is a limit to the possible amounts of shear, with this limit being essentially the same for all directions of shearing. We note that, from (15) and Figure 2, this locking shear decreases with decreasing values of J. This is consistent with the smaller value of J reflecting a stiffer tissue.

The shear stress for the model (11) is

$$\overline{T}_{12} \equiv \frac{T_{12}}{\mu} = \kappa + \gamma s(c + \kappa s)\frac{J(\lambda^2 - 1)}{J - (\lambda^2 - 1)^2},\tag{16}$$

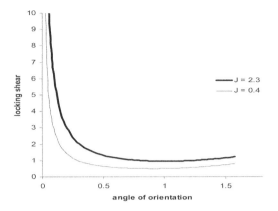

Figure 2.   Locking shear versus angle of orientation.

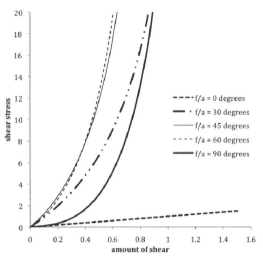

Figure 3.   Shear stress versus amount of shear for various fibre angles.

where here, and henceforth, the superposed bar denotes stresses normalized with respect to the shear modulus μ and we have introduced the dimensionless parameter $\gamma = E/\mu$. It follows from (16) that the shear stress becomes unbounded as the amount of shear approaches the locking shear value. This behaviour provides an idealisation of the rapid stiffening produced by collagen fibres in soft tissues under shear load.

The effect of the fibre angle on the shear stress response will now be investigated. We choose $J = 2.3$ and set $\gamma = 20$, a value used by Destrade *et al.* (2008) reflecting experimental data of Ning *et al.* (2006) for brainstems of 4 week old pigs. A plot of the shear stress versus amount of shear for different angles of orientation is given in Figure 3. The striking feature of the plots in Figure 3 is that there are essentially two distinct modes of shear response: the response for fibres with angles of orientation close to $0°$ and the response for all other angles. This mirrors the previously observed dependence of the locking shear on the angle of orientation. We note from (16) that a $0°$ orientation is equivalent to the neo-Hookean *isotropic* response mode for the shear stress with a linear dependence on the amount of shear. Finally we note that, *for small amounts of shear* $\kappa < 0.2$, the $90°$ orientation stress response is most like the linear isotropic response corresponding to a fibre angle of $0°$.

In general, the shear modulus can be defined as

$$s_m \equiv \frac{dT_{12}}{d\kappa}(\kappa = 0).\tag{17}$$

It follows from (16) that for the model of interest here

$$s_m = \mu \left(1 + \frac{\gamma}{2}\sin^2 2\theta\right).\tag{18}$$

Thus in the range $0 \leq \theta \leq \pi/2$, the maximum shear modulus occurs at $\theta = \pi/4$ with value $\mu(1 + \gamma/2)$ and the minima occur at $\theta = 0, \pi/2$ with values μ.

5   REMARKS ON EXPERIMENTAL DATA

There are few experimental data on the simple shear of rubber in the literature and even less data available for soft tissue (the most widely cited data for biological materials are those of Dokos *et al.* (2002) for passive myocardial tissue). This reflects the difficulty in performing simple shear experiments in comparison to the standard material characterisation tests of simple and biaxial tension and pure shear. An immediate practical issue that arises in designing and performing experiments based on the deformation field (1) is maintaining the constant dimension of the block in the plane of shear. On assuming that this issue can be resolved (a particularly elegant solution to this problem for rubber can be found, for example, in Brown 2006), there are other difficulties that need to be overcome and these are particularly acute for soft tissue.

Although both rubbers and soft biological tissues exhibit a severe strain-stiffening effect, they are distinguished by the fact that this occurs at much lower strain values for soft tissue than for rubberlike materials (see, e.g., Holzapfel 2005). In the context of simple shear experiments, this can be easily seen by comparing the plots of shear stress in Andreev & Burlakova (2007) with those in Dokos *et al.* (2002). In one of the few simple shear experiments on elastomers reported in the literature, Andreev & Burlakova (2007) sheared a layer of a rubber-like polymer (plastisol) of thickness 5 mm between platens of size 7.5 cm $\times$ 4 cm, which yields a ratio of the dimensions in the plane of shearing of 1:15. This aspect ratio was used presumably to ensure homogeneity of deformation for the range of shear considered. Amounts of shear up to 1.2 were studied and it is easily seen from their Figure 2 that the shear stresses remain bounded with a maximum of only 7 kPa. In contrast, Figure 6 of Dokos *et al.* (2002) shows that some of the shear stresses reported for passive ventricular myocardium are increasing very rapidly at amounts of shear of the order 0.5. Thus close to the values of the locking shears for soft tissue, the sizes of the shear and normal components of the traction on the inclined faces for soft tissue, necessary to maintain the homogeneous deformation (1), are likely to be an order of magnitude *greater* than those for rubber for the same amount of shear. In practice, however, tractions are never applied to the inclined faces and consequently homogeneity of the deformation is lost. This loss of homogeneity of deformation has long been recognized as a problem when rubbers are sheared (see, for example, Sommer & Yeoh 2001). To circumvent this difficulty rectangular specimens of rubber are usually cut so that the X_2 dimension is *smaller* than the other two, a ratio of at least 1:4 being recommended. We note that the experiments of Dokos *et al.* (2002) were performed on *cuboid* samples. There is no guidance in the literature as to the *range of shear stress* for which homogeneity is maintained for blocks proportioned in this way, although it seems implicit in the data and experimental protocol of Andreev & Burlakova (2007) that a range bounded above by the value of the shear modulus s_m defined in (17) is appropriate. This will be assumed here.

It was observed in the last section that there are essentially only two modes of shear stress response in the simple shear of fibre-reinforced materials modeled by (11), one of which is the neo-Hookean isotropic shear stress response corresponding to $\theta = 0°$. Since it follows from (4) and (13) that $W_4(\theta = 0°) = 0$, one finds that the shear traction on the inclined faces for fibre-reinforced materials for this angle of orientation is the same as the corresponding neo-Hookean *isotropic* shear traction given by

$$S_{iso} = \mu\kappa/1 + \kappa^2 . \tag{19}$$

Thus it would be expected that for specimens of aspect ratio of at least 1:4, homogeneity will be maintained for shear stresses up to the order of the shear modulus which, from (18), has the value μ. The shear stress response is qualitatively the same for all other fibre-orientations. Dokos *et al.* (2002) only sheared specimens for which $\theta = 0, 90°$ and consequently the shearing of specimens for which $\theta = 90°$ will only be considered further here. For this angle, it is shown in Horgan & Murphy (2011a) that again the shear traction necessary to maintain simple shear is given by the isotropic relation (19). Thus, again, it would be expected that for specimens of aspect ratio of at least 1:4, homogeneity will be maintained for shear stresses up to the order of the shear modulus s_m, which from (18) again has the value μ. The important difference between the two shearing modes is that the *amount of shear* κ for which homogeneity is preserved for $\theta = 90°$ is *significantly smaller* than the allowable range of shear for $\theta = 0°$. This is most easily seen from the plots of shear stress given in the last section. Recall that the plotted shear stress is normalized with respect to μ so that the maximum allowable shear stress value

is 1. It is expected that homogeneity is preserved up to this stress value for properly proportioned specimens. It is seen from Figure 3 that the allowable amount of shear when $\theta = 90°$ is in the approximate range $(0, 0.4)$ whereas for $\theta = 0°$ the corresponding range is $(0, 1)$.

6 CONCLUDING REMARKS

This paper has focused attention on the effect of fibre orientation on the shear stress response of fibre-reinforced soft biomaterials in simple shear. Results on the corresponding normal and hydrostatic stresses are described in Horgan & Murphy (2011a, b). The constitutive model used reflects the stretch induced stiffening of collagen fibres with increased shear loading. The corresponding strain-energy densities are of logarithmic form in the anisotropic invariant and are a viable alternative to the exponential forms usually used in biomechanical modelling of soft tissues. The results obtained are relevant to the development of accurate shear test protocols for the determination of constitutive properties of soft biomaterials.

ACKNOWLEDGMENTS

The work of COH was supported by the US National Science Foundation under Grant CMMI 0754704. This research was completed while this author held a Science Foundation Ireland E. T. S. Walton Fellowship at Dublin City University.

REFERENCES

Andreev, V. G. & Burlakova, T. A., 2007 Measurement of shear elasticity and viscosity of rubberlike materials. *Acoustical Physics* 53, 44–47.

Brown, R. 2006. *Physical testing of rubber*. New York, NY: Springer.

Destrade, M., Gilchrist, M. D., Prikazchikov, D. A. & Saccomandi, G., 2008 Surface instability of sheared soft tissues. *J. Biomech. Eng.* 130, 061007:1–6.

Dokos, S., LeGrice, I. J., Smaill, B. H., Kar, J. & Young, A. A. 2000 A triaxial-measurement shear-test device for soft biological tissues. *J. Biomech. Eng.* 122, 471–478.

Dokos, S., Smaill, B. H., Young, A. A. & LeGrice, I. J. 2002 Shear properties of passive ventricular myocardium. *Am. J. Physiol. Heart Circ. Physiol.* 283, H2650–H2659.

Gardiner, J. C. & Weiss, J. A. 2001 Simple shear testing of parallel-fibered planar soft tissues. *J. Biomech. Eng.* 123, 170–175.

Gent, A. N. 1996 A new constitutive relation for rubber. *Rubber Chem. Technol.* 69, 59–61.

Holzapfel, G. A. 2005 Similarities between soft biological tissues and rubberlike materials. In: *Constitutive models for rubber IV, Proceedings of the 4th European conference on "Constitutive Models for Rubber", (ECCMR 2005)*, Stockholm, Sweden (eds. P.-E. Austrell & L. Kari) pp. 607–617. Lisse, Balkema.

Horgan, C. O. & Murphy, J. G. 2010 Simple shearing of incompressible and slightly compressible isotropic nonlinearly elastic materials. *J. Elasticity* 98, 205–221.

Horgan, C. O. & Murphy, J. G. 2011a Simple shearing of soft biological tissues. *Proc. R. Soc. Lond. A* 467, 760–777.

Horgan, C. O. & Murphy, J. G. 2011b On the normal stresses in simple shearing of fiber-reinforced nonlinearly elastic materials. *J. Elasticity* (in press).

Horgan, C. O. & Polignone, D. A. 1995 Cavitation in nonlinearly elastic solids: a review. *Appl. Mech. Reviews* 48, 471–485.

Horgan, C. O. & Saccomandi, G. 2003 A description of arterial wall mechanics using limiting chain extensibility constitutive models. *Biomech. Model. Mechanobiol.* 1, 251–266.

Horgan, C. O. & Saccomandi, G. 2005 A new constitutive model for fiber-reinforced incompressible nonlinearly elastic solids. *J. Mech. Phys. Solids* 53, 1985–2015.

Horgan, C. O. & Saccomandi, G. 2006 Phenomenological hyperelastic strain-stiffening constitutive models for rubber. *Rubber Chem. Technol.* 79, 152–169.

Humphrey, J. D. 2002 *Cardiovascular solid mechanics*. New York, NY: Springer.

Lin, H. T., Dorfmann, A. L. & Trimmer, B. A. 2009 Soft-cuticle biomechanics: a constitutive model of anisotropy for caterpillar integument. *J. Theor. Biol.* 256, 447–457.

Merodio, J. & Ogden, R. W. 2005 Mechanical response of fiber-reinforced incompressible nonlinear elastic solids. *Int. J. Nonlinear Mech.* 40, 213–227.

Merodio, J., Saccomandi, G. & Sgura, I. 2006 The rectilinear shear of fiber-reinforced incompressible non-linearly elastic solids. *Int. J. Nonlinear Mech.* 41, 1103–1115.

Ning, X., Zhu, Q., Lanir, Y. & Margulies, S. S. 2006 A transversely isotropic viscoelastic constitutive equation for brainstem undergoing finite deformation. *J. Biomech. Eng.* 128, 925–933.

Ogden, R. W. & Saccomandi, G. 2007 Introducing mesoscopic information into constitutive equations for arterial walls. *Biomech. Model. Mechanobiol.* 6, 333–344.

Polignone, D. A. & Horgan, C. O. 1993a Cavitation for incompressible anisotropic nonlinearly elastic spheres. *J. Elasticity* 33, 27–65.

Polignone, D. A. & Horgan, C. O. 1993b Effects of material anisotropy and inhomogeneity on cavitation for composite incompressible anisotropic nonlinearly elastic spheres. *Int. J. Solids Struct.* 30, 3381–3416.

Sadovsky, A.V., Baldi, P. F. & Wan, F. Y. M. 2007 A theoretical study of the *in vivo* mechanical properties of angiosperm roots: constitutive theories and methods of parameter estimation. *J. Eng. Materials Tech.* 129, 483–487.

Schmid, H., Nash, M. P., Young, A. A. & Hunter, P. J. 2006 Myocardial material parameter estimation-a comparative study for simple shear. *J. Biomech. Eng.* 128, 742–750.

Sommer, J. G. & Yeoh, O. H. 2001 Tests and specifications. In *Engineering with rubber: how to design rubber components* (ed A. N. Gent), pp. 307–366. Cincinnati, OH: Hanser-Gardner Publications.

Taber, L. A. 2004 *Nonlinear theory of elasticity: applications in biomechanics.* Singapore: World Scientific.

Constitutive Models for Rubber VII – Jerrams & Murphy (eds)
© 2012 Taylor & Francis Group, London, ISBN 978-0-415-68389-0

Three-dimensional carbon black aggregate reconstruction from two orthogonal TEM images

J. López-de-Uralde, M. Salazar, A. Santamaría, A. Zubillaga & P.G. Bringas
DeustoTech, University of Deusto, Bilbao, Spain

T. Guraya, A. Okariz & E. Gómez
EUITIB, University of the Basque Country, Bilbao, Spain

Z. Saghi
Department of Materials Science and Metallurgy, University of Cambridge, UK

ABSTRACT: Current morphological characterization techniques extract the basic properties of Carbon Black aggregates from single bi-dimensional images, which are just planar projections of the real structures. Although widely used, this approach provides limited and inaccurate characterization of the aggregates. Several techniques have been proposed to achieve a full three-dimensional characterization of nano-structures, such as electron tomography and FIB-SEM. However, the capability of these techniques to characterize Carbon Black aggregates is still under study. In this work, we present a novel solution that aims to improve the characterization of Carbon Black aggregates by providing a fast, flexible and robust reconstruction method. The proposed method takes advantage of genetic algorithms to obtain a complete three-dimensional model of the aggregates from two orthogonal TEM images. Results were compared to an electron tomography reconstruction obtained from a full tilt series.

1 INTRODUCTION

Filled polymers are widely employed in many application areas such as paper, plastics, rubber and paint industries, where their polymer properties (e.g. density, conductivity, mechanical properties) and rheological properties are enhanced by means of mixing them with some filler. Different fillers have different effects on the material's chemical nature, particle size and shape, and therefore result in quite complex geometrical objects to which specific reinforcing properties can be associated. This is the case of carbon black where carbon black spherical units form three-dimensional aggregated structures s whose particular shape has an important impact on the reinforcing properties of the polymer. It is well known that stiffness, tensile strength, compression and tear strength, fatigue and wear resistance or dynamic performance significantly increase through these filler additions, but the mechanisms responsible for these effects are still under discussion (Kohls, & Beaucage 2002, Fukahori 2003).

Filler characterization is essential in order to state the relationship between the morphological characteristics of the fillers and their reinforcement effect on the polymer: the employment of Finite Elements Analysis (FEA), for example, requires a great amount of precise data. To this extent, it's clear that a greater precision in the filler characterization and a more complete characterization will lead to remarkable improvements in predictive models and FEA.

The filler characterization is standardized, and microscopic techniques are used for the characterization of the filler: a TEM/AIA procedure is well defined in the corresponding standard (ASTM D3849 – 07). This standard describes the procedure to characterize the carbon black's aggregates of each grade, but important parameters related to the reinforcing capabilities of the grade such as the aggregate's superficial area are derived from measurements based on TEM images, which are two-dimensional projections of the aggregate. It's clear that improvement of the morphological characterization is therefore an important issue, and that the three-dimensional characterization of the fillers could be a step forward.

Over the last years there have been several attempts to carry out a 3D reconstruction of fillers (Gruber et al. 1994). Electron tomography and FIB-SEM techniques are nowadays the most promising opportunities for achieving a complete 3D reconstruction of nanostructures. In the tomography area, Kohjiya et al. (2005, 2006 and 2008) have reported some 3D reconstructions of different fillers in rubber. The main difference between both techniques, apart from the microscopy technique itself (TEM for tomography and SEM for FIB-SEM) is that in the first case images are taken in a set of different tilts of the sample with respect to the incident electron beam and the volume is reconstructed later by means of some complex mathematic algorithms (Frank 2005), while in the FIB-SEM technique layers of the nanometer order are successively eroded from the surface of the sample employing a Focused Ion Beam (FIB) and a new SEM image of the surface is obtained, giving rise to a stack of images of the different layers of the sample, which can be reconstructed into a volume in a relatively easy way.

Both techniques are time-consuming and require a big amount of data (image stacks), and although after the volume reconstruction morphologic parameters such as the aggregate's volume or superficial area can be determined, the exact location and size of each of the carbon units in the aggregate can not be determined. Further work must be carried out upon these reconstructions in order to extract this kind of information. Monte Carlo methods could be an effective way of deducing the exact structure of the reconstructed volume (Eguzkitza et al., in prep.).

In this work we explore the possibility of reconstructing the same 3D TEM-images of the fillers by means of a less time demanding technique and, at the same time, obtaining exhaustive information about the morphology of the filler. A 3D model of a carbon black's aggregate will be obtained from two TEM images of the same aggregate at different angles. This model will be exhaustive enough to give all the details about the aggregate and its structure: the number, relative position and size of the carbon units and the volume and superficial area of the complete aggregate.

This technique is based on a genetic algorithm (Barricelli 1954), which applies a set of genetic operators to a group of possible solutions –called population–, in search for the fittest ones at each iteration –or generation–. These operators may include random mutations and crossover methods (i.e. forming an instance from two or more parents) (Koza 1995).

The result of this operation only provides us with the position and size of the particles in the aggregate. Therefore, in order to obtain more substantial information, a reconstruction process is usually achieved through the use of a three-dimensional volumetric field (a voxelization grid in which each of its elements, called voxels, contains the volumetric information at a given point). Furthermore, for the estimation of the superficial area, isosurface generation algorithms such as Marching Cubes (Lorensen & Cline 1987) are commonly used.

2 ALGORITHM'S DESCRIPTION

The algorithm proposed in this work can be divided into three main parts. The core is the 3D genetic algorithm where particles are randomly generated in pursue of an accurate three-dimensional aggregate structure determination. Initially, we need to process two TEM images captured at 90 degrees difference as shown in Figure 1 to obtain their height maps. Later we project the particles with the aim of obtaining similar height maps to the originals. Once we have the definition of

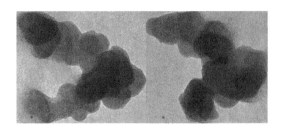

Figure 1. Original TEM images at −45° and +45°.

the aggregate structure, that is, the position and size of its particles, we generate a 3D model with two main objectives: the visualization of the particles, and the estimation of their volume and superficial area.

2.1 Image processing

First, the aggregate from the two images is segmented with an enhanced version of a binarization algorithm we have previously published (López-de-Uralde et al. 2011). Afterwards, the images are transformed into height maps.

2.2 Segmentation

In order to be able to segment the aggregate we first have to binarize a noise-free image. To this extent, a Gaussian smoother is applied. This 2D convolution operator eliminates noise at the price of losing a bit of detail (Pajares & de la Cruz 2007). Then, a threshold for aggregate-background discrimination is estimated using Otsu's method (Otsu 1975). This threshold is adjusted to be more adequate for TEM images. We generate a binary image considering that pixels with a value below the threshold correspond to background and pixels above it are part of the aggregate area.

Moreover, we improve the edge quality by dilating and eroding it with a disk shape morphological structuring element.

2.2.1 Height map

A height map is a representation of the height of each point on a surface in a two-dimensional matrix. It is typically represented by an image indicating the surface height at each point by its corresponding pixel with its colour. In our case, this map represents the number of overlapping particles in each point of the initial image taken with a microscope. Our height map, as can be seen in Figure 2, is generated by the following steps.

2.2.1.1 Image pre-processing

First, the image is resized with a bicubic filter to reduce the processing time of all subsequent steps.

Then, with the aim of reducing the image noise, we apply the Median filter on the image obtained with the microscope.

To end this process, it is required to reverse the resulting image from the previous step, obtaining

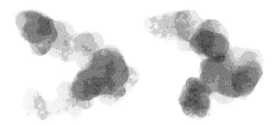

Figure 2. Height maps at −45° and +45°.

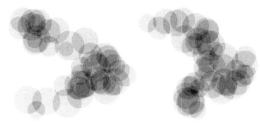

Figure 3. Particle projections at −45° and +45°.

a predominantly black background with an aggregate whose accumulation of material in each pixel is directly related to the amount of white of that pixel.

2.2.1.2 Background removal

At this stage the objective is to completely eliminate the background colour, both the background itself and the white increase in the aggregate caused by the background material. In this way, the average value of the background pixels is calculated using the mask in the segmented image.

The final step of this section results in the aggregate obtained without any background. This will require painting black all the pixels marked as background by the segmentation image in the original image. It is also necessary to subtract the average value of the background obtained in the previous steps in all pixels of the aggregate image that the segmented image marks as not-background.

2.2.1.3 Particle value detection

To build the height map, the average amount of material of a single particle is needed. To do this we must normalize the heights from the above process by converting the maximum height into pure white and minimal height into pure black. By this step we get similar levels of white in all images.

Finally, to obtain the value of a single particle, it is necessary to recount the number of existing heights in the image and calculate a value for each level indicating the presence of this height in the image. With this value it is possible to discriminate the heights that do not have much presence in the image and keep the lowest height that exceeds a particular threshold.

2.2.1.4 Height map generation

Having calculated the height value of a single particle, we pass to the calculation of the height map. To do this we have to populate the array of heights with the value obtained by dividing the amount of white in that pixel by the amount of white that a single particle should have, therefore obtaining the number of overlapping particles in each pixel of the image stored in the height matrix.

The calculation of the value of a single particle is overly dependent on the original image, the original particles distribution and the segmented image. For that reason, the obtained value may be to be too low, generating too many particles at each point. To solve this problem it is necessary to make a normalization of the depth of the map. To make such standardization

it is necessary to establish the maximum number of overlapped particles.

2.3 3D genetic algorithm

This algorithm is based on the theory of evolution, but is simplified in the way that there are only mutations and not crossover operators (i.e. mixing two intermediate solutions).

The goal that drives this evolution process is to obtain similar height maps from the TEM images projecting particles at −45 degrees and +45 degrees as shown in Figure 3. As they are orthogonal planes, the projection of a particle only requires to obviate one axis (e.g. (3, 8, 7) corresponds to (3, 8) and (7, 8)).

First, the population is initialized. Then, an iterative process starts where each instance of the population suffers one mutation. Later, the mutated instances are compared with the previous ones by means of the defined fitness function and the best 30 instances are kept. Finally, when the best instance reaches the desired fitness a result is established and the algorithm terminates.

2.3.1 Initialization

The first step of the genetic algorithm is to initialize the population. Ten empty instances are created. Each instance will be an intermediate solution containing a list of particles that will form an aggregate.

2.3.2 Mutations

We have defined five types of mutations that are applied during the mutation process. One and only one of them is chosen randomly:

– Adding a particle: this mutation generates a three-dimensional point randomly. Its radius is established randomly between a minimum and maximum radius manually defined previously.
– Removing a particle: this mutation eliminates a particle.
– Moving a particle: this mutation changes the position of a particle.
– Moving a particle a little: this mutation changes the position of a particle only one pixel.
– Resizing a particle: this mutation changes the radius of a particle.

It is worth pointing out that when creating, moving or resizing a particle there are several constraints. Firstly, we ensure that the sphere is inside the square

Figure 4. Sphere and volumetric mixed with isosurface representations of the aggregate at 0°.

Figure 5. Original TEM image at 0° and its tilted tomographic reconstruction.

prism formed by the two images. Secondly, we check that the centre of the particle is inside the masks of the two projections.

Moreover, when applying a moving or resizing mutation to an instance, the particle involved is chosen randomly.

2.3.3 *Selection*

To select between the instances of the population, we have defined a fitness function. This function computes the differences between the height maps of the two projections and their corresponding projections of the instance particles.

Furthermore, an instance is penalized for the portion of mask that is not filled by particles. Additionally, isolated particles harm the fitness function result proportionally to their radius and distance to the closest particle.

2.3.4 *Termination check*

There are two termination conditions: the best instance reaches a given fitness or a sufficiently good result is observed by manual inspection.

2.4 *Three-dimensional reconstruction and visualization*

The particle data obtained in previous steps enables us to faithfully recreate the original Carbon Black aggregate in a three-dimensional space. The resulting model not only provides us with a visual depiction of the aggregate (which can be used to further validate previous results), but also introduces additional volumetric and superficial information.

We provide the user with different representations. Each of them provides a different view of the model data so it can be perceived and understood more easily. In Figure 4 all the three representations are shown.

– Sphere-based representation: The simplest of the three, it utilizes spheres to display every particle of the aggregate. In this way it provides a fast and easy way for the graphic hardware to draw the surface of the aggregate. The user can check with this representation if the current best instance of the genetic is good enough.
– Volumetric representation: In this representation, only the volumetric information is shown. This is achieved through the creation of a voxelization grid, whose individual elements (called voxels), have a

different value depending on the distance to the closest particle centre. This value is later translated into a colour through the use of an interpolable colour palette and drawn on screen. Furthermore, in order to show the inner section of the aggregate, additional clipping planes are introduced for the user to control.
– Isosurface representation: Taking into account the voxelization grid values established in the previous representation, a triangulated mesh is generated through our enhanced implementation of the Marching Cubes algorithm. By doing this, we obtain a surface that more accurately matches the original one.

3 METHODOLOGY

In order to get the three-dimensional characterization of a carbon black aggregate, a stack of TEM images at different tilt angles has been obtained. The carbon black has been supplied by Cabot Corp. and is a CSX 691 commercial grade. Some specimens have been prepared to be analysed according to the ASTM standard procedure (ASTM D3849 – 07) employed for carbon black grades characterization. The microscope is a JEM-2200FS/CR transmission electron microscope (JEOL, Japan) equipped with an ULTRASCAN 4000 SP (4008×4008 px) and a cooled slow scan CCD camera (Gatan, UK). Tilted images have been taken in the $\pm60°$ interval at every 1.5°, see Figure 5 for the image corresponding to 0° tilt.

The complete image stack has been used for a tomography reconstruction by means of the IMOD free software (Mastronarde & McIntosh 1996, IMOD Home Page). Once the whole tomogram is computed, the segmentation of the reconstructed volume gives rise to the carbon black aggregate's volume as can be seen on Figure 5. And the values for the superficial area and volume of this reconstructed volume are determined with the 3dmod program of the IMOD package.

At the same time, two images from the full image stack have been chosen for the calculation of the three-dimensional model deduced by means of the genetic algorithm presented in this paper. These images correspond to $\pm45°$ tilted ones in the stack employed for the tomographic reconstruction.

After the images are processed as described in the previous section, the height map of each image is calculated and the 3D genetic algorithm is employed to calculate a 3D model according to these height maps.

The designed genetic algorithm predicts the position and size of the particles in three dimensions. An initial population of 10 individuals is created. Each individual contains a list of particles with their positions and radius, and the algorithm works with two projections in the $\pm45°$ corresponding to that list of particles: it calculates a height map that estimates the quality of each individual at this stage and after the mutations that give rise to new individuals. To evaluate the quality of each individual we used a fitness function. This function returns a value calculated by comparison of the individuals' local height map and the previously calculated height map of the original image.

At each iteration, new descendants (individuals created by mutations) are generated using the entire population as well as 10 additional "children" of the 10 best individuals of the previous iteration. Then, the best 30 individuals of the population are selected and the rest are discarded from the population. We obtained this value of the population by an empirical validation with different configurations, maximizing the trade-off between accuracy and efficiency because a larger population can avoid the appearance of local maximum. However, such a population requires more overhead processing.

The possible mutations that may appear, as we aforementioned, include creation, removal, repositioning and rescaling of the particles. Both in the rescale and creation, it is mandatory to set a radius, whose maximum and minimum values are set manually.

Once there is a solution, the validation of the results is performed by extracting the volume from the voxelization grid and the superficial area from the reconstructed isosurface. These values are later compared to the same values obtained in the tomographic reconstruction of the TEM images.

In addition, we have performed another validation with an image taken at 0°. With this purpose we have segmented it and generated its height map. Then we have projected the particles of the solution shown through this paper into the 0° plane. This projection is not as straightforward as the ones at −45° and +45°. In this case, a change to the coordinate system by a rotation is needed. This third image, like the tomographic reconstruction is only used with validation purposes, so the number of images needed for the reconstruction is still two.

4 RESULTS

Values for the superficial area and volume of the re-constructed tomogram are $5.93 \cdot 10^5$ nm^2 and $1.69 \cdot 10^7$ nm^3 respectively. It must be pointed out that in order to perform these measurements, a segmentation task must be carried out on the orthoslices

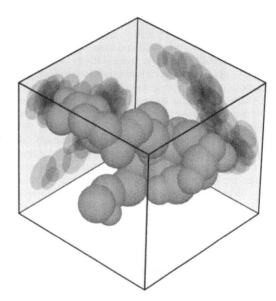

Figure 6.   Reconstructed aggregate with its projections.

obtained after reconstruction. This manual intervention along with artefacts inherent to the reconstruction process itself (Frank 2005) induces errors in the quantification that should be kept in mind.

On the other hand, our algorithm has calculated the same parameters with only two TEM images and nearly no human aid, giving rise to a superficial area of $4.45 \cdot 10^5$ nm^2 and a volume of $1.21 \cdot 10^7$ nm^3. In the Figure 6 is shown their three-dimensional reconstruction.

It is obvious that further work must be carried out in order to improve the algorithm and draw final conclusions, but these results show a great concordance and confirm the validity of the proposed method.

5 CONCLUSIONS

We have presented a new and fast method for the three-dimensional reconstruction of carbon black aggregates. This study has led us to insightful conclusions; however, the research in this field is far from being completed. Since the global impact of carbon black aggregates on the mechanical properties has still a high degree of uncertainty, further study is required. In addition, it would be very interesting to validate it on more complex aggregates and other types of nanofillers such as nanoclays, graphene and so forth. In order to improve the algorithm, the following changes will be implemented in future work.

The projection method should be changed so that the centre of the particle projects more density than the outer area. Additionally, aggregate density is uniform across the aggregate so the fitness function should take this into account for particle intersections.

Different genetic techniques like crossover, drastic mutations and keeping randomly some elements despite having a bad fitness value. We expect to avoid getting stuck in local maxima.

These modifications will make the algorithm more computationally expensive, so we strongly suggest the study of a method to accelerate it by means of General-Purpose computation on Graphics Processing Units (GPGPU). In doing so, the higher resolution speed will increase dramatically enabling us to analyse more aggregates of the same sample.

Additionally, we propose the development of advanced skeletonization methods to correctly assess the internal structure of the carbon black aggregates. Since the exact location of each individual particle has already been calculated, the new skeletonization algorithm could achieve a faithful estimation of the internal structure of the aggregate. Subsequently, the resulting three-dimensional skeleton would provide several new features that, once processed by machine learning algorithms, could lead to a better classification of carbon black aggregates.

Finally, we suggest the creation of a User Interface based on Augmented Reality (AR) devices to properly visualize the different layers of three-dimensional information. Furthermore, by employing a stereoscopic device (e.g. a Head-Mounted Display) the new Human-Computer Interaction model should be able to provide a higher level of depth perception, allowing the user to easily understand the structure of the carbon black aggregates (even those with complex ramifications).

ACKNOWLEDGEMENTS

This work has been supported by the University of the Basque Country with the NUPV09/03 incentive for investigation, the project DFV10/04 of the Regional Government of Biscay and the project UEGV09/C19 of the Basque Government. We also want to thank the Electron Microscopy Platform Service of CIC Biogune for the TEM images in the Electronic Microscopy.

REFERENCES

ASTM D3849 – 07, 2008. Standard Test Method for Carbon Black-Morphological Characterization of Carbon Black Using Electron Microscopy.

Barricelli, N. 1954. Esempi numerici di processi di evoluzione. Methodos: 45–68.

Frangakis, A. S. & Hegerl R. 2001. Noise reduction in electron tomographic reconstructions using nonlinear anisotropic diffusion. *Journal of structural biology* 135(3): 239–250.

Frank J. 2005. *Electron tomography. Methods for three-dimensional visualization of structures in the cell*, Springer.

Fukahori, Y. 2003. The mechanics and mechanism of the carbon black reinforcement of elastomers. *Rubber Chemical Technology* 76: 548–566.

Gruber T. & Zerda, T.W. 1994. 3D morphological characterization of carbon black aggregates using transmision electron microscopy. *Rubber Chemical Technology* 67: 280–287.

IMOD Home page, http://bio3d.colorado.edu/imod/

Kohjiya, S. et al. 2005. Three-dimensional nanostructure of in situ silica in natural rubber as revealed by 3D-TEM/electron tomography. *Polymer* 46: 4440–4446.

Kohjiya, S. et al. 2006. Visualization of carbon black networks in rubbery matrix by skeletonisation of 3D image. *Polymer* 47: 3298–3301.

Kohjiya, S. et al. 2008. Visualization of nanostructure of soft matter by 3D-TEM: Nanoparticles in a natural rubber matrix. *Progress in Polymer Science* 33: 979–997.

Kohls D. J. & Beaucage G. 2002. *Current Opinion in Solid State and Material Sciences*, 6: 183–194.

Koza, J.R. 1995. Survey of genetic algorithms and genetic programming. *Microelectronics Communications Technology Producing Quality Products Mobile and Portable Power Emerging Technologies:* 589.

López-de-Uralde, J. et al. 2011. Automatic morphological categorisation of carbon black nano-aggregates. *Database and Expert Systems Applications*: 185–193.

Lorensen, W. E, & Cline, H. E. 1987. Marching cubes: A high resolution 3D surface construction algorithm. *ACM Siggraph Computer Graphics* 21(4): 163–169.

Mastronarde, D.N. 1997. Dual-axis tomography: an approach Ruth alignment methods that preserve resolution, *Journal of Structural Biology*, 120: 343–352.

Otsu, N. 1975. A threshold selection method from gray-level histograms, *IEEE Transactions On Systems Man And Cybernetics* 9(1): 62–66.

Pajares, G. & de la Cruz, J. 2007. *Visión por Computador.* Madrid: Ra-Ma Publishers.

Finite element analysis and design of rubber specimen for mechanical test

C.S. Woo, W.D. Kim & H.S. Park
Korea Institute Machinery & Materials, Daejeon, Korea

ABSTRACT: The material properties of rubber were determined by the experiments of simple tension, simple compression tension, pure shear, equi-biaxial tension test. In simple tension test, dumbbell specimen is generally used to obtain a state of pure tensile strain. It is shown that a narrow strip specimen for length is over 10 times of the width can be also used in. In simple compression test, the effect of the friction force between the specimen and the platen is investigated. The test device with the tapered platen is proposed to overcome the effect of friction. It is verified by experimental and finite element analysis results. In pure shear tests, it is shown that the width of specimen must be at least 10 times of the height. Specimen and equipment of equi-biaxial tension was development a state of strain equivalent to pure compression.

1 INTRODUCTION

The objective of the testing described herein is to define and to satisfy the input requirements of mathematical material models that exist in structural, non-linear finite element analysis software. The material modeling of hype-elastic properties in rubber is generally characterized by the strain energy function. These functions are based on the assumption that the rubbery material is isotropic and elastic. The strain energy functions have been represented either in term of the strain invariants that are functions of the stretch ratios, or directly in terms of the principal stretch. Successful modeling and design of rubber components relies on both the selection of an appropriate strain energy function and an accurate determination of material constants in the function. Material constants in the strain energy functions can be determined from the curve fitting of experimental stress-strain data. There are several different types of experiments, including simple tension, simple compression, equi-biaxial, pure shear tests. In general, a combination of simple tension, equi-biaxial, and pure shear tests are used to determine the material constants. Although the test procedures and the configuration of test specimens are given a full explanation in standards, they can still be modified for the purpose of easy application of test equipment and a minimization of experimental errors.

In this study, the test errors due to the strip specimen used in the tension test are predicted and evaluated using finite element analysis and experiments. In the compression test, the effect of the friction force between the specimen and the platen is investigated, and a new test device with a tapered platen is proposed to overcome the effect of friction. It was turned out that the relationship of the stress and strain using the tapered platen was in fairly agreement with the pure compressive state. Also, specimen and equipment of equi-biaxial tension was development a state of strain equivalent to pure compression. In the pure shear test, a proper specimen shape is suggested to obtain a state of pure shear strain.

2 EXPERIMENT

2.1 Simple tension test

Simple tension experiments are very popular for rubber material. There are several standards for the testing of rubber in tension. However, the experimental requirements for analysis are somewhat different than most standardized test methods.

The most significant requirement is that in order to achieve a state of pure tensile strain, the specimen be much longer in direction of stretching than in the width and thickness dimensions. Two types of test specimens are generally used, which are the dumbbells and the strip specimens. A dumbbell specimen is generally used to prevent specimen failure in the clamped region. The experiment with the dumbbell specimen needs an extensometer. Particularly, a non-contacting device such as a laser extensometer is required in the environmental test within the chamber. If the material constants in the relatively small strain range are to be determined, then the stress-strain curve can be obtained using a strip specimen without the extensometer. The elongation can be measured by the crosshead displacement. In this case, experimental errors are inevitable because the uniform rectangular shape of the specimen cannot be preserved after deformation. When these errors are small enough to neglect, the strip specimen is conveniently used with a universal test apparatus. These errors can be predicted and evaluated using non-linear finite element analyses and experiments for dumbbell and strip specimens. A deformed shape of the dumbbell specimen is shown

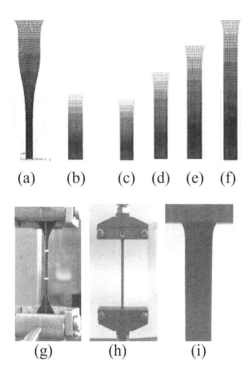

(a)   (b)   (c)   (d)   (e)   (f)

(g)   (h)   (i)

Figure 1. Simple tension test using (a) deformed shape of dumbbell specimen (b) gage length (c) deformed shape of 1:4 strip specimen (d) 1:6 (e) 1:8 (f) 1:10 (g) dumbbell specimen (h) strip specimen (g) clamped edge in strip specimen.

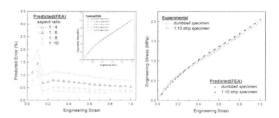

Figure 2. Predicted error between dumbbell and strip specimens with various aspects ratios (length/width) up to 100% strain by finite element analysis.

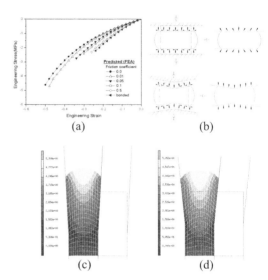

(a)   (b)

(c)   (d)

Figure 3. Simple compression test (a) friction effects on stress-strain curves (b) free body diagrams of flat and tapered platens (c) deformed shapes with 5% in taper angle and friction coefficient 0.05 (d) 10% in taper angle and 0.1 in friction coefficient.

in Fig. 1(a) and (f). The shape within the gauge length of the specimen is uniform during deformation with uniform strain distribution. Figs. 1(g) and (i) show the deformed shape and the clamped edge of the strip specimen, respectively. In this case, the pure states of stress and strain cannot be achieved because of the clamped region. As the length of the strip increases, the effect of the clamped region will decrease. According to the results of finite element analyses, it is shown that the error in stress magnitude between the dumbbell and strip specimens decreases as the length increases. This error is smaller than one percent, when the length of a strip specimen is 10 times longer than the width as shown in Fig. 2. Experimental and predicted results within the 100 percent strain level of the simple tension tests using the dumbbell and the strip specimens are shown in Fig. 2. Even though there exist some differences in the stress-strain response between the prediction and the experiment, they can be neglected when considering the testing error and the inherent scatter of the rubber properties.

2.2 Simple compression test

A compression test is in many ways easier to carry out than a tension test, and in view of the large number of applications of rubber in compression, should be more often used. The compression test piece detailed in ASTM D575 is a cylindrical shape with 28.6 ± 0.1 mm diameter and 12.5 ± 0.5 mm thick.

In order to use the compression test data for determining material constants in the strain energy functions, the pure state of compressive strain is required. The pure compression state is, however, quite difficult to achieve due to the friction between the specimen and the compression platens. If friction exists, then the stress and strain distributions are not uniform and barreling occurs. Even though the friction is very small, below 0.1, there is some difference between the cases of friction and perfect slippage. It is impossible to maintain perfect slippage during testing even if a lubricant is used. The effect of the friction on stress-strain curves is shown in Fig. 3(a). Fig. 3(b) shows the force diagram with the flat and tapered platens whose slop is the same as the coefficient of friction. Because the combined force of friction and compression is applied in the direction of restraining the deformation of specimen, barreling and deviation in load-deformation occurs. By employing the tapered platens in the test, the direction of the combined force can be made vertical, thus achieving the effect of eliminating the friction forces. Figs 3(c) and (d) show the

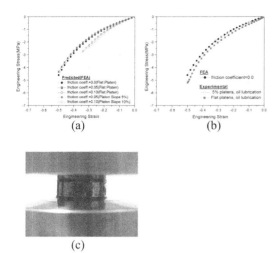

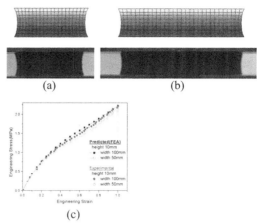

(a)   (b)

(c)

Figure 5. Deformed shape of pure shear test (a) aspect ratio is 5:1 and (b) 10:1 (c) stress-strain curves.

(c)

Figure 4. Stress-strain curves in simple compression test (a) predicted (b) experiments (c) taper platen.

deformed shapes by finite element analyses with the tapered platens of 5% and 10%, corresponding to the coefficients of friction of 0.05 and 0.1, respectively. The amount of barreling for the tapered platens has been reduced, compared to the flat platens. Fig. 4(a) shows the compressive stress-strain curves predicted by the finite element analysis using the flat and the tapered platens. The results show that the predicted curves with the tapered platens are in a close agreement with the predicted curve with flat platens in the case of perfect slippage. The stress-strain curves during compression are shown in Fig. 4(b) for various friction conditions and platen types. The results of the newly designed platens show good correlation between the theoretical and the predicted results. The compression specimen using the tapered platens with a lubricant strained without any barreling as shown in Fig. 4(c).

2.3 Pure shear test

A shear strain state is a more important mode of deformation for engineering applications than tension. There are pure shear and simple shear in shear deformation mode. The quad lap simple shear test piece is standardized. But, pure shear test is not yet standardized. There are two difficulties in the simple shear test. The first difficulty is making the specimen. This may require either bonded to the rigid supports during vulcanization or molded blocks are adhered with a high modulus adhesive. Secondly, the low shear strain range is limited because the rigid plates are bent on straining. Alternatively, pure shear test can be developed high strain range than simple shear test. If the material is incompressible and the width of the specimen is longer than the height, a pure shear state exists in the specimen at 45 degrees angle to the stretching direction. Aspect ratio of the specimen is most significant in pure shear test because the specimen is perfectly constrained in the horizontal direction. Fig. 5 shows

the deformed shapes by the finite element analysis at 100 percent stretching for the aspect ratio of 5:1 and 10:1. Stress-strain curves obtained from the tests are shown Fig. 5(c), compared to those predicted by the finite element analysis. Even though there exist some differences in the stress-strain responses between the experiment and the analysis, fairly good correlations are observed. A better agreement can be seen for the aspect ratio of 10:1, compared to 5:1. The differences are attributed to the specimen slippage from the clamp edges, leading to the inadequate states of pure shear strain. Therefore, it is necessary to design a gripping device to prevent specimen slippage, in order to improve the test accuracy.

2.4 Equi-biaxial tension test

For incompressible or nearly incompressible materials, equi-biaxial tension of a specimen creates a state equivalent to pure compression. Although the actual experiment is more complex than the simple compression test, a pure state of strain can be achieved which will result in more accurate material model. Finite element analysis of the specimen is required to determine the appropriate geometry of the clamping point in Fig. 6(a). The equi-biaxial strain state may be achieved by radial stretching a circular disc in Fig. 6(b). Once again, a non-contacting strain measuring device must be used such that strain is measured away from the clamp edges in Fig. 6(c).

2.5 Strain range effect

A stress softening behavior is observed at any strain range below a level to which the rubber has been previously stretched. This stress softening in rubber is called the Mullins effect. The effect of pre-stressing is due to the physical breakdown or the reformation of the rubber network structures. One example of this behavior is shown in Fig. 7(a) and (b) where a filled

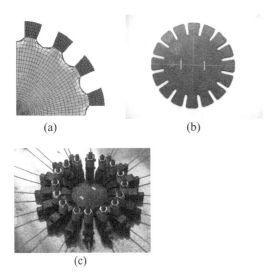

(a) (b)

(c)

Figure 6. Specimen of equi-biaxial tension test (a) finite element analysis of specimen (b) Configuration of specimen (c) equi-biaxial tension test using a laser extensometer.

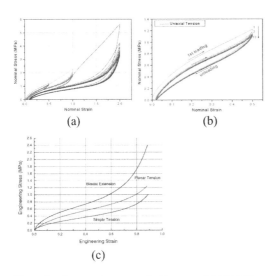

(a) (b)

(c)

Figure 7. Stress-stain curves of rubber material (a) Mullins effect (b) 10 repetition (c) reduced data setprepared for analysis.

natural rubber is strained to 50% strain for 10 repetitions followed by straining to 200% for 10 repetitions. Stress-strain curves are generally stabilized after 10 cycles for natural rubber. Therefore, in order to predict the behavior of the rubber components using the finite element analysis, the rubber constants must be determined from the stabilized cyclic stress-strain curve. The stress-strain curve varies significantly depending on the cyclic strain levels. A typical set of 3 stress-strain curves appropriate for input into fitting routines are shown in Fig. 7(c).

Table 1. Mooney-Rivlin 2-term fit summary for multiple modes.

case	simple tension	pure shear	equi-biaxial	C10	C01	shear modulus (G)
1	O	X	X	0	0.775	1.550
2	O	O	X	0.202	0.497	1.398
3	O	X	O	0.612	0.021	1.266
4	O	O	O	0.629	0.013	1.285

3 RUBBERY MATERIAL PROPERTIES

3.1 Strain energy function

In the FEA of rubbery materials, material models can be characterized by strain energy functions have been represented either in term of the strain invariants which are functions of the stretch ratios or directly in terms of the stretch ratios themselves. Such a material is called hyper-elastic material. The earliest model of nonlinear elasticity is Mooney-Rivlin may be given as follows:

$$W = C_{10}(I_1 - 3) + C_{01}(I_2 - 3) \tag{1}$$

Although, it shows a good agreement with tensile test data up to 100% strains, it has been found inadequate in describing the compression mode of deformation. Moreover, the Mooney-Rivlin model fails to account for the stiffening of the material at large strains. It is important to note that a Mooney-Rivlin model will give a straight line, not the nonlinear plot. This model obtained by fitting tensile data is quite inadequate in other modes of deformation, especially compression.

Ogden proposed the energy function as separable functions of principal stretches. This model gives a good correlation with test data in simple tension up to 700%.

$$W = \sum_{n=1}^{N} \frac{\mu_n}{\alpha_n}(\lambda_1^{\alpha_n} + \lambda_2^{\alpha_n} + \lambda_3^{\alpha_n} - 3) \tag{2}$$

The model accommodates non-constant shear modulus and slightly compressible material behavior. Ogden model has become quite popular recently, and has been successfully applied to the analysis of industrial products.

3.2 Fit multiplemodes

We performed the curve fitting with simple tension, pure shear test and equi-biaxial test data. Mooney-Rivlin 2-term fits that uses progressively more information as the basis for the curve fitting.

The table above summarizes the coefficient calculated in each case. Clearly the multiple modes have the biggest influence and capture better behavior. As shown in table 1, using only tensile data over predicts stress in the multiple modes and would be a poor fit for balloon inflation.

4 CONCLUSIONS

The improvement of test method and error prediction of test was investigated using finite element analysis and experiments. The validity of a simple tension test using a narrow strip specimen was evaluated by the finite element analysis. When the ratio of length to width of the specimen was greater than 10, the difference between the narrow strip and the dumbbell specimen could be reduced to 1.0%. Even though there were some differences in stress-strain responses between the two types of the simple tension specimen, they can be neglected considering the testing error and the inherent scatter of rubber properties.

The compression test method using the tapered platen with a slope equivalent to the friction coefficient was proposed to eliminate the effect of friction. It turned out that the stress-strain curve using the tapered platen was in close agreement with the curve using the flat platen with perfect slippage. The experimental results suggested that the compression test using the tapered platens was more reliable for achieving the state of pure compression than that using the flat platens. Also, specimen and equipment of equi-biaxial tension was development a state of strain equivalent to pure compression.

The pure shear specimens must have a length that is 10 times the height to achieve pure strain states in lateral direction and decrease the problem of grip slippage.

In order to predict the behavior of the rubber components using the finite element analysis, the rubber constants must be determined from the stabilized cyclic stress-strain curve and multiple modes have the biggest influence and capture better behavior.

ACKNOWLEDGEMENT

This work was supported by the Industrial Strategic technology development program (10037360, A Multidimensional Design Technology Considering Perceived Quality (BSR) Based on Reliability) funded by the Ministry of Knowledge Economy (MKE, Korea).

REFERENCES

Brown. R. P. (1996), Physical Testing of Rubber, Rapra Technology Ltd, Shewbury.

Frederick, R. E. (1982), Science and technology of Rubber, Rubber Division of American Chemical Society.

Gent, A. N. (1992), Engineering with Rubber – How to design rubber components, Oxford University Press, New York.

Lake, G.J. (1997), Fatigue and Fracture of Elastomers, Rubber Chemistry & Technology, 68, 435–460.

Mal, A. K. & Singh, S. J. (1990), Deformation of Elastic Solids, Prentice Fall PTR.

Mullin, L. (1969), Rubber Chemistry & Technology, Vol. 42, pp. 339–362.

Takeuchi. K. & Nagakawa. M. (1993), Int'l Polymer Sci. Vol. 20, No. 10, pp. 64–69.

A new machine for accurate fatigue and crack growth analysis of rubber compounds

A. Favier

01dB-Metravib, Limonest, France

ABSTRACT: 01dB-Metravib is well-known for its range of Dynamic Mechanical Analysis (DMA) instruments including high specifications (high force, high strain, high frequency, . . .) which presents unusual capabilities for dynamic analysis of rubber and elastomer compounds. The company enlarges its range of testing products by introducing a new machine: DMA+300. This instrument is specifically designed for fatigue and crack growth tests analysis. Goal of this new instrument is to characterize with accuracy the crack growth rate and tearing energy of elastomers materials. This work is introducing the key points of the instrument.

1 INTRODUCTION

Resistance to tearing is a key factor for elastomer in many applications: (tire, dampers, seals, . . .), and there is today a need for accurate capability of crack propagation characterization.

The principle of the crack growth test consists in initiating a crack on an elastomer film and following up the growth of the crack while applying on the specimen some excitation. Results of these test are crack growth rate, rearing energy, as a function of dynamic and static excitation (stress or strain controlled), temperature, oxygen rate, . . .

2 DESCRIPTION OF THE INSTRUMENT

2.1 *Mechanical frame*

DMA+300 consists of a floor standing mechanical frame, control and acquisition electronics and a computer station equipped with dedicated software.

The dynamic mechanical excitation is delivered by an original electrodynamic actuator developed by 01dB-Metravib that includes a very performing guiding and anti-rotation system maintaining the excitation within the specimen plane. Its 300N force range was defined based on the elastomer specimen geometry in order to offer an optimum analysis range.

The dynamic and static components of the specimen strain, as well as the applied force, are measured respectively at each end of the specimen.

The excitation signal is programmed using a new software program, devoted to fatigue and crack growth tests: MULTITEST.

The excitation signal is controlled by controlling either the strain or the stress.

The floor standing mechanical test frame was de-signed so as to grant the operator easy and

Figure 1.   DMA+300 instrument.

comfortable access to the specimen, to the cutting system and to the crack growth follow up optical system.

2.2 *Specimen mounting device*

The specimen is initially mounted on a specific mounting template according to a very strict protocol. This ensures that the jaws are very accurately positioned on the same plane and parallel to the specimen plane. Additionally, this makes possible to handle the specimen, while installing on the instrument, with absolutely no risk of degradation, for example by applying unwanted preload and torsion on the sample.

The specimen used on the DMA+300 can either be pure shear or tensile mode. Practically, concerning

Figure 2. Thermal chamber and crack initiation tools included in front door.

Figure 3. Optical measurement system.

crack growth tests, the pure shear is the preferential mode, because the results are independent of the crack length in this kind of excitation.

2.3 Temperature and oxygen controlled chamber

Thermal conditioning is ensured by a dedicated thermal chamber operating through forced convection technology. The chamber was designed for an operating temperature range well beyond the requirements commonly encountered for elastomer analysis, i.e., from $-150°C$ to $500°C$. The thermal chamber includes a large-sized window that allows observing the entire surface of the specimen. The optical quality of this window allows a highly accurate control and optical analysis and measurement of the specimen.

The test can also be carried out by setting the specimen to a specific oxygen rate environment. Oxygen is a key factor in the propagation rate of the crack, which makes important to control this parameter with accuracy. The oxygen rate is measured at the entrance of the thermal chamber using a dedicated probe. The oxygen level can be controlled from 10 ppm up to 20% in the chamber

2.4 Crack initiation

The philosophy of DMA+300 consists in initiating the crack, and to follow its propagation. The way the crack is initiated is particularly important and can influence the test results. In order the have a good reproducibility a specific cutting system has been added inside the chamber. The cutting system is integrated inside the front door of the thermal chamber, with external operating manual controls.

A cut can then be made in the specimen to initiate a crack, very precisely and repeatedly, without opening the chamber, hence without disturbing the temperature regulation or the gas mix. This operating mode is also ensuring the safety of the operator.

The cutting system includes two blades, located respectively on the left and right of the specimen. It offers the possibility to generate a crack on each side

of the sample, to make some reproducibility measurement on the propagation during the same test and same sample. The cutting system allows to control the vertical position of the crack, as well as the depth of the initiated crack.

2.5 Optical measurement system

The optical measurement system includes: a binocular microscope, a ring light system, a mechanical translation system driving the motions of the microscope along X, Y and Z axes and driving electronics interfaced with the test software.

Micrometric translation plates driven by a stepbystep engine are used to drive and control the microscope displacements very precisely.

When the crack growth test is started, the microscope is placed in front of the specimen. After the cutting phase, the operator takes a geometric reference shot that will represent the origin of the crack growth follow-up.

The microscope remains in this position during the entire crack growth test to allow for as many crack propagation measurements as necessary during the test. The operator is tracking the crack end position using the joystick. Then the operator selection of the position generates automatic acquisition and calculation of the crack rate.

2.6 Software dedicated to fatigue and crack growth test

Test driving and acquisition are achieved by a new software developed by 01dB-Metravib engineers in close collaboration with research engineers of the rubber and tire industry. The same software is running both fatigue and crack growth tests on the instrument. Thanks to its multiple harmonics control and measurements capability, it is possible to control different excitation modes (waveforms): sine, Haversine, pulse, triangle, square, etc. It is also possible to apply customized waveforms imported from ASCII files.

The goal is to reproduce on the instrument, an excitation which is as close as possible from the real life of the product, in order to get the most meaningful results.

The excitation frequency can be programmed from 2 Hz to 1 kHz.

The test can be controlled using different parameters: displacement, strain, force or stress amplitude, excitation frequency or strain rate. It is also possible to work at constant energy, which is useful for comparison of different compounds.

3 CRACK GROWTH TEST CAMPAIGN

A crack growth test campaign is composed of different sequences that are specific of this type of analysis:

– Accommodation,
– Characterization
– Crack generation
– Crack growth.

3.1 Accommodation

The accommodation sequence consists in applying a certain number of excitation cycles to the specimen in order to stabilise it. This operation is only a conditioning phase for the specimen and does not yield any analysis results.

The accommodation sequence always occurs before the characterisation sequence, and is normally done at the highest strain of the rest of the campaign.

This sequence is done at defined temperature and oxygen control, which can be kept constant for the complete campain

3.2 Characterization

The characterisation sequence consists in applying a strain sweep (with imposed frequency or strain rate) to the specimen, and get the tearing energy (G, in J/m^2), i.e., the energy delivered by area unit.

The plot of G versus the amplitude of the imposed displacement is particularly interesting to characterise since it allows comparing the crack growth behaviour of different materials under similar conditions, i.e., at identical tearing energies.

In a crack growth test campaign, it is possible to define test with tearing energy settings. In such case, the dynamic excitation amplitude to be applied is automatically calculated from the G curve obtained during the characterisation sequence.

Characterization is done under temperature and oxygen controlled conditions.

3.3 Crack generation

After the characterization, the crack is generated in the specimen. One crack can be initiated on each side of the specimen, with control position and depth, by using the specific tools.

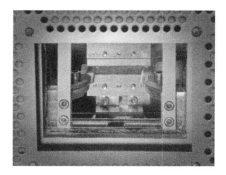

Figure 4. Crack generation in specimen.

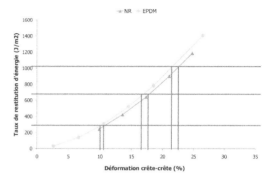

Figure 5. G (J/m^2) versus Strain sweep (%) for NR and EPDM.

3.4 Crack growth

The crack growth sequence consists in following up the crack growth, under given stress conditions. To do so, crack length measurements are carried out periodically.

The crack length measurement consists in stopping the dynamic test and using the microscope to read the geometric coordinates of the crack tip.

During the crack growth sequence, the user can perform a crack length measurement at any time or do it on predefined schedule.

Successive acquisitions of the crack tip coordinates throughout the test allow getting the crack growth curve (length in mm, crack growth rate in nm/number of cycles).

4 RESULTS

01dB-Metravib has set up early 2011 a collaboration with LRCCP in order to make some studies on the DMA+300. Very first results are presenting a comparison of to compounds based on NR and EPDM

– Temperature: Room temperature (23°C ± 2°C)
– 25 Hz, sinus waveform excitation
– Pure shear samples geometry

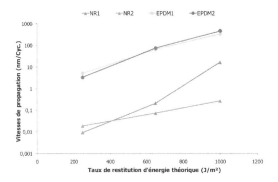

Figure 6. Crack growth rate (nm/cycle) versus G (J/m²) for NR and EPDM sample.

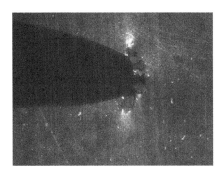

Figure 7. Spreading of the crack for NR2 sample.

4.1 *Characterization*

Selection of G used for crack growth test: $G = 250\,\text{J/m}^2$, $650\,\text{J/m}^2$ and $1000\,\text{J/m}^2$.

4.2 *Crack growth results*

The results display an excellent reproducibility for the EPDM sample, and its higher crack growth rate compared to the NR material. For the NR sample, especially the $G = 1000\,\text{J/m}^2$ is displaying significant differences.

This can be explained by the spread of the crack into several cracks for NR2 sample. Then the crack rate of the main crack is affected by the multiple crack, which explains the smaller rate.

Another points is that the crack has also propagated in the thickness of the samples (front and back length of the crack were different).

This should lead in several improvement in the testing processes of future experiment:

- Reducing the thickness of the samples to avoid propagation in the thickness of sample.
- Reinitiating a new crack in the samples when multiple cracks appear.

5 CONCLUSIONS

DMA+300 is proposing a new testing capability to characterize the crack growth of elastomer specimens. It makes possible to control with high accuracy, the excitation settings (dynamic and static amplitude, waveforms) and the temperature and oxygen environment of the specimen. The different tools for specimen handling, crack growth initiation, crack measurements, are key parameters for the reliability and good reproducibility of the results

Beyond this specificity, DMA+300 is a very versatile test machine that can be used to characterize properties of elastomers compounds, such as Payne effect, Mullins effect, frequency and temperature dependence, fatigue . . .

Constitutive Models for Rubber VII – Jerrams & Murphy (eds)
© 2012 Taylor & Francis Group, London, ISBN 978-0-415-68389-0

Biaxial fracture testing of rubber compounds

F. Caimmi, R. Calabrò, C. Marano & M. Rink
Dipartimento di Chimica, Materiali e Ingegneria Chimica "Giulio Natta" Politecnico di Milano, Milan, Italy

ABSTRACT: In this work the fracture toughness of natural rubber compounds under biaxial tensile loading condition was evaluated determining the J-integral by finite element analysis. To this aim a sound constitutive equation from tests under simple well-identified stress states such as tensile and plane strain test configurations was first determined for the studied materials. The models were then validated through the numerical simulation of both tensile tests under more complex stress states and of fracture tests in pure shear test configuration. The method was then applied to biaxial fracture tests previously performed.

1 INTRODUCTION

In a previous work (Marano et al. 2010) tensile tests under biaxial loading conditions were performed using a central notched cross-shaped specimen to study the fracture behaviour of carbon black-filled natural rubber compounds. The test consisted of two steps: a drawing step was initially performed loading the specimen in the direction parallel to the notch plane, up to different draw ratios, and then the specimen was loaded in the direction normal to the notch plane up to fracture. The fracture mechanics approach was applied to evaluate fracture toughness as a function of the draw-ratio set in the drawing step, showing a dependence of toughness on the draw ratio, at least for low draw ratios. As a fracture mechanics parameter a stress intensity factor was determined considering the central part of the cross-shaped specimen as a biaxially loaded infinite plate. This approach has several shortcomings. First, the part of the specimen under biaxial load is not easily identified and the stress at the boundaries is, thus, not precisely known. Further, the high strains and the highly non-linear behaviour of the material imply strong limitations in applying the stress intensity factor, whose use can therefore lead to misleading results.

The present work aims at determining an energy based parameter, such as J-integral, for this test configuration by means of numerical simulation via Finite Element (FE) method. In order to obtain sensible data, several steps are needed, namely: identification of a constitutive law, validation of its predictive capabilities and finally calculation of the J-integral at fracture from the experimental load data.

In the first part of this work a sound constitutive equation was determined, for each of the studied materials, from tests under simple well identified stress states such as uniaxial and plane strain tensile configurations. In this work it was *a priori* decided to focus on hyper-elastic material models to describe rubber

mechanical response; among the various models available, see *e.g.* (Holzapfel 2000), Ogden model was selected due to its ability to carefully reproduce the experimental data.

In the second part tensile tests under more complex stress states, that is the tensile loading of a cross-shaped specimen, and fracture tests in pure shear configuration (Boggio 2010) were carried out and the experimental results compared to numerical simulation predictions for model validation. It is worth to highlight that non-linear material models bring up a number of issues. In particular, in the material parameters identification step, the non-linear optimization problem to be solved in order to identify the material constants may have multiple solutions, depending also on the number of experimental tests used for fitting; such a multiplicity can lead to very different numerical solutions when these solutions are used to solve some given boundary value problem (Ogden et al., 2004). The experimental validation step on some further configuration is therefore mandatory in order to obtain sound results and to be confident in the actual predictive capabilities of the identified law.

The FE method was then applied to the tests previously performed in (Marano et al. 2010).

2 EXPERIMENTAL STUDY

2.1 *Materials*

Natural Rubber (NR0) and carbon black-filled natural rubber (NR50), kindly provided by Bridgestone, containing 50 g of carbon black N330 per hundred grams of rubber (phr), which corresponds to a volume fraction of 0.21. Plates having a thickness, B, between 1.5 and 3 mm were compression moulded at 160 °C and 8 MPa for 15 min, so as to assure complete sulphur vulcanization.

2.2 Tests for materials' constitutive equation determination

Tests in uniaxial and plane strain tensile configurations were performed to determine the constitutive equations of the studied materials. The tests were carried out at 23°C on a INSTRON 1185 dynamometer, at a nominal strain rate of 1.5 min⁻¹. A video-extensometer was used for strain measurement both in the loading and in the transverse direction.

Dumbbell specimens were used for uniaxial tensile test.

The plane strain tensile stress state was obtained subjecting rectangular specimens 30 mm high and having 3×98 mm² section subjected to uniaxial loading.

2.3 Tests for model validation

Test under biaxial tensile configuration were performed on a displacement-controlled biaxial dynamometer developed at LaBS, Politecnico di Milano, and made available for this research. Cross-shaped specimens, shown in Figure 1, were used. Parts in dark grey are made of reinforced rubber, which is vulcanized together with the rubber compound during the compression moulding process. During testing, deformation of the reinforced part of the specimen can be neglected. The tests were carried out at 23°C, drawing the sample at 30 mm/min in direction 1 and 2 simultaneously. Samples surface was sprayed with a target pattern in order to obtain a deformation field through the digital image correlation technique.

Fracture tests in pure shear test configuration were performed using a notched version of the specimen described in 2.2 for plane strain test. Notches 30 mm length were introduced with a cutter and sharpened with a razor blade. To ensure that crack propagates along the notch plane a groove was cut on both sides of the PS specimen. The ratio of the actual specimen thickness on the notch plane to that in the un-grooved region was about 0.6.

The tests were carried out under the same conditions as for the tests in 2.2.

Material fracture toughness was evaluated as follows:

$$J = \eta \left(\frac{a}{W} \right) \frac{U}{B_c} (W - a) \tag{1}$$

where η = dimensionless factor which for ungrooved pure shear specimen is equal to 1 (Hocine et al. 2003), (Kim & Joe, 1989); B_c = the thickness of the grooved sample; W = the specimen width; a = the crack length; U = the fracture input energy.

Pure shear fracture tests are thoroughly described in (Boggio et al. 2011).

2.4 Fracture test

The numerical simulation was applied to the fracture tests reported in (Marano et al. 2010), which are shortly described in the following. Tensile tests under biaxial

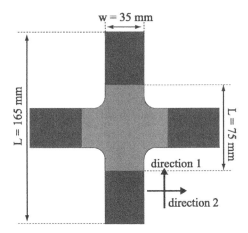

Figure 1. Test specimen geometry (dark gray: reinforced rubber).

loading conditions were performed using the cross-shaped specimen described in 2.3 in which a 4 mm long central notch was introduced with a sharp razor blade. The test was video-recorded to determine fracture initiation time. To measure a local (near tip) draw ratio, hereafter denoted simply as λ_2, a grid was drawn on the biaxially loaded zone of the specimen. (Marano et al. 2010). The test consisted of two steps: the sample was first drawn, at 100 mm/min, along a direction parallel to the notch plane (direction 2), up to a given draw ratio λ_2 (in the range 1–2.5) and then loaded perpendicularly to the notch plane (direction 1), at 30 mm/min, up to fracture. Test procedure and load trends are schematically depicted in Figure 2. The load measured in direction 2, P_2 was let to relax down to a fairly constant value before loading in direction 1.

Note that this is a pure Mode I fracture test, due to specimen symmetry and loading configuration. In the studied materials molecular orientation induced by deformation is expected to give rise to strength anisotropy and to a dependence of toughness on λ_2, at variance with what one would anticipate if the materials were isotropic linear elastic.

3 MODELING

3.1 Constitutive model

As mentioned earlier, at this stage of the research work, the description of material behaviour was deliberately restrained to the class of hyper-elastic materials. In this class, Ogden's celebrated hyper-elastic material model was selected (Ogden, 1972). Among the various modifications of the model proposed in literature, the one implemented in the FE software ABAQUS (Dassault Systèmes, 2010) was selected for obvious convenience, as ABAQUS is the software used for FE simulation.

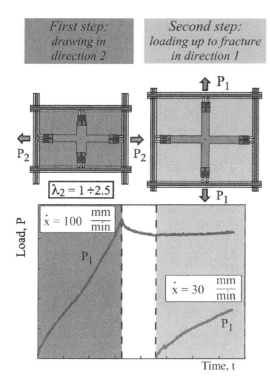

First step:
drawing in
direction 2

Second step:
loading up to fracture
in direction 1

P_1

P_2 | | | | P_2

$\boxed{\lambda_2 = 1 \div 2.5}$

P_1

$\dot{x} = 100 \; \dfrac{mm}{min}$

P_1

$\dot{x} = 30 \; \dfrac{mm}{min}$

P_1

Load, P

Time, t

Figure 2. Procedure of the fracture test on cross-shaped sample.

In ABAQUS Ogden's model is expressed in terms of a hyper-elastic strain energy function as

$$\Psi = \sum_{i=1}^{N} \frac{2\mu_i}{\alpha_i}\left(\bar{\lambda}_1^{\alpha_i^2} + \bar{\lambda}_2^{\alpha_i^2} + \bar{\lambda}_3^{\alpha_i^2} - 3\right) +$$
$$\sum_{i=1}^{N} \frac{1}{D_i}(J-1)^{2i} \qquad (2)$$

where the first summation represents the deviatoric part of the mechanical response and the second the volumetric one. λ_j ($j = 1, 2, 3$) are the principal stretches, while $\bar{\lambda}_j$ are the deviatoric principal stretches *i.e.*

$$\bar{\lambda}_j = J^{-1/3}\lambda_j \qquad (3)$$

with J being the volume ratio, *i.e.* the determinant of the deformation gradient.

In equation (2) N is the number of terms in the summations, and is a fit parameter. μ_i, α_i and D_i are material constants to be identified; the μ_i are related to G, the shear modulus of the material in the reference configuration (*i.e.* $\sum_i \mu_i = G$) while the D_i are related to volumetric expansion.

To model NR0, material incompressibility was assumed and thus all of the D_i were set to zero. As to NR50, experimental data obtained on the lateral contraction of plane strain specimens indicated that the such material was compressible, thus to obtain a good fit a volumetric term was inserted in the model,

i.e. D_1 was not set to zero, while for $i > 1$ the D_i were set to zero also for NR50. With such a choice, D_1 can be directly related to bulk modulus, K, in the reference configuration by the relation $K = 2/D_1$ (Dassault Systèmes, 2010).

Identification of material constants can be performed directly by ABAQUS using the least-square method (Dassault Systèmes, 2010). ABAQUS internal algorithms were used for NR50. With NR0 ABAQUS was unable to obtain fits corresponding to thermodynamically stable materials over the whole range of experimental deformations considered in this work. Thus an *ad-hoc* identification routine was written; the routine minimizes the square of the Euclidean distance between the experimental stress-strain response (simultaneously for plane strain and uniaxial tests) and the one calculated at varying material parameters. The optimization was constrained by imposing that the tangent stiffness matrix be positive for $\lambda = 1, 2, 3, 4$ and 5 for uniaxial loading, plane strain loading and biaxial loading. The need to impose the positive definiteness at various levels of strain comes from the fact that for Ogden materials the tangent stiffness matrix depends on deformation, as it can be readily verified by calculating it from (2). While this procedure does not rigorously grant the obtainment of a stable material for all possible stress states and for every possible level of deformation from the identification routine, in this work it was enough to give a stable material for biaxial, uniaxial and plane strain stress states at every level of nominal deformation among those reached during the experiments.

3.2   FE models

In this work two finite element models were used and implemented in the ABAQUS FE code (Dassault Systèmes, 2010). The first model was used to reproduce the behaviour of pure shear fracture specimens. It is a 3D model reproducing only one quarter of the specimen; the need to use a 3D model arise from the presence of the groove in the specimens. Velocity is prescribed on the top part of the boundary, while symmetry conditions are imposed on the other faces. Simulations were run up to the experimentally determined fracture time. Notches were modelled as region with unconstrained boundary displacements. 20-nodes brick element with reduced integration and mixed formulation (C3D20RH in ABAQUS, helpful in the case of incompressible materials) were used to model most of the specimen. In a region having a radius of 5 mm around the crack front fully integrated elements were used. Near the crack tip typical element edge dimension was 0.05 mm. A total of about 7000 elements were used in the model.

The second one is the model for the cross-shaped specimen. It is a plane stress 2D model of the specimen in 1; both reinforced rubber and unreinforced rubber were modelled. Reinforced rubber was modelled as isotropic liner elastic material with modulus 720 MPa (measured from uniaxial tensile tests), while unreinforced rubber was modelled with the Ogden model previously described. Only a quarter of the specimen was modelled due to symmetry. At the boundary (edges of the cross arms) a velocity corresponding to the crosshead displacement rate was prescribed in the case of un-notched biaxial specimens used for validation, while, for easiness, a stress corresponding to the fracture initiation load was prescribed in the case of fracture specimens. 8-noded rectangular element with reduced integration and mixed formulation were used for most of the model; elements with standard displacements formulation were used to model the reinforced rubber regions. In the tip region fully integrated elements were used. A total of 15500 were used in the model.

The J-integral was determined as a line integral using the ABAQUS built-in routine, thus avoiding the need to obtain accurate stress fields at the tip.

4 RESULTS AND DISCUSSION

4.1 Constitutive law identification

Stress-strain curves obtained in uniaxial tensile and plane strain tests together with the identified constitutive laws are reported for NR0 and NR50 in figures 3 (a) and (b) respectively.

Very good agreement exists between the identified law and the experimental data.

For the NR0 material it has been found that a three terms Ogden strain energy was needed to find a good fit to the experimental data. In the case of NR50 two terms gave a better results than three terms. As recalled earlier, in this case a single volumetric term corresponding to a Poisson's ratio of 0.4 was needed to obtain a good fit (see 3.1); the value was determined from lateral strain measurements and supplied directly to the identification routine.

4.2 FE models validation

In Figure 4 some representative predictions from the FE models of the pure shear fracture specimens are given in terms of load-displacement traces recorded up to the fracture load. Excellent agreement was obtained for NR0 predictions while the agreement for NR50 is slightly less satisfactory. In particular, the initial compliance is not correctly reproduced nor is the strain hardening behaviour, which begins to show itself at crosshead displacements of about 11 mm.

Prediction for the un-notched cross-shaped specimen are shown in Figures 5 and 6. For NR0 the agreement is more than satisfactory in terms of predictions of both the load-displacement trace (Fig. 5a) and the strain field (an example of the comparison between the prediction and some measurements is given in Figure 6).

In the case of NR50 (Fig. 5b) a good agreement could be obtained only at low forces. The FE model clearly fails to catch the abrupt increase in stiffness shown by the experiments. Whether this is due to a

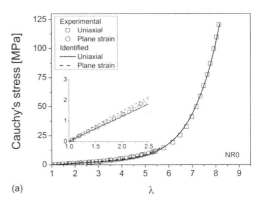

(a)

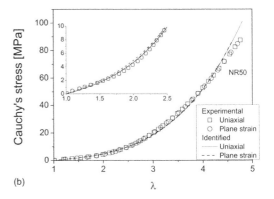

(b)

Figure 3. Experimental and identified stress-strain behaviour in terms of true stress/stretch ratio. Insets show the initial region of the plot. (a) NR0. (b) NR50.

problem with the FE model of the cross-shaped specimen (arising at high loads and therefore unnoticed for NR0) or to a problem with the identified constitutive model is at present unclear. As to this last hypothesis, it is worth remembering that the larger the number of experimental data available for fitting, the better are the predictive capabilities of Ogden's model (Ogden et al. 2004). It is thus possible that better results could be obtained by increasing the number of experimental stress states used as inputs for the identification procedure. Further investigation is currently underway.

4.3 J-integral determination

Due to the unsatisfactory results obtained with the validation of the NR50 models, J-integral calculations were performed only for NR0. The results from such calculations are shown in Figure 7.

A strong decrease of J with the draw ratio can be observed thus confirming the qualitative trends obtained in (Marano et. al. 2010) and showing the sensitivity of rubber toughness to deformation induced structural anisotropy.

Toughness at $\lambda_2 = 1$ is about 30 kJ/m^2. From pure shear fracture tests on the same material (Boggio 2010) a value of 23 ± 1 kJ/m^2 was obtained at similar nominal strain rate. Further tests are underway to asses the repeatability of the cross-shaped fracture tests.

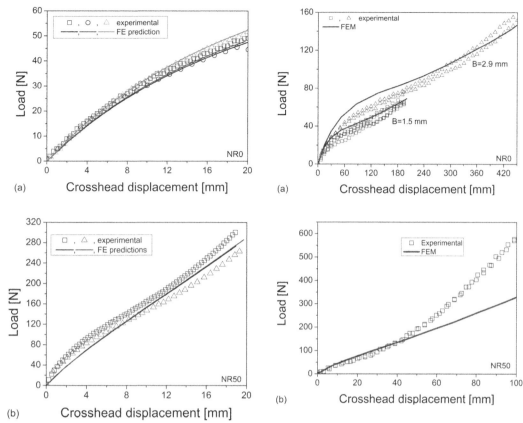

Figure 4. Load-displacement traces for pure-shear fracture specimens. Various experimental samples are shown (symbols) together with the respective FE prediction. Corresponding grey-tones indicate results for the same sample. (a) NR0. (b) NR50.

Figure 5. Load-displacement traces for the un-notched cross-shaped specimens. Various experimental samples are shown (symbols) together with the respective FE prediction (a) NR0. Samples with different thickness are shown (b) NR50. Samples thickness is 1.5 mm.

5 CLOSING REMARKS

An *ad-hoc* identification and validation technique to identify Ogden's model parameters was developed and applied to predict the fracture toughness of natural rubber compounds under a special biaxial loading condition.

The identification of Ogden's parameters from plane strain and uniaxial tests seemed straightforward. Anyway it was possible to find a set of parameters actually able to predict the response of the material under very different loading conditions only for NR0. In the case of NR50 the predictions were much less satisfying. The reasons for such a disagreement may lay in every step behind the validation one:

– Ogden's model was actually unable to catch the behaviour of the carbon-black-filled material;
– the number of stress-states used for identification was insufficient;
– the intrinsic problems arising in non-linear optimization prevented the finding of a parameter set corresponding to the actual material behaviour.

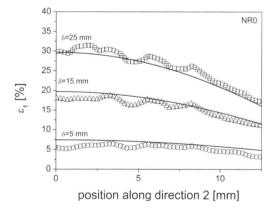

Figure 6. Nominal strain along direction 1, ε_1, vs. position for NR0 cross-shaped unnotched samples. (solid lines: calculated via FE; symbols: measured via DIC). Reference system origin is in the symmetry center of the specimen. δ = crosshead displacement. Specimen thickness $B = 1.5$ mm.

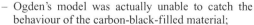

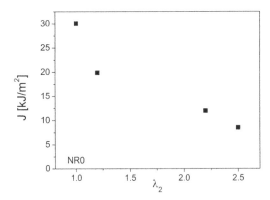

Figure 7. Natural rubber fracture toughness as a function of the transverse draw ratio for cross-shaped fracture specimens.

Nonetheless the technique was successfully applied to calculate from the experimental data a fracture parameter for natural rubber (NR0) as a function of the applied draw ratio, confirming the actual dependence on transverse strain of natural rubber fracture toughness.

ACKNOWLEDGMENTS

This work was supported by Bridgestone Technical Center Europe, Roma. The authors thank the Bridgestone technical staff for the helpful discussion.

F.C. gratefully acknowledges partial support by Italy's Consorzio Interuniversitario Nazionale per la Scienza e la Tecnologia dei Materiali (INSTM) through grant ID/DB234.

REFERENCES

Boggio, M. 2010. Ph.D. Thesis. Fracture Behaviour of Carbon Black Filled Natural Rubber; Politecnico di Milano, Milano, Italy
Boggio, M., Marano, C., Rink, M. 2011. Time-Dependence of Fracture Beahviour of Carbon Balck Filled Natural Rubber. 7th Europen Conference on Constitutive Models for Rubber (ECCMR7), Dublin, Ireland.
Dassault Systèmes Simulia Corp. 2010. ABAQUS/Analysis User's Manual, version 6.10. Providence, RI, USA.
Holzapfel G.A., *Nonlinear Solid Mechanics* ([n.p]: Wiley, 2000).
Marano C., Calabrò R. & Rink M. 2010. Effect of Molecular Orientation on the Fracture Behavior of Carbon Black-Filled Natural Rubber Compounds. *Journal of Polymer Science: Part B: Polymer Physics*, 48: 1509–1515.
Ogden R.W., Large Deformation Isotropic Elasticity: On the Correlation of Theory and Experiment for Compressible Rubberlike Solids, *Proceedings of the Royal Society of London. Series A, Mathematical and Physical Sciences*, 328 (1972), 567–83.
Ogden R.W., Saccomandi G. & Sgura I., Fitting hyperelastic models to experimental data, *Computational Mechanics*, 34 (2004), 484–502.

Constitutive Models for Rubber VII – Jerrams & Murphy (eds)
© 2012 Taylor & Francis Group, London, ISBN 978-0-415-68389-0

Dynamic analysis and test investigation on rubber anti-vibration component

R.K. Luo, W.J. Mortel, S. Sewell, D. Moore & J. Lake
Department of Engineering and Technology, Trelleborg IAVS, Leicester, UK

ABSTRACT: There is very little literature regarding rubber dynamic simulations. In many engineering applications the dynamic analysis is normally not required. However there is more demanding requirement for quantitative specification of dynamic behaviour of rubber-to-metal bonded anti-vibration components. This case study is based on an engine installation where there is a need for understanding of the products complex stiffness and damping. Initially the model has been verified by quasi-static tests, followed by a dynamic test programme within a test laboratory. A three-dimensional finite element model has been developed to predict the natural frequencies and mode shapes for the product. At same time a steady state dynamic response from a damped harmonic input is also evaluated. Compared with the test result, it is concluded that the simulation is reliable and can be used in the proper design stage when dynamic characteristics are needed.

1 INTRODUCTION

There is very little literature regarding rubber dynamic simulations. The dynamic performance of rubber anti-vibration components is strongly influenced by their own characteristics and service environment. In recent years the tendency has been to produce such components not only for safety but also for improved dynamic response in a complicated loading conditions. This has led to design of the components to a more challenging standard than before. First they should meet the reasonable stiffness and the service life requirements, see Luo etc. Secondly the dynamic resonance should be avoided from the frequencies of the external dynamic loading. In the rubber industry computer simulation and laboratory tests are employed simultaneously to obtain the best design. This paper reports the predictions from the simulation and validation against the test results. The approach is the same as used for previous dynamic analysis for rail vehicles and impact on composite structures, also see Luo etc.

2 FINITE ELEMENT MODEL

Finite element analysis has been used to predict the free-vibration and steady state dynamic responses. The finite element calculations have been performed using the finite element code Abaqus. A picture of the model is shown in Figure 1. Two materials have been used in the component, one is a mild steel and the other is the rubber with nominal 65 hardness. The component is subjected to a pre-compression force and then a dynamic loading along the shear direction is applied (parallel to the flat surface). Three-dimensional finite elements with full integration have been used for both steel and rubber. For the rubber, elements with extra

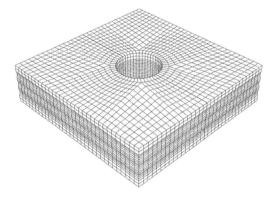

Figure 1. The finite element model.

pressure freedom were used. The total degrees of freedom is 79,000. The bottom surface of the model is fixed. The shrinkage of the component is not modelled as the other sandwich mount [see reference, Luo 2011] due to much thinner rubber section. The dynamic shear movement is applied to the top surface. Total CPU run time is about 8 hours and 20 minutes using intel Xeon 2.0 GHz with 8 GB Ram.

3 FREE VIBRATION ANALYSIS

The most straightforward type of dynamic analysis is the determination of natural frequencies and mode shapes which are independent of any loading. This type of calculation can give considerable insight into the dynamic behaviour of the component. The equations of motion for the free vibration of an undamped system are

$$[M]\{\ddot{\delta}\} + [K]\{\delta\} = \{0\} \tag{1}$$

in which $\{0\}$ is a zero vector. The problem of vibration analysis consists of determining the conditions under which equation (1) will permit motions to occur. A solution exists if a determinant is zero. That is

$$\|[K] - \omega^2[M]\| = 0 \qquad (2)$$

Equation (2) is called the frequency equation of the system. Expanding the determinant will give an algebraic equation of the nth degree in the frequency parameter ω^2 for a system having n degrees of freedom. The n roots of this equation ($\omega_1^2, \omega_2^2, \omega_3^2, \ldots, \omega_n^2$) represent the frequencies of the n modes of vibration which are possible in the system. The natural frequencies and mode shapes of the component are the dynamic behaviour of the structure itself. This characteristic is of great consequence in the study of the dynamic response of the system to external forces. The Lanczos method is used for free vibration analysis. Lanczos eigensolver is a powerful tool for extraction of the extreme eigenvalues, and the corresponding eigenvectors of a sparse symmetric generalized eigenproblem has been implemented in many applications. The Lanczos procedure consists of a set of Lanczos "runs," in each of which a set of iterations called steps is performed. For each Lanczos run the following spectral transformation is applied:

$$[M]\{K\} - \sigma[M]^{-1}[M]\{\emptyset\} = \theta[M]\{\emptyset\} \qquad (3)$$

where σ is the shift, θ is the eigenvalue, and $\{\emptyset\}$ is the eigenvector. This transformation allows rapid convergence to the desired eigenvalues. The eigenvectors of the symmtrized problem (2) and the transformed problem (3) are identical, while the eigenvalues of the original problem and the transformed problem are related in the following manner:

$$\omega^2 = \frac{1}{\theta} + \sigma \qquad (4)$$

More details can be seen in reference.

4 STEADY STATE DYNAMIC RESPONSE

Steady-state dynamic analysis provides the steady-state amplitude and phase of the response of a system subjected to harmonic excitation at a given frequency. The general dynamic equations can be written as

$$[M]\{\ddot{\delta}\} + [C]\{\dot{\delta}\} + [K]\{\delta\} = \{F\} \qquad (5)$$

where $\{F\}$ is the harmonic load vector, which can be expressed as a function of both sine term and cosine term. The damping effect can be included as factors of both mass and stiffness of the component. An assumption can be made that the damping is proportional to the mass and the stiffness. The expressions (Rayleigh damping) is shown below:

$$[M]\{\ddot{\delta}\} + [C]\{\dot{\delta}\} + [K]\{\delta\} = \{F\} \qquad (5)$$

where α and β are proportional constants.

Figure 2.   First mode shape of the component.

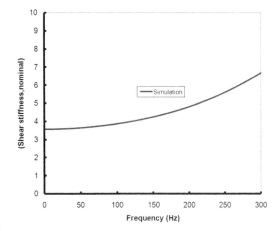

Figure 3.   Response of the harmonic dynamic loading.

Usually such analysis is done as a frequency sweep, by applying the loading at a series of different frequencies and recording the response.

5 RESULTS AND DISCUSSIONS

The first mode shape of the component is shown in Figure 2. The deformation is the shear mode, which is easily excited by an external shear loading. The fundamental frequency is 73 Hz. Hence in an engineering applications an exciting frequency from an external loading should be kept well away from this fundamental frequency. Figure 3 shows the spectral response of the component in the form of dynamic shear stiffness. There is only a small increase in stiffness over the first 50 Hz. The stiffness is doubled when external loading increased to 300 Hz, which indicates the hysteresis effect will contribute a much greater proportion of the complex stiffness as the dynamic frequency increases. These values are very important for engineering design and applications.

A dynamic test was carried out in a laboratory and the response curve of the complex stiffness is shown in Figure 4. Compared with the simulation the two curves

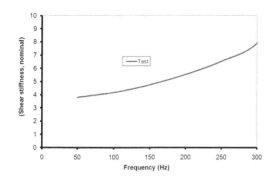

Figure 4. Comparison between the simulation and the test.

are matched very well. It is therefore concluded that the simulation is reliable and can be used in the proper design stage when dynamic characteristics are needed.

ACKNOWLEDGEMENT

The authors would like to thank Trelleborg IAVS to allow the publication of this paper.

REFERENCES

Ogden, R.W. 1984. Non-linear elastic deformations. Chichester: Ellis Horwood Limited

Luo, R.K., Gabbitas B.L. & Brickle B.V., 1994, Fatigue life evaluation of a railway vehicle bogie using an integrated dynamic simulation, Journal of Rail and Rapid Transit, 208, 123–132

Luo, R.K., Gabbitas B.L. & Brickle B.V., 1994, An integrated dynamic simulation of metro vehicles in a real operating environment, Vehicle System Dynamics, 23, 334–345

Luo, R.K., Gabbitas B.L. & Brickle B.V., 1996, Fatigue design in railway vehicle bogies based on dynamic simulation, Vehicle System Dynamics, 25, 449–459

Luo, R.K., Gabbitas B.L. & Brickle B.V., 1996, Dynamic stress analysis of an open-shaped railway bogie frame, Engineering Failure Analysis, 1(3), 53–64

Luo, R.K., Green E.R. & Morrison C.J., 1999, Impact damage analysis of composite plates, International Journal of Impact Engineering, 4(22), 435–448

Luo, R.K. & Wu, W.X., 2006, Fatigue failure analysis of anti-vibration rubber spring, Engineering Failure Analysis, Vol. 13, No. 1, 110–116

Luo, R.K., Mortel, W.J., & Wu, X.P., 2009, Fatigue failure investigation on anti-vibration springs, Engineering Failure Analysis, Vol. 16, No. 5, 1366–1378

Luo, R.K., Mortel, W.J., Cook P.W. and Lake, J., 2011, Computer simulation and experimental investigation of offset sandwich mount, Plastics, Rubber & Composites, 4(40), 155–160

ABAQUS User Manual, 2010, Dassult Systems, USA

Constitutive Models for Rubber VII – Jerrams & Murphy (eds)
© 2012 Taylor & Francis Group, London, ISBN 978-0-415-68389-0

Author index

Printed and bound by CPI Group (UK) Ltd, Croydon, CR0 4YY

21/10/2024

01777095-0006